ELECTRICAL MOTOR CONTROLS

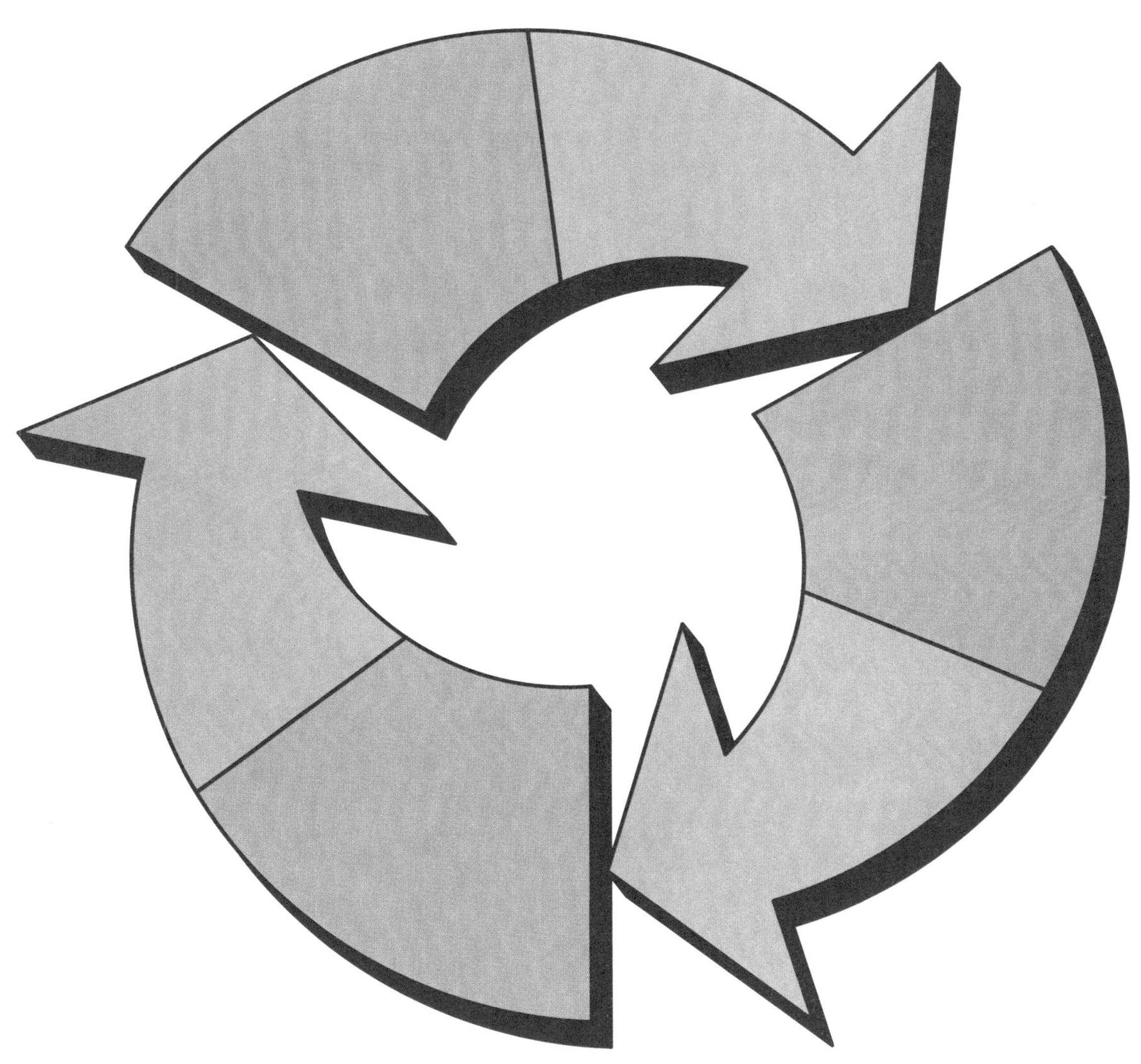

AMERICAN TECHNICAL PUBLISHERS, INC.
HOMEWOOD, ILLINOIS 60430

Gary Rockis
Glen A. Mazur

1 2 3 4 5 6 7 8 9 – 97 – 9 8 7 6 5 4 3 2

Printed in the United States of America

Library of Congress Cataloging-in-Publication Data

Rockis, Gary.
Electrical motor controls / Gary Rockis, Glen A. Mazur.
p. cm.
Includes index.
ISBN 0-8269-1671-6 (hard)
1. Electric controllers. 2. Electric motors--Electronic control.
I. Mazur, Glen. II. Title
TK2851.R63 1997
621.31′042--DC20

96-31168
CIP

The authors and publisher are grateful to the following companies and organizations for providing technical information and assistance. Companies preceded by an asterisk (*) have provided photographs that were used on the cover.

- ABB Power T&D Company Inc.
- Advanced Assembly Automation Inc.
- AEMC Instruments
- American Precision Industries, Motion Technologies Group
- ARO Fluid Products Div., Ingersol-Rand
- Atlas Technologies Inc.
- Baldor Electric Co.
- Banner Engineering Corp.
- Bergey Windpower Co., Inc.
- Boeing Commercial Airplane Group
- Carlo Gavazzi Inc. Electromatic Business Unit
- Cincinnati Milacron
- Clausing Industrial
- Clippard Instrument Laboratory, Inc.
- Cutler-Hammer
- Danfoss Electronic Drives
- DoALL Company
- Eagle Signal Industrial Controls
- Electrical Apparatus Service Association, Inc.
- Fanuc Robotics North America
- Fluke Corporation
- The Foxboro Company
- Furnas Electric Co.
- GE Motors & Industrial Systems
- *General Electric Company
- Giddings & Lewis, Inc.
- Gould Inc.
- Gould Instrument Systems, Inc.
- Greenlee Textron Inc.
- Guardian Electric Mfg. Co.
- Heidelberg Harris, Inc.
- *Honeywell's MICRO SWITCH Division
- Humphrey Products Company
- Ideal Industries, Inc.
- Ircon, Inc.
- Kay-Ray/Sensall, Inc.
- Klein Tools, Inc.
- Leeson Electric Corporation
- Lennox Industries Inc.
- Milwaukee Electric Tool Corporation
- Namco Controls Corporation
- Omron Electronics, Inc.
- Panduit Corp.
- Park Detroit
- Products Unlimited
- Ridge Tool Company
- *Rockwell Automation, Allen-Bradley Company, Inc.
- Rofin Sinar
- Ruud Lighting, Inc.
- SEW-EURODRIVE, Inc.
- Siemens Corporation
- Sprecher + Schuh
- Square D Company
- SSAC, Inc.
- Thermometrics Inc.
- *Xycom, Inc.

CONTENTS

INTRODUCTION

Electrical Motor Controls provides comprehensive coverage of the control devices used in contemporary industrial electrical systems. Beginning with basics such as tools, instruments, safety, electrical symbols, and line diagrams, each chapter builds on previous chapters. The 17 chapters detail motor starters, solenoids, control devices, motor circuits, power distribution systems, programmable controllers, reduced-voltage starting, and accelerating and decelerating methods. Special emphasis is placed on preventive maintenance and the development of troubleshooting skills.

CAD-drawn line art and electronic text have been combined with photographs from over 50 major manufacturers and organizations to present technical information in a direct-to-the-point, heavily-illustrated, colorful format designed to represent the broad range of uses for electricity, and specifically for the control of electricity, in different industries today.

Review Questions follow each chapter. Review Question Answers are located in the back of this textbook. The Appendix contains many useful tables, charts, formulas, etc. The Glossary defines over 400 technical terms related to motor control and control devices. *Electrical Motor Controls Workbook* provides Tech-Cheks and Worksheets for each major concept presented in the textbook. Answers to workbook problems are in the *Instructor's Guide*. Transparencies are available to enhance instruction.

The Publisher

Electrical Motor Controls by Gary Rockis and Glen Mazur

Technology Utilized:

Editorial	**American Technical Publishers, Inc.** Homewood, IL
Technical Edit	WordPerfect® for Windows 5.1
Copy Edit	WordPerfect® for Windows 5.1
Typesetting	Ventura™ Publisher 4.1.1
Text	10.5 pt Times
Heads	11.5 pt Times Bold
Captions	9 pt Helvetica
Art	AutoCAD®, AutoScript®, B-Coder™, Corel Ventura™ 5, Designer™ 3.1, and Micrografx® Picture Publisher 5.0
Film	Linotronic® 330 Imagesetter
Printing	**The Banta Book Group** Menasha, WI
Imposition	OptiCopy Camera
Plates	Burgess Plate Framer
Paper	50 # OptiMatte
Press	Baker Perkins 8 Unit
Endsheets	80 # Standard White
Cover	10 pt Lexotone
Press	Heidelberg SpeedMaster
Binding	**Welsh Bindery** New Berlin, WI
Adhesive	Harris Universal Binder
Case	Kolbus-Case/In-Line

Chapter 1

Electrical Tools, Instruments, and Safety

The proper tools must be selected for each job. Tools must be organized and readily available for use. Consult the operator's manual for correct operation and use before using a new tool or instrument. Electrical power can be dangerous. An electrician must be aware of the dangers associated with electrical power and the potential hazards on the job. Safe work habits and proper procedures minimize the possibility of an accident.

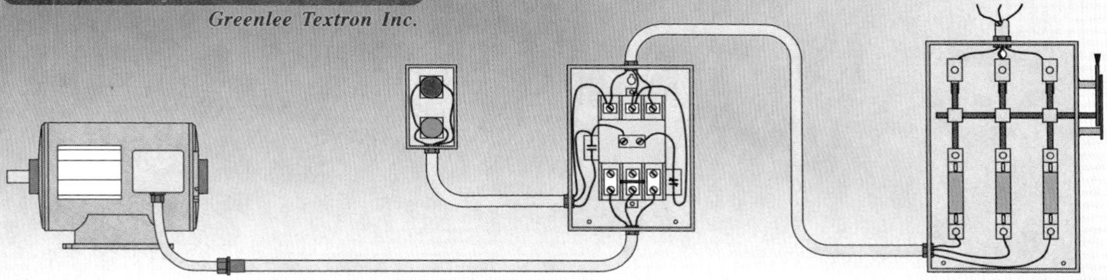

Greenlee Textron Inc.

TOOLS

Various hand tools are used by electricians for the maintenance, troubleshooting, and installation of electrical equipment. Different tools are designed for the efficient and safe completion of a specific job. Proper use of tools is required for safe and efficient electrical work.

Screwdrivers

A *screwdriver* is a hand tool with a tip designed to fit into a screw head for turning. Electricians use screwdrivers in many installation, troubleshooting, and maintenance activities to secure various threaded fasteners. Various types of screwdrivers are available. The two main types of screwdrivers are the flathead and Phillips. Flathead and Phillips screwdrivers are available as standard, offset, and screwholding. See Figure 1-1.

Greenlee Textron Inc.

Hydraulic punch drivers from Greenlee Textron Inc. allow fast hole punching for easy installation of conduit in electrical enclosures.

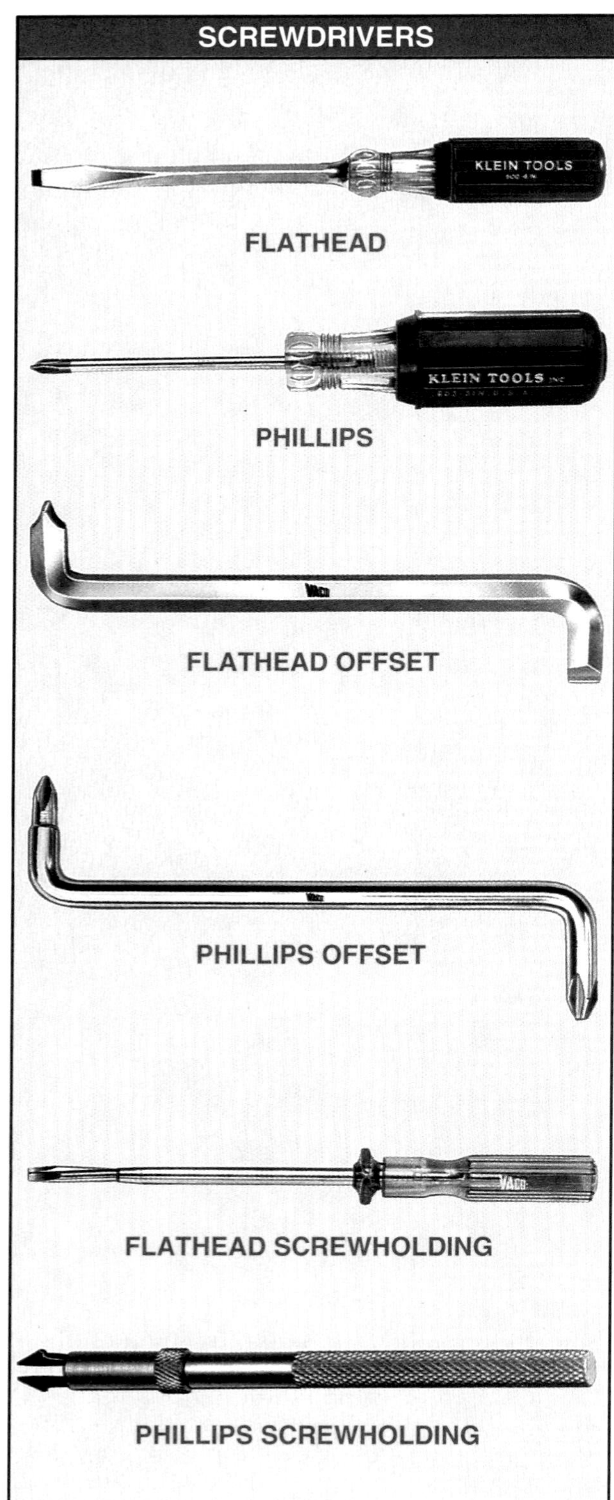

Klein Tools, Inc.

Figure 1-1. Electricians use screwdrivers in many installation, troubleshooting, and maintenance activities to secure various threaded fasteners.

Standard screwdrivers are used for the installation and removal of threaded fasteners. Offset screwdrivers provide a means for reaching difficult screws. A screwholding screwdriver is used to hold screws in place when working in tight spots. Once started, the screw is released and tightened with a standard screwdriver. Screwdrivers are available with square shanks to which a wrench can be applied for the removal of stubborn screws. They may also have a thin shank to reach and drive screws in deep, counter-bored holes.

When using a screwdriver, ensure that the tip fits the slot of the screw snugly and does not project beyond the screw head. Never use a screwdriver as a cold chisel or punch. Do not use a screwdriver near live wires. Never expose a screwdriver to excessive heat. Redress a worn tip with a file to regain a good straight edge. Discard a screwdriver that has a worn or broken handle.

Pliers

Pliers are hand tools with opposing jaws for gripping and/or cutting. Pliers are used by electricians for various gripping, turning, cutting, positioning, and bending operations. The most common pliers include slip-joint, tongue-and-groove, long-nose, diagonal-cutting, side-cutting, end-cutting, and locking pliers. See Figure 1-2.

General Electric Company

Screwdrivers are used on electrical devices to add or remove components and make adjustments to the device.

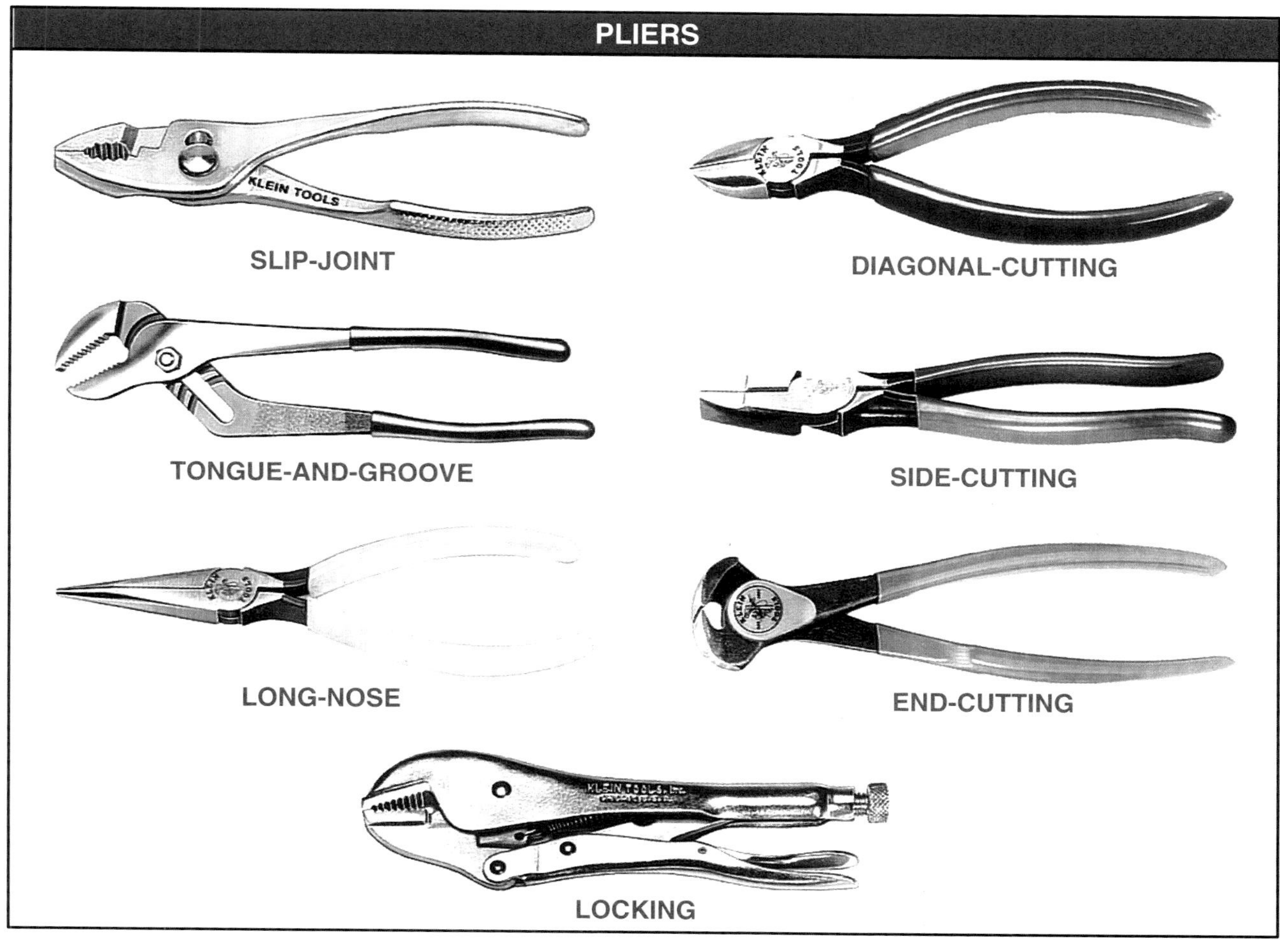

Klein Tools, Inc.

Figure 1-2. Pliers are used by electricians for various gripping, turning, cutting, positioning, and bending operations.

Slip-joint pliers are used to tighten box connectors, lock nuts, and small-sized conduit couplings. Tongue-and-groove pliers are used for a wide range of applications involving gripping, turning, and bending. Their adjustable jaws provide a wide range of sizes. Long-nose pliers are used for bending and cutting wire and positioning small components. Larger diagonal-cutting pliers are used for cutting cables and wires too difficult for side-cutting pliers.

Side-cutting (lineman's) pliers are used for cutting cable, removing knockouts, twisting wire, and deburring conduit. End-cutting pliers are used for cutting wire, nails, rivets, etc. close to the workpiece. Locking pliers are used to lock on to any workpiece. The pliers may be adjusted to lock at any size with any desired amount of pressure.

Wrenches

A *wrench* is a hand tool with jaws at one or both ends that are designed to turn bolts, nuts, or pipes. Common wrenches include socket, combination, adjustable, hex key, and pipe. See Figure 1-3.

Socket wrenches are used to tighten a variety of items, such as hex head lag screws, bolts, and various electrical connectors. Adjustable wrenches are used to tighten items, such as hex head lag screws, bolts, and larger-sized conduit couplings. Hex key wrenches are used for tightening hex head bolts. A combination wrench is a hand tool with an open-end wrench on one end and a closed-end (box) wrench on the other. Pipe wrenches may be straight, offset, strap, or chain. Pipe wrenches are used to tighten and loosen larger pipes and conduit.

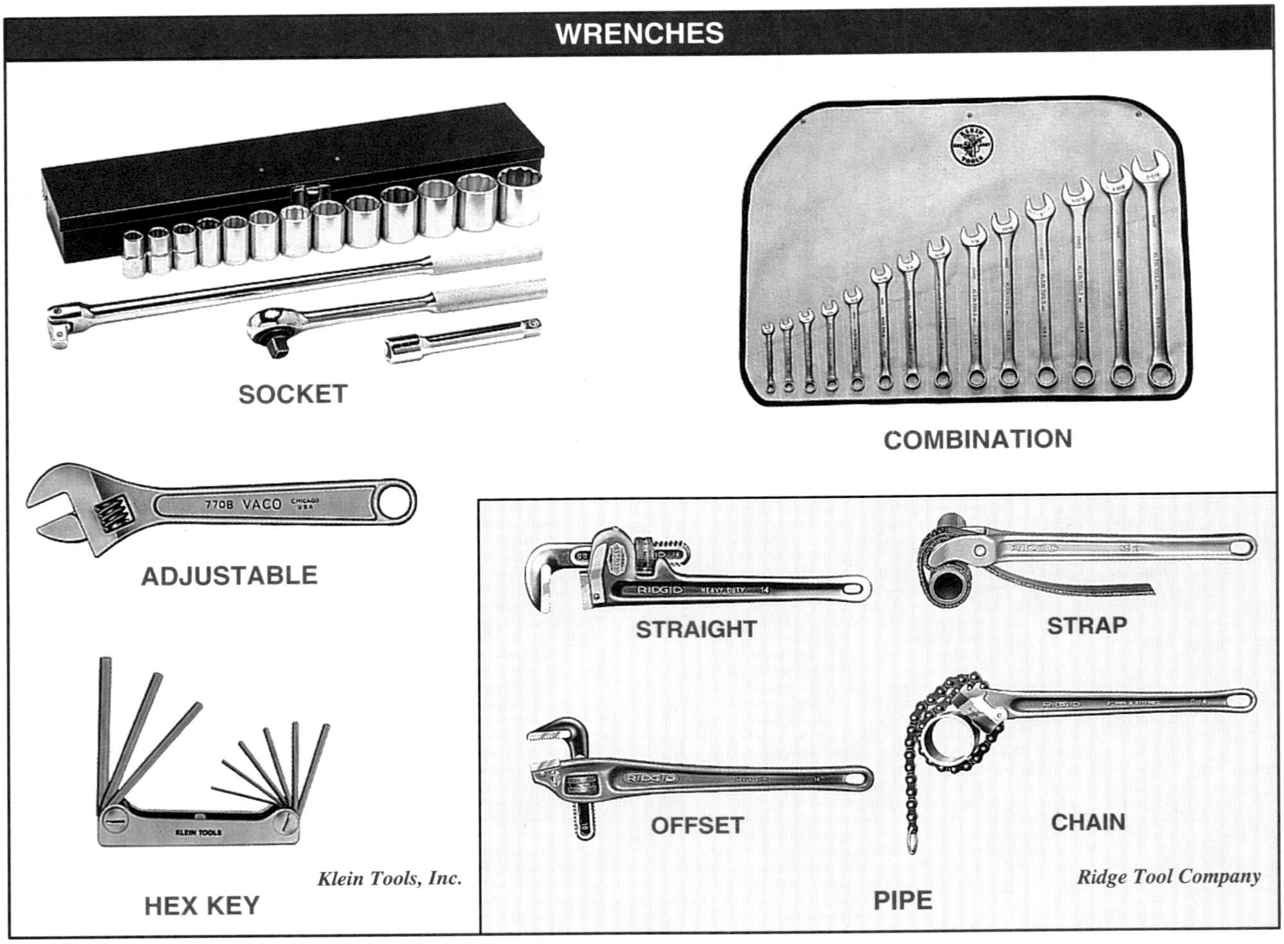

WRENCHES

SOCKET

COMBINATION

ADJUSTABLE

HEX KEY

Klein Tools, Inc.

STRAIGHT

STRAP

OFFSET

CHAIN

PIPE

Ridge Tool Company

Figure 1-3. Common wrenches include socket, combination, adjustable, hex key, and pipe.

When using wrenches, never use a pipe extension or other form of "cheater" to increase the leverage of the wrench. Select a wrench with an opening which exactly fits the nut. Too large an opening can spread the jaws of an open-end wrench and can batter the points of a box or socket wrench. Care should be taken in selecting inch wrenches for inch fasteners and metric wrenches for metric fasteners. If possible, always pull on a wrench handle. The safest wrench is a box or socket wrench because they cannot slip and injure the worker. Always use a straight handle rather than an offset handle if conditions permit.

Slip-joint pliers are useful in many applications, including gripping, twisting, turning, and fastening.

Hammers

A *hammer* is a striking or splitting tool with a hardened head fastened perpendicular to the handle. Common hammers include the electrician's, ball peen, and sledge hammer. See Figure 1-4.

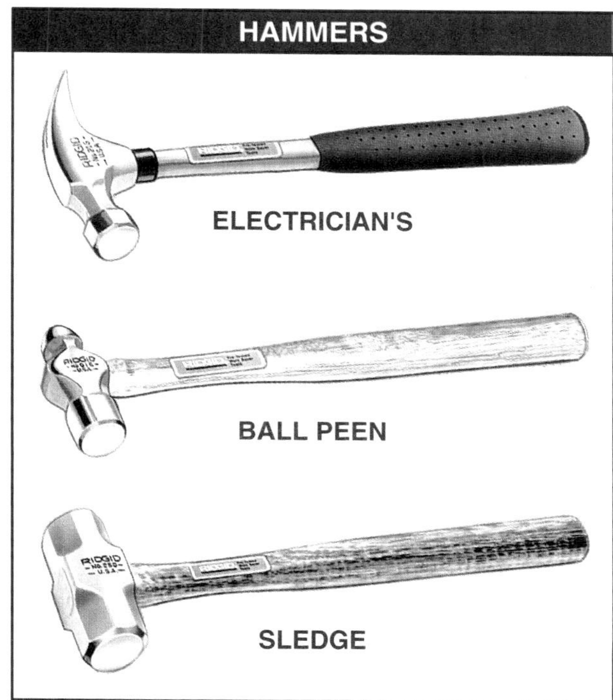

Ridge Tool Company

Figure 1-4. Common hammers include the electrician's, ball peen, and sledge hammer.

An electrician's hammer is used to mount electrical boxes and drive nails. An electrician's hammer may also be used to determine the height of receptacle boxes because most hammers are 12″ in length from head to end of handle, or can be so marked. Ball peen hammers of the proper size are designed for striking chisels and punches. They may also be used for riveting, shaping, and straightening unhardened metal. Medium-sized sledge hammers (5 lb) are used for driving stakes and other heavy-duty pounding.

Pipe Vises

A *vise* is a portable or stationary clamping device used to firmly hold work in place. Pipe vises include the yoke and chain vises. See Figure 1-5. A pipe vise has a hinge at one end and a hook at the other. This enables the vise to be opened so that the pipe does not need to be threaded through the vise jaws to be worked on. A yoke pipe vise is bolted to a workbench. A clamp kit vise is a vise that contains a clamp for mounting. A clamp kit vise can be temporarily mounted for light-duty work without drilling holes. The yoke pipe vise and chain vise are available in portable workbench models.

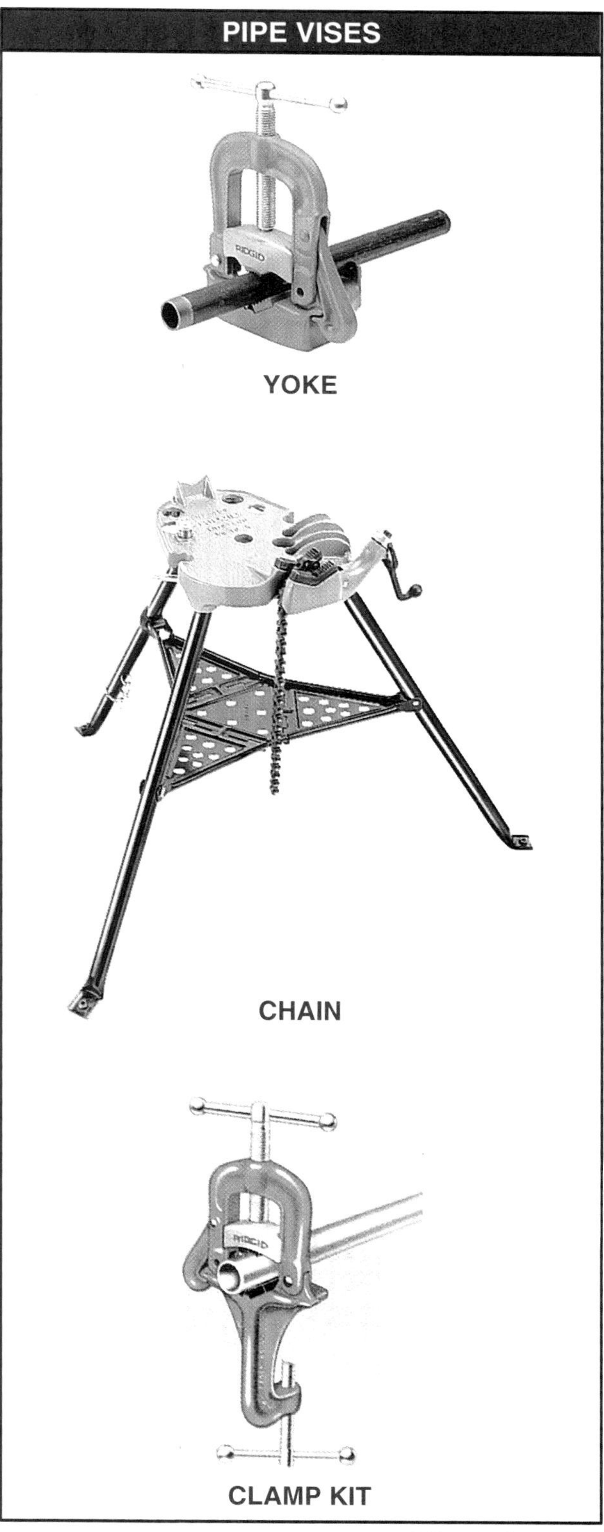

Ridge Tool Company

Figure 1-5. A vise is a portable or stationary clamping device used to firmly hold work in place.

Miscellaneous

Other tools are used by electricians when doing electrical work. These tools include nut drivers, knives, strippers, fish tape, saws, etc. An electrician must know the proper operation of all tools before use. See Figure 1-6.

Organization

Tools should be marked so that they can be easily identified as belonging to an individual or to a department in a company. To be effective, tools must be available when needed.

MISCELLANEOUS TOOLS . . .

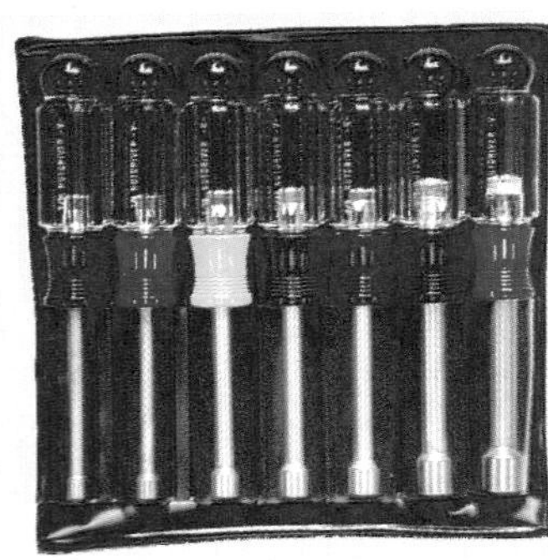

Used to tighten positive-sized hex head nuts and screws.

NUT DRIVER SET

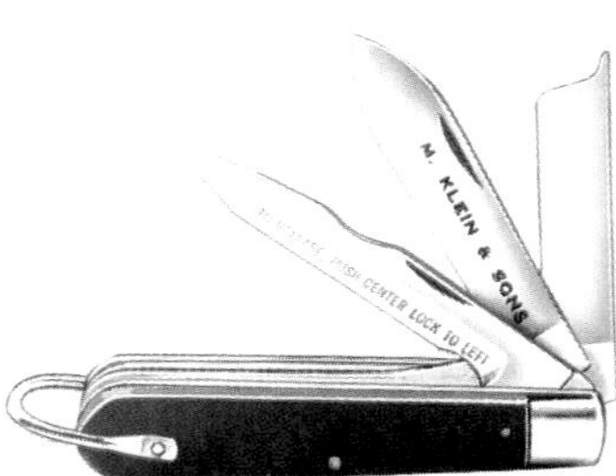

Klein Tools, Inc.

Removes insulation from cables and service conductors.

ELECTRICIAN'S KNIFE

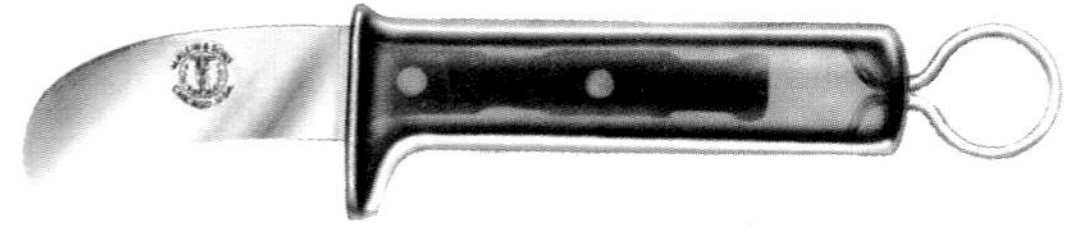

Klein Tools, Inc.

Removes insulation from cables and service conductors.

SKINNING KNIFE

Greenlee Textron Inc.

Removes insulation from small diameter wire, secures crimp type connectors, and cuts small bolts to length.

COMBINATION WIRE STRIPPER, BOLT CUTTER, AND CRIMPER KNIFE

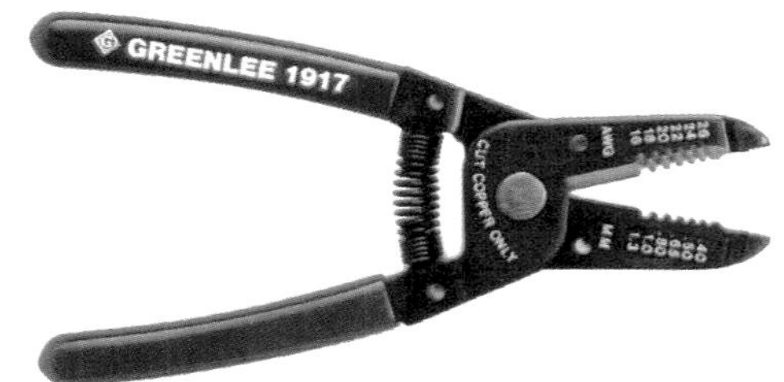

Greenlee Textron Inc.

Removes insulation from small diameter wire.

WIRE STRIPPER

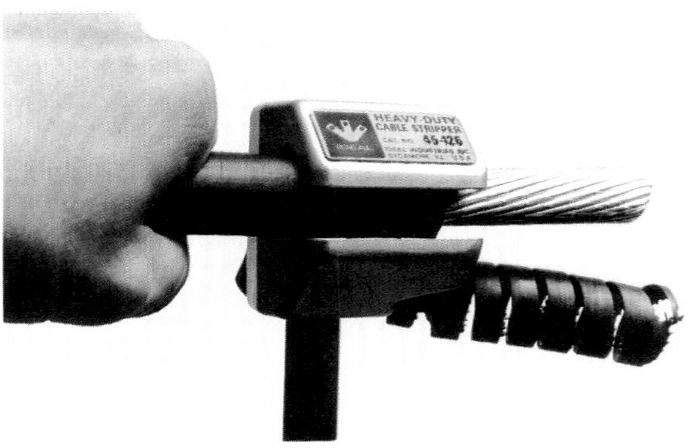

Ideal Industries, Inc.

Removes insulation from heavy-duty cables.

CABLE STRIPPER

Figure 1-6 continued . . .

. . . MISCELLANEOUS TOOLS . . .

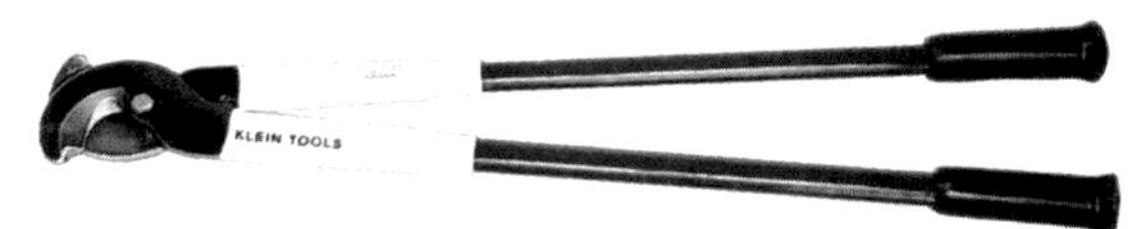

Provides shear-type cable cutting for relatively large diameter cables. Makes a clean, even cut for ease in fitting lugs and terminals.

Klein Tools, Inc.

CABLE CUTTER

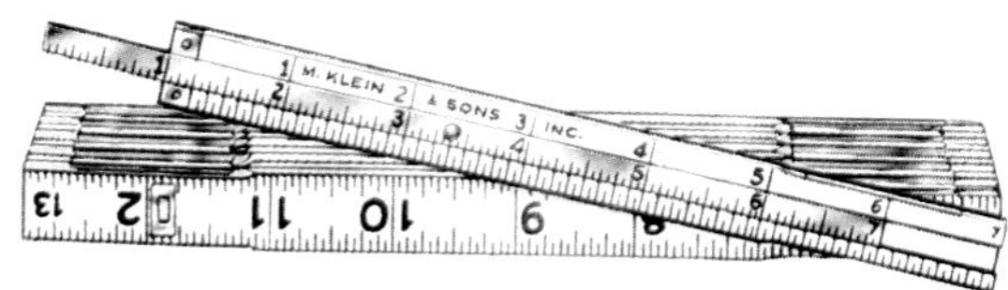

Klein Tools, Inc.

Works better than tapes for measuring layout work.

FOLDING RULE

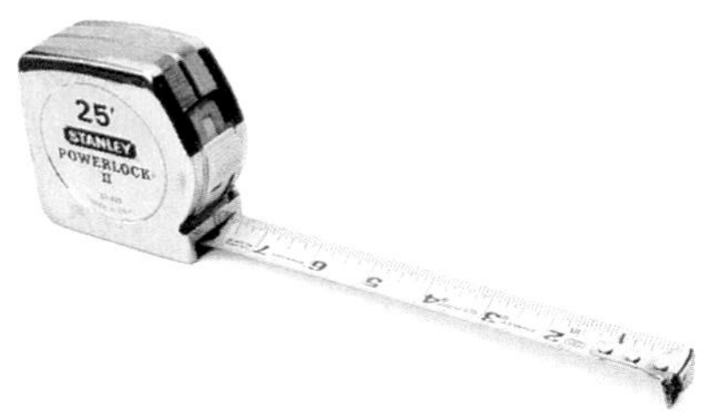

Used for rapid layout in measurements. Should be as wide as possible for easy extension.

POWER RETURN STEEL TAPE

Milwaukee Electric Tool Corporation

Ideal for drilling concrete to install fasteners.

HEAVY-DUTY HAMMER DRILL

Useful in leveling control panels and conduit bends.

POCKET TORPEDO LEVEL

Used to level long conduit runs and bus bar installations.

24-INCH LEVEL

Greenlee Textron Inc.

Useful to pull wire through conduit and "fish" wires around obstructions in walls.

FISH TAPE WITH HOLDER

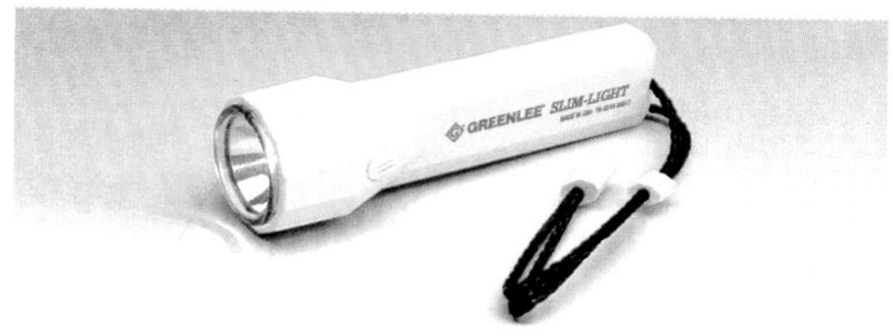

Greenlee Textron Inc.

Used for emergency lighting and inspection.

FLASHLIGHT

Figure 1-6 continued . . .

. . . MISCELLANEOUS TOOLS . . .

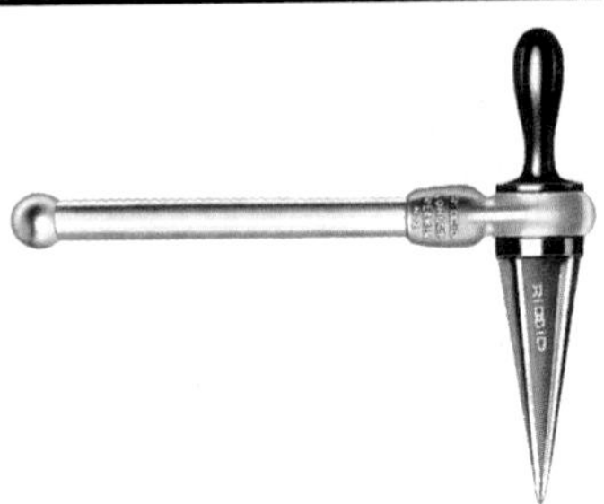

Ridge Tool Company

Used to deburr/remove rough edges from inside conduit.

REAMING TOOL

Greenlee Textron Inc.

Used to pull larger cables and wires into place.

POWER WIRE PULLER

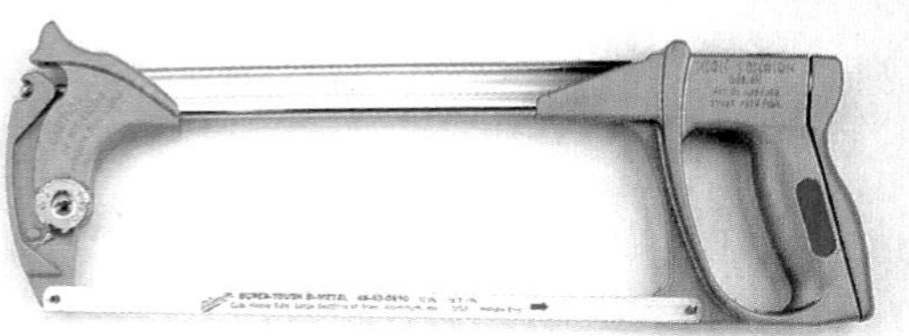

Milwaukee Electric Tool Corporation

Cuts heavy cable, pipe, and conduit.

HACKSAW

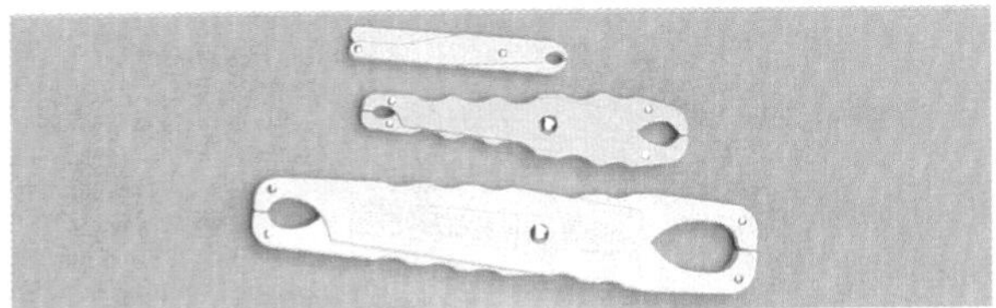

Greenlee Textron Inc.

Safely removes cartridge type fuses.

FUSE PULLER

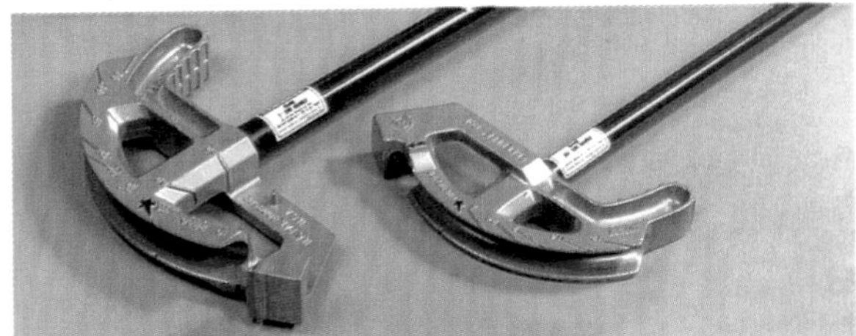

Greenlee Textron Inc.

Used to bend rigid conduit into a variety of shapes. Available for different sized conduits.

RIGID CONDUIT HICKEY

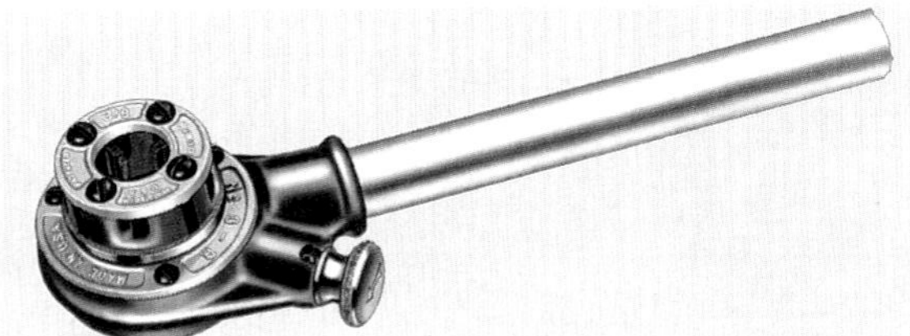

Ridge Tool Company

Used to thread rigid conduit pipe at a variety of locations on the job site.

HAND THREADER

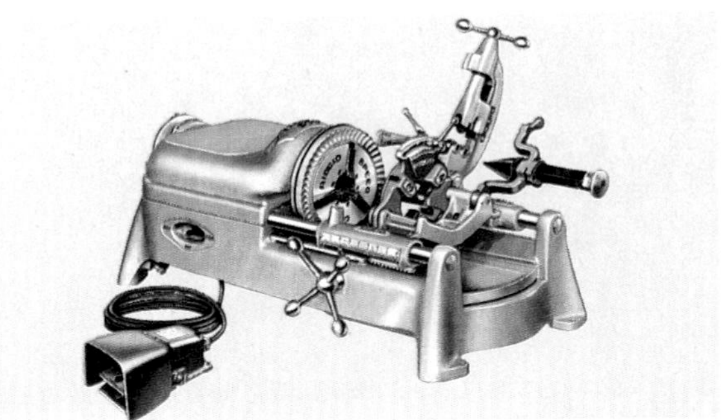

Ridge Tool Company

Used to thread rigid conduit pipe at a more centralized location.

POWER THREADER WITH STAND

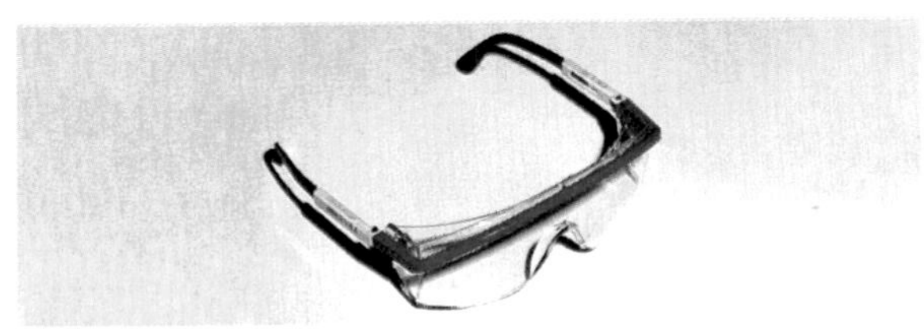

Protects eyes from flying debris. Required on all construction sites.

SAFETY GLASSES

Figure 1-6 continued . . .

. . . MISCELLANEOUS TOOLS

Greenlee Textron Inc.

Punches holes in metal enclosures and cabinets through the use of wrenches or sockets.

KNOCK OUT PUNCH

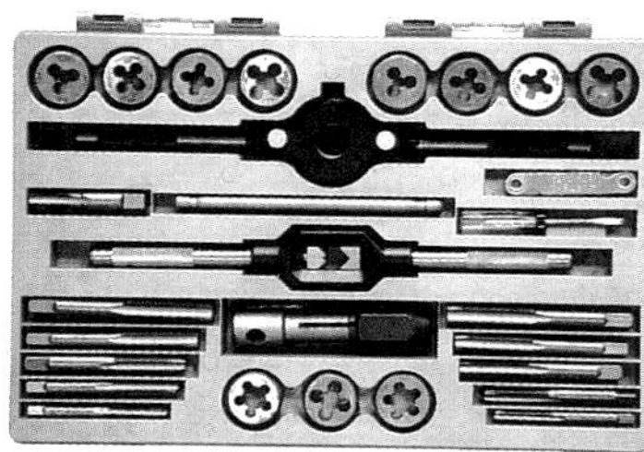

Provides threading for control panel mounting, bolts, nuts, and steel rod.

TAP AND DIE SET

Greenlee Textron Inc.

Allows for wires to be pulled more easily through conduit when applied to the outside of electrical wires.

WIRE LUBRICANT

Protects against impact, chemicals, and high voltage. Required on all construction sites.

SAFETY HARD HAT

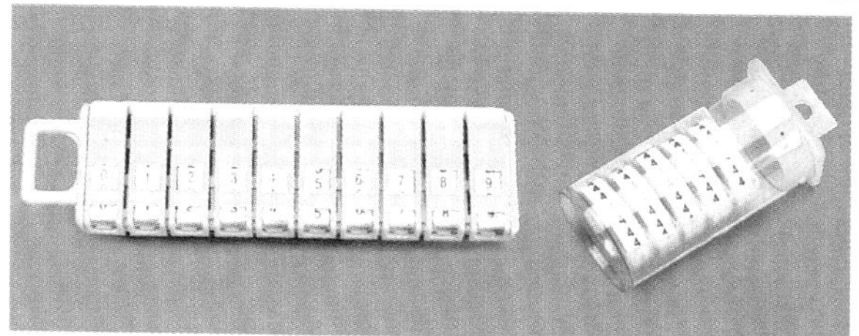

Greenlee Textron Inc.

Properly identifies wires and matching terminal ends.

WIRE MARKERS

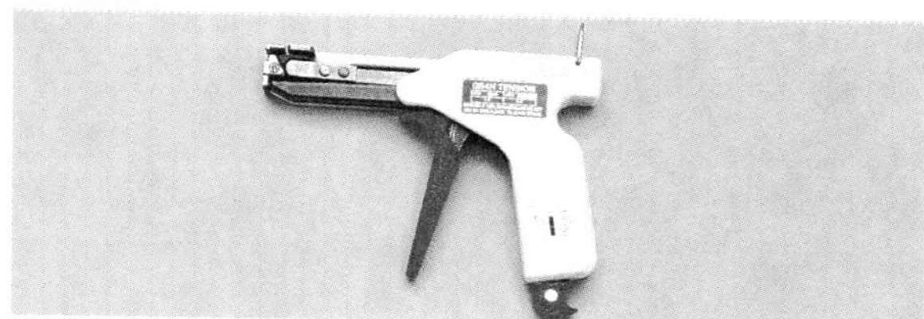

Panduit Corp.

Secures wires in wire runs and harnesses.

TIE-RAP GUN

Provides insulation to electrical connections once repairs and/or installation have been made.

ELECTRICIAN'S TAPE

Provides lockout protection for electrical equipment. It may also be used on a tool box.

PADLOCK

Figure 1-6. Other tools are used by electricians when doing electrical work.

Tools must not be damaged by abuse. An organized tool system provides both a central location and a means of protection for tools.

Electrical tools can be organized in several ways depending on where and how frequently they are used. A pegboard may be appropriate if the tools are used at a repair bench. An electrician's leather pouch may be used if tools are used only at a construction site. A portable toolbox is normally best when tools are used at a bench and on a jobsite.

Pegboards. Pegboard is available in 4′ × 8′ sheets at most lumber yards. Heavy-duty tempered board is normally best for tools. Once the pegboard is mounted, outlines of the tools can be made to maintain tool inventory. The outlines may be painted on the board or cut out of self-adhesive vinyl paper.

Tool Pouches. Tool pouches are used to safely transport and store many small electrical hand tools and instruments. A tool pouch is normally made of heavy-duty leather. Pouches vary in design and size. Tool pouches are chosen based on specific needs and comfort. Some pouches hold only a few tools, while others hold a wide selection. The pouch selected depends on the work required. See Figure 1-7.

TOOL POUCH

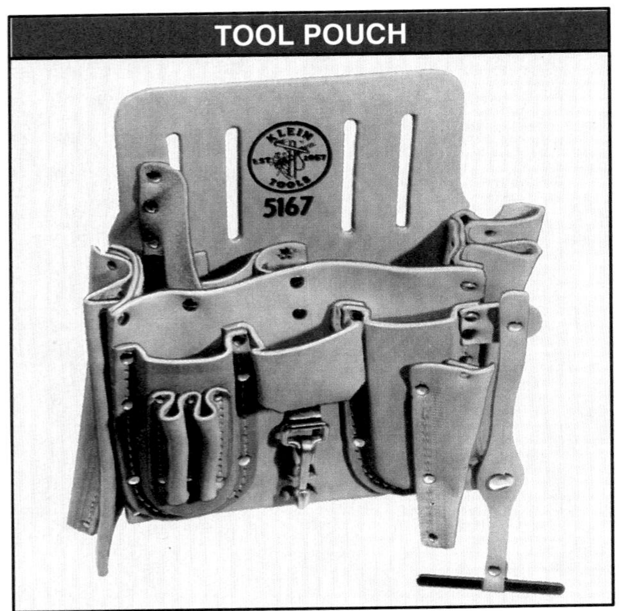

Klein Tools, Inc.

Figure 1-7. Tool pouches are used to conveniently and safely transport and store many small electrical hand tools and instruments.

Toolboxes are used to transport, store, and secure electrical tools and instruments.

Toolboxes, Chests, and Cabinets. Toolboxes, chests, and cabinets are used by many electricians to store, organize, and carry tools. A well-designed toolbox can be locked and helps keep tools clean and dry. Toolboxes may have a lift-out tray or may contain lever-operated trays that open automatically as the cover is lifted.

Tool chests are more substantial than toolboxes. A tool chest may have from two to 10 drawers. A tool cabinet is used to organize and store a variety of tools and materials in one location. Tool cabinets are always mounted on casters. Tool chests can be added on top of the cabinet. A list of all tools should be kept in a toolbox, chest, or cabinet to ensure a complete inventory after each job. See Figure 1-8.

An organized tool system ensures that clean, dry tools can be found when and where they are needed. Several manufacturers produce chemicals which can be placed in a toolbox, chest, or cabinet to keep the tools from rusting.

Tool Safety

The proper tool must be used for the job. Tool safety requires tool knowledge. The correct size must be used. Use good quality tools and use them for the job they were designed to accomplish.

Hand Tools. Learn how to use the tools properly. Learn the safe way of working with each tool. Do not force a tool or use tools beyond their capacity. The right tool does the job faster and safer. The time spent looking for the right tool and the expense of purchasing it are less costly than a serious accident.

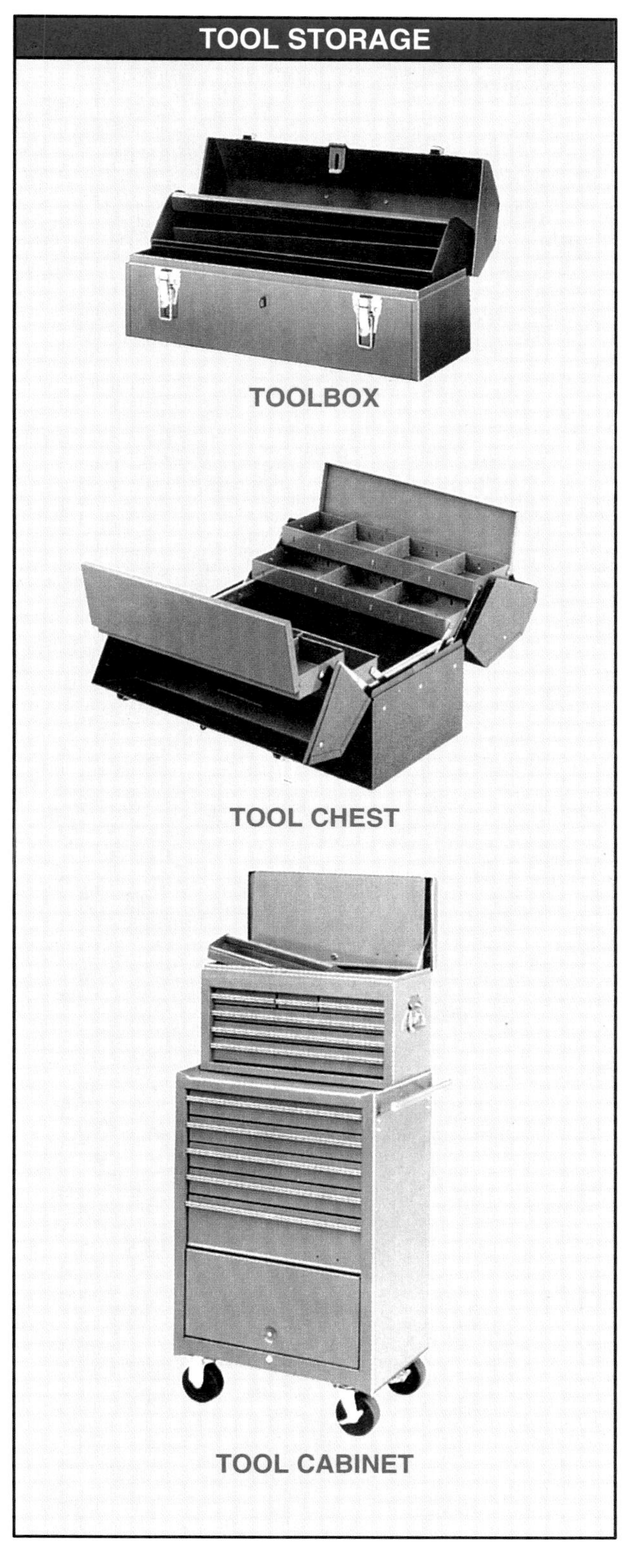

Klein Tools, Inc.

Figure 1-8. Toolboxes, chests, and cabinets are used by many electricians to store, organize, and carry tools.

Keep tools in good condition. Periodic checks on tools help to keep them in good condition. Always inspect a tool before using it. Do not use a tool which is in poor or faulty condition. Use only safe tools. Tool handles should be free of cracks and splinters and should be fastened securely to the working part. Damaged tools are dangerous and less productive than those in good working condition. Repair or replace a tool immediately when inspection shows a dangerous condition.

Cutting tools should be sharp and clean. Dull tools are dangerous. The extra force exerted in using dull tools often results in losing control of the tool. Dirt or oil on a tool may cause it to slip while being used and cause injury.

Keep tools in a safe place. Even good tools can be dangerous when left in the wrong place. Many accidents are caused by tools falling off ladders, shelves, and scaffolds that are being moved.

Each tool should have a designated place in a toolbox, chest, or cabinet. Do not carry tools in pockets unless the pocket is designed for that tool. Keep pencils in the pocket designed for them.

Keep sharp-edged tools away from the edge of a bench or work area. Brushing against the tool may cause it to fall and injure a leg or foot. Carry edged and sharply pointed tools with the cutting edge or point down and outward from the body.

Power Tools. Do not attempt to use any power tools without knowing their principles of operation, methods of use, and safety precautions. Obtain authorization from supervisors before using power tools.

All power tools should be grounded unless they are approved double-insulated. Power tools must have a 3-wire conductor cord. A three-prong plug connects into a grounded electrical outlet (receptacle). Approved receptacles may be locking or nonlocking. Consult OSHA (Occupational Safety Health Administration) and local codes for proper grounding specifications. See Figure 1-9.

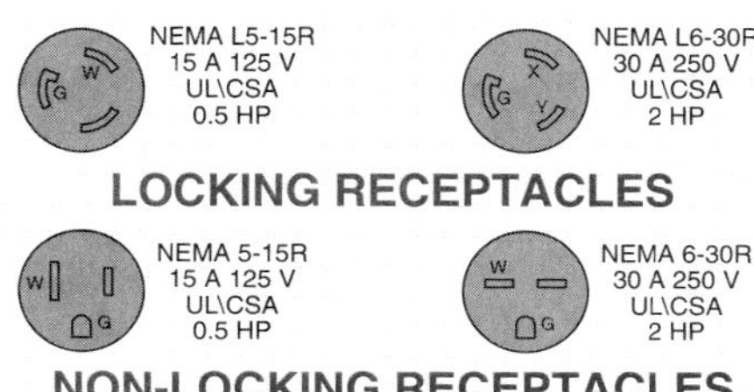

Figure 1-9. Approved receptacles may be locking or non-locking.

It is dangerous to use an adapter to plug a three-prong plug into a two-hole outlet unless a separate ground wire or strap is connected to an approved ground. The ground ensures that any short circuit trips the circuit breaker or blows the fuse. **Warning:** An ungrounded power tool can kill.

Double-insulated tools have two prongs and have a notation on the specification plate that they are double insulated. Electrical parts in the motor of a double-insulated tool are surrounded by extra insulation to help prevent electrical shock. For this reason, the tool does not have to be grounded. Both the interior and exterior should be kept clean of grease and dirt that might conduct electricity.

Safety rules must be followed when using power tools. Electrical power tool safety rules include:

- Know and understand all manufacturer's safety recommendations
- Read the owner's manual before using the tool
- Ensure that all safety guards are properly in place and in working order
- Wear safety goggles and a dust mask when required
- Ensure that the material to be worked is free of obstructions and securely clamped
- Ensure that the switch is in the OFF position before connecting a tool to a power source
- Keep attention focused on the work
- A change in sound during tool operation normally indicates trouble. Investigate immediately
- Power tools should be inspected and serviced by a qualified repair person at regular intervals as specified by the manufacturer or by OSHA
- Inspect electrical cords to see that they are in good condition
- Shut OFF the power when work is completed. Wait until all movement of the tool stops before leaving a stationary tool or laying down a portable tool
- Clean and lubricate all tools after use
- Remove all defective power tools from service. Alert others to the situation
- Take extra precautions when working on damp or wet surfaces. Use additional insulation to prevent any body part from coming into contact with a wet or damp surface
- Always work with at least one other coworker in hazardous or dangerous locations

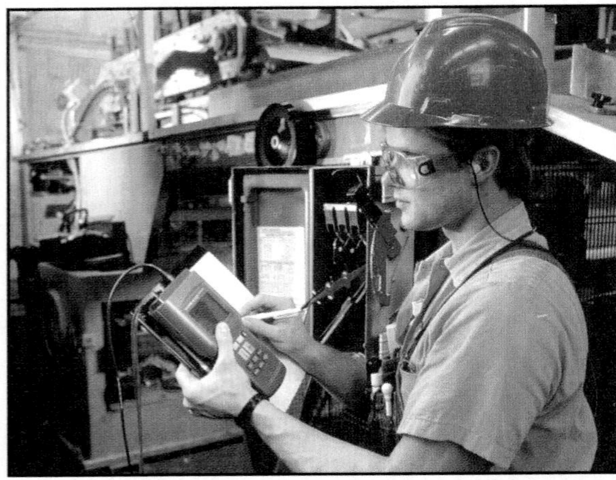

Fluke Corporation

Proper test and personal safety equipment must be used when troubleshooting live circuits.

ELECTRICAL INSTRUMENTS

Electrical instruments are used by electricians to aid in the taking of various electrical measurements. Electrical instruments include scopes, meters, probes, testers, etc. See Figure 1-10. Always consult the owner's manual and fully understand all functions of an instrument before using it. Care must be taken when using electrical instruments because damage to the instrument or personal injury may occur.

Reading Analog Displays

An *analog display* is an electromechanical device that indicates readings on a meter by the mechanical motion of a pointer. Analog displays use scales to display measured values. Analog scales may be linear or nonlinear. A *linear scale* is a scale that is divided into equally spaced segments. A *nonlinear scale* is a scale that is divided into unequally spaced segments. See Figure 1-11.

Analog scales are divided using primary divisions, secondary divisions, and subdivisions. A primary division is a division with a listed value. A secondary division is a division that divides primary divisions in halves, thirds, fourths, fifths, etc. A subdivision is a division that divides secondary divisions in halves, thirds, fourths, fifths, etc.

ELECTRICAL INSTRUMENTS . . .

Greenlee Textron Inc.

Measures AC/DC voltages up to 600 V.

VOLTAGE TESTER

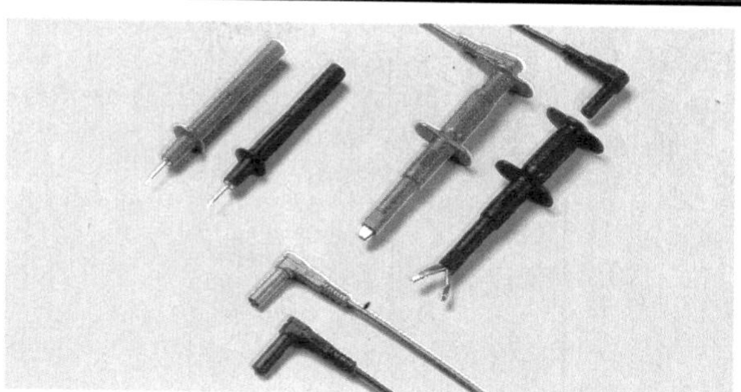

Fluke Corporation

Used with test equipment and control circuits in isolation problems.

TEST LEAD SET

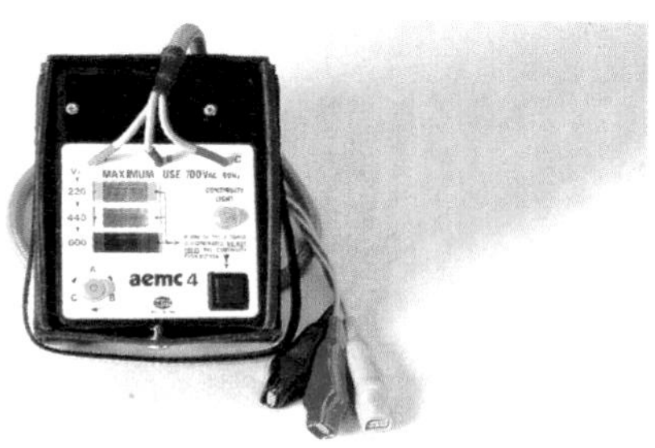

AEMC Instruments

Checks phase sequence (rotation) on the incoming line, voltage up to 600 VAC, and continuity.

PHASE SEQUENCE, VOLTAGE, AND CONTINUITY TESTER

Greenlee Textron Inc.

Verifies proper installation of 3-wire outlets.

POLARIZED RECEPTACLE TESTER

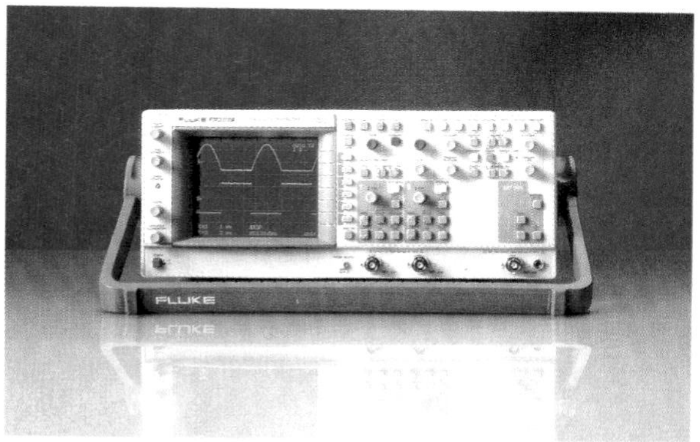

Fluke Corporation

Troubleshoots and verifies outputs and inputs of electronic devices.

OSCILLOSCOPE

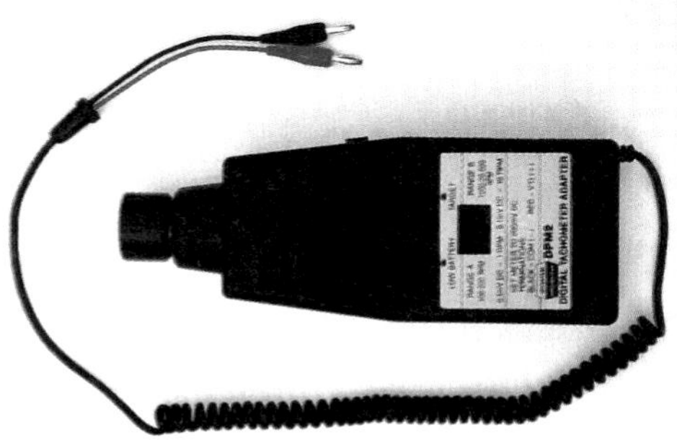

Used with a multimeter to record the rotational speed of motors.

DIGITAL TACHOMETER ADAPTER

Figure 1-10 continued . . .

. . . ELECTRICAL INSTRUMENTS

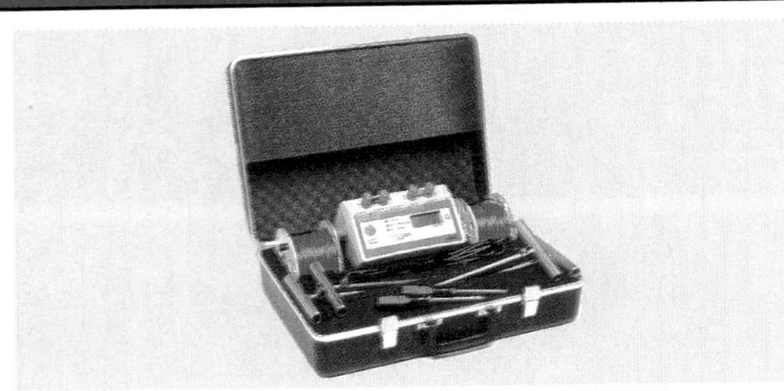

AEMC Instruments

Tests resistance of manufactured grounds and solid resistivity for corrosion control.

GROUND RESISTANCE TESTER

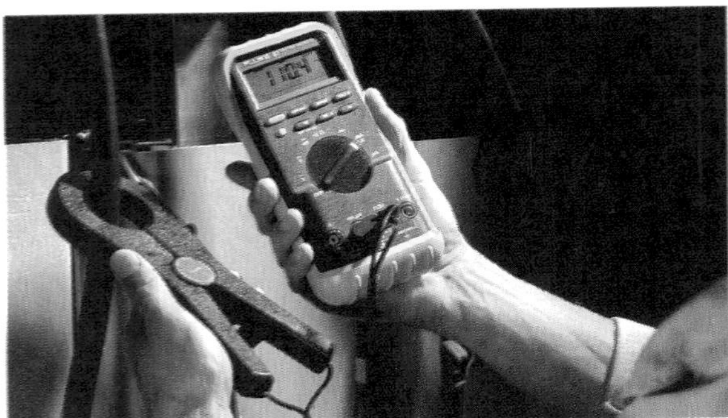

Fluke Corporation

Measures voltage, current, and resistance. Ammeter attachment allows AC/DC current to be measured without breaking circuit.

INDUSTRIAL VOM WITH CLAMP-ON AMMETER

Greenlee Textron Inc.

Determines proper phasing of 3ϕ systems.

PHASE SEQUENCE INDICATOR

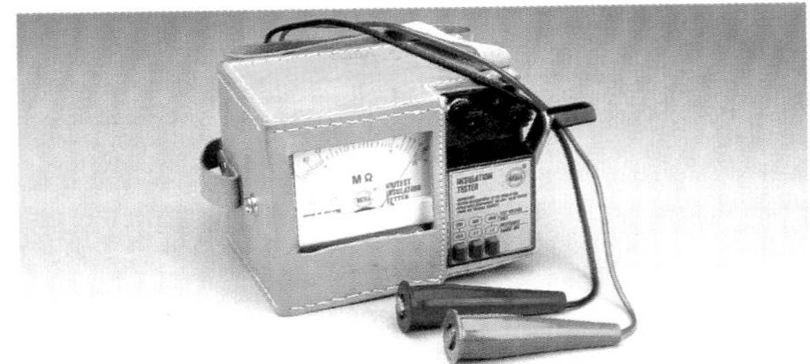

Greenlee Textron Inc.

Tests the insulation properties of motors, transformers, cables, generators, and related equipment.

MEGOHMMETER

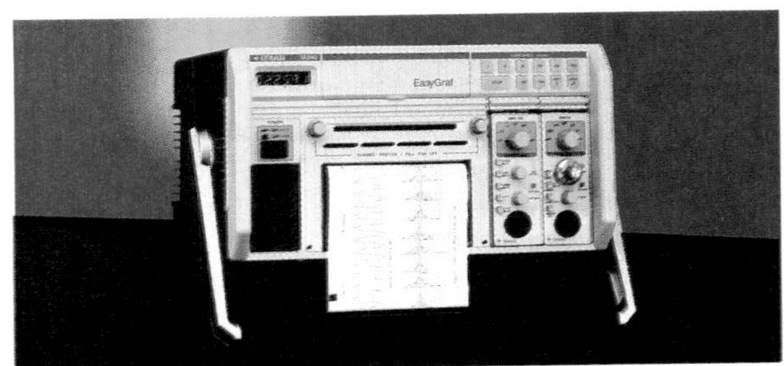

Gould Instrument Systems, Inc.

Monitors voltage and current variations over extended periods of time.

STRIP CHART RECORDER

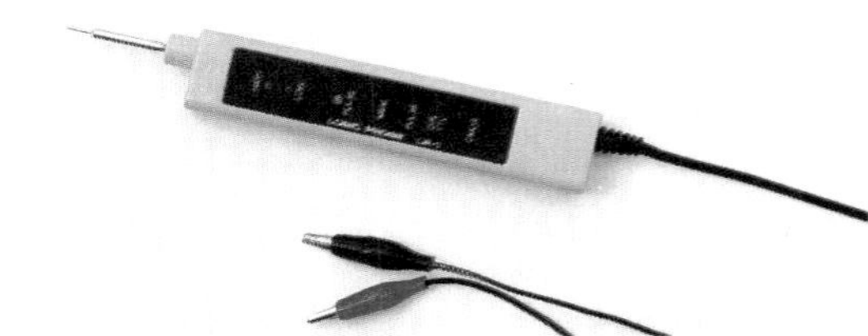

Troubleshoots and verifies outputs and inputs of digital logic circuits.

DIGITAL LOGIC PROBE

AEMC Instruments

Determines 3ϕ coil relationships and continuity in motors and gyro windings.

MOTOR POLARITY TESTER

Figure 1-10. Electrical instruments are used by electricians to aid in the taking of various electrical measurements.

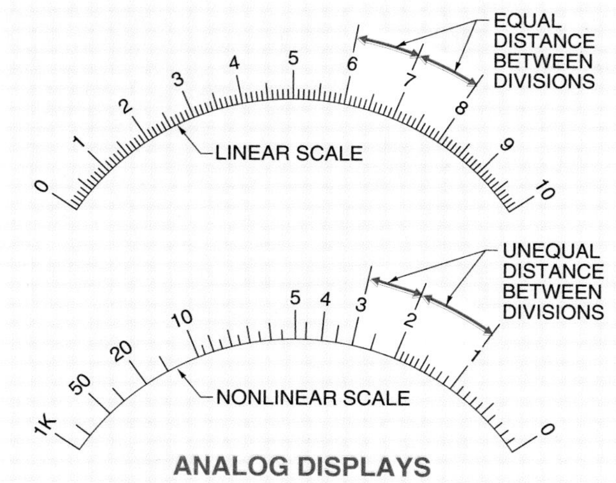

Figure 1-11. An analog display is an electromechanical device that indicates readings on a multimeter by the mechanical motion of a pointer.

Secondary divisions and subdivisions do not have listed numerical values. When reading an analog scale, add the primary, secondary, and subdivision readings. See Figure 1-12.

To read an analog scale, apply the procedure:

1. Read the primary division.
2. Read the secondary division if the pointer moves past a secondary division. *Note:* This may not occur with very low readings.
3. Read the subdivision if the pointer is not directly on a primary or secondary division. Round the reading to the nearest subdivision if the pointer is not directly on a subdivision. Round the reading to the next highest subdivision if rounding to the nearest subdivision is unclear.

Add the primary division, secondary division, and subdivision readings to obtain the analog reading.

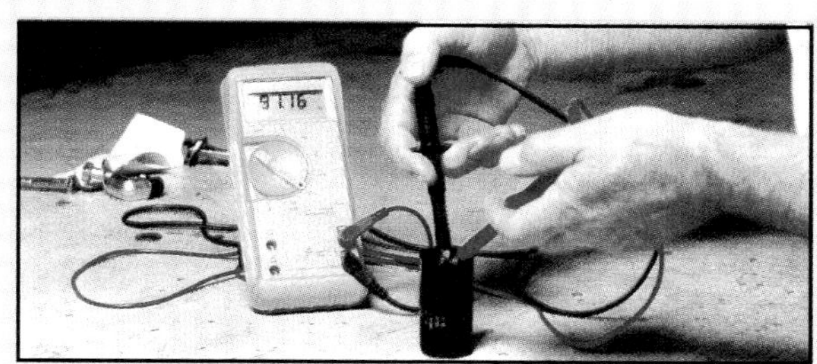

Fluke Corporation

Care must be taken when testing some components such as capacitors, because they store an electrical charge and can damage test equipment and/or cause an electrical shock.

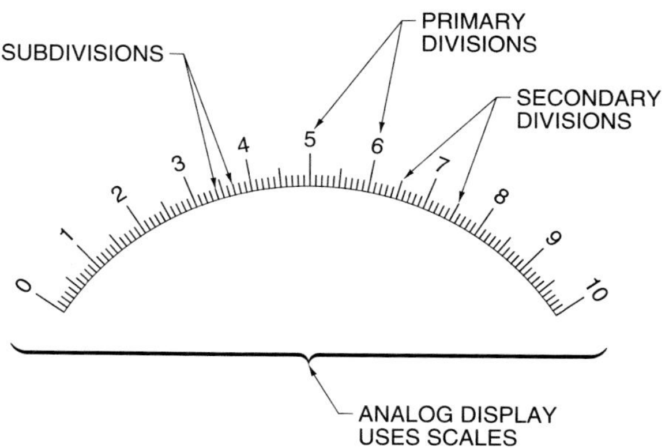

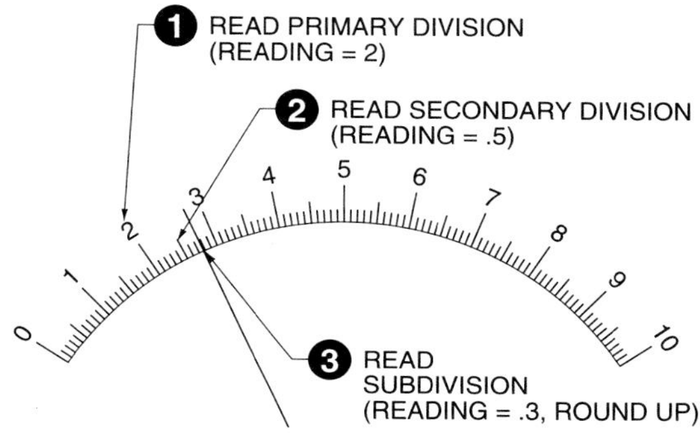

	READING
1 PRIMARY DIVISION	2
2 SECONDARY DIVISION	.5
3 SUBDIVISION	.3
METER READING	2.8

READING ANALOG SCALES

Figure 1-12. When reading an analog scale, add the primary, secondary, and subdivision readings.

Reading Digital Displays

A *digital display* is an electronic device that displays readings on a meter as numerical values. Digital displays help eliminate human error when taking readings by displaying exact values measured. Errors occur when reading a digital display if the displayed prefixes, symbols, and decimal points are not properly applied.

Digital displays display values using either a light-emitting diode (LED) or a liquid crystal display (LCD). LED displays are easier to read and use more power than LCD displays. Most portable digital meters use an LCD.

The exact value on a digital display is determined from the numbers displayed and the position of the decimal point. A range switch determines the placement of the decimal point.

Typical voltage ranges on a digital display are 3 V, 30 V, and 300 V. The highest possible reading with the range switch on 3 V is 2.999 V. The highest possible reading with the range switch on 30 V is 29.99 V. The highest possible reading with the range switch on 300 V is 299.9 V. Accurate readings are obtained by using the range that gives the best resolution without overloading. See Figure 1-13.

Bar Graph

A *graph* is a diagram that shows a variable in comparison to other variables. Most digital displays include a bar graph to show changes and trends in a circuit. A *bar graph* is a graph composed of segments that function as an analog pointer. The displayed bar graph segments increase as the measured value increases and decrease as the measured value decreases. Reverse the polarity of test leads if a negative sign is displayed at the beginning of the bar graph. See Figure 1-14.

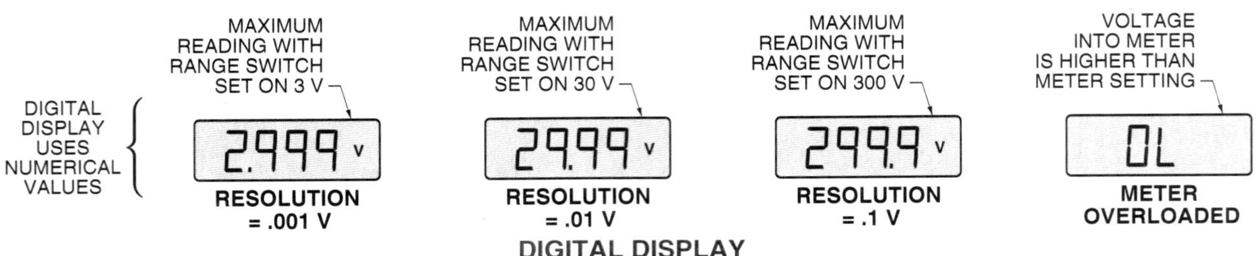

Figure 1-13. Digital displays display values using either a light-emitting diode (LED) or a liquid crystal display (LCD).

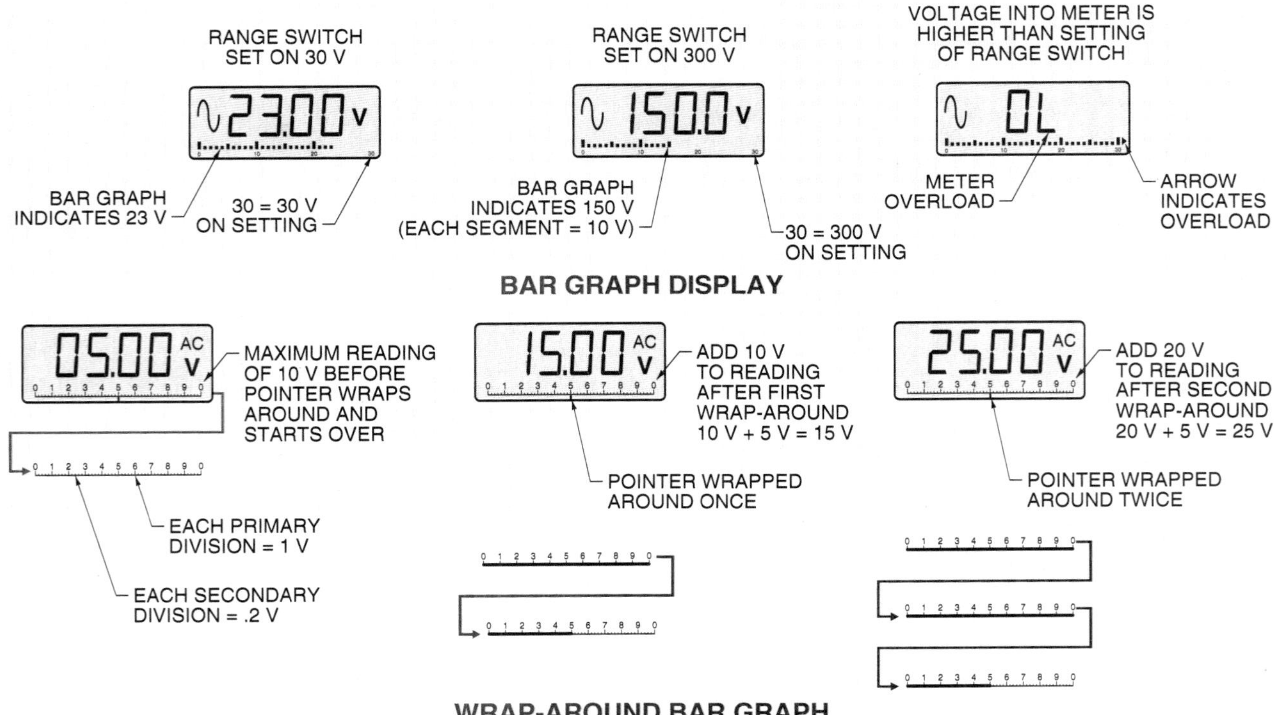

Figure 1-14. A bar graph is composed of segments that function as an analog pointer.

A *wrap-around bar graph* is a bar graph that displays a fraction of the full range on the graph. The pointer wraps around and starts over when the limit of the bar graph is reached.

Using a Bar Graph. A bar graph reading is updated 30 times per second. A digital reading is updated 4 times per second. The bar graph is used when quickly changing signals cause the digital display to flash or when there is a change in the circuit that is too rapid for the digital display to detect.

For example, mechanical relay contacts may bounce open when exposed to vibration. Contact bounce causes intermittent problems in electrical equipment. Frequency and severity of contact bounce increase as a relay ages.

A contact's resistance changes momentarily from zero to infinity and back when a contact bounces open. *Infinity* is an unlimited number or amount. A digital display cannot indicate contact bounce because most digital displays require more than 250 milliseconds (ms) to update their displays. The quick response of bar graphs enables detection of most contact bounce problems. The contact bounce is displayed by the movement of one or more segments the moment the contact opens.

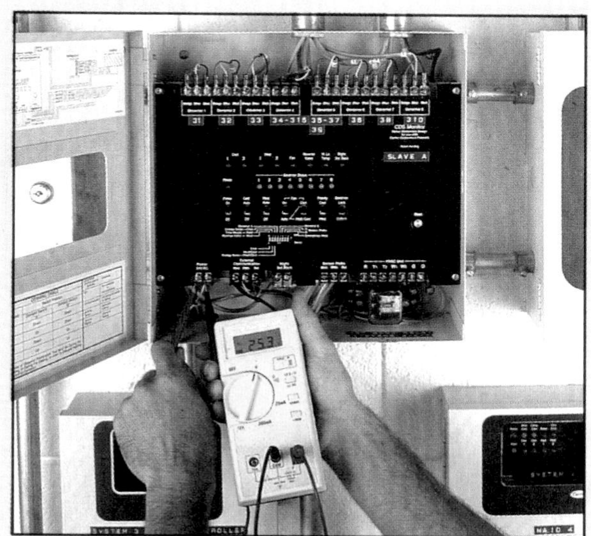

Greenlee Textron Inc.

The 93-3000 digital multimeter from Greenlee Textron Inc. measures AC/DC current and voltage, resistance, and continuity in electrical circuits.

Ghost Voltages

A meter set to measure voltage may display a reading before the meter is connected to a powered circuit. The displayed voltage is a ghost voltage that appears as changing numbers on a digital display or as a vibrating analog display. A *ghost voltage* is a voltage that appears on a meter not connected to a circuit.

Ghost voltages are produced by the magnetic fields generated by current-carrying conductors, fluorescent lighting, and operating electrical equipment. Ghost voltages enter a meter through the test leads because test leads not connected to a circuit act as antennae for stray voltages. See Figure 1-15.

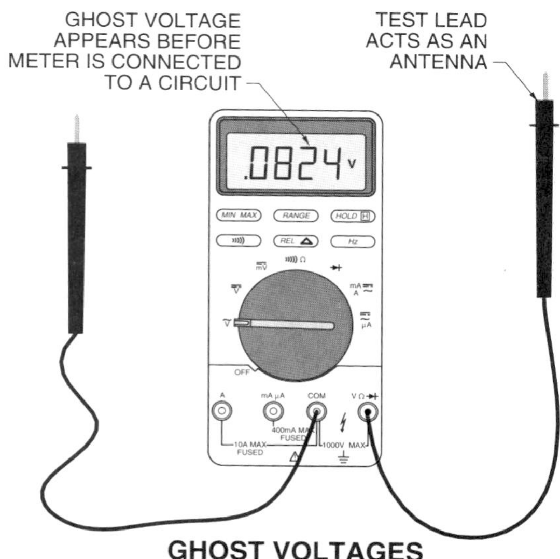

Figure 1-15. Ghost voltages are produced by the magnetic fields generated by current-carrying conductors, fluorescent lighting, and operating electrical equipment.

Ghost voltages do not damage a meter. Ghost voltages may be misread as circuit voltages when a meter is connected to a circuit that is believed to be powered. A circuit that is not powered can also act as an antenna for stray voltages. To ensure true circuit voltage readings, connect a meter to a circuit for a long enough time so that the meter displays a constant reading.

Informational Outputs

An *informational output* is any output device that displays data about the circuit or operation. Information must be monitored and maintained as systems, circuits, or processes increase in size and complexity.

Output devices such as gauges give a visual indication of what is currently happening. Output devices, such as recorders, maintain records of what has happened over a period of time.

Informational devices are classified as outputs because they can be used as the final output load in a circuit. These devices, however, are usually used along with other outputs, such as motors and heating elements. Informational outputs can also be used in any part of the circuit or system, including the input and decision sections.

Informational outputs for electrical quantities measure volts, amperes, watts, hertz, ohms, mhos, etc. See Figure 1-16.

Informational outputs for nonelectrical quantities measure pressure, speed, temperature, air velocity, pressure differential, relative humidity, etc. See Figure 1-17.

An automobile dashboard is an example of how informational outputs have increased as system complexity has increased. Old dashboards displayed automobile speed, engine temperature, battery condition, seat belt condition, and light status. Today's dashboards display all these conditions as well as outside temperature, estimated miles remaining, "lights suggested" indicator, "door open" indicator, plus extensive engine monitoring indicators that show when service is required.

INFORMATIONAL OUTPUTS – ELECTRICAL QUANTITIES

Device		Measures	Unit of Measure	Typical Uses
VOLTMETER	14.7 V	Amount of electrical pressure in a circuit	Volts (V)	Gives a voltage reading in applications such as battery chargers, power supplies, and power distribution systems
AMMETER	1.5 10 A	Amount of electron flow in a circuit	Amperes (A)	Indicates amount of current a load, circuit, or process is using
WATTMETER	16.8 W	Amount of electrical power in a circuit	Watts (W)	Gives a power reading in applications such as amplifiers, heating elements, and power distribution systems
FREQUENCY METER	57.1 Hz	Number of electrical cycles per second	Hertz (Hz)	Indicates AC power line frequency in applications such as variable-speed motor drives
OHMMETER	10.08 kΩ	Resistance of electrical flow	Ohms (Ω)	Indicates a load, circuit, or component resistance before power is applied to a circuit
CONDUCTIVITY METER	0 10 20 30 40 50 60 70 80 90 100 • HI LO •	Ability to conduct electricity	Mhos (usually mhos/cm), or microsiemens (usually μS/cm)	Measures the ability of a solution (usually water) to conduct electricity

Figure 1-16. Informational outputs for electrical quantities measure volts, amperes, watts, hertz, ohms, mhos, etc.

INFORMATIONAL OUTPUTS – NONELECTRICAL QUANTITIES				
Device		**Measures**	**Unit of Measure**	**Typical Uses**
PRESSURE GAUGE	500 1000 1500 2000 PSI	Amount of fluid (air or liquid) pressure in a system	Pounds per square inch (psi), kilograms per centimeter (kg/cm), or bar (bar)	Monitors HVAC systems, pollution control systems, fluid systems, and machine conditions
TACHOMETER	1000 RPM	Speed of a rotating object	Revolutions per minute (rpm)	Monitors speed of moving objects, motors, gears, engines, and machine parts
TEMPERATURE METER	478 °C	Amount of heat in an object or area	Degrees Fahrenheit (°F) or Degrees Celsius (°C)	Monitors temperature of products, machines, fluids, processes, and areas
ANEMOMETER	12 m/sec	Air velocity or force (distance traveled per unit of time)	Feet per minute (fpm) or meters per second (mps)	Monitors flow of air in HVAC systems
MANOMETER	0 1 2 3 4 5 6 7	Pressure differential between two points in a system	Inches water column (in H_2O), pounds per square inch (psi), or centimeters (cm)	Monitors pressure drop across air filters, dampers, and refrigeration coils
HYGROMETER	0 10 20 30 40 50 60 70 80 90 100 % RELATIVE HUMIDITY	Relative humidity	Percent (0% to 100%) relative humidity (%RH)	Monitors freezers, HVAC systems, storage bins, computer rooms, libraries, and warehouses
pH METER	7.08 pH	Acidity or alkalinity of a solution	0 pH to 14 pH	Monitors cooling towers, process steam, feedwater, pulp and paper operations, and wastewater treatment
VIBRATION METER	2.19 IN/SEC	Amount of imbalance in a machine or system	Velocity (in/s, cm/s) or displacement (in., mm)	Monitors machines and motors for excess vibration Indicates when a machine is not properly loaded or when alignment is required
FLOWMETER	CAL RATE ON TOTAL 3.67	Amount of fluid moving in a system	Gallons per minute (gpm) or standard cubic feet per minute (scfm)	Indicates that gas or liquid is moving and monitors rate of movement
COUNTER	654321	Number of devices moving past a given location	Numerical	Maintains production values, parts used, and inventory

Figure 1-17. Informational outputs for nonelectrical quantities measure pressure, speed, temperature, air velocity, pressure differential, relative humidity, etc.

Resistance Measurement

Ensure that no voltage is present in the circuit or component under test before taking resistance measurements. Low voltage applied to a meter set to measure resistance causes inaccurate readings. High voltage applied to a meter set to measure resistance causes meter damage. Check for voltage using a voltmeter. See Figure 1-18.

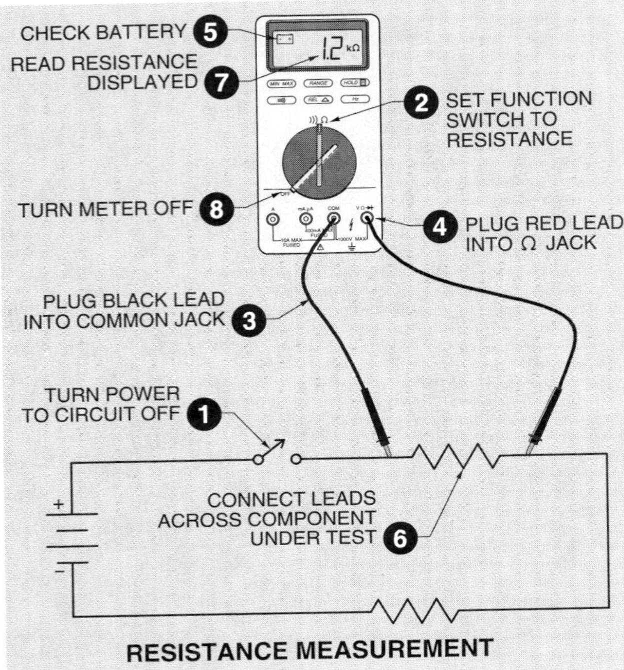

Figure 1-18. Ensure that no voltage is present in the circuit or component under test before taking resistance measurements.

To measure resistance using an ohmmeter, apply the procedure:

1. Check that all power is OFF in the circuit or component under test.
2. Set the function switch to the resistance position, which is marked Ω on digital meters.
3. Plug the black test lead into the common jack.
4. Plug the red test lead into the resistance jack.
5. Ensure that the meter batteries are in good condition. The battery symbol is displayed when the batteries are low. Digital meters are zeroed by an internal circuit.
6. Connect the meter test leads across the component under test. Ensure that contact between the test leads and the circuit is good. Dirt, solder flux, oil, and other foreign substances greatly affect resistance readings.
7. Read the resistance displayed on the meter. Check the circuit schematic for parallel paths. Parallel paths with the resistance under test cause reading errors. Do not touch exposed metal parts of the test leads during the test. Resistance of your body can cause reading errors.
8. Turn the meter OFF after measurements are taken to save battery life.

AC Voltage Measurement

Exercise caution when measuring AC voltages over 24 V. See Figure 1-19.

Warning: Ensure that no body part contacts any part of the live circuit, including the metal contact points at the tip of the test leads.

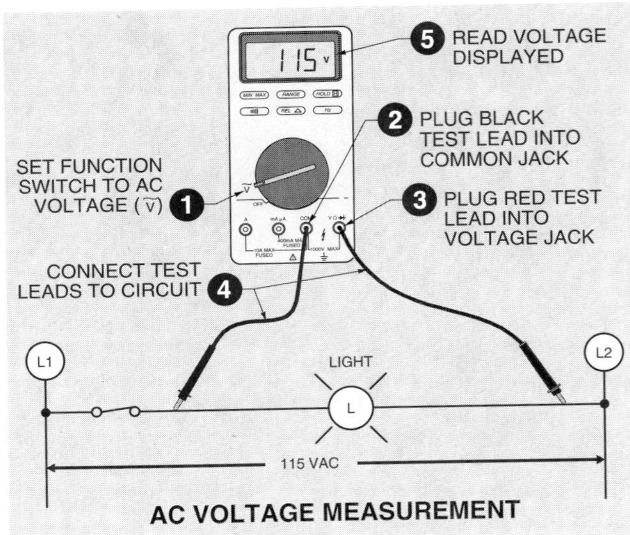

Figure 1-19. Exercise caution when measuring AC voltages over 24 V.

To measure AC voltages with a voltmeter, apply the procedure:

1. Set the function switch to AC voltage. Set the meter on the highest voltage setting if the voltage in the circuit is unknown.

2. Plug the black test lead into the common jack.
3. Plug the red test lead into the voltage jack.
4. Connect the test leads to the circuit. The position of the test leads is arbitrary. Common industrial practice is to connect the black test lead to the grounded (neutral) side of the AC voltage.
5. Read the voltage displayed on the meter.

DC Voltage Measurement

Exercise caution when measuring DC voltages over 60 V. See Figure 1-20.

Warning: Ensure that no body part contacts any part of a live circuit, including the metal contact points at the tip of the test leads.

To measure DC voltages with a voltmeter, apply the procedure:

1. Set the function switch to DC voltage. Select a setting high enough to measure the highest possible circuit voltage if the meter has more than one voltage position or if the circuit voltage is unknown.
2. Plug the black test lead into the common jack.
3. Plug the red test lead into the voltage jack.
4. Discharge any capacitors.
5. Connect the meter test leads in the circuit. Connect the black test lead to circuit ground and the red test lead to the point at which the voltage is under test. Reverse the black and red test leads if a negative sign appears in front of the reading on a digital meter.
6. Read the voltage displayed on the meter.

Fluke Corporation

A Model 36 true rms clamp meter from Fluke Corporation is used when measuring high currents.

Measuring Current – Clamp-On Ammeters

Clamp-on ammeters take current readings without opening a circuit. The jaws of a clamp-on ammeter are opened and encircled around the conductor under test. Readings are taken on bare or insulated conductors. When the jaws close, a current reading is indicated on the meter's scale. The reading indicates the amount of current drawn by loads connected to the conductor.

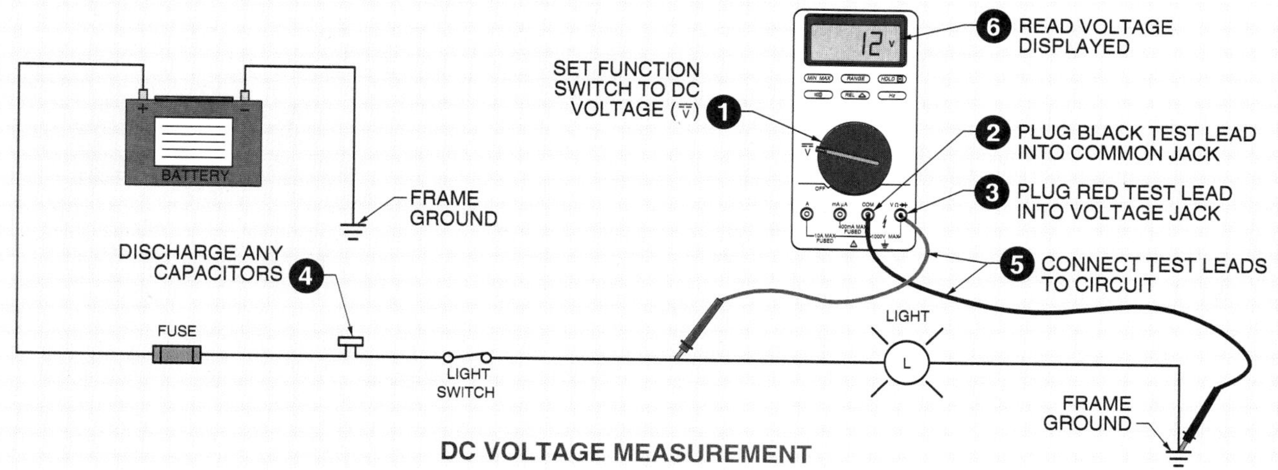

Figure 1-20. Exercise caution when measuring DC voltages over 60 V.

Clamp-On Ammeter – AC Measurement

Clamp-on ammeters measure the current in a circuit by measuring the strength of the magnetic field around a conductor. Care must be taken to ensure that the meter does not pick up stray magnetic fields. Whenever possible, separate the conductors under test from other surrounding conductors by a few inches. See Figure 1-21.

To measure AC using a clamp-on ammeter, apply the procedure:

1. Set the function switch to AC current. Select the proper setting to measure the highest possible circuit current if the meter has more than one current position or if the circuit current is unknown.
2. Plug the current probe accessory into the meter when using a multimeter that requires a current probe.
3. Open the jaws by pressing against the trigger.
4. Enclose one conductor in the jaws. Ensure that the jaws are completely closed before taking readings.
5. Read the current displayed on the meter.

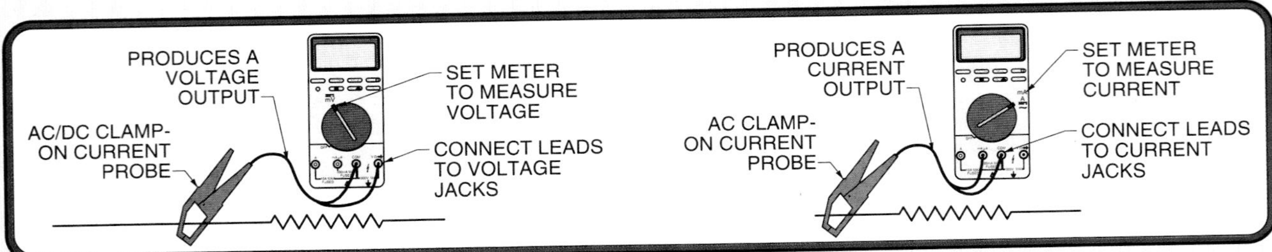

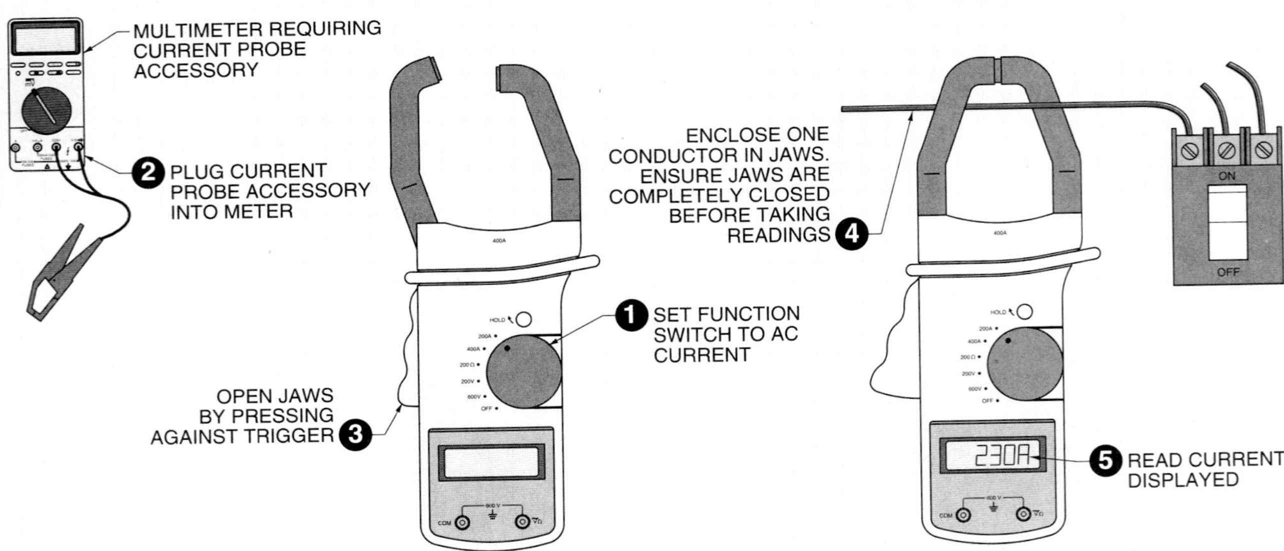

CLAMP-ON AMMETER – AC MEASUREMENT

Figure 1-21. Clamp-on ammeters measure the current in a circuit by measuring the strength of the magnetic field around a conductor.

ELECTRICAL SAFETY

OSHA and state safety laws have helped to provide safe working areas for electricians. Individuals can work safely on electrical equipment with today's safeguards and recommended work practices. In addition, an understanding of the principles of electricity is gained. Ask supervisors when in doubt about a procedure. Report any unsafe conditions, equipment, or work practices as soon as possible.

Fuses

Before removing any fuse from a circuit, be sure the switch for the circuit is open or disconnected. When removing fuses, use an approved fuse puller and break contact on the hot side of the circuit first. When replacing fuses, install the fuse first into the load side of the fuse clip, then into the line side.

GFCIs

A *ground fault circuit interrupter (GFCI)* is an electrical device which protects personnel by detecting potentially hazardous ground faults and quickly disconnecting power from the circuit. A potentially dangerous ground fault is any amount of current above the level that may deliver a dangerous shock. Any current over 8 mA is considered potentially dangerous depending on the path the current takes, the amount of time exposed to the shock, and the physical condition of the person receiving the shock.

Therefore, GFCIs are required in such places as dwellings, hotels, motels, construction sites, marinas, receptacles near swimming pools and hot tubs, underwater lighting, fountains, and other areas in which a person may experience a ground fault.

A GFCI compares the amount of current in the ungrounded (hot) conductor with the amount of current in the neutral conductor. If the current in the neutral conductor becomes less than the current in the hot conductor, a ground fault condition exists. The amount of current that is missing is returned to the source by some path other than the intended path (fault current). A fault current as low as 4 mA to 6 mA activates the GFCI and interrupts the circuit. Once activated, the fault condition is cleared and the GFCI manually resets before power may be restored to the circuit. See Figure 1-22.

GFCI protection may be installed at different locations within a circuit. Direct-wired GFCI receptacles provide a ground fault protection at the point of installation. GFCI receptacles may also be connected to provide GFCI protection at all other receptacles installed downstream on the same circuit. GFCI CBs, when installed in a load center or panelboard, provide GFCI protection and conventional circuit overcurrent protection for all branch-circuit components connected to the CB.

Plug-in GFCIs provide ground fault protection for devices plugged into them. These plug-in devices are often used by personnel working with power tools in an area that does not include GFCI receptacles.

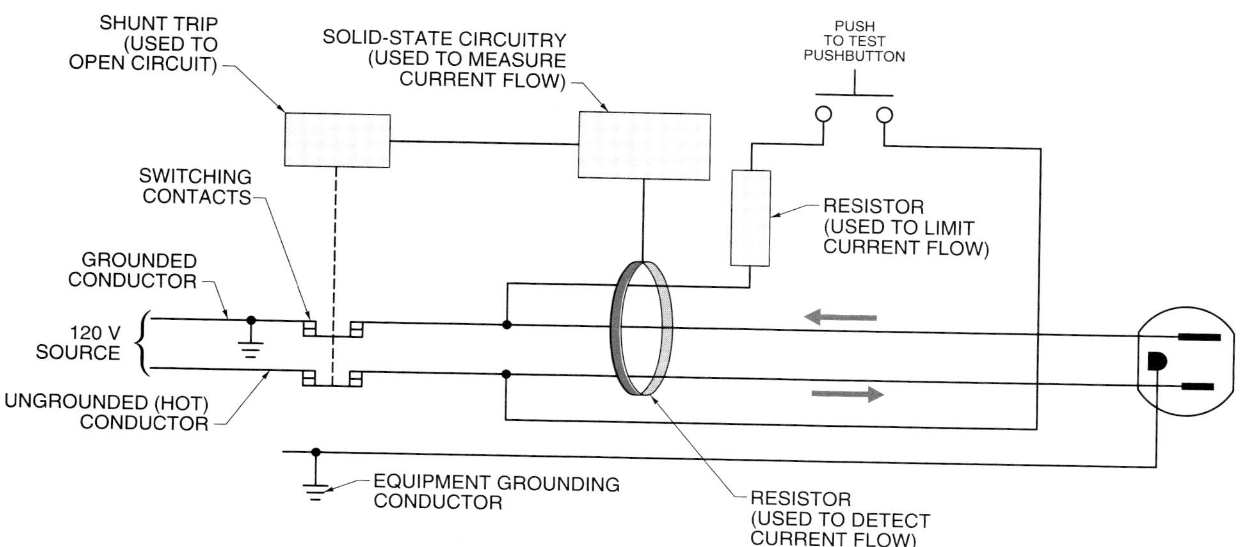

Figure 1-22. A GFCI compares the amount of current in the ungrounded (hot) conductor with the amount of current in the neutral conductor.

Electrical Shock

Electrical shock occurs when a person comes in contact with two conductors of a circuit or when the body becomes part of the electrical circuit. In either case, a severe shock can cause the heart and lungs to stop functioning. Also, severe burns may occur where current enters and exits the body.

Prevention is the best medicine for electrical shock. Respect all voltages, have a knowledge of the principles of electricity, and follow safe work procedures. Do not take chances. All electricians should be encouraged to take a basic course in CPR (cardiopulmonary resuscitation) so they can aid a coworker in emergency situations.

Always make sure portable electric tools are in safe operating condition. Make sure there is a third wire on the plug for grounding in case of shorts. The fault current should flow through the third wire to ground instead of through the operator's body to ground if electric power tools are grounded and if an insulation breakdown occurs.

Lockout/Tagout

Electrical power must be removed when electrical equipment is inspected, serviced, or repaired. To ensure the safety of personnel working with the equipment, power is removed and the equipment must be locked out and tagged out.

Per OSHA standards, equipment is locked out and tagged out before any preventive maintenance or servicing is performed. *Lockout* is the process of removing the source of electrical power and installing a lock which prevents the power from being turned ON. *Tagout* is the process of placing a danger tag on the source of electrical power which indicates that the equipment may not be operated until the danger tag is removed. See Figure 1-23.

A danger tag has the same importance and purpose as a lock and is used alone only when a lock does not fit the disconnect device. The danger tag shall be attached at the disconnect device with a tag tie or equivalent and shall have space for the worker's name, craft, and other required information. A danger tag must withstand the elements and expected atmosphere for as long as the tag remains in place. A lockout/tagout is used when:

- Servicing electrical equipment that does not require power to be ON to perform the service
- Removing or bypassing a machine guard or other safety device
- The possibility exists of being injured or caught in moving machinery
- Clearing jammed equipment
- The danger exists of being injured if equipment power is turned ON

Panduit Corp.

Figure 1-23. Equipment must be locked out and tagged out before preventive maintenance or servicing is performed.

Lockouts and tagouts do not by themselves remove power from a circuit. An approved procedure is followed when applying a lockout/tagout. Lockouts and tagouts are attached only after the equipment is turned OFF and tested to ensure that power is OFF. The lockout/tagout procedure is required for the safety of workers due to modern equipment hazards. OSHA provides a standard procedure for equipment lockout/tagout. OSHA's procedure is:

1. Prepare for machinery shutdown.
2. Machinery or equipment shutdown.
3. Machinery or equipment isolation.
4. Lockout or tagout application.
5. Release of stored energy.
6. Verification of isolation.

Warning: Personnel should consult OSHA Standard 29CFR1910.147 for industry standards on lockout/tagout.

A lockout/tagout shall not be removed by any person other than the person that installed it, except in an emergency. In an emergency, the lockout/tagout may be removed only by authorized personnel. The authorized personnel shall follow approved procedures. A list of company rules and procedures are given to any person that may use a lockout/tagout. Always remember:

- Use a lockout and tagout when possible
- Use a tagout when a lockout is impractical. A tagout is used alone only when a lock does not fit the disconnect device
- Use a multiple lockout when individual employee lockout of equipment is impractical
- Notify all employees affected before using a lockout/tagout
- Remove all power sources including primary and secondary
- Measure for voltage using a voltmeter to ensure that power is OFF

Lockout Devices. Lockout devices are lightweight enclosures that allow the lockout of standard control devices. Lockout devices are available in various shapes and sizes that allow for the lockout of ball valves, gate valves, and electrical equipment such as plugs, disconnects, etc.

Lockout devices resist chemicals, cracking, abrasion, and temperature changes. They are available in colors to match ANSI pipe colors. Lockout devices are sized to fit standard industry control device sizes. See Figure 1-24.

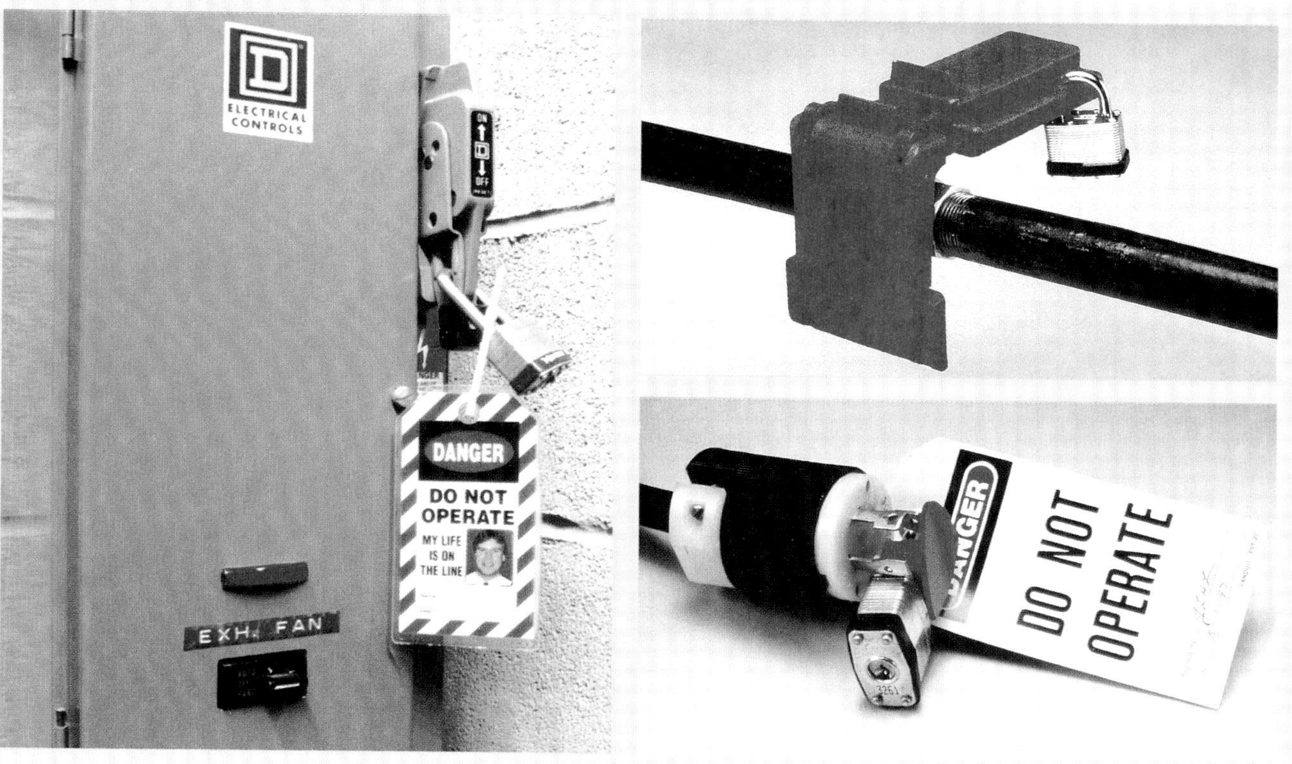

Panduit Corp.

Figure 1-24. Lockout devices are available in various shapes and sizes that allow for the lockout of standard control devices.

Locks used to lock out a device may be color-coded and individually keyed. The locks are rust-resistant and are available with various size shackles.

Danger tags provide additional lockout and warning information. Various danger tags are available. Danger tags may include warnings such as "Do Not Start," "Do Not Operate," or may provide space to enter worker, date, and lockout reason information. Tag ties must be strong enough to prevent accidental removal and must be self-locking and nonreusable.

Lockout/tagout kits are also available. A lockout/tagout kit contains items required to comply with the OSHA lockout/tagout standards. Lockout/tagout kits contain reusable danger tags, tag ties, multiple lockouts, locks, magnetic signs, and information on lockout/tagout procedures. See Figure 1-25. Be sure the source of electricity remains open or disconnected when returning to work whenever leaving a job for any reason or whenever the job cannot be completed the same day.

Figure 1-25. Lockout/tagout kits comply with OSHA lockout/tagout standards.

Clothing and Personal Protective Equipment

Clothing should fit snugly to avoid danger of becoming entangled in moving machinery or creating a tripping or stumbling hazard. See Figure 1-26.

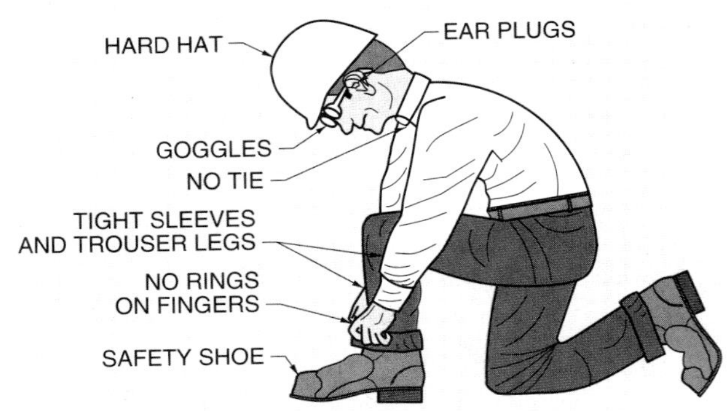

Figure 1-26. Clothing should fit snugly to avoid danger of becoming entangled in moving machinery or creating a tripping or stumbling hazard.

Recommended safe work clothes include:

- Thick-soled work shoes for protection against sharp objects such as nails. Wear work shoes with safety toes if the job requires. Make sure the soles are oil resistant if the shoes are subject to oils and grease
- Rubber boots for damp locations
- A hat or cap. Wear an approved safety helmet (hard hat) if the job requires

Confine long hair or keep hair trimmed and avoid placing the head in close proximity to rotating machinery. Do not wear jewelry. Gold and silver are excellent conductors of electricity.

FIRE SAFETY

The chance of fire is greatly decreased by good housekeeping. Keep rags containing oil, gasoline, alcohol, shellac, paint, varnish, or lacquer in a covered metal container. Keep debris in a designated area away from the building. Sound an alarm if a fire occurs. Alert all workers on the job and then call the fire department. After calling the fire department, make a reasonable effort to contain the fire.

Fire Extinguishers

Always read instructions before using a fire extinguisher. Always use the correct fire extinguisher for the class of fire. See Figure 1-27. Fire extinguishers are normally red. Fire extinguishers may be located on a red background so they can be easily located.

Be ready to direct firefighters to the fire. Inform them of any special problems or conditions that exist, such as downed electrical wires or leaks in gas lines.

Report any accumulations of rubbish or unsafe conditions that could be fire hazards. Also, if a portable tool bin is used on the job, a good practice is to store a CO_2 extinguisher in it.

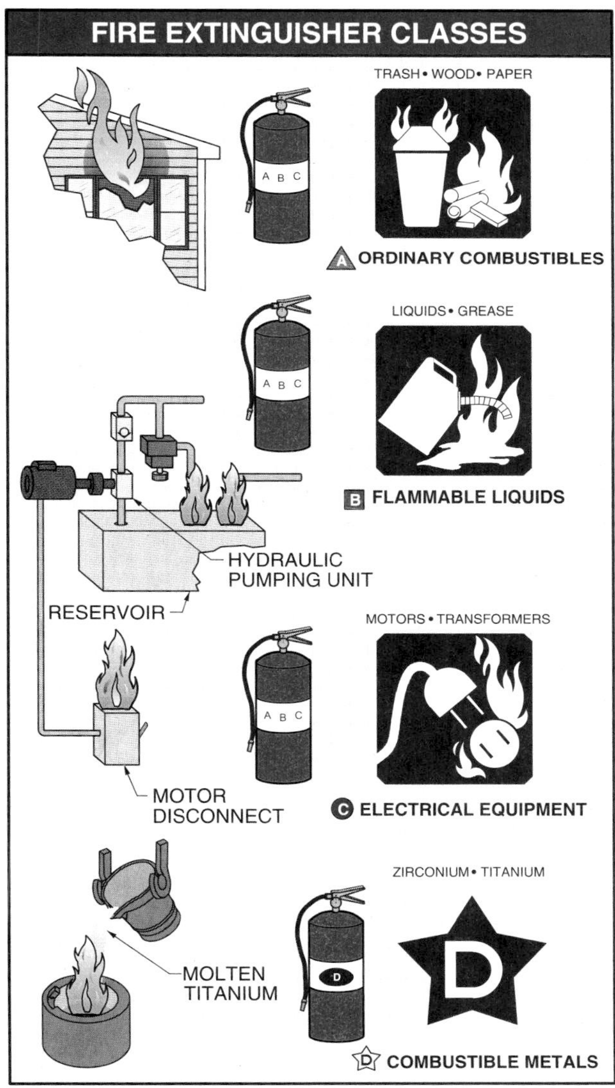

Figure 1-27. Always use the correct fire extinguisher for the class of fire.

In-Plant Training

A select group of personnel (if not all personnel) should be acquainted with all extinguisher types and sizes available in a plant or work area. Training should include a tour of the facility indicating special fire hazard operations.

In addition, it is helpful to periodically practice a dry run, discharging each type of extinguisher. Such practice is essential in learning how to activate each type, knowing the discharge ranges, realizing which types are affected by winds and drafts, familiarizing oneself with discharge duration, and learning of any precautions to take as noted on the nameplate.

Panduit Corp.

Panduit's Lockout/Tagout Training Program is designed to protect employees and help plants comply with OSHA lockout/tagout general training requirements.

Extinguisher Maintenance Tips

Inspect extinguishers at least once a month. It is common to find units that are missing, damaged, or used. Consider contracting for such a service. Contract for annual maintenance with a qualified service agency. Never attempt to make repairs to extinguishers. This is the chief cause of dangerous shell ruptures.

Hazardous Locations

The use of electrical equipment in areas where explosion hazards are present can lead to an explosion and fire. This danger exists in the form of escaped flammable gases such as naphtha, benzine, propane, and others. Coal, grain, and other dust suspended in air can also cause an explosion. Article 500 of The National Electrical Code® (NEC®) covers hazardous locations. Any hazardous location requires the maximum in safety and adherence to local, state, and federal guidelines and laws, as well as in-plant safety rules. Hazardous locations are indicated by Class, Division, and Group. See Appendix.

1. What are three of the different methods used to organize electrical tools?
2. What are 10 of the basic rules to follow for proper and safe tool usage?
3. Why should all power tools be grounded before working with them?
4. Why is it dangerous to use a screwdriver around a live electrical wire?
5. Is it safe to use a pipe extension to increase the leverage of a wrench?
6. What tool should be used for bending and cutting small diameter wire?
7. What tool should be used for cutting cable and twisting wire?
8. What tool should be used for removing insulation from small diameter wire?
9. What tool should be used to pull wire through conduit?
10. What tool should be used to bend conduit?
11. What tool should be used to remove cartridge type fuses?
12. What is an analog display?
13. What is a digital display?
14. What type of switch determines the placement of the decimal?
15. How do digital displays display values?
16. What is a bar graph?
17. What is a ghost voltage?
18. What is an informational output device?
19. What are four electrical information output devices?
20. What are four non-electrical informational output devices?
21. What instrument can be used to measure voltage in a circuit?
22. What instrument can be used to measure current in a circuit?
23. What is the procedure for measuring AC voltage?
24. What is the purpose of a GFCI?
25. What are three situations where a lockout/tagout is used?
26. When should rubber boots be used?
27. What is a Class A fire?
28. What is a Class B fire?
29. What is a Class C fire?
30. What is a Class D fire?

2

Chapter

Electrical Symbols and Line Diagrams

All trades have a certain language to enable rapid and efficient transfer of information. This language may include symbols, drawings or diagrams, words, phrases, or abbreviations. Control language is communicated using line diagrams. Line diagrams show the logic of an electrical circuit. Electrical control circuits may be manual, automatic, or magnetic.

Sprecher + Schuh

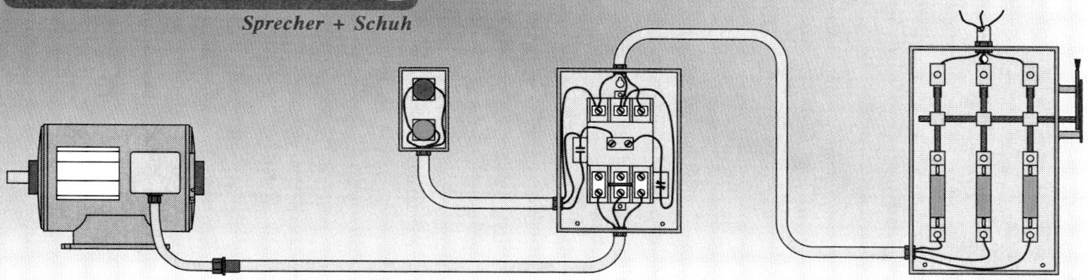

CONTROL LANGUAGE

All trades and professions have a certain language that enables the rapid and efficient transfer of information and ideas. This language may include symbols, drawings or diagrams, words, phrases, or abbreviations. Line (ladder) diagrams use industrial electrical symbols to provide the information necessary to understand the operation of any electrical control system. Industrial electrical symbols illustrate the electrical devices within a circuit. See Appendix.

Furnas Electric Co.

Pilot lights manufactured by Furnas Electric Co. may contain incandescent, neon, or LED lamps and are used to indicate the status of a circuit.

Line (Ladder) Diagrams

Control language is communicating by using line (ladder) diagrams. See Figure 2-1. A *line (ladder) diagram* is a diagram which shows, with single lines and symbols, the logic of an electrical circuit or system of circuits and components. A line diagram indicates the location of electrical devices within a circuit.

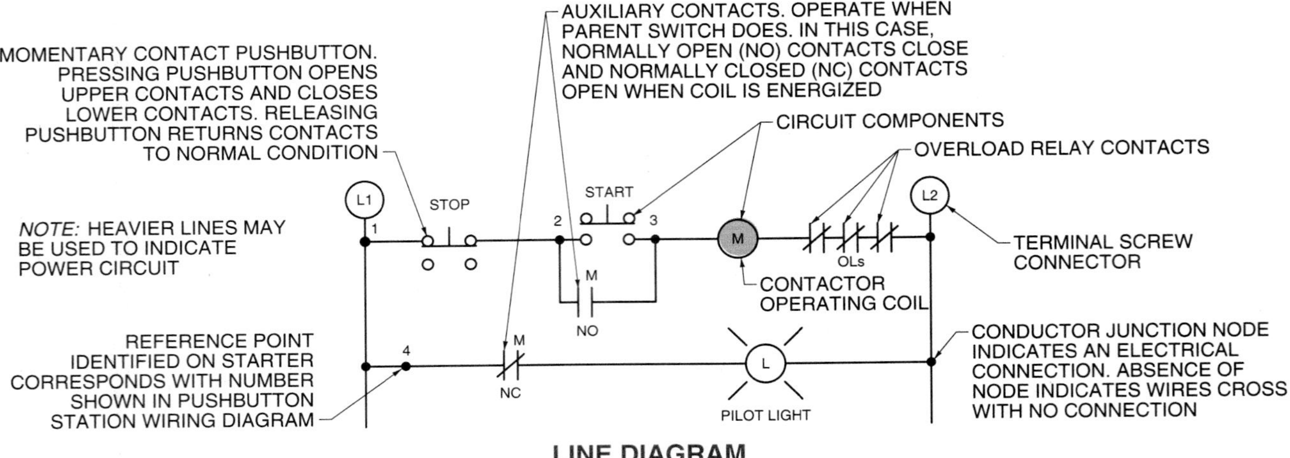

Figure 2-1. A line (ladder) diagram consists of a series of symbols interconnected by lines to indicate the flow of current through the various devices.

Circuit design and modifications to existing circuits are possible using line diagrams. A line diagram shows the power source and how current flows through the various parts of the circuit, such as pushbuttons, contacts, coils, and overloads. A line diagram shows the circuitry necessary for the operation of the controller. A line diagram is not intended to show the physical relationship of the various devices in the controller. Line diagrams are much easier to draw than pictorial drawings.

Manual Control Circuits

A *manual control circuit* is any circuit that requires a person to initiate an action for the circuit to operate. A line diagram may be used to illustrate a manual control circuit of a pushbutton controlling a pilot light. See Figure 2-2. In the line diagram, the lines labeled L1 and L2 represent the power circuit. The voltage of the power circuit is normally indicated on the circuit near these lines. In this circuit, the voltage is 115 V, but may be 230 V, 460 V, 575 V, or 2300 VAC. A DC control voltage may be marked with a negative (–) or positive (+) sign and may be 90 V, 120 V, 180 V, 240 V, 500 V, or 550 VDC.

The black nodes on a circuit indicate an electrical connection. The wires cross each other and are not electrically connected if a node is not present.

Line diagrams are read from left (L1) to right (L2). In this circuit, pressing pushbutton 1 (PB1) allows current to pass through the closed contacts of PB1, through pilot light 1 (PL1) and on to L2. This forms a complete circuit which activates PL1. Releasing PB1 opens contacts PB1, stopping the current flow to the pilot light. This turns OFF the pilot light.

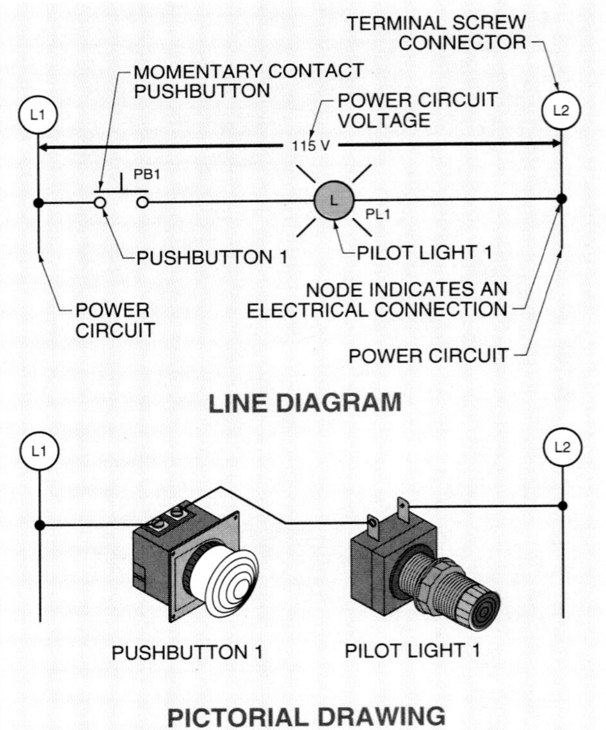

Figure 2-2. A line diagram may be used to illustrate a manual control circuit of a pushbutton controlling a pilot light.

A line diagram may be used to illustrate the control and protection of a 1ϕ motor using a manual starter with overload protection. See Figure 2-3. The manual starter is represented in the line diagram by the set of normally open (NO) contacts S1 and by the overload contacts OL1. The line diagram is drawn for ease of reading and does not indicate where the devices are physically located. For this reason, the overloads are shown between the motor and L2 in the line diagram but are physically located in the manual starter.

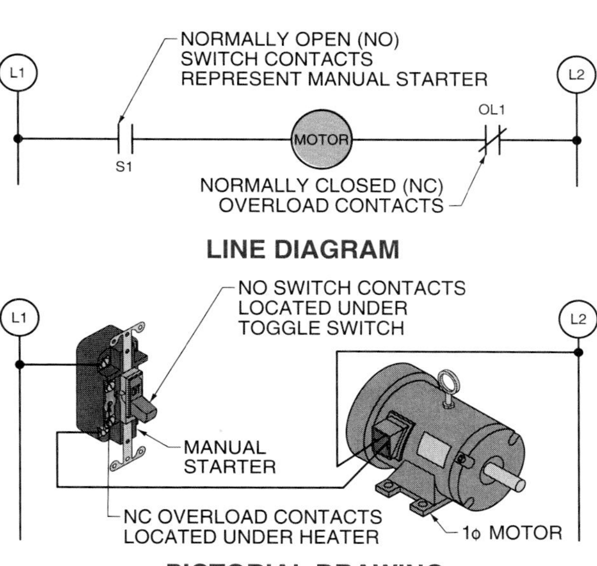

Figure 2-3. A line diagram may be used to illustrate the control and protection of a 1ϕ motor using a manual starter with overload protection.

In this circuit, current passes through contacts S1, the motor, the overloads, and on to L2 when the manual starter (S1 NO contacts) is closed. This starts the motor. The motor runs until contacts S1 are opened, a power failure occurs, or the motor experiences an overload. In the case of an overload, the OL1 contacts open and the motor stops. The motor cannot be restarted until the overload is removed and the overload contacts are reset to their normally closed (NC) position. *Note:* For consistency, the overload symbol is always drawn in a line diagram after the motor. In the actual circuit, the overload is located before the motor. It does not matter that the overload is shown after the motor in the line diagram because the overload is a series device and opens the motor control circuit in either position.

Automatic Control Circuits

Automatically-controlled devices have replaced many jobs that were once performed manually. As a part of automation, control circuits are designed to replace manual devices. Any manual control circuit may be converted to automatic operation. For example, an electric motor on a sump pump can be turned ON and OFF automatically by adding an automatic control device such as a float switch. See Figure 2-4. This control circuit is used in basements to control a sump pump to prevent flooding. The float switch senses the change in water level and automatically starts the pump which removes the water when it reaches a predetermined level.

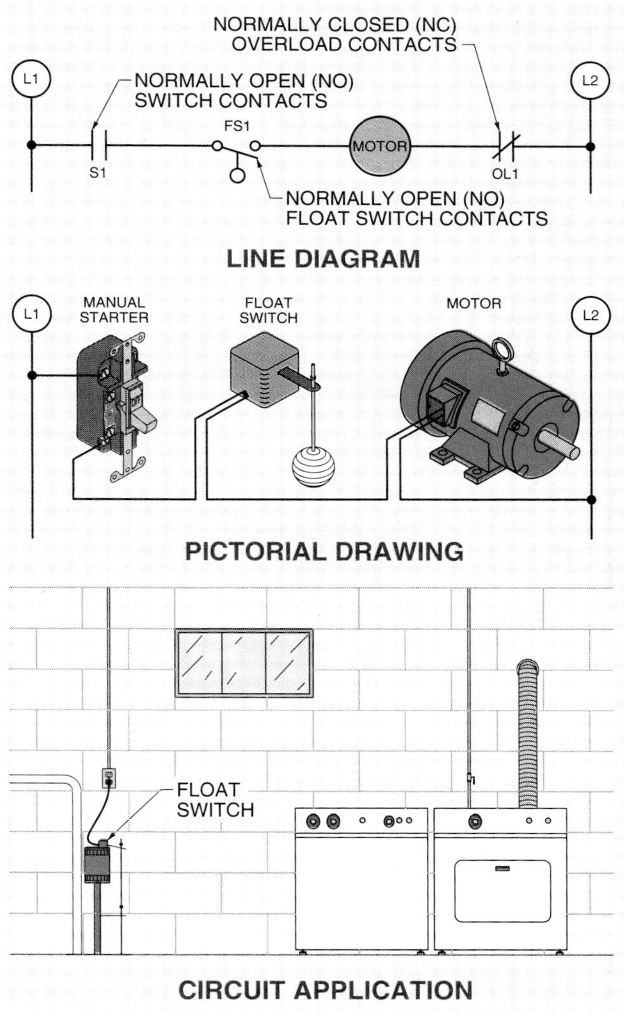

Figure 2-4. An electric motor on a sump pump can be turned ON and OFF automatically by using an automatic control device such as a float switch.

In this circuit, float switch contacts FS1 determine if current passes through the circuit when switch contacts S1 are closed. Current passes through contacts S1, FS1, the motor, its overload, and on to L2 when the float switch contacts FS1 are closed. This starts the pump motor. The pump motor pumps water until the water level drops enough to open contacts FS1 and shut OFF the pump motor. A motor overload, power failure, or the manual opening of contacts S1 stops the pump motor from automatically pumping water when it reaches the predetermined level.

Control devices, such as float switches, are normally designed with NO and NC contacts. Variations in the application of the float switch are possible because the NO and NC contacts can close or open contacts when changes in liquid level occur. For example, the NC contacts of the float switch may be used for a pump operation to maintain certain levels of water in a livestock water tank. See Figure 2-5. The NC contacts close and start the pump motor when the water level drops, due to evaporation or drinking. The float opens the NC contacts, shutting the pump motor OFF when the water level rises to a predetermined level.

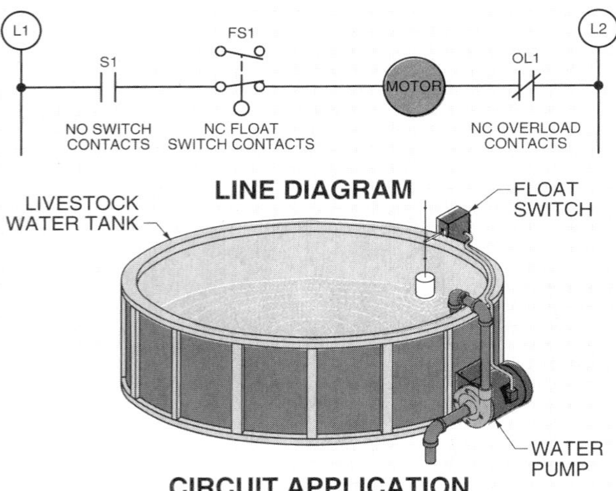

Figure 2-5. The NC contacts of a float switch may be used for a pump operation to maintain a certain level of water in a livestock water tank.

In this circuit, when contacts S1 and float switch contacts FS1 are closed, current passes through them, the motor, the overloads, and on to L2. This starts the pump motor. The pump motor pumps water until the water level rises high enough to open contacts FS1 and shut OFF the pump. A motor overload, a power failure, or the manual opening of contacts S1 stops the pump from automatically filling the tank to a predetermined level. In certain circuits, overloads that automatically reset themselves may be used. In other situations, such as flooding, it is considered less serious to let the motor burn out than to shut the operation down.

Magnetic Control Circuits

Although manual controls are compact and sometimes less expensive than magnetic controls, industrial and commercial installations often require that electrical control equipment be located in one area while the load device is located in another. Solenoids, contactors, and magnetic motor starters are used for remote control of devices.

Solenoids. A *solenoid* is an electric output device that converts electrical energy into a linear mechanical force. A solenoid consists of a frame, plunger, and coil. See Figure 2-6. A magnetic field is set up in the frame when the coil is energized by an electric current passing through it. This magnetic field causes the plunger to move into the frame. The result is a straight line force, normally as a push or pull action.

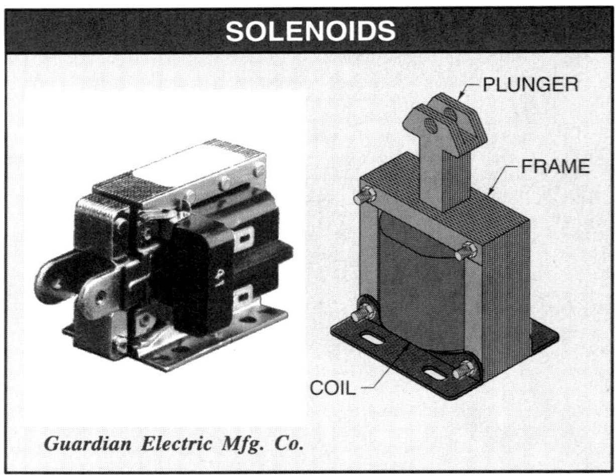

Figure 2-6. A solenoid is an electric output device that converts electrical energy into a linear mechanical force.

A solenoid may be used to control a door lock which is opened only when a pushbutton is pressed. See Figure 2-7. In this circuit, pressing pushbutton 1 allows an electric current to flow through the solenoid, creating a magnetic field. The magnetic field, depending on the solenoid construction, causes the plunger to push or pull. In this circuit, the door may open as long as the pushbutton is pressed. The door is locked when the pushbutton is released. This circuit provides security access to a building or room.

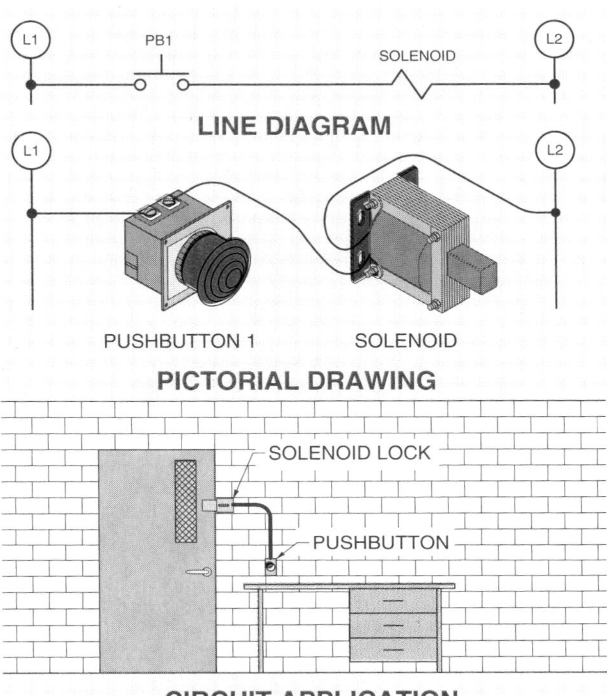

Figure 2-7. A solenoid may be used to control a door lock which is opened only when a pushbutton is pressed.

General Electric Company

A contactor may be used to control fractional horsepower motors that contain overload protection.

A solenoid may also be used to control a holding clamp. A part may be held in position or released based on the decision of the pushbutton operator.

Contactors. A *contactor* is a control device that uses a small control current to energize or de-energize the load connected to it. A contactor does not include overload protection. A contactor is constructed and operates similarly to a solenoid. See Figure 2-8. A contactor has a frame, plunger, and coil like a solenoid. The action of the plunger, however, is directed to close sets of contacts. The closing of the contacts allows electrical devices to be controlled from remote locations.

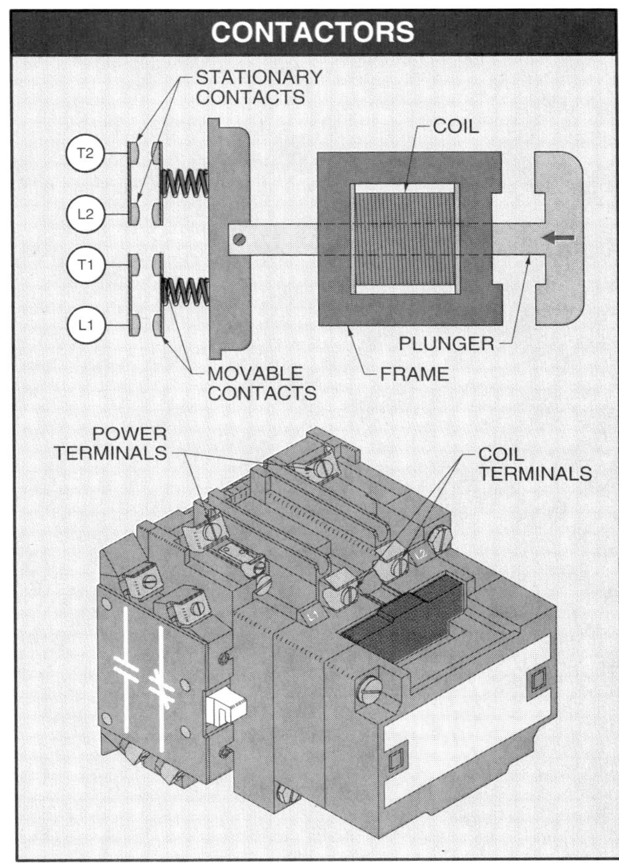

Figure 2-8. A contactor is a control device that uses a small control current to energize or de-energize the load connected to it.

A line diagram may be used to show the electrical operation of a contactor. See Figure 2-9. In this circuit, pressing pushbutton 1 (PB1) allows current to pass through the switch contacts, the contactor coil (C1), and on to L2. This energizes the contactor coil (C1). The activation of C1 closes the power contacts of the contactor. The contactor contains power contacts which close each time the control circuit is activated. These contacts are not normally shown in the line diagram.

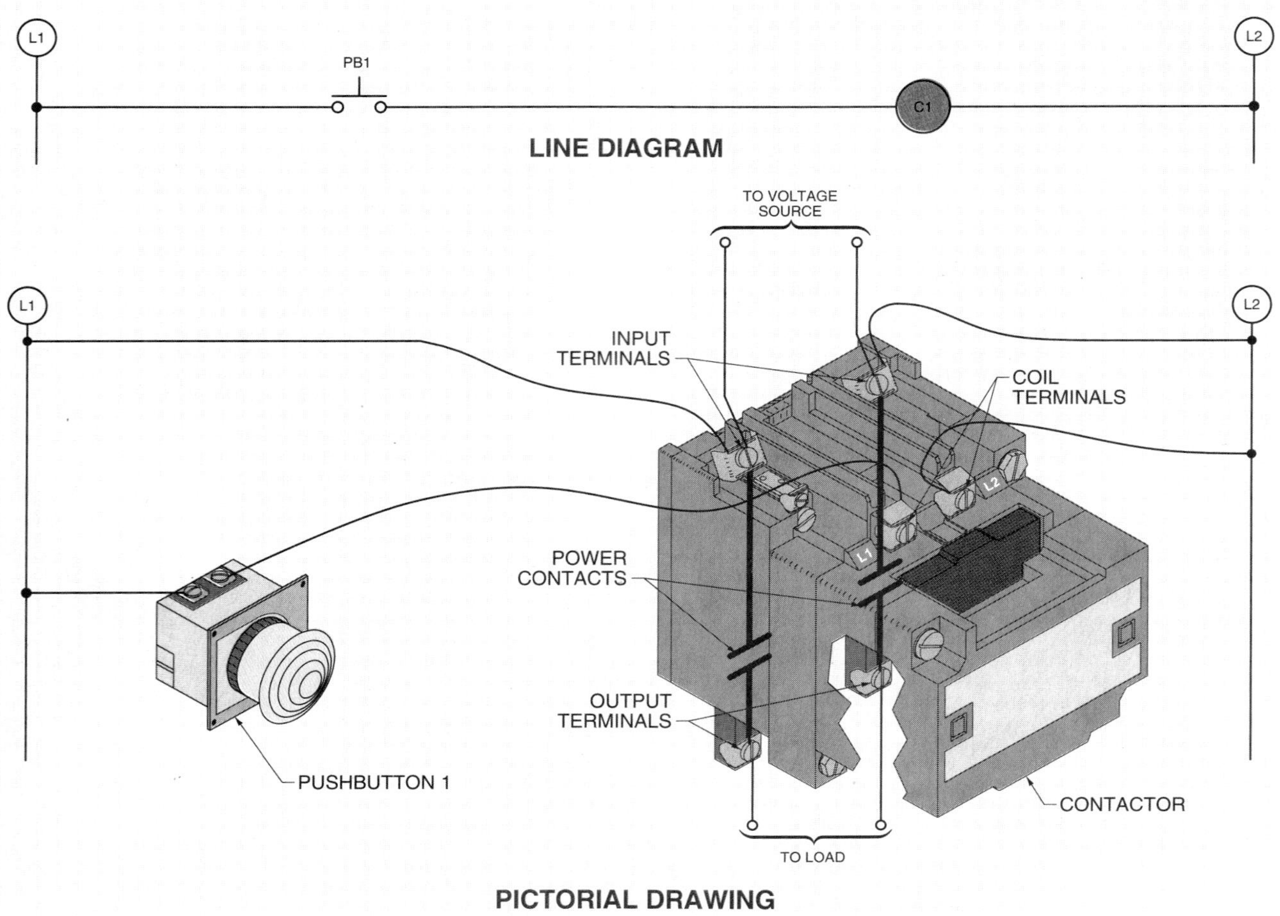

L1
PB1
C1
L2

LINE DIAGRAM

TO VOLTAGE SOURCE
L1
L2
INPUT TERMINALS
COIL TERMINALS
POWER CONTACTS
OUTPUT TERMINALS
PUSHBUTTON 1
CONTACTOR
TO LOAD

PICTORIAL DRAWING

Figure 2-9. A line diagram may be used to show the electrical operation of a contactor.

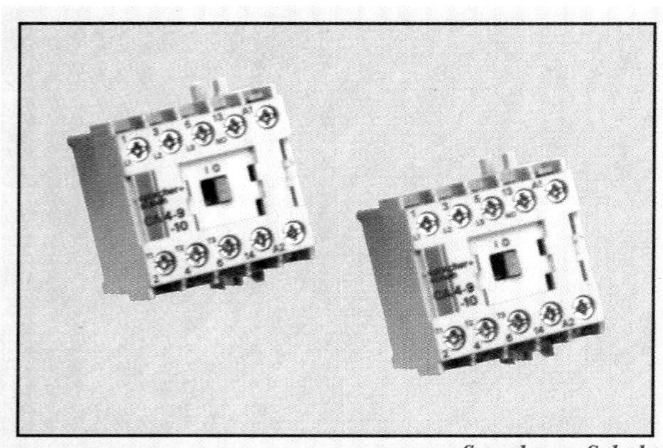

Sprecher + Schuh

The contacts in Sprecher + Schuh's CA4 Series miniature contactors have an electrical life of 700,000 operations, while the AC magnet system has a mechanical life of 10,000,000 operations.

Releasing PB1 stops the flow of current to the contactor coil and de-energizes the coil. The power contacts return to their NO condition when the coil de-energizes. This shuts OFF the lights, heaters, or other loads connected to the contacts.

This circuit works well in turning ON and OFF various loads remotely. It does, however, require someone to hold the contacts closed if the coil must be continuously energized. Auxiliary contacts may be added to a contactor to form an electrical holding circuit and to eliminate the necessity of someone holding the pushbutton continuously. See Figure 2-10. The auxiliary contacts are attached to the side of the contactor and are opened and closed with the power contacts as the coil is energized or de-energized. These contacts are shown on the line diagram because they are part of the control circuit.

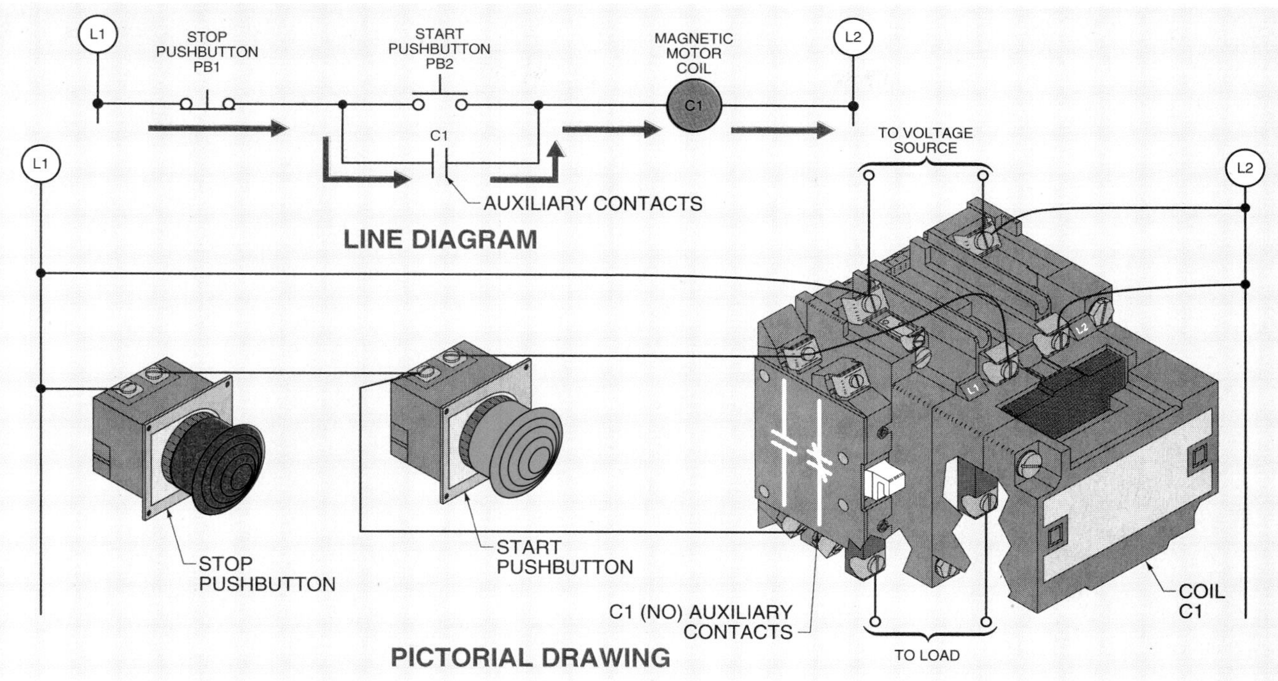

Figure 2-10. Auxiliary contacts may be added to a contactor to form an electrical holding circuit.

A line diagram may be used to show the logic of the electrical holding circuit. In this circuit, pressing the start pushbutton (PB2) allows current to pass through the closed contacts of the stop pushbutton (PB1), through the closed contacts of the start pushbutton, through coil C1, and on to L2. This energizes coil C1. With coil C1 energized, auxiliary contacts C1 close and remain closed as long as the coil is energized. This forms a continuous electrical path around the start pushbutton (PB2) so that even if the start pushbutton is released, the circuit remains energized because the coil remains energized.

The circuit is de-energized by a power failure or by pressing the NC stop pushbutton (PB1). In either case, the current flow to coil C1 stops and the coil is de-energized, causing auxiliary contacts C1 to return to their NO position. The start pushbutton (PB2) must be pressed to re-energize the circuit.

Magnetic Motor Starters. A *magnetic motor starter* is an electrically-operated switch (contactor) that includes motor overload protection. Magnetic motor starters are used to start and stop motors. They are identical to contactors except that they have overloads attached to them. See Figure 2-11.

The overloads have heaters (located in the power circuit) which sense excessive current flow to the motor. The heaters open the NC contacts (located in the control circuit) when the overload becomes dangerous to the motor.

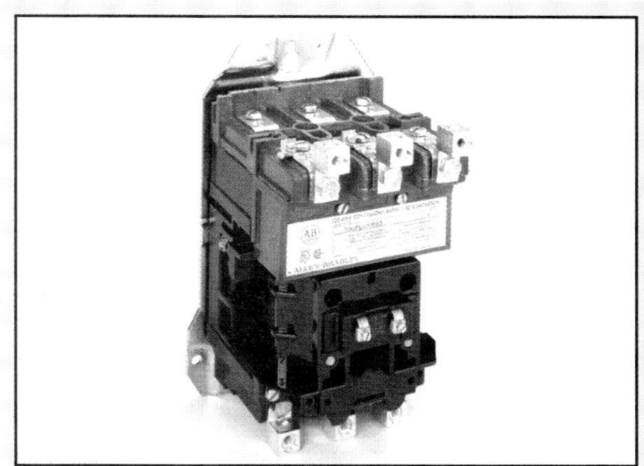

Rockwell Automation, Allen-Bradley Company, Inc.
Bulletin 500FL contactors from Allen-Bradley are electrically-held contactors designed to switch current to incandescent, fluorescent, and mercury-vapor lamps, as well as capacitors and other non-motor loads.

MAGNETIC MOTOR STARTERS

MOUNTING PLATE
POWER TERMINALS
COIL TERMINALS
AUXILIARY CONTACTS
HEATERS
OVERLOAD CONTACTS
RESET BUTTON

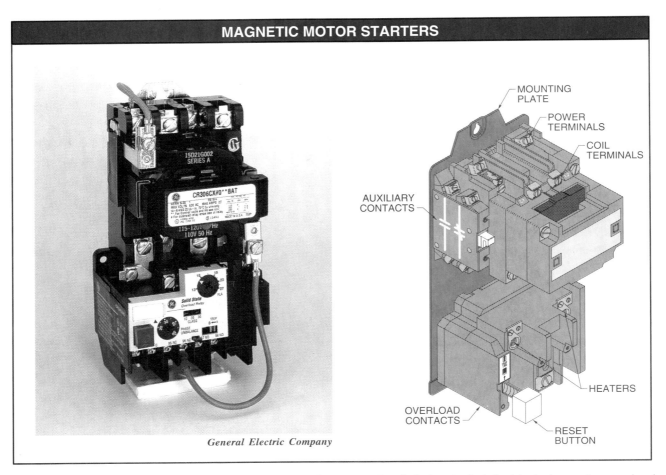

General Electric Company

Figure 2-11. A magnetic motor starter is an electrically-operated switch (contactor) that includes motor overload protection.

Products Unlimited

Definite purpose starters with ratings from 2 A to 88 A and overloads that are ambient compensated from –25°C to 55°C are available from Products Unlimited.

A line diagram may be used to illustrate the use of a magnetic starter with overloads. See Figure 2-12. The only difference in this line diagram and the line diagram of a contactor circuit is the addition of a motor starter coil (M). In this circuit, pressing the start pushbutton allows current to pass through coil M1 and overload OL1. This energizes coil M1. With coil M1 energized, auxiliary contacts M1 close and the circuit remains energized even if the start pushbutton is released.

This circuit is de-energized if the stop pushbutton is pressed, a power failure occurs, or any one of the overloads sense a problem in the power circuit. Coil M1 de-energizes, causing auxiliary contacts M1 to return to their NO condition if one of these situations occurs. The overload must be removed, the overload device reset, and the start pushbutton pressed to restart the motor when it stops due to an overload.

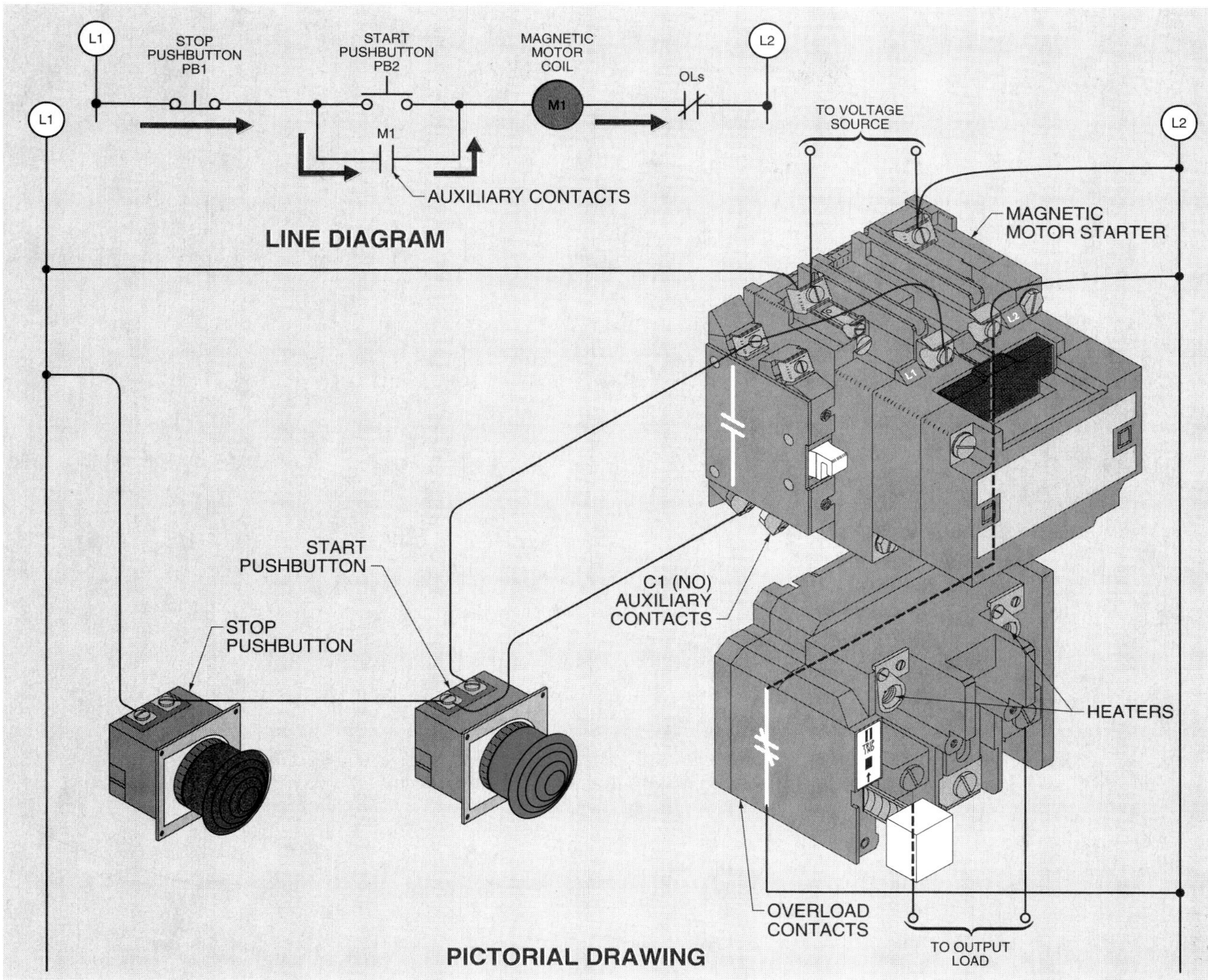

Figure 2-12. A line diagram may be used to illustrate the use of a magnetic starter with overloads.

1. What is the function of the line diagram?
2. How are wires that are electrically connected illustrated on a line diagram?
3. Where are the overload contacts drawn in a line diagram?
4. Are normally open (NO) or normally closed (NC) contacts used when a float switch is used to maintain a predetermined level?
5. How is a solenoid illustrated in a line diagram?
6. How are coils illustrated in a line diagram?
7. What are auxiliary contacts?
8. Are normally open (NO) or normally closed (NC) contacts used when using auxiliary contacts to maintain an electrical holding circuit?
9. What is a magnetic motor starter?
10. How can the power be removed from a magnetic motor starter coil after the start pushbutton is pressed in a line diagram that uses auxiliary contacts to form a holding circuit around the start pushbutton?
11. What is a manual control circuit?
12. What are the three major parts of a solenoid?
13. How are magnetic motor starters similar to contactors?
14. What is control language?
15. How are line diagrams read?

3 Chapter

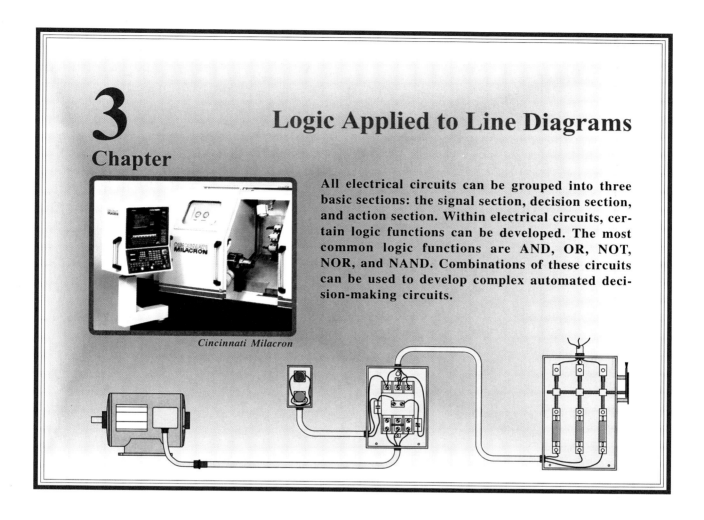

Logic Applied to Line Diagrams

All electrical circuits can be grouped into three basic sections: the signal section, decision section, and action section. Within electrical circuits, certain logic functions can be developed. The most common logic functions are AND, OR, NOT, NOR, and NAND. Combinations of these circuits can be used to develop complex automated decision-making circuits.

Cincinnati Milacron

BASIC RULES OF LINE DIAGRAMS

The electrical industry has established a universal set of symbols and rules on how line diagrams (circuits) are laid out. By applying these standards, an electrician establishes a working practice common to all electricians.

One Load Per Line

No more than one load should be placed in any one circuit line between L1 and L2. A pilot light can be connected into a circuit with a single-pole switch. See Figure 3-1. In this circuit, the power lines are drawn vertically on sides of the drawing. The lines are marked L1 and L2. The space between L1 and L2 represents the voltage of the control circuit. This voltage appears across pilot light PL1 when switch S1 is closed. Current flows through S1, and PL1 glows because the voltage between L1 and L2 is the proper voltage for the pilot light.

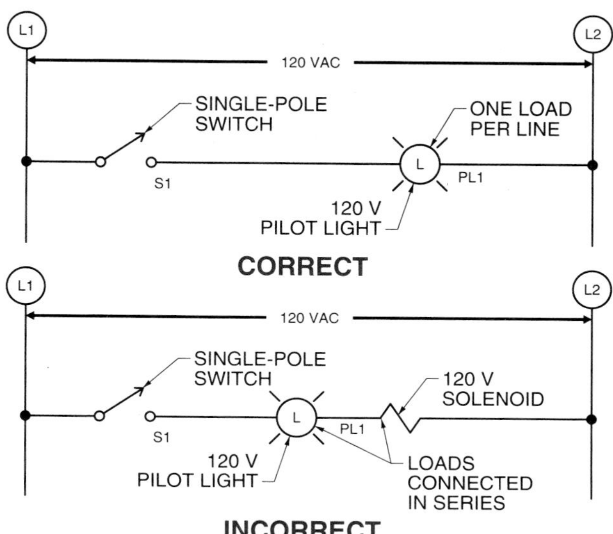

Figure 3-1. No more than one load should be placed in any one circuit line between L1 and L2.

Two loads must not be connected in one line of a line diagram. If the two loads are connected in series, then the voltage between L1 and L2 must divide across both loads when S1 is closed. The result is that neither device receives the entire 120 V necessary for proper operation.

The load that has the highest resistance drops the highest voltage. The load that has the lowest resistance drops the lowest voltage.

Loads must be connected in parallel when more than one load must be connected in the line diagram. See Figure 3-2. In this circuit, there is only one load for each line between L1 and L2 even though there are two loads in the circuit. The voltage from L1 and L2 appears across each load for proper operation of the pilot light and solenoid. This circuit has two lines, one for the pilot light and one for the solenoid.

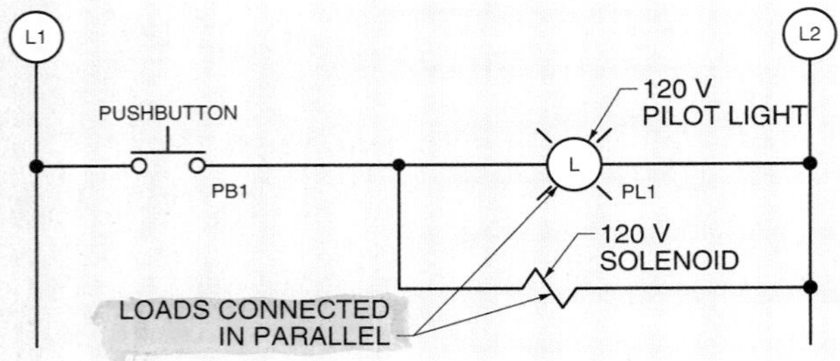

Figure 3-2. Loads must be connected in parallel when more than one load must be connected in the line diagram.

Load Connections

A *load* is the electrical device in the line diagram that uses the electrical power from L1 to L2. Control relays, solenoids, and pilot lights are loads that are connected directly or indirectly to L2. See Figure 3-3.

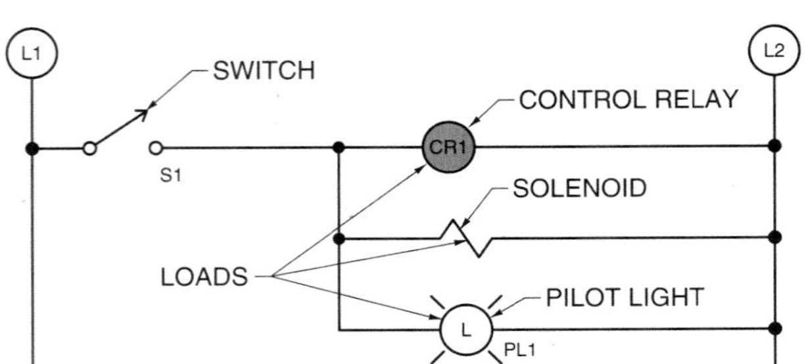

Figure 3-3. Control relays, solenoids, and pilot lights are loads that are connected directly or indirectly to L2.

Magnetic motor starter coils are connected to L2 indirectly through normally closed (NC) overload contacts. See Figure 3-4. An overload contact is normally closed and opens only if an overload condition exists in the motor. The number of NC overload contacts between the starter coil and L2 depends on the type of starter and power that is used in the circuit.

Anywhere from one to three NC contacts are shown between the starter and L2 in all line diagrams. To avoid confusion, it is common practice to draw one set of NC contacts and mark these contacts all overloads (OLs).

An overload marked this way indicates that the circuit is correct for any motor or starter used. The electrician knows to connect all the NC overload contacts that the starter is designed for in series.

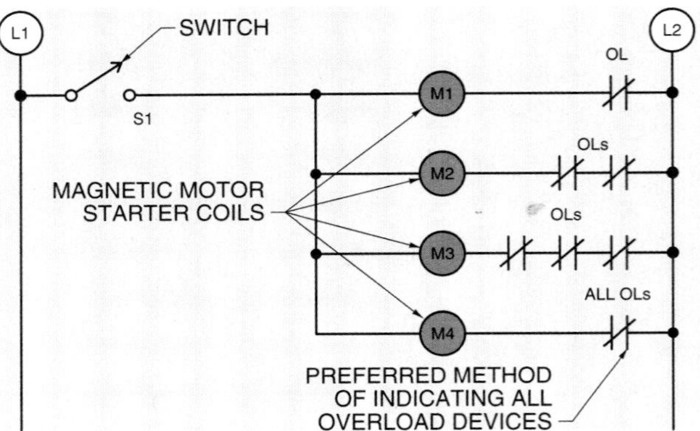

Figure 3-4. Magnetic motor starter coils are connected to L2 indirectly through NC overload contacts.

Control Device Connections

Control devices are connected between L1 and the operating coil (or load). Operating coils of contactors and starters are activated by control devices such as pushbuttons, limit switches, and pressure switches. See Figure 3-5.

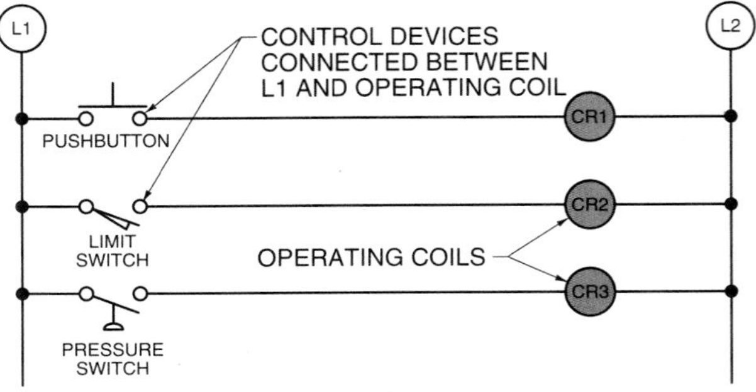

Figure 3-5. Control devices are connected between L1 and the operating coil.

Each line includes at least one control device. The operating coil is ON all the time if no control device is included in a line. A circuit may contain as many control devices as is required to make the operating coil function as specified. These control devices may be connected in series or parallel when controlling an operating coil.

Two control devices (a flow switch and a temperature switch) can be connected in series to control a coil in a magnetic motor starter. See Figure 3-6. The flow switch and temperature switch must close to allow current to pass from L1 through the control device, the magnetic starter coil, and the overloads to L2. Two control devices (a pressure switch and a foot switch) can be connected in parallel to control a coil in a magnetic motor starter. Either the pressure switch or the foot switch can be closed to allow current to pass from L1 through the control device, the magnetic starter coil, and the overloads to L2. Regardless of how the control devices are arranged in a circuit, they must be connected between L1 and the operating coil (or load). The contacts of the control device may be either normally open (NO) or normally cloased (NC). The contacts used and the way the control devices are connected into a circuit (series or parallel) determines the function of the circuit.

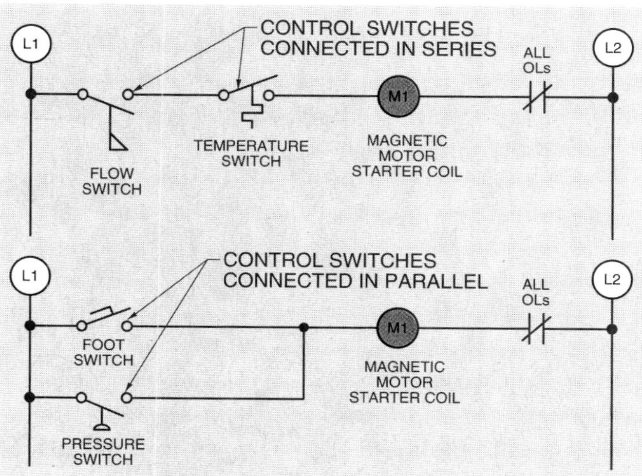

Figure 3-6. Two control devices may be connected in series or parallel to control a coil in a magnetic motor starter.

Line Number Reference

Each line in a line diagram should be numbered starting with the top line and reading down. See Figure 3-7.

Line 1 connects PB1 to the solenoid to complete the path from L1 to L2. Line 2 connects PS1 to the solenoid to complete the path from L1 to L2. PB1 and PS1 are marked as two separate lines even though they control the same load, because either the pushbutton or the pressure switch completes the path from L1 to L2. Line 3 connects a foot switch and a temperature switch to complete the path from L1 to L2. The foot switch and temperature switch both appear in the same line, because it takes both the foot switch and the temperature switch to complete the path to the pilot light.

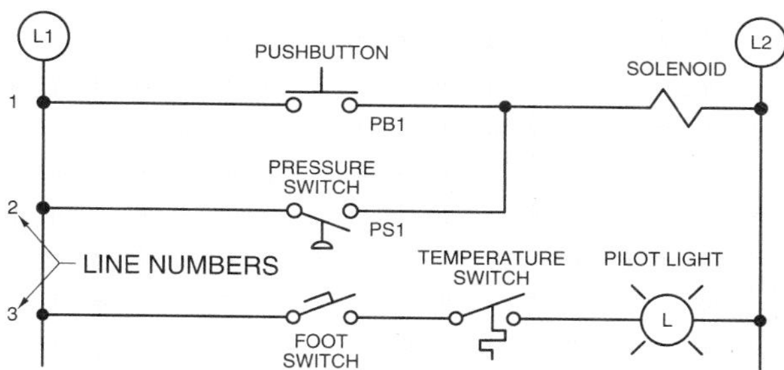

Figure 3-7. Each line in a line diagram should be numbered starting with the top line and reading down.

Numbering each line simplifies the understanding of the function of a circuit. The importance of this numbering system becomes clear as circuits become more complex and lines are added.

Numerical Cross-Reference Systems

Numerical cross-reference systems are required to trace the action of a circuit in complex line diagrams. Common rules help to quickly simplify the operation of complex circuits.

Furnas Electric Co.

Electrical switching devices must be installed inside the proper enclosure for protection against the environment.

Numerical Cross-Reference System (NO Contacts). Relays, contactors, and magnetic motor starters normally have more than one set of auxiliary contacts. These contacts may appear at several different locations in the line diagram. Numerical cross-reference systems quickly identify the location and type of contacts controlled by a given device. A numerical cross-reference system consists of numbers in parenthesis to the right of the line diagram. NO contacts are represented by line numbers. See Figure 3-8.

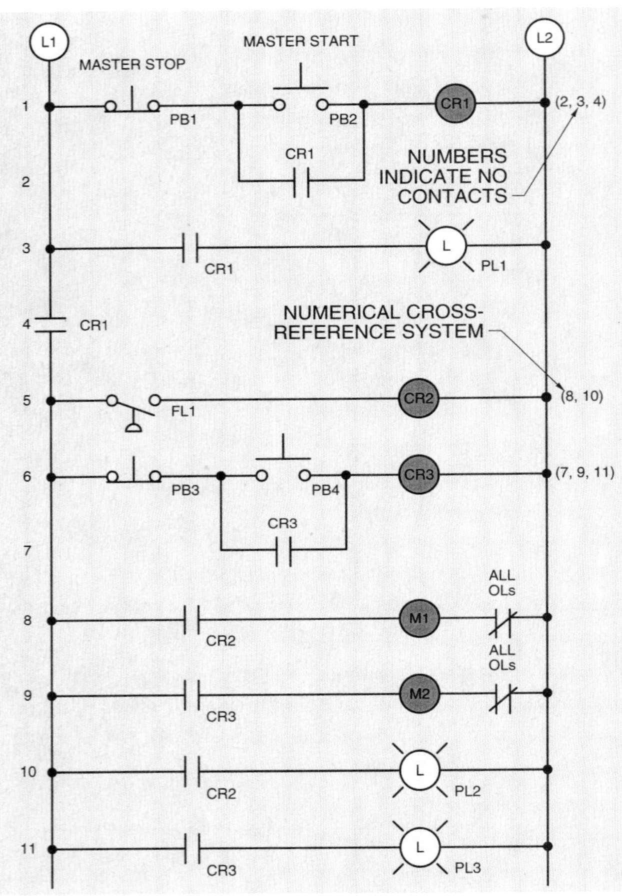

Figure 3-8. The location of NO contacts controlled by a device is determined by the numbers on the right side of the line diagram.

In this circuit, pressing master start pushbutton PB2 energizes control relay coil CR1. Control relay coil CR1 controls three sets of NO contacts. This is shown by the numerical codes (2, 3, 4) on the right side of the line diagram. Each number indicates the line in which the NO contacts are located.

In line 2, the NO contacts form the holding circuit (memory) for maintaining the coil CR1 after master start pushbutton PB2 is released. In line 3, the NO contacts energize pilot light PL1, indicating that the circuit has been energized. In line 4, the NO contacts allow the remainder of the circuit to be activated by connecting L1 to the remainder of the circuit. The numerical cross-reference system shows the location of all contacts controlled by coil CR1 as well as the effect each has on the operation of the circuit.

In line 5, control relay CR2 energizes if float switch FL1 closes. Control relay CR2 closes the NO contacts located in lines 8 and 10 as indicated by the numerical codes (8, 10). The magnetic motor starter controlled by coil M1 is energized when the NO contacts of line 8 close. Pilot light PL2 turns ON indicating the motor has started when the NO contacts in line 10 close.

In line 6, several NO contacts located in lines 7, 9, and 11 are used to control other parts of the circuit through control relay CR3. In line 7, the NO contacts form the memory circuit for maintaining the circuit to control relay coil CR3 after pushbutton PB4 is released. The NO contacts in line 9 close, energizing the magnetic motor starter controlled by coil M2 when coil CR3 is energized. Simultaneously, the NO contacts in line 11 close, causing pilot light PL3 to light as an indicator that the motor has started.

Heidelberg Harris, Inc.

The use of standard numbering and cross-reference systems enables efficient troubleshooting of complex electrical systems.

The numerical cross-reference system allows the simplification of complex line diagrams. Each NO contact must be clearly marked because each set of NO contacts are numbered according to the line in which they appear.

Numerical Cross-Reference System (NC Contacts). In addition to NO contacts, there are also NC contacts in a circuit. To differentiate between NO and NC, NC contacts are indicated as a number which is underlined. See Figure 3-9. For example, lines 9 and 11 contain devices which control NC contacts in lines 12 and 13 as indicated by the underlined numbers (12, 13) to the right of the line diagram.

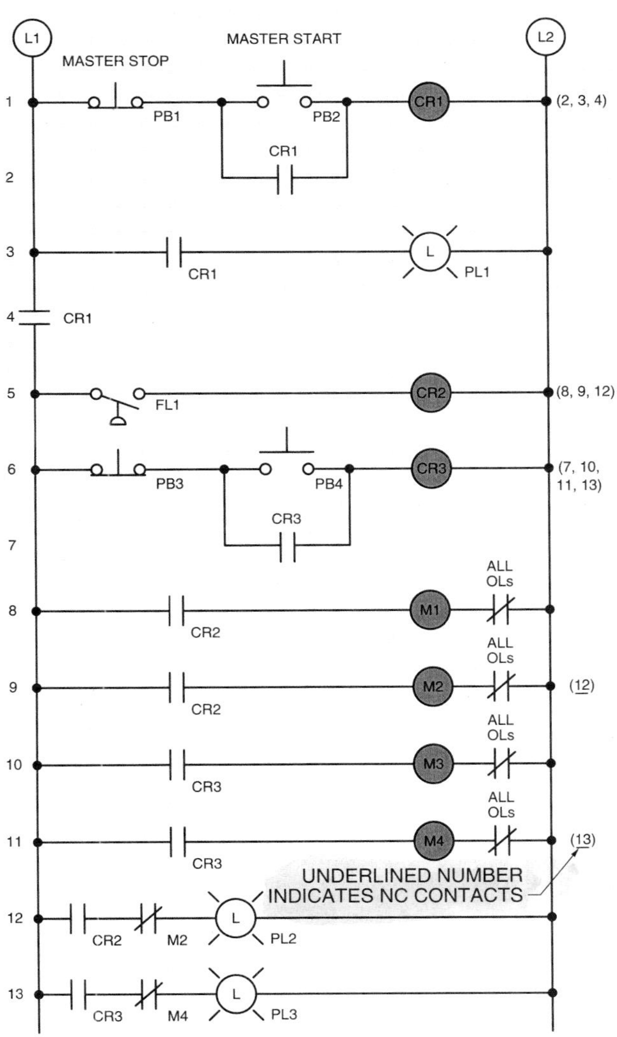

Figure 3-9. NC contacts are indicated by numbers which are underlined to distinguish them from NO contacts.

In this circuit, pressing master start pushbutton PB2 energizes control relay coil CR1. Control relay coil CR1 controls three sets of NO contacts. This is shown by the numerical codes (2, 3, 4) on the right side of the line diagram. In line 2, the NO contacts form the holding circuit (memory) for maintaining the coil CR1 after master start pushbutton PB2 is released. In line 3, the NO contacts energize pilot light PL1, indicating that the circuit has been energized. In line 4, the NO contacts allow the remainder of the circuit to be activated by connecting L1 to the remainder of the circuit.

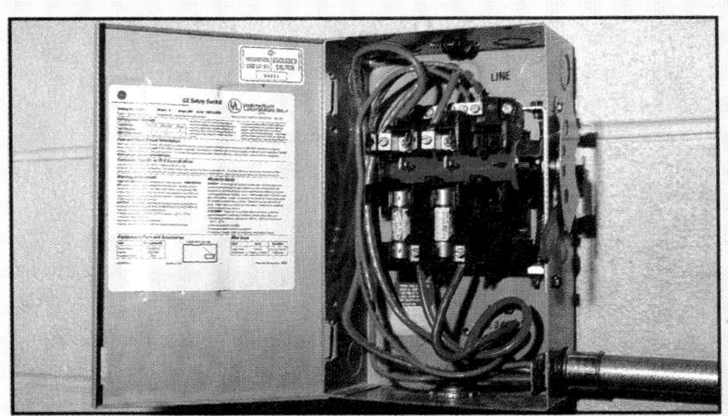

The use of wire reference numbers helps identify the wires within an electrical system.

In line 5, when control relay coil CR2 is energized, the NO contacts in lines 8, 9, and 12 close. Closing these contacts energizes coils M1 and M2 and completes the circuit going to the pilot light in line 12. The pilot light in line 12 does not glow because the NC contacts controlled by coil M2 in line 12 are opened at the same time that the NO contacts are closed, leaving the circuit open. With the NO contacts of CR2 closed and the NC contacts of M2 open, the light stays OFF unless something happens to shut down line 9, which contains coil M2. For example, if coil M2 represents a safety cooling fan protecting the motor controlled by M1, the light would indicate the loss of cooling.

A similar sequence of events took place when line 6 was energized. In this case, pressing pushbutton PB4 energizes control relay coil CR3, which closes NO contacts in line 7, 10, 11, and 13. A memory circuit is formed in line 7, coils M3 and M4 are energized in lines 10 and 11, and part of the circuit to pilot light PL3 in line 13 is completed. Because coil M4 is energized, the NC contacts it controls open in line 13, forming a similar alarm circuit to the one in line 12.

If coil M4 in line 11 drops out for any reason, the NC contacts in line 13 return to their NC position and the pilot light alarm signal in line 13 is turned ON. This circuit could be used where it is extremely important for the operator to know when something is not functioning.

Wire-Reference Numbers

Each wire in a control circuit is assigned a reference point (number) on a line diagram to keep track of the different wires that connect the components in the circuit. Each reference point is assigned a reference number. Reference numbers are normally assigned from the top left to the bottom right. This numbering system can apply to any control circuit such as single-station, multistation, or reversing circuits. See Figure 3-10.

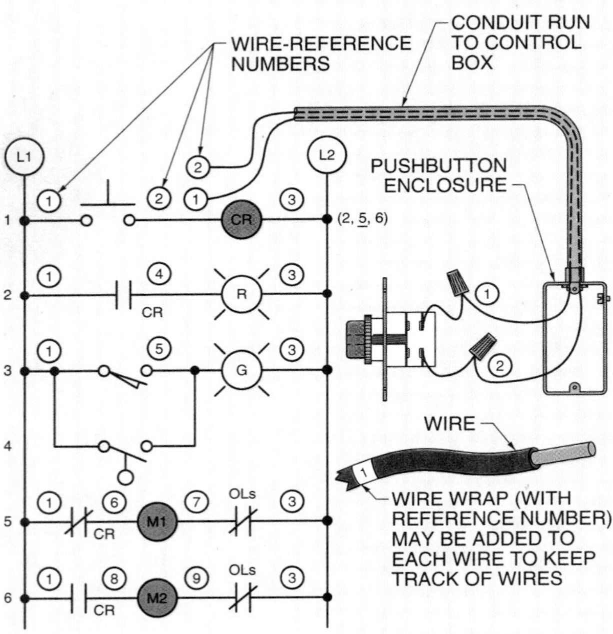

Figure 3-10. Each wire in a control circuit is assigned a reference point on a line diagram to keep track of the different wires that connect the components in the circuit.

Any wire that is always connected to a point is assigned the same number. The wires that are assigned a number vary from 2 to the number required by the circuit. Any wire that is prewired when the component is purchased is normally not assigned a reference number. Different numbering systems can be used. The exact numbering system varies for each manufacturer or design engineer. One common method used is to circle the wire-reference numbers. Circling the wire-reference numbers helps separate them from other numbering systems.

Manufacturer's Terminal Numbers

Manufacturers of electrical relays, timers, counters, etc., include numbers on the terminal connection points. These terminal numbers are used to identify and separate the different component parts (coil, NC contacts, etc.) included on the individual pieces of equipment. Manufacturer's terminal numbers are often added to a line diagram after the specific equipment to be used in the control circuit is identified. See Figure 3-11.

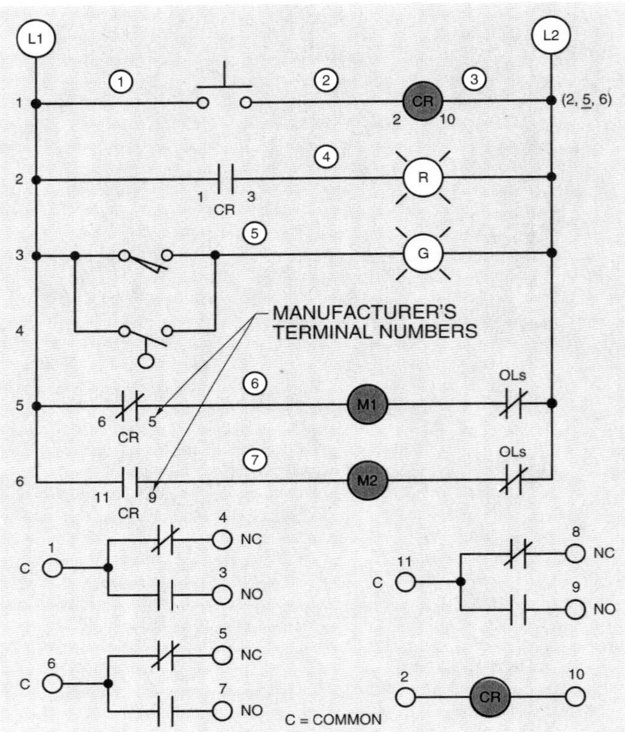

Figure 3-11. Manufacturers include terminal numbers to identify and separate the different component parts included on individual pieces of equipment.

Cross-Referencing Mechanically-Connected Contacts

Control devices such as limit switches, flow switches, temperature switches, liquid level switches, and pressure switches normally have more than one set of contacts operating when the device is activated. These devices normally have at least one set of NO contacts and one set of NC contacts which operate simultaneously. For all practical purposes, the multiple contacts of these devices normally do not control other devices in the same lines of a control circuit.

The two methods used to illustrate how contacts found in different control lines belong to the same control switch are the dashed line method and the numerical cross-reference method. See Figure 3-12.

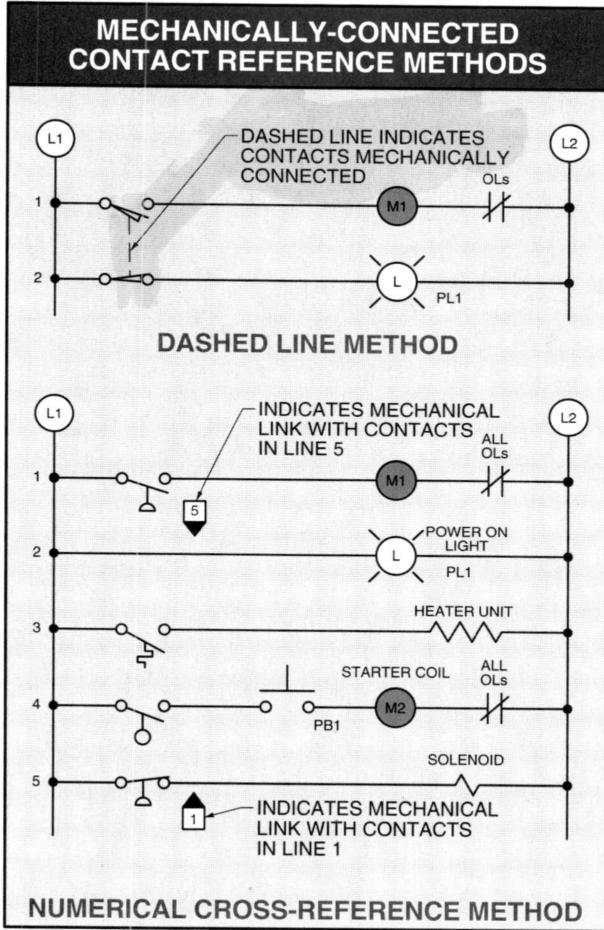

Figure 3-12. Contacts found in different control lines that belong to the same control switch are illustrated using the dashed line or numerical cross-reference methods.

In the dashed line method, the dashed line between the NO and NC contacts indicates that both contacts move from the normal position when the arm of the limit switch is moved. In this circuit, pilot light PL1 is ON, and motor starter coil M1 is OFF. After the limit switch is actuated, pilot light PL1 turns OFF, and the motor starter coil M1 turns ON.

The dashed line method works well when the control contacts are close together and the circuit is relatively simple. If a dashed line must cut across many lines, the circuit becomes hard to follow.

The numerical cross-reference method is used on complex line diagrams where a dashed line cuts across several lines. In this circuit, a pressure switch with a NO contact in line 1 and an NC contact in line 5 is used to control a motor starter and a solenoid. The NO and NC contacts of the pressure switch are simultaneously actuated when a predetermined pressure is reached. A solid arrow pointing down is drawn by the NO contact in line 1 and is marked with a 5 to show the mechanical link with the contact in line 5. A solid arrow pointing up is drawn by the NC contact in line 5 and is marked with a numeral 1 to show the mechanical linkage with the contact in line 1. This cross-reference method eliminates the need for a dashed line cutting across lines 2, 3, and 4. This makes the circuit easier to follow and understand. This system may be used with any type of control switch found in a circuit.

LINE DIAGRAMS – SIGNALS, DECISIONS, AND ACTION

The concept of control is to accomplish specific work in a predetermined manner. A circuit must respond as designed, without any changes. To accomplish this consistency, all control circuits are composed of three basic sections: the signals, the decisions, and the action sections. See Figure 3-13. Complete understanding of these sections enable easy understanding of any existing industrial control circuit, as well as those which are being created by industry as it becomes more mechanized and automated.

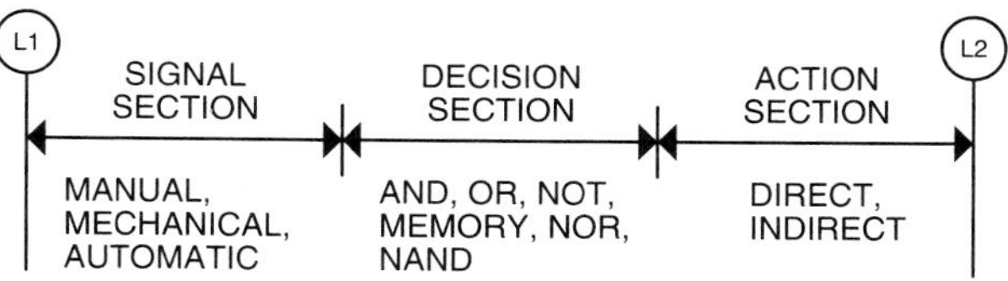

Figure 3-13. All control circuits are composed of signals, decisions, and action sections.

Signals

A signal starts or stops the flow of current by closing or opening the control device's contacts. Current is allowed to flow through the control device if the contacts are closed. Current is not allowed to flow through the control device if the contacts are opened. Pushbuttons, limit switches, flow switches, foot switches, temperature switches, and pressure switches may be used as the signal section of a control circuit.

All signals depend on some condition that must take place. This condition can be manual, mechanical, or automatic. A *manual condition* is any input into the circuit by a person. Foot switches and pushbuttons are control devices that respond to a manual condition. A *mechanical condition* is any input into the circuit by a mechanically moving part. A limit switch is a control device that responds to a mechanical condition. When a moving object, such as a box, hits a limit switch, the limit switch normally has a lever, roller, ball, or plunger actuator which causes a set of contacts to open or close. An *automatic condition* is any input which responds automatically to changes in a system. Flow switches, temperature switches, and pressure switches respond to automatic conditions. These devices automatically open and close sets of contacts when a change in the flow of a liquid is created, a change in temperature is sensed, or pressure varies. The signal accomplishes no work by itself, it merely starts or stops the flow of current in that part of the circuit.

Decisions

The decision section of a circuit determines what work is to be done and in what order the work is to occur. The decision section of a circuit adds, subtracts, sorts, selects, and redirects the signals from the control devices to the load. For the decision part of the circuit to perform a definite sequence, it must perform in a logical manner. The way the control devices are connected into the circuit gives the circuit logic. The basic logic functions are AND, OR, NOT, NOR, and NAND logic. The decision section of the circuit accepts informational inputs (signals), makes logical decisions based on the way the control devices are connected into the circuit, and provides the output signal that controls the load.

Action

Once a signal is generated and the decision has been made within a circuit, some action (work) should result. In most cases it is the operating coil in the circuit which is responsible for initiating the action. This action is direct when devices such as motors, lights, and heating elements are turned ON as a direct result of the signal and the decision. This action is indirect when the coils in solenoids, magnetic starters, and relays are energized. The action is indirect because the coil energized by the signal and the decision may energize a magnetic motor starter which actually starts the motor. Regardless of how this action takes place, the load causes some action (direct or indirect) in the circuit and, for this reason, is the action section of the circuit.

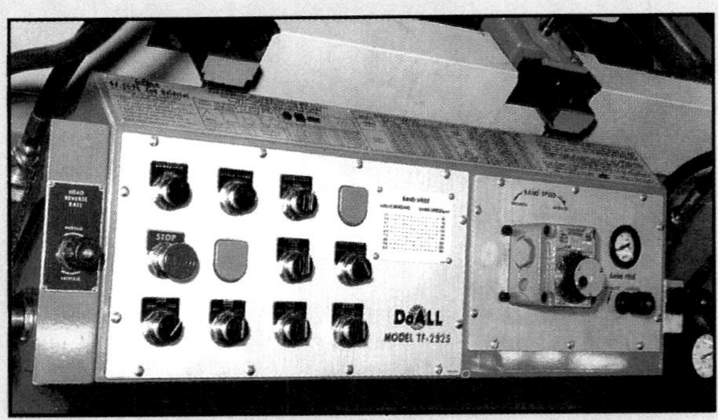

DoAll Company

The DoALL Model TF-2525 band saw includes manual, mechanical, and automatic inputs.

LOGIC FUNCTIONS

Control devices such as pushbuttons, limit switches, and pressure switches are connected into a circuit so that the circuit can function in a predetermined manner. All control circuits are basic logic functions or combinations of logic functions. Logic functions are common to all areas of industry. This includes electricity, electronics, hydraulics, pneumatics, math, and other routine activities. See Appendix. Logic functions include AND, OR, AND/OR, NOT, NOR, and NAND.

AND Logic

AND logic is used in industry when two pushbuttons are connected in series to control a solenoid. See Figure 3-14.

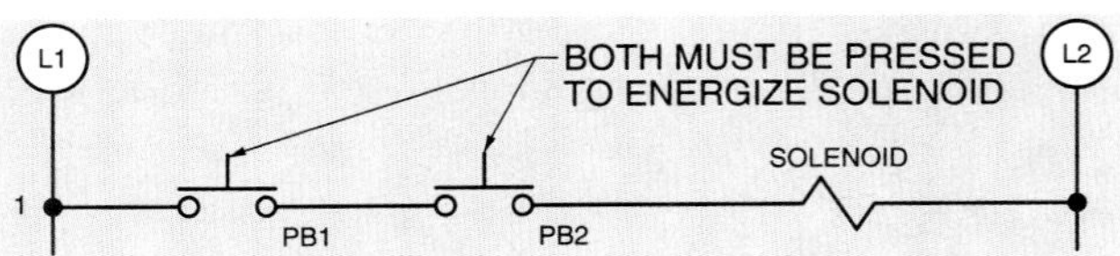

Figure 3-14. In AND logic, the load is ON if both of the control signal's contacts are closed.

PB1 and PB2 must be pressed before the solenoid is energized. The logic function that makes up the decision section of this circuit is AND logic. The reason for using the AND function could be to build in safety for the operator of this circuit.

If the solenoid were operating a punch press or shear, the pushbuttons could be spaced far enough apart so that the operator would have to use both hands to make the machine operate. This ensures that the operator's hands are not near the machine when it is activated. With AND logic, the load is ON only if all the control signal contacts are closed. As with any logic function, the signals may be manually, mechanically, or automatically controlled. Any control device such as limit switches, pressure switches, etc., with NO contacts can be used in developing AND logic. The NO contacts of each control device must be connected in series for AND logic.

A simple example of AND logic takes place whenever an automobile that has an automatic transmission is started. The ignition switch must be turned to the start position and the transmission selector must be in the park position before the starter is energized. Before the action (load ON) in the automobile circuit can take place, the control signals (manual) must be performed in a logical manner (decision).

OR Logic

OR logic is used in industry when a pushbutton and a temperature switch are connected in parallel. See Figure 3-15. In this circuit, the load is a heating element that is controlled by two control devices.

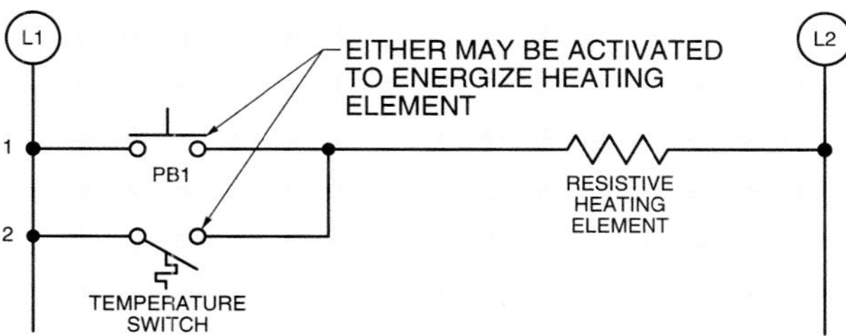

Figure 3-15. In OR logic, the load is ON if any one of the control signal's contacts is closed.

The logic of this circuit is OR logic because the pushbutton or the temperature switch energizes the load. The temperature switch is an example of an automatic control device that turns the heating element ON and OFF to maintain the temperature setting for which the temperature switch is set. The manually-controlled pushbutton could be used to test or turn ON the heating element when the temperature switch contacts are open.

In OR logic, the load is ON if any one of the control signal's contacts is closed. The control devices are connected in parallel. Series and parallel refer to the physical relationship of each control device to other control devices or components in the circuit. This series and parallel relationship is only part of what determines the logic function of any circuit.

An example of OR logic is in a dwelling that has two doorbells (pushbuttons) controlling one bell. The bell (load) may be energized by pressing (signal ON) either the front or the back pushbutton (control device). Here, as in the automobile circuit, the control devices are connected to respond in a logical manner.

AND/OR Logic Combination

The decision section of any circuit may contain one or more logic functions. See Figure 3-16. In this circuit, both pressure and flow must be present in addition to the pushbutton or the foot switch being engaged to energize the starter coil (load). This provides the circuit with the advantage of both AND logic and OR logic. The machine is protected because both pressure and flow must exist before it is started, and there is a choice between using a pushbutton or a foot switch for final operation. The action taking place in this circuit is energizing a coil in a magnetic motor starter. The signal inputs for this circuit have to be two automatic and at least one manual.

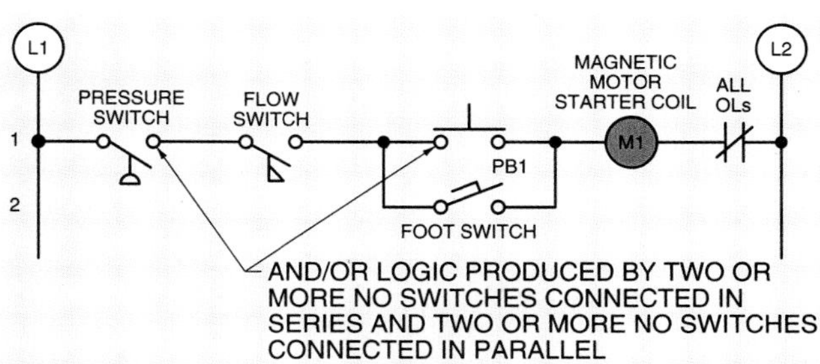

Figure 3-16. The decision section of any circuit may contain one or more logic functions.

Each control device responds to its own input signal and has its own decision-making capability. When multiple control devices are used in combination with other control devices making their own decisions, a more complex decision can be made through the combination of all control devices used in the circuit. All industrial control circuits consist of control devices capable of making decisions in accordance with the input signals received.

NOT Logic

NOT logic has an output if the control signal is OFF. For example, replacing NO contacts on a pushbutton with NC contacts energizes the solenoid and pilot light without pressing the pushbutton. See Figure 3-17. Pressing the pushbutton in this circuit de-energizes the loads. There must not be a signal if the loads are to remain energized. With NOT logic, the output remains ON only if the control signal contacts remain closed.

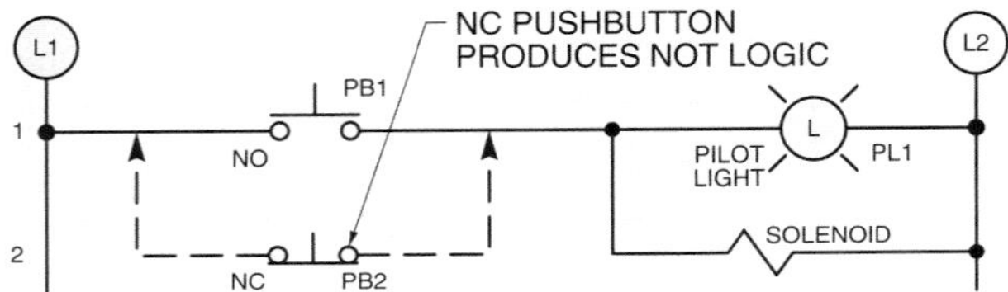

Figure 3-17. In NOT logic, the load is ON only if the control signal contacts are closed.

An example of NOT logic is the courtesy light in a refrigerator. The light is ON if the control signal is OFF. The control signal is the door of the refrigerator. Any time the door is open (signal OFF), the load (courtesy light) is ON. The condition that controls the signal can be manual, mechanical, or automatic. With the refrigerator door, the condition is mechanical.

NOR Logic

NOR logic is an extension of NOT logic in that two or more NC contacts in series are used to control a load. See Figure 3-18. In this circuit, additional operator safety is provided by adding several emergency stop pushbuttons (NOT logic) to the control circuit. Pressing any emergency stop pushbutton de-energizes the load (coil M1). By incorporating NOR logic, each machine may be controlled by one operator, but any operator or supervisor can have the capability of turning OFF all the machines on the assembly line to protect individual operators or the entire system. With the knowledge of NOR logic, the electrician can readily add stop pushbuttons by wiring them in series to perform their necessary function.

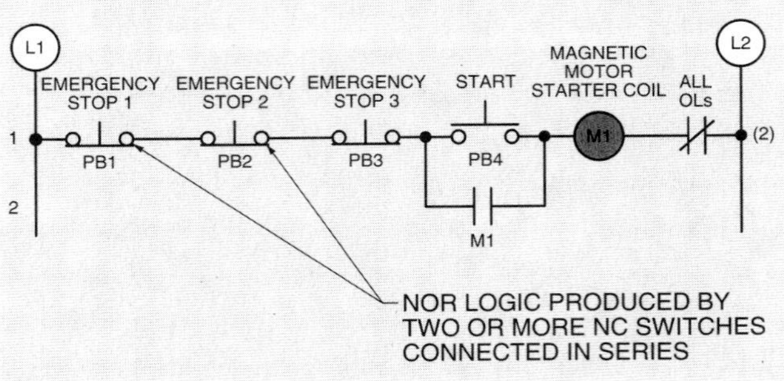

Figure 3-18. NOR logic is an extension of NOT logic in that two or more NC contacts in series are used to control a load.

NAND Logic

NAND logic is an extension of NOT logic in which two or more NC contacts are connected in parallel to control a load. See Figure 3-19. In this circuit, two interconnecting tanks are filled with a liquid. When pushbutton PB3 is pressed, coil M1 is energized and auxiliary contacts M1 close until both tanks are filled. Both tanks fill to a predetermined level because the float switch in tank 1 and tank 2 do not open until both tanks are full. Every NOT must be open (signal OFF) to stop the filling process based on the input of the float switches. NOR logic is also present in this circuit because the emergency stops (NOT) at tank 1 or tank 2 may stop the process if an operator at either of the tanks see a problem.

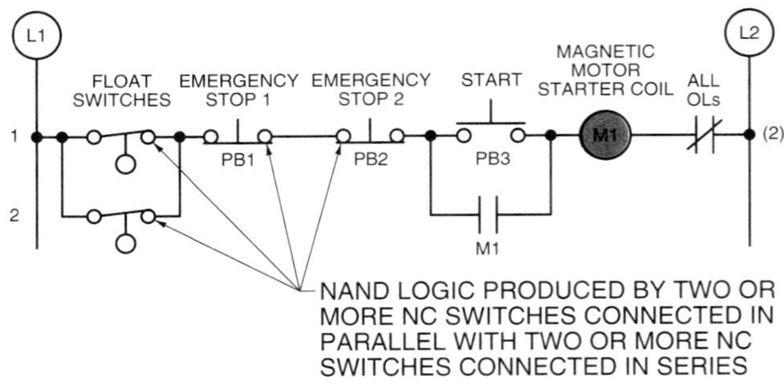

Figure 3-19. NAND logic is an extension of NOT logic in which two or more NC contacts are connected in parallel to control a load.

An example of a NAND circuit is the courtesy light in an automobile. In an automobile, the light is ON if the control signal is OFF. This circuit is different from a refrigerator door in that an automobile may have two or more doors which must not be open for the load to be OFF. Every NOT (signal OFF) energizes the load.

Memory

Many of today's industrial circuits require their control circuits not only to make logic decisions such as AND, OR, and NOT, but also to be capable of storing, memorizing, or retaining the signal inputs to keep the load energized even after the signals are removed. A switch that controls house lights from only one location is an example of a memory circuit. When it is ON, it remains ON until it is turned OFF, and remains OFF until it is turned ON. It performs a memory function because the output corresponds to the last input information until new input information is received to change it. In the case of the house light switch, the memory circuit was accomplished by a switch that mechanically stays in one position or another.

In industrial control circuits, it is more common to find pushbuttons with return spring contacts than those which mechanically stay held in one position. Auxiliary contacts are added to give circuits with pushbuttons memory. See Figure 3-20. Once coil M1 of the magnetic motor starter is energized, it causes coil contacts M1 to close and remain closed (memory) until the coil is de-energized.

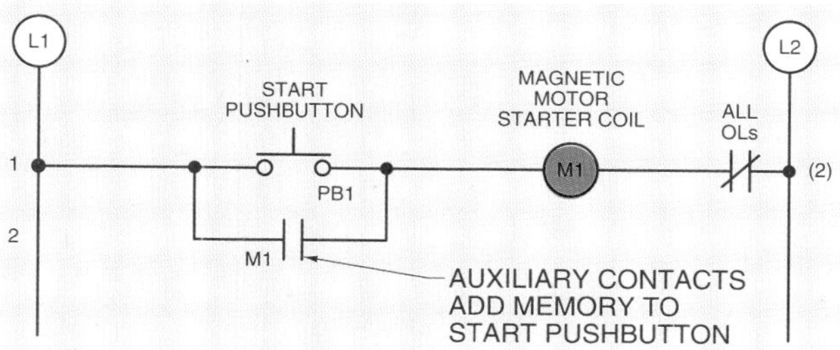

Figure 3-20. Auxiliary contacts are added to give circuits with pushbuttons memory.

NOT logic may be added to memory logic to create a common start/stop control circuit. See Figure 3-21. When stop pushbutton PB1 is activated, current to coil M1 stops and contacts M1 open, returning the circuit to its original condition.

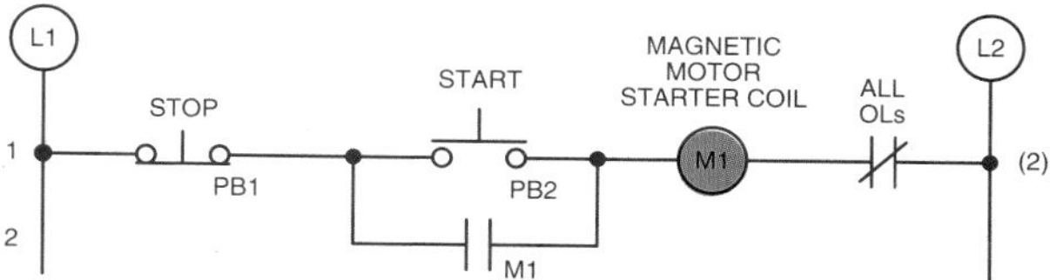

Figure 3-21. A common start/stop control circuit is created by adding the NOT logic of a stop pushbutton to the memory logic of magnetic coil contacts.

COMMON CONTROL CIRCUITS

Various control circuits are commonly used in commercial and industrial electrical circuits. An electrician must understand the entire circuit operation to begin wiring or troubleshooting the circuit.

Heidelberg Harris, Inc.

The JF-35 combination folder from Heidelberg Harris includes different control circuits combined into one common circuit.

Start/Stop Stations Controlling Magnetic Starters

A load is often required to be started and stopped from more than one location. See Figure 3-22. In this circuit, the magnetic motor starter may be started or stopped from two locations. Additional stop pushbuttons are connected in series (NOR logic) with the existing stop pushbuttons.

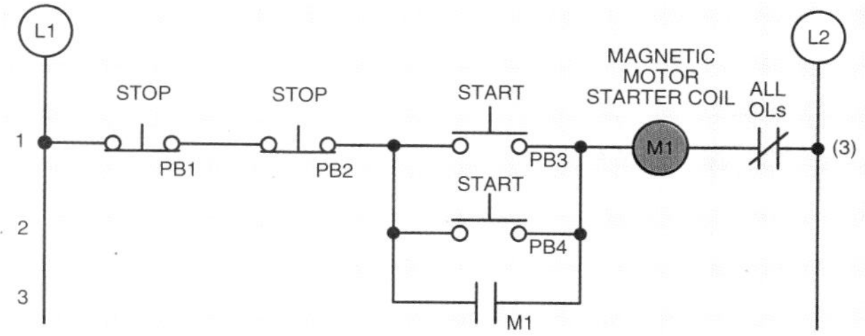

Figure 3-22. Two stop pushbuttons connected in series and two start pushbuttons connected in parallel are used to control a motor from two locations.

Additional start pushbuttons are connected in parallel (OR logic) with the existing start pushbuttons. Pressing any one of the start pushbuttons (PB3 or PB4) causes coil M1 to energize. This causes auxiliary contacts M1 to close, adding memory to the circuit until coil M1 is de-energized. Coil M1 may be de-energized by pressing stop pushbuttons PB1 or PB2, by an overload which would activate the OLs, or by a loss of voltage to the circuit. In the case of an overload, the overload has to be removed and the circuit overload devices reset before the circuit would return to normal starting condition.

Two Magnetic Starters Operated by Two Start/Stop Stations with Common Emergency Stop

In almost all electrical systems, several devices can be found running off a common supply voltage. Two start/stop stations may be used to control two separate magnetic motor starter coils with a common emergency stop protecting the entire system. See Figure 3-23. Pressing start pushbutton PB3 causes coil M1 to energize and seal in auxiliary contacts M1. Pressing start pushbutton PB5 causes coil M2 to energize and seal in auxiliary contacts M2. Once the entire circuit is operational, emergency stop pushbutton PB1 can shut down the entire circuit or the individual stop pushbuttons PB2 or PB4 can de-energize the coils in their respective circuits. Each circuit is overload protected and does not affect the other when one magnetic motor starter experiences a problem.

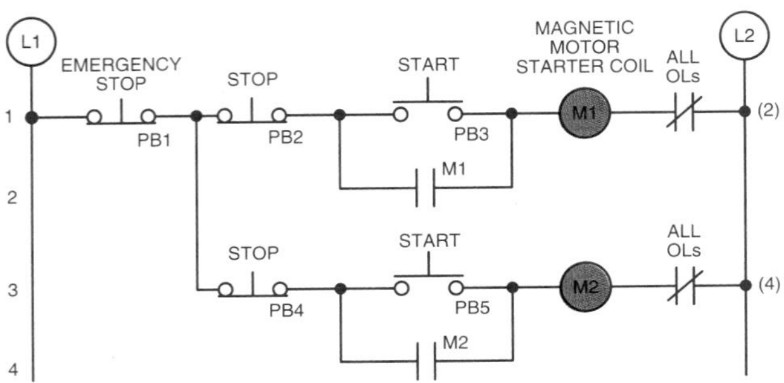

Figure 3-23. Two start/stop stations are used to control two separate magnetic motor starter coils with a common emergency stop protecting the entire system.

Start/Stop Station Controlling Two or More Magnetic Starters

Steel mills, paper mills, bottling plants, and canning plants are industries which require simultaneous operation of two or more motors. In each industry, products or materials are spread out over great lengths which must be started together to prevent product separation or stretching. To accomplish this, two motors can be started almost simultaneously from one location. See Figure 3-24.

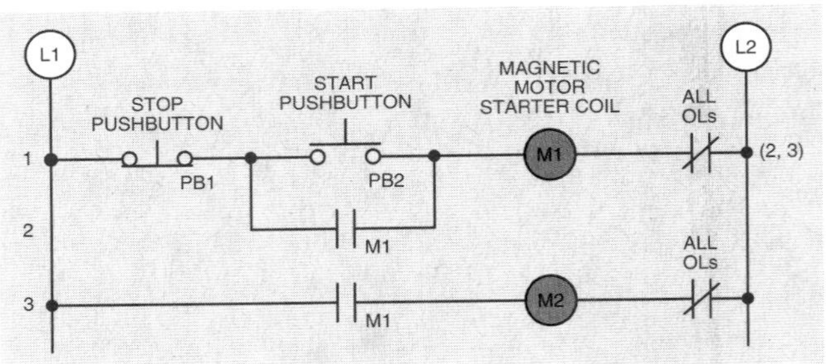

Figure 3-24. Two motors can be started almost simultaneously from one location to prevent product separation or stretching.

In this circuit, pressing start pushbutton PB2 energizes coil M1 and seals in both sets of auxiliary contacts M1. *Note:* It is acceptable to have more than one set of auxiliary contacts controlled by one coil. When both sets of contacts close, the first set of M1 contacts (line 2) provides memory for the start pushbutton and completes the circuit to energize coil M1.

Furnas Electric Co.

Pushbuttons, selector switches, and indicator lights may be combined to create an operator station to control any industrial process.

The second set of M1 contacts (line 3) completes the circuit to coil M2, energizing coil M2. The motors associated with these magnetic motor starters start almost simultaneously because both coils energize almost simultaneously. Pushing the stop pushbutton breaks the circuit (line 1), de-energizing coil M1. When coil M1 drops out, both sets of auxiliary contacts are deactivated. The motors associated with these magnetic motor starters stop almost simultaneously because both coils de-energize almost simultaneously. An overload in magnetic motor starter M2 affects only the operation of coil M2. The entire circuit is shut down if an overload exists in motor starter M1. The entire circuit stops because de-energizing coil M1 also affects both sets of auxiliary contacts M1. This protection might be used where a machine, such as an industrial drill, would be damaged if the cooling liquid pump shut OFF while the drill was still operating.

Pressure Switch with Pilot Light Indicating Device Activation

Pilot lights are manufactured in a variety of colors, shapes, and sizes to meet the needs of industry. The illumination of these lights signals an operator that any one of a sequence of events may be taking place. A pilot light may be used with a pressure switch to indicate when a device is activated. See Figure 3-25.

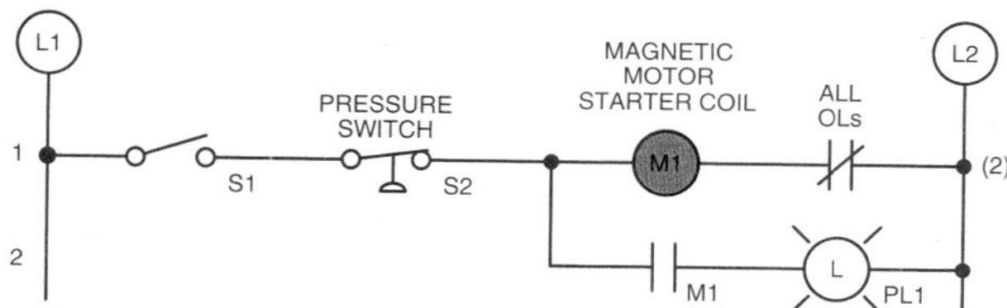

Figure 3-25. A pilot light is used with a pressure switch to indicate when a device is activated.

Square D Company

The variety of selector switches and indicator lights manufactured by Square D Company enables the design of many custom control circuits.

In this circuit, pressure switch S2 has automatic control over the circuit when switch S1 is closed. When the pressure to switch S2 drops, the switch closes and activates coil M1 which controls the magnetic starter of the compressor motor, starting the compressor. At the same time, contacts M1 close and pilot light PL1 turns ON. The compressor continues to run and the pilot light stays ON as long as the motor runs. When pressure builds sufficiently to open pressure switch S2, coil M1 de-energizes and the magnetic motor starter drops out, stopping the compressor motor. The pilot light goes out because contact M1 controlled by coil M1 opens. The pilot light is ON only when the compressor motor is running. This circuit might be used in a garage to let the owner know when the air compressor is ON or OFF.

A pilot light may be used with a start/stop station to indicate when a device is activated. See Figure 3-26. In this circuit, pressing start pushbutton PB2 energizes coil M1, causing auxiliary contacts M1 to close. Closing contacts M1 provides memory for start pushbutton PB2 and maintains an electrical path for the pilot light. As long as coil M1 is energized, the pilot light stays ON. Pressing stop pushbutton PB1 de-energizes coil M1, opening contacts M1 and turning OFF the pilot light. An overload in this circuit also de-energizes coil M1, opening contacts M1 and turning OFF the pilot light. A circuit like this can be used as a positive indicator that some process is taking place. The process may be in a remote place such as in a pump well or in another building.

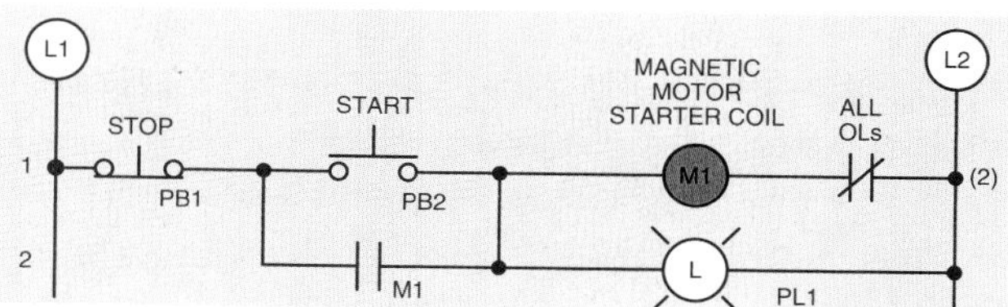

Figure 3-26. A pilot light is used with a start/stop station to indicate when a device is activated.

Start/Stop Station with Pilot Light Indicating No Device Activation

Pilot lights may be used to show when an operation is stopped as well as when it is started. NOT logic is used in a circuit when a pilot light is used to show that an operation has stopped. NOT logic is established by placing one set of NC contacts in series with a device. See Figure 3-27.

Figure 3-27. NOT logic is used to indicate when a device is not operating.

In this circuit, pressing start pushbutton PB2 energizes coil M1, causing both sets of auxiliary contacts M1 to energize. NO contacts M1 (line 2) close, providing memory for PB2, and NC contacts M1 (line 3) open, disconnecting pilot light PL1 from the line voltage, causing the light to turn OFF. Pressing stop pushbutton PB1 de-energizes coil M1, causing both sets of contacts to return to their normal positions. NO contacts M1 (line 2) return to their NO position, and NC contacts M1 (line 3) return to their NC position, causing the pilot light to be reconnected to the line voltage and causing it to turn ON. The pilot light is ON only when the coil to the magnetic motor starter is OFF. A bell or siren could be substituted for the pilot light to serve as a warning device. A circuit like this is used to monitor critical operating procedures such as a cooling pump for a nuclear reactor. When the cooling pump stops, the pilot light, bell, or siren immediately calls attention to the fact that the process has been stopped.

Cutler-Hammer

Conveyor systems may use a sequence control circuit to prevent material pileup.

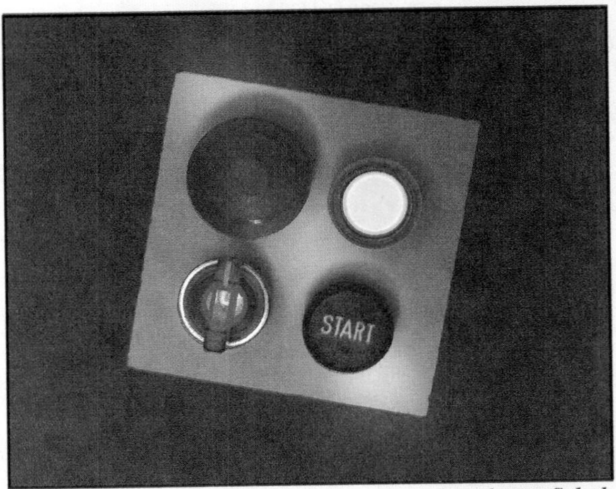

Sprecher + Schuh

Pushbuttons, selector switches, and indicator lights control and provide indication of the status of motor control circuits.

Pushbutton Sequence Control

Conveyor systems often require one conveyor system to feed boxes or other materials onto another conveyor system. A circuit is needed to prevent the pileup of material on the second conveyor if the second conveyor is stopped. A sequence control circuit does not let the first conveyor operate unless the second conveyor has started and is running. See Figure 3-28.

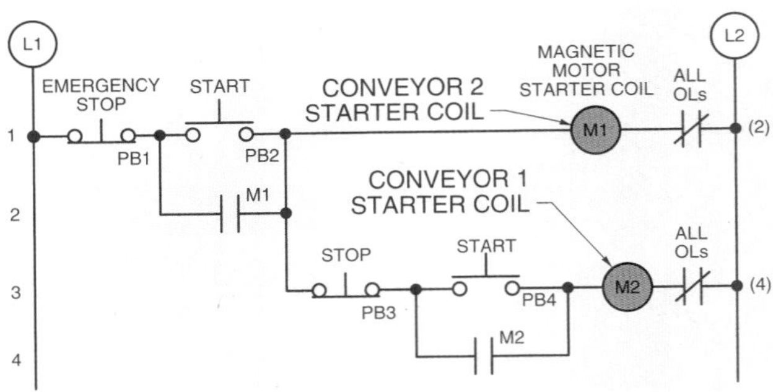

Figure 3-28. A sequence control circuit does not let the first conveyor operate unless the second conveyor has started and is running.

In this circuit, pressing start pushbutton PB2 energizes coil M1 and causes auxiliary contacts M1 to close. With auxiliary contacts M1 closed, PB2 has memory and provides an electrical path to allow coil M2 to be energized when start pushbutton PB4 is pressed.

With start pushbutton PB4 pressed, coil M2 energizes and closes contacts M2, providing memory for start pushbutton PB4 so that both conveyors run. Conveyor 1 (coil M2) cannot start unless conveyor 2 (coil M1) is energized. Both conveyors shut down if an overload occurs in the circuit with coil M1 or if stop pushbutton PB1 is pressed. Only conveyor 1 shuts down if conveyor 1 (coil M2) experiences an overload. A problem in conveyor 1 does not affect conveyor 2.

Jogging with a Selector Switch

Jogging is the frequent starting and stopping of a motor for short periods of time. Jogging is used to position materials by moving the materials small distances each time the motor starts. A selector switch is used to provide a common industrial jog/run circuit. See Figure 3-29. The selector switch (two-position switch) is used to manually open or close a portion of the electrical circuit. In this circuit, the selector switch determines if the circuit is a jog circuit or run circuit. With selector switch S1 in the open (jog) position, pressing start pushbutton PB2 energizes coil M1, causing the magnetic motor starter to operate. Releasing start pushbutton PB2 de-energizes coil M1, causing the magnetic motor starter to stop. With selector switch S1 in the closed (run) position, pressing start pushbutton PB2 energizes coil M1, closing auxiliary contacts M1, providing memory so that the magnetic starter operates and continues to operate until stop pushbutton PB1 is pressed.

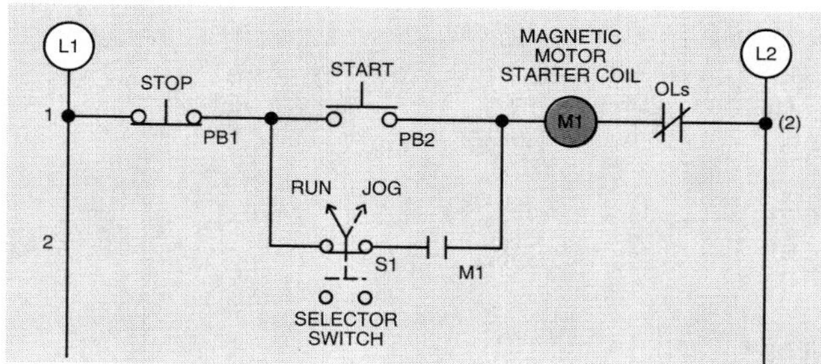

Figure 3-29. A selector switch is used to provide a common industrial jog/run circuit.

When stop pushbutton PB1 is pressed, coil M1 de-energizes and all circuit components return to their original condition. The overloads may also open the circuit and must be reset after the overload is removed to return the circuit to normal operation. This circuit may be found where an operator may run a machine continuously for production, but may stop it at any time for small adjustments or repositioning.

1. How many electrical loads can be placed in the control circuit of any one line between L1 and L2?
2. How are the loads connected if more than one load must be connected in a line diagram?
3. All loads are connected directly or indirectly to which power line?
4. Where are control devices connected in a line diagram?
5. How is each line in a line diagram marked to distinguish that line from all other lines?
6. How are NO and NC contacts identified when using the numerical cross-reference system?
7. In what order are wire-reference numbers normally assigned?
8. What are the two methods used to illustrate how contacts found in different control lines belong to the same control switch, such as a limit switch?
9. What are the three basic sections of all control circuits?
10. What is the signal(s) section of a control circuit?
11. What is the decision section of a control circuit?
12. What is the action section of a control circuit?
13. What is AND logic as applied to control circuits?
14. What is OR logic as applied to control circuits?
15. What is NOT logic as applied to control circuits?
16. What is NOR logic as applied to control circuits?
17. What is NAND logic as applied to control circuits?
18. What is memory logic as applied to control circuits?
19. When additional stops are added to a control circuit, should they be connected in series or parallel?
20. When additional starts are added to a control circuit, should they be connected in series or parallel?
21. What is jogging?

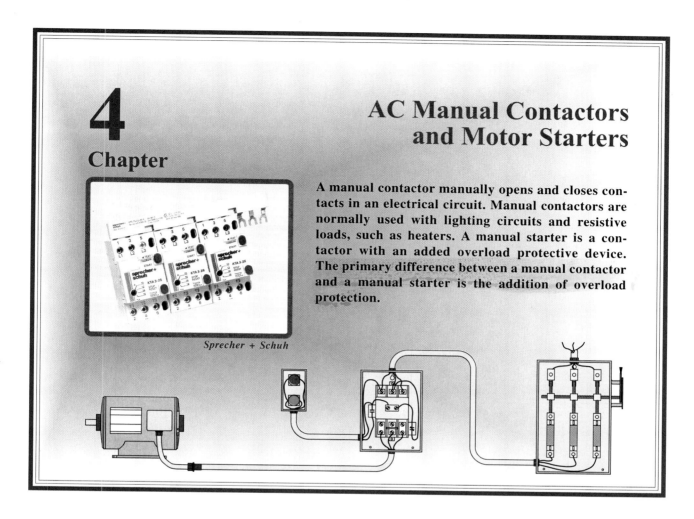

4 Chapter

AC Manual Contactors and Motor Starters

A manual contactor manually opens and closes contacts in an electrical circuit. Manual contactors are normally used with lighting circuits and resistive loads, such as heaters. A manual starter is a contactor with an added overload protective device. The primary difference between a manual contactor and a manual starter is the addition of overload protection.

Sprecher + Schuh

MANUAL SWITCHING

The starting and stopping of electric motors was first introduced in the late 1800s and was accomplished through the use of knife switches. See Figure 4-1. Knife switches were eventually discontinued as a means of connecting and disconnecting line voltage directly to motor terminals for three basic reasons. First, the open knife switch had exposed (live) parts which presented an extreme electrical hazard to the operator. In addition, any applications where dirt or moisture were present made the open switch concept extremely vulnerable to problems. Second, the speed of opening and closing contacts was determined solely by the operator. Considerable arcing and pitting of the contacts led to rapid wear and eventual replacement if the operator did not open or close the switch quickly. Finally, most knife switches were made of soft copper which required replacement after repeated arcing, heat generation, and mechanical fatigue.

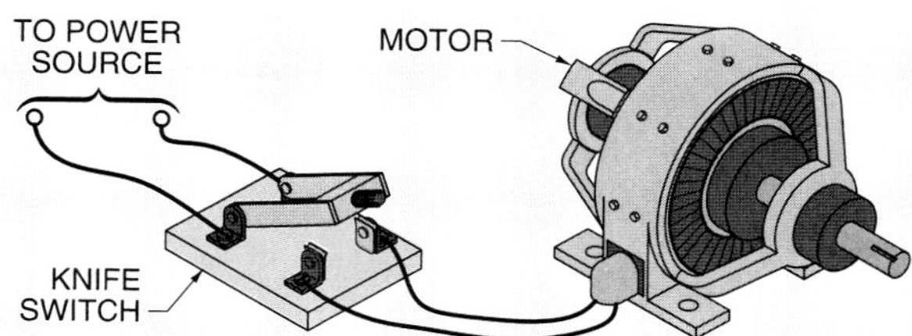

Figure 4-1. Knife switches were the first devices used to start and stop electric motors.

Mechanical Improvements

As industry demanded more electric motors at the turn of the century, improvements were made to knife switches to make them more acceptable as a controller. First, the knife switch was enclosed in a steel enclosure to protect the switch. An insulated external handle was added to protect the operator. Second, an operating spring was attached to the handle to ensure quick opening and closing of the knife blade.

The switch handle was designed so that once the handle was moved a certain distance, the tension on the spring forced the contacts to open or close at the same continuous speed each time it was operated. See Figure 4-2.

Figure 4-2. A knife switch is enclosed in a steel enclosure, has an insulated external handle, and includes an operating spring for improved operation and safety.

Even with these improvements, the blade and jaw mechanism of a knife switch had a short mechanical life when the knife switch was used as a direct controller. The knife switch mechanism was discontinued as a means of direct control for motors because of the short life. Knife switches are currently used as electrical disconnects. A *disconnect* is a device used only periodically to remove electrical circuits from their supply source. The mechanical life of the knife switch mechanism is not of major concern because a disconnect is used infrequently.

MANUAL CONTACTORS

A *manual contactor* is a control device that uses pushbuttons to energize or de-energize the load connected to it. See Figure 4-3. A manual contactor manually opens and closes contacts in an electrical circuit. Manual contactors cannot be used to start and stop motors because they have no overload protection built into them. Manual contactors are normally used with lighting circuits and resistive loads, such as heaters. A fuse or circuit breaker is normally included in the same enclosure with the manual contactor.

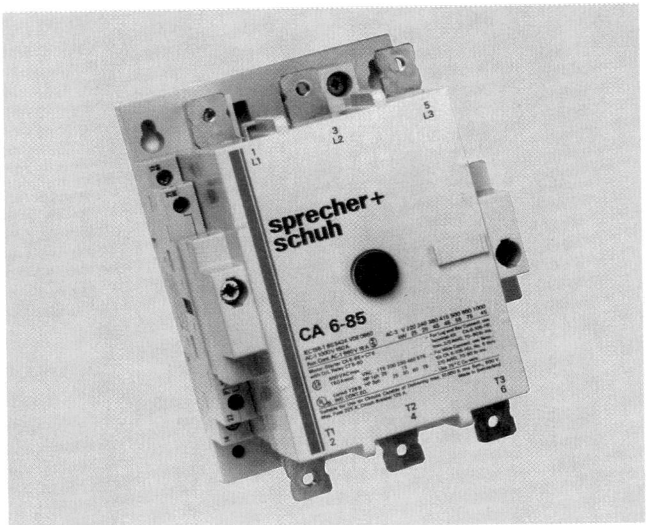

Sprecher + Schuh

Figure 4-3. A manual contactor uses pushbuttons to energize or de-energize the load connected to it.

Double-Break Contacts

Double-break contacts have replaced the knife switch as a direct controller. *Double-break contacts* are contacts that break the electrical circuit in two places. See Figure 4-4.

Rockwell Automation, Allen-Bradley Company, Inc.

Contactors include power contacts that are designed to switch the large currents drawn by high-power loads.

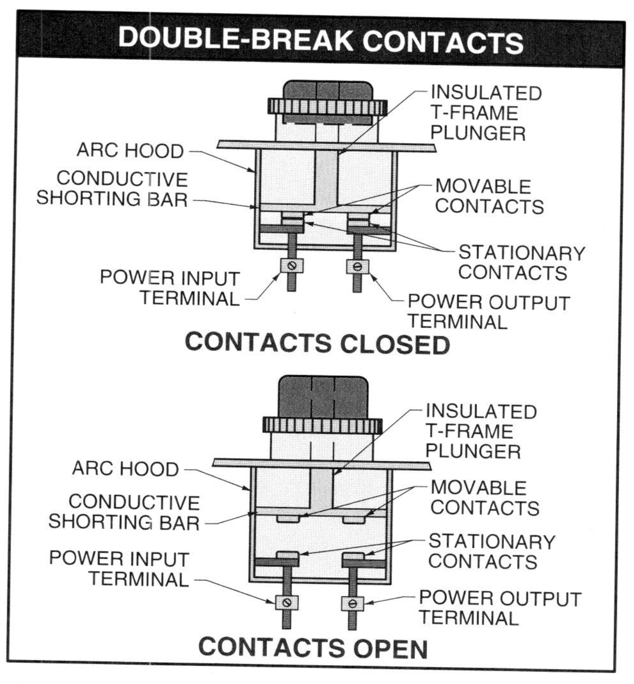

Figure 4-4. Double-break contacts break the electrical circuit in two places.

Double-break contacts allow devices to be designed that have a higher contact rating (current rating) in a smaller space than devices designed with single-break contacts. With double-break contacts, the movable contacts are forced against the two stationary contacts to complete the electrical circuit when a set of normally open (NO) double-break contacts are energized. The movable contacts are pulled away from the stationary contacts and the circuit is open when the manual contactor is de-energized. The procedure is reversed when normally closed (NC) double-break contacts are used.

A 3ϕ manual contactor has three sets of NO double-break contacts. One set of NO double-break contacts is used to open and close each phase in the circuit. The movable contacts are located on an insulated T-frame and are provided with springs to soften their impact. The T-frame is activated by a pushbutton mechanism. The mechanical linkage consistently and quickly makes or breaks the circuits similar to that of a disconnect. See Figure 4-5.

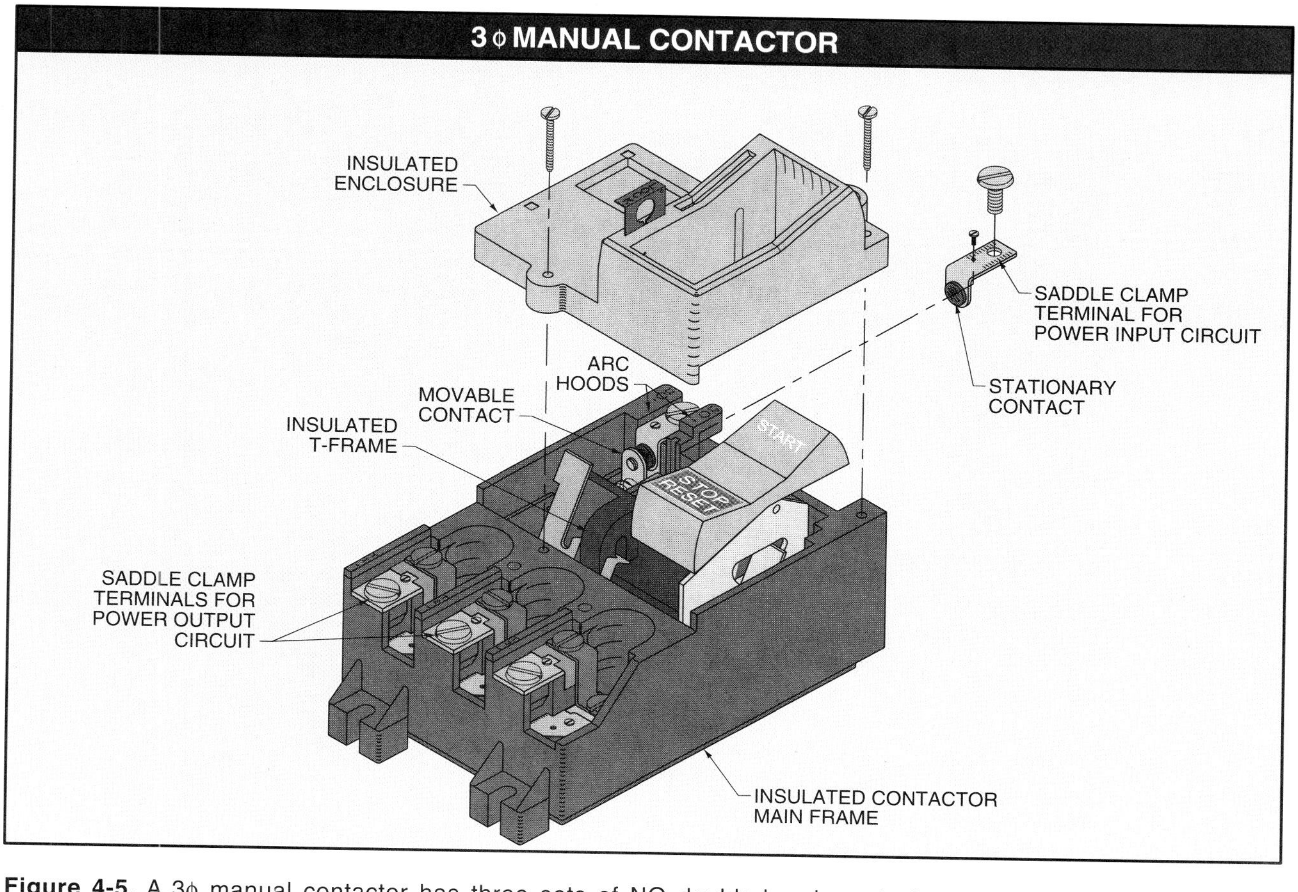

Figure 4-5. A 3ϕ manual contactor has three sets of NO double-break contacts.

The movable contacts have no physical connection to external electrical wires. The movable contacts move into the arc hoods and bridge the gap between the set of fixed contacts to make or break the circuit. All physical electrical connections are made indirectly to the fixed contacts through saddle clamps.

Contact Construction

Metal alloys have helped contactors become popular control devices. In the past, a major problem with knife switches was that they were constructed from soft copper. Today, most contacts are made of a low-resistance silver alloy. Silver is alloyed (mixed) with cadmium or cadmium oxide to make an exceptionally arc-resistant material which has good conductivity. In addition, the silver alloy has good mechanical strength, enabling it to endure the continual wear encountered by many opening and closings. Another advantage of silver-alloy contacts is that the oxide (rust) that forms on the metal is an excellent conductor of electricity. Even when the contacts appear dull or tarnished, they are still capable of operating normally. See Figure 4-6.

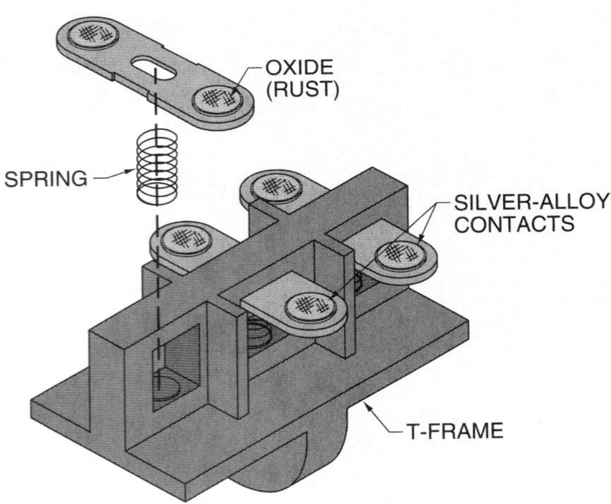

Figure 4-6. The oxide that forms on silver-alloy contacts is an excellent conductor of electricity.

Manual contactors directly control power circuits. Power circuit wiring is shown on a wiring diagram. A *wiring diagram* is a diagram that shows the connection of an installation or its component devices or parts. A wiring diagram shows, as closely as possible, the actual connection and placement of all component devices or parts in a circuit, including power circuit wiring. An understanding of a wiring diagram is required because an electrician may be required to make changes in power circuits as well as in control circuits. See Figure 4-7.

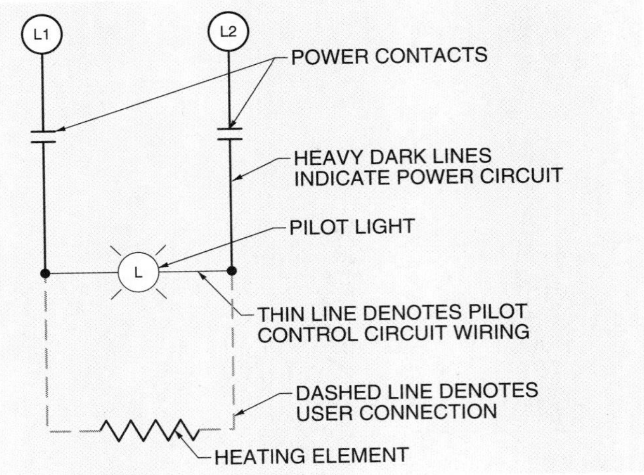

Figure 4-7. A wiring diagram shows the connection of an installation or its component devices or parts.

The wiring diagram for a single-pole manual contactor and pilot light shows the power contacts and how they are connected to the load. Like the line diagram, the power circuit is indicated through heavy dark lines and the control circuit is indicated by thin lines. In this circuit, current passes from L1 through the pilot light on to L2, causing the pilot lamp to glow when power contacts in L1 and L2 close. At the same time, current passes from L1 through the heating element and on to L2, causing the heating element to be activated. The pilot light and heating element are connected in parallel with each other.

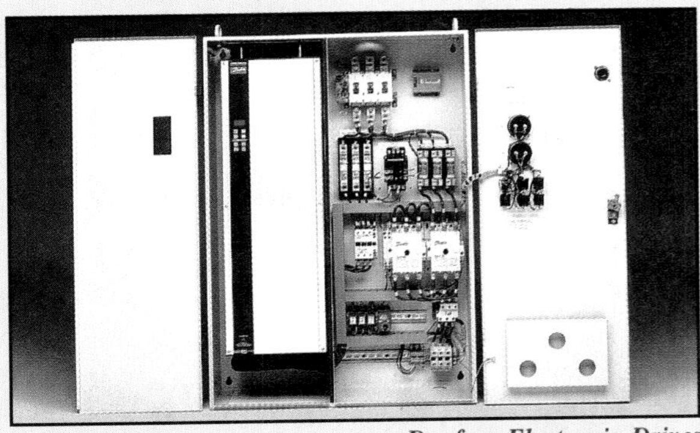

Danfoss Electronic Drives

Contacts are placed inside motor control cabinets along with other components used to control a system.

Wiring diagrams may be complex. For example, the wiring diagram for a dual-element heater with pilot lights contains various circuit paths. In this circuit, the low-heat heating element is operated when the low contacts in L1 and L2 are closed so that a connection is made to the low and common terminals of the heater. This allows the low-heat heating element to be energized. See Figure 4-8.

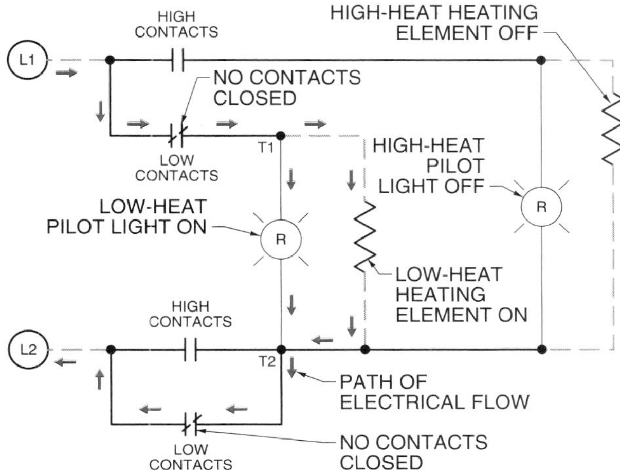

Figure 4-8. In the wiring diagram for a dual-element heater with pilot lights, the low-heat heating element is operated when the low contacts in L1 and L2 are closed so that a connection is made to the low and common terminals of the heater.

To operate the high-heat heating element, the high contacts in L1 and L2 are closed so that a connection is made to the high and common terminals of the heater. This allows the high-heat heating element to be energized. A low-heat pilot light and high-heat pilot light turn ON to indicate each condition because each pilot light is in parallel with the appropriate heating element. See Figure 4-9.

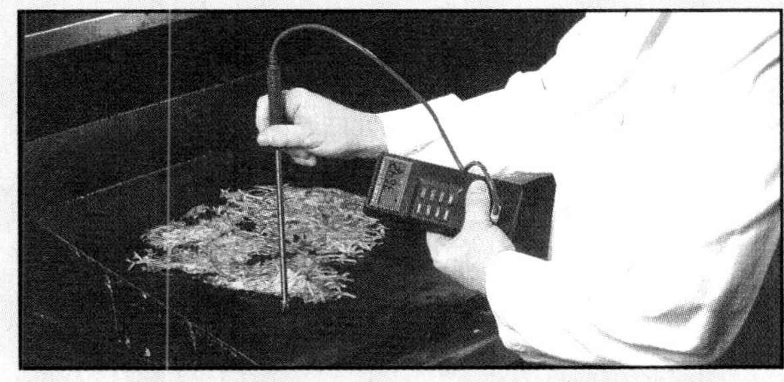

Fluke Corporation

The actual temperature output of heating elements can be measured using a temperature probe.

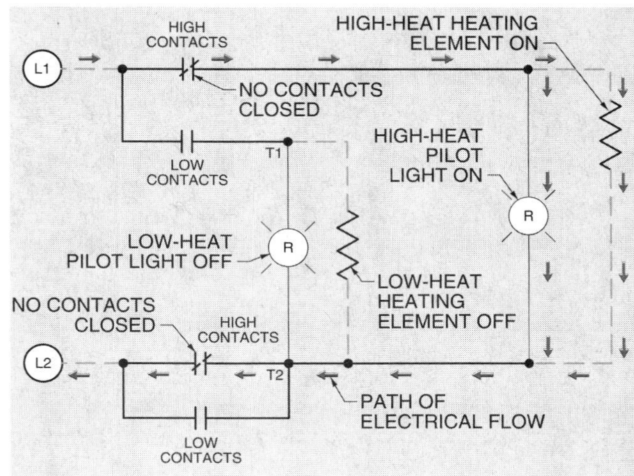

Figure 4-9. In the wiring diagram for a dual-element heater with pilot lights, the high-heat heating element is operated when the high contacts in L1 and L2 are closed so that a connection is made to the high and common terminals of the heater.

One problem that may arise with a dual-element start is that someone may try to energize both sets of elements at the same time. This causes serious damage to the heater. To prevent this problem from occurring, most manual contactors are equipped with a mechanical interlock. *Mechanical interlocking* is the arrangement of contacts in such a way that both sets of contacts cannot be closed at the same time. Mechanical interlocking can be established by a mechanism that forces open one set of contacts while the other contacts are being closed. Another method is to provide a blocking bar or holding mechanism that does not allow the first set of contacts to close until the second set of contacts open. An electrician can determine if a device is mechanically interlocked by consulting the wiring diagram information provided by the manufacturer. This information is normally packaged with the equipment when it is delivered or is glued to the inside of the enclosure.

MANUAL STARTERS

A *manual starter* is a contactor with an added overload protection device. Manual starters are used only in electrical motor circuits. The primary difference between a manual contactor and a manual starter is the addition of overload protection. See Figure 4-10.

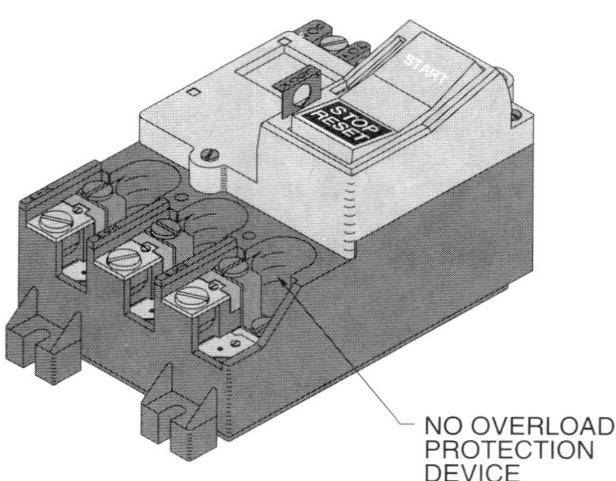

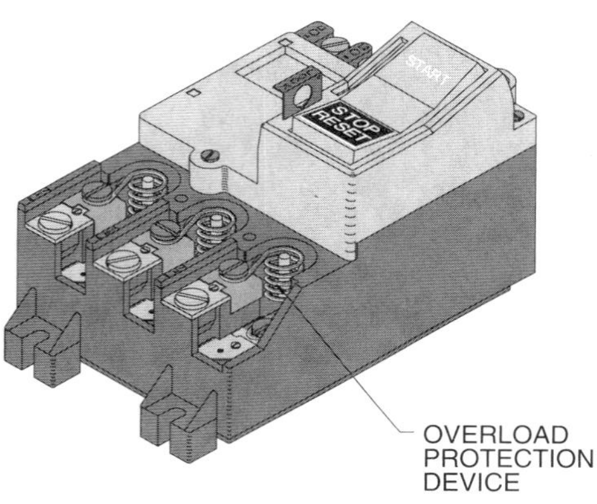

Figure 4-10. A manual starter is a contactor with an added overload protection device.

The overload protection device must be added because the National Electrical Code® (NEC®) requires that the control device shall not only turn a motor ON and OFF, but shall also protect the motor from destroying itself under an overloaded situation, such as a locked rotor. A *locked rotor* is a condition when a motor is loaded so heavily that the motor shaft cannot turn. A motor with a locked rotor draws excessive current and burns up if not disconnected from the line voltage. The overload device senses the excessive current and opens the circuit.

Overload Protection

A motor goes through three stages during normal operation. The three stages include resting, starting, and operating under load. See Figure 4-11. A motor at rest requires no current because the circuit is open. A motor that is starting draws a tremendous inrush current (normally six to eight times the running current) when the circuit is closed. Fuses or circuit breakers must have a sufficiently high ampere rating to avoid opening the circuit immediately by the large inrush current required by a motor when starting.

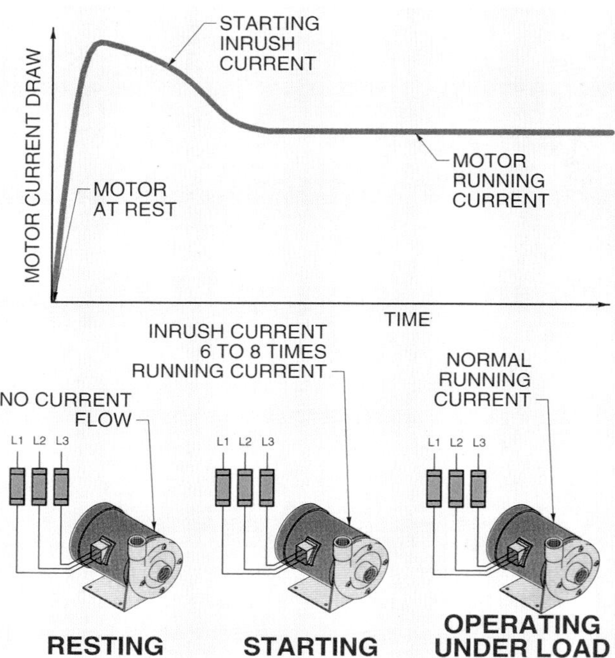

Figure 4-11. The three stages a motor goes through during normal operation include resting, starting, and operating under load.

A motor may encounter an overload while running. While it may not draw enough current to blow the fuses or trip circuit breakers, it is large enough to produce sufficient heat to burn up the motor. The intensive heat concentration generated by excessive current in the windings causes the insulation to fail and burn up the motor. It is estimated that for every 1°C rise over normal ambient temperatures, ratings for insulation can reduce the life expectancy of a motor almost a year per degree. *Ambient temperature* is the temperature of the air surrounding the motor. The normal rating for many motors is about 40°C (104°F).

Fuses or circuit breakers must protect the circuit against very high currents of a short circuit or a ground. An overload relay is required that does not open the circuit while the motor is starting, but opens the circuit if the motor gets overloaded and the fuses do not blow. See Figure 4-12.

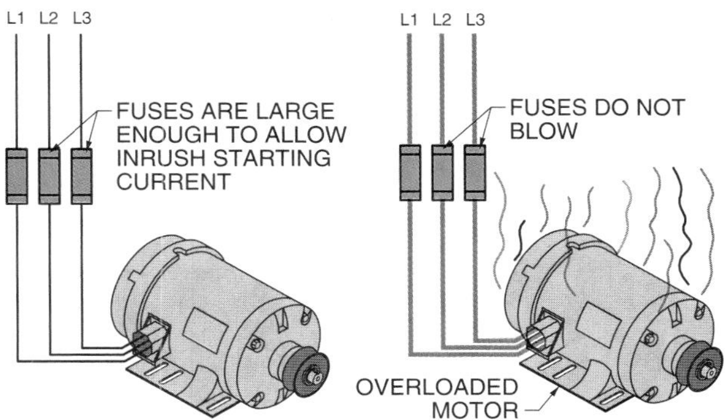

Figure 4-12. An overload relay does not open the circuit while the motor is starting, but opens the circuit if the motor gets overloaded and the fuses do not blow.

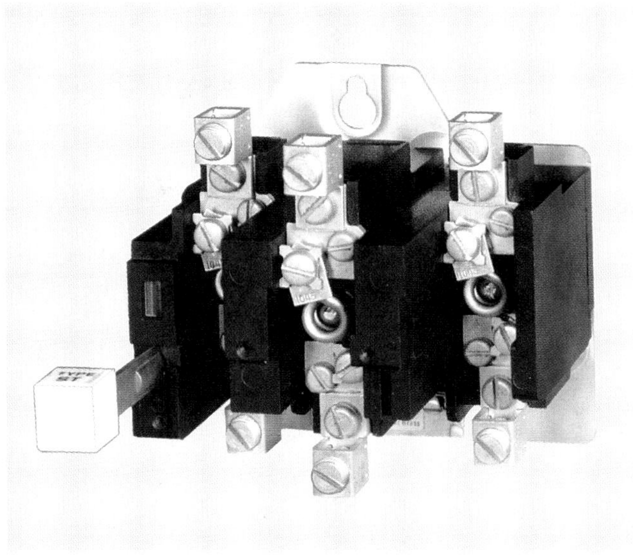

Cutler-Hammer

Figure 4-13. A heater coil is a sensing device used to monitor the heat generated by excessive current and the heat created through ambient temperature rise.

To meet motor protection needs, overload relays are designed to have a time delay to allow harmless, temporary overloads without disrupting the circuit. Overload relays must also have a trip capability to open the circuit if mildly dangerous currents that could result in motor damage continue over a period of time. All overload relays have some means of resetting the circuit once the overload is removed.

Melting Alloy Overloads. One of the most popular methods of providing overload protection is by using a melting alloy overload relay. Heat is the end product which destroys a motor. To be effective, an overload relay must measure the temperature of the motor by monitoring the amount of current being drawn. The overload relay must indirectly monitor the temperature conditions of the motor because the overload relay is normally located at some distance from the motor.

A *heater coil* is a sensing device used to monitor the heat generated by excessive current and the heat created through ambient temperature rise. Many different types of heater coils are available. The operating principle of each is the same. A heater coil converts the excess current drawn by the motor into heat which is used to determine whether the motor is in danger. See Figure 4-13.

Most manufacturers rely on a eutectic alloy in conjunction with a mechanical mechanism to activate a tripping device when an overload occurs. A *eutectic alloy* is a metal that has a fixed temperature at which it changes directly from a solid to a liquid state. This temperature never changes and is not affected by repeated melting and resetting.

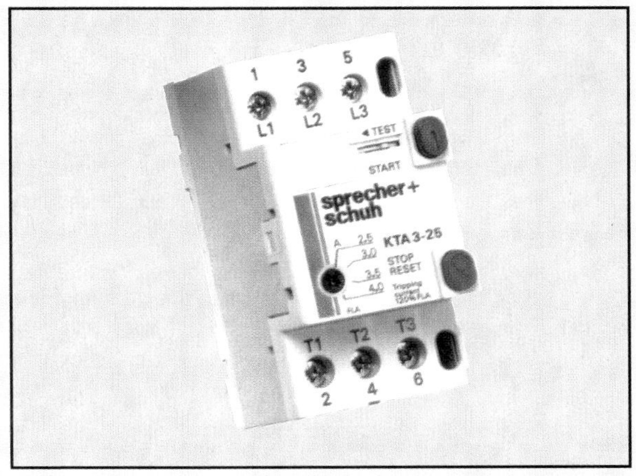

Sprecher + Schuh

The KTA3 line of motor circuit controllers from Sprecher + Schuh combines the functions of motor short circuit protection, thermal overload protection, switching, and signaling in one small package.

Most manufacturers use a ratchet wheel and eutectic alloy tube combination to activate a trip mechanism when an overload occurs. The eutectic alloy tube consists of an outer tube and an inner shaft connected to a ratchet wheel. The ratchet wheel is held firmly in the tube by the solid eutectic alloy. The inner shaft and ratchet wheel are locked into position by a pawl (locking mechanism) so that the wheel cannot turn when the alloy is cool. See Figure 4-14. Excessive current applied to the heater coil melts the eutectic alloy. This allows the ratchet wheel to turn freely.

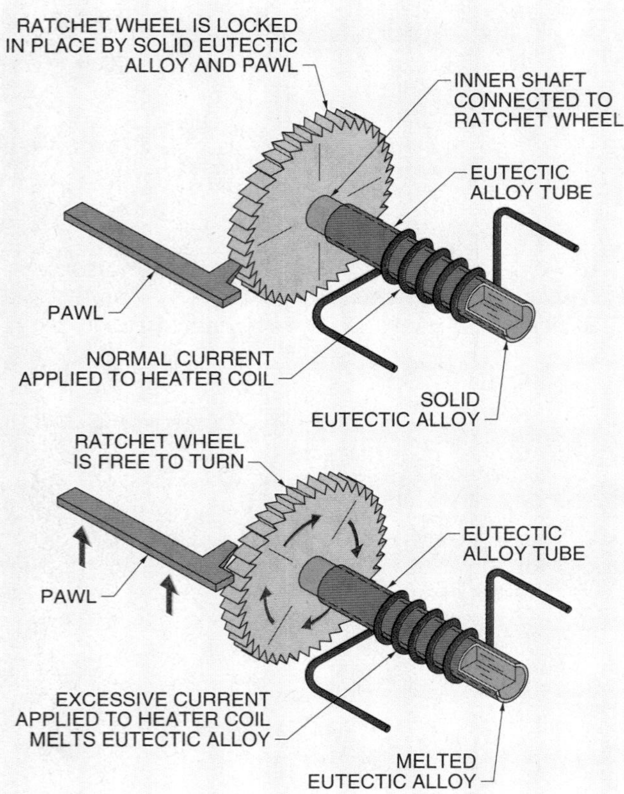

Figure 4-14. Most manufacturers use a ratchet wheel and eutectic alloy combination to activate a trip mechanism when an overload occurs.

The main device in an overload relay is the eutectic alloy tube. The compressed spring tries to push the NC contacts open when motor current conditions are normal. The pawl is caught in the ratchet wheel and does not let the spring push up to open the contacts. See Figure 4-15.

The heater coil heats the eutectic alloy tube when an overload occurs. The heat melts the alloy which allows the ratchet wheel to turn. The spring pushes the reset button up which opens the contacts to the voltage coil of the contactor. The contactor opens the circuit to the motor which stops the current flow through the heater coil. The heater coil cools, which solidifies the alloy and ratchet wheel.

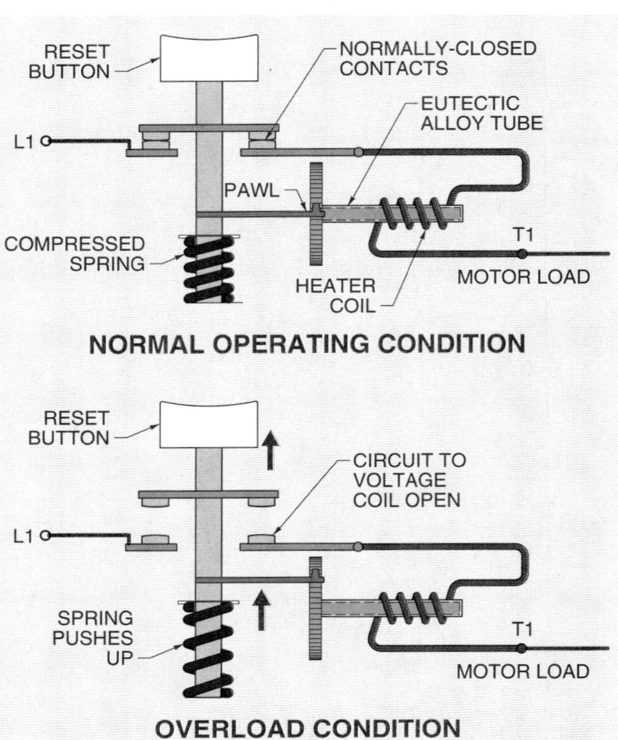

Figure 4-15. In a manual starter overload relay, the compressed spring tries to push the NC contacts open under normal operating conditions.

The stop pushbutton is used to reset the overload relay on a manual starter.

Resetting Overload Devices. The cause of the overload must be found before resetting the overload relay. The relay trips on resetting if the overload is not removed. Once the overload is removed, the device can be reset. The reset button is pushed, forcing the pawl across the ratchet wheel until the contacts are closed and the spring and ratchet wheel are returned to their original condition. The start pushbutton can then be pressed to start the motor. See Figure 4-16.

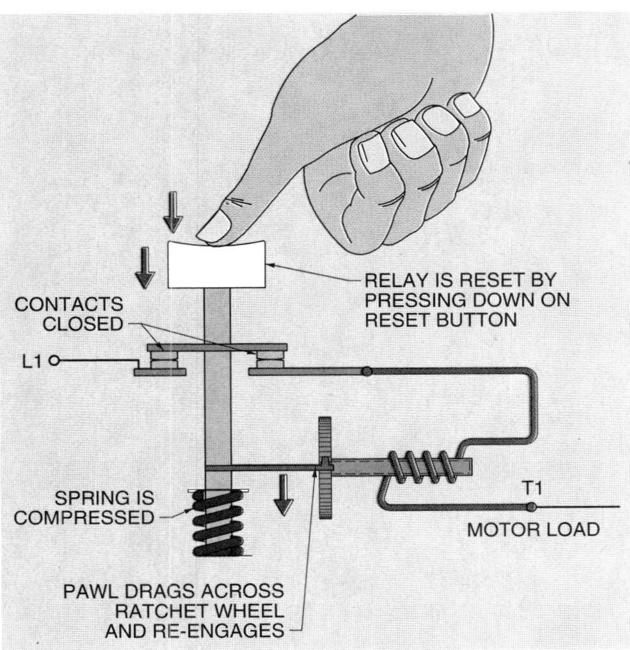

Figure 4-16. The overload relay is reset by pressing the reset button which forces the pawl across the ratchet wheel until the contacts are closed and the spring and ratchet wheel are returned to their original condition.

Nothing requires replacement or repair when an overload device trips. Once the cause of the overload is removed, the reset button may be pressed. Normally, a few minutes should be allowed to cool the eutectic alloy.

The same basic overload relay is used with all sizes of motors. The only difference is that the heater coil size is changed. For small horsepower motors, a small heater coil is used. For large horsepower motors, a large heater coil is used. Check the NEC® for appropriate selection of overload heater sizes.

Selecting AC Manual Starters

Electricians are often required to select manual starters for new installations or replace ones which have been severely damaged due to an electrical fire or explosion. In either case, the electrician must specify certain characteristics of the starter to obtain the proper replacement. See Figure 4-17. Manual starters are selected based on phasing, number of poles, voltage consideration, starter size, and enclosure type. Starter sizes are given in general motor protection tables. General motor protection tables indicate motor protection device sizes based on motor horsepower, current, fuse, and wire size. See Appendix.

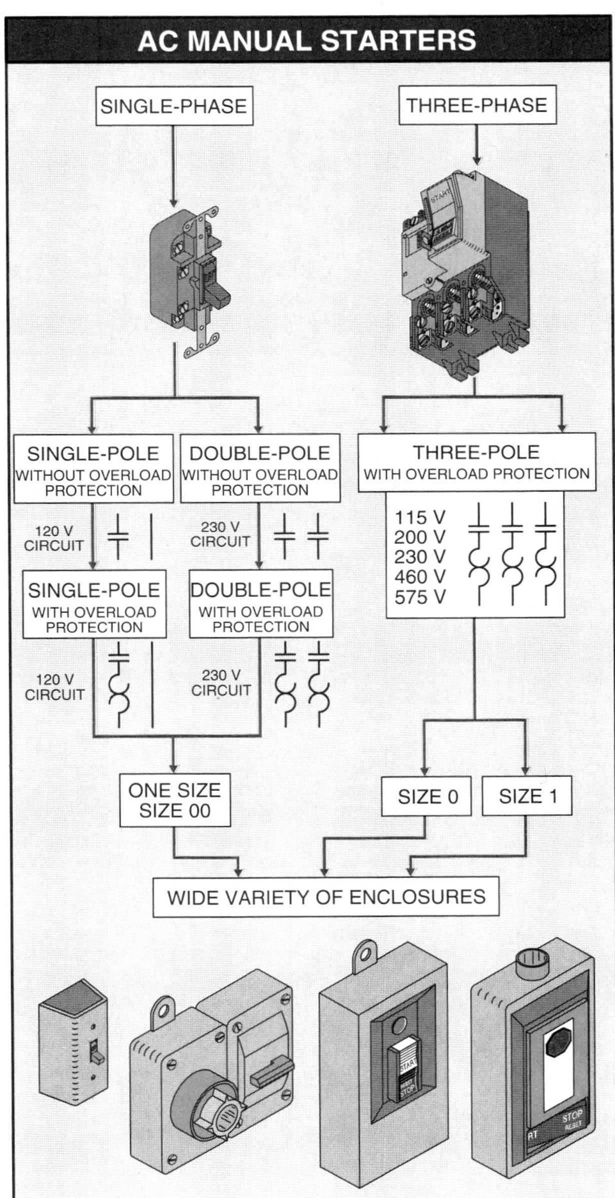

Figure 4-17. Manual starters are selected based on phasing, number of poles, voltage consideration, starter size, and enclosure type.

Phasing. AC manual contactors can be divided into 1ϕ and 3ϕ contactors. See Figure 4-18. A 120 V, 1ϕ power source has one hot wire (ungrounded conductor) and one neutral wire (grounded conductor). A 230 V, 1ϕ power source has two hot wires: L1, L2 (ungrounded conductors), and no neutral. A 3ϕ power source has three hot wires: L1, L2, L3, and no neutral.

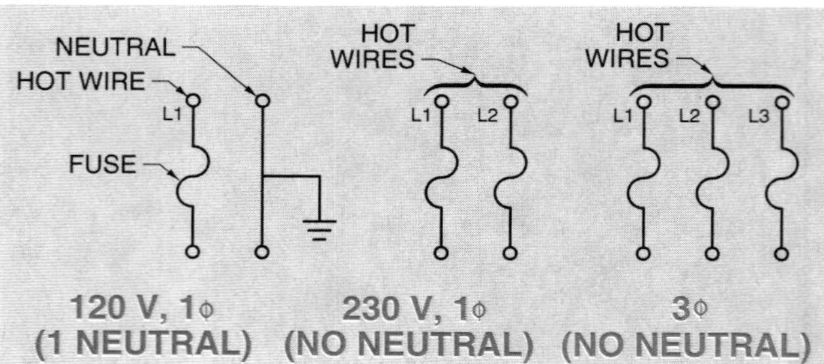

Figure 4-18. AC manual contactors can be divided into 1ϕ and 3ϕ contactors.

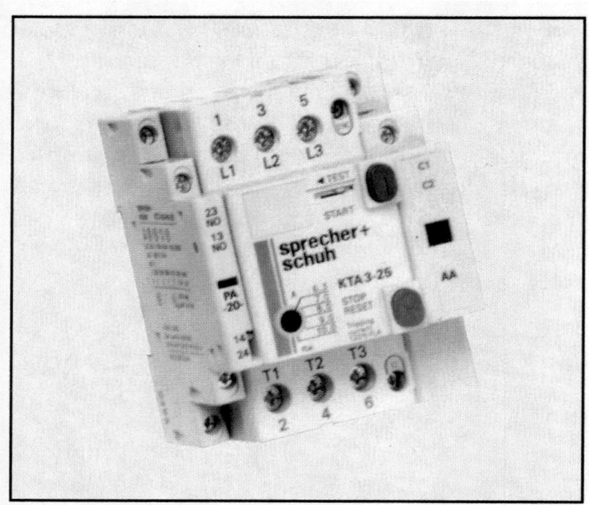

Sprecher + Schuh

Auxiliary contacts, trip indicators, and undervoltage and shunt release modules may be added to Sprecher + Schuh's KTA3 motor circuit controllers to enhance the performance of the devices.

Single-phase manual starters are available as single-pole and double-pole devices because the NEC® requires that each ungrounded conductor (hot wire) be open when disconnecting a device. A single-pole device is used on 120 V circuits and a double-pole device is used on 230 V circuits.

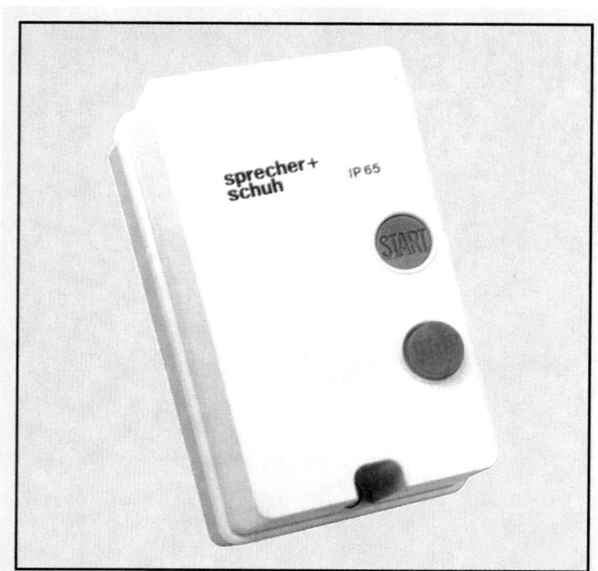

Sprecher + Schuh

Sprecher + Schuh's compact starters are housed in high impact F4 watertight enclosures that are designed to protect the starters and save space.

Single-phase manual starters have limited horsepower ratings because of their physical size. They may be used as starters for motors of 1 HP maximum. Single-phase manual starters are available in only one size for all motors rated at 1 HP or less. The size established for 1ϕ starters is size 00. Single-phase manual contactors and starters are used for 1ϕ, 1 HP and under motors where low-voltage protection is not needed. They are also used for 1ϕ that do not require a high frequency of operation.

Three-phase manual starters are physically larger than 1ϕ manual starters and may be used on motors up to 10 HP. Three-phase manual contactors are normally pushbutton-operated instead of toggle-operated as in 1ϕ starters.

Motor circuits require a manual starter which has overloads. Contactors, however, can be used in certain applications, such as in lighting circuits, without overload devices. In those cases, the fuse provides the overload protection.

Three-phase devices are designed with three-pole switching because 3ϕ devices have three hot wires which must be disconnected. Three-phase, like 1ϕ, uses butt-type contacts and have quick-make and quick-break mechanisms. Three-phase contactors and starters are designed to be used on circuits from 115 V up to and including 575 V.

Three-phase starters are used for 3ϕ, 7.5 HP and under motors operating at 208/230 V, or 10 HP and under operating at 380/575 V. Three-phase starters are also used for 3ϕ motors where low-voltage protection is not needed, motors that do not require a high frequency of operation, and motors that do not need remote operation by pushbuttons or limit switches.

Enclosures. Enclosures provide mechanical and electrical protection for the operator and the starter. See Appendix. Although the enclosures are designed to provide protection in a variety of situations (water, dust, oil, and hazardous locations), the internal electrical wiring and physical construction of the starter remain the same.

Consult the NEC® and local codes to determine the proper selection of an enclosure for a particular application. For example, NEMA Type 1 enclosures are intended for indoor use primarily to provide a degree of protection against human contact with the enclosed equipment in locations where unusual service conditions do not exist.

Manual Starter Applications. Manual motor starters are used in applications such as air compressors, conveyor systems, and drill presses. See Figure 4-19. In most applications, the manual starter provides the means of turning ON and OFF the device while providing motor overload protection.

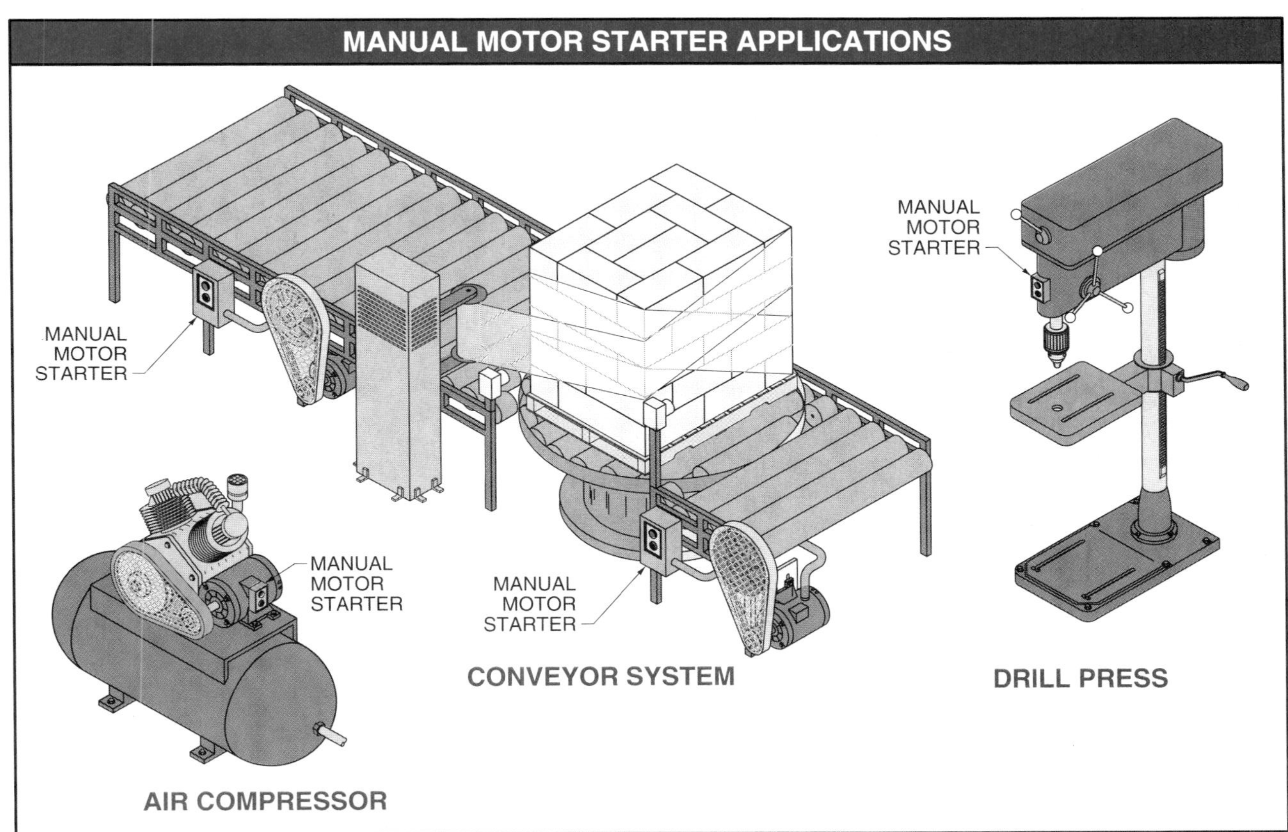

Figure 4-19. Manual motor starters are used in applications such as air compressors, conveyor systems, and drill presses.

1. What are the disadvantages of using a knife switch for starting and stopping electric motors?
2. What are knife switches used for today?
3. What is a NO double-break contact?
4. What is the function of arc hoods?
5. Why is a silver alloy used on switching contacts?
6. How does a wiring diagram differ from a line diagram?
7. What do the heavy dark lines indicate on a wiring diagram?
8. What is the function of a mechanical interlock?
9. How does a manual starter differ from a manual contactor?
10. Why must a motor be protected by an overload relay and a fuse or breaker?
11. What is ambient temperature?
12. What are the basic requirements of an overload relay?
13. How does a basic overload relay determine an overload?
14. How is an overload relay reset?
15. What type of device is required to switch a 120 V circuit?
16. What type of device is required to switch a 230 V circuit?
17. What type of device is required to switch a 3ϕ circuit?
18. Why are contactors and starters placed in an enclosure?
19. What is the purpose of a NEMA Type 1 enclosure?
20. What type of motor starter is used in a hazardous dust location?

5 Chapter

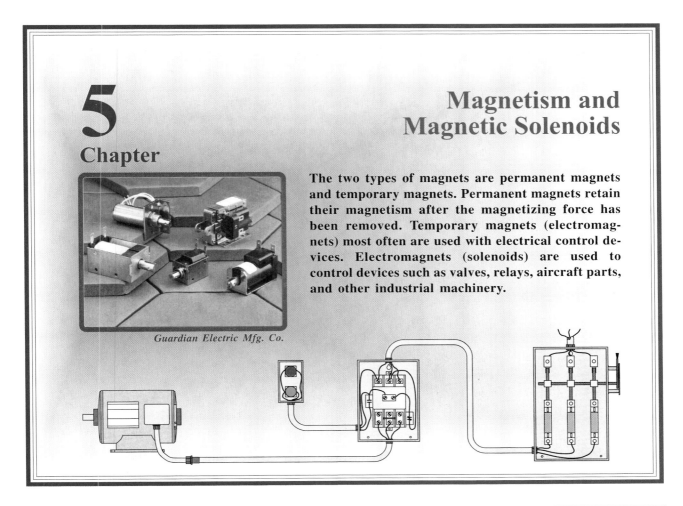

Magnetism and Magnetic Solenoids

Guardian Electric Mfg. Co.

The two types of magnets are permanent magnets and temporary magnets. Permanent magnets retain their magnetism after the magnetizing force has been removed. Temporary magnets (electromagnets) most often are used with electrical control devices. Electromagnets (solenoids) are used to control devices such as valves, relays, aircraft parts, and other industrial machinery.

MAGNETISM

Magnetism was first discovered by the Greeks when they noticed that a certain type of stone attracted bits of iron. This stone was first found in Asia Minor in the province of Magnesia. The stone was named magnetite after this province.

Magnets

A *magnet* is a substance that attracts iron and produces a magnetic field. Magnets are either permanent or temporary. *Permanent magnets* are magnets which can retain their magnetism after the magnetizing force has been removed. Permanent magnets have a high retentivity because they can retain the residual magnetism (leftover magnetism) once the magnetizing force has been removed. Permanent magnets include natural magnets (magnetite) and manufactured magnets. See Figure 5-1.

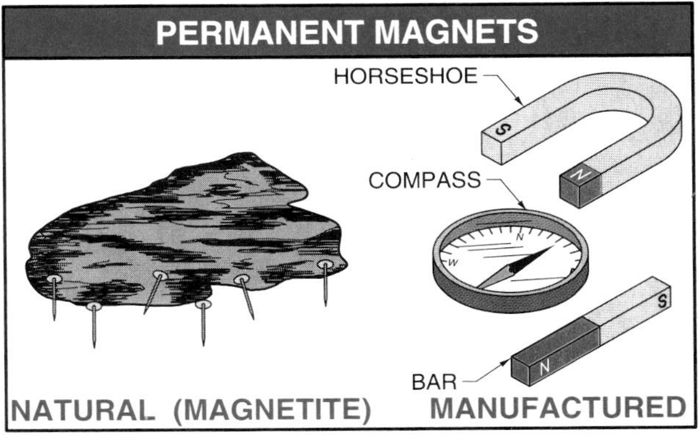

Figure 5-1. Permanent magnets include natural magnets (magnetite) and manufactured magnets.

Temporary magnets are magnets which have extreme difficulty in retaining any magnetism after the magnetizing force has been removed. See Figure 5-2. Temporary magnets have a low retentivity. Very little residual magnetism (leftover magnetism) remains once the magnetizing force has been removed.

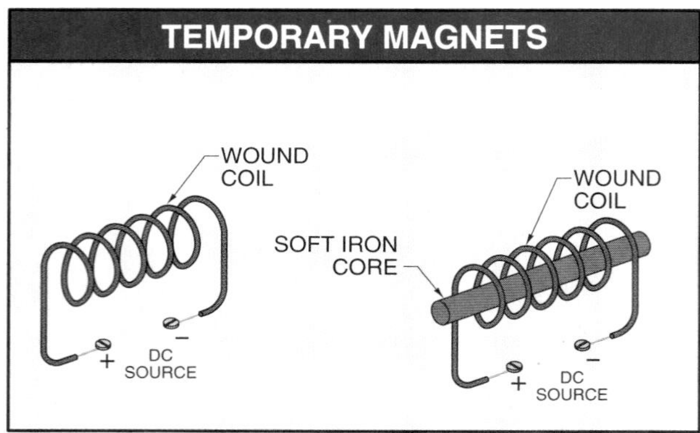

Figure 5-2. Temporary magnets are magnets which have extreme difficulty in retaining any magnetism after the magnetizing force has been removed.

Molecular Theory of Magnetism

The *molecular theory of magnetism* states that all substances are made up of an infinite number of molecular magnets that can be arranged in either an organized or disorganized manner. See Figure 5-3. The material is demagnetized if it has disorganized molecular magnets. The material is magnetized if it has organized molecular magnets.

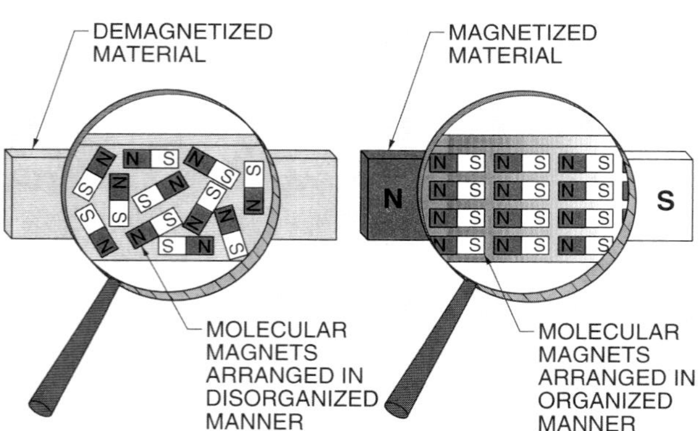

Figure 5-3. The molecular theory of magnetism states that all substances are made up of an infinite number of molecular magnets that can be arranged in either an organized or disorganized manner.

The molecular theory of magnetism explains how certain materials react to magnetic fields when used as control devices. For example, it explains why hard steel is used for permanent magnets, while soft iron is used for temporary magnets found in control devices.

The dense molecular structure of hard steel does not easily disorganize once a magnetizing force has been removed. Hard steel is difficult to magnetize and demagnetize, making it a good permanent magnet.

The loose molecular structure of soft iron can be magnetized and demagnetized easily. Soft iron is ideal for temporary magnets used in control devices because it does not retain residual magnetism very easily.

Electricity and Magnetism Relationship

In 1819, Hans C. Oersted, a Danish physicist, discovered that a magnetic field is created around an electrical conductor when electricity flows through the conductor. See Figure 5-4. He tried several experiments to increase the strength of the magnetic field, and discovered that the three ways to increase the strength of the magnetic field produced by a conductor are to increase the amount of current by increasing the voltage, increase the number of coils, and insert an iron core through the coils. See Figure 5-5.

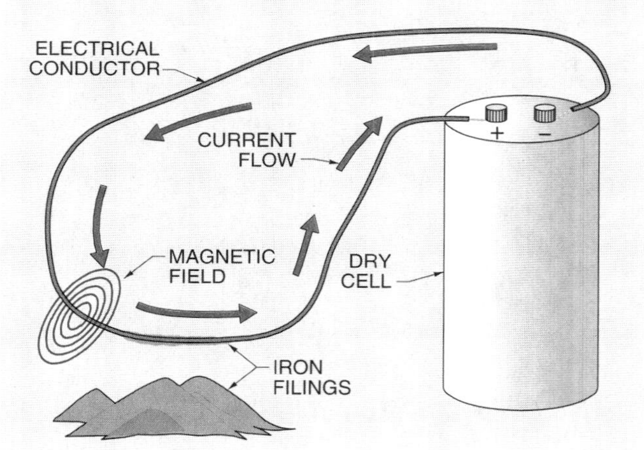

Figure 5-4. In 1819, Hans C. Oersted, a Danish physicist, discovered that a magnetic field is created around an electrical conductor when electricity flows through the conductor.

These experiments led to the development of a huge control industry which depends on magnetic coils to convert electrical energy into usable magnetic energy. A magnetic coil device used extensively in the control industry is the solenoid.

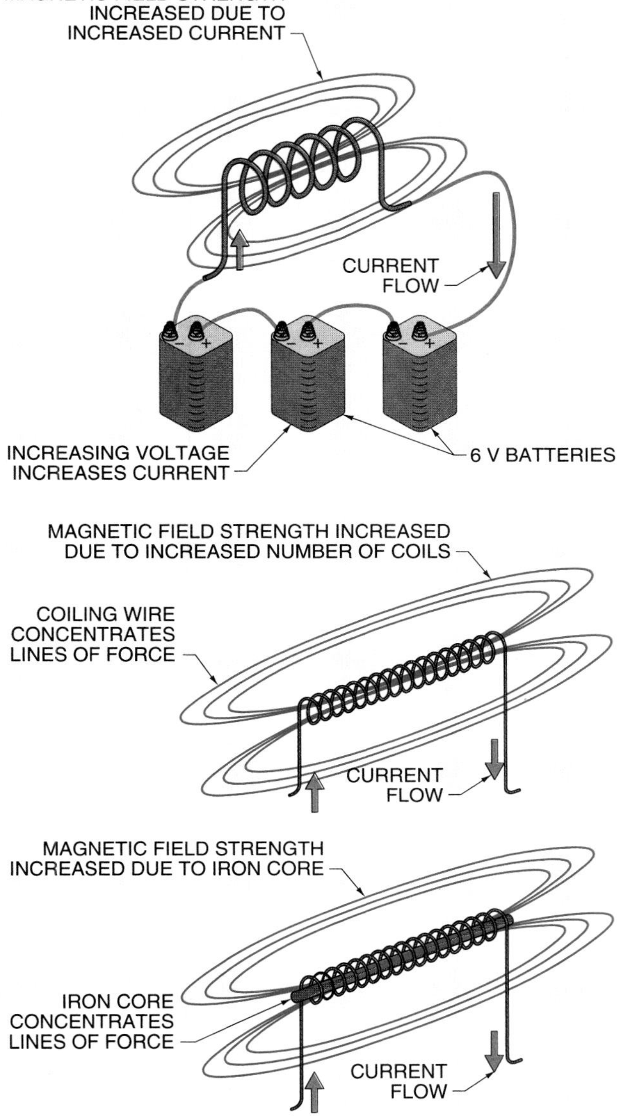

Figure 5-5. The strength of a magnetic field produced by a conductor may be increased by increasing the amount of current by increasing the voltage, increasing the number of coils, and inserting an iron core through the coils.

SOLENOIDS

A *solenoid* is an electric output device that converts electrical energy into a linear mechanical force. The magnetic attraction of a solenoid may be used to transmit force. Solenoids may be combined with a moving armature which transmits the force created by the solenoid into useful work.

Solenoid Configurations

Solenoids are configured in various ways for different applications and operating characteristics. The five solenoid configurations include clapper, bell-crank, horizontal-action, vertical-action, and plunger. See Figure 5-6.

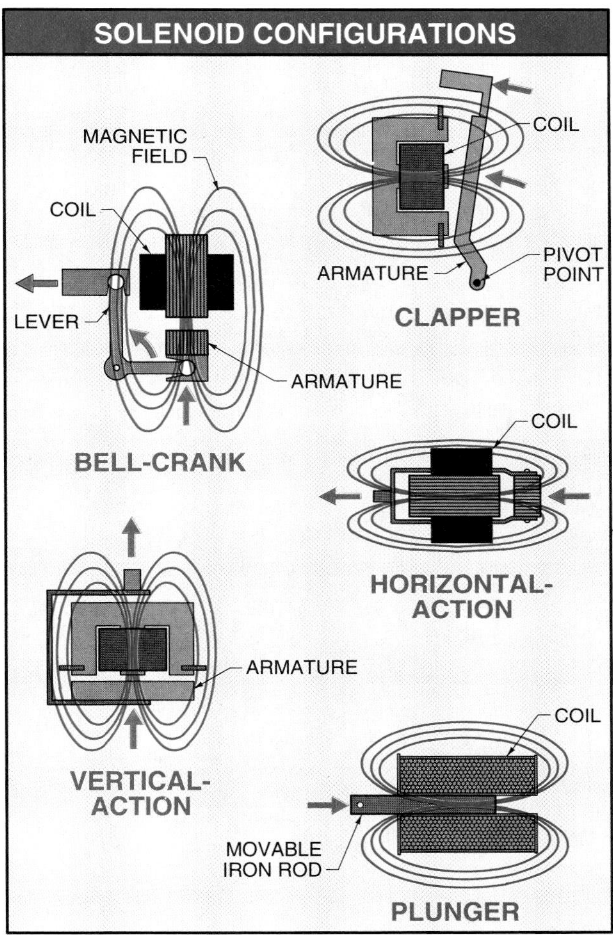

Figure 5-6. The five solenoid configurations include clapper, bell-crank, horizontal-action, vertical-action, and plunger.

A clapper solenoid has the armature hinged on a pivot point. As voltage is applied to the coil, the magnetic effect produced pulls the armature to a closed position so that it is picked up (sealed in). A bell-crank solenoid uses a lever attached to the armature to transform the vertical action of the armature into a horizontal motion. The use of the lever allows the shock of the armature to be absorbed by the lever and not transmitted to the end of the lever. This is beneficial when a soft but firm motion is required in certain controls.

A horizontal-action solenoid is a direct-action device. The movement of the armature moves the resultant force in a straight line. Horizontal-action solenoids are one of the most common configurations. A vertical-action solenoid also uses a mechanical assembly but transmits the vertical action of the armature in a straight line motion as the armature is picked up.

A plunger solenoid contains only a moving iron cylinder. A movable iron rod placed within the electrical coil tends to equalize or align itself within the coil when current passes through the coil. The current causes the rod to center itself so that the rod ends line up with the ends of the solenoid, if the rod and solenoid are of equal length.

In a plunger solenoid, a spring is used to move the rod a short distance from its center in the coil. The rod moves against the spring tension to recenter itself in the coil when the current is turned ON. The spring returns the rod to its off-center position when the current is turned OFF. The motion of the rod is used to operate any number of mechanical devices. See Figure 5-7.

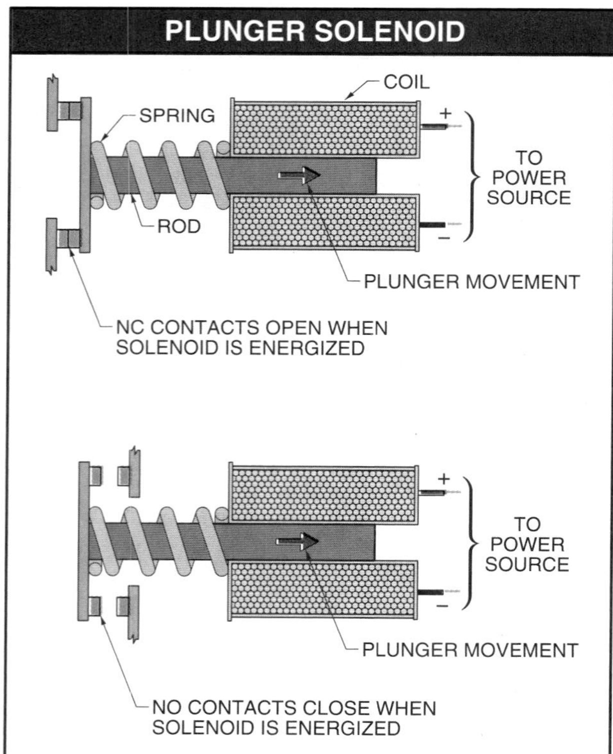

Figure 5-7. In a plunger solenoid, a spring is used to move the rod a short distance from its center in the coil. The rod moves against the spring tension to recenter itself when the current is turned ON.

Solenoid Construction

Solenoids are constructed of many turns of wire wrapped around a magnetic laminate assembly. Passing electric current through the coil causes the armature to be pulled toward the coil. When appropriate, devices may be attached to the solenoid to accomplish tasks like opening and closing contacts.

Eddy Currents. *Eddy current* is unwanted current induced in the metal structure of a device due to the rate of change in the induced magnetic field. Strong eddy currents are generated in solid metal when used with alternating current. In AC solenoids, the magnetic assembly and armature consist of a number of thin pieces of metal laminated together. The thin pieces of metal reduce the eddy currents produced by transformer action in the metal. See Figure 5-8. Eddy currents are confined to each lamination, thus reducing the intensity of the magnetic effect and subsequent heat build up. For DC solenoids, a solid core is acceptable because the current is in one direction and continuous.

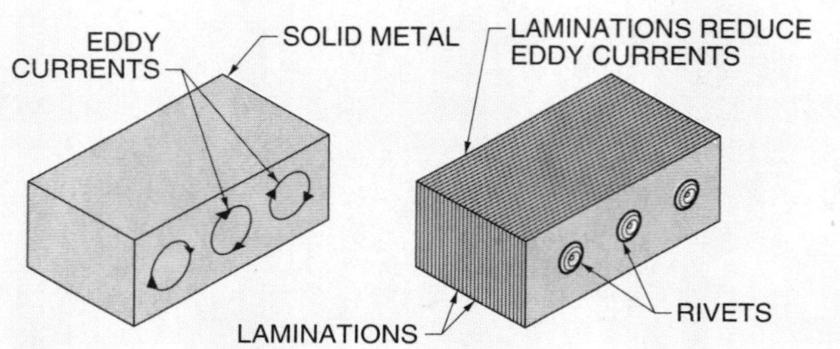

Figure 5-8. In AC solenoids, the magnetic assembly and armature consist of a number of thin pieces of metal laminated together.

Clippard Instrument Laboratory, Inc.

Clippard produces intrinsically safe solenoid-operated valves for use in hazardous locations.

Armature Air Gap. To avoid chattering, solenoids are designed so that the armature is attracted to its sealed-in position so that it completes the magnetic circuit as completely as possible. To ensure this, both the faces on the magnetic laminate assembly and those on the armature are machined flat to a very close tolerance.

As the coil is de-energized, some magnetic flux (residual magnetism) always remains and could be enough to hold the armature in the sealed position. To eliminate this possibility, a small air gap is always left between the armature and magnetic laminate assembly to break the magnetic field and allow the armature to drop away freely when de-energized. See Figure 5-9.

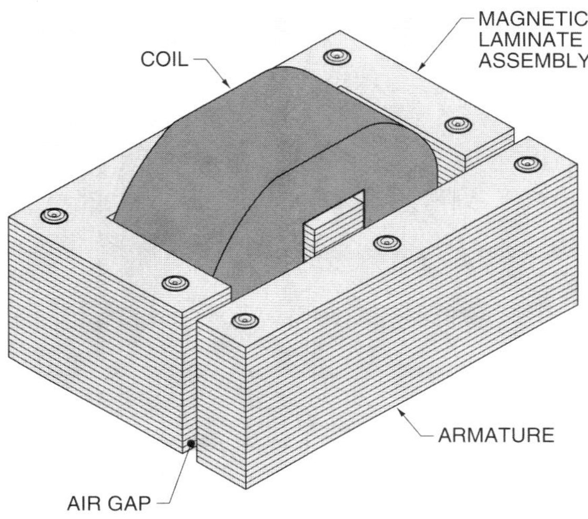

Figure 5-9. A small air gap is always left in the magnetic laminate assembly to break the magnetic field and allow the armature to drop away freely after being de-energized.

Shading Coil. A *shading coil* is a single turn of conducting material (normally copper or aluminum) mounted on the face of the magnetic laminate assembly or armature. See Figure 5-10. A shading coil sets up an auxiliary magnetic field which is out of phase with the main coil magnetic field so that it helps hold in the armature as the main coil magnetic field drops to zero in an AC circuit.

The magnetic field generated by alternating current periodically drops to zero. This makes the armature drop out or chatter. The out-of-phase attraction of the shading coil adds enough pull to the unit to keep the armature firmly seated. Without the shading coil, excessive noise, wear, and heat builds up on the armature faces, reducing the armature's life expectancy. See Figure 5-11.

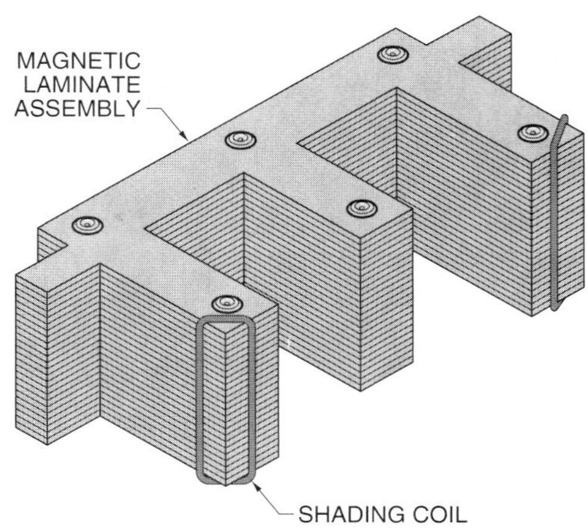

Figure 5-10. A shading coil sets up an auxiliary magnetic field which is out of phase with the main coil magnetic field.

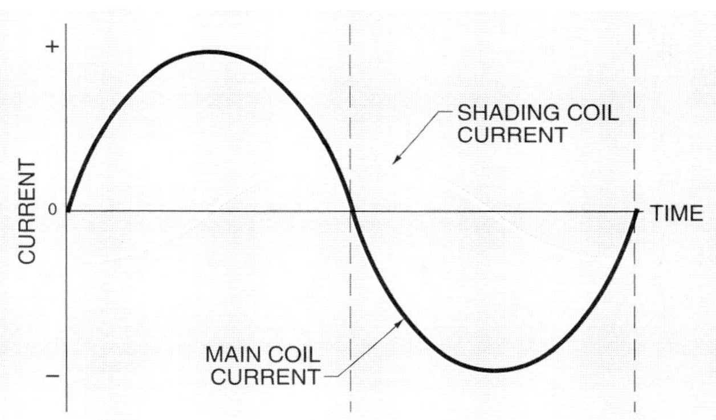

Figure 5-11. The shading coil current and magnetic field are out of phase with the main coil current and magnetic field.

SOLENOID CHARACTERISTICS

The two primary characteristics of a solenoid are the amount of voltage applied to the coil and the amount of current allowed to pass through the coil. Solenoid voltage characteristics include pick-up voltage, seal-in voltage, and drop-out voltage. Solenoid current characteristics include coil inrush current as well as sealed current.

Coils

Magnetic coils are normally constructed of many turns of insulated copper wire wound on a spool. The mechanical life of most coils is improved by encapsulating them in an epoxy resin or glass-reinforced alkyd material. See Figure 5-12. In addition to increasing mechanical strength, these materials greatly increase the moisture resistance of the magnetic coil. Coils are replaced instead of repaired because these devices are encapsulated and cannot be repaired when they go bad.

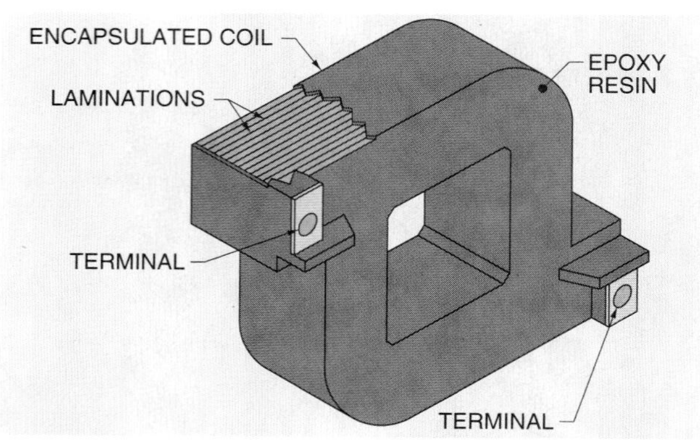

Figure 5-12. The mechanical life of most coils is improved by encapsulating them in an epoxy resin or glass-reinforced alkyd material.

Coil Inrush and Sealed Currents. Solenoid coils, like motors, tend to draw more current when first energized than is required to keep them running. See Figure 5-13. In a solenoid coil, the inrush current is approximately six to 10 times the sealed current. After the solenoid has been energized for some time, the coil becomes hot, causing the coil current to fall and stabilize at approximately 80% of its value when cold. The reason for such a high inrush current is that the basic opposition to current flow when a solenoid is energized is only the resistance of the copper coil. Upon energizing, however, the armature begins to move iron into the core of the coil. This large amount of iron in the magnetic circuit greatly increases the impedance (alternating current opposition to current flow) of the coil and thus decreases the current through the coil. The heat produced by the coil further reduces current flow because copper wire increases in resistance when hot. This limits some current flow.

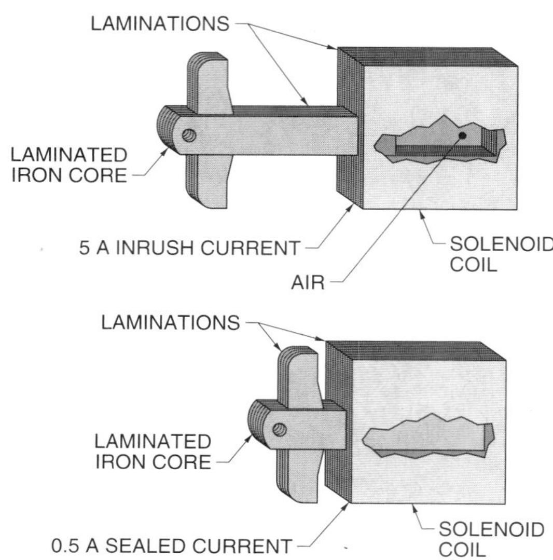

Figure 5-13. Solenoid inrush current is approximately six to 10 times the sealed current.

Coil Inrush and Sealed Current Ratings. Magnetic coil data is normally given in volt amperes (VA). For example, a solenoid with a 120 V coil rated at 600 VA inrush and 60 VA sealed has an inrush current of 5 A ($^{600}/_{120}$ = 5 A), and a sealed current of 0.5 A ($^{60}/_{120}$ = 0.5 A). The same solenoid with a 480 V coil draws only 1.25 A ($^{600}/_{480}$ = 1.25 A) inrush current, and 0.125 A ($^{60}/_{480}$ = 0.125 A) sealed current. The VA rating helps determine the starting and energized current load drawn from the supply line.

Cincinnati Milacron

Solenoids are used on machine tools to control the flow of lubricant and/or coolant.

Coil Voltage Characteristics. *Pick-up voltage* is the minimum control voltage which causes the armature to start to move. *Seal-in voltage* is the minimum control voltage required to cause the armature to seal against the pole faces of the magnet. *Drop-out voltage* is the voltage which exists when the voltage has reduced sufficiently to allow the solenoid to open. Most solenoids have a seal-in voltage that is less than the pick-up voltage.

The bell-crank solenoid configuration has characteristics which allow its design to have a lower seal-in voltage than pick-up voltage. This allows the device to definitely seal in if it has enough voltage to be picked up.

Voltage Variation Effects

Voltage variations are one of the most common causes of solenoid failure. Precautions must be taken in selecting the proper coil for a solenoid. Excessive or low voltage must not be applied to a solenoid coil.

High Voltage. A coil draws more than its rated current if the voltage applied to the coil is too high. Excessive heat is produced which causes early failure of the coil insulation. The magnetic pull is also too high and causes the armature to slam in with excessive force. This causes the magnetic faces to wear rapidly and reduce the expected life of the solenoid.

Low Voltage. Low voltage on the coil produces low coil current and reduced magnetic pull. The solenoid may pick up but does not seal in when the applied voltage is greater than the pick-up voltage, but less than the seal-in voltage. The greater pick-up current (six to 10 times sealed current) quickly heats up and burns out the coil because it is not designed to carry a high continuous current. The armature also chatters, which creates a great deal of noise and increases the wear of the magnetic faces.

SELECTING PROPER SOLENOIDS

Solenoids are selected by analyzing the work to be done and selecting the correct solenoid to perform the work. Solenoid application rules are also considered when selecting solenoids.

Solenoid Application Rules

Solenoids are selected for an application based on the loading conditions which give the optimum performance. Rules to determine good solenoid application include:

- Obtain complete data on load requirements. Both the ultimate life of the solenoid and the life of its linkage depend on the loading of the solenoid. The solenoid may not seal correctly, resulting in coil noise, overheating, and eventual burnout if overloaded. An accurate estimate of the required force at specific inch strokes (lb load v. in. travel) is required to prevent solenoid overload.
- Allow for possible low-voltage conditions of the power supply. Some allowance must be made for low-voltage conditions of the power supply because the pull of the solenoid varies as the square of the voltage (4 lb at 10 V, 16 lb at 20 V, etc.). Solenoids should be applied in accordance with their recommended load. This rating is based on the amount of force the solenoid can develop with 85% of the rated voltage applied to the coil.
- Use the shortest possible stroke. Shorter strokes produce faster operating rates, require less power, produce greater force, and decrease coil heating. Any decrease in heating increases the life expectancy of the coil. The greater force allows a small, low-rated, low-cost solenoid to be used. Also, less destructive mechanical energy is normally available from shorter strokes. This decrease in destructive energy or impact force helps to reduce solenoid wear.
- Never use an oversized solenoid. Use of an oversized solenoid is inefficient, resulting in higher initial cost, a physically larger unit, and greater power consumption. Any energy not expended in useful work must be absorbed by the solenoid in the form of impact force, because energy produced by a solenoid is constant regardless of the load. This results in reduced mechanical life and subjects the linkage mechanism to unnecessary strain.

Solenoid Selection Methods

After reviewing the basic rules of selecting proper solenoids, specific parameters are analyzed. Solenoids are selected based on the outcome(s) required.

Push or Pull. A solenoid may push or pull depending on the application. In the case of a door latch, the unit must pull. In a clamping jig, the unit must push.

Length of Stroke. The length of the stroke is calculated after determining whether the solenoid must push or pull. For example, a door latch requires ½″ maximum stroke length.

Required Force. Manufacturer's specification sheets are used to determine the correct solenoid based on the required force. See Figure 5-14. For example, an A 100 solenoid is used for an application that requires a horizontal force of 2.7 lb. Select a solenoid from the Horizontal Force @ 85% Voltage* column with the next highest force if the required force is not given.

Duty Cycle. Solenoid characteristic tables are also used to check the duty cycle requirements of the application against the duty cycle information given for the solenoid. For example, an A 101 solenoid is required for an application requiring 190 operations per minute.

Uniform Force Curve. Manufacturers provide specification curves to help determine the overall operating characteristics of a solenoid. The force curve of a solenoid must meet the load throughout its length of travel. See Figure 5-15. For example, an A 100 solenoid may be used in an application requiring 2.7 lb of horizontal force over a stroke length of approximately ⅝″.

½″ PULL SOLENOID CHARACTERISTICS

Solenoid	Seated Force*		Plunger Weight*	Shipping Weight*	VA 100% Voltage Seated	½″ Maximum Stroke		
	85% Voltage	100% Voltage				Horizontal Force 85% Voltage*	VA 100% Voltage	Duty Cycle 50% Time ON**
A 100	7	9	.2	1.3	40	2.7	230	240
A 101	9	12	.3	1.5	50	4.0	322	190
A 102	11	15	.3	1.7	50	4.7	420	180
B 100	11	15	.4	2.3	60	6.2	520	200
B 101	13	18	.5	2.6	70	9.6	790	109

* in lb
** in ops/min

Figure 5-14. Manufacturer's specification sheets are used to determine the correct solenoid based on the required force.

Guardian Electric Mfg. Co.

Solenoids are available in a variety of sizes and designs to meet various application needs.

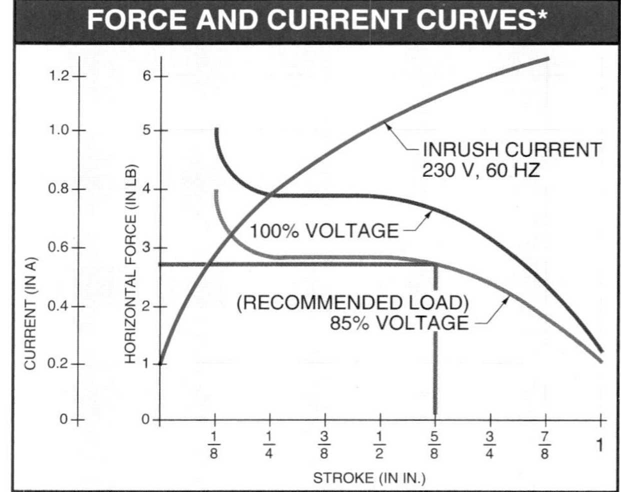

Figure 5-15. Manufacturers provide specification curves to determine operating characteristics of solenoids.

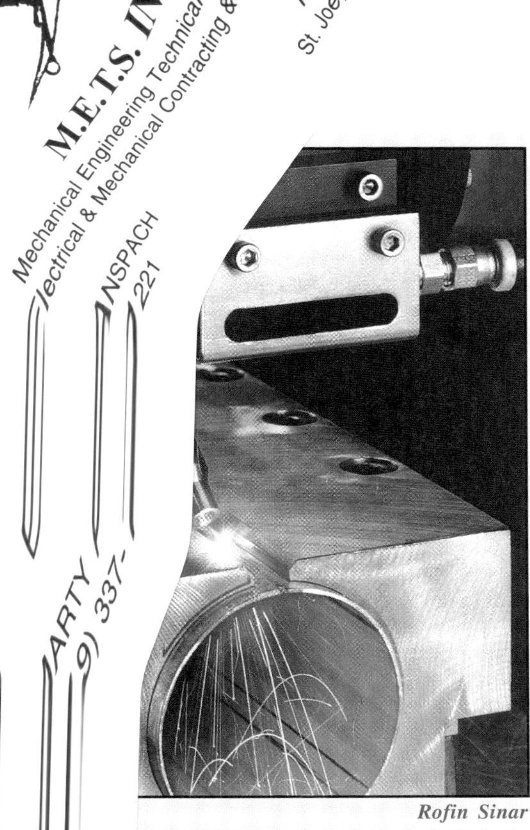

Rofin Sinar

Sole... to control the flow of gas during tube weld-
ing ...

Mou... ufacturers provide letter or number
code... the solenoid mount. See Figure 5-
16. I... , an A solenoid is selected for a door
latcl... because the door latch application
requ... mounting solenoid.

C...	Mounting
...OID MOUNTING CODE	
	End
	Right side
	Throat
	None (for thru-bolts)
	Left side
	Both sides

Fig... anufacturers provide letter or number
cod... te the solenoid mount.

Vo... ...g. Manufacturers provide letter or
nu... to indicate the voltages that are avail-
abl... ...n solenoid. See Figure 5-17. For ex-
am... ...lenoid may be used for an application
th... 115 V coil.

SOLENOID VOLTAGE RATINGS	
Number	**Voltage (in V)**
2 A	115
3 A	230
4 A	460
5 A	575

Figure 5-17. Manufacturers provide letter or number codes to indicate the voltages that are available for a given solenoid.

Other Considerations. Additional background information may be helpful in obtaining the proper solenoid in other situations. Most manufacturers provide a specification order sheet that has space for additional information to help select the correct solenoid for an application. See Figure 5-18.

Heidelberg Harris, Inc.

Large complex systems use solenoids to control product positioning as well as other functions.

LINEAR SOLENOID DESIGN DATA SHEE

(1) Type: Push ______ Pull ______

(2) Stroke: ______inches

(3) Force: ______lb at start ______lb at end of stroke (for h

(4) Duty Cycle: Continuous ______ Intermittent ______

(5) Type of Mounting: Horizontal ______ Vertical ______

(6) Cycles Per Second: ______

(7) Voltage: DC ______ Min ______ Max

AC ______ Min ______ Max

(8) Ambient Temperature: ______

(9) Body Size: Length ______

Diameter ______

(10) Electrical Connection:

____

(11) Environmental Conditions:

Dust ______ Water ______ Oil ______

Other ______________________________

Description of Solenoid Application Considered: ____________

____

____

Figure 5-18. Most manufacturers provide a specification order sheet that has s to help select the correct solenoid for an application.

Rofin Sinar

Solenoid operating environments must be considered when selecting a solenoid for an application.

SOLENOID APPL

Solenoids are common dustrial control circu various solenoids are u filling processes. See

Solenoids can be fo ment. In residential e found in doorbells, w kitchen appliances.

Hydraulics/Pneumat

Hydraulic and pneumati iently switched by electr position, double-solenoid a standard hydraulic cyl

When PB1 is momentarily pressed, the valve spool in the solenoid valve is shifted so that the fluid is directed to the back side of the cylinder piston, causing it to move forward. The piston continues to travel to its full extension or until it stalls against the work it is performing.

When PB2 is momentarily pressed, the valve spool in the solenoid valve shifts to the opposite direction so that the fluid is directed to the front side of the cylinder, causing the piston to move backward. *Note:* Coils and pushbuttons should be wired so that they are interlocked. Failure to do so results in coil burn-out. Reversal can be made to take place at any point in the piston cycle with the piston still moving. The unit cannot, however, be stopped in either direction part-way through a stroke and held there.

ARO Fluid Products Div., Ingersol-Rand

The Alpha™ Series Sub-base valve manufactured by ARO has an operating pressure of 0 to 150 psi and actuator options of solenoid/spring, solenoid/solenoid, pilot/spring, pilot/pilot, or solenoid/pilot.

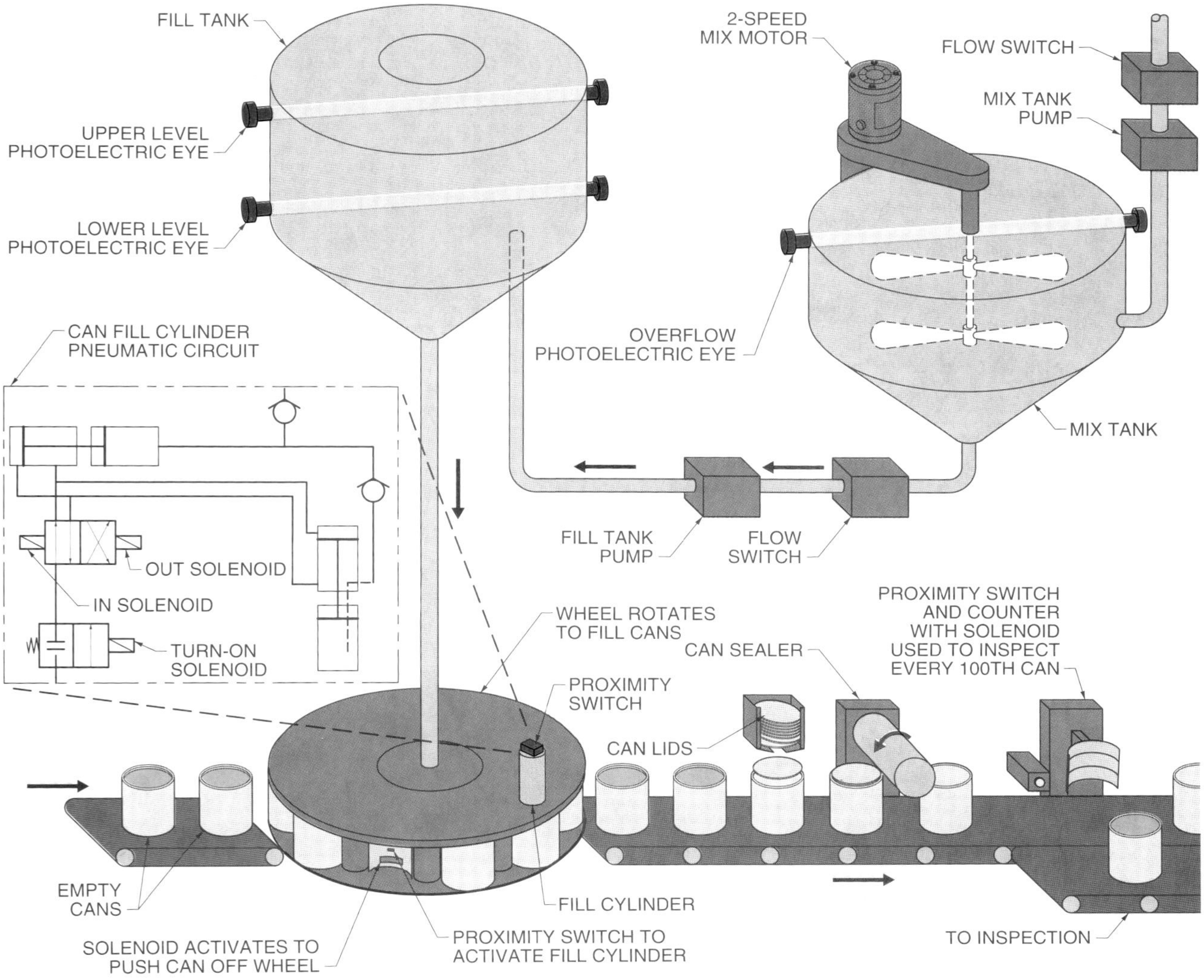

Figure 5-19. Solenoids are commonly used in commercial and industrial control circuit applications.

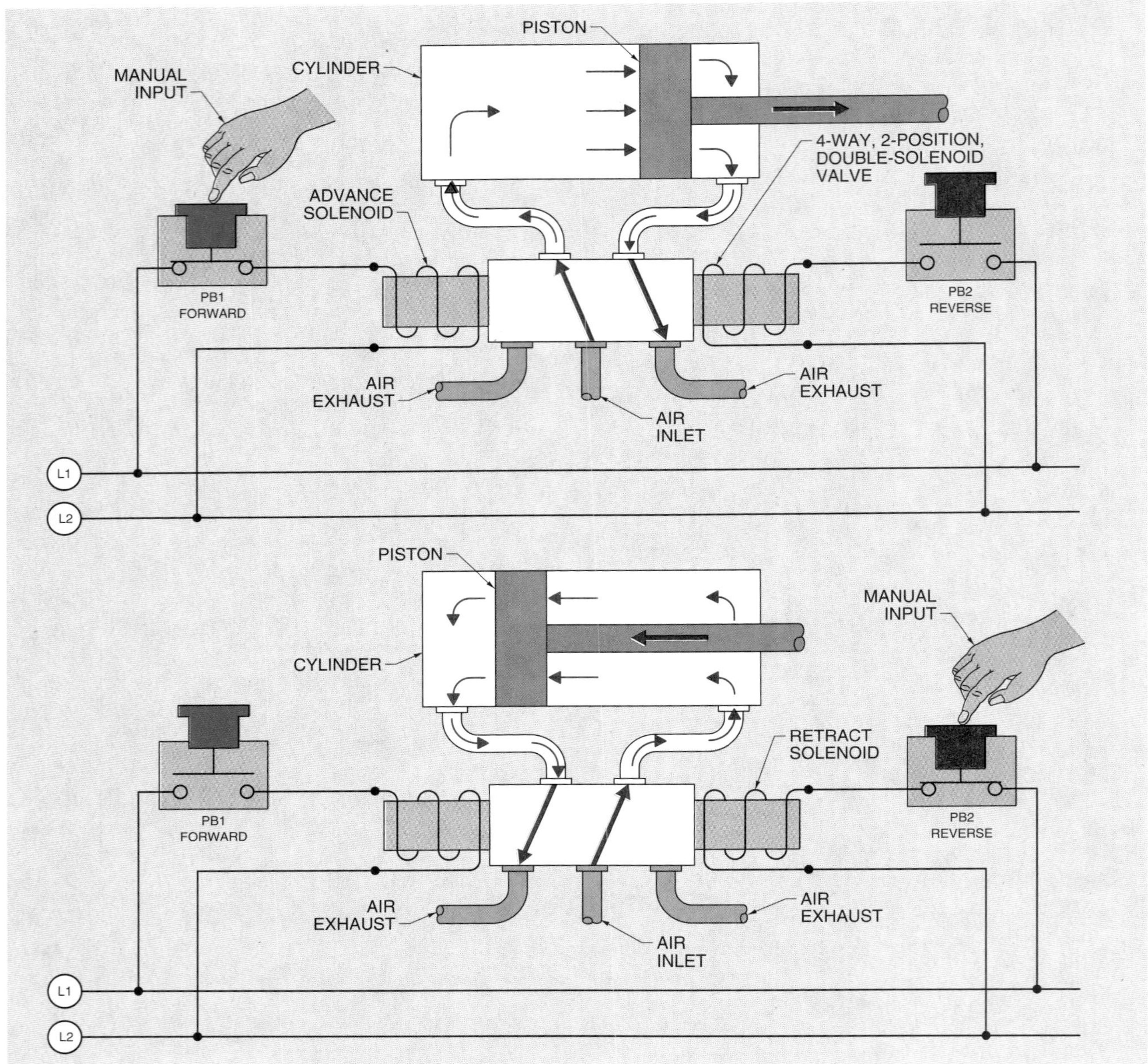

Figure 5-20. Four-way, 2-position, double-solenoid valves can be used to electrically control a standard hydraulic or pneumatic cylinder.

Refrigeration

Direct-acting, two-way valves are common in refrigeration equipment. See Figure 5-21. Two-way (shutoff) valves have one inlet and one outlet pipe connection. These units may be constructed as normally open (NO), where the valve is open when de-energized and closed when energized, or they may be constructed as normally closed (NC), where the valve is closed when de-energized and open when energized.

A number of different solenoids may be used in a typical refrigeration system. See Figure 5-22. The liquid line solenoid valves could be operated by 2-wire or 3-wire thermostats. The hot gas solenoid valve remains closed until the defrost cycle and then feeds the evaporator with hot gas for the defrost operation.

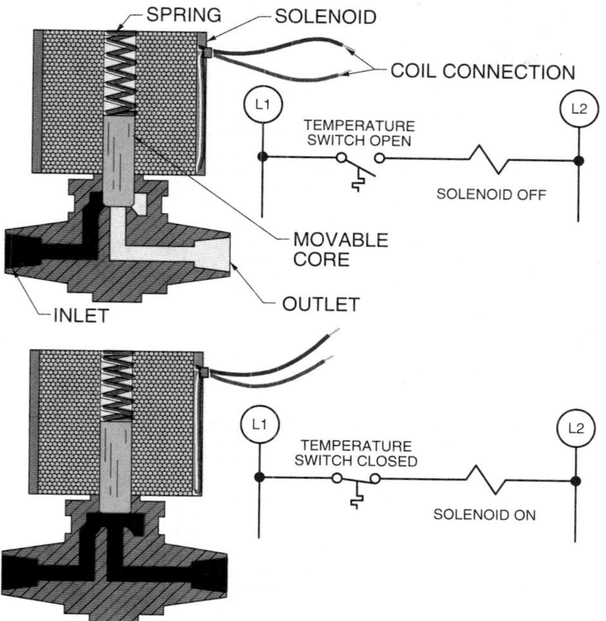

Figure 5-21. Direct-acting, 2-way valves open and close lines in a refrigeration system.

Humphrey Products Company

310/410 Series 3-way and 4-way solenoid-operated valves manufactured by Humphrey Products Company are used for pilot operation of larger valves, control of actuators, and single- and double-acting cylinders from $^3/_4''$ to $1^1/_2''$ bore.

Combustion

Different solenoids may be used in an oil-fired single burner system. See Figure 5-23. The solenoids are crucial in the start-up and normal operating functions of the system.

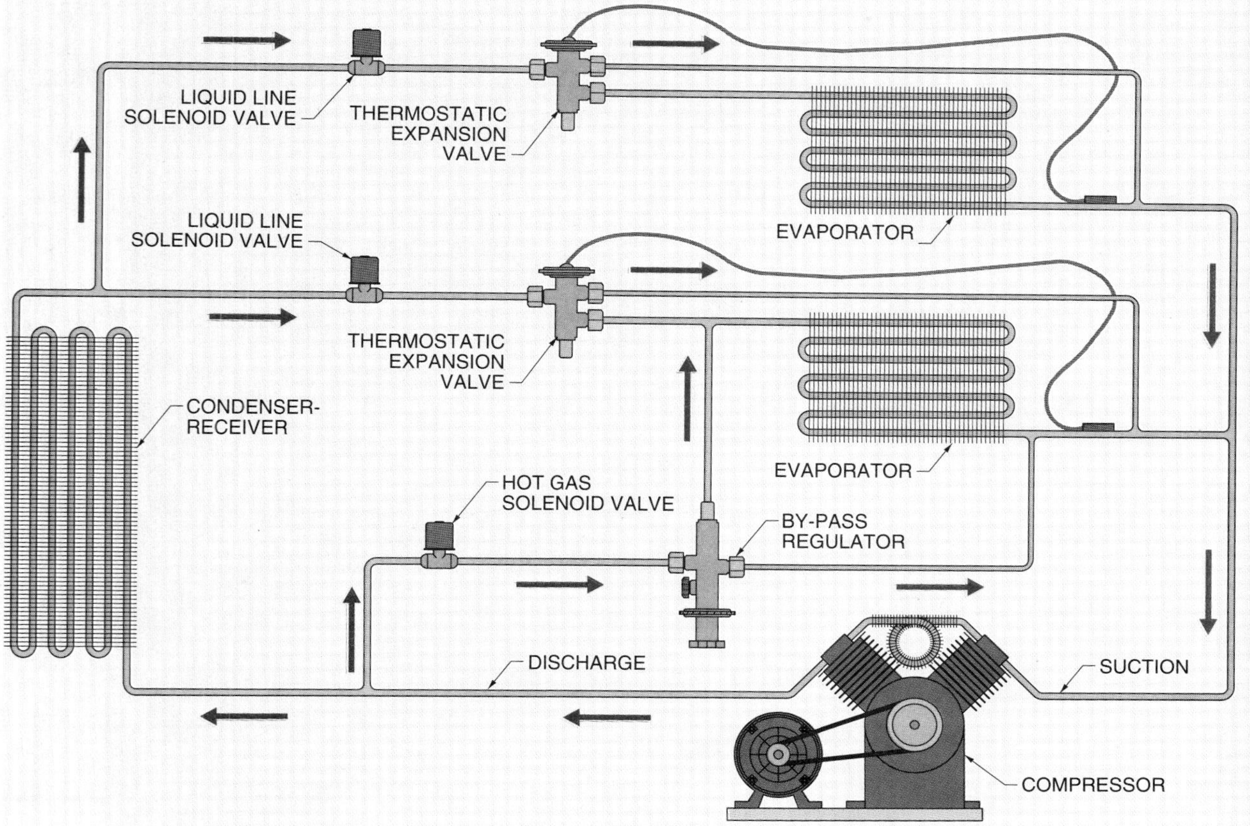

Figure 5-22. Refrigeration systems use solenoid valves to stop, start, and redirect the flow of refrigerant.

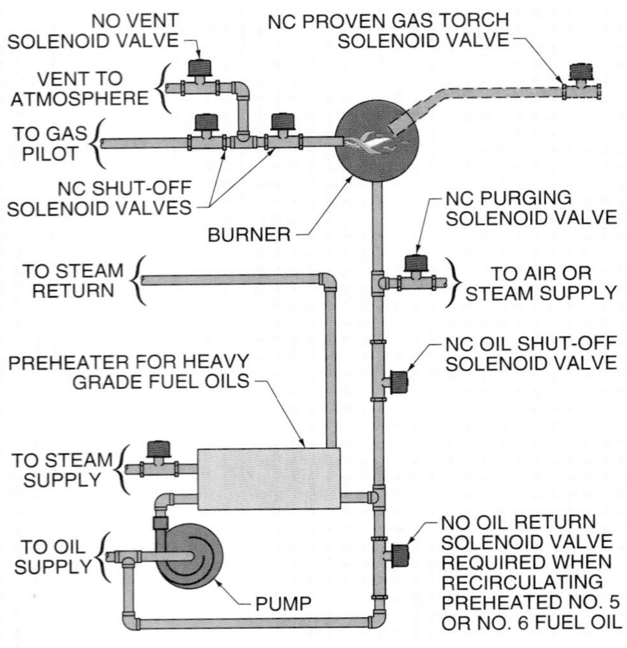

Figure 5-23. Different solenoids are used for the safe operation of an oil-fired single burner system.

Safety

A solenoid can be attached to a manual starter to provide low-voltage protection. See Figure 5-24. Low-voltage protection is accomplished by a continuous-duty solenoid which is energized whenever the line voltage is present. The solenoid de-energizes, opening the starter contacts, when the line voltage is lost. The contacts do not automatically close when the line voltage is restored because of a latching mechanism. The device must be manually reset to close the contacts.

General Purpose

In addition to commercial and industrial use, solenoids are used for general-purpose applications. Typical general-purpose applications include the use of solenoids in products such as printing calculators, copy machines, and airplanes. See Figure 5-25.

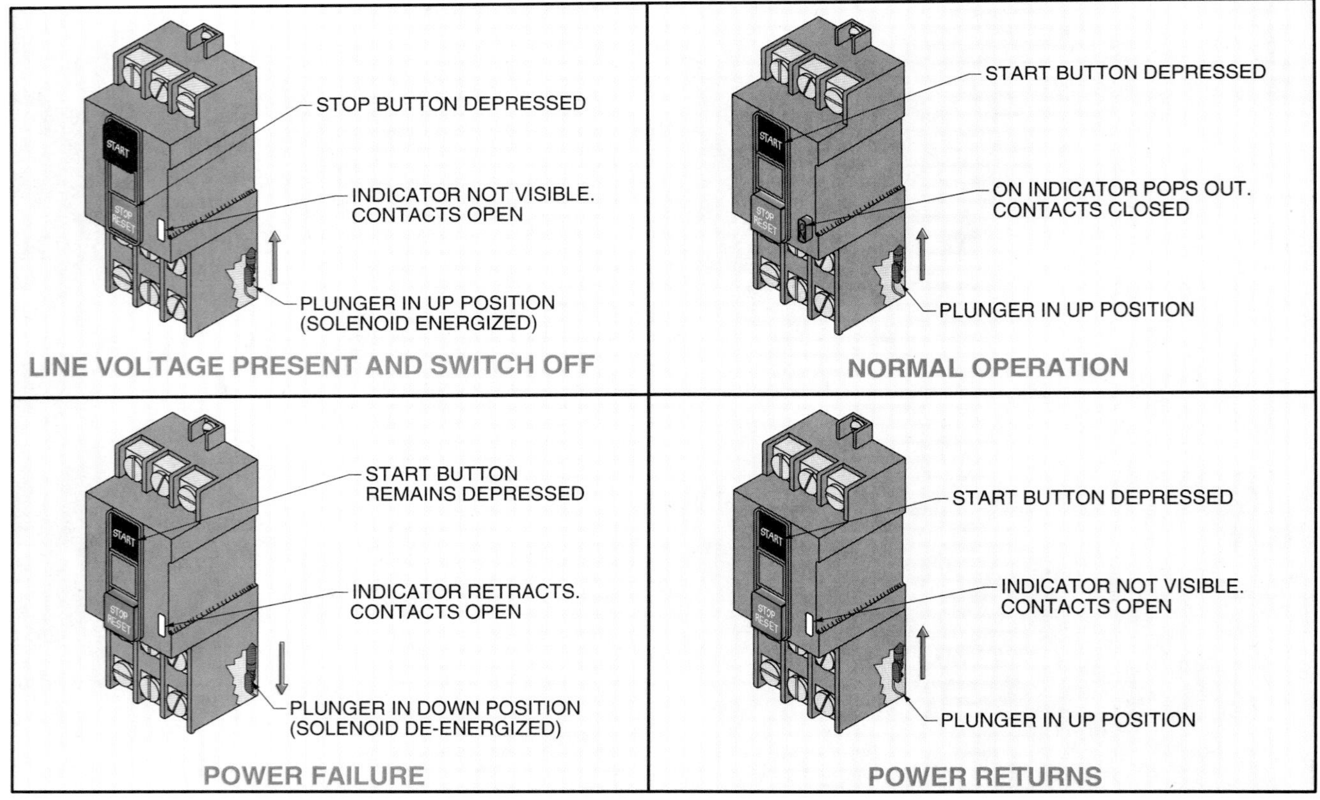

Figure 5-24. A solenoid can be attached to a manual starter to provide low-voltage protection.

Boeing Commercial Airplane Group

Figure 5-25. General purpose solenoids are used in printing calculators, copy machines, and airplanes.

Fluke Corporation

Fluke meters may be used to troubleshoot solenoids used in aircraft.

TROUBLESHOOTING SOLENOIDS

Solenoids fail due to coil burnout or mechanical damage. Manufacturer charts are used to help in the determination of the cause of solenoid failure. See Figure 5-26.

Incorrect Voltage

The voltage applied to a solenoid should be ±10% of the rated solenoid value. The voltage is measured directly at the valve when the solenoid is energized. A voltmeter is set on an AC voltage range for AC solenoids. The voltmeter is set on a DC voltage range for DC solenoids. The range setting must be greater than the applied voltage. See Figure 5-27.

A solenoid overheats when the voltage is excessive. The heat destroys the insulation on the coil wire and burns out the solenoid. The solenoid has difficulty moving the spool inside the valve when the voltage is too low. The slow operation causes the solenoid to draw its high inrush current longer. Longer high inrush current causes excessive heat.

Incorrect Frequency

Solenoids are available with frequency ratings of 50 Hz or 60 Hz. Solenoids with a frequency rating of 50 Hz are used for export applications. Solenoids with a frequency rating of 60 Hz are used for domestic applications. Imported machines may have solenoid valves with a frequency rating of 50 Hz. A solenoid may operate if the frequency is not correct, but may have a higher failure rate and may produce noise.

Transients

In most industrial applications, the power supplying a solenoid comes from the same power lines that supply electric motors and other solenoids. High transient voltages are placed on the power lines as these inductive loads are turned ON and OFF. Transient voltages may damage the insulation on the solenoid coil, nearby contacts, and other loads. The transient voltages may be suppressed by using snubber circuits. A *snubber circuit* suppresses noise and high voltage on the power lines.

SOLENOID FAILURE CHARACTERISTICS		
Problem	**Possible Causes**	**Comments**
Failure to operate when energized	Complete loss of power to solenoid	Normally caused by blown fuse or control circuit problem
	Low voltage applied to solenoid	Voltage should be at least 85% of solenoid's rated value
	Burned out solenoid coil	Normally evident by pungent odor caused by burnt insulation
	Shorted coil	Normally a fuse is blown and continues to blow when changed
	Obstruction of plunger movement	Normally caused by a broken part, misalignment, or the presence of a foreign object
	Excessive pressure on solenoid plunger	Normally caused by excessive system pressure in solenoid-operated valves
Failure to operate spring-return solenoids when de-energized	Faulty control circuit	Normally a problem of the control circuit not disengaging the solenoid's hold or memory circuit
	Obstruction of plunger movement	Normally caused by a broken part, misalignment, or the presence of a foreign object
	Excessive pressure on solenoid plunger	Normally caused by excessive system pressure in solenoid-operated valves
Failure to operate electrically-operated return solenoids when de-energized	Complete loss of power to solenoid	Normally caused by a blown fuse or control circuit problem
	Low voltage applied to solenoid	Voltage should be at least 85% of solenoid's rated value
	Burned out solenoid coil	Normally evident by pungent odor caused by burnt insulation
	Obstruction of plunger movement	Normally caused by broken part, misalignment, or presence of a foreign object
	Excessive pressure on solenoid plunger	Normally caused by excessive system pressure in solenoid-operated valves
Noisy operation	Solenoid housing vibrates	Normally caused by loose mounting screws
	Plunger pole pieces do not make flush contact	An air gap may be present causing the plunger to vibrate. These symptoms are normally caused by foreign matter
Erratic operation	Low voltage applied to solenoid	Voltage should be at least 85% of the solenoid's rated voltage
	System pressure may be low or excessive	Solenoid size is inadequate for the application
	Control circuit is not operating properly	Conditions on the solenoid have increased to the point where the solenoid cannot deliver the the required force

Figure 5-26. Manufacturer's charts are used to help in the determination of the cause of solenoid failure.

Environmental Conditions

A solenoid must operate within its rating and not be mechanically damaged or damaged by the surrounding atmosphere. A solenoid coil is subject to heat during normal operation. This heat comes from the combination of fluid flowing through the valve, the temperature rise from the coil when energized, and the ambient temperature of the solenoid.

Solenoid Troubleshooting Procedure

A voltmeter and ohmmeter are required when troubleshooting a solenoid. See Figure 5-28.

To troubleshoot a solenoid, apply the procedure:

1. Turn electrical power to solenoid or circuit OFF.
2. Measure the voltage at the solenoid to ensure the power is OFF.

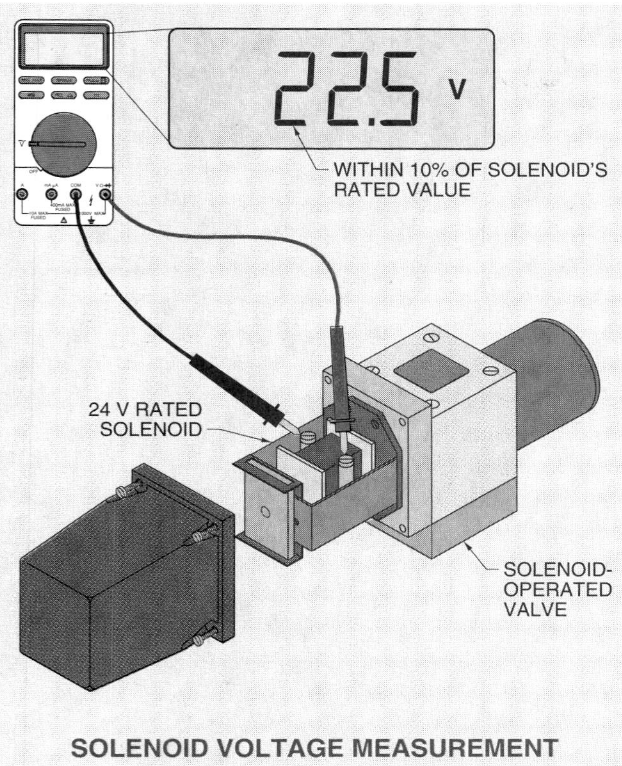

Figure 5-27. The voltage applied to a solenoid should be ±10% of the rated solenoid value.

3. Remove the solenoid cover and visually inspect the solenoid. Look for a burnt coil, broken parts, or other problems. Replace the coil when burnt. Replace the broken parts when available. Replace the valve, contactor, starter, or solenoid-operated device when the parts are not available. *Note:* Determine the fault before installing a new coil when a solenoid has failed due to a burnt or shorted coil. The new coil burns out if the fault is not corrected. Always observe solenoid operation after a solenoid is replaced.
4. Disconnect the solenoid wires from the electrical circuit when no obvious problem is observed.
5. Check the solenoid continuity. Connect the meter leads to the solenoid wires with all power turned OFF. The meter should indicate a resistance reading of æ15% of the coil's normal reading.

Unknown readings are obtained by testing a good solenoid. A low or zero reading indicates a short or partial short. Replace the solenoid if there is a short. No movement of the needle on an analog meter or infinity resistance on a digital meter indicates the coil is open and defective. Replace the solenoid if the open is not obvious. Note: Set the ohmmeter to R × 100 if the solenoid uses a small gauge wire. The small wire has high resistance.

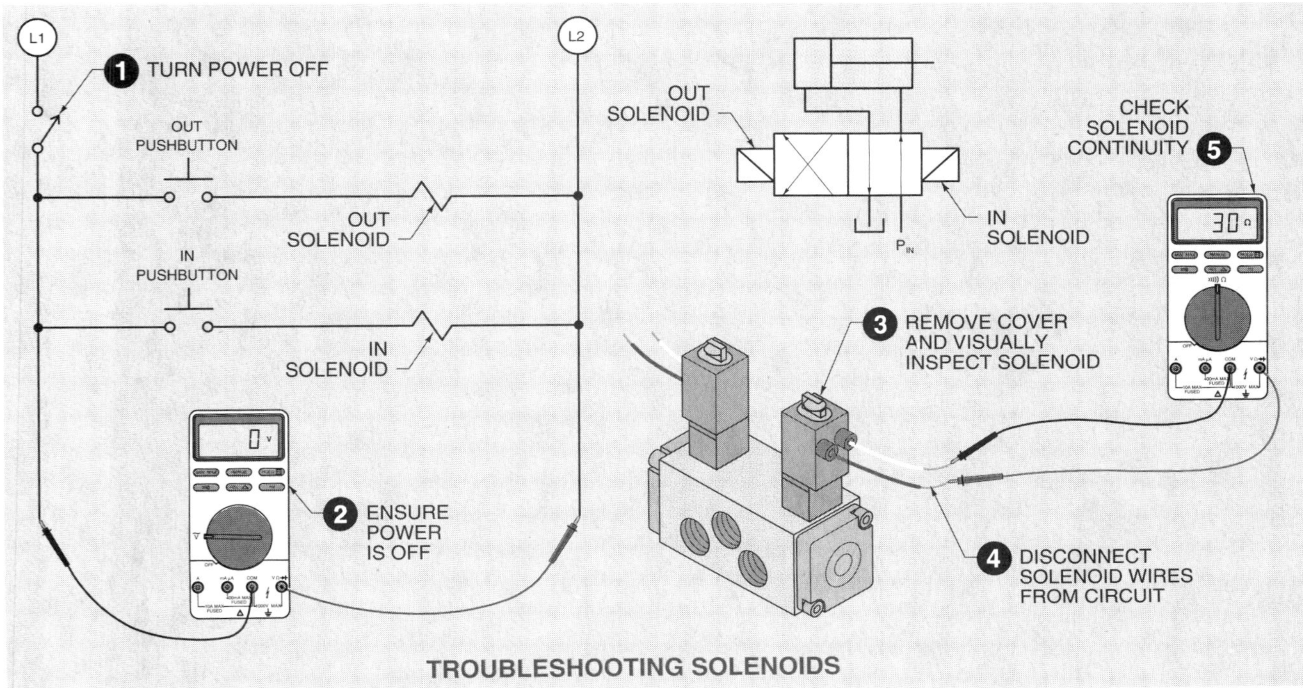

Figure 5-28. A voltmeter and ohmmeter are required when troubleshooting a solenoid.

1. What are the two main types of magnets?
2. Why is soft iron more easily magnetized than hard steel?
3. How can the strength of an electromagnet be increased?
4. What are the different types of solenoids used to transmit force?
5. Why is the armature of a solenoid made from a number of thin laminated pieces instead of a solid piece?
6. What is the purpose of having a small air gap on the armature face?
7. What function does the shading coil perform?
8. Why does a solenoid coil have a much higher inrush current than sealed current?
9. What is the effect of a higher-than-rated voltage applied to a solenoid coil?
10. What is the effect of a lower-than-rated voltage applied to a solenoid coil?
11. What are the four basic rules followed when selecting a solenoid?
12. What are the requirements that must be determined when selecting the proper solenoid?
13. What are the common applications of solenoids?
14. What are the problems that may develop in a solenoid?
15. Solenoids are available with what frequency ratings?
16. What function does a snubber circuit perform?
17. What are the two primary characteristics of a solenoid?
18. What are the two primary causes of solenoid failure?
19. Shading coils are normally made of what material?
20. Magnetic coils are normally made of what kind of wire?

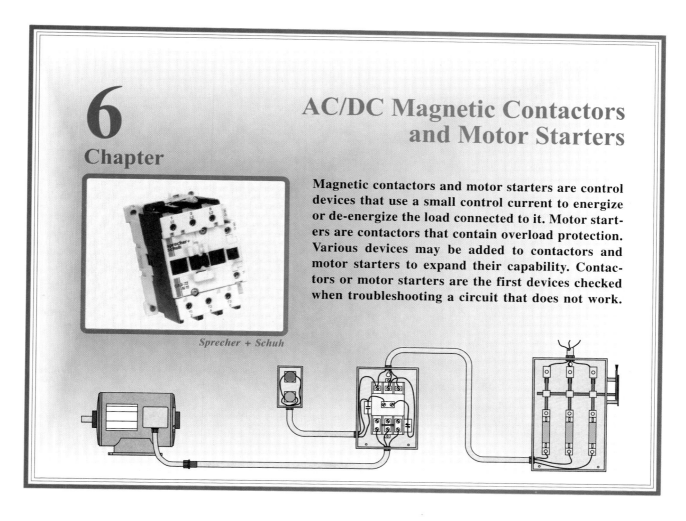

6 Chapter

AC/DC Magnetic Contactors and Motor Starters

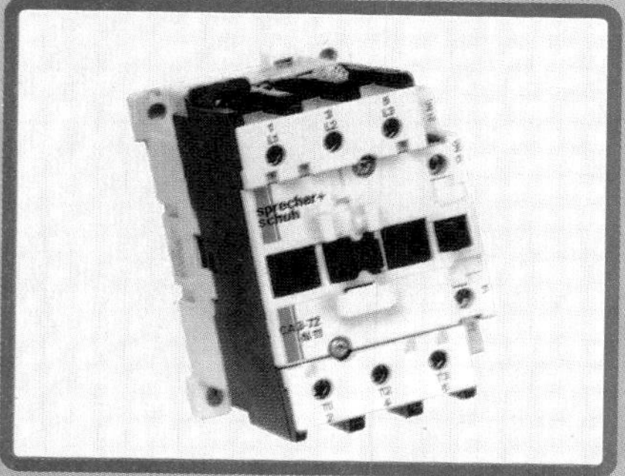

Sprecher + Schuh

Magnetic contactors and motor starters are control devices that use a small control current to energize or de-energize the load connected to it. Motor starters are contactors that contain overload protection. Various devices may be added to contactors and motor starters to expand their capability. Contactors or motor starters are the first devices checked when troubleshooting a circuit that does not work.

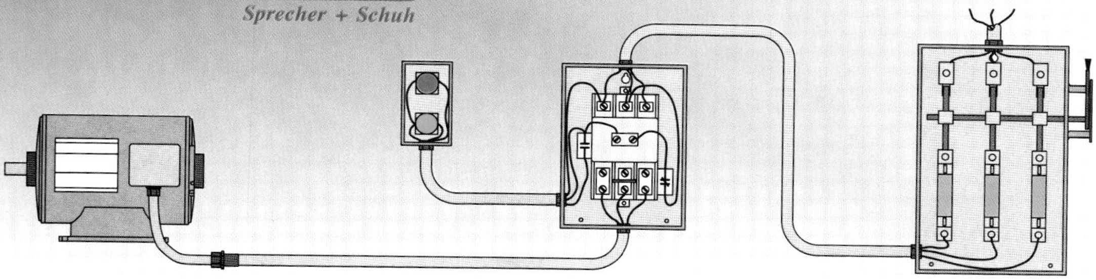

CONTACTORS

A *contactor* is a control device that uses a small control current to energize or de-energize the load connected to it and has no overload protection. Contactors may be operated manually or magnetically. Contactors are devices for establishing and interrupting an electrical power circuit repeatedly. Contactors are used to make and break the electrical power circuit to such loads as lights, heaters, transformers, and capacitors. See Figure 6-1.

Contactor Construction

Solenoid action is the principle operating mechanism for a magnetic contactor. The linear action of a solenoid is used to open and close sets of contacts instead of pushing and pulling levers and valves. See Figure 6-2. The use of solenoid action rather than manual input is an advantage of a magnetic contactor over a manual contactor. Remote control and automation, which are impossible with manual contactors, can be designed into a system using magnetic contactors.

Products Unlimited

Contactors have higher current ratings than relays because they are used to control high-power loads.

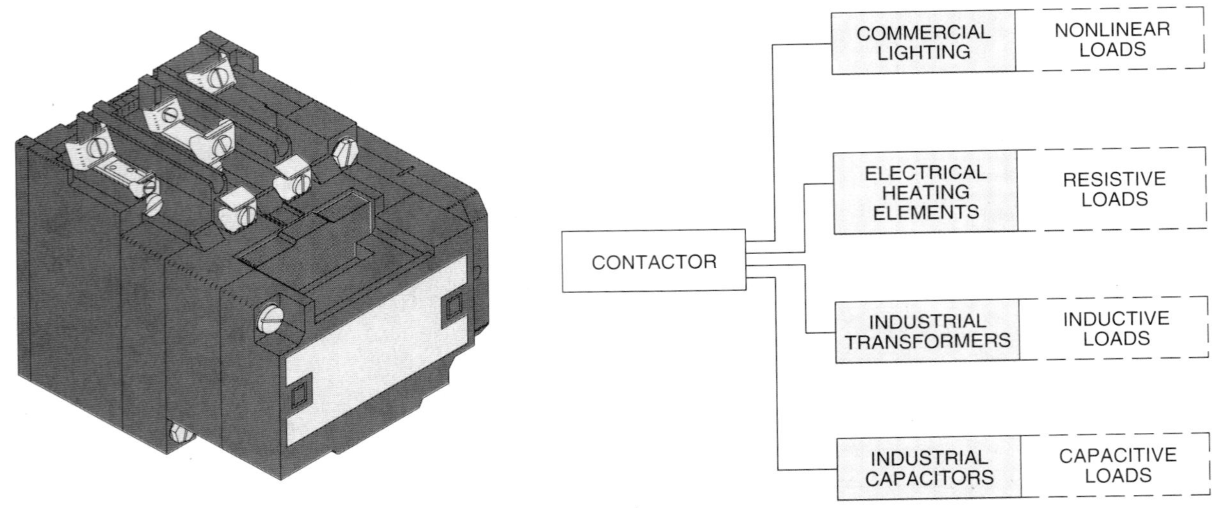

Figure 6-1. Contactors are used to make and break the electrical power circuit to lights, heaters, transformers, and capacitors.

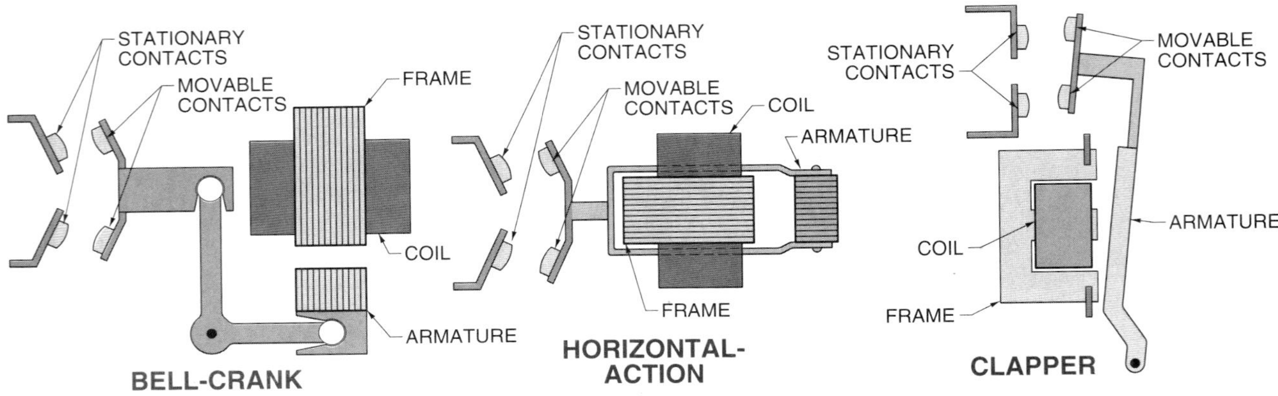

Figure 6-2. Solenoid action is the principle operating mechanism for magnetic contactors.

Contactor Wiring

Control circuits are often referred to by the number of conductors used in the control circuit, such as 2-wire and 3-wire control. Two-wire control involves two conductors to complete the circuit. Three-wire control involves three conductors to complete the circuit.

Two-Wire Control. Two-wire control has two wires leading from the control device to the starter or contactor. See Figure 6-3. The control device could be a thermostat, float switch, or other contact device. When the contacts of the control device close, they complete the coil circuit of the contactor, causing it to energize. This connects the load to the line through the power contacts. The contactor coil is de-energized when the contacts of the control device open. This de-energizes coil M1 which opens the contacts that control the load. The contactor functions automatically in response to the condition of the control device without the attention of an operator.

A 2-wire control circuit provides low-voltage release, but not low-voltage protection. In the event of a power loss in the control circuit, the contactor de-energizes (low-voltage release), but also re-energizes if the control device remains closed when the circuit has power restored. Low-voltage protection cannot be provided in this circuit because there is no way for the operator to be protected from the circuit once it has been re-energized.

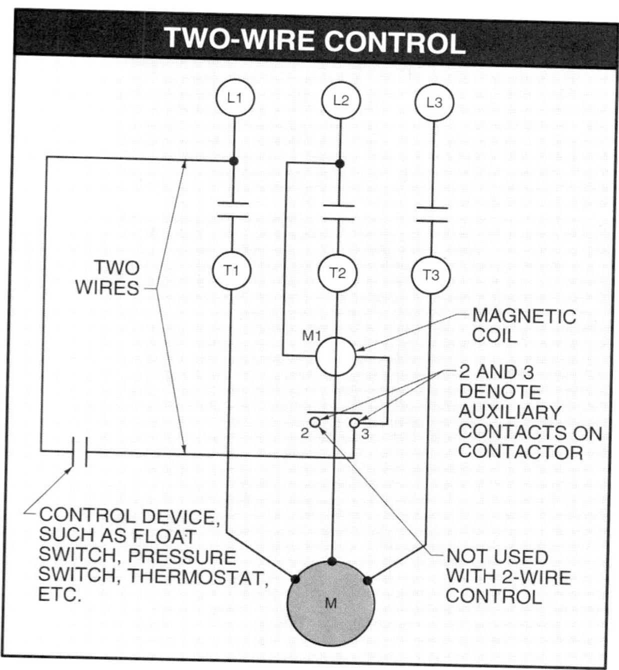

Figure 6-3. In 2-wire control, two wires lead from the control device to the starter or contactor.

Caution must be exercised in the use and service of 2-wire control circuits because of the lack of low-voltage protection. Two-wire control is normally used for remote or inaccessible installations, such as pumping stations, water or sewage treatment, air conditioning or refrigeration systems, and process line pumps where an immediate return to service after a power failure is required.

Three-Wire Control. Three-wire control has three wires leading from the control device to the starter or contactor. See Figure 6-4. The circuit uses a momentary contact stop pushbutton (NC) wired in series with a momentary contact start pushbutton (NO) wired in parallel to a set of contacts which forms a holding circuit interlock (memory).

When the NO start pushbutton is pressed, current flows through the stop pushbutton (NC), through the momentarily closed start pushbutton, through magnetic coil M1, the overloads, and on to L2. This causes the magnetic coil to energize. When energized, the auxiliary holding circuit interlock contacts (memory) close, sealing the path through to the coil circuit even if the start pushbutton is released.

Pressing the stop pushbutton (NC) opens the circuit to the magnetic coil, causing the contactor to de-energize. A power failure also de-energizes the contactor. The interlock contacts (memory) reopen when the contactor de-energizes. This opens both current paths to the coil through the start pushbutton and the interlock.

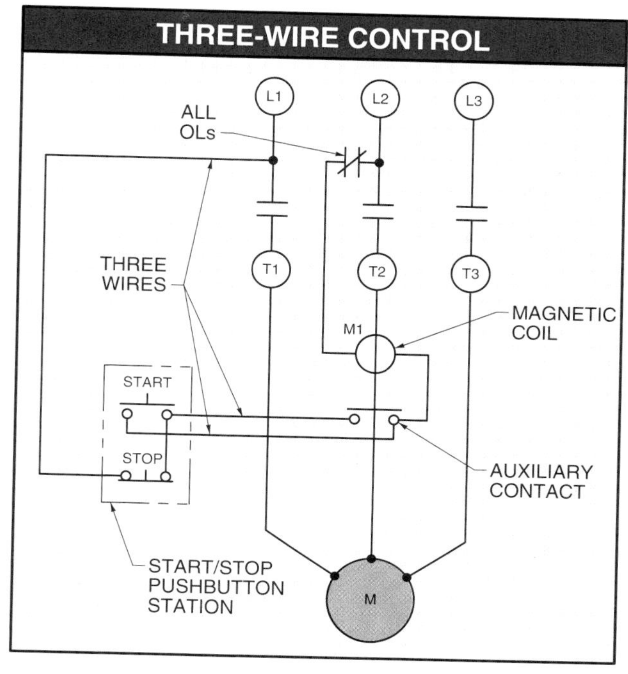

Figure 6-4. In 3-wire control, three wires lead from the control device to the starter or contactor.

Three-wire control provides low-voltage release and low-voltage protection. The coil drops out at low or no voltage and cannot be reset unless the voltage returns and the operator presses the start pushbutton.

AC and DC Contactors

AC contactor assemblies may have several sets of contacts. DC contactor assemblies typically have only one set of contacts. See Figure 6-5. In 3ϕ AC contactors, all three power lines must be broken. This creates the need for several sets of contacts. For multiple contact control, a T-bar assembly allows several sets of contacts to be activated simultaneously. In a DC contactor, it is necessary to break only one power line.

AC contactor assemblies are made of laminated steel, while DC assemblies are solid. Laminations are unnecessary in a DC coil because the current travels in one direction at a continuous rate and does not create eddy-current problems. The other major differences between AC and DC contactors are the

General Electric Company

Rockwell Automation, Allen-Bradley Company, Inc.

Figure 6-5. AC contactor assemblies may have several sets of contacts. DC contactor assemblies typically have only one set of contacts.

electrical and mechanical requirements necessary for suppressing the arcs created in opening and closing contacts under load.

Arc Suppression

Arc suppression is required on contactors and motor starters. An *arc suppressor* is a device that dissipates the energy present across opening contacts. Without arc suppression, contactors and motors may require maintenance prematurely and could result in excessive downtime.

Opening Contact Arc. A short period of time (a few thousandths of a second) exists when a set of contacts is opened under load during which the contacts are neither fully in touch with each other nor completely separated. See Figure 6-6.

As the contacts continue to separate, the contact surface area decreases, increasing the electrical resistance. With full-load current passing through the increasing resistance, a substantial temperature rise is created on the surface of the contacts. This temperature rise is often high enough to cause the contact surfaces to become molten and emit ions of vaporized metal into the gap between the contacts. This hot ionized vapor permits the current to continue to flow in the form of an arc, even though the contacts are completely separated. The arcs produce additional heat, which if continued, can damage the contact surfaces. The sooner the arc is extinguished, the longer the life expectancy of the contacts.

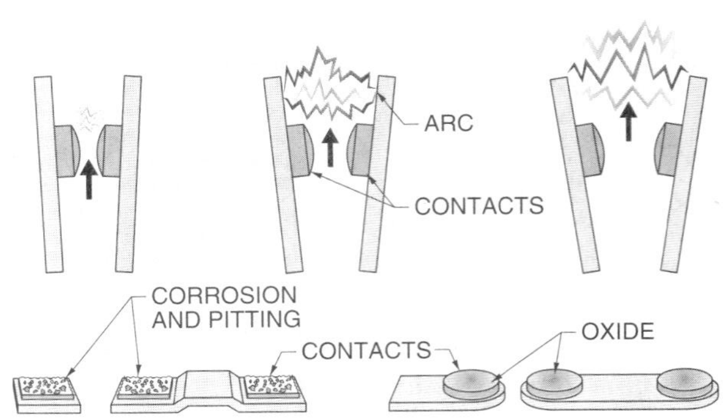

Figure 6-6. An electrical arc is created between contacts as they are opened. Prolonged arcing may result in damage to contact surfaces.

DC Arc Suppression. DC arcs are considered the most difficult to extinguish because the continuous DC supply causes current to flow constantly and with great stability across a much wider gap than does an AC supply of equal voltage. To reduce arcing in DC circuits, the switching mechanism must be such that the contacts separate rapidly and with enough air gap to extinguish the arc as soon as possible on opening.

It is also necessary in closing DC contacts to move the contacts together as quickly as possible to avoid some of the same problems encountered in opening them. DC contactors are larger than AC contactors to allow for the additional air gap. In addition, the operating characteristics of DC contactors are faster than AC contactors.

One disadvantage in rapid closing of DC contactors is that the contacts must be buffered to eliminate contact bounce due to excessive closing force. Contact bounce may be minimized through the use of certain types of solenoid action and springs attached under the contacts to absorb some of the shock.

AC Arc Suppression. An AC arc is self-extinguishing when a set of contacts is opened. In contrast to a DC supply of constant voltage, the AC supply has a voltage which reverses its polarity every $\frac{1}{120}$ of a second when operated on a 60 Hertz (Hz) line frequency. See Figure 6-7. The alternation allows the arc to have a maximum duration of no more than a half-cycle. During any half-cycle, the maximum arcing current is reached only once in that half-cycle.

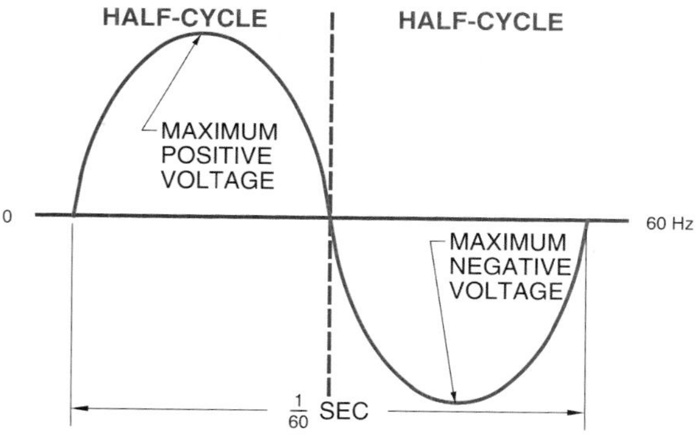

Figure 6-7. The maximum arcing current is reached only once during any half-cycle of AC voltage.

The contacts can be separated more slowly and the gap length may be shortened because an AC arc is self-extinguishing. This short gap keeps the voltage across the gap and the arc energy low. With low gap energy, ionizing gases cool more rapidly, extinguishing the arc and making it difficult to restart. AC contactors need less room for operation and run cooler which increases contact life.

Arc at Closing. Arcing may also occur on AC and DC contactors when they are closing. The most common arcing occurs when the contacts come close enough that an arc is able to bridge the open space between the contacts.

Arcing also occurs if a whisker or rough edge of the contact touches first and melts, causing an ionized path which allows current to flow. In either case, the arc lasts until the contact surfaces are fully closed. Contactor design for both AC and DC devices are quite similar. The contactor should be designed so that the contacts close as rapidly as possible without bouncing to minimize the arc at each closing.

Arc Chutes. An *arc chute* is a device that confines, divides, and extinguishes arcs drawn between contacts opened under load. See Figure 6-8. Arc chutes are used to contain large arcs and the gases created by them. Arc chutes employ the de-ion principle which confines, divides, and extinguishes the arc for each set of contacts.

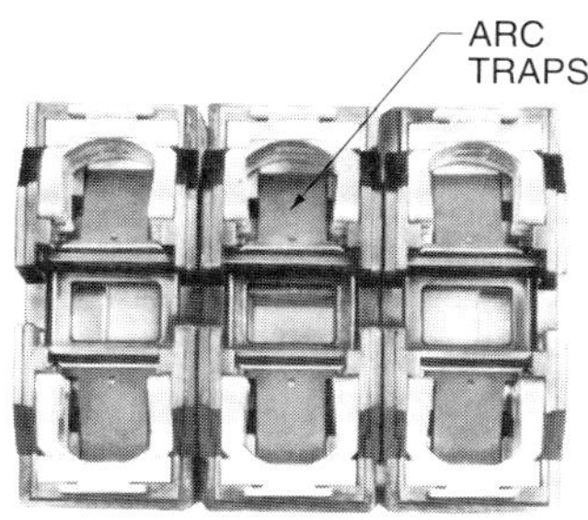

Cutler-Hammer

Figure 6-8. Arc chutes and arc traps are used to confine, divide, and extinguish arcs drawn between contacts opened under load.

Arcs may also be extinguished by using special arc traps and arc-quenching compounds. This circuit breaker technique attracts, splits, and quickly cools arcs as well as vents ionized gases. Vertical barriers between each set of contacts, as well as arc covers, confine arcs to separate chambers and quickly quench them.

DC Magnetic Blowout Coils. When a DC circuit carrying large amounts of current is interrupted, the collapsing magnetic field (flux) of the circuit current may induce a voltage which helps sustain the arc. Action must be taken to quickly limit the damaging effect of the heavy current arcs because a sustained electrical arc may melt the contacts, weld them together, or severely damage them.

One way to stop the arc quickly is to move the contacts some distance from each other as quickly as possible. The problem is that the contactor has to be large enough to accommodate such a large air gap.

Magnetic blowout coils are used to reduce the distance required and yet quench arcs quickly. Magnetic blowout coils provide a magnetic field which blows out the arc similarly to blowing out a match.

A magnetic field is created around the current flow whenever a current flows through a conductive medium (in this case ionized air). The direction of the magnetic field (flux) around the conductor is determined by wrapping the right or left hand around the conductor. When the thumb on the right hand points in the direction of conventional current flow, the wrapping fingers point in the direction of the resulting magnetic flux. When the thumb on the left hand points in the direction of electron current flow, the wrapping fingers point in the direction of the resulting magnetic flux. See Figure 6-9.

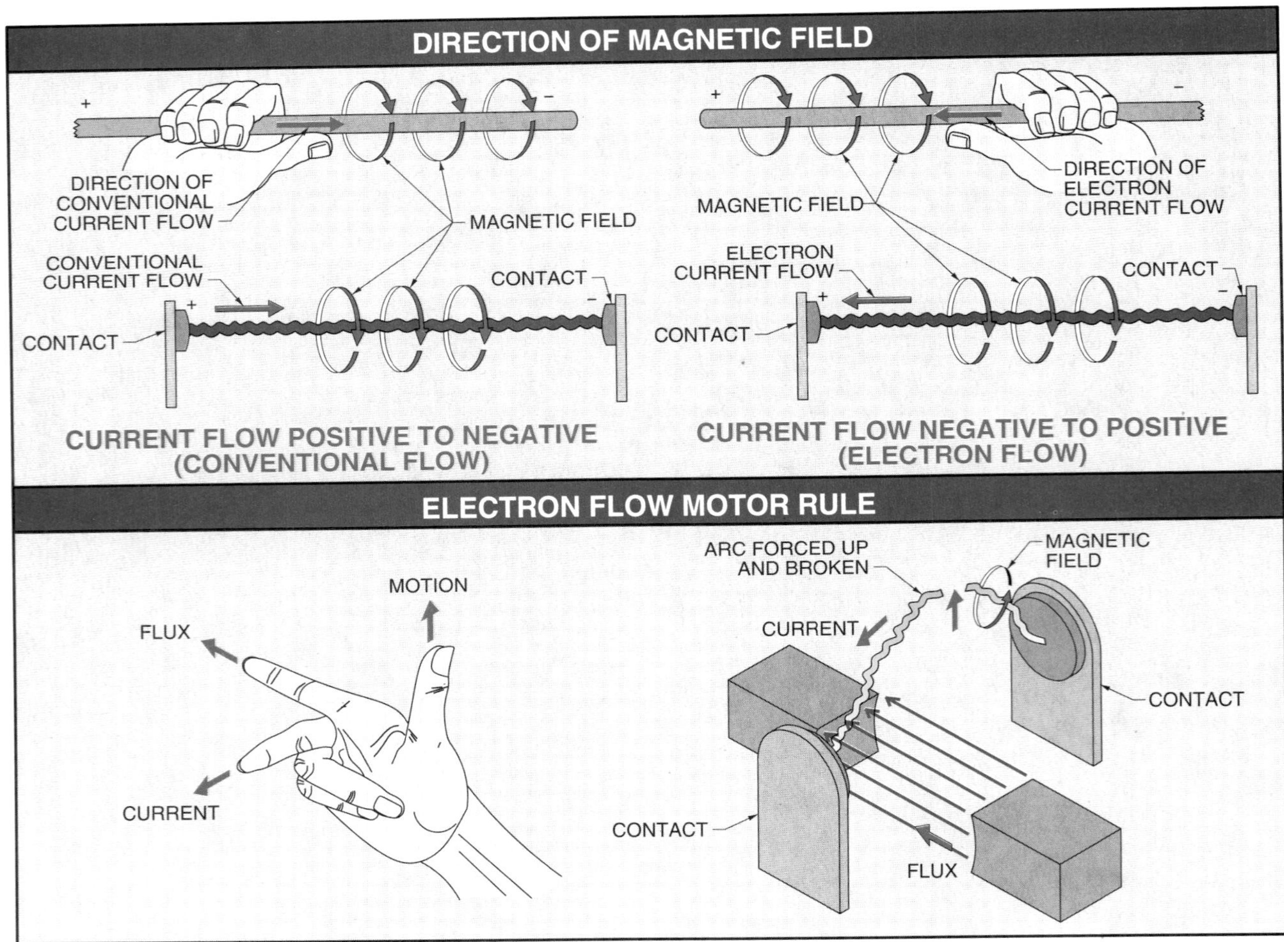

Figure 6-9. The direction of the magnetic field around the conductor is determined by wrapping the right or left hand around the conductor. The electron flow motor rule indicates the motion of an arc cutting through flux lines.

The electron flow motor rule states that when a current-carrying conductor (represented by the middle finger) is placed in a parallel magnetic field (represented by the index finger), the resulting force or movement is in the direction of the thumb. This action occurs because the magnetic field around the current flow opposes the parallel magnetic field above the current flow. This makes the magnetic field above the current flow weaker, while aiding the magnetic field below the current flow, making the magnetic field stronger. The net result is an upward push which quickly elongates the arc current so that it breaks (blows out). An electromagnetic blowout coil is often referred to as a puffer because of its blowout ability. See Figure 6-10.

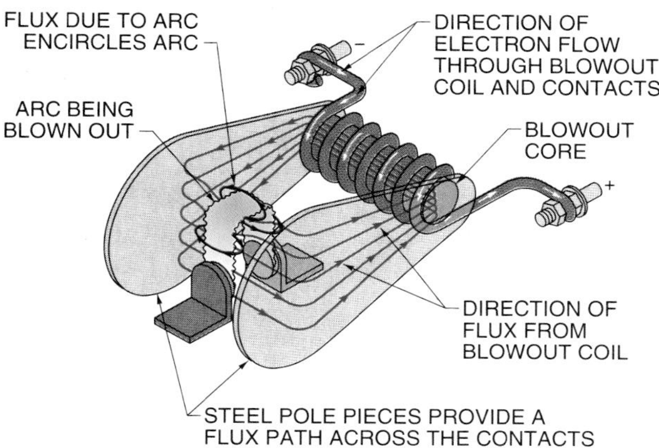

Figure 6-10. Electromagnetic blowout coils rapidly extinguish DC arcs.

Contact Construction

Contact design and materials depend on the size, current rating, and application of the contactor. Double-break contacts are normally made of a silver cadmium alloy. Single-break contacts in large contactors are frequently made of copper because of the low cost.

Single-break copper contacts are designed with a wiping action to remove the copper oxide film which forms on the copper tips of the contacts. The wiping action is necessary because copper oxide formed on the contact when not in use is an insulator and must be eliminated for good circuit conductivity.

In most cases, the slight rubbing action and burning that occur during normal operation keep the contact surfaces clean for proper service. Copper contacts that seldom open or close, or those being replaced, should be cleaned to reduce contact resistance. High contact resistance often causes serious heating of the contacts.

General Purpose AC/DC Contactor Sizes and Ratings

Magnetic contactors, like manual starters, are rated according to the size and type of load by the National Electrical Manufacturers Association (NEMA). Tables are used to indicate the number/size designations and establish the current load carried by each contact in a contactor. See Figure 6-11. The rating is for each contact individually, not for the entire contactor. For example, a Size 0, 3-pole contactor rated at 18 A is capable and rated for switching three separate 18 A loads simultaneously.

60 Hz AC CONTACTOR STANDARD NEMA RATINGS

Size	8 Hr Open Rating*	Power Rating**				
		3φ			1φ	
		200 V	230 V	230/460 V	115 V	230 V
00	9	1½	1½	2	⅓	1
0	18	3	3	5	1	2
1	27	7½	7½	10	2	3
2	45	10	15	25	3	7½
3	90	25	30	50	—	—
4	135	40	50	100	—	—
5	270	75	100	200	—	—
6	540	150	200	400	—	—
7	810	—	300	600	—	—
8	1215	—	450	900	—	—
9	2250	—	800	1600	—	—

* in A ** in HP

DC CONTACTOR STANDARD NEMA RATINGS

Size	8 Hr Open Rating*	Power Rating**		
		115 V	230 V	550 V
1	25	3	5	—
2	50	5	10	20
3	100	10	25	50
4	150	20	40	75
5	300	40	75	150
6	600	75	150	300
7	900	110	225	450
8	1350	175	350	700
9	2500	300	600	1200

* in A ** in HP

Figure 6-11. Tables indicate the number/size designations and establish the current load carried by each contact in a contactor.

Contactor dimensions vary greatly. The range is from inches to several feet in length. Contactors are selected based on type, size, and voltage available. Contactors are also available in a wide variety of enclosures. See Figure 6-12. The wide variety of enclosures offer protection ranging from the most basic protection to high levels of protection required in hazardous locations where any spark caused by the closing or opening of the contact could cause an explosion.

MAGNETIC MOTOR STARTERS

A *magnetic motor starter* is a contactor that includes overload protection. Motor starters include overload relays that detect excessive current passing through a motor. Motor starters are used to switch all types and sizes of motors. Motor starters are available in sizes that can switch loads of a few amperes to several hundred amperes. See Figure 6-13.

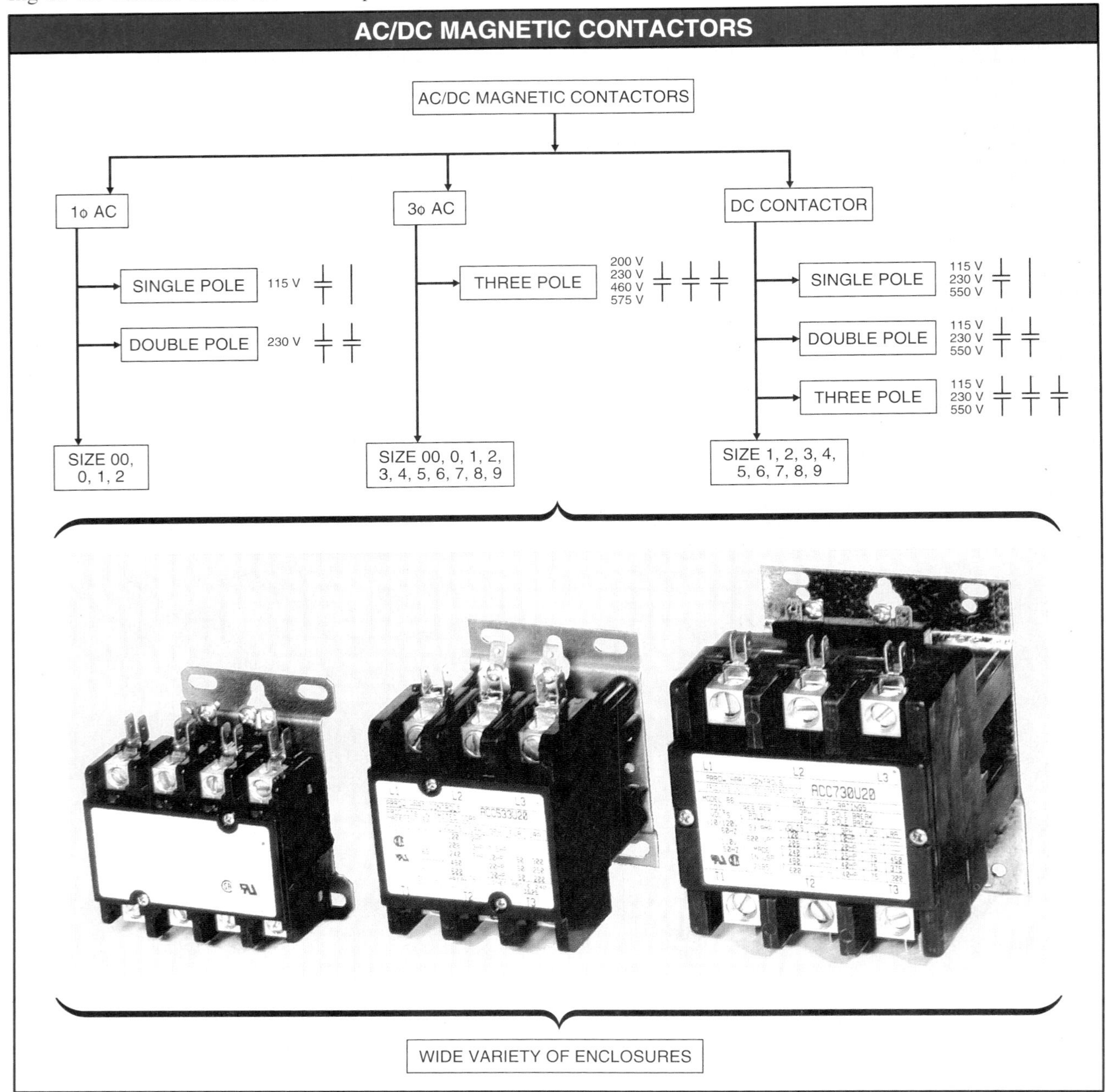

Products Unlimited

Figure 6-12. Contactor dimensions vary from inches to several feet in length.

Furnas Electric Co.

Figure 6-13. A magnetic motor starter is a contactor with overload protection added.

Overload Protection

The main difference between the sensing device for a manual starter and a magnetic starter is that a manual overload opens the power contacts on the starter. The overload device on a magnetic starter opens a set of contacts to the magnetic coil, de-energizing the coil and disconnecting the power. Overload devices include melting alloy, magnetic, and bimetallic overload relays.

Melting Alloy Overload Relays. The melting alloy overload relays used in magnetic starters are similar to the melting alloy overload relays used in manual starters. They consist of a heater coil, eutectic alloy, and mechanical mechanism to activate a tripping device when an overload occurs.

Magnetic Overload Relays. Magnetic overload relays provide another means of monitoring the amount of current drawn by a motor. A magnetic overload relay operates through the use of a current coil. At a specified overcurrent value, the current coil acts as a solenoid, causing a set of NC contacts to open. This causes the circuit to open and protect the motor by disconnecting it from power. See Figure 6-14.

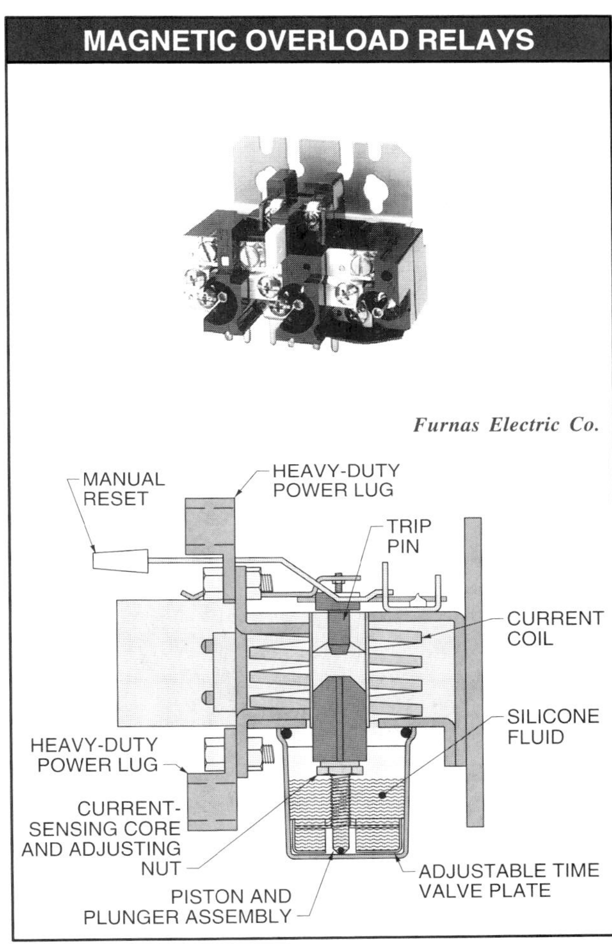

Figure 6-14. Magnetic overload relays use a current coil which, at a specified overcurrent value, acts like a solenoid and causes a set of NC contacts to open.

Magnetic overload relays are used in special applications, such as steel mill processing lines or other heavy-duty industrial applications where holding a specified level of motor current is required. A magnetic overload relay is also ideal for special applications, such as slow acceleration motors, high-current inrush motors, or any use where normal time/current curves of thermal overload relays do not provide satisfactory operation. This flexibility is made possible because the magnetic unit may be set for either instantaneous or inverse time-tripping characteristics. The device may also offer independent adjustable trip time and trip current.

Magnetic overload relays are extremely quick to reset because they do not require a cooling-off period before being reset. Magnetic overload relays are much more expensive than thermal overload relays.

Bimetallic Overload Relays. In certain applications, such as walk-in meat coolers, remote pumping stations, and some chemical process equipment, overload relays which reset automatically to keep the unit operating up to the last possible moment may be required. A *bimetallic overload relay* is an overload relay which resets automatically. Bimetallic overload relays operate on the principle of the bimetallic strip. A bimetallic strip is made of two pieces of different metal. The dissimilar metals are permanently joined by lamination. Heating the bimetallic strip causes it to warp because metals expand and contract at different rates. The warping effect of the bimetallic strip is used as a means for separating contacts. See Figure 6-15.

Once the tripping action has taken place, the bimetallic strip cools and reshapes itself. In certain devices, such as circuit breakers, a trip lever needs to be reset to make the circuit operate again. In other devices, such as bimetallic overload relays, the device automatically resets the circuit when the bimetallic strip cools and reshapes itself.

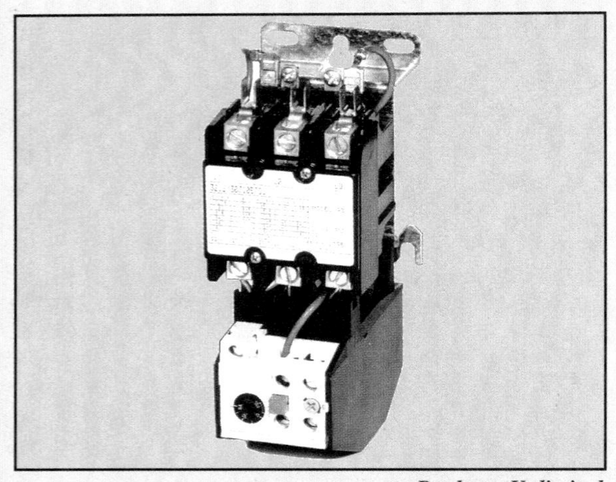

Products Unlimited

The ACV Series definite purpose starters from Products Unlimited feature IEC-rated adjustable overload blocks, phase loss and unbalanced load protection, and trip indication and circuit test buttons.

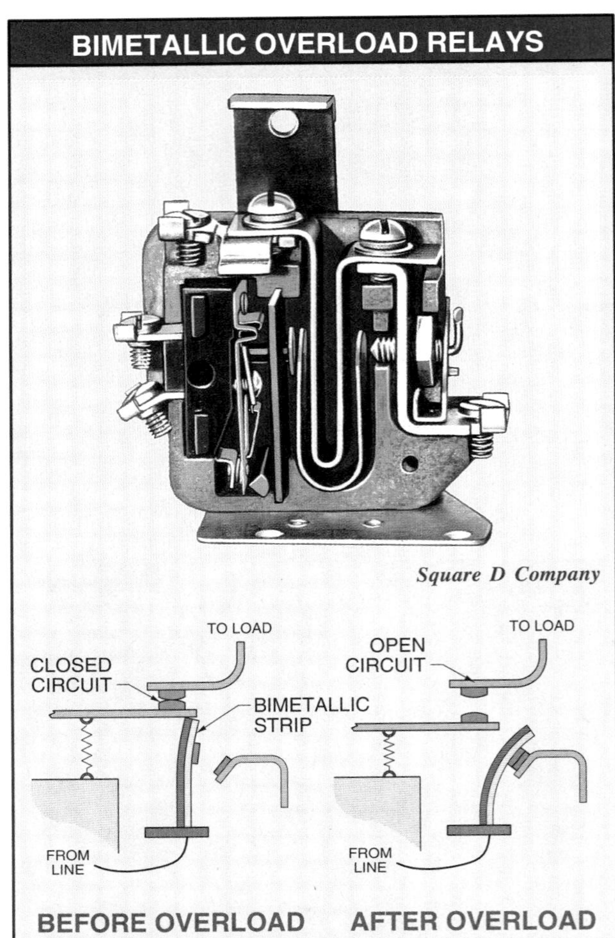

Figure 6-15. The warping effect of the bimetallic strip is used as a means for separating contacts.

The motor restarts even when the overload has not been cleared, and trips and resets itself again at given intervals. Care must be exercised in the selection of a bimetallic overload relay because repeated cycling eventually burns out the motor. The bimetallic strip may be shaped in the form of a U. The U-shape provides a uniform temperature response.

Trip Indicators. Many overload devices have a trip indicator built into the unit to indicate to the operator that an overload has taken place within the device. See Figure 6-16. A red metal indicator appears in a window located above the reset button when the overload relay has tripped. The red indicator informs the operator or electrician why the unit is not operating and that it potentially is capable of returning with an automatic reset.

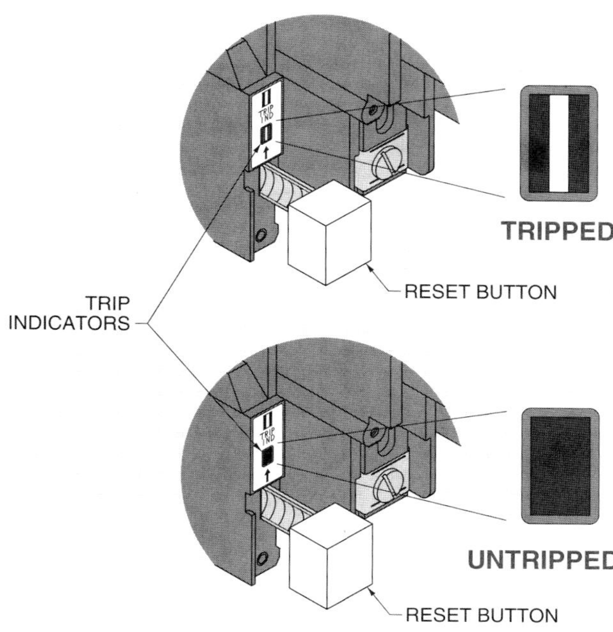

Figure 6-16. Trip indicators indicate that an overload has taken place within the device.

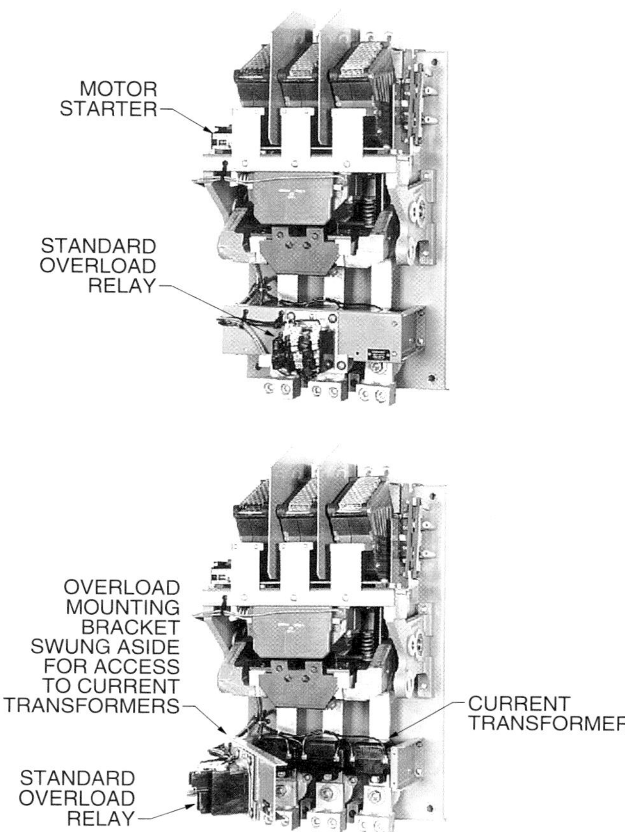

Figure 6-17. Standard overload relays may be used on very large starters by using current transformers with specific reduction ratios.

Overload Current Transformers. Large horsepower motors have currents that exceed the values of standard overload relays. To make the overload relays larger would greatly increase their physical size and would become a space problem in relation to the magnetic motor starter. To avoid such a conflict, current transformers are used to reduce the current in a fixed ratio. See Figure 6-17. The current transformer determines the amount of current going to the motor, but reduces the current to a lower value for the overload relay. For example, if 50 A were going to a motor, only 5 A would be going to the overload relay through the use of the current transformer. Standard current transformers are normally rated in primary and secondary rated current such as 50/5 or 100/5.

Because the ratio is always the same, an increase to the motor also increases the current to the overload relay. If the correct current transformer and overload relay combination are selected, the same overload protection can be provided to a motor as if the overload relay were actually in the load circuit. The overload relay contacts open and the coil to the magnetic motor starter is de-energized when excessive current is sensed. This shuts the motor OFF. Several different current transformer ratios are available to make this type of overload protection easy to provide.

Overload Heater Sizes

Each motor must be sized according to its own unique operating characteristics and applications. Thermal overload heaters are selected based on the full-load current rating (FLC), service factor (SF), and ambient temperature (surrounding air temperature) of the motor when it is operating.

Full-Load Current Rating. Selection of thermal overload heaters is based on the full-load current shown on the motor nameplate or in the motor manufacturer's specification sheet. See Figure 6-18. The current value reflects the current to be expected when the motor is running at specified voltages, speeds, and normal torque characteristics. Heater manufacturers develop current charts indicating which heater should be used with each FLC.

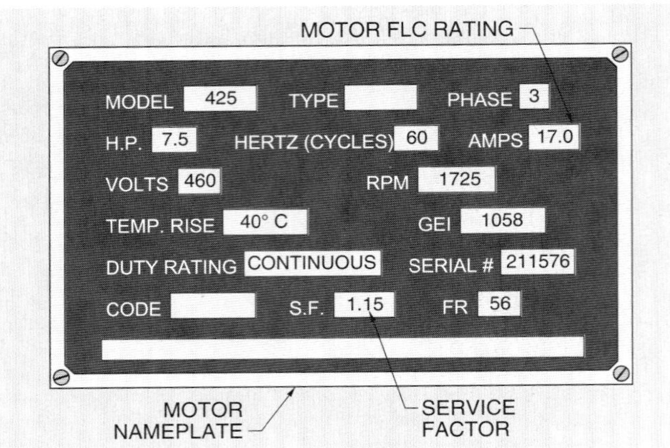

Figure 6-18. Selection of thermal overload heaters is based on the full-load current shown on the motor nameplate.

Service Factor. In most motor applications, there are times when the motor must produce more than its rated horsepower for a short period of time without damage. A *service factor (SF)* is a number designation that represents the percentage of extra demand that can be placed on the motor for short intervals without damaging the motor. Common service factors range from 1.00 to 1.25, indicating that the motor can produce 0% to 25% extra demand over that for which it is normally rated. A 1.00 SF indicates that the motor cannot produce more power than it is rated for and to do so would result in damage. An SF of 1.25 indicates that the motor can produce up to 25% more power than it is rated for, but for short periods of time.

The excessive current which can be safely handled by a given motor for short periods of time is approximated by multiplying the SF by the FLC. For example, if a motor is rated at an FLC of 10 A with an SF of 1.15, the excess short-term current equals 11.5 A ($10 \times 1.15 = 11.5$ A). The motor could handle an additional 1.5 A for a short period of time.

Ambient Temperature. A thermal overload relay operates on the principle of heat. Sufficient heat is generated by the excessive current to melt a metal alloy, produce movement in a current coil, or warp a bimetallic strip and allow the device to trip when an overload takes place. The temperature surrounding a thermal overload relay must be considered because the relay is sensitive to heat from any source. The ambient temperature is a factor when considering moving a thermal overload relay from a refrigerated meat packing plant to a location near a blast furnace.

Overload relay devices are normally rated to trip at a specific current when surrounded by an ambient temperature of 40°C (104°F). This standard ambient temperature is acceptable for most control applications. Compensation must be provided for higher or lower ambient temperatures.

Overload Heater Selection

Overload heater coils for continuous-duty motors are selected from manufacturer's tables based on the motor's nameplate FLC for maximum motor protection and compliance with Section 430-32 of the NEC®. The class, type, and size information of a magnetic starter are found on the nameplate on the face of the starter. See Figure 6-19. The phase, SF, and FLC of the motor are determined from the motor nameplate. Common applications use 40°C as the ambient temperature. Questionable ambient temperatures should be measured at the job site or determined by some other method.

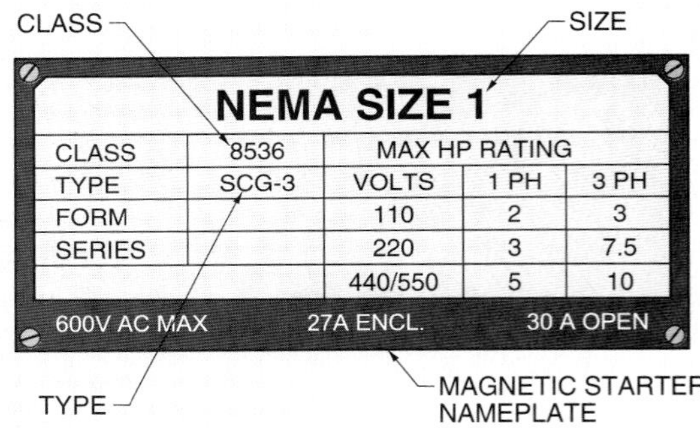

Figure 6-19. The nameplate of a magnetic starter includes the class, type, and size of the starter.

Rockwell Automation, Allen-Bradley Company, Inc.

A motor starter nameplate includes motor starter information and is used when installing a motor starter.

Always refer to the manufacturer's instructions on thermal overload relay selection to see if any restrictions are placed on the class of starter required. See Figure 6-20. For example, unless a class 8198 starter is used, motors with SFs of 1.15 to 1.25 may use 100% of the motor FLC for thermal overload selection.

Motor and controller in _same ambient temperature:_

a. All starter classes, except Class 8198

1. For 1.15 to 1.25 service factor motors use 100% of motor full-load current for thermal unit selection.
2. For 1.0 service factor motors use 90% of motor full-load current for thermal unit selection.

b. Class 8198 only:

CLASS RESTRICTION

1. For 1.0 service factor motors use 100% of motor full-load current for thermal unit selection.
2. For 1.15 to 1.25 service factor motors use 110% of motor full-load current for thermal unit.

Figure 6-20. Manufacturer's instructions on thermal overload relay selection detail restrictions that are placed on classes of starters.

Manufacturer's Heater Selection Charts. Manufacturers provide charts for use in selecting proper thermal overload heaters. The correct chart must be used for the appropriate size starter. See Figure 6-21. See Appendix. This information is also found within the enclosure of many motor starters.

Rockwell Automation, Allen-Bradley Company, Inc.

The Bulletin 825 smart motor manager from Allen-Bradley is a programmable, electronic overload protection relay with communication capability that provides motor protection by detecting thermal overload, phase loss, stalling, ground fault, short circuit, and underload.

THERMAL UNIT CURRENT RATINGS

Motor Full-Load Current (Amps)			Thermal Unit Number
1 Unit	**2 Units**	**3 Units**	
0.29 – 0.31	0.29 – 0.31	0.28 – 0.30	B0.44
0.32 – 0.34	0.32 – 0.34	0.31 – 0.34	B0.51
0.35 – 0.38	0.35 – 0.38	0.35 – 0.37	B0.57
0.39 – 0.45	0.39 – 0.45	0.38 – 0.44	B0.63
0.46 – 0.54	0.46 – 0.54	0.45 – 0.53	B0.71
0.51 – 0.61	0.51 – 0.61	0.54 – 0.59	B0.81
0.62 – 0.66	0.62 – 0.66	0.60 – 0.64	B0.92
0.67 – 0.73	0.67 – 0.73	0.64 – 0.72	B1.03
0.74 – 0.81	0.74 – 0.81	0.73 – 0.80	B1.16
0.82 – 0.94	0.82 – 0.94	0.81 – 0.90	B1.30
0.95 – 1.05	0.95 – 1.05	0.91 – 1.03	B1.45
1.06 – 1.22	1.06 – 1.22	1.04 – 1.14	B1.67
1.23 – 1.34	1.23 – 1.34	1.15 – 1.27	B1.88
1.35 – 1.51	1.35 – 1.51	1.28 – 1.43	B2.10
1.52 – 1.71	1.52 – 1.71	1.44 – 1.62	B2.40
1.72 – 1.93	1.72 – 1.93	1.63 – 1.77	B2.65
1.94 – 2.14	1.94 – 2.14	1.78 – 1.97	B3.00
2.15 – 2.40	2.15 – 2.40	1.98 – 2.32	B3.30
2.41 – 2.72	2.41 – 2.72	2.33 – 2.51	B3.70
2.73 – 3.15	2.73 – 3.15	2.52 – 2.99	B4.15
3.16 – 3.55	3.16 – 3.55	3.00 – 3.42	B4.85
3.56 – 4.00	3.56 – 4.00	3.43 – 3.75	B5.50
4.01 – 4.40	4.01 – 4.40	3.76 – 3.98	B6.25
4.41 – 4.88	4.41 – 4.88	3.99 – 4.48	B6.90
4.89 – 5.19	4.89 – 5.19	4.49 – 4.93	B7.70
5.20 – 5.73	5.20 – 5.73	4.94 – 5.21	B8.20
5.74 – 6.39	5.74 – 6.39	5.22 – 5.84	B9.10
6.40 – 7.13	6.40 – 7.13	5.85 – 6.67	B10.2
7.14 – 7.90	7.14 – 7.90	6.68 – 7.54	B11.5
7.91 – 8.55	7.91 – 8.55	7.55 – 8.14	B12.8
8.56 – 9.53	8.56 – 9.53	8.15 – 8.72	B14.0
9.54 – 10.6	9.54 – 10.6	8.73 – 9.66	B15.5
10.7 – 11.8	10.7 – 11.8	9.67 – 10.5	B17.5
11.9 – 13.2	11.9 – 12.0	10.6 – 11.3	B19.5
13.3 – 14.9	—	11.4 – 12.0	B22.0
15.0 – 16.6	—	—	B25.0
16.7 – 18.0	—	—	B28.0
Following Selections for Size 1 Only			
—	11.9 – 13.2	—	B19.5
—	13.3 – 14.9	11.4 – 12.7	B22.0
—	15.0 – 16.6	12.8 – 14.1	B25.0
16.7 – 18.9	16.7 – 18.9	14.2 – 15.9	B28.0
19.0 – 21.2	19.0 – 21.2	16.0 – 17.5	B32.0
21.3 – 23.0	21.3 – 23.0	17.6 – 19.7	B36.0
23.1 – 25.5	23.1 – 25.5	19.8 – 21.9	B40.0
25.6 – 26.0	25.6 – 26.0	22.0 – 24.4	B45.0
—	—	24.5 – 26.0	B50.0

Figure 6-21. Manufacturers provide charts for use in selecting proper overload heaters.

For example, a thermal unit number B2.40 is the correct overload heater for controlling a 3ϕ motor with an FLC of 1.50 A. Column three is used because all three phases of the 3ϕ motor must have thermal overload protection. The heater must provide protection of approximately 1.5 A (1.44 – 1.62) based on the motor FLC. Manufacturers have different numbers to relate to their specific heaters, but the selection procedure is similar.

Checking Selections. Article 430 of the NEC® indicates that a motor must be protected up to 125% of its FLC. Because the minimum full load of A B 2.40 overload device is 1.44 A, the device trips at 125% of this value or 1.8 A (1.44 × 1.25 = 1.8 A). Dividing the minimum trip current (1.8 A) by the FLC of the motor (1.5 A) and multiplying by 100% determines if this range is acceptable. ($^{1.8}/_{1.5}$ × 100% = 120%). The heater selection is correct because the trip current is less than the NEC® limit of 125%.

Ambient Temperature Compensation. As ambient temperature increases, less current is needed to trip overload devices. As ambient temperature decreases, more current is needed to trip overload devices. Most heater manufacturers provide special overload heater selection tables that provide multipliers to compensate for temperature changes above or below the standard temperature of 40°C. The multipliers ensure that the increase or decrease in temperature does not affect the proper protection provided by the overload relay. See Figure 6-22.

THERMAL UNIT SELECTION

Controller Class	Continuous-Duty Motor Service Factor	Melting Alloy and Non-Compensated Bimetallic Relays — Ambient Temperature of Motor		
		*	**	***
		Full-Load Current Multiplier		
All Classes except 8198	1.15 – 1.25	1.0	0.9	1.05
	1.0	0.9	0.8	.95
Class 8198	1.15 – 1.25	1.1	1.0	1.15
	1.0	1.0	0.9	1.05

* same as controller ambient

** constant 10°C (18°F) higher than controller ambient

*** constant 10°C (18°F) lower than controller ambient

Figure 6-22. Special overload heater selection tables provide multipliers to compensate for ambient temperatures above or below the standard temperature of 40°C.

Rockwell Automation, Allen-Bradley Company, Inc.

NEMA Size 6 full-voltage starters with SMP-3™ overload relays have a continuous ampere rating of 540 A and a maximum rating of 200 HP at 230 V.

For example, a multiplier of 0.9 is required for an ambient temperature increase of 10°C to 50°C. Multiplying the motor FLC (1.5 A) by the correction factor (0.9) determines the compensated overload heater current rating of 1.35 A (1.5 A × 0.9 = 1.35 A). Using the heater selection chart, the acceptable current range is 1.28 A to 1.43 A. A B2.10 heater is required based on the increase in ambient temperature. This is one size smaller than the heater required (B2.40) at a 40°C ambient temperature.

The temperature surrounding an overload heater is 30°C if the ambient temperature is decreased 10°C. The correction multiplier is 1.05 for a 10°C decrease in ambient temperature. The corrected current is 1.575 A using an FLC of 1.5 A (1.5 A × 1.05 = 1.575 A). In the selection guide, a range of 1.44 to 1.62 is acceptable. In this case, the same size heater would be used. Always consult manufacturer's specifications and tables for proper sizing.

In rare instances, such as older installations or severely damaged equipment, it may be impossible to determine a motor's FLC from its nameplate. Manufacturers provide charts listing approximate FLCs based on average motor FLCs. See Figure 6-23.

AMPERE RATINGS OF 3φ, 60 Hz, AC INDUCTION MOTORS

HP	rpm Speed	220 V	230 V	380 V	460 V	575 V	2200 V
¼	1800	1.09	.95	.55	.48	.38	—
	1200	1.61	1.40	.81	.70	.56	
	900	1.84	1.60	.93	.80	.64	
⅓	1800	1.37	1.19	.69	.60	.48	—
	1200	1.83	1.59	.92	.80	.64	
	900	2.07	1.80	1.04	.90	.72	
½	18800	1.98	1.72	.99	.86	.69	—
	1200	2.47	2.15	1.24	1.08	.86	
	900	2.74	2.38	1.38	1.19	.95	
¾	1800	2.83	2.46	1.42	1.23	.98	—
	1200	3.36	2.92	1.69	1.46	1.17	
	900	3.75	3.26	1.88	1.63	1.30	
1	3600	3.22	2.80	1.70	1.40	1.12	—
	1800	4.09	3.56	2.06	1.78	1.42	
	1200	4.32	3.76	2.28	1.88	1.50	
	900	4.95	4.30	2.60	2.15	1.72	
1½	3600	5.01	4.36	2.64	2.18	1.74	—
	1800	5.59	4.86	2.94	2.43	1.94	
	1200	6.07	5.28	3.20	2.64	2.11	
	900	6.44	5.60	3.39	2.80	2.24	
2	3600	6.44	5.60	3.39	2.80	2.24	—
	1800	7.36	6.40	3.87	3.20	2.56	
	1200	7.87	6.84	4.14	3.42	2.74	
	900	9.09	7.90	4.77	3.95	3.16	
3	3600	9.59	8.34	5.02	4.17	3.34	—
	1800	10.8	9.40	5.70	4.70	3.76	
	1200	11.7	10.2	6.20	5.12	4.10	
	900	13.1	11.4	6.90	5.70	4.55	
5	3600	15.5	13.5	8.20	6.76	5.41	—
	1800	16.6	14.4	8.74	7.21	5.78	
	1200	18.2	15.8	9.59	7.91	6.32	
	900	18.3	15.9	9.60	7.92	6.33	
7½	3600	22.4	19.5	11.8	9.79	7.81	—
	1800	24.7	21.5	13.0	10.7	8.55	
	1200	25.1	21.8	13.2	10.9	8.70	
	900	26.5	23.0	13.9	11.5	9.19	
10	3600	29.2	25.4	15.4	12.7	10.1	—
	1800	30.8	26.8	16.3	13.4	10.7	
	1200	32.2	28.0	16.9	14.0	11.2	
	900	35.1	30.5	18.5	15.2	12.2	
15	3600	41.9	36.4	22.0	18.2	14.5	—
	1800	45.1	39.2	23.7	19.6	15.7	
	1200	47.6	41.4	25.0	20.7	16.5	
	900	51.2	44.5	26.9	22.2	17.8	
20	3600	58.0	50.4	30.5	25.2	20.1	—
	1800	58.9	51.2	31.0	25.6	20.5	
	1200	60.7	52.8	31.9	26.4	21.1	
	900	63.1	54.9	33.2	27.4	21.9	

HP	rpm Speed	200V	230 V	380 V	460 V	575 V	2200 V
25	3600	69.9	60.8	36.8	30.4	24.3	—
	1800	74.5	64.8	39.2	32.4	25.9	
	1200	75.4	65.6	39.6	32.8	26.2	
	900	77.4	67.3	40.7	33.7	27.0	
30	3600	84.4	73.7	44.4	36.8	29.4	—
	1800	86.9	75.6	45.7	37.8	30.2	
	1200	90.6	78.8	47.6	39.4	31.5	
	900	94.1	81.8	49.5	40.9	32.7	
40	3600	111.0	96.4	58.2	48.2	38.5	—
	1800	116.0	101.0	61.0	50.4	40.3	
	1200	117.0	102.0	61.2	50.6	40.4	
	900	121.0	105.0	63.2	52.2	41.7	
50	3600	138.0	120.0	72.9	60.1	48.2	—
	1800	143.0	124.0	75.2	62.2	49.7	
	1200	145.0	126.0	76.2	63.0	50.4	
	900	150.0	130.0	78.5	65.0	52.0	
60	3600	164.0	143.0	86.8	71.7	57.3	—
	1800	171.0	149.0	90.0	74.5	59.4	
	1200	173.0	150.0	91.0	75.0	60.0	
	900	177.0	154.0	93.1	77.0	61.5	
75	3600	206.0	179.0	108.0	89.6	71.7	—
	1800	210.0	183.0	111.0	91.6	73.2	
	1200	212.0	184.0	112.0	92.0	73.5	
	900	222.0	193.0	117.0	96.5	77.5	
100	3600	266.0	231.0	140.0	115.0	92.2	—
	1800	271.0	236.0	144.0	118.0	94.8	
	1200	275.0	239.0	145.0	120.0	95.6	
	900	290.0	252.0	153.0	126.0	101.0	
125	3600	—	292.0	176.0	146.0	116.0	—
	1800		293.0	177.0	147.0	117.0	23.6
	1200		298.0	180.0	149.0	119.0	24.2
	900		305.0	186.0	153.0	122.0	24.8
150	3600	—	343.0	208.0	171.0	137.0	—
	1800		348.0	210.0	174.0	139.0	29.2
	1200		350.0	210.0	174.0	139.0	29.9
	900		365.0	211.0	183.0	146.0	30.9
200	3600	—	458.0	277.0	229.0	164.0	—
	1800		452.0	274.0	226.0	181.0	34.8
	1200		460.0	266.0	230.0	184.0	35.5
	900		482.0	2.79.0	241.0	193.0	37.0
250	3600	—	559.0	338.0	279.0	223.0	—
	1800		568.0	343.0	284.0	227.0	57.5
	1200		573.0	345.0	287.0	229.0	58.5
	900		600.0	347.0	300.0	240.0	60.5
300	1800	—	678.0	392.0	339.0	271.0	69.0
	1200		684.0	395.0	342.0	274.0	70.0
400	1800	—	896.0	518.0	448.0	358.0	91.8
500	1800	—	1110.0	642.0	555.0	444.0	116.0

Figure 6-23. Most manufacturers provide charts for approximating full-load current when motor nameplate information is not available.

These charts should be used only as a last resort. This technique is not suggested as a standard procedure because the average rating could be higher or lower for a specific motor and, therefore, selection on this basis always involves risk. For fully reliable motor protection, select heat coils based on the motor's FLC rating as shown on the motor nameplate. The FLC of the motor stated on charts should be used in the selection of the heater, using the same procedure as if it were the motor nameplate information. These charts provide approximately the same information that may be found on the motor nameplate, but should be used only if motor nameplate information is not available.

Inherent Motor Protectors

Inherent motor protectors are overload devices located directly on or in a motor to provide overload protection. Certain inherent motor protectors base the sensing element on the amount of heat generated or the amount of current consumed by the motor. The device directly or indirectly (using contactors) trips a circuit which disconnects the motor from the power circuit based on what it senses. Bimetallic thermodiscs and thermistor overload devices are inherent motor protectors.

Bimetallic Thermodiscs. A bimetallic thermodisc operates on the same principle as a bimetallic strip. The differences between these devices are the shape of the device and its location. A thermodisc has the shape of a miniature dinner plate and is located within the frame of the motor. See Figure 6-24. The bimetallic thermodisc warps and opens the circuit when the motor is overloaded. Bimetallic thermodiscs are normally used on small horsepower motors to disconnect the motor directly from the power circuit. Bimetallic thermodiscs may be tied into the control circuit of a magnetic contactor coil where they can be used as indirect control devices.

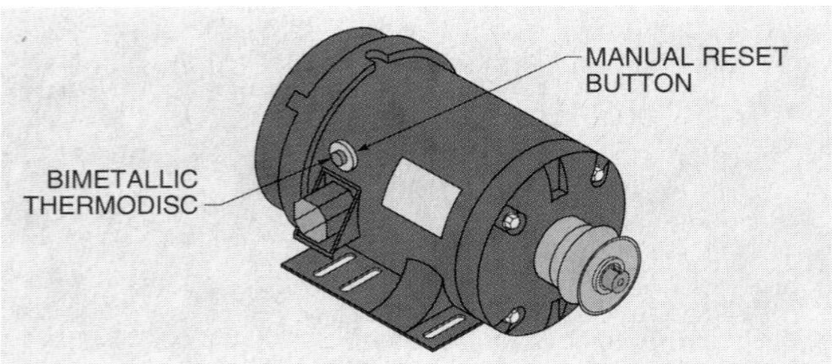

Figure 6-24. Bimetallic thermodiscs are normally used on small horsepower motors to directly disconnect the motor from the power circuit.

Always ensure power to the motor is turned OFF before resetting a manual-reset thermodisc. This prevents a potential hazard when the motor restarts.

Thermistor Overload Devices. A thermistor-based overload is a sophisticated form of inherent motor protection. A thermistor overload device combines a thermistor, solid-state relay, and contactor into a custom-built overload protector. See Figure 6-25.

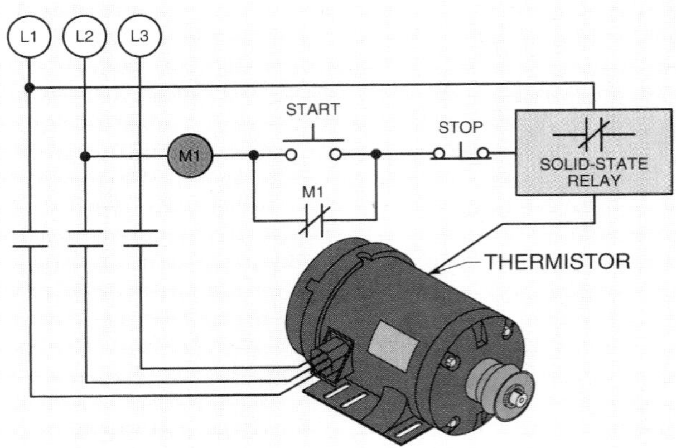

Figure 6-25. A thermistor overload device combines a thermistor, solid-state relay, and contactor into a custom-built overload protector.

A thermistor is similar to a resistor in that its resistance changes with the amount of heat applied to it. As the temperature increases, the resistance of the thermistor decreases and the amount of current passing through the thermistor increases. The changing signal must be amplified before it can do any work, such as triggering a relay, because the thermistor is a low-power device (normally in the thousandths of ampere range). When a thermistor overload device is amplified, a relay may open a set of contacts in the control circuit of a magnetic starter, de-energizing the power circuit of the motor.

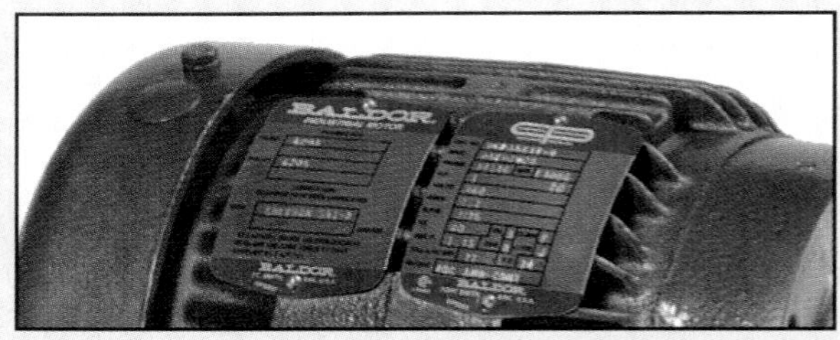

Baldor Electric Co.

The nameplate of a motor is used to select the proper overload protection device.

The major drawback to thermistor overload devices is that they require a close coordination between the user and the manufacturer to customize the design. Custom work costs more than standard off-the-shelf overload protectors. Custom work is uneconomical except for special and high-priced motors requiring extensive protection.

Electronic Overload Protection

An *electronic overload* is a device that has built-in circuitry to sense changes in current and temperature. Because electronic devices may include amplifiers, small changes can be responded to before mechanical devices can be activated. An electronic overload monitors the current in the motor directly by measuring the current in the power lines of the motor.

CONTACTOR AND MAGNETIC MOTOR STARTER MODIFICATIONS

Certain devices may be added to basic contactors or starters to expand their capability. These devices include additional electrical contacts, power poles, pneumatic timers, transient suppression modules, and control circuit fuse holders. See Figure 6-26.

Additional Electrical Contacts

Most contactors and starters have the ability to control several additional electrical contacts by adding them to existing auxiliary contacts. They may be used as extra auxiliary contacts. Both NO and/or NC contacts may be wired to control additional loads. NC contacts are used to turn additional loads ON any time the contactor or starter is OFF, as well as provide electrical interlocking. NO contacts are used to turn additional loads ON any time the contactor or starter is ON.

Power Poles

In certain cases, additional power poles (contacts capable of carrying a load) may be added to a contactor. The power poles are available with NO or NC contacts. Normally, only one power pole unit with one or two contacts is added per contactor or starter.

In certain cases with large-sized contactors or starters, it may be necessary to replace the coil to handle the additional load created in energizing the additional poles. Most power poles are factory or field installed.

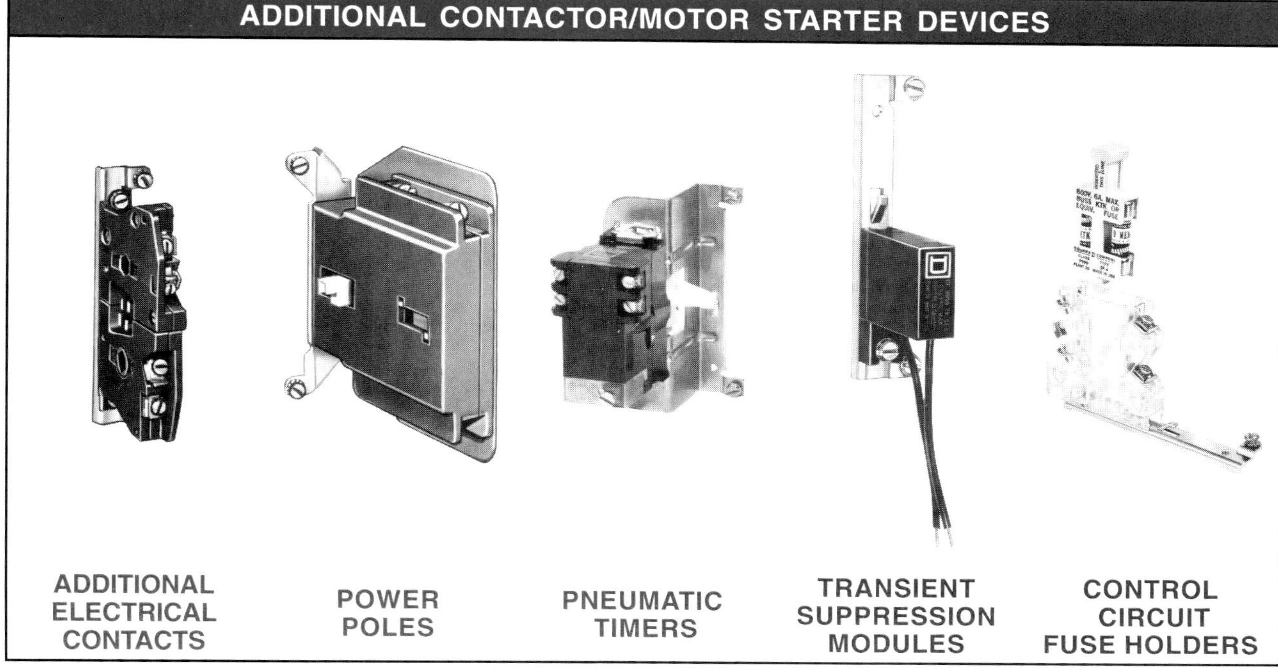

Square D Company

Figure 6-26. The devices that may be added to basic contactors or magnetic motor starters to expand their capability include additional electrical contacts, power poles, pneumatic timers, transient suppression modules, and control circuit fuse holders.

Pneumatic Timers

A mechanically-operated pneumatic timer can be mounted on some sizes of contactors and starters for applications requiring the simultaneous operation of a timer and a contactor. The use of mechanically-operated timers results in considerable savings in panel space over a separately-mounted timer. Available in time delay after de-energization (OFF-delay) or time delay after energization (ON-delay), the timer attachment has an adjustable timing period over a specified range.

Most manufacturers provide units that are field convertible from ON-delay to OFF-delay (or vice-versa) without additional parts. They are ordered either fixed or variable. Most timers mount on the side of the contactor and are secured firmly. One single-pole, double-throw contact is provided.

Transient Suppression Modules

Transient suppression modules are designed to be added where the transient voltage generated when opening the coil circuit interferes with the power operation of nearby components and solid-state control circuits. Modules normally consist of resistance/capacitance (RC) circuits and are designed to suppress the voltage transients to approximately 200% of peak coil supply voltage.

In certain cases, a voltage transient is generated when switching the integral control transformer that powers the coil control circuit. The transient suppression module, when used with devices wired for common control, is connected across the 120 V transformer secondary. The transient suppression module is not connected across the control coil.

Control Circuit Fuse Holders

Control circuit fuse holders can be attached to contactors or starters when either one or two control circuit fuses may be required. The fuse holder helps satisfy the NEC® requirements in Section 430-72.

INTERNATIONAL STANDARDS

The International Electrotechnical Commission (IEC), headquartered in Geneva, Switzerland, is primarily associated with equipment used in Europe. The National Electrical Manufacturers Association (NEMA), headquartered in Washington, DC, is primarily associated with equipment used in North America. IEC and NEMA rate contactors and motor starters. This causes confusion because ratings are different for the same horsepower. IEC devices are smaller in size for the equivalent-rated contactor. See Figure 6-27.

IEC AND NEMA DEVICE COMPARISON

Considerations	IEC	NEMA
Size	Smaller per horsepower rating than NEMA	Larger per horsepower rating than IEC
Cost	Lower price per horsepower	Higher price per horsepower
Performance	Normally, an electrical life of 1,000,000 operations is acceptable	Electrical life typically is 2.5 to 4 times higher for equivalent IEC device
Applications	Application sensitive with greater knowledge and care necessary	Application easier with fewer parameters to consider
Overloads	Fixed heaters that are adjustable to match different motors at same horsepower. Heaters are not field changeable	Field-changeable heaters allow adjustment to motors of different horsepowers
Additional Information	Reset/stop dual function operation mechanism typical	Reset only mechanism typical
	Hand/auto reset typical	Hand reset only typical
	Typically designed for use with fast-acting, current-limiting European fuses	Designed for use with domestic time delay fuses and circuit breakers

Figure 6-27. The difference between IEC and NEMA devices is based on size, cost, performance, applications, and overloads.

IEC devices are built with materials required for average applications. NEMA devices are built for a high level of performance in a variety of applications. IEC devices are less expensive, but more application sensitive. NEMA devices are more costly, but less application sensitive. IEC devices are commonly used in original equipment manufacturer (OEM) machines where machine specifications are known and do not change. NEMA devices are commonly used where machine requirements and specifications may vary.

TROUBLESHOOTING CONTACTORS AND MOTOR STARTERS

Contactors or motor starters are the first device checked when troubleshooting a circuit that does not work or has a problem. Contactors or motor starters are checked first because they are the point where the incoming power, load, and control circuit are connected. Basic voltage readings are taken at a contactor or motor starter to determine where the problem lies. The same basic procedure used to troubleshoot a motor starter works for contactors because a motor starter is a contactor with added overload protection.

Check the tightness of all terminals and busbar connections when troubleshooting control devices. Loose connections in the power circuit of contactors and motor starters cause overheating. Overheating leads to equipment malfunction or failure. Loose connections in the control circuit cause control malfunctions. Loose connections of grounding terminals lead to electrical shock and cause electromagnetic-generated interference.

Troubleshoot the power circuit and the control circuit if the control circuit does not correctly operate a motor. The two circuits are dependent on each other, but are considered two separate circuits because they are normally at different voltage levels and always at different current levels. See Figure 6-28.

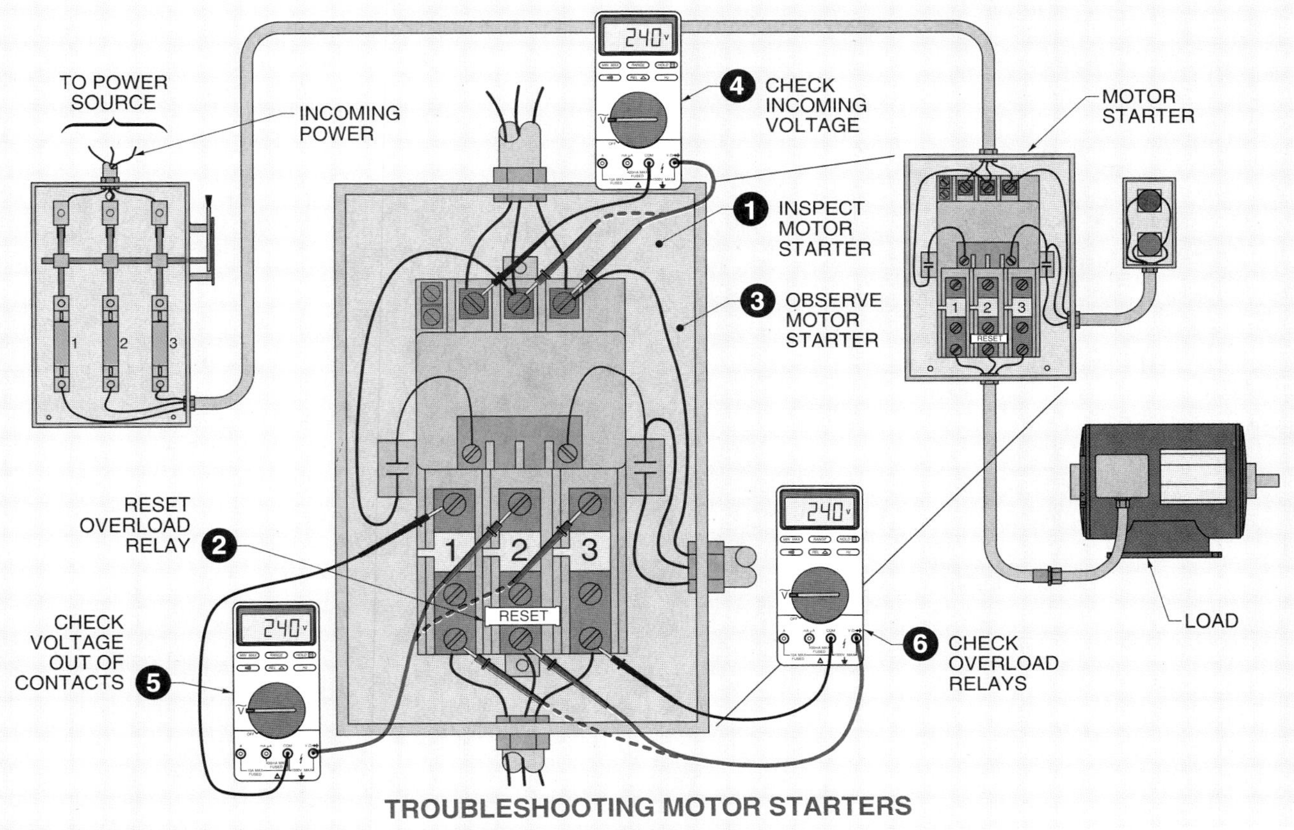

Figure 6-28. The contactor or motor starter is the first device checked when troubleshooting a circuit that does not work or has a problem.

To troubleshoot a motor starter, apply the procedure:

1. Inspect the motor starter and overload assembly. Service or replace motor starters that show heat damage, arcing, or wear. Replace motor starters that show burning.
2. Reset the overload relay if there is no visual indication of damage. Replace the overload relay if there is visual indication of damage.
3. Observe the motor starter for several minutes if the motor starts after resetting the overload relay. The overload relay continues to open if an overload problem continues to exist.
4. Check the voltage going into the starter if resetting the overload relay does not start the motor. Check circuit voltage ahead of the starter if the voltage reading is 0 V. The voltage is acceptable if the voltage reading is within 10% of the motor's voltage rating. The voltage is unacceptable if the voltage reading is not within 10% of the motor's voltage rating.
5. Energize the starter and check the starter contacts if the voltage into the starter is present and at the correct level. The starter contacts are good if the voltage reading is acceptable. Open the starter, turn the power OFF, and replace the contacts if there is no voltage reading.
6. Check the overload relay if voltage is coming out of the starter contacts. Turn the power OFF and replace the overload relay if the voltage reading is 0 V. The problem is downstream from the starter if the voltage reading is acceptable and the motor is not operating.

Troubleshooting Guides

A troubleshooting guide is used when troubleshooting contactors and motor starters. The guide states a problem, its possible cause(s), and corrective action(s) that may be taken. See Figure 6-29.

CONTACTOR AND MOTOR STARTER TROUBLESHOOTING GUIDE...

Problem	Possible Cause	Corrective Action
Humming noise H-M-M-M-M	Magnet pole faces misaligned	Realign. Replace magnet assembly if realignment is not possible.
	Too low voltage at coil	Measure voltage at coil. Check voltage rating of coil. Correct any voltage that is 10% less than coil rating.
	Pole face obstructed by foreign object, dirt, or rust	Remove any foreign object and clean as necessary. Never file pole faces.
Loud buzz noise	Shading coil broken	Replace coil assembly.
Controller fails to drop out POWER IN POWER OUT	Voltage to coil not being removed	Measure voltage at coil. Trace voltage from coil to supply looking for shorted switch or contact if voltage is present.
	Worn or rusted parts causing binding	Clean rusted parts. Replace worn parts.
	Contact poles sticking	Check for burning or sticky substance on contacts. Replace burned contacts. Clean dirty contacts.
	Mechanical interlock binding	Check to ensure interlocking mechanism is free to move when power is OFF. Replace faulty interlock.
Controller fails to pull in POWER IN POWER OUT	No coil voltage	Measure voltage at coil terminals. Trace voltage loss from coil to supply voltage if voltage is not present.
	Too low voltage	Measure voltage at coil terminals. Correct voltage level if voltage is less than 10% of rated coil voltage. Check for a voltage drop as large loads are energized.
	Coil open	Measure voltage at coil. Remove coil if voltage is present and correct but coil does not pull in. Measure coil resistance for open circuit. Replace if open.

Figure 6-29 continued...

...CONTACTOR AND MOTOR STARTER TROUBLESHOOTING GUIDE		
Problem	**Possible Cause**	**Corrective Action**
	Coil shorted	Shorted coil may show signs of burning. The fuse or breakers should trip if coil is shorted. Disconnect one side of coil and reset if tripped. Remove coil and check resistance for short if protection device does not trip. Replace shorted coil. Replace any coil that is burned.
	Mechanical obstruction	Remove any obstructions.
Contacts badly burned or welded 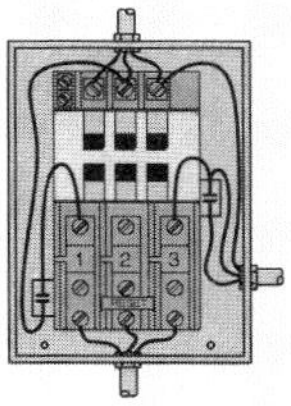	Too high inrush current	Measure inrush current. Check load for problem if higher-than-rated load current. Change to larger controller if load current is correct but excessive for controller.
	Too fast load cycling	Change to larger controller if load cycled ON and OFF repeatedly.
	Too large overcurrent protection device	Size overcurrent protection to load and controller.
	Short circuit	Check fuses or circuit breakers. Clear any short circuit.
	Insufficient contact pressure	Check to ensure contacts are making good connection.
Nuisance tripping 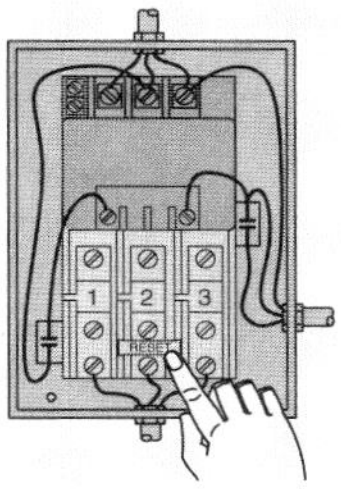	Incorrect overload size	Check size of overload against rated load current. Size up if permissible per NEC®.
	Lack of temperature compensation	Correct setting of overload if controller and load are at different ambient temperatures.
	Loose connections	Check for loose terminal connection.

Figure 6-29. A troubleshooting guide used when troubleshooting contactors and motor starters states a problem, its possible cause(s), and corrective action(s) that may be taken.

1. What is a contactor?
2. What is 2-wire control?
3. What is low-voltage release?
4. What is 3-wire control?
5. What is low-voltage protection?
6. What are the differences between AC and DC contactors?
7. Why is arc suppression needed?
8. Why is it harder to extinguish an arc on contacts passing DC than on contacts passing AC?
9. What are the methods used to reduce an arc across a contact?
10. How are AC and DC contactors rated?
11. What is a magnetic motor starter?
12. What are the different types of overloads used with a magnetic motor starter?
13. What is the function of overloads?
14. What is the purpose of a current transformer?
15. What general factors must be considered when selecting overloads?
16. What is the service factor rating of a motor?
17. What is ambient temperature?
18. What is ambient temperature compensation?
19. What is an inherent motor protector?
20. What are some of the optional devices that may be added to contactors and magnetic motor starters?

7 Chapter

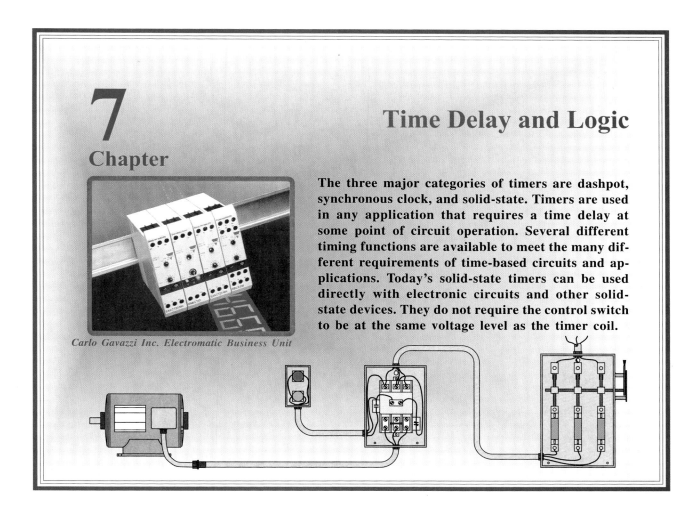

Time Delay and Logic

The three major categories of timers are dashpot, synchronous clock, and solid-state. Timers are used in any application that requires a time delay at some point of circuit operation. Several different timing functions are available to meet the many different requirements of time-based circuits and applications. Today's solid-state timers can be used directly with electronic circuits and other solid-state devices. They do not require the control switch to be at the same voltage level as the timer coil.

Carlo Gavazzi Inc. Electromatic Business Unit

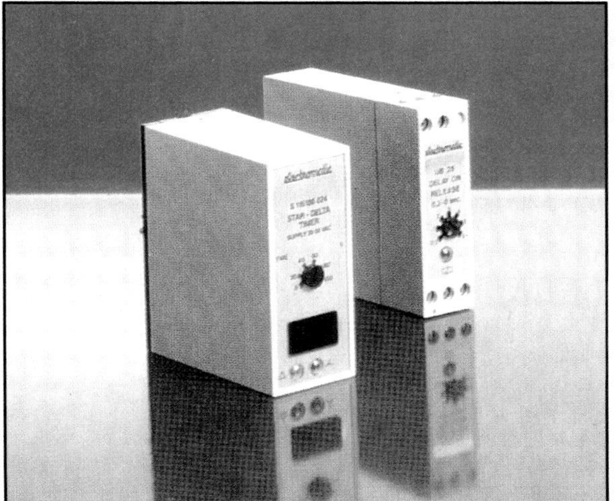

Carlo Gavazzi Inc. Electromatic Business Unit

The S- and U-housing timers manufactured by Carlo Gavazzi Inc. include LEDs to indicate when the timer is timing and when the timer is timed out.

TIMERS

The three major categories of timers are dashpot, synchronous clock, and solid-state. See Figure 7-1. The dashpot timer is the oldest industrial timer and is rarely used in new installations. Synchronous clock timers have been installed in millions of control applications over the last 50 years and, even though their use is declining, they remain the timer of choice for many applications. Today, solid-state timers are by far the most common timers used in most timing applications. Although each device accomplishes its task in a different way, all timers have the common ability to introduce some degree of time delay into a control circuit.

Dashpot Timers

A *dashpot timer* is a timer that provides time delay by controlling how rapidly air or liquid is allowed to pass into or out of a container through an orifice (opening) that is either fixed in diameter or variable. See Figure 7-2. If the piston of a hand-operated tire pump is forced down, the piston rapidly moves down if the valve opening is unrestricted. However, if the valve opening is restricted, the travel time of the piston increases. The smaller the opening, the longer the travel time.

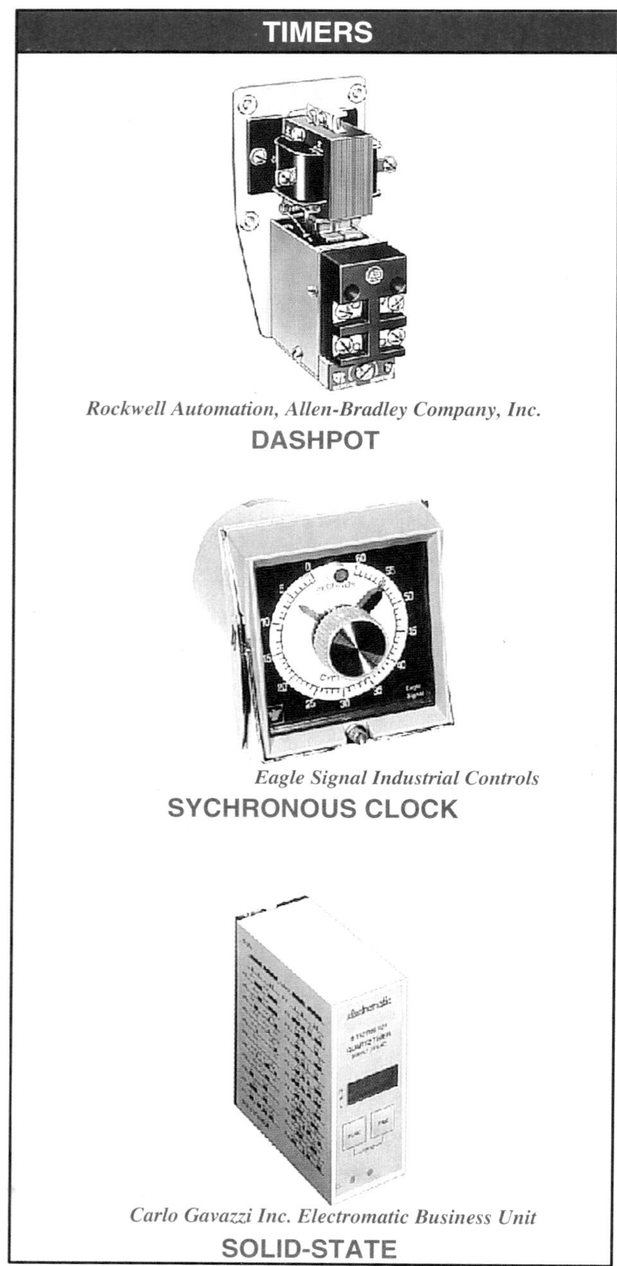

Figure 7-1. The three major categories of timers are dashpot, synchronous clock, and solid-state.

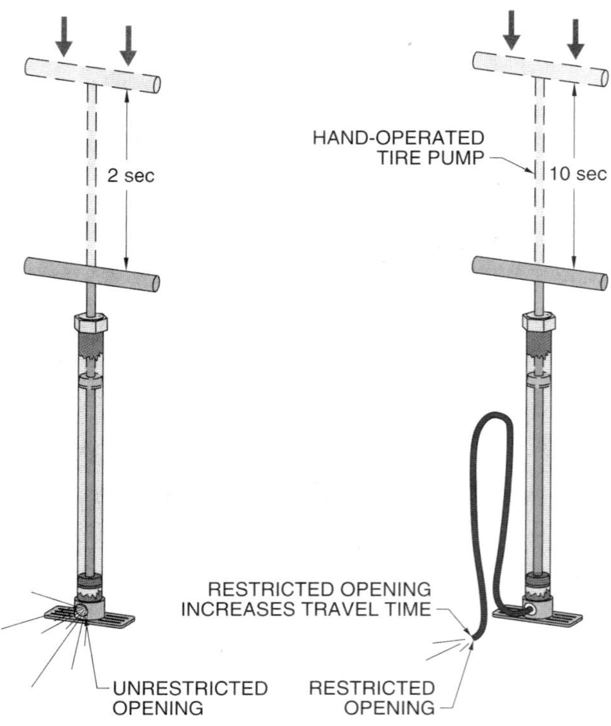

Figure 7-2. A dashpot timer provides time delay by controlling how rapidly air or liquid is allowed to pass into or out of a container.

Synchronous Clock Timers

A *synchronous clock timer* is a timer that opens and closes a circuit depending on the position of the hands of a clock. See Figure 7-3. The timer may have one or more contacts through which the circuit may be opened or closed.

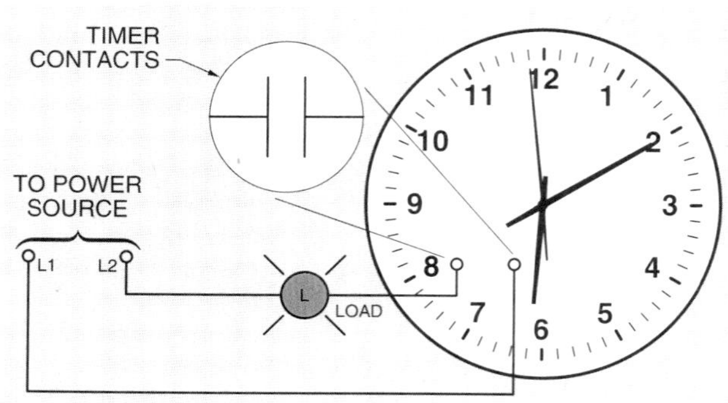

Figure 7-3. A synchronous clock timer opens and closes a circuit depending on the position of the hands of a clock.

The time delay is provided by the speed at which the clock hands move around the perimeter of the face of the clock. In this case, the contacts are closed once every 12 hours. The timer is operated by a synchronous clock motor. Synchronous clock motors are AC-operated and maintain their speed based on the frequency of the AC power line which feeds them. Synchronous clock timers are accurate timers because power companies maintain strict tolerance on line frequency.

Solid-State Timers

A *solid-state timer* is a timer whose time delay is provided by solid-state electronic devices enclosed within the timing device. The solid-state timing circuit provides a very accurate timing function at the most economical cost. Solid-state timers can control timing functions ranging from a fraction of a second to hundreds of hours. Most solid-state timers are designed as plug-in modules for quick replacement.

Solid-state timers can replace dashpot and synchronous timers in most applications. Solid-state timers are less susceptible to outside environmental conditions because they are often encapsulated in epoxy resin for protection. However, because they are encapsulated and cost less, they are normally impossible to repair. See Figure 7-4.

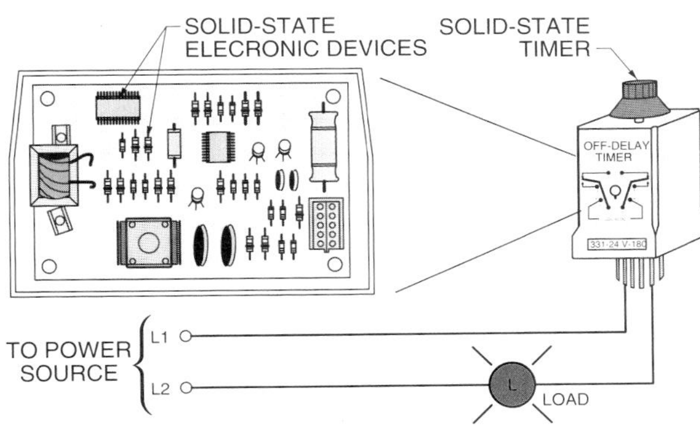

Figure 7-4. A solid-state timer has a time delay provided by solid-state electronic devices enclosed within the timing device.

TIMING FUNCTIONS

Several different timing functions are available to meet the many different requirements of time-based circuits and applications. ON-delay and OFF-delay timing functions were the only two timing functions available when dashpot and synchronous timers were the only timers used.

When solid-state timers became available, they offered ON-delay, OFF-delay, one-shot, and recycle functions. Today's solid-state timers offer dozens of special timing functions in addition to the four basic timing functions, because solid-state timing circuits can be easily modified. Several of the special timing functions are normally combined into one multiple-function timer.

ON-Delay

An *ON-delay (delay on operate) timer* is a device that has a preset time period that must pass after the timer has been energized before any action occurs on the timer contacts. Once activated, the timer may be used to turn a load ON or OFF, depending on the way the timer contacts are connected into the circuit. The load energizes after the pre-programmed time delay when an NO timer contact is used. The load de-energizes after the preset time delay when an NC timer contact is used.

ON-delay timer contacts do not change position until the set time period passes after the timer receives power. See Figure 7-5. After the preset time has passed, the timer contacts change position.

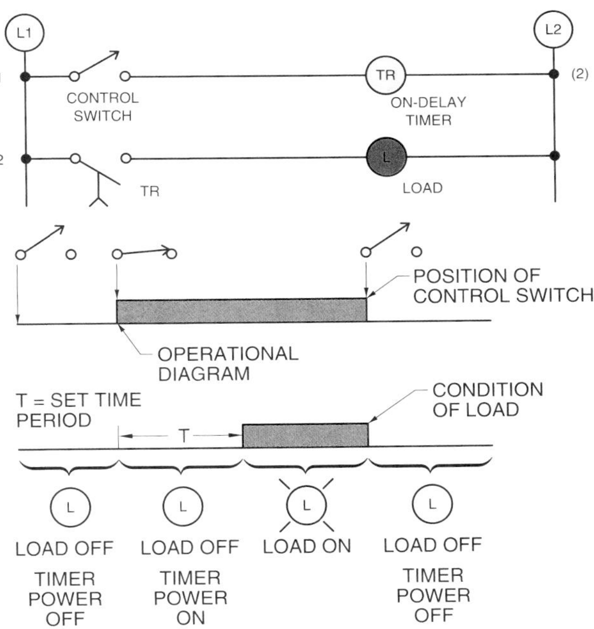

Figure 7-5. ON-delay timer contacts do not change position until the set time period passes after the timer receives power.

In the ON-delay timer circuit, the NO contacts close and energize the load. The load remains energized as long as the control switch remains closed. The load de-energizes the second the control switch is opened. An operational diagram is used to show timer operation. In the operational diagram, the top line shows the position of the control switch and the bottom line shows the condition of the load.

ON-Delay (timed-closed). An ON-delay (timed-closed) function may be illustrated using two balloons. See Figure 7-6. The solenoid plunger forces air out of balloon A through orifice B and into balloon C when control switch S1 is closed. Contacts TR1 close, energizing the circuit to the load after balloon C is filled. This energizes the load. The ON-delay function takes 5 sec if it takes 5 sec for balloon C to fill.

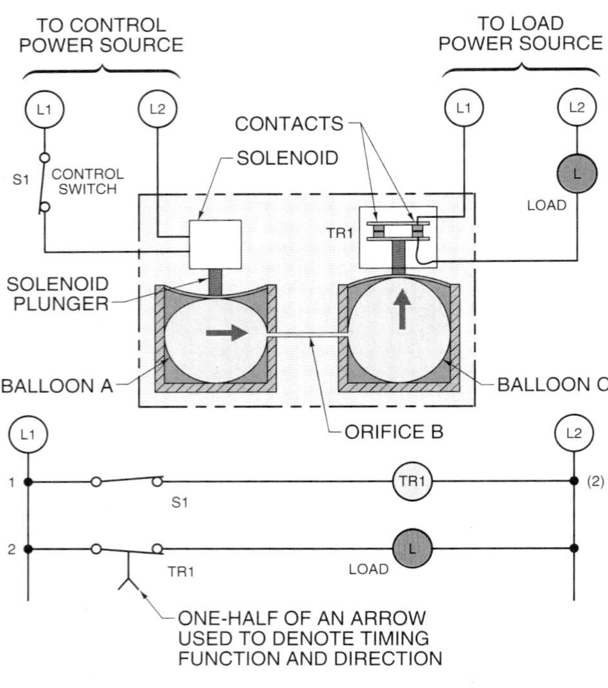

Figure 7-6. With an ON-delay (timed-closed) function, the contacts close after the timing cycle is complete.

One-half of an arrow is used to indicate the direction of time delay of the NO timing contacts in ON-delay timers. The half arrow points in the direction of ON delay. The operational diagram should be used if an arrow is not used with an ON-delay timer.

ON-Delay (timed-open). An ON-delay timer could be designed to open or close a circuit after a predetermined time delay. With an ON-delay (timed-open) function, the balloon forces the contacts open after the timing cycle is complete. See Figure 7-7. With control switch S1 closed, the solenoid plunger forces air from balloon A to balloon C through orifice B. After 5 sec, contacts TR1 open the circuit to the load and the load is de-energized. One-half of an arrow is shown in the line diagram. The arrow indicates that the NC contacts open after the ON-delay function has taken place. This pneumatically-operated timing function is how dashpot timers operate. A synchronous clock timer or solid-state timer could be substituted for the pneumatic timer. A pneumatic timer is the easiest to understand in terms of mechanical and timing operation.

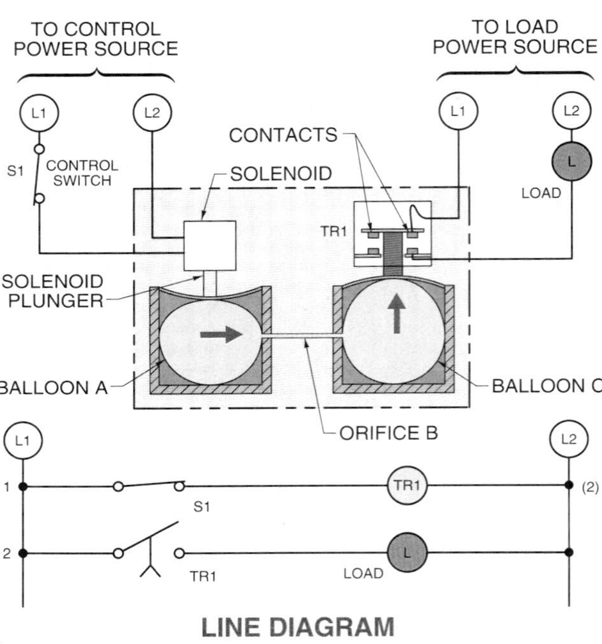

Figure 7-7. With an ON-delay (timed-open) function, the contacts open after the timing cycle is complete.

OFF-Delay

An *OFF-delay (delay on release) timer* is a device that does not start its timing function until the power is removed from the timer. See Figure 7-8. In this circuit, a control switch is used to apply power to the timer. The timer contacts change immediately and the load energizes when power is first applied to the timer.

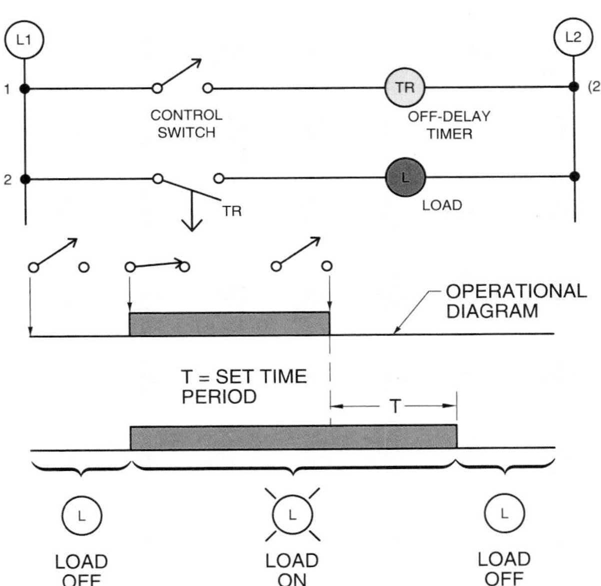

Figure 7-8. An OFF-delay (delay on release) timer is a device that does not start its timing function until the power is removed from the timer.

The timer contacts remain in the changed position and the time period starts when power is removed from the timer. The timer contacts return to their normal position and the load is de-energized on expiration of the set time period.

OFF-Delay (timed-open). An OFF-delay (timed-open) contact circuit may be used to provide cooling in a projector once the bulb has been turned OFF but has not had time to cool down. See Figure 7-9.

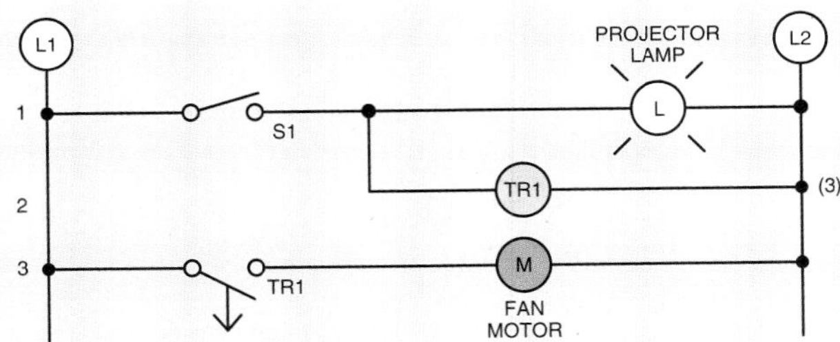

Figure 7-9. An OFF-delay (timed-open) contact circuit may be used to provide cooling in a projector once the bulb has been turned OFF but has not had time to cool down.

In this circuit, closing switch S1 turns ON the projector bulb and activates timer coil TR1. With timer TR1 energized, NO contacts TR1 immediately close, energizing the fan motor which controls the cooling of the projector.

The projector bulb and the cooling fan remain ON as long as switch S1 stays closed. When switch S1 is opened, the projector bulb turns OFF and power is removed from the timer. Contacts TR1 remain closed for a predetermined OFF delay and then open, causing the cooling fan to turn OFF. This OFF-delay timed-open circuit is generally set to adequately cool the projector equipment before it shuts OFF. This circuit could also be used for large cooling fan motors by replacing the fan motor in the control circuit with a motor starter. The motor starter could be used to control any size motor.

OFF-Delay (timed-closed). An OFF-delay (timed-closed) contact circuit may be used to provide a pumping system with backspin protection and surge protection on stopping. See Figure 7-10.

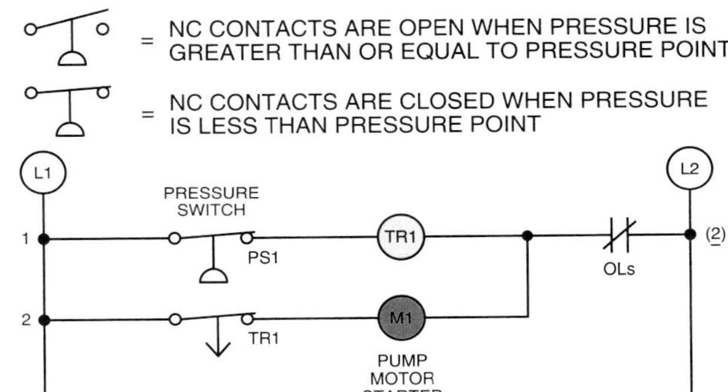

Figure 7-10. An OFF-delay (timed-closed) contact circuit may be used to provide a pumping system with backspin protection and surge protection on stopping.

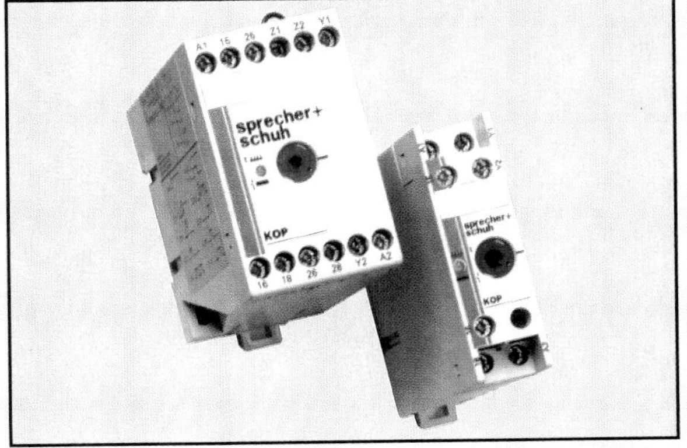

Sprecher + Schuh

Sprecher + Schuh's KOP Series of electronic timing relays are available in a variety of output functions for all types of industrial control applications.

Surge protection is often necessary when a pump is turned OFF and a high column of water is stopped by a check valve. The force of the sudden stop may cause surges which operate the pressure switch contacts, subjecting the starter to chattering. *Backspin* is the backward turning of a centrifugal pump when the head of water runs back through the pump just after it has been turned OFF. Starting the pump during backspin might damage the pump motor.

To eliminate the damage resulting from surges and backspin, actuating pressure switch PS1 causes timer relay coil TR1 to energize. With TR1 energized, the NC contacts of TR1 immediately open and cause coil M1 to de-energize, shutting OFF the pump motor. Even if pressure switch PS1 recloses, the coil and motor remain OFF for a predetermined time (to allow surges and backspin to clear) and then restart once the OFF-delay function has taken place. This OFF-delay function prevents the system from operating until TR1 has timed out and its NC contacts return to their normally-closed position.

A comparison chart may be used to compare the operation of ON-delay and OFF-delay timing functions and contacts. See Figure 7-11. To help compare timing functions, instantaneous relay contacts are also included. Some manufacturers also use abbreviations in their catalogs to describe the type of contacts used.

SSAC, Inc.

The TS1 Series delay ON make solid-state timers from SSAC, Inc. contain no moving parts or contacts to arc, wear, and eventually fail.

TIMER COMPARISON

Abbreviation	Meaning	Function	Symbols
NOTC	Normally open, timed-closed	ON-delay (timed-closed) contact – Timer contact normally open. Timed-closed on timer energization. Opens immediately on timer de-energization	
NCTO	Normally closed, timed-open	ON-delay (timed-open) contact – Timer contact normally closed. Timed-open on timer energization. Closes immediately on timer de-energization	
NOTO	Normally open, timed-open	OFF-delay (timed-open) contact – Timer contact normally open. Closes immediately on timer energization. Timer contact times open on timer de-energization	
NCTC	Normally closed, timed-closed	OFF-delay (timed-closed) contact – Timer contact normally closed. Opens immediately on timer energization. Timer contact times closed on timer de-energization	
NO	Normally open	Instantaneous contact – Normally open. Contact closes immediately on relay energization. Opens immediately on relay de-energization	
NC	Normally closed	Instantaneous contact – Normally closed. Contact opens immediately on relay energization. Closes immediately on relay de-energization	

Figure 7-11. A comparison chart may be used to compare the operation of ON-delay and OFF-delay timing functions and contacts.

One-Shot

A *one-shot (interval) timer* is a device in which the contacts change position immediately and remain changed for the set period of time after the timer has received power. See Figure 7-12. After the set period of time has passed, the contacts return to their normal position.

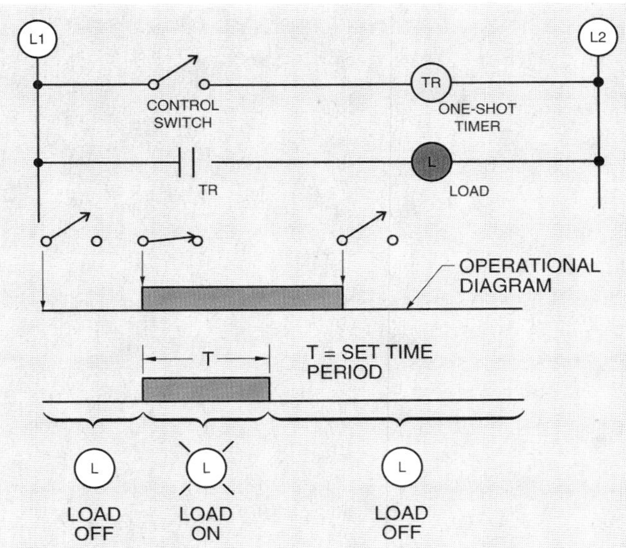

Figure 7-12. A one-shot (interval) timer has contacts which change position and remain changed for the set period of time after the timer has received power.

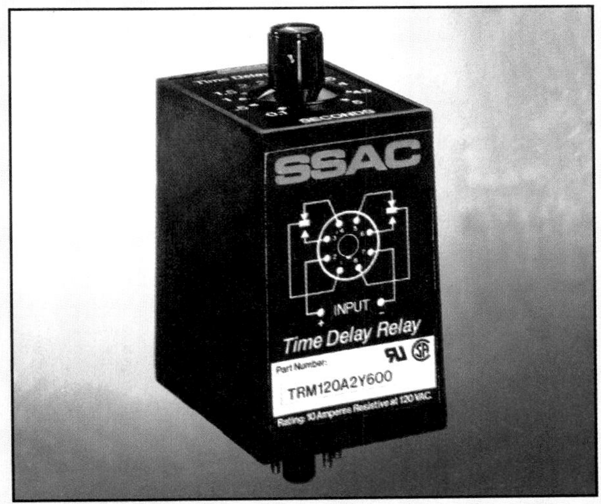

SSAC, Inc.

The TRM Series delay ON make time delay relays from SSAC, Inc. are available with a time adjustment knob and 24 V to 230 V operation.

One-shot timers are used in applications in which a load is ON for only a set period of time. One-shot timer applications include coin-operated games, dryers, car washes, and other machines. One-shot timing functions have not been available as long as ON-delay and OFF-delay timing functions because the one-shot timing function became available only after solid-state timers were available. For this reason, and the fact that so many other timing functions are now available, no standard symbol was established for any other timer contacts except ON-delay and OFF-delay. Today, the basic normally open (NO) and normally closed (NC) contacts, along with the timer type and/or operational diagram, are used with all timers that are not ON-delay or OFF-delay.

Recycle

A *recycle timer* is a device in which the contacts cycle open and closed repeatedly once the timer has received power. The cycling of the contacts continues until power is removed from the timer. See Figure 7-13. In the recycle timer circuit, the closing of the control switch starts the cycling function. The load continues to turn ON and OFF at regular time intervals as long as the control switch is closed. The cycling function stops when the control switch is opened.

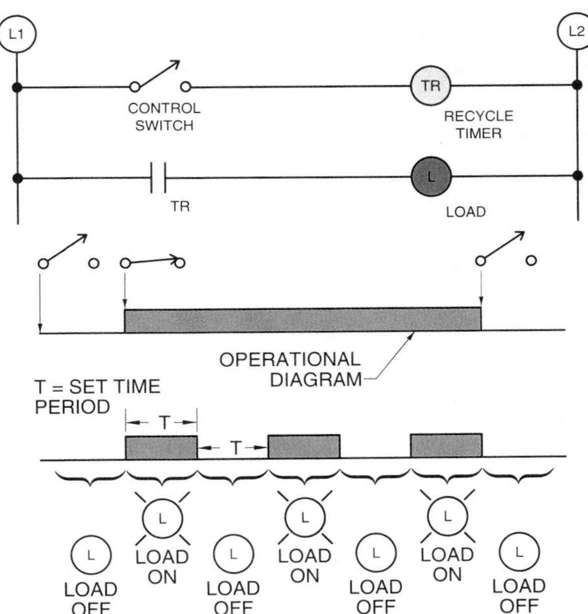

Figure 7-13. A recycle timer is a device in which the contacts cycle open and closed repeatedly once the timer has received power.

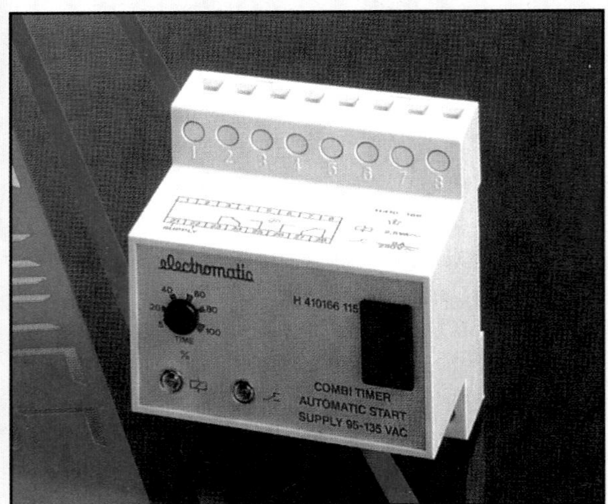

Carlo Gavazzi Inc. Electromatic Business Unit

The Type H 410 multi-function timer from Carlo Gavazzi, Inc. has four selectable functions (delay on operate, interval timer, and symmetrical recycler with ON- or OFF-time first) and four selectable time ranges (0.15 sec to 800 sec).

Recycle timers may be symmetrical or asymmetrical. A *symmetrical recycle timer* operates with equal ON and OFF time periods. An *asymmetrical recycle timer* is a timer which has independent adjustments for the ON and OFF time periods. Asymmetrical timers always have two different time adjustments.

Multiple Function

ON-delay, OFF-delay, one-shot, and recycle timers are considered monofunction timers. That is, they perform only one timing function, such as ON-delay or OFF-delay. Multiple function timers are solid-state timers that can perform many different timing functions. Multiple function timers are normally programmed for different timing functions by the placement of DIP (dual in-line package) switches located on the timer. See Figure 7-14.

In this timer, four DIP switches are used to set the timer range and function. The first two DIP switches set the time range from .8 sec to 60 min. The last two DIP switches set the timer function. The timer can be set for an ON-delay, one-shot, or recycle time function. The recycle time function can be set to start with the OFF time occurring first or the ON time occurring first.

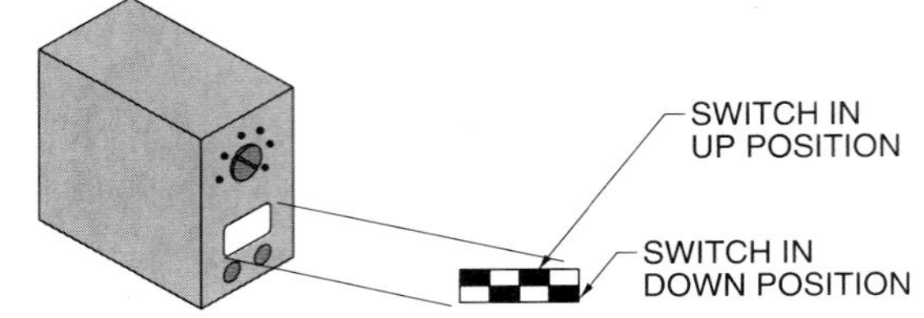

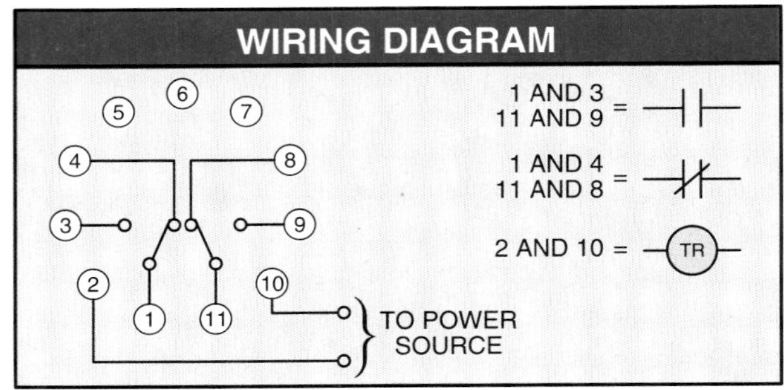

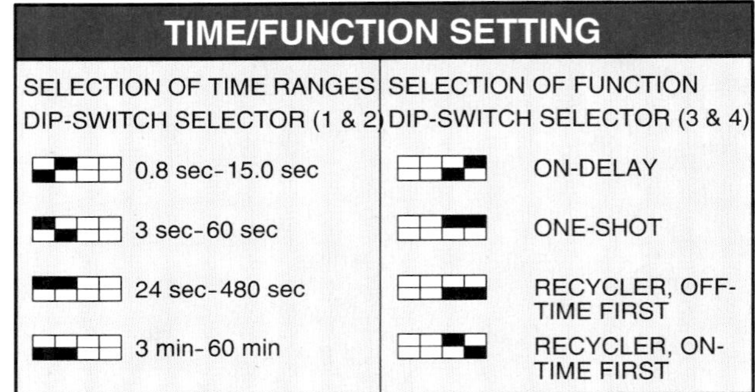

TIME/FUNCTION SETTING

SELECTION OF TIME RANGES DIP-SWITCH SELECTOR (1 & 2)	SELECTION OF FUNCTION DIP-SWITCH SELECTOR (3 & 4)
0.8 sec-15.0 sec	ON-DELAY
3 sec-60 sec	ONE-SHOT
24 sec-480 sec	RECYCLER, OFF-TIME FIRST
3 min-60 min	RECYCLER, ON-TIME FIRST

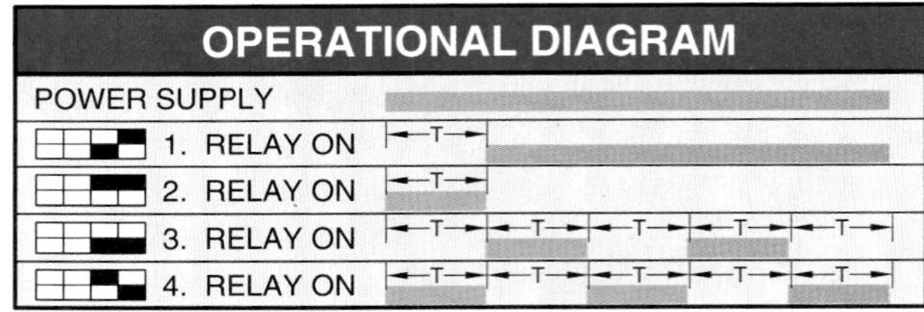

Figure 7-14. A multiple function timer may use the placement of DIP switches to determine the type of timing function and timer setting.

In addition to some (or all) of the basic timing functions, many multiple function timers can also be programmed for special timing functions. See Figure 7-15. In this multiple function timer, many timing functions can be programmed with a time range from .15 sec to 220 hr. This timer includes standard time functions such as an ON-delay (program setting 1) and special timing functions, such as a combination of both ON-delay and OFF-delay (program setting 5).

1 2 3 4 5 6 7 8 9 — 9 DIP SWITCH SELECTOR

OPERATIONAL DIAGRAM

1 2 3 4	POWER SUPPLY
	1. RELAY ON
	2. RELAY ON
	3. RELAY ON
	4. RELAY ON
	5. RELAY ON
	6. RELAY ON
	7. RELAY ON
	8. RELAY ON
	9. RELAY ON

TIME SETTING

SELECTION OF TIME RANGE
DIP SWITCH SELECTOR (5- 9)

5 6 7 8 9	5 6 7 8 9
0.15 sec-1.50 sec	2.6 min-26.0 min
0.3 sec-3.0 sec	5 min-50 min
0.6 sec-6.0 sec	10 min-100 min
1.2 sec-12.0 sec	20 min-200 min
2.5 sec-25.0 sec	40 min-400 min
5 sec-50 sec	1.4 hr-14.0 hr
10 sec-100 sec	2.8 hr-28.0 hr
20 sec-200 sec	5.5 hr-55 hr
40 sec-400 sec	11 hr-110 hr
1.3 min-13.0 min	22 hr-220 hr

Figure 7-15. Multiple function timers can be programmed for many timing functions of various times.

WIRING DIAGRAMS

In the past, when only dashpot, synchronous clock, and the first solid-state timers were used, timing functions were controlled by applying and removing power from the timer's coil (or circuit). This meant that the control switch had to be rated for the same type and level of voltage as the timer coil. This is still true when using dashpot, synchronous clock, and basic solid-state timers. Today, solid-state timers are available that use different methods of controlling the timer. The advantage of the new solid-state timers is that they can be used directly with electronic circuits, other solid-state devices (photoelectric, proximity, temperature sensors), and do not require the control switch to be at the same voltage level as the timer coil.

Supply Voltage Controlled

A *supply voltage timer* is a timer that requires the control switch to be connected so that it controls power to the timer coil. See Figure 7-16. In this circuit, the control switch is connected in series with the timer coil. The advantage of this control method is that it is exactly the same as most electrical control circuits used over the years. The disadvantage is that if a standard 115 VAC timer is used, the control switch has to switch the 115 VAC. This means that the control switch has to be installed using standard AWG #14 Copper wire and properly enclosed for safety.

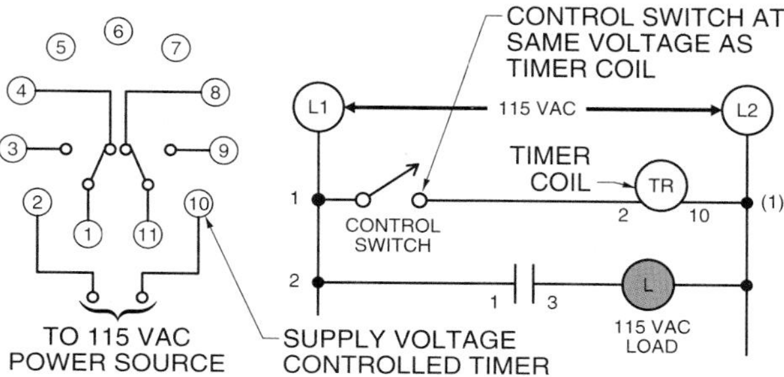

Figure 7-16. A supply voltage timer requires the control switch to be connected so that it controls power to the timer coil.

Contact-Controlled

A *contact-controlled timer* is a timer that does not require the control switch to be connected in line with the timer coil. See Figure 7-17. In a contact-controlled timer, the timer supplies the voltage to the circuit in which the control switch is placed. The voltage of the control circuit is normally less than 24 VDC.

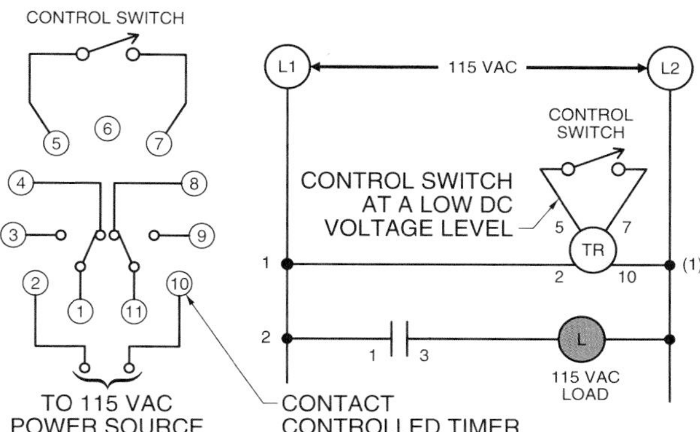

Figure 7-17. A contact-controlled timer does not require the control switch to be connected in line with the timer coil.

The advantage of this control method is that the control switch can be wired using low-voltage wire (AWG #16, #18, #20). The control switch contacts can be small and normally require less than a 100 mA rating because the timer control circuit requires little current to pass through the control switch. The disadvantage of this control method is that many electricians are unfamiliar with connecting the control switch outside a standard 115 VAC control circuit. The timer's low-voltage circuit (pins 5 and 7) is often connected to the timer's 115 VAC circuit (pins 2 and 10). This results in the destruction of the timer.

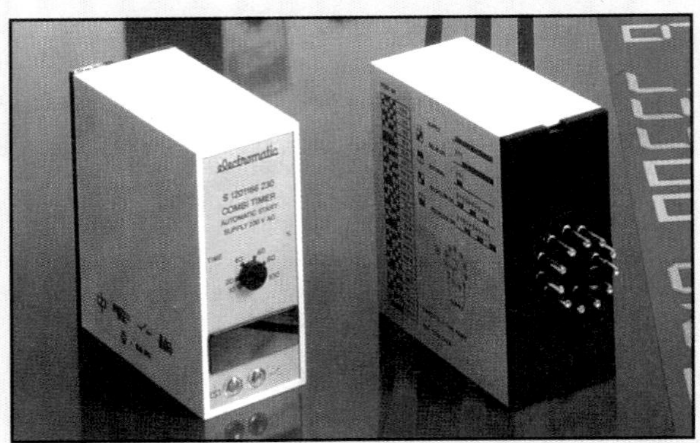

Carlo Gavazzi Inc. Electromatic Business Unit

The Type S 1201 multi-function, plug-in time relay from Carlo Gavazzi, Inc. has 20 selectable time ranges up to 220 hr and four selectable modes of operation.

Transistor-Controlled

Transistor-controlled timers are like contact-controlled timers. A *transistor-controlled timer* is controlled by an external transistor from a separately powered electronic circuit. See Figure 7-18. Today's industrial control circuits are often connected to DC electronic circuits that include a transistor as their output. Such devices include solid-state temperature controls, counters, computers, and other control devices. A timer that uses a transistor as the control input can be connected to almost any electronic circuit that includes a transistor output. The advantages and disadvantages of a transistor-controlled timer are the same as the contact-controlled timer. Any transistor-controlled timer can use a contact as the control device. However, not all contact-controlled timers can use a transistor as the control device. Check the manufacturer's specification sheet when using any unfamiliar timer.

Sensor-Controlled

A sensor-controlled timer is like contact-controlled and transistor-controlled timers that include an additional output from the timer. A *sensor-controlled timer* is a timer controlled by an external sensor in which the timer supplies the power required to operate the sensor. See Figure 7-19. The advantage of a sensor-controlled timer is that by supplying power out of the timer itself, no external power supply is required to operate the control sensor. Sensor-controlled timers are generally used with photoelectric and proximity controls, but may be used with any control that meets the specifications of the timer.

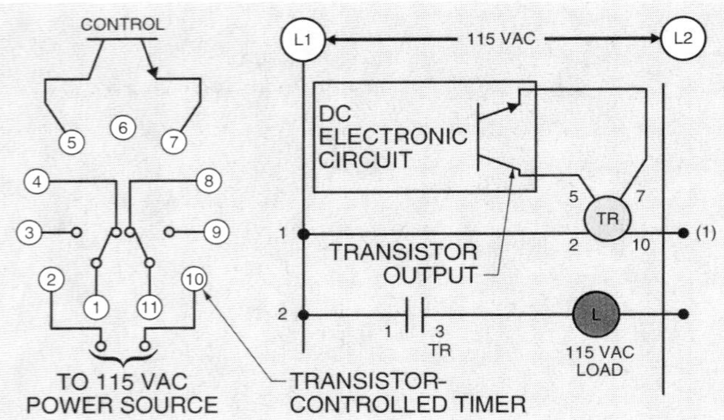

Figure 7-18. A transistor-controlled timer is controlled by an external transistor from a separately powered electronic circuit.

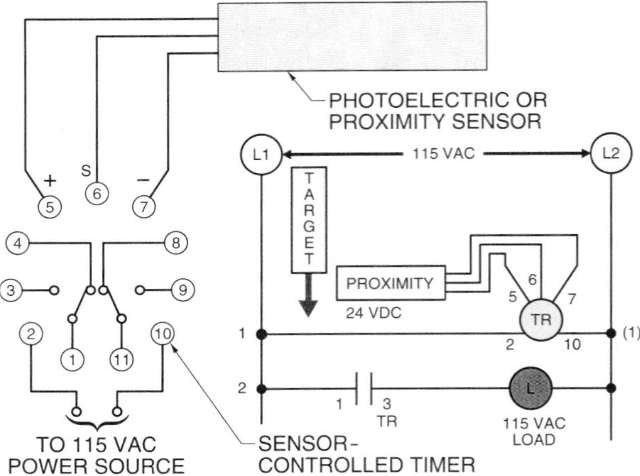

Figure 7-19. A sensor-controlled timer is controlled by an external sensor in which the timer supplies the power required to operate the sensor.

MULTIPLE CONTACT TIMERS

In the past, synchronous clock timer manufacturers included instantaneous and time-delay contacts on their timers to meet the many different application requirements of timers. The timers may be used for numerous applications when they have both types of contacts. These timers are still often used and have been updated to include solid-state timing circuits and can be converted from ON-delay to OFF-delay timing functions.

Wiring Diagrams

Wiring diagrams are required on multiple contact timers because several different connection points must be located and wired to the timer for it to perform properly. See Figure 7-20. The wiring diagram for a timer is normally located on the back of the timer. By quickly surveying the diagram, the timer clutch coil, the synchronous motor (or solid-state circuit), and the timer pilot light can be located.

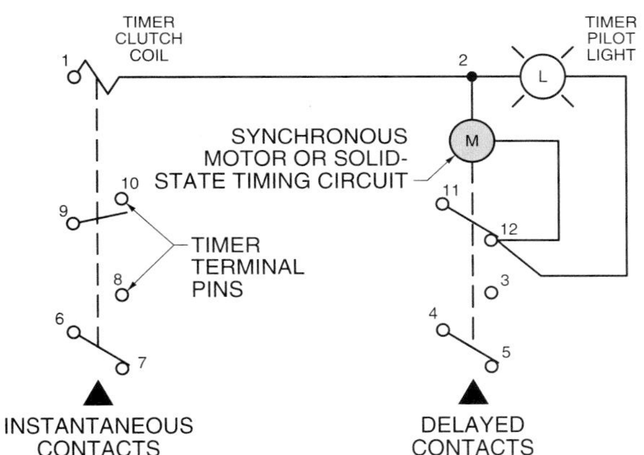

Figure 7-20. Wiring diagrams simplify locating connection points and wiring multiple contact timers.

In this timer, the timer clutch coil engages and disengages the motor. This is similar to those found in an automobile. The timer clutch coil engages and disengages contacts 9 and 10, and 6, 7, and 8. When the clutch is engaged, contacts 9 and 10, and 6 and 8 close instantaneously. Contacts 6 and 7 open simultaneously. In other words, the timer clutch controls two NO contacts and one NC contact instantaneously. The timing motor controls contacts 11 and 12, and 3, 4, and 5 through a time delay. When the motor times out, contacts 4 and 5, and 11 and 12 open and contacts 3 and 4 close.

Contacts 11 and 12 open slightly later than contacts 4 and 5 after the motor times out. The timer pilot light, wired in parallel with the motor, indicates when the motor is timing. Contacts 4 and 5, and 11 and 12 close and contacts 3 and 4 open when the timer is reset.

Wiring Motor-Driven Timers. A motor-driven timer can be wired into L1 and L2 to provide power to the circuit. See Figure 7-21. In this circuit, current flows from L1 through the timer clutch coil and on to L2. Current also flows from terminal 1 through the closed contacts 11 and 12, feeding the parallel circuit provided by the timer motor and timer pilot light and then on to L2.

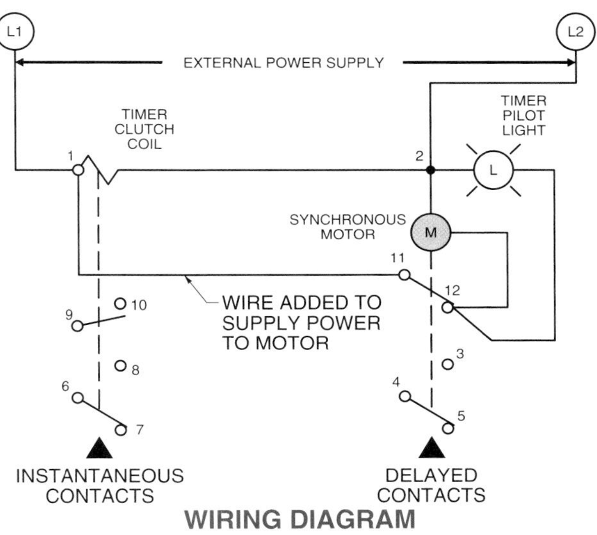

Figure 7-21. A motor-driven timer can be wired into L1 and L2 of a multiple contact timer to provide power.

Operation of Motor-Driven Timers. A timer activated by a limit switch has effects on various loads wired into the circuit. This timer can be controlled by a manual, mechanical, or automatic input. The timer may be used to achieve control using a sustained mechanical input. See Figure 7-22. This input (limit switch) must remain closed to energize the timer clutch coil and power the timer motor. The timer is connected for ON-delay, requiring input power to close the clutch, starting the timer. The contacts, both instantaneous and time delay, are connected to four loads marked A, B, C, and D. These loads may be any load, such as a solenoid, magnetic starter, light, etc. A code is added above each load to illustrate the sequence during reset, timing, and when timed out.

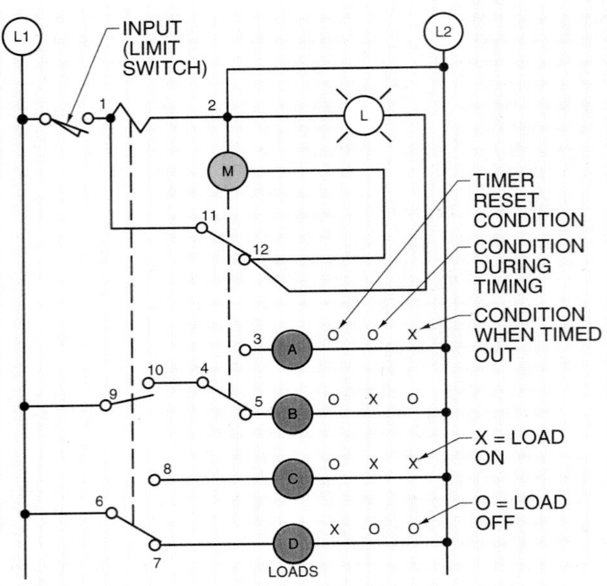

Figure 7-22. Multiple contact timers may be used to achieve control using a sustained mechanical input.

The code is used to indicate the condition (ON or OFF) of each load during the three stages of the timer. The three stages include the reset condition (no power applied to timer), the timing condition (time at which the timer is timing, but not timed out), and the timed-out condition. An O indicates when the load is de-energized and an X indicates when the load is energized.

Loads C and D are relay type responses, using only the instantaneous contacts. Loads A and B use the combined action of the instantaneous and delay contacts to achieve the desired sequence. In this circuit, the timing motor is wired through delay contacts 11 and 12 to ensure motor cutoff after the timer times out. This is required because the limit switch remains closed after timing out. The limit switch otherwise would have to be opened to reset the timer. This also means that a loss of plant power resets the timer, because the clutch opens when power is lost.

TIMER APPLICATIONS

Timers are used in any application that requires a time delay at some point of circuit operation. The timer used (ON-delay, OFF-delay, etc.) depends on the application. The timer selected (dashpot, solid-state, etc.) depends on cost, expected usage, type of equipment, operating environment, and personal preference. In general, monofunction solid-state timers are the best choice for most applications in which the circuit is not likely to change in function or time range. For applications that require changing timing functions and/or large time range fluctuations, multiple-function programmable timers are the best choice. Even if most of the machines and equipment in a plant use monofunction timers, multiple-function timers are often stocked in the maintenance shop because they can be used to replace many different types of timers.

ON-Delay Timer Applications

ON-delay timers are the most common timer in use. ON-delay timers may be used to monitor a patient's breathing. See Figure 7-23. In this application, the timer is used to sound an alarm if a patient does not take a breath within 10 sec. The circuit includes a low-pressure actuated switch built into a patient monitoring system. Pressure switches are available that can activate electrical contacts at pressures less than 1 psi. Pressure switches that react to pressures less than 1 psi are rated in inches of water column (in. WC). For example, the pressure switch used in this application may be rated at 4″ WC. Approximately 27 in. of WC is equal to 1 psi.

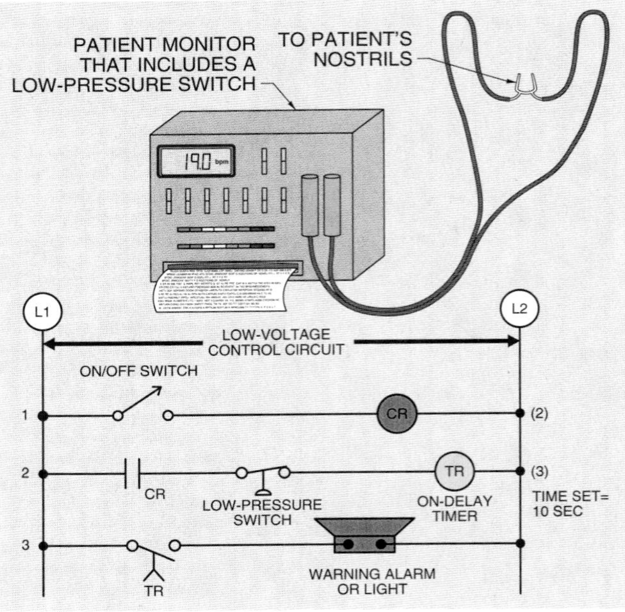

Figure 7-23. ON-delay timers may be used to monitor a patient's breathing.

The circuit is turned ON by the ON/OFF switch once the patient is connected to the monitor. If the patient does not take a breath, the timer starts timing and continues timing until the patient takes a breath (which resets the timer) or the timer times out. If the timer times out, the timer contacts close which sounds a warning.

OFF-Delay Timer Applications

OFF-delay timers are used in applications that require a load to remain energized even after the input control has been removed. See Figure 7-24. In this circuit, the OFF-delay timer is used to keep the water flowing for 1 min after the emergency shower pushbutton is pressed and released.

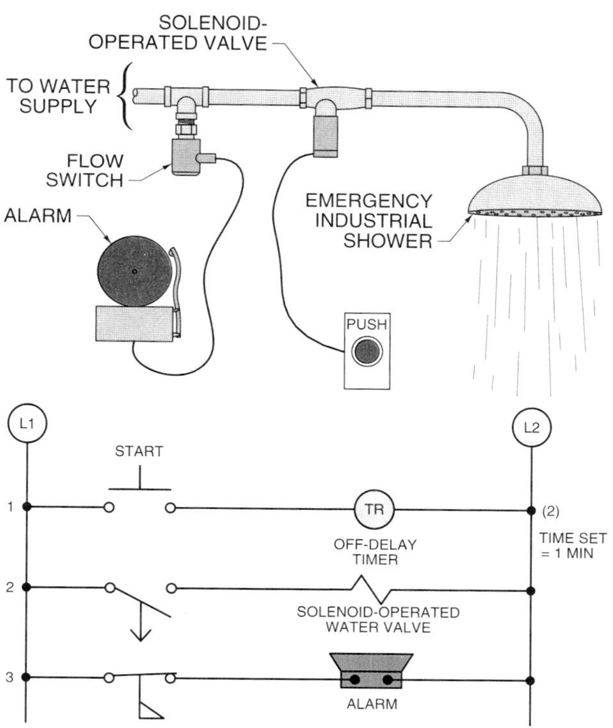

Figure 7-24. OFF-delay timers are used in applications that require a load to remain energized even after the input control has been removed.

After the shower pushbutton is pressed, the timer contacts close and the solenoid-operated valve starts the flow of water. The water flows even if the pushbutton is released. A flow switch is used to indicate when water is flowing. The flow switch sounds the alarm that can be used to bring help. The flow switch also sounds the alarm if there is a water break anywhere downstream from the switch.

One-Shot Timer Applications

One-shot timers are used in applications that require a fixed-timed output for a set period of time. See Figure 7-25. In this application, a one-shot timer is used to control the amount of time that plastic wrap is wound around a pallet of cartons.

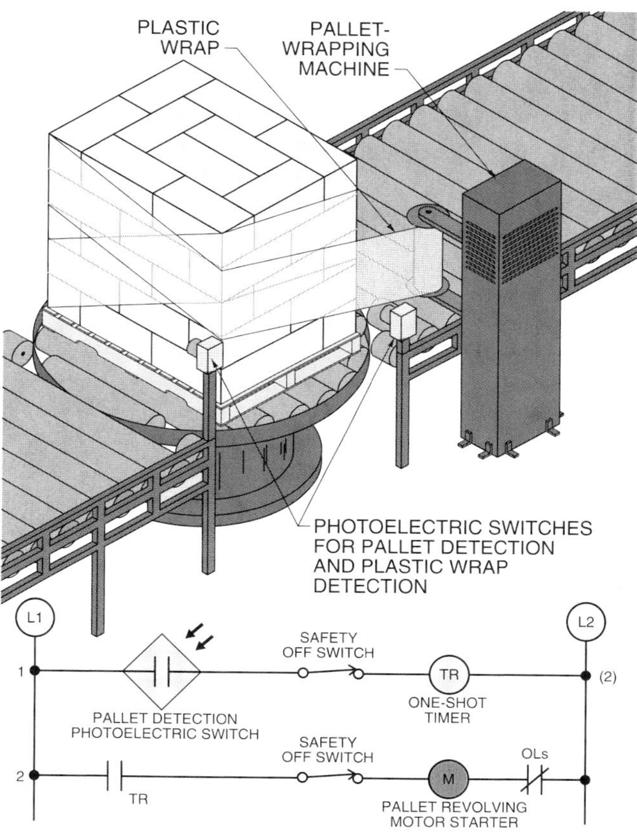

Figure 7-25. One-shot timers are used in applications that require a fixed-timed output for a set period of time.

A photoelectric switch detects a pallet entering the plastic wrap machine. The photoelectric switch energizes the one-shot timer. The one-shot timer contacts close, starting the wrapping process. The wrapping process continues for the setting of the timer. A second photoelectric switch could be used to detect that the plastic wrap is actually being applied. This is helpful in indicating a tear in the plastic or an empty roll.

Recycle Timer Applications

Recycle timers are used in applications that require a fixed ON and OFF time period. See Figure 7-26. In this application, a recycle timer is used to keep a product automatically mixed.

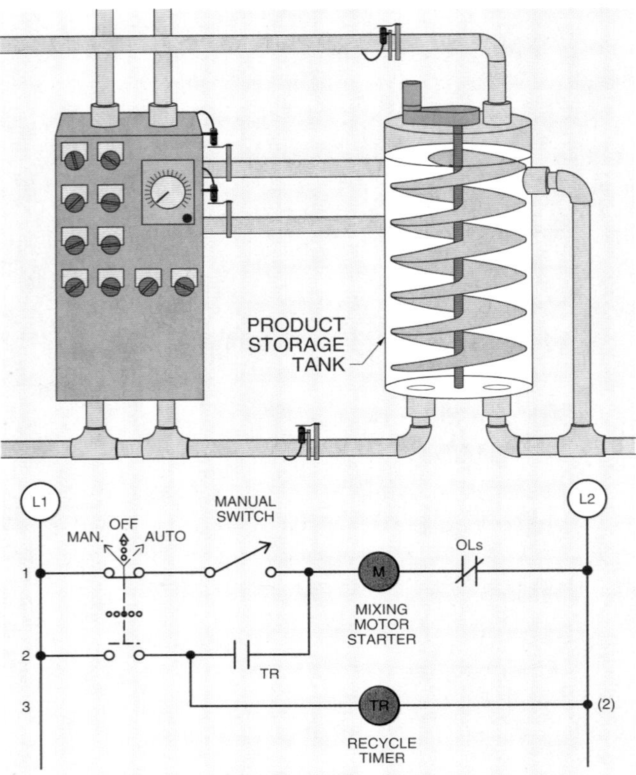

Figure 7-26. Recycle timers are used in applications that require a fixed ON and OFF time period.

Power is applied to the timer when the three-position selector switch is placed in the automatic position. The timer starts recycling for as long the selector switch is in the automatic position. The recycling timer turns the mixing motor ON and OFF at the set time. In this application, an asymmetrical timer works best. The ON time (mixer motor ON) is set less than the OFF time (mixer motor OFF). For example, the timer may be set to mix the product for 5 min every 2 hr.

Multiple Contact Timer Applications

Multiple contact timers are used in applications where a sustained input is used to control a circuit. See Figure 7-27. In this circuit, the cartons coming down the conveyor belt are to be filled with detergent. Each carton must be filled with the same amount of detergent and the process must be automatic. To accomplish this, the timer circuit is used to control the time it takes to fill one carton. A limit switch sustained input is used to detect the carton. A motor drives the conveyor belt and a solenoid opens or closes the hopper full of detergent. The limit switch could be replaced with a photoelectric or proximity switch.

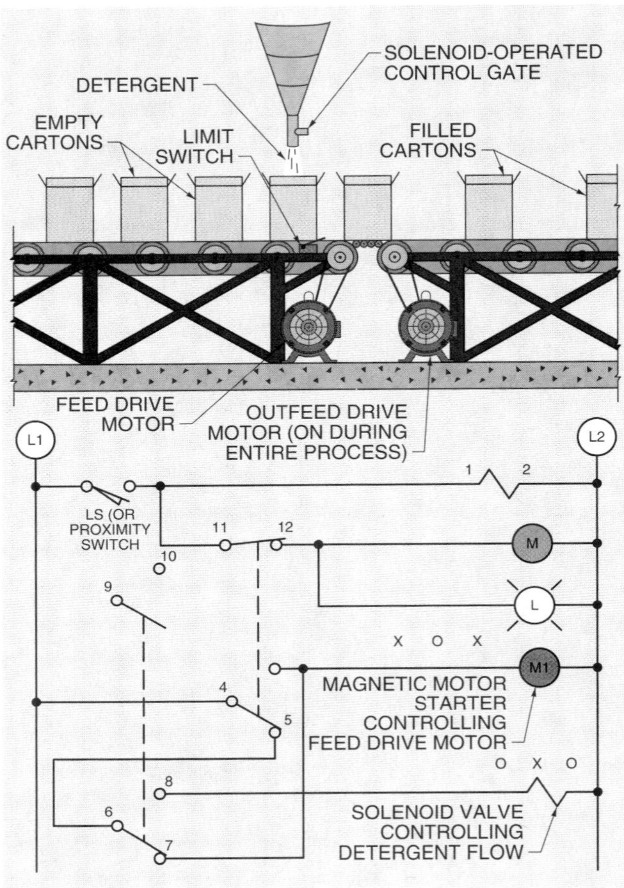

Figure 7-27. Multiple contact timers are used in applications where a sustained input controls a circuit.

As the cartons are coming down the conveyor, the feed drive motor is ON and the solenoid valve is OFF (no detergent fill). As a carton contacts the limit switch, the feed drive motor shuts OFF. This stops the conveyor and energizes the solenoid. The solenoid opens the control gate on the hopper of detergent. After the timer times out, the feed drive motor turns ON and the solenoid valve closes. This removes the filled carton which opens the limit switch; resetting the timer. The code for each load is added above the respective load. The code illustrates the desired sequence of operation of the loads.

Memory could be added to a circuit if the circuit requires a momentary input, such as a pushbutton to initiate timing. See Figure 7-28. In this circuit, a pushbutton is used as the input signal. As with any memory circuit, a NO instantaneous contact must be connected in parallel with the pushbutton if memory is to be added into the circuit. To accomplish this, a NO instantaneous contact 9 and 10 is connected in parallel with the pushbutton so that power is maintained when the control switch is released. A separate reset (stop) switch is used to reset the timer. After the timer has timed out, the timer resets only if the reset pushbutton is pressed. The sequence of each load is illustrated by the code.

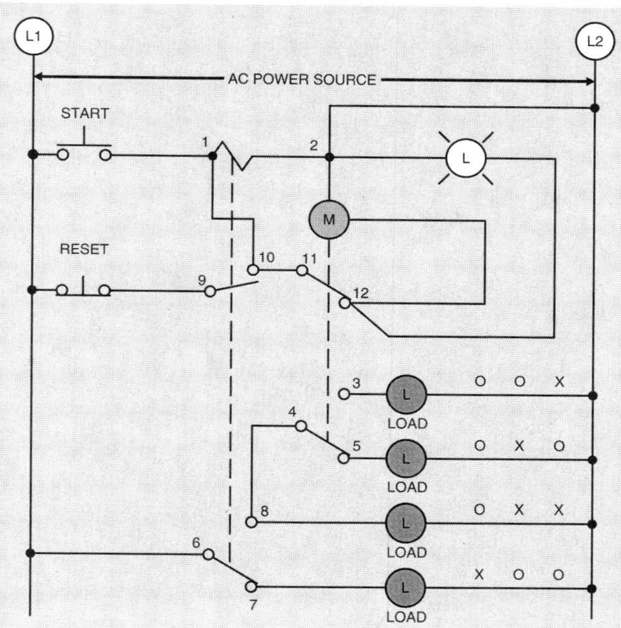

Figure 7-28. Memory could be added to a circuit if the circuit requires a momentary input, such as a pushbutton to initiate timing.

These diagrams are not in pure line diagram form. Line diagrams are standards used by all manufacturers. Wiring diagrams are the actual diagrams matching the logic of the line diagram to the manufacturer's product designed to perform that logic. A wiring diagram is used because this is the actual diagram found on the timer.

TROUBLESHOOTING TIMING CIRCUITS

Troubleshooting timing circuits is a matter of checking power to the timer and checking for proper timer contact operation. The timer is replaced if there is a problem with any part of the timer. The most common problem is contact failure. Measure the current at the contacts if the contacts failed prematurely. The current must not exceed the rating of the contacts. See Figure 7-29. To troubleshoot timers, apply the procedure:

1. Measure the voltage of the control circuit. The voltage must be within the specification range of the timer. Correct the voltage problem if the the voltage is not within the specification range of the timer. A common problem is an overloaded control transformer that is delivering a low-voltage output.
2. Measure the voltage at the timer coil. The voltage should be the same as the voltage of the control circuit. Check the control switch if the voltage is not the same. Check the voltage at the timer contacts if the voltage is correct.
3. Measure the voltage into the timer contacts. The voltage must be within the range of the load the timer is controlling. Correct the voltage problem if the voltage is not within the range.
4. Measure the voltage out of the timer contacts if the voltage into the timer contacts is correct. The voltage should be the same as the voltage of the control circuit. Check the timer contact connection points and wiring for a bad connection or corrosion if the voltage is not the same.

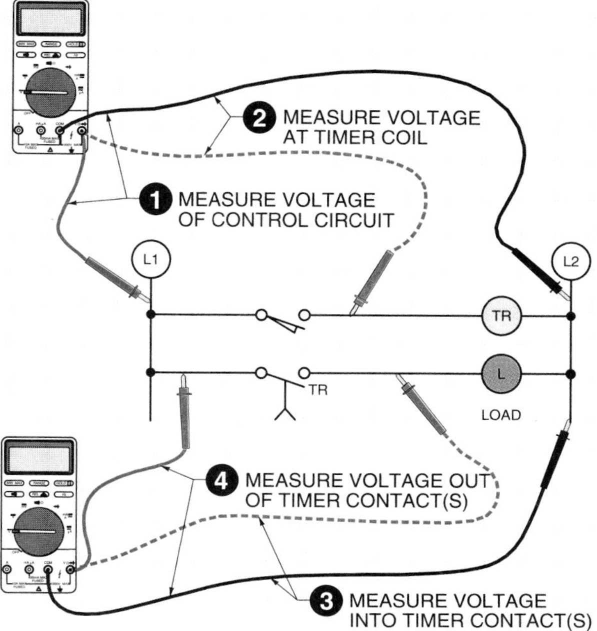

Figure 7-29. Troubleshooting timing circuits is a matter of checking power to the timer and checking for proper timer contact operation.

1. What are the three major categories of timers?
2. How does a dashpot timer develop a time delay?
3. How does a synchronous timer develop a time delay?
4. How does a solid-state timer develop a time delay?
5. What is an ON-delay timer?
6. What is another name for ON-delay?
7. What is an OFF-delay timer?
8. What is another name for OFF-delay?
9. What is a one-shot timer?
10. What is another name for one-shot?
11. What is a recycle timer?
12. How is the control switch connected when a supply voltage timer is used?
13. What is the advantage of using a contact-controlled timer?
14. Can a transistor-controlled timer use a contact as the control device?
15. What type of timer provides power to allow photoelectric and proximity control of the timer?
16. When using an X or O code to indicate the condition of a load, how is a de-energized load coded?
17. When using an X or O code to indicate the condition of a load, how is an energized load coded?
18. What type of recycle timer operates with equal ON and OFF time periods?
19. What does an operational diagram for a multiple function timer show?
20. What is the most common problem in timing circuits?

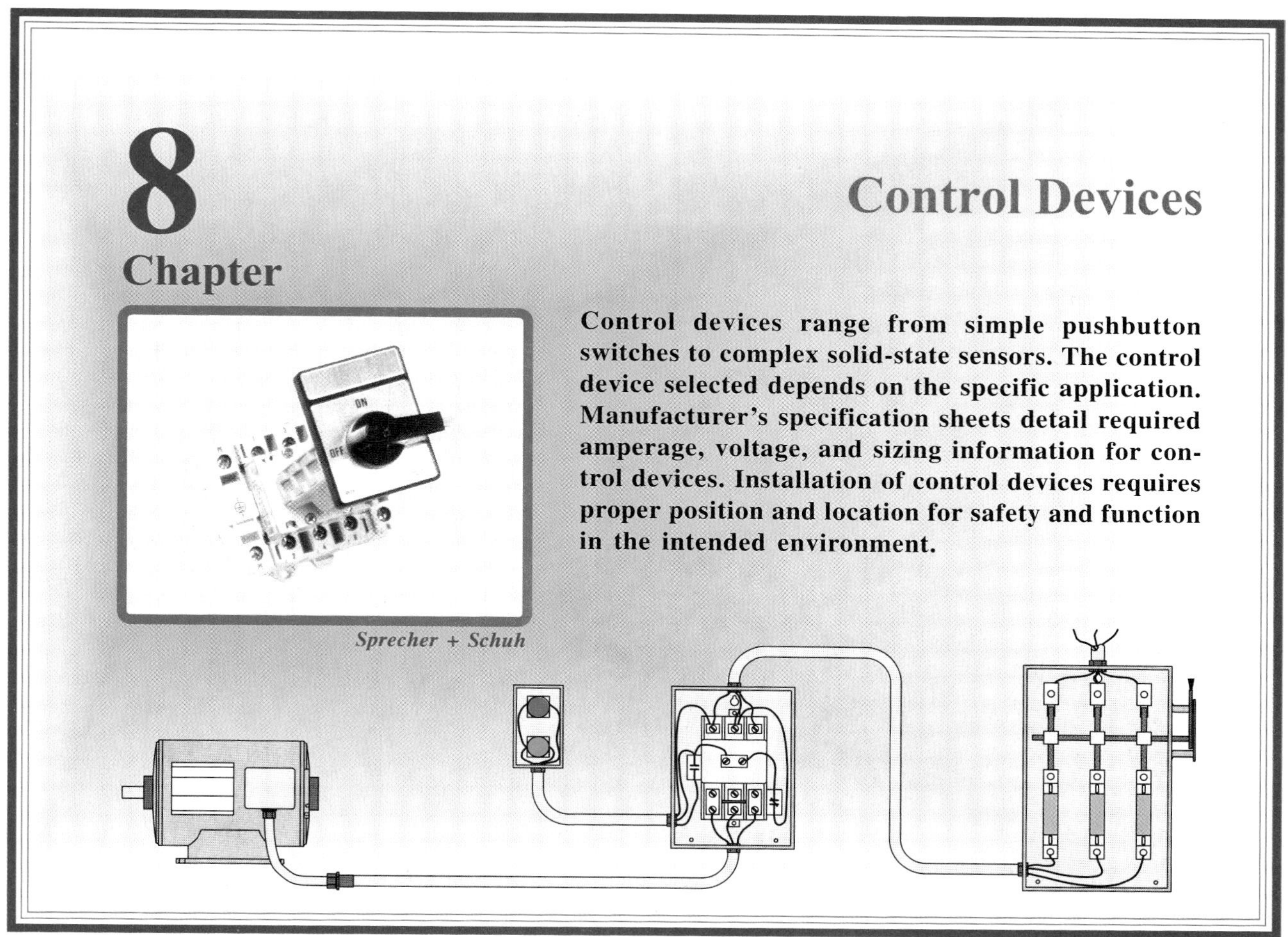

8 Chapter

Control Devices

Control devices range from simple pushbutton switches to complex solid-state sensors. The control device selected depends on the specific application. Manufacturer's specification sheets detail required amperage, voltage, and sizing information for control devices. Installation of control devices requires proper position and location for safety and function in the intended environment.

Sprecher + Schuh

INDUSTRIAL PUSHBUTTONS

Pushbuttons are the most common control switches used on industrial equipment. Almost all industrial machines and processes have a manually-controlled position, even if the machine or process is designed to operate automatically. An industrial pushbutton consists of a legend plate, an operator, and one or more contact blocks (electrical contacts). See Figure 8-1.

Legend Plates

A *legend plate* is the part of a switch that includes the written description of the switch's operation. A legend plate indicates the pushbutton's function in the circuit. Legend plates are available indicating common circuit operations such as start, stop, jog, up, down, ON, OFF, reset, run, or are available blank. The lettering on legend plates is normally uppercase for clarity and visibility. Legend plates are also available in different colors. The color red is normally used for such circuit functions as stop, OFF, and emergency stop. The color black with white lettering is used for most other circuit functions. However, different colored legend plates can be used along with colored operators to highlight different circuit functions. When color is used, red is normally used to indicate a stop or OFF function, green is used to indicate an ON or open function, and amber is used to indicate a manual override or reset function.

Operators

An *operator* is the device that is pressed, pulled, or rotated by the individual operating the circuit. An operator activates the pushbutton's contacts. Operators are available in many different colors, shapes, and sizes. Standard pushbutton operators include the flush, half-shrouded, extended, and jumbo mushroom button. The operator used depends on the application. See Figure 8-2.

Furnas Electric Co.

Figure 8-1. An industrial pushbutton consists of a legend plate, an operator, and one or more contact blocks (electrical contacts).

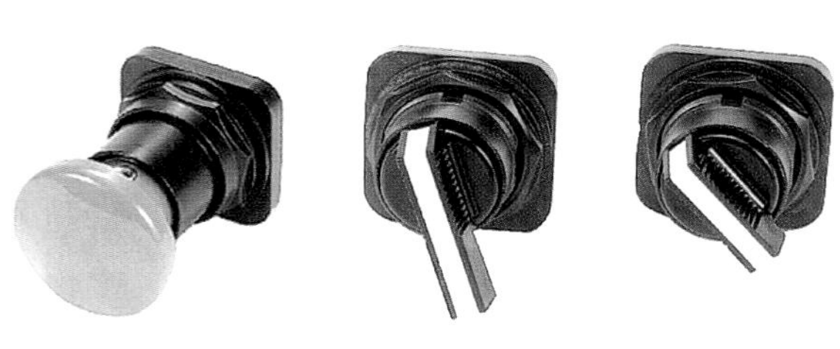

Furnas Electric Co.

Figure 8-2. An operator is the device that is pressed, pulled, or rotated by the individual operating the circuit.

Flush Button Operators. A *flush button operator* is a pushbutton with a guard ring surrounding the button which prevents accidental operation. The flush button operator is the most common operator used in applications in which accidental turn-ON may create a dangerous situation.

Half-Shrouded Button Operators. A *half-shrouded button operator* is a pushbutton with a guard ring which extends over the top half of the button. This helps prevent accidental operation, but allows for easier operation with the thumb. The half-shrouded button operator is used where avoiding accidental operation is preferred, but where the operator may be wearing gloves. Wearing gloves makes depressing a flush button operator difficult.

Extended Button Operators. An *extended button operator* is a pushbutton that has the button extended beyond the guard. An extended button operator is easily accessible and the color of the operator may be seen from all angles. The extended button operator is the most common operator used in applications in which an accidental start is not dangerous, such as when turning on lights.

Jumbo Mushroom Button Operators. A *jumbo mushroom button operator* is a pushbutton that has a large curved operator extending beyond the guard. A jumbo mushroom button operator is easily seen because of its large size. It can be operated from any angle and is used in applications that require fast operation such as emergency stops, motor stops, and valve shut-offs.

Contact Blocks

A *contact block* is the part of the pushbutton that is activated when the operator is pressed. A contact block includes the switching contacts of the pushbutton. Contact blocks include normally open (NO), normally closed (NC), or both NO and NC contacts. The most common contact block includes one NO and one NC contact. NO contacts make the circuit when the pushbutton operator is pressed and are used mainly for start or ON functions. NC contacts break the circuit when the pushbutton operator is pressed and are used mainly for stop or OFF functions. See Figure 8-3.

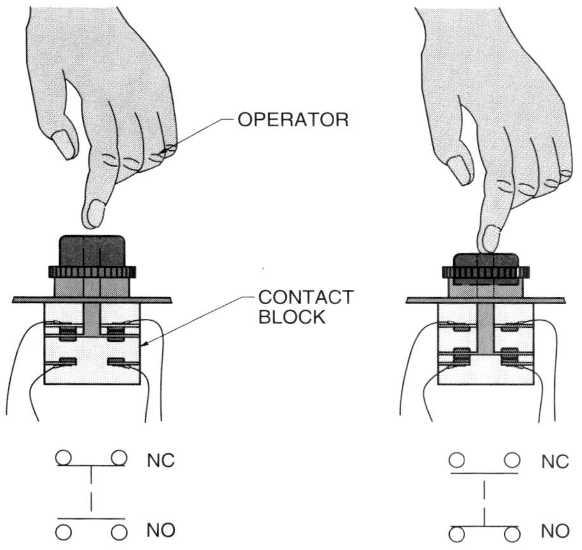

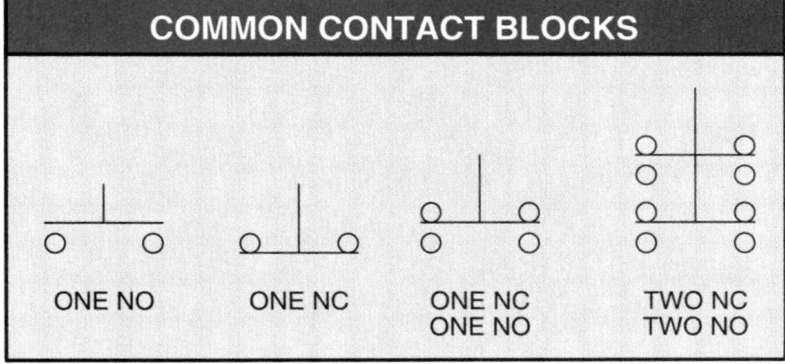

Figure 8-3. Contact blocks include normally open (NO), normally closed (NC), or both NO and NC contacts.

Pushbuttons are housed in pushbutton stations. A *pushbutton station* is an enclosure that protects the pushbutton, contact block, and wiring from dust, dirt, water, or corrosive fluids. Enclosures are available in various sizes and with a number of punched holes for mounting the operators. Every basic NEMA enclosure is available because pushbutton stations need to be mounted where they are conveniently operated.

A pushbutton must be placed in the proper enclosure for continuous and safe operation. Pushbuttons are often required to operate in environments where dust, dirt, oil, vibration, corrosive material, extreme variations of temperature and humidity, as well as other damaging factors are present. Always match the correct components and enclosure with the environment in which they operate. See Figure 8-4.

Cutler-Hammer

Pushbuttons manufactured by Cutler-Hammer are designed for space-saving installation and heavy-duty and corrosive industrial applications.

SELECTOR SWITCHES

A *selector switch* is a switch with an operator that is rotated (instead of pushed) to activate the electrical contacts. Selector switches select one of several different circuit conditions. They are normally used to select either two or three different circuit conditions. However, selector switches are available that have more than three positions.

Sprecher + Schuh

The operator of a selector switch is rotated to control the operation of an electrical circuit.

ENCLOSURES					
Type	**Use**	**Service Conditions**	**Tests**	**Comments**	**Type**
1	Indoor	No unusual	Rod entry, rust resistance		
3	Outdoor	Windblown dust, rain, sleet, and ice on enclosure	Rain, external icing, dust, and rust resistance	Do not provide protection against internal condensation or internal icing	1
3R	Outdoor	Falling rain and ice on enclosure	Rod entry, rain, external icing, and rust resistance	Do not provide protection against dust, internal condensation, or internal icing	
4	Indoor/outdoor	Windblown dust and rain, splashing water, hose-directed water, and ice on enclosure	Hosedown, external icing, and rust resistance	Do not provide protection against internal condensation or internal icing	4
4X	Indoor/outdoor	Corrosion, windblown dust and rain, splashing water, hose-directed water, and ice on enclosure	Hosedown, external icing, and corrosion resistance	Do not provide protection against internal condensation or internal icing	4X
6	Indoor/outdoor	Occasional temporary submersion at a limited depth			
6P	Indoor/outdoor	Prolonged submersion at a limited depth			
7	Indoor locations classified as Class I, Groups A, B, C, or D, as defined in the NEC®	Withstand and contain an internal explosion of specified gases, contain an explosion sufficiently so an explosive gas-air mixture in the atmosphere is not ignited	Explosion, hydrostatic, and temperature	Enclosed heat-generating devices shall not cause external surfaces to reach temperatures capable of igniting explosive gas-air mixtures in the atmosphere	7
9	Indoor locations classified as Class II, Groups E or G, as defined in the NEC®	Dust	Dust penetration, temperature, and gasket aging	Enclosed heat-generating devices shall not cause external surfaces to reach temperatures capable of igniting explosive gas-air mixtures in the atmosphere	9
12	Indoor	Dust, falling dirt, and dripping noncorrosive liquids	Drip, dust, and rust resistance	Do not provide protection against internal condensation	12
13	Indoor	Dust, spraying water, oil, and noncorrosive coolant	Oil explosion and rust resistance	Do not provide protection against internal condensation	

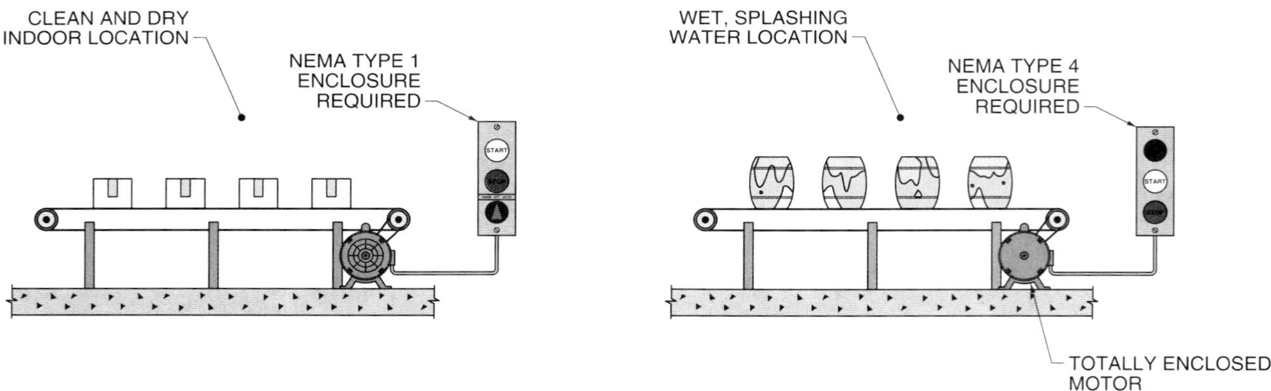

Figure 8-4. A pushbutton station is an enclosure that protects the pushbutton, contact block, and wiring from dust, dirt, water, or corrosive fluids.

Two-Position Selector Switches

A two-position selector switch allows the operator to select one of two circuit conditions. For example, a two-position selector switch may be used to place a heating circuit in the manual (HAND) or automatic (AUTO) condition. Only the manual control switch can turn the heating contactor ON or OFF when the selector switch is placed in the HAND position. The heating contactor controls the high-power heating elements. Only the temperature switch can turn the heating contactor ON or OFF when the selector switch is placed in the AUTO position. Circuit conditions controlled by two-position selector switches include ON/OFF, left/right, manual/automatic, up/down, slow/fast, run/stop, forward/reverse, jog/run, and open/close conditions. See Figure 8-5.

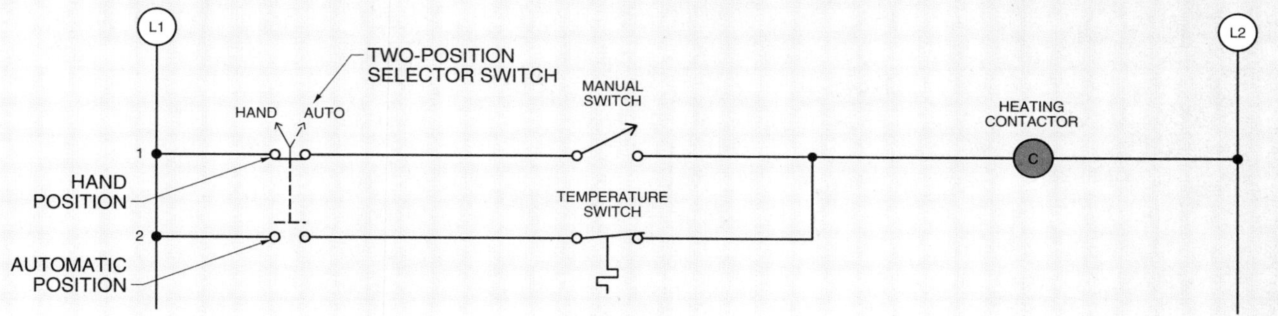

Figure 8-5. A two-position selector switch allows the operator to select one of two circuit conditions.

Three-Position Selector Switches

A three-position selector switch allows the operator to select one of three circuit conditions. For example, a three-position selector switch may be used to place a heating circuit in the manual, automatic, or OFF position. The OFF position is added for safety. In the OFF position, the heating contactor (or other machine being controlled) cannot be energized by the manual or automatic switch. Circuit conditions controlled by three-position selector switches include manual/OFF/automatic, heat/OFF/cool, forward/OFF/reverse, jog/OFF/run, slow/stop/fast, and up/stop/down conditions. See Figure 8-6.

Furnas Electric Co.

Selector switches and pushbuttons, when properly enclosed in a NEMA 7 or 9 enclosure, reduce the risk of ignition of hazardous atmospheres.

Truth Tables

Contact position on a selector switch may be illustrated using truth tables (target tables) or solid lines, dashed lines, and a series of small circles. See Figure 8-7. In truth tables, each contact on the line diagram is marked A, B, etc. and each position of the selector switch is marked 1, 2, etc. The truth table is made and positioned near the switch to illustrate each position and each contact.

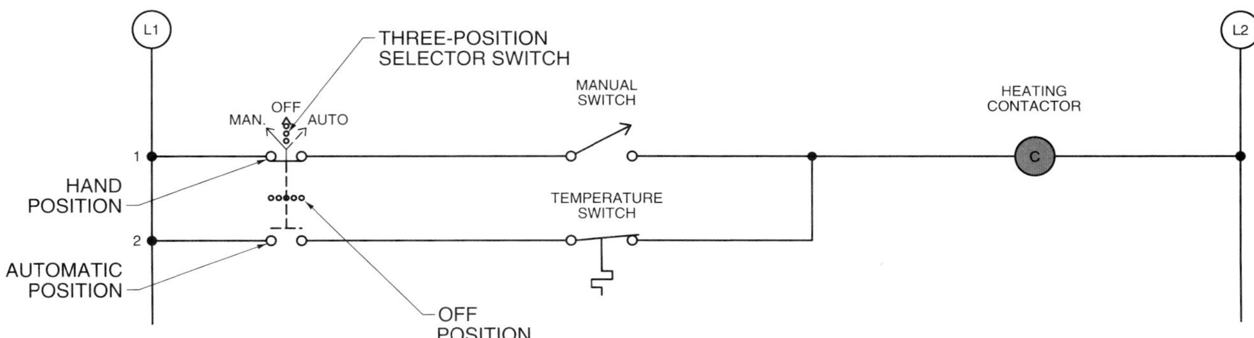

Figure 8-6. A three-position selector switch allows the operator to select one of three circuit conditions.

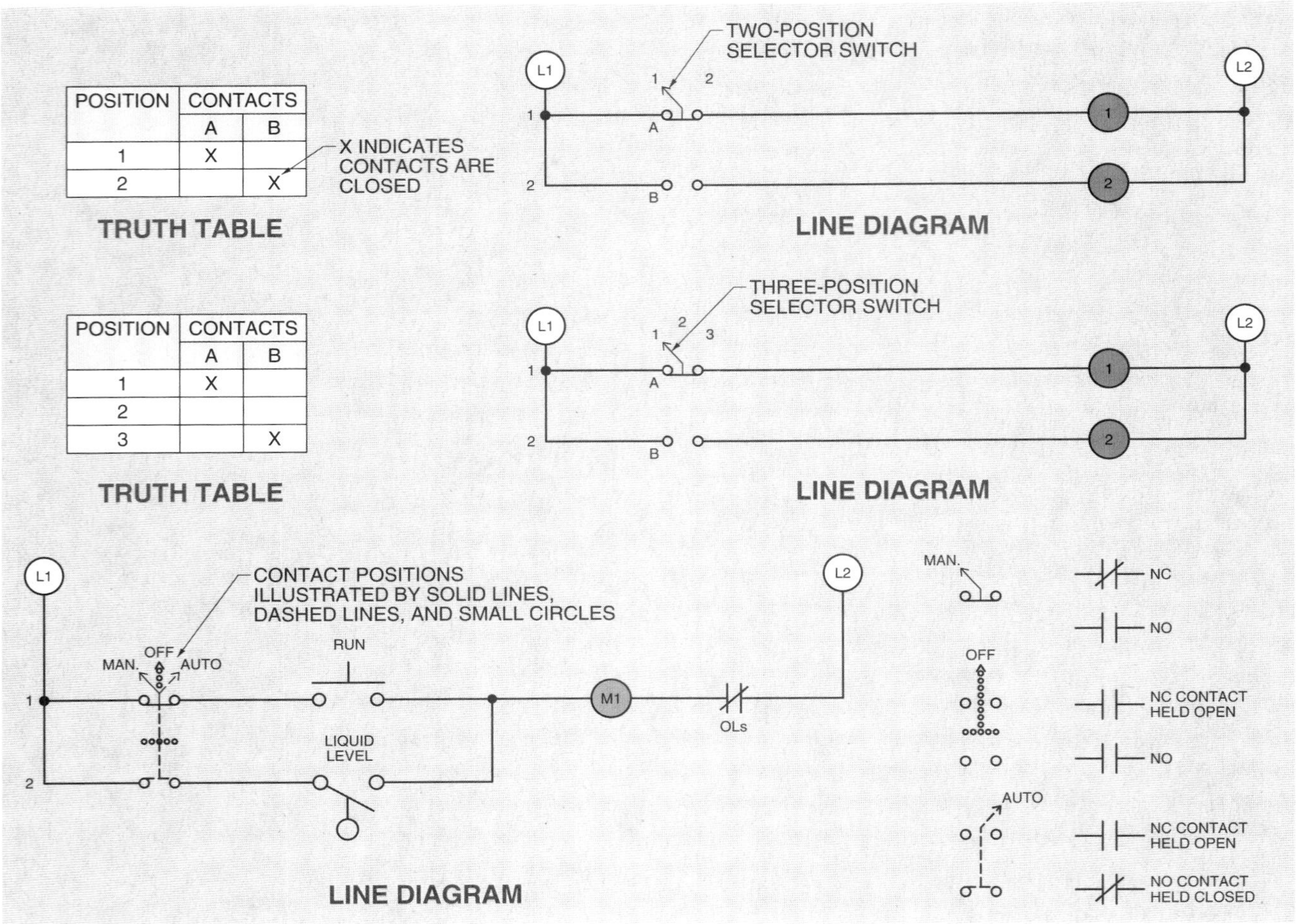

POSITION	CONTACTS	
	A	B
1	X	
2		X

X INDICATES CONTACTS ARE CLOSED

TRUTH TABLE

TWO-POSITION SELECTOR SWITCH

L1 L2 1 2 A B

LINE DIAGRAM

POSITION	CONTACTS	
	A	B
1	X	
2		
3		X

TRUTH TABLE

THREE-POSITION SELECTOR SWITCH

L1 L2 1 2 3 A B

LINE DIAGRAM

CONTACT POSITIONS ILLUSTRATED BY SOLID LINES, DASHED LINES, AND SMALL CIRCLES

L1 L2 MAN. OFF AUTO RUN LIQUID LEVEL M1 OLs

MAN. — NC

NO

OFF — NC CONTACT HELD OPEN

NO

AUTO — NC CONTACT HELD OPEN

NO CONTACT HELD CLOSED

LINE DIAGRAM

Figure 8-7. Contact position on a selector switch may be illustrated using truth tables (target tables) or solid lines, dashed lines, and a series of small circles.

Square D Company

Pushbutton stations are designed to allow the used of interchangeable elements to meet various control application needs.

An X is placed in the table if a contact is closed in any position. The table is easily read as to what contacts are closed in what positions. Truth tables illustrate the selector switch contacts more clearly than the method of using solid and dashed lines and small circles when a selector switch has more than two contacts or more than three positions.

JOYSTICKS

A *joystick* is an operator that selects one to eight different circuit conditions by shifting from the center position into one of the other positions. The most common joysticks can move from the center position into one of four different positions (up, down, left, or right).

The advantage of a joystick is that a person may control many operations without removing their hand from the joystick and do not have to take their eyes off the operation performed by the circuit.

The most common circuit conditions controlled by a joystick is in controlling a hoist (or crane) in the raise, lower, left, right, or OFF position. See Figure 8-8. In the hoist application, two reversing motors move the hoist and pulleys. One forward and reversing motor starter controls the hoist drive motor, and another forward and reversing motor starter controls the pulley motor. The joystick can turn only one motor starter ON at a time.

Two methods are used to indicate which position the joystick must be placed in to operate the contacts.

In the first method, a dot is placed in the symbol of the joystick to indicate the position the joystick must be in to switch the contacts. The NO contacts close and the NC contacts open when the contacts are switched.

In the second method, a truth table is used to indicate which contacts are switched in each position. In the truth table, an X indicates when the contact is closed. Truth tables are normally given in manufacturer's catalogs showing joystick operation, and a dot in the symbol is normally used on the line diagram to indicate its position.

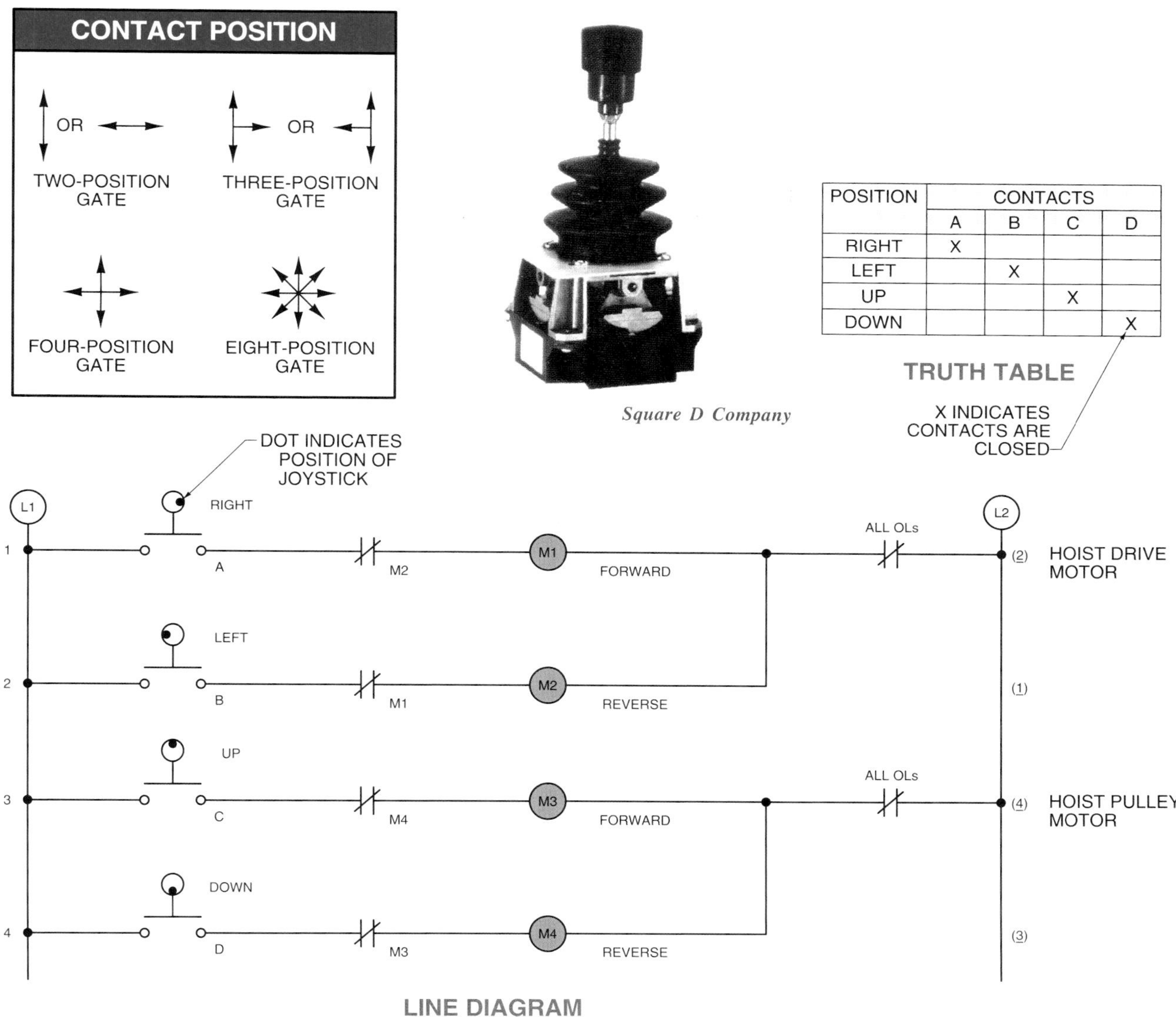

POSITION	CONTACTS			
	A	B	C	D
RIGHT	X			
LEFT		X		
UP			X	
DOWN				X

Figure 8-8. A joystick is used to control many different circuit operations from one location.

LIMIT SWITCHES

A *limit switch* is a mechanical input that requires physical contact of the object with the switch actuator. The physical contact is obtained from a moving object that comes in contact with the limit switch. The mechanical motion physically opens or closes a set of contacts within the limit switch enclosure. The contacts start or stop the flow of current in the electrical circuit. The contacts start, stop, operate in forward, operate in reverse, recycle, slow, or speed an operation. See Figure 8-9.

For example, a limit switch is used to automatically turn ON a light in a refrigerator or prevent a microwave oven from operating with the door open. In a washing machine, a limit switch is used to automatically turn OFF the washer if the load is not balanced. In an automobile, limit switches are used to automatically turn ON lights when a door is opened, and prevent overtravel of automatically-operated windows. In industry, limit switches are used to limit the travel of machine parts, sequence operations, detect moving objects, monitor an object's position, and provide safety (detect guards in place).

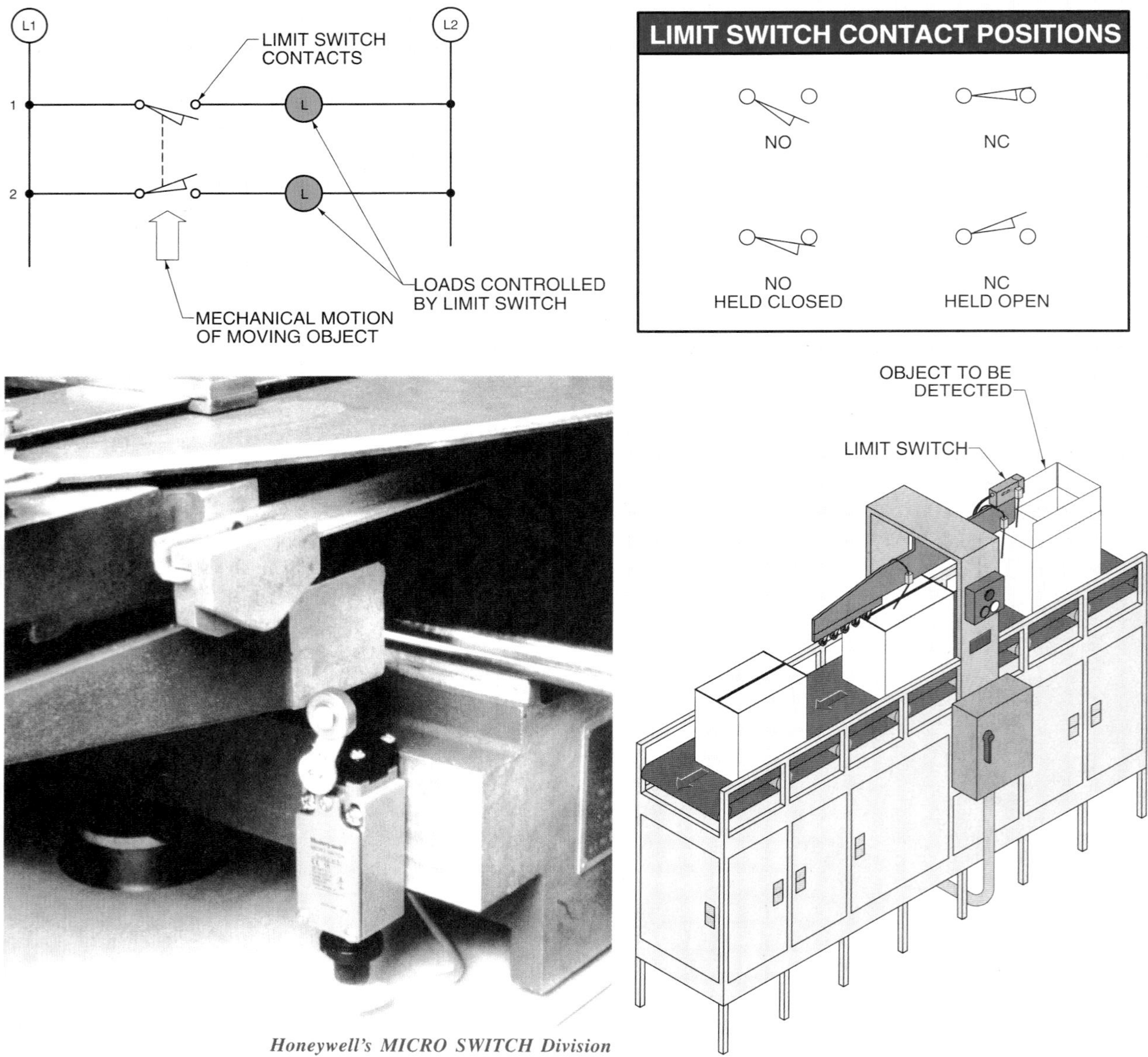

Figure 8-9. Limit switches are used to convert a mechanical motion into an electrical signal.

Limit switch contacts are normally snap-acting, which quickly change position to minimize arcing at the contacts. Limit switch contacts may be NO, NC, or any combination of NO and NC contacts. Most limit switches include one NO contact and one NC contact. Contacts are rated for the maximum current and voltage they can safely control.

Limit switch contacts must be connected to the proper polarity. See Figure 8-10. There is no arcing between the contacts when the contacts energize and de-energize the load as long as the contacts are at the same polarity. Arcing or welding of the contacts may occur from a possible short circuit if the contacts are connected to opposite polarity. Contacts must be selected according to proper voltage and current size according to the load that is listed in the manufacturer's specifications.

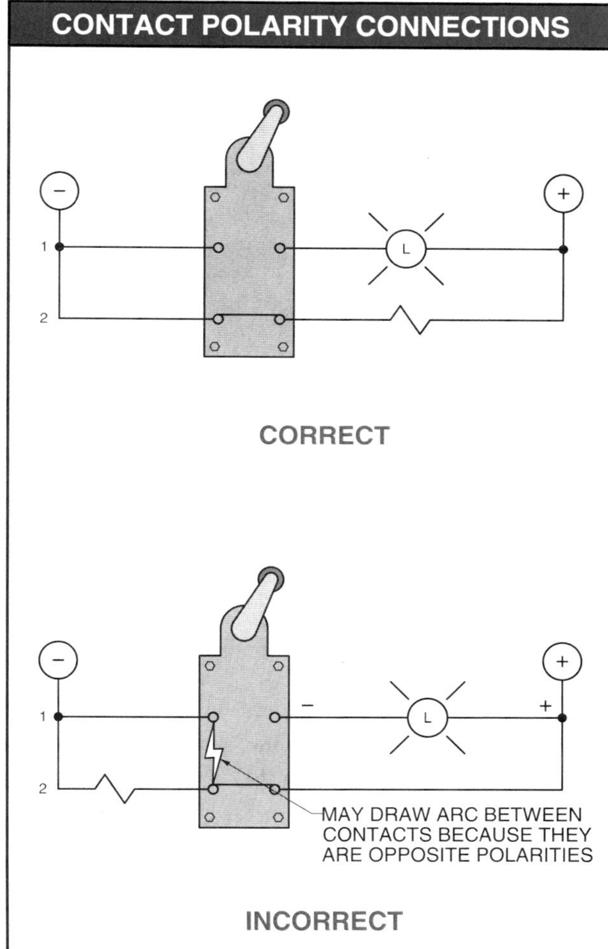

Figure 8-10. Arcing or welding of the contacts may occur from a possible short circuit if the contacts are connected to opposite polarity.

A relay, contactor, or motor starter must be used to interface the limit switch with the load if the load current exceeds the contact rating. See Figure 8-11.

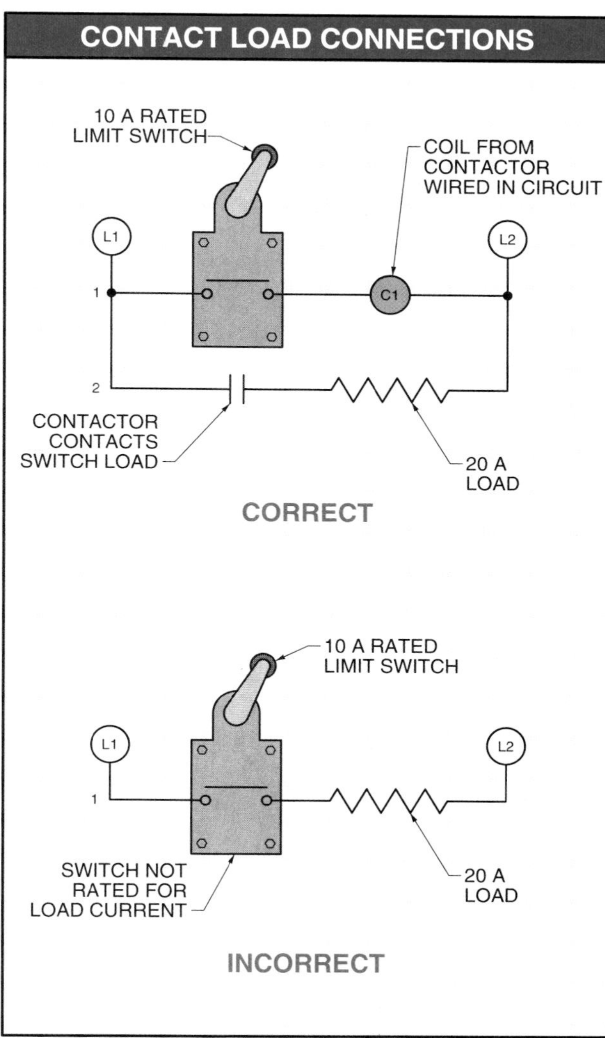

Figure 8-11. A relay, contactor, or motor starter must be used to interface the limit switch with the load if the load current exceeds the contact rating.

Actuators

An *actuator* is the part of a limit switch that transfers the mechanical force of the moving part to the electrical contacts. See Figure 8-12. Small limit switches are available with one fixed actuator. The fixed actuator may be any one of several different types, but is not removable nor interchangeable. Large limit switches are available with a knurled shaft that allows different actuators to be attached. The actuator used depends on the application.

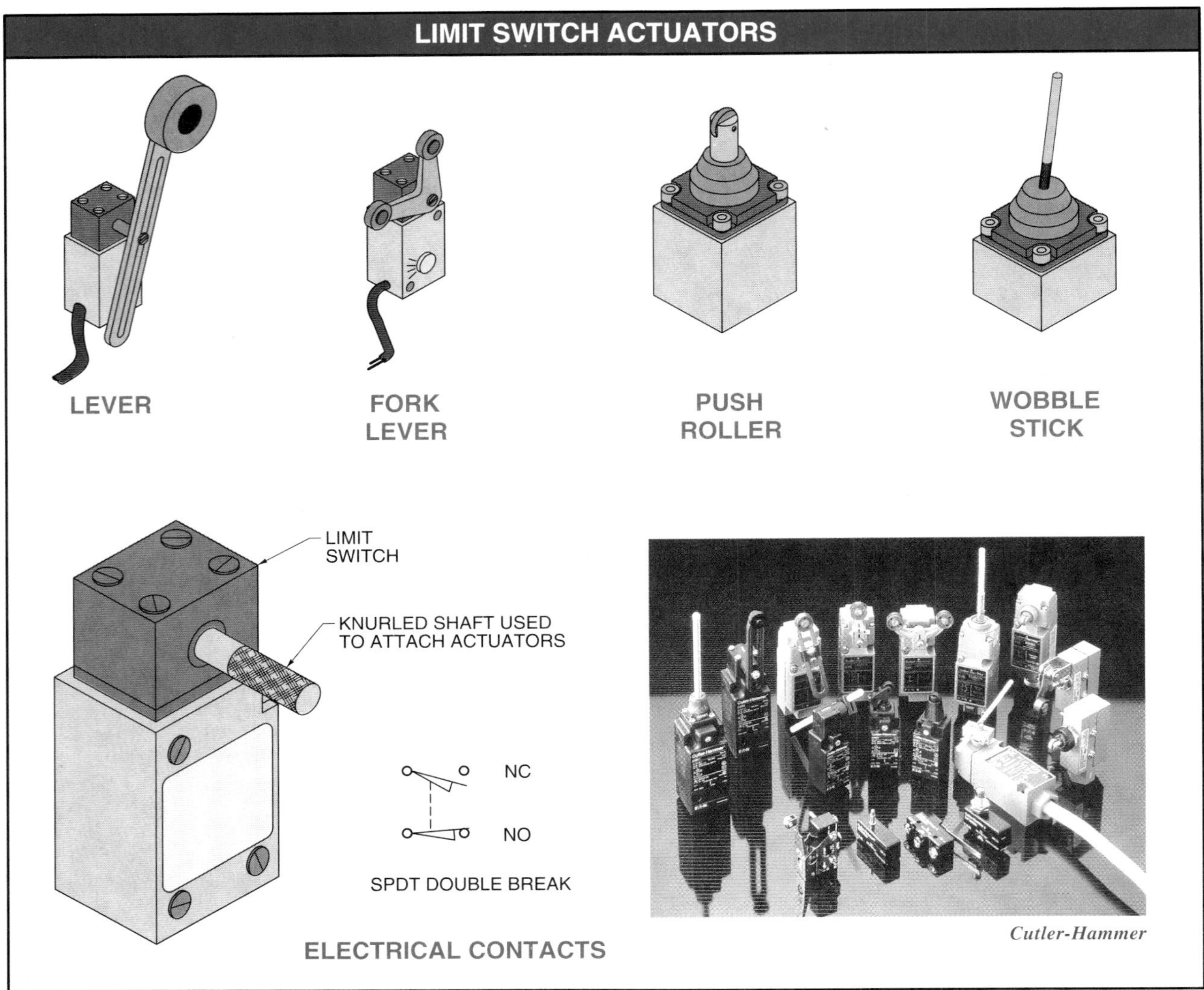

Figure 8-12. An actuator is the part of a limit switch that transfers the mechanical force of the moving part to the electrical contacts.

The basic actuators used on limit switches include lever, fork lever, push roller, and wobble stick. Most manufacturers offer several variations in addition to the basic actuators.

Levers. A *lever actuator* is an actuator operated by means of a lever which is attached to the shaft of the limit switch. The lever actuator includes a roller on the end which helps prevent wear. The length of the lever may be fixed or adjustable. The adjustable lever is used in applications in which the length of the arm, or actuator travel, may require adjustment. A lever actuator may be operated from either direction, but is normally used in applications in which the actuating object is moving in only one direction.

Fork Levers. A *fork-lever actuator* is an actuator operated by either one of two roller arms. Fork-lever actuators are used where the actuating object travels in two directions. A typical application is a grinder that automatically alternates back and forth.

Push Rollers. A *push-roller actuator* is an actuator operated by direct forward movement into the limit switch. A direct thrust with very limited travel is accomplished. push-roller actuators are commonly used to prevent overtravel of a machine part or object. The switch contacts stop the forward movement of the object when the machine part comes in contact with the limit switch.

Namco Controls Corporation

Namco Snaplock® limit switches are ideal for lighter force applications and are available with top-mounted plunger and push-roller actuators.

Wobble Sticks. A *wobble-stick actuator* is an actuator operated by means of any movement into the switch, except a direct pull. The wobble-stick actuator normally has a long arm that may be cut to the required length. Wobble-stick actuators are used in applications that require detection of a moving object from any direction.

Limit Switch Installation

Limit switches are actuated by a moving part. Limit switches must be placed in the correct position in relationship to the moving part. Limit switches should not be operated beyond the manufacturer's recommended travel specifications. See Figure 8-13.

Limit switch contacts do not operate if the actuating object does not force the limit switch actuator to move far enough. Limit switch contacts operate, but may return to their normal position if the actuating object forces the actuator of the limit switch to move too far. Overtravel may also damage the limit switch or force it out of position.

A rotary cam-operated limit switch must be installed according to manufacturer recommendations. A push-roller actuator should not be allowed to snap back freely. The cam should be tapered to allow a slow release of the lever. This helps to eliminate roller bounce, switch wear, and allows for better repeat accuracy. See Figure 8-14.

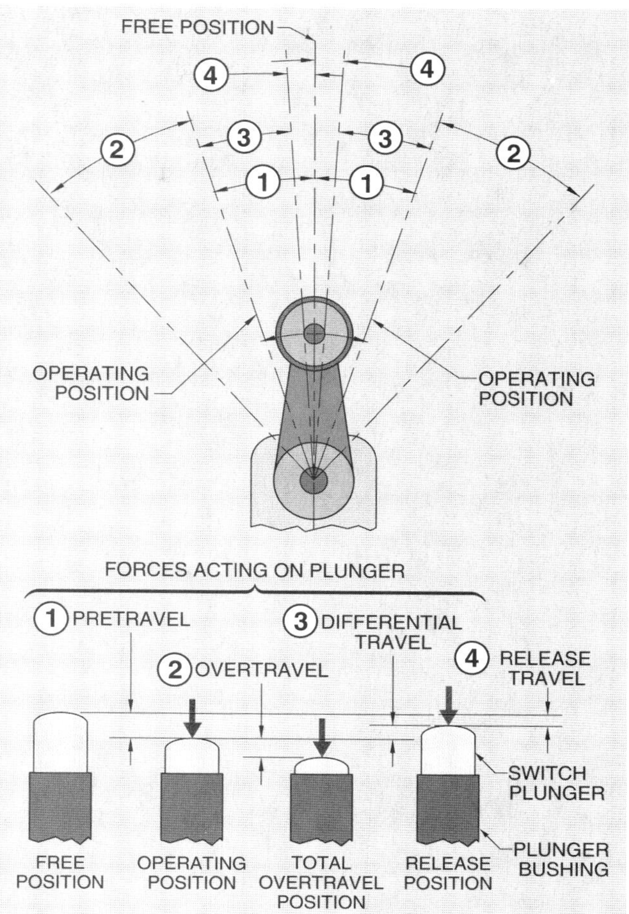

Figure 8-13. Limit switches should not be operated beyond the manufacturer's recommended travel specifications.

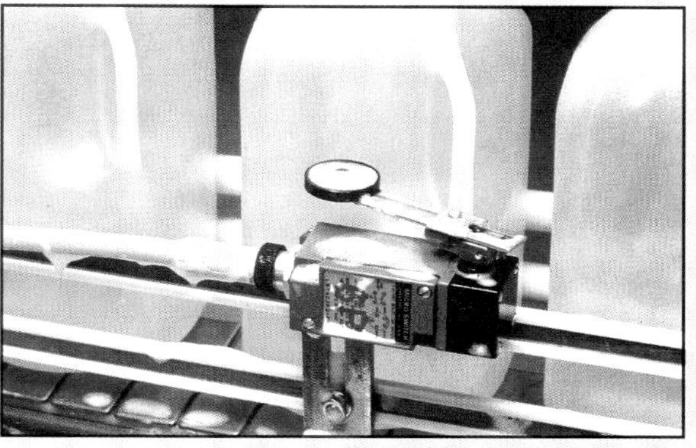

Honeywell's MICRO SWITCH Division

The HDLS side rotary-actuated limit switches from Honeywell have excellent sealing capability for withstanding harsh-duty food and beverage washdowns and severe machine tool environments.

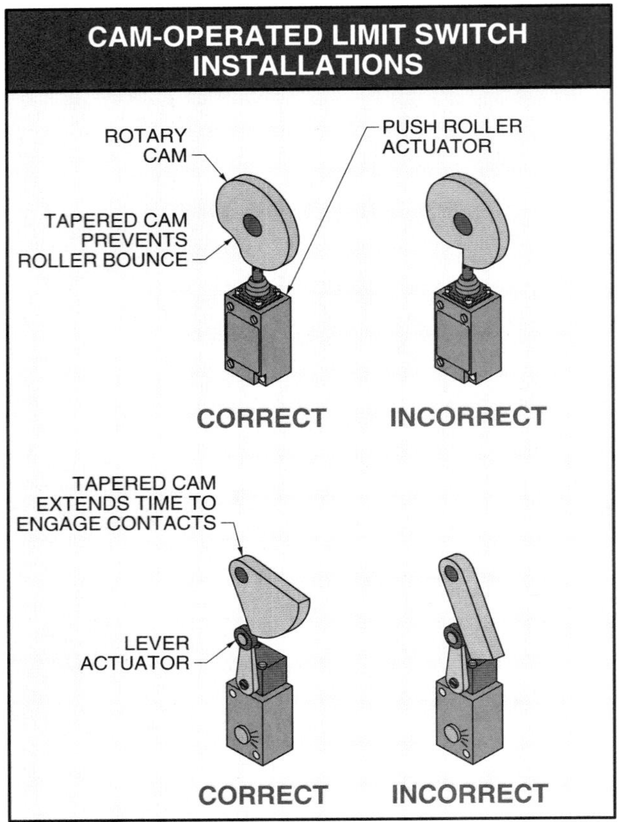

Figure 8-14. A cam-operated limit switch must be installed to prevent severe impact and allow a slow release of the lever.

Honeywell's MICRO SWITCH Division

The CLS Series cable-pull limit switches from Honeywell detect through a manual reset feature which identifies a broken or slackened cable and direct-acting contacts to enhance reliability and increase safety.

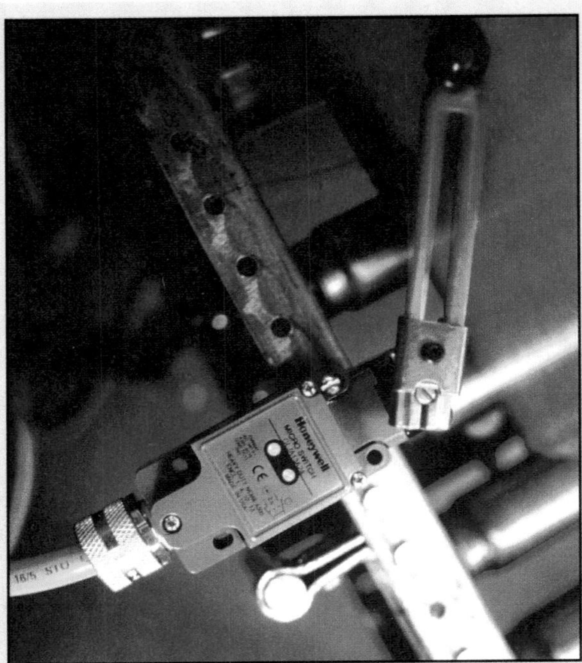

Honeywell's MICRO SWITCH Division

The GLS (Global Limit Switch) Series from Honeywell is available with a full range of actuator heads that make it ideal for material handling equipment, conveyors, machine tools, transfer lines, process equipment, textile machinery, and construction equipment.

Limit switches installed where relatively fast motions are involved must be installed so that the limit switch's lever does not receive a severe impact. The cam should be tapered to extend by the time it takes to engage the electrical contacts. This prevents wear on the switch and allows the contacts a longer closing time. This ensures that the circuit is complete.

Limit switches using push-roller actuators must not be operated beyond their travel in emergency conditions. A lever actuator is used instead of a push-roller actuator in applications where an override may occur. See Figure 8-15. A lever actuator has an extended range which prevents damage to the switch and mounting in case of an overtravel condition. A limit switch should never be used as a stop. A stop plate should always be added to protect the limit switch and its mountings from any damage due to overtravel.

Limit switches are designed to be used as an automatic controller that is mechanically-activated. Care should be taken to avoid any human error. Limit switches should be mounted so that an operator cannot accidentally activate the limit switch. See Figure 8-16.

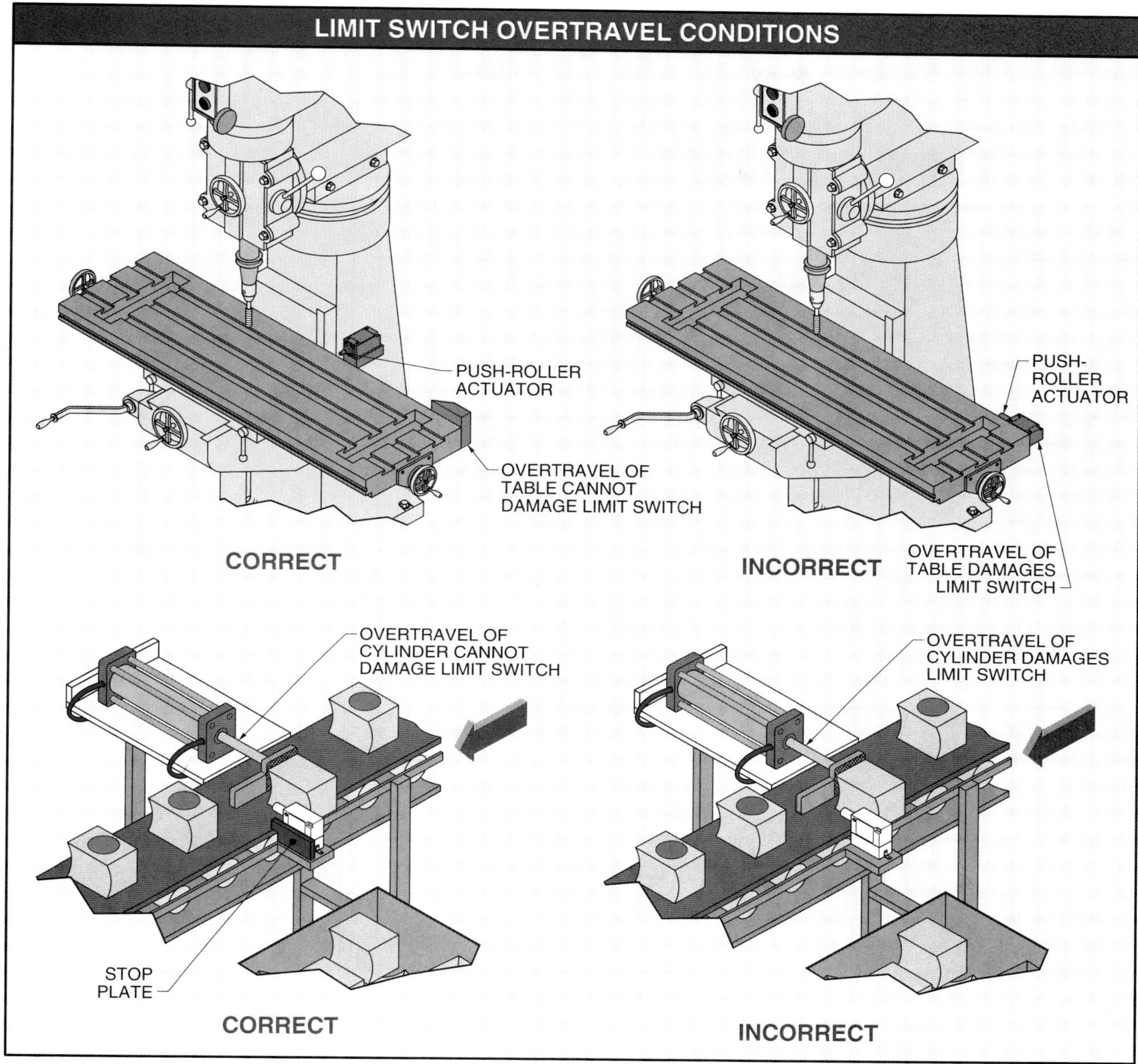
LIMIT SWITCH OVERTRAVEL CONDITIONS
PUSH-ROLLER ACTUATOR
OVERTRAVEL OF TABLE CANNOT DAMAGE LIMIT SWITCH
CORRECT
PUSH-ROLLER ACTUATOR
INCORRECT
OVERTRAVEL OF TABLE DAMAGES LIMIT SWITCH
OVERTRAVEL OF CYLINDER CANNOT DAMAGE LIMIT SWITCH
STOP PLATE
CORRECT
OVERTRAVEL OF CYLINDER DAMAGES LIMIT SWITCH
INCORRECT

Figure 8-15. Limit switches using push-roller actuators must not be operated beyond their travel limit.

The atmosphere and surroundings of the limit switch must be considered when mounting a limit switch. A limit switch must be mounted in a location where machining chips or other materials do not accumulate. These could interfere with the operation of the limit switch and cause circuit failure. This applies to splashing or submerging the limit switch in oils, coolants, or other liquids. Heat levels above the specified limits of the switch must be avoided. Always position a limit switch to avoid any excessive heat.

Foot Switches

A *foot switch* is a control switch that is operated by a person's foot. A foot switch is used in applications that require a person's hands to be free, or an additional control point. Foot switch applications include sewing machines, drill presses, lathes, and other similar machines. Most foot switches have two positions, a toe-operated position and an OFF position.

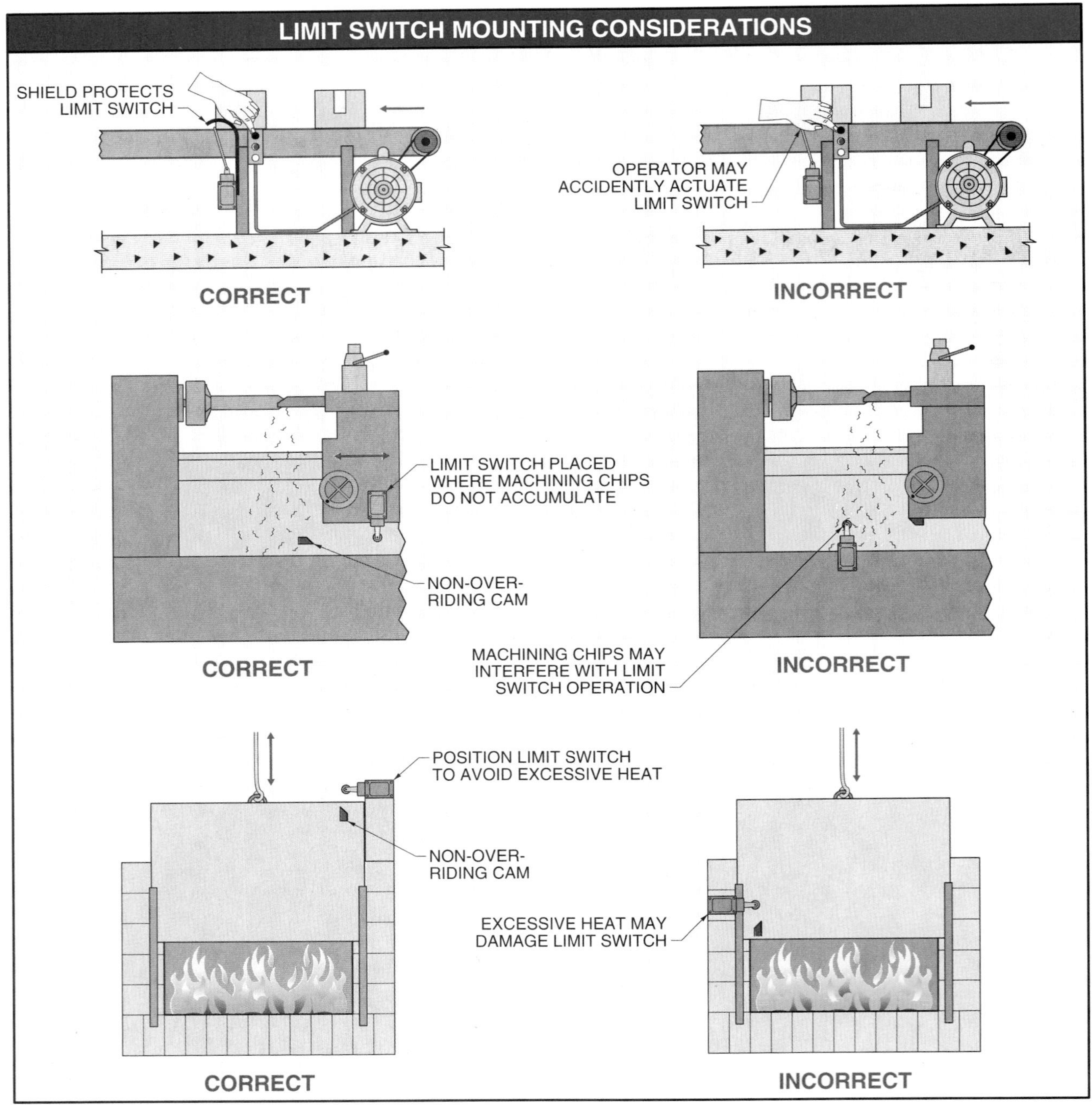
LIMIT SWITCH MOUNTING CONSIDERATIONS
SHIELD PROTECTS LIMIT SWITCH
CORRECT
OPERATOR MAY ACCIDENTLY ACTUATE LIMIT SWITCH
INCORRECT
LIMIT SWITCH PLACED WHERE MACHINING CHIPS DO NOT ACCUMULATE
NON-OVER-RIDING CAM
CORRECT
MACHINING CHIPS MAY INTERFERE WITH LIMIT SWITCH OPERATION
INCORRECT
POSITION LIMIT SWITCH TO AVOID EXCESSIVE HEAT
NON-OVER-RIDING CAM
EXCESSIVE HEAT MAY DAMAGE LIMIT SWITCH
CORRECT
INCORRECT

Figure 8-16. Limit switches should be mounted to avoid accidental activation, accumulated materials, and excessive heat.

The OFF position is normally spring-loaded so that the switch automatically returns to the OFF position when released. Foot switches with three positions include a pivot on a fulcrum to allow toe or heel control. Like the two-position foot switch, the three-position foot switch is normally spring-loaded so that the switch automatically returns to the OFF position when released. See Figure 8-17.

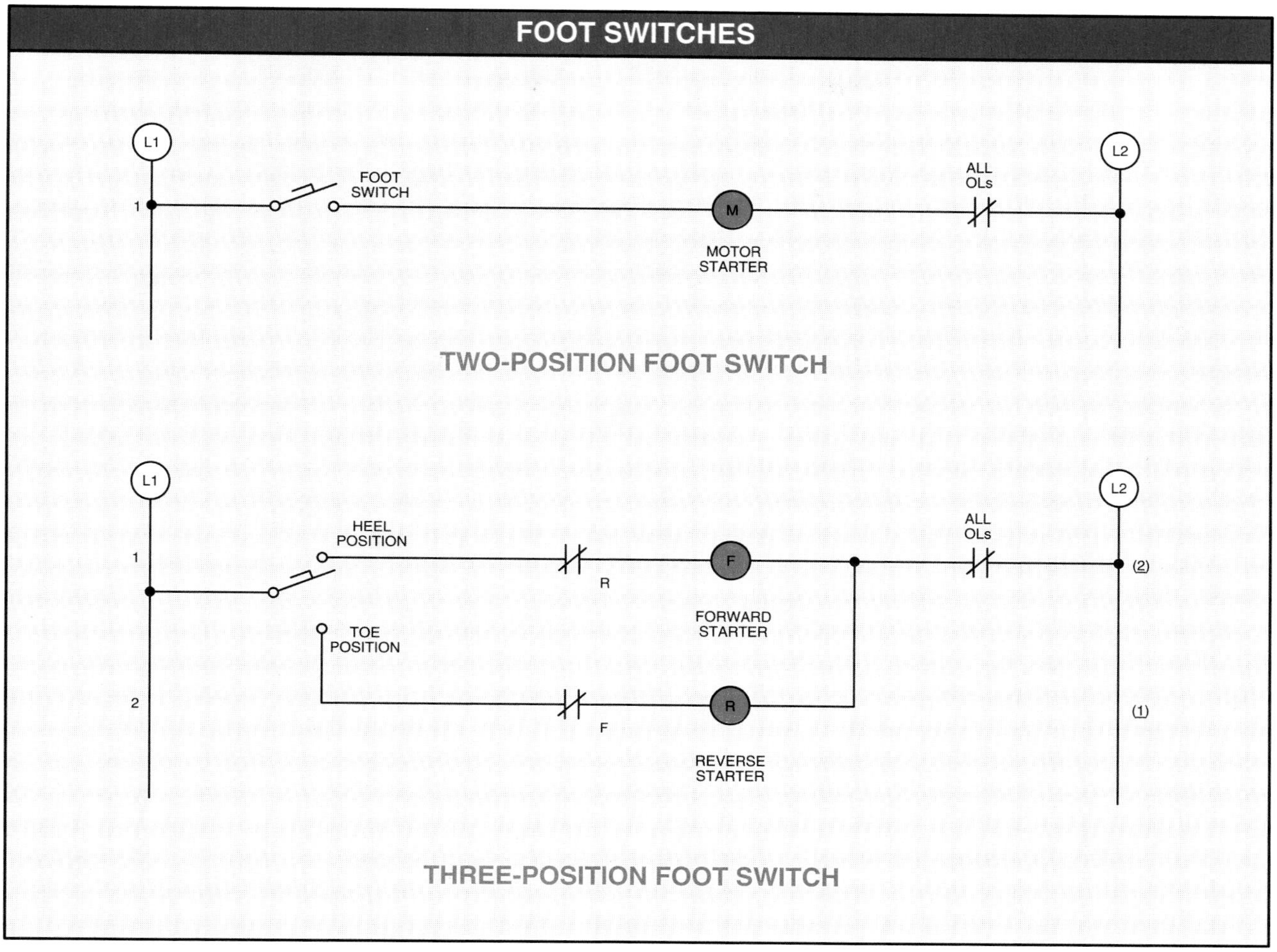

FOOT SWITCHES

L1
FOOT SWITCH
1
M
MOTOR STARTER
ALL OLs
L2

TWO-POSITION FOOT SWITCH

L1
HEEL POSITION
1
R
F
FORWARD STARTER
ALL OLs
L2
(2)
TOE POSITION
2
F
R
REVERSE STARTER
(1)

THREE-POSITION FOOT SWITCH

Figure 8-17. A foot switch is used to allow hands-free control or an additional control point.

Ruud Lighting, Inc.

The Luma low-voltage track lighting system from Ruud is available with a photocontrol/time clock option which turns the lights ON at dusk and keeps them ON for a set time period.

DAYLIGHT SWITCHES

A *daylight switch* is a switch that automatically turns lamps ON at dusk and OFF at dawn. Daylight switches are used to control outdoor lamps such as street lights and signs. They are also used to provide safety and security around buildings and other areas that require lighting at night.

Daylight switches use a sensor that changes resistance with a change in light intensity. The higher the light source, the lower the sensor's resistance. Current flows through the relay coil when the sensor's resistance is low. The NC contacts open and the lamp is turned OFF when the relay coil energizes.

The sensor's resistance increases as the light source decreases. The increased resistance reduces the flow of current to the point that the relay de-energizes. This causes the NC contacts to close and the lamp to turn ON. See Figure 8-18.

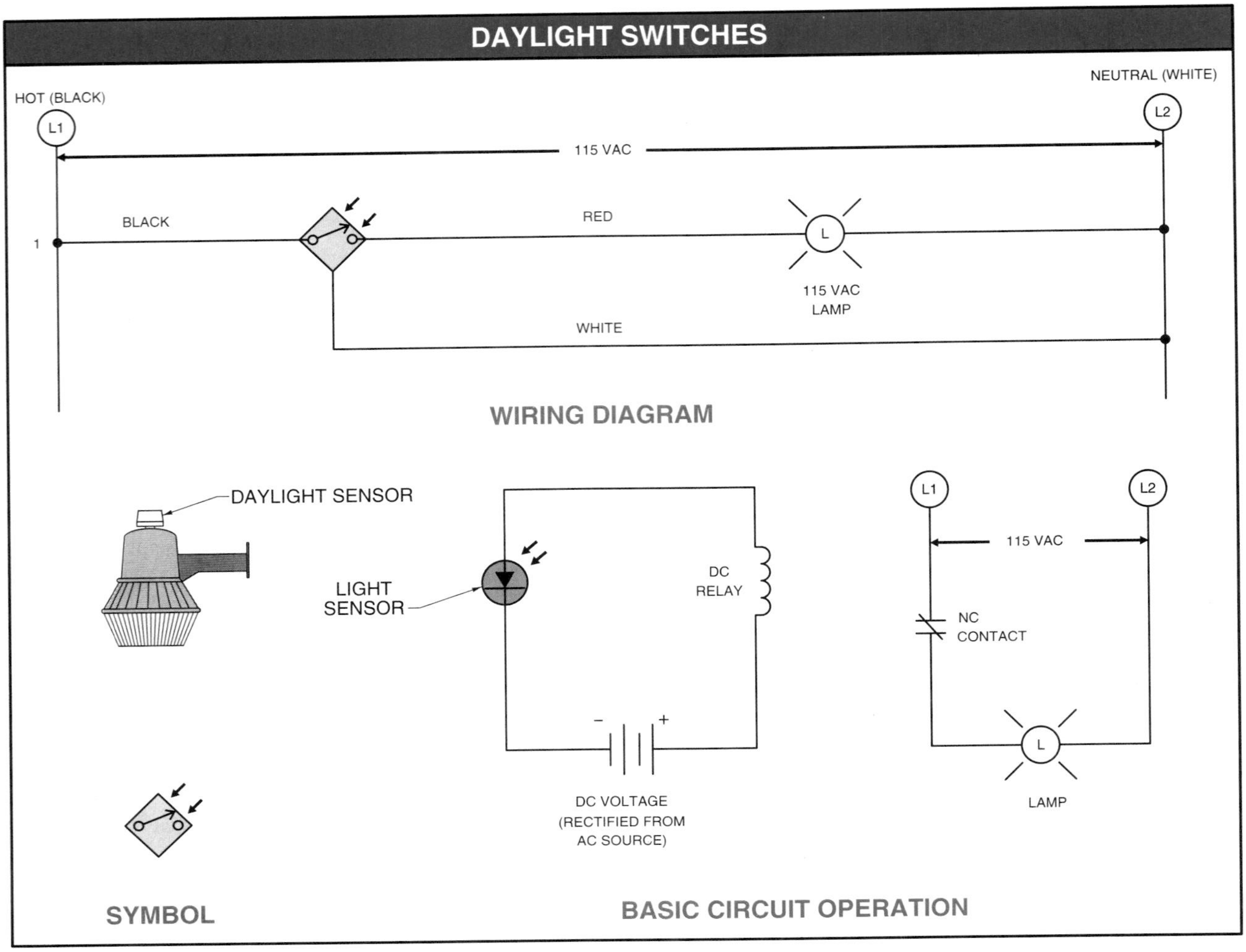

DAYLIGHT SWITCHES

HOT (BLACK)
L1
NEUTRAL (WHITE)
L2
115 VAC
1
BLACK
RED
L
115 VAC
LAMP
WHITE

WIRING DIAGRAM

DAYLIGHT SENSOR
LIGHT SENSOR
DC RELAY
DC VOLTAGE (RECTIFIED FROM AC SOURCE)
L1
L2
115 VAC
NC CONTACT
L
LAMP

SYMBOL

BASIC CIRCUIT OPERATION

Figure 8-18. A daylight switch is a switch that automatically turns lamps ON at dusk and OFF at dawn.

The sensor of the switch must be positioned so that artificially-produced light from the lamp that is being controlled (as well as other lamps), does not fall on the sensor. Most daylight switches have an adjustment to reduce or increase the amount of light falling on the sensor. Most daylight switches include an approximate 30 sec time delay to prevent nuisance switching caused by automobile headlights.

Ruud Lighting, Inc.

Daylight switches increase the security and convenience of landscape lighting.

PRESSURE SWITCHES

Pressure is force exerted over a surface divided by its area. The exerted force always produces a deflection or change in the volume or dimension on the area to which it is applied. Pressure is expressed in pounds per square inch (psi). Low pressures are expressed in inches of water column (wc). One psi equals 27.68″ wc.

A *pressure switch* is a control switch that detects a set amount of force and activates electrical contacts when the set amount of force is reached. The contacts may be activated by positive, negative (vacuum), or differential pressures. Differential pressure switches are connected to two different system pressures. See Figure 8-19.

NC or NO contacts are used depending on the application. NC contacts are used to maintain system pressure. The closed contacts energize a pump motor until system pressure is reached.

When system pressure is reached, the contacts open and the pump motor is turned OFF. NO contacts are used to signal an overpressure condition. An alarm is sounded when the open contacts close. The alarm remains ON until the pressure is reduced.

Pressure switches use different sensing devices to detect the amount of pressure. The pressure switch used depends on the application and system pressure. Most pressure switches use either a diaphragm, bellows, or piston-sensing device.

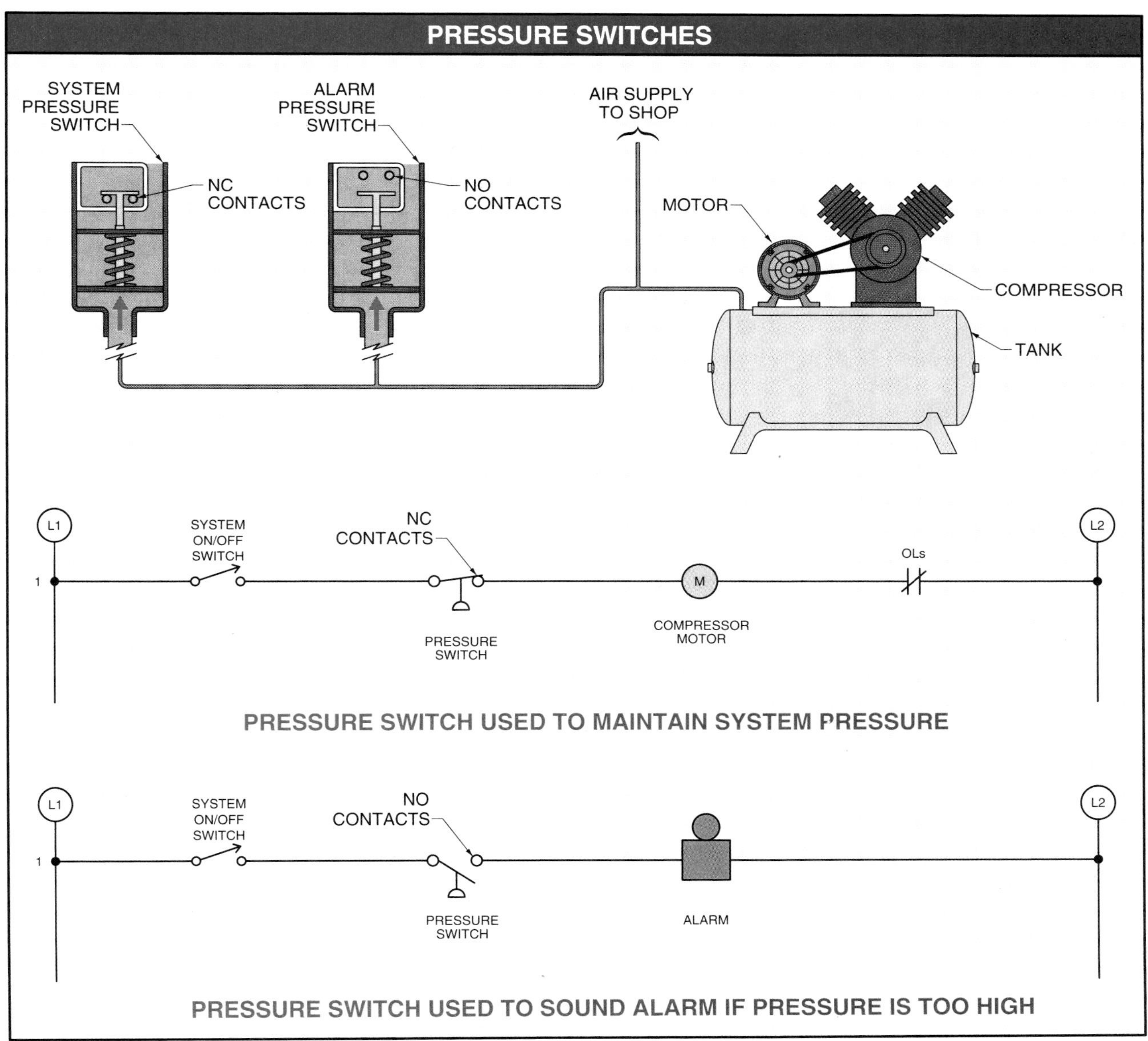

Figure 8-19. A pressure switch is a control switch that detects a set amount of force and activates electrical contacts when the set amount of force is reached.

Diaphragm-sensing devices are used for low pressure applications. Piston-sensing devices are used for high pressure applications. See Figure 8-20.

Diaphragms

A *diaphragm* is a deflecting mechanism that moves when a force (pressure) is applied. One side of the diaphragm is connected to the pressure to be detected (source pressure) and the other side is vented to the atmosphere.

The diaphragm moves against a spring switch mechanism which operates electrical contacts when the source pressure increases. The spring tension is adjustable to allow for different pressure settings. A diaphragm pressure switch is used with pressures of less than 200 psi, but some are designed to detect several thousand pounds of pressure.

Bellows

A *bellows* is a cylindrical device with several deep folds which expand or contract when pressure is applied. One end of the bellows is closed and the other end is connected to the source pressure. The expanding bellows moves against a spring switch mechanism which operates electrical contacts when the source pressure increases.

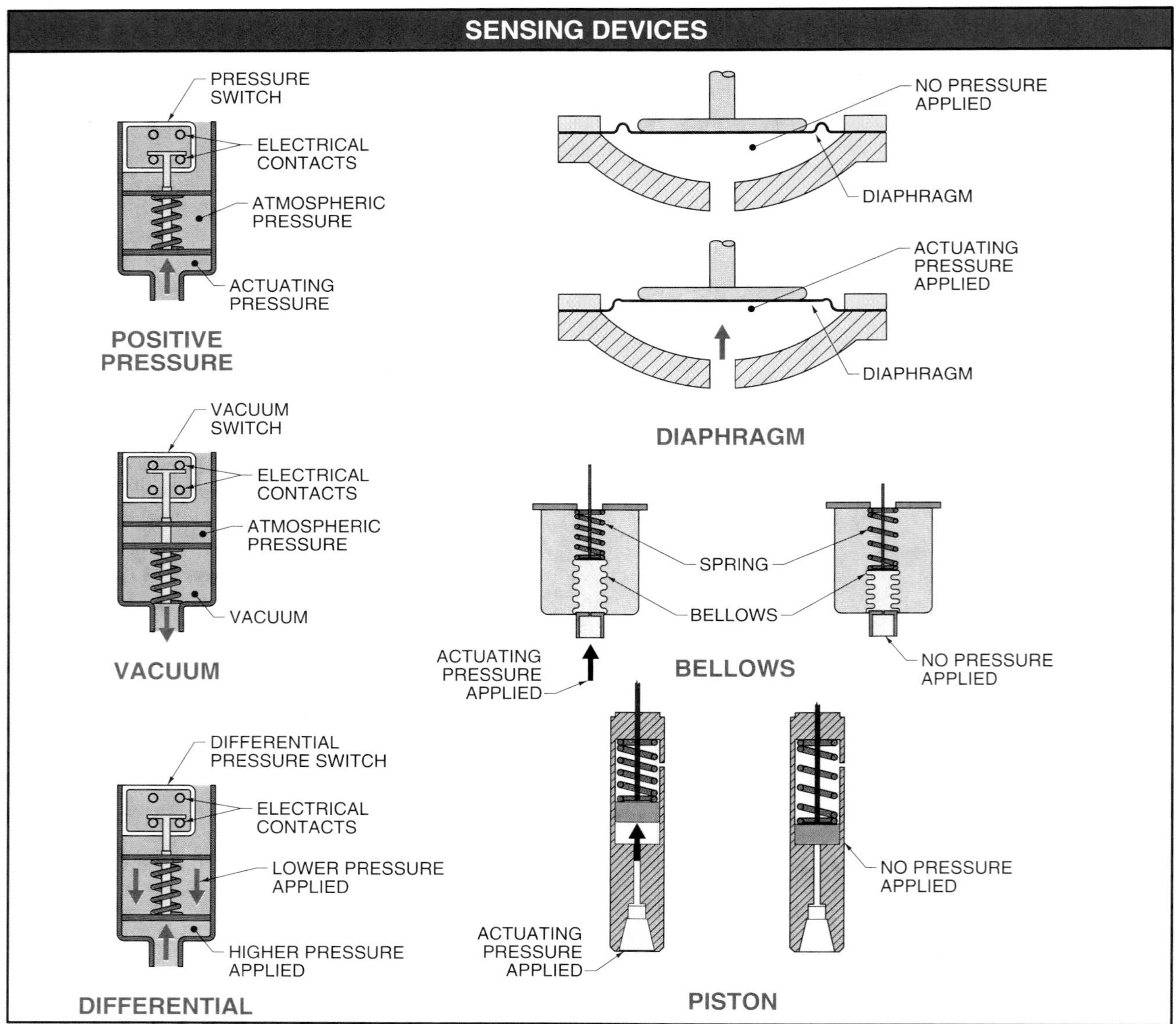

Figure 8-20. Pressure switches use different sensing devices to detect the amount of pressure.

The spring tension is adjustable to allow for different pressure settings. A bellows pressure switch is used with pressures of up to 500 psi, but some are designed for higher pressures.

Pistons

A *piston* is a cylinder that is moved back and forth in a tight-fitting chamber by the pressure applied in the chamber. A piston-operated pressure switch uses a stainless steel piston moving against a spring tension to operate electrical contacts. The piston moves a switch mechanism which operates electrical contacts when the source pressure increases.

The spring tension is adjustable to allow for different pressure settings. Piston-operated pressure switches are designed for high-pressure applications of 10,000 psi or more.

Dead Band

When a change in pressure occurs causing the diaphragm to move far enough to actuate the switch contacts, some of the pressure must be removed before the switch resets for another cycle. *Dead band (differential)* is the amount of pressure that must be removed before the switch contacts reset for another cycle after the setpoint has been reached and the switch has been actuated. See Figure 8-21.

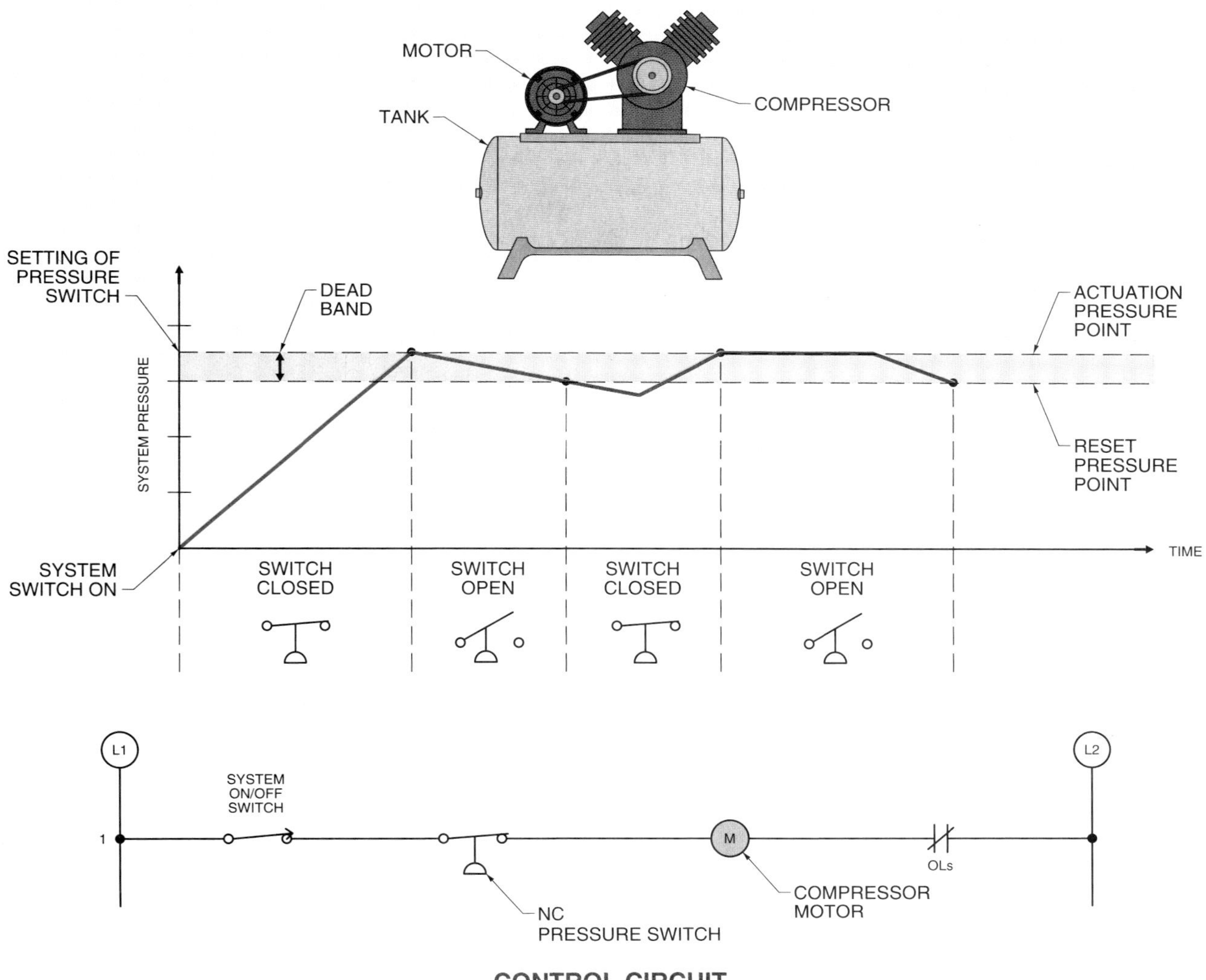

Figure 8-21. Dead band is the amount of pressure that must be removed before the switch contacts reset for another cycle after the setpoint has been reached and the switch has been actuated.

Dead band is inherent in all pressure, temperature, level, flow, and most automatically-actuated switches. Dead band is not a fixed amount, but is different at each setpoint. Dead band is minimum when the setpoint is at the low end of the switch range. Dead band is maximum when the setpoint is at the high end of the switch range.

Dead band may be beneficial or detrimental. Without a dead band range, or too small of one, electrical contacts chatter ON and OFF as a pressure switch approaches the setpoint. However, a large dead band is detrimental in applications that require the pressure to be maintained within a very close range. Different switches have different dead band ratings. Always check the amount of listed dead band when using pressure switches in different applications.

Pressure Switch Applications

Most pressure switches are used to maintain a predetermined pressure in a tank or reservoir. Pressure switches may also be used to sequence the return of pneumatic or hydraulic cylinders. See Figure 8-22. The two-position, 4-way, directional control valve solenoid is energized when the operator presses the start pushbutton. This changes the directional control valve from the spring position to the solenoid-actuated position. The cylinder advances because the flow of pressure is changed in the cylinder. The control relay energizes and its NO contacts close because it is in parallel with the solenoid. This adds memory to the circuit. The operator releases the pushbutton and the cylinder continues to advance. The cylinder advances until the preset pressure is reached on the pressure switch, which signals the return of the cylinder by de-energizing the solenoid and relay. The de-energizing of the solenoid returns the directional control valve to the spring position. The return of the directional control valve to the spring position reverses the flow in the cylinder which returns it. The return of the cylinder depends on the setting of the pressure switch. The setting of the pressure switch depends on the application of the cylinder. This setting may be low for packing fragile materials or high for forming metals.

The advantage of using a pressure switch over a pushbutton for returning the cylinder is that the load always receives the same amount of pressure before the cylinder returns. An emergency stop could be added to the control circuit for the manual return of the cylinder.

A low-range pressure switch may be used with a metal tubing arrangement in a fluidic sensor. See Figure 8-23. In this application, a constant low-pressure stream of air is directed at a sheet of material through the metal tubing. As long as the material in process is present, the air stream is deflected. The stream of air is sensed in the receiver tube if the material breaks. The pressure switch would signal corrective action through the control relay. The fluidic sensor has certain advantages over using a photoelectric sensor. For example, the air stream flowing over the material in normal operation may perform a second function such as cooling, cleaning, or drying the material. A fluidic sensor is also inexpensive compared to a photoelectric sensor because the only cost is the metal tube and the pressure switch.

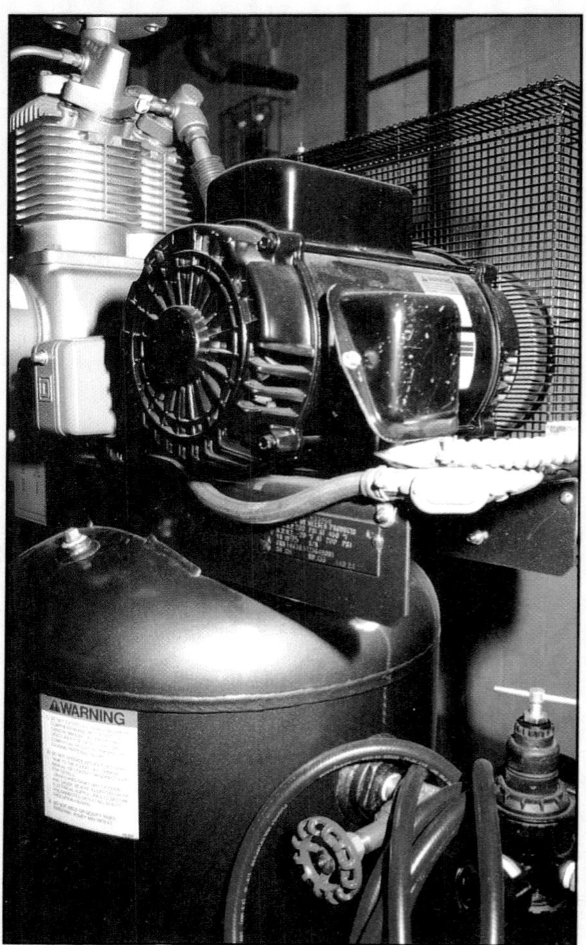

Pressure switches are used in air compressor circuits to cycle the motor ON and OFF to maintain the proper tank pressure.

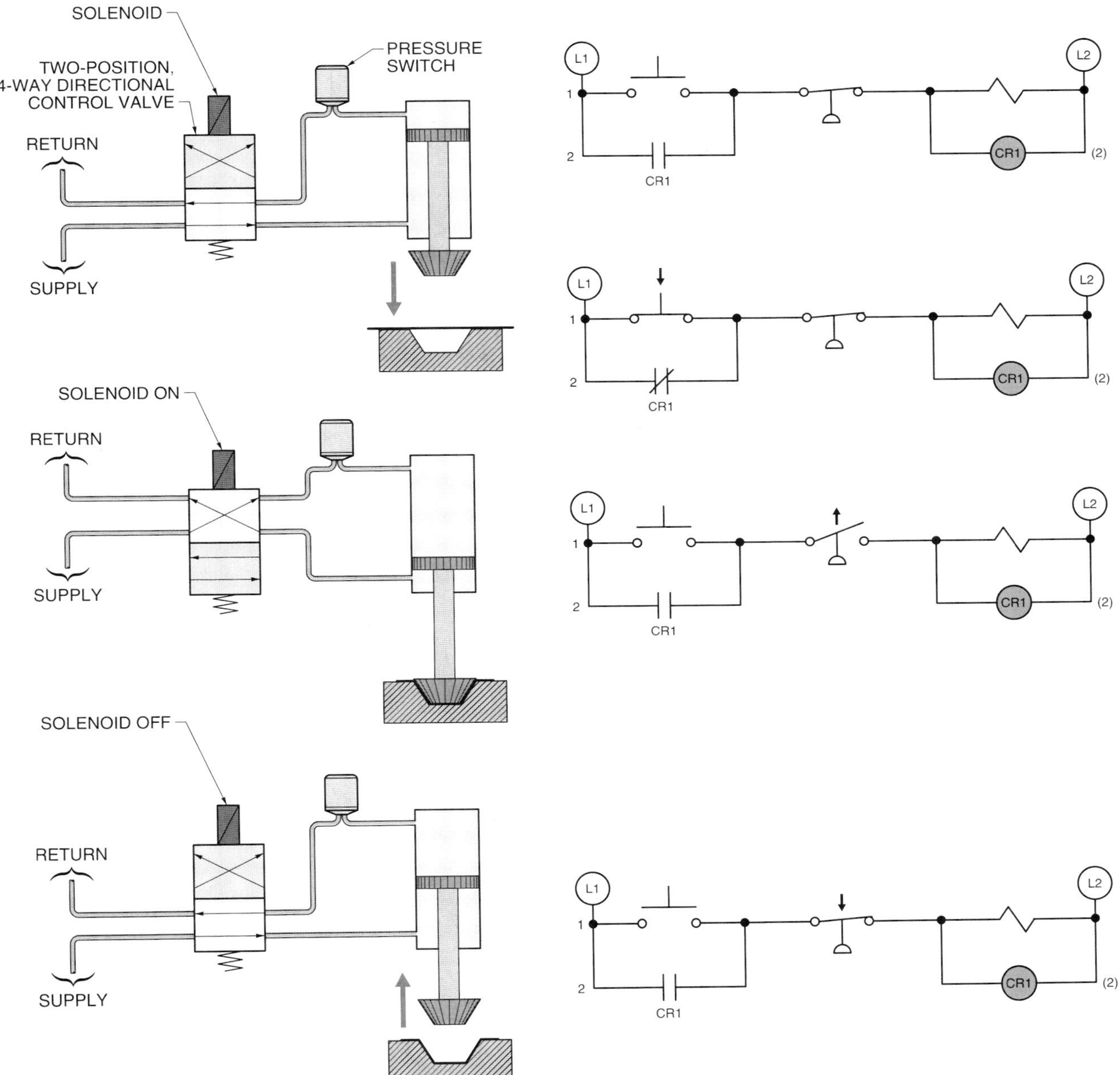

Figure 8-22. Pressure switches are used to maintain a predetermined pressure in a tank or reservoir or sequence the return of pneumatic or hydraulic cylinders.

TEMPERATURE SWITCHES

Temperature switches are control devices that react to heat intensity. Temperature switches are used in heating systems, cooling systems, fire alarm systems, process control systems, and equipment/circuit protection systems. In most applications, temperature switches react to rising or falling temperatures. Cooling systems, alarm systems, and protection systems use temperature switches that react to rising temperatures. Heating systems use temperature switches that react to falling temperatures.

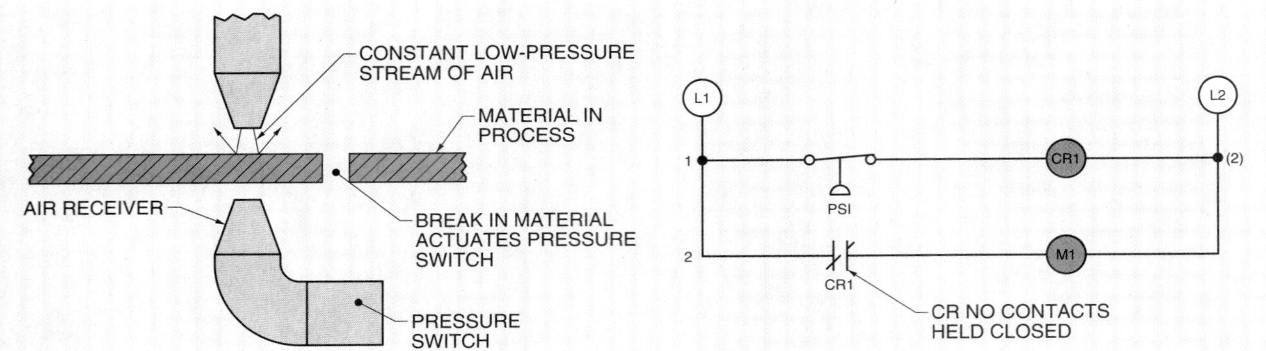

Figure 8-23. A low-range pressure switch may be used with a metal tubing arrangement in a fluidic sensor to determine the presence of a material in process.

A heating system maintains a set temperature when the ambient temperature drops without additional heat. In a heating system, as ambient temperature drops, the switch contacts close and turn ON the heat-producing device. The heat-producing device may be an electric coil, gas furnace, heat pump, or any device that produces heat. See Figure 8-24.

Lennox Industries Inc.

Temperature switches are wired into furnace control circuits to protect the system from overheating and control the firing of burners and the operation of heating elements.

The temperature switch energizes a heating contactor. The heating contactor energizes the heat-producing coils. By having the temperature switch control a contactor, the high current required by the heating coils does not pass through the contacts of the temperature switch. This reduces the size of the required temperature switch and increases the life of the contacts. A temperature switch may also sound an alarm if the temperature rises too high.

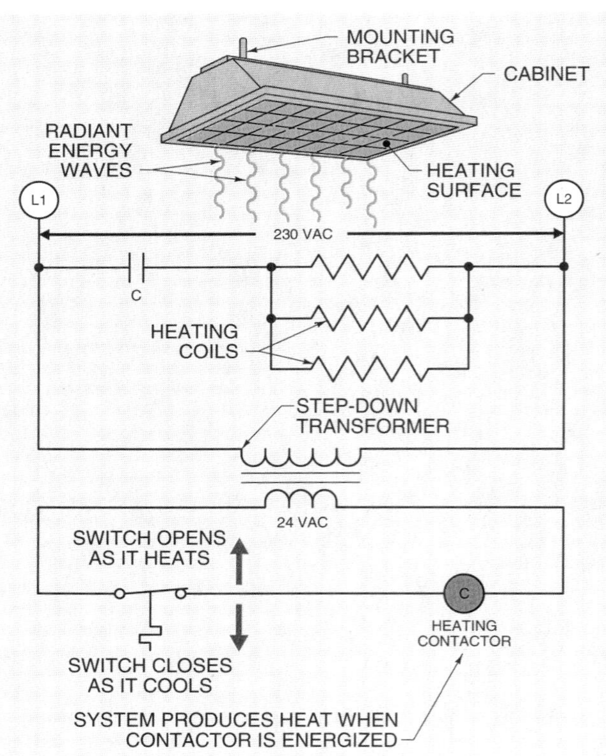

Figure 8-24. In a heating system, heat is produced when the temperature switch contacts cool and close.

A cooling system maintains a set temperature when the ambient temperature rises without additional cooling. In a cooling system, as the ambient temperature rises, the switch contacts close and turn ON a cooling device. The cooling device may be a standard air conditioning unit, cooling tower, radiator, or any device that is used to cool. See Figure 8-25.

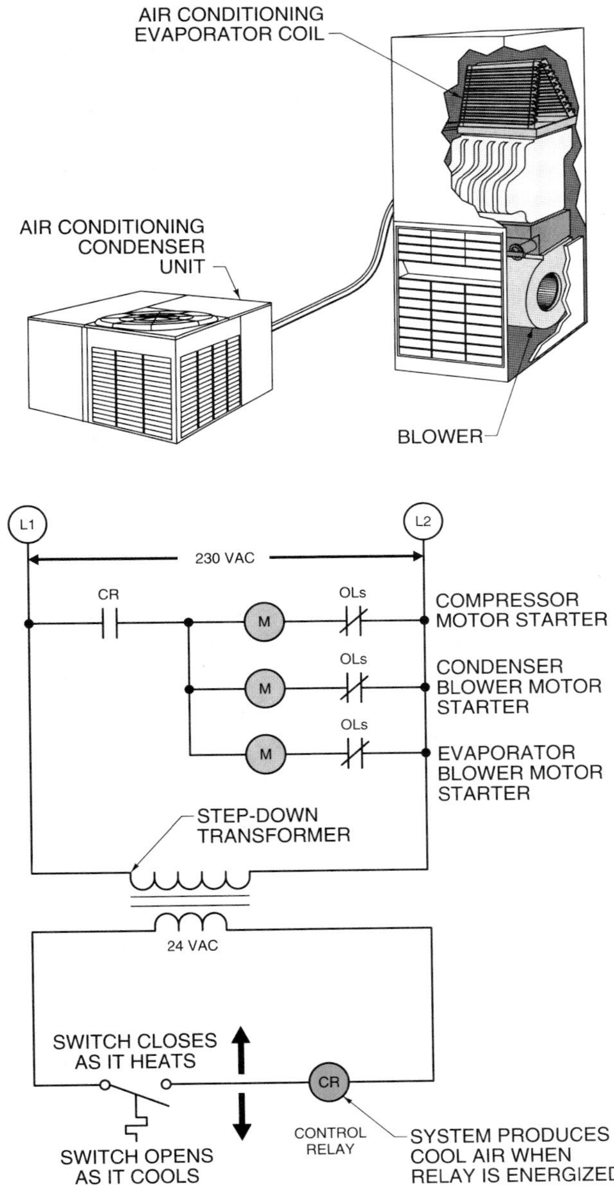

Figure 8-25. In a cooling system, cool air is produced when the temperature switch contacts heat and close.

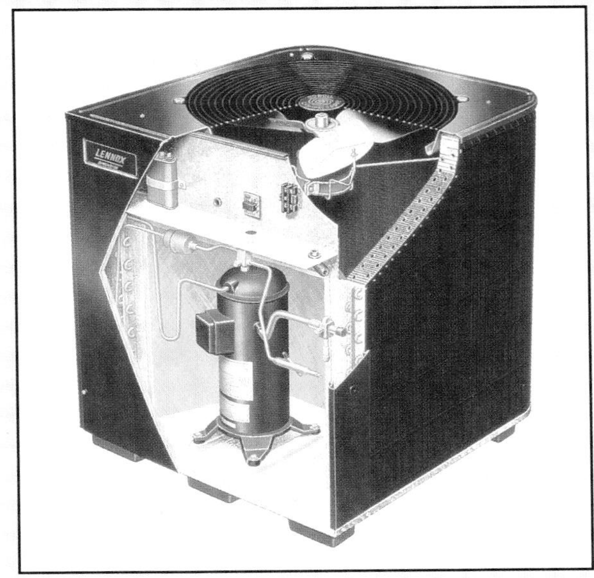

Lennox Industries Inc.

Temperature switches are used to turn ON a cooling system when the temperature in a building rises to a predetermined setpoint.

The temperature switch energizes a control relay. The control relay energizes the motor starters. The motor starters control the motors that produce cooling and provide overload protection for the motors. The control relay may also open a valve that allows water (or coolant) to circulate over the heated area.

Temperature switches may be activated by several different sensors. The different sensors used to activate electrical contacts in response to temperature changes include bimetallic, capillary tube, thermistor, and thermocouple. See Figure 8-26.

Bimetallic

A *bimetallic sensor* is a sensor that bends or curls when the temperature changes. It is made of two different metals bonded together. The two metals expand at different rates when heated. This causes the metal strip to bend into an arc which is used to open and close electrical contacts. To improve performance, most bimetallic sensors use a strip wound into a coil. The coil moves a bulb filled with liquid mercury. The mercury moves across a set of electrical contacts, completing an electrical circuit, when the coil expands.

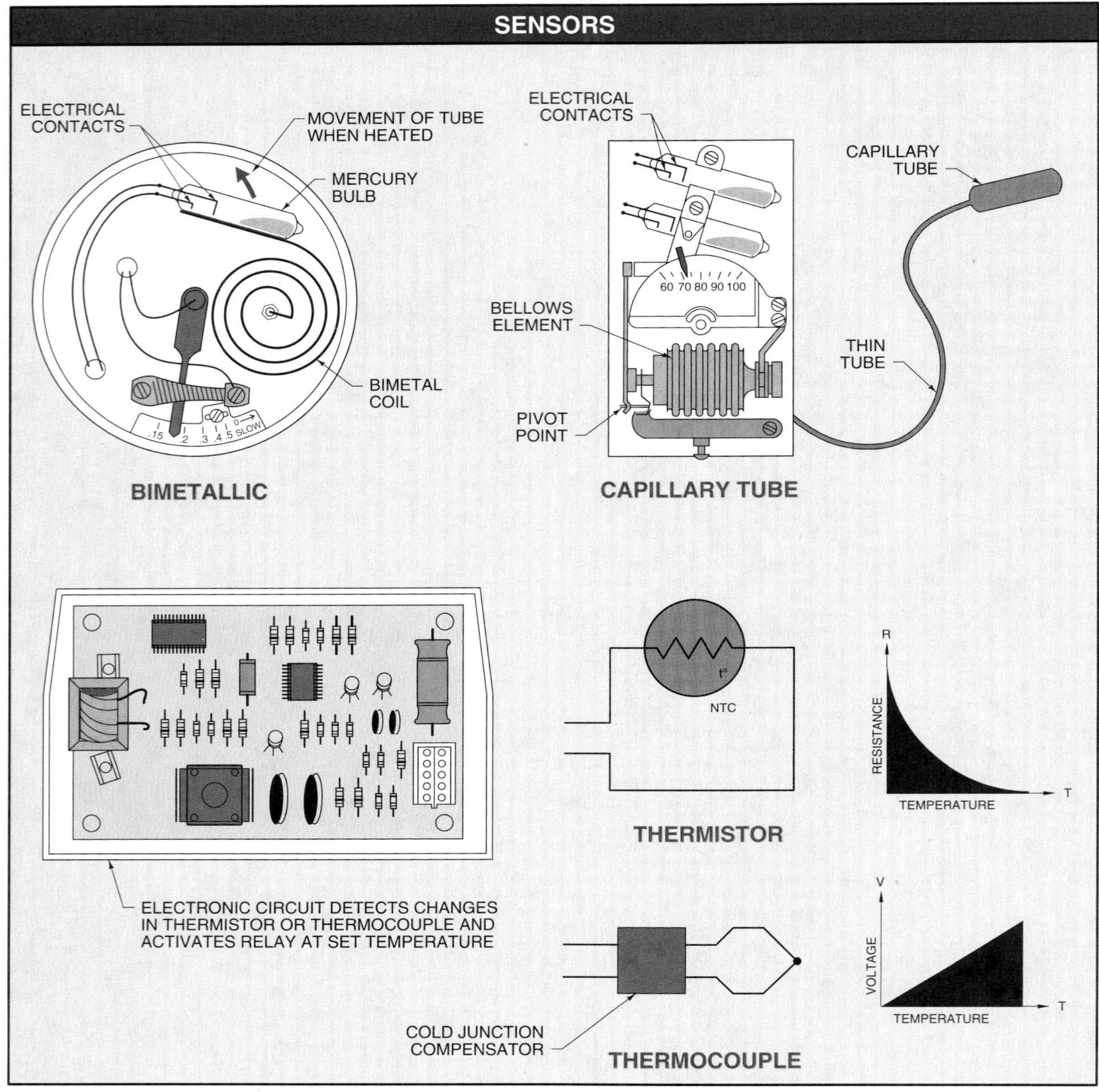

Figure 8-26. The different sensors used to activate electrical contacts in response to temperature changes include bimetallic, capillary tube, thermistor, and thermocouple.

Capillary Tubes

A *capillary tube sensor* is a sensor that changes internal pressure with a change in temperature. A capillary tube sensor uses a tube filled with a temperature-sensitive liquid. The pressure in the tube changes in proportion to the temperature surrounding the tube. As the temperature rises, the pressure in the tube increases. As the temperature decreases, the pressure in the tube decreases. The pressure change inside the tube is transmitted to a bellows inside the temperature control through a capillary tube. The movement of the bellows activates electrical contacts, completing an electrical circuit.

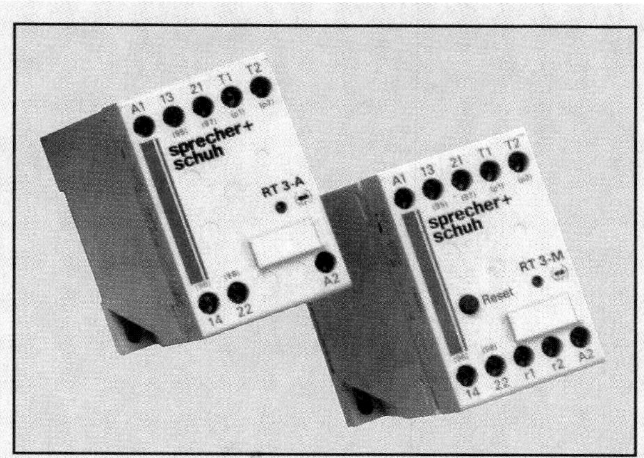

Sprecher + Schuh

Series RT3 thermistor protection relays from Sprecher + Schuh are designed for use in applications where exact temperature monitoring is critical.

Thermistors

A *thermistor* is a temperature-sensitive resistor that changes its electrical resistance with a change in temperature. Thermistors have either a positive temperature coefficient (PTC) or a negative temperature coefficient (NTC). PTC thermistors increase in resistance when the temperature increases. NTC thermistors increase in resistance when the temperature decreases. The thermistor is placed at the point where the temperature is to be measured, and is connected to an electronic circuit which is set to respond to the changes in resistance. The electronic circuit activates a relay at the set temperature, completing an electrical circuit.

Liquid, Air, or Surface Temperature Control Applications. Thermistors are available which monitor the temperature of liquid, air, or a surface temperature. See Figure 8-27. Any one or all may be used depending on the application.

A thermistor is used to monitor and control liquids in many applications. This is one of the most common process controls used in industry. The temperature of many liquids must be at a set point before a mixing or fill process can start. In other applications, a process may have to be stopped if the liquid is too cool or too hot. Thermistors are available in a wide range of temperatures with extreme ranges from −400°F to 3200°F (−240°C to 1760°C).

Thermistors that sense air temperatures are available in a wide range of temperatures, with the most common in the −30°F to 150°F range. These sensors are used mostly in heating and air conditioning control to maintain a desired temperature of a room or a storage unit.

Thermistors that sense surface temperatures are normally designed to attach to a metallic surface. They can be used to detect an ice build-up in an air conditioning system or heat build-up in many other processes. The advantage of surface temperature sensors is that they are easy to install on pipes without having to open the system.

Thermocouples

A *thermocouple* is a temperature sensor of two dissimilar metals joined at the end where heat is to be measured, which produces a voltage output at the other end proportional to the measured temperature. Thermocouples work on the principle that two different metals connected together produce a voltage when the junction is heated. A cold junction compensator is required to monitor this voltage change. The cold junction compensator establishes a reference point between the measuring point and a given point. The cold junction compensator is connected to an electronic circuit, which is set to respond to the changes in voltage. The electronic circuit activates a relay at the set temperature.

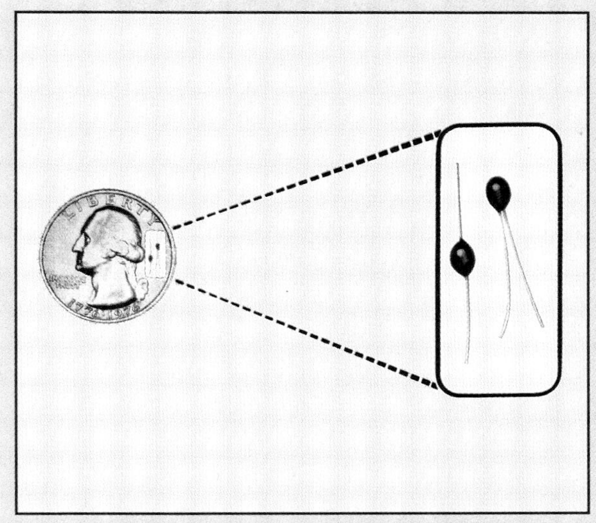

Thermometrics Inc.

NTC thermistors from Thermometrics Inc. are suited for applications such as gas chromatography, gas flow measurement, thermal conductivity analysis, and liquid level sensing.

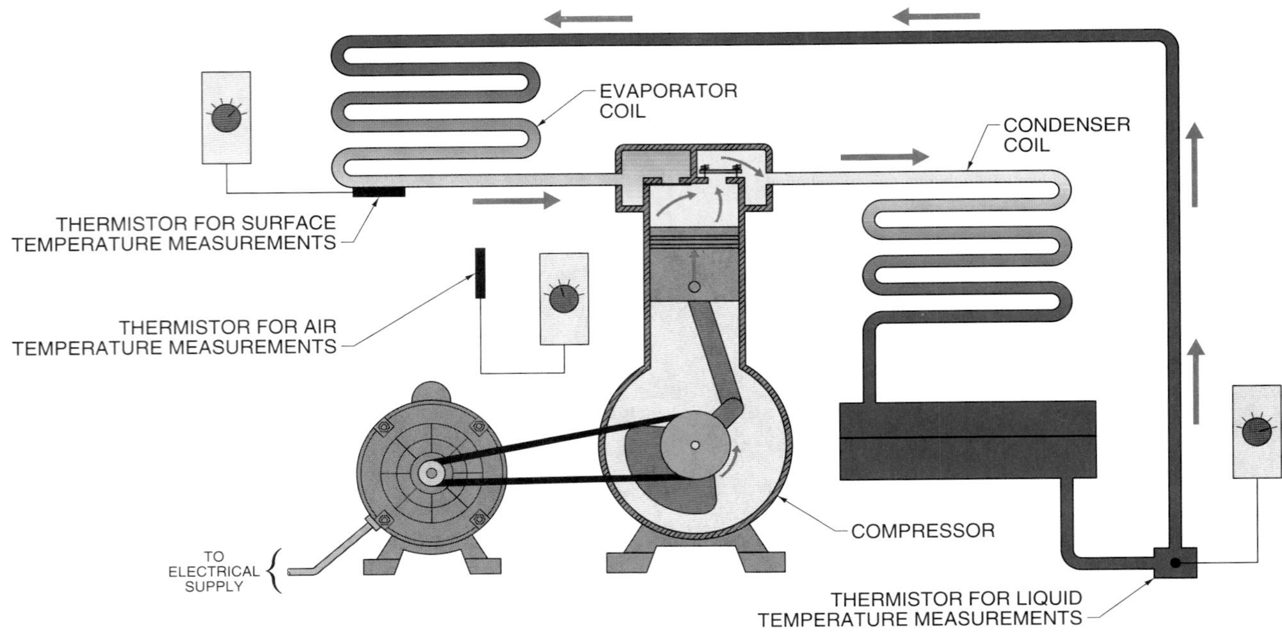

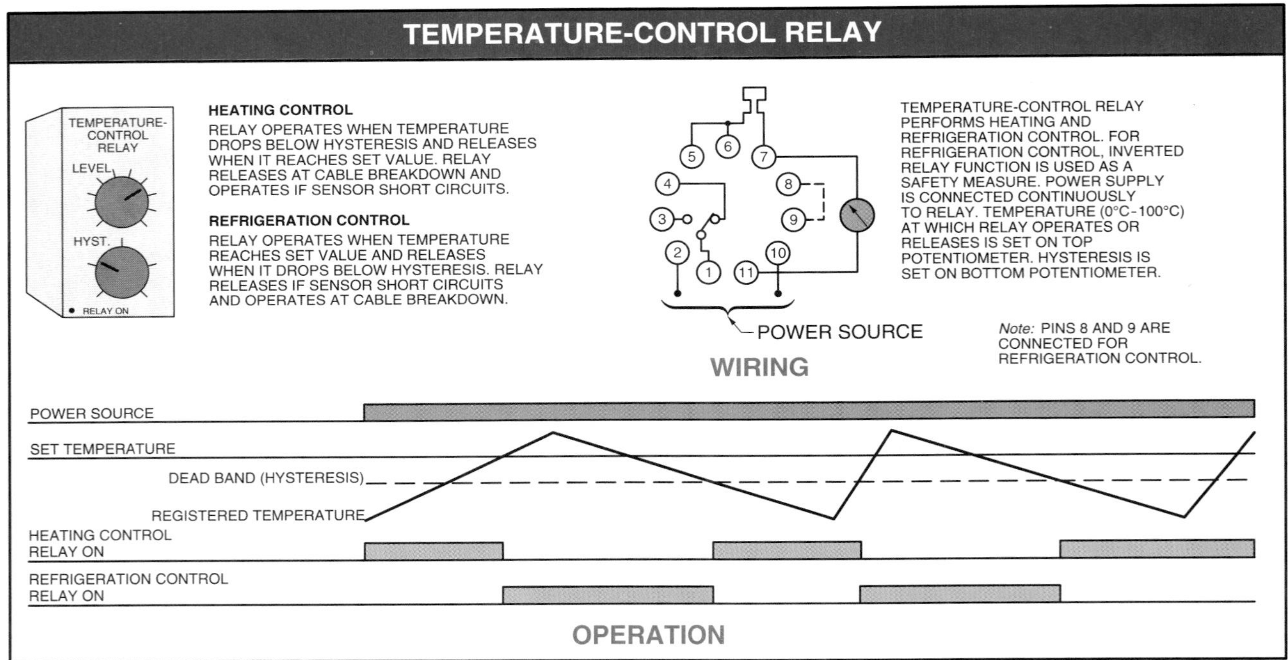

Figure 8-27. Thermistors are available that monitor the temperature of liquid, air, or a surface temperature.

FLOW SWITCHES

Flow is the travel of fluid in response to a force caused by pressure or gravity. The fluid may be air, water, oil, or some other gas or liquid. Most industrial processes depend on fluids flowing from one location to another. Problems may occur if the flow is stopped or slowed. Flow may be stopped by a frozen pipe, clogged pipe, or an improperly closed valve (manual or automatic).

A *flow switch* is a control switch that detects the movement of a fluid. See Figure 8-28.

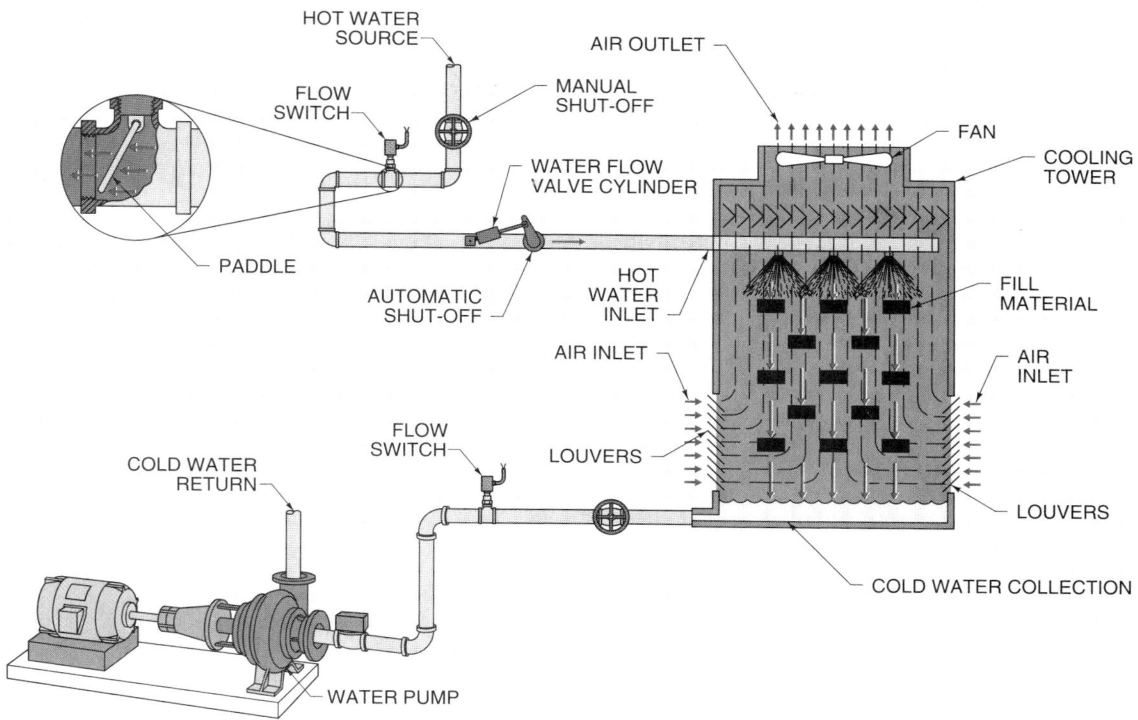

Figure 8-28. A flow switch is a control switch that detects the movement of a fluid.

Applications that use flow switches to detect the presence or absence of flow include:

- Boilers
- Cooling lines
- Air compressors
- Fluid pumps
- Food processing systems
- Machine tools
- Sprinkler systems
- Water treatment systems
- Heating processes
- Refrigeration systems
- Chemical processing and refining

Flow switches use different methods to detect if the fluid is flowing. The methods used to detect if the fluid are flowing include the paddle and transmitter/receiver methods. In the paddle method, a paddle extends into the pipe or duct. The paddle moves and actuates electrical contacts when the fluid flow is sufficient to overcome the spring tension on the paddle. The spring tension is adjustable on many flow switches, allowing for different flow rate adjustments.

In the transmitter/receiver method, a transmitter sends a signal through the pipe. The receiver picks up the transmitted signal. The strength of the signal changes when the product is flowing. One common transmitter/receiver method uses a sound transmitter to produce sound pulses through the fluid. Moving solids or bubbles in the fluid reflect a distorted sound back to the receiver. The transmitter/receiver unit is adjustable for detecting different flow rates.

Both NO and NC electrical contacts are used with flow switches. See Figure 8-29. In some applications, a flow of fluid indicates a problem. For example, in an automatic sprinkler system used as fire protection, the flow of water indicates a problem. In this application, an NO contact on the flow switch could be used to sound an alarm. When a fire (or high heat) opens the sprinkler head, the water starts flowing through the pipe. The flow of water closes the NO contacts and the alarm sounds. The alarm sounds as long as the water is flowing.

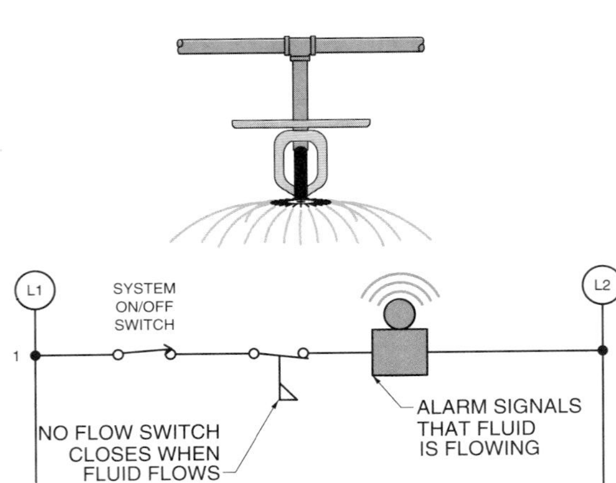

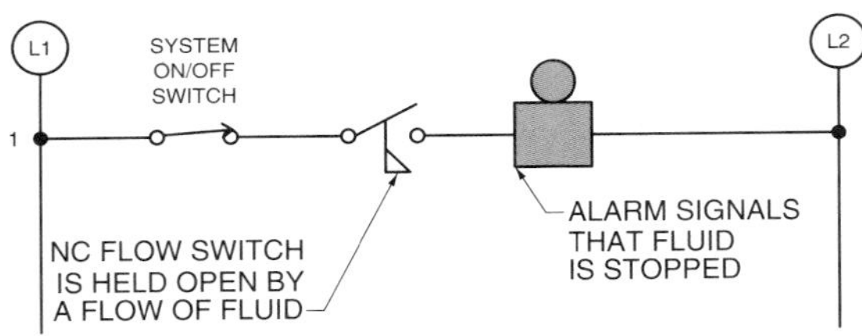

Figure 8-29. Both NO and NC electrical contacts are used with flow switches.

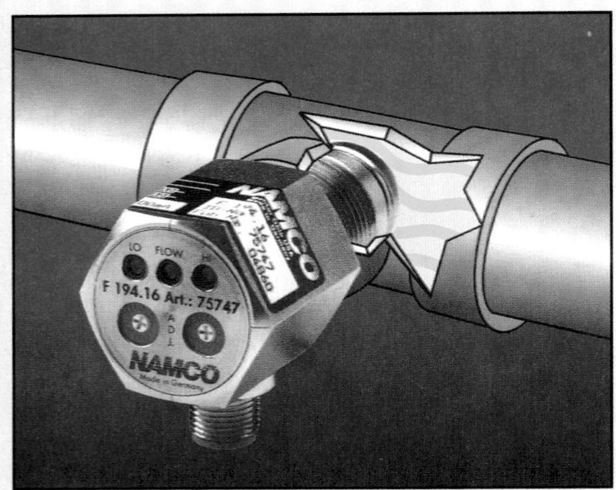

Namco Controls Corporation

The electronic flow monitors manufactured by Namco Controls Corporation are self-contained sensors that detect variation in media flow based on thermal conduction.

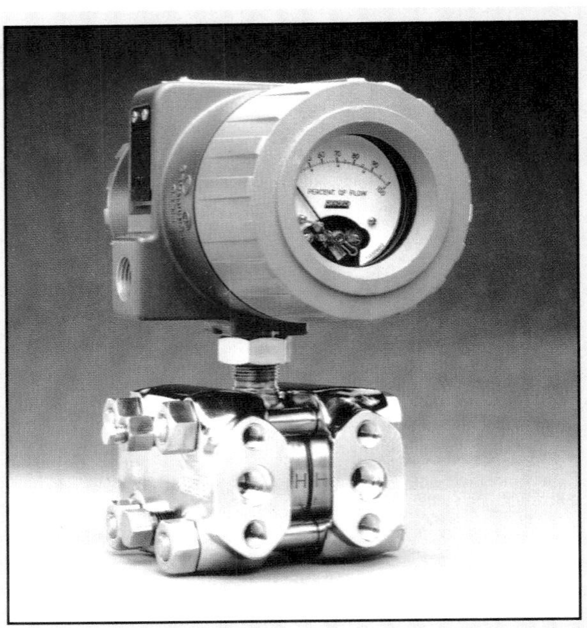

The Foxboro Company

The electronic differential pressure transmitter from Foxboro is commonly used for flow measurement.

An NC contact is used to signal when a fluid is not flowing. The NC contact may be used to sound an alarm if fluid stops flowing. When fluid is flowing, the NC contacts are held open by the fluid flow.

A flow switch may also be used to detect air flow across the heating elements of an electric heater. See Figure 8-30. The heating elements burn out if sufficient air flow is not present. The flow switch is used as an economical way to turn the heater OFF any time there is not enough air flow. This circuit can also be applied to an air conditioning or refrigeration system. In this circuit, the flow switch is used to detect insufficient air flow over the refrigeration coils. The restricted air flow is normally caused by the icing of the coils, which blocks the air flow. In this case, the flow switch would automatically start the defrost cycle of the refrigeration unit.

Flow switches may also be used to detect the proper air flow in a ventilation system. See Figure 8-31. The ventilation system may be directing dangerous gases away from the operator. Poisonous gases could overcome the operator or damage could occur to the process involved if there is insufficient air flow.

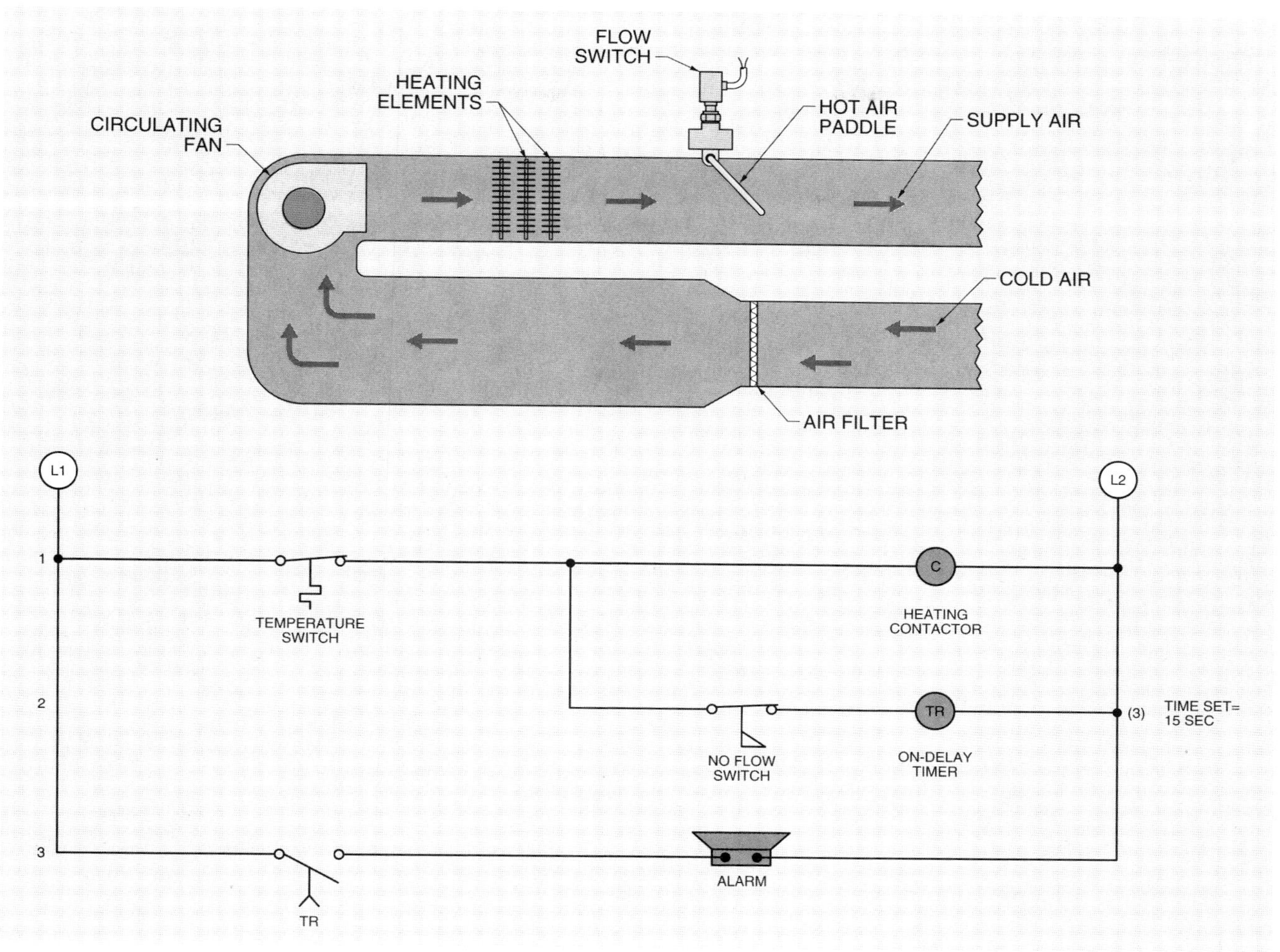

FLOW SWITCH
HEATING ELEMENTS
CIRCULATING FAN
HOT AIR PADDLE
SUPPLY AIR
COLD AIR
AIR FILTER
L1
L2
1
TEMPERATURE SWITCH
C
HEATING CONTACTOR
2
NO FLOW SWITCH
TR
ON-DELAY TIMER
(3)
TIME SET= 15 SEC
3
TR
ALARM

Figure 8-30. A flow switch may be used to determine if sufficient air is flowing across the heating elements of an electric heater.

L1
L2
STOP
START
1
M
ALL OLs
(2)
BLOWER MOTOR STARTER
2
M
3
ALARM
NO FLOW SWITCH
BLOWER
FLOW SWITCH
VAPORS

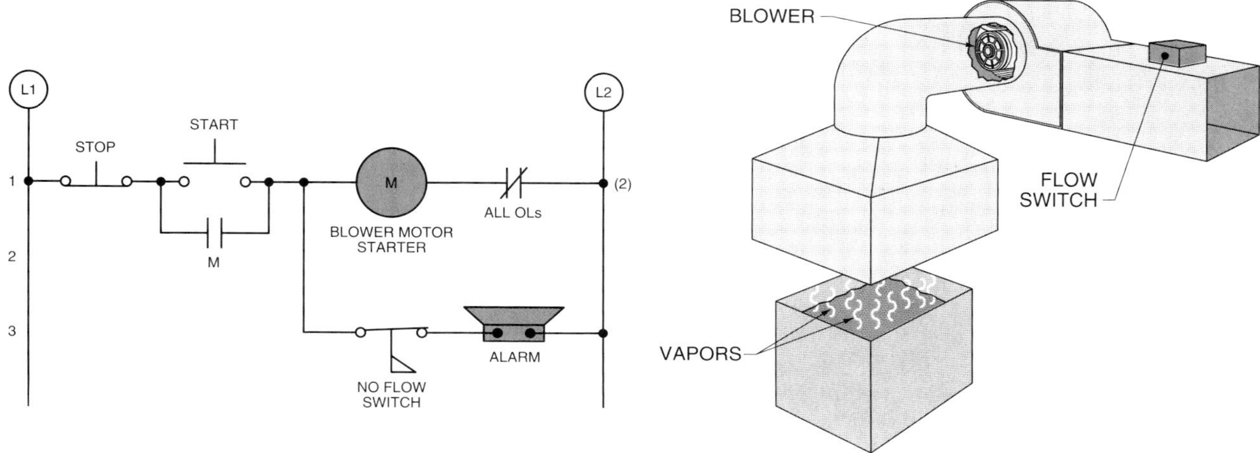

Figure 8-31. A flow switch may be used to maintain a critical ventilation process.

Flow switches are often used to protect the large motion picture projector used in theaters where poor air flow would cause heat build-up and reduce the life expectancy of the expensive bulbs used in projectors. Air flow may be restricted from a large draft caused by high winds outside of the building or by clogged air filters in the intake system.

A flow switch may be used to advance a clogged filter based on restricted air flow. See Figure 8-32. The flow switch is used to start a gear-reduced motor that slowly advances the roll of filter material until sufficient air flow is present.

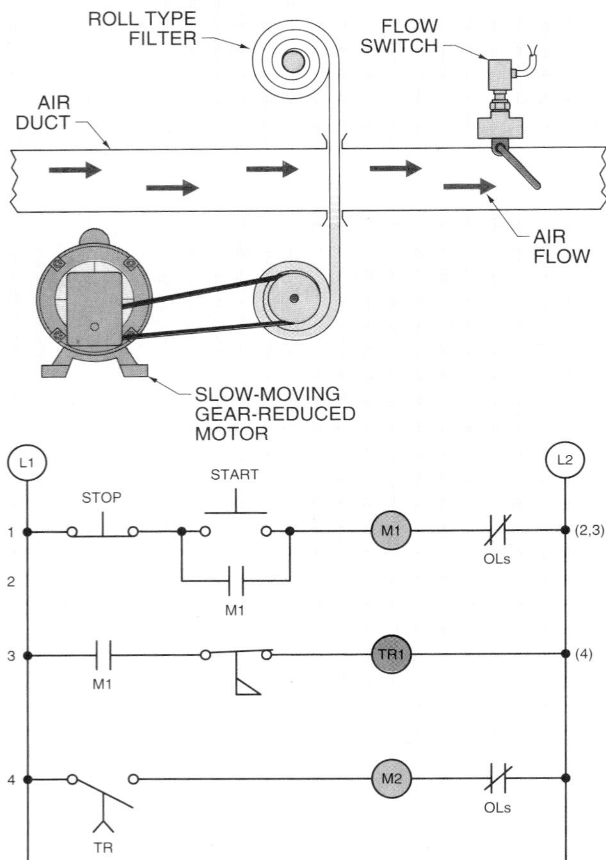

Figure 8-32. A flow switch may be used to advance a clogged filter based on restricted air flow.

SMOKE/GAS SWITCHES

Smoke/gas switches detect vapor. A *vapor* is a gas that can be liquefied by compression without lowering the temperature. A *smoke switch (smoke detector)* is a switch that detects a set amount of smoke caused by smoldering or burning material and activates a set of electrical contacts. See Figure 8-33. A smoke switch is used as an early-warning device in fire protection systems. They are available in self-contained units such as the common units used in most houses, or as industrial units used as part of a large fire protection system.

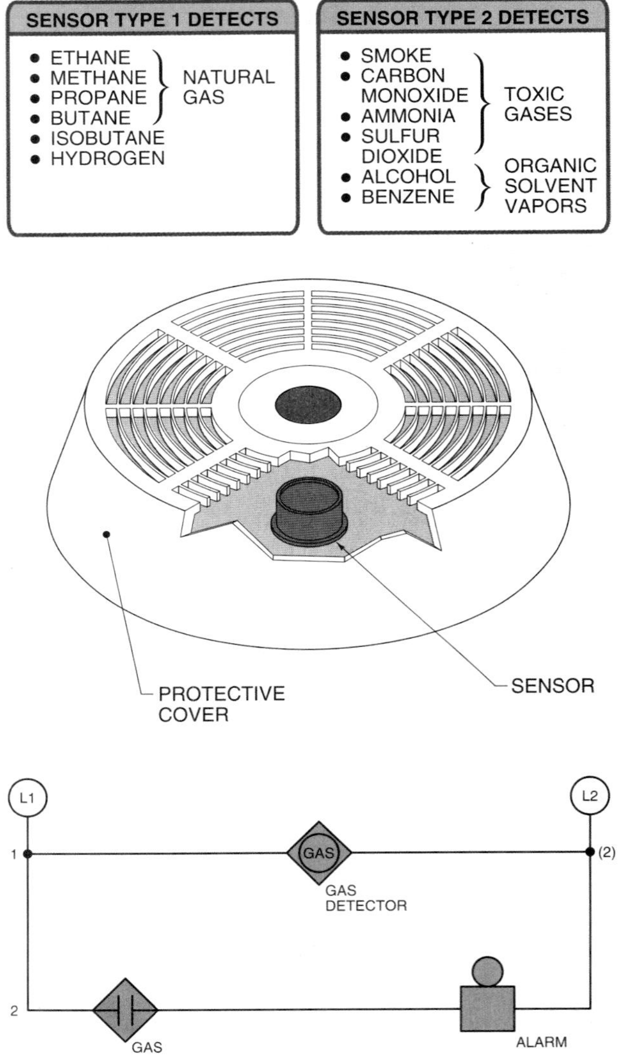

Figure 8-33. A smoke switch (smoke detector) is a switch that detects a set amount of smoke caused by smoldering or burning material and activates a set of electrical contacts.

A *gas switch (gas detector)* is a switch that detects a set amount of a specified gas and activates a set of electrical contacts. Many gas detectors have interchangeable sensor units that can detect different groups of gases. For example, one common gas sensor is designed to detect gases such as propane and butane, but not carbon monoxide and smoke. This gas detector is used in applications such as detecting leaks in areas that have (or fill) propane tanks. Another sensor is designed to detect toxic gases such as carbon monoxide and ammonia. This gas detector is used in applications such as detecting high carbon monoxide levels in an area with operating combustion engines.

LEVEL SWITCHES

In most industrial plants there are tanks, vessels, reservoirs, and other containers in which process water, waste water, raw materials, or product must be stored or mixed. The level of the product must be controlled. A *level switch* is a switch that detects the height of a liquid or solid (gases have no level) inside a tank.

Systems that use level switches include: milk, water, oil, beer, wine, solvents, plastic granules, coal, grains, sugar, chemicals, and many other product processing systems. Different level switches are used to detect each product. Factors that determine the correct level switch to use for an application include:

- Motion – Turbulence causes some level switches to chatter ON and OFF or actuate falsely.
- Corrosiveness – Level switches are made of different materials such as stainless steel, copper, plastic, etc. Always use a level switch made of a material that is compatible with the product to be detected.
- Density – All solid materials have a certain density. Capacitive level switches are designed to detect different amounts of density.
- Physical state – The physical state of a liquid depends on its type, temperature, and condition. Any liquid or solid may be detected if the correct level switch is used. Different level switches are designed to operate at different temperatures and to detect different types and thicknesses of liquids.
- Movement – A moving product may require a special level switch. A product that is stationary for a long period may cause certain mechanical level switches to stick.
- Conductivity – Some level switches with metal probes placed in a liquid depend on the liquid to be a conductor for proper operation.
- Abrasiveness – Non-contact level switches should be used with abrasive products.
- Sensing distance – Some level switches are designed to detect short distances and others are designed to detect long distances.

All level switches are designed to detect a certain range of materials. Some level switches can only detect liquids, others can detect both liquids and solids. Some level switches must come in direct contact with the product to be detected, others do not have to make contact. The level switch used depends on the application, cost, life expectancy, and product to be detected. The different level switches include mechanical, magnetic, conductive probe, capacitive, and optical.

Mechanical

Mechanical level switches were the first level switches used and are still one of the most common. *Mechanical level switches* are level switches that use a float which moves up and down with the level of the liquid and activates electrical contacts at a set height. See Figure 8-34. Mechanical level switches may be used with many different liquids because the float is the only part of the switch that is in contact with the liquid. Mechanical level switches work well with water (even dirty water) and any other liquid that dries without leaving a crust. Mechanical limit switches are not used with paint because the paint builds up on the float as the paint dries. The dried paint weighs the float down and affects the operation of the switch.

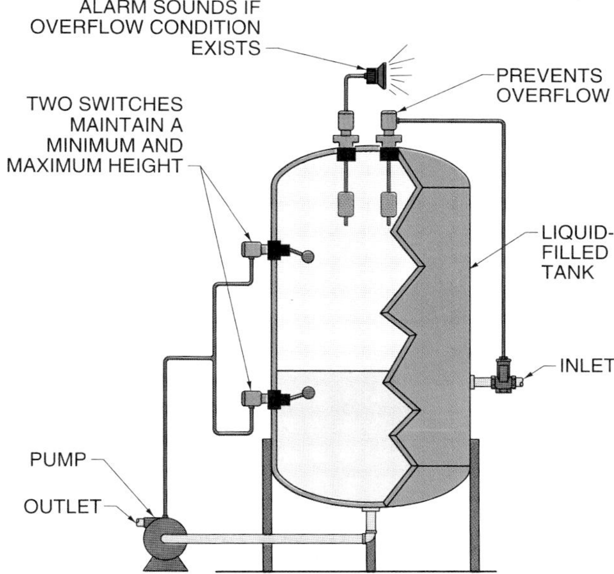

Figure 8-34. A mechanical level switch uses a float which moves up and down with the level of the liquid and activates electrical contacts at a set height.

One of the most common applications of a mechanical level switch is in sump pumps found in most houses with basements. In a sump application, the level switch turns ON a pump when the level reaches a set height.

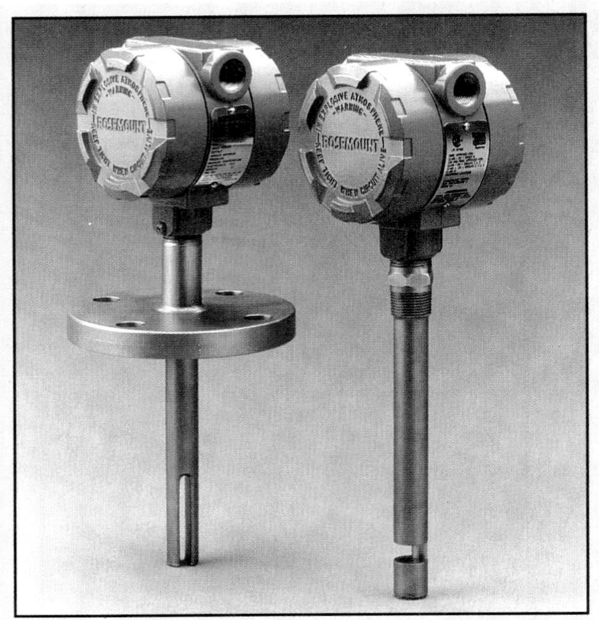

Kay-Ray/Sensall, Inc.

Sensall's Sapphire™ liquid-level sensors use state-of-the-art ultrasonic and electronic technologies to produce a reliable and technically advanced level switch.

Magnetic

A *magnetic level switch* is a switch that contains a float, a moving magnet, and a magnetically-operated reed switch to detect the level of a liquid. The float moves with the level of the liquid. A permanent magnet inside the float moves up and down with the liquid. The magnet passes alongside a magnetically-operated reed switch as it moves with the level of the liquid. The reed switch contacts change position when the magnetic field is present. The reed switch contacts return to their normal position when the magnetic field moves away. An advantage of using a magnetic level switch is that several individual switches may be placed on one housing. See Figure 8-35.

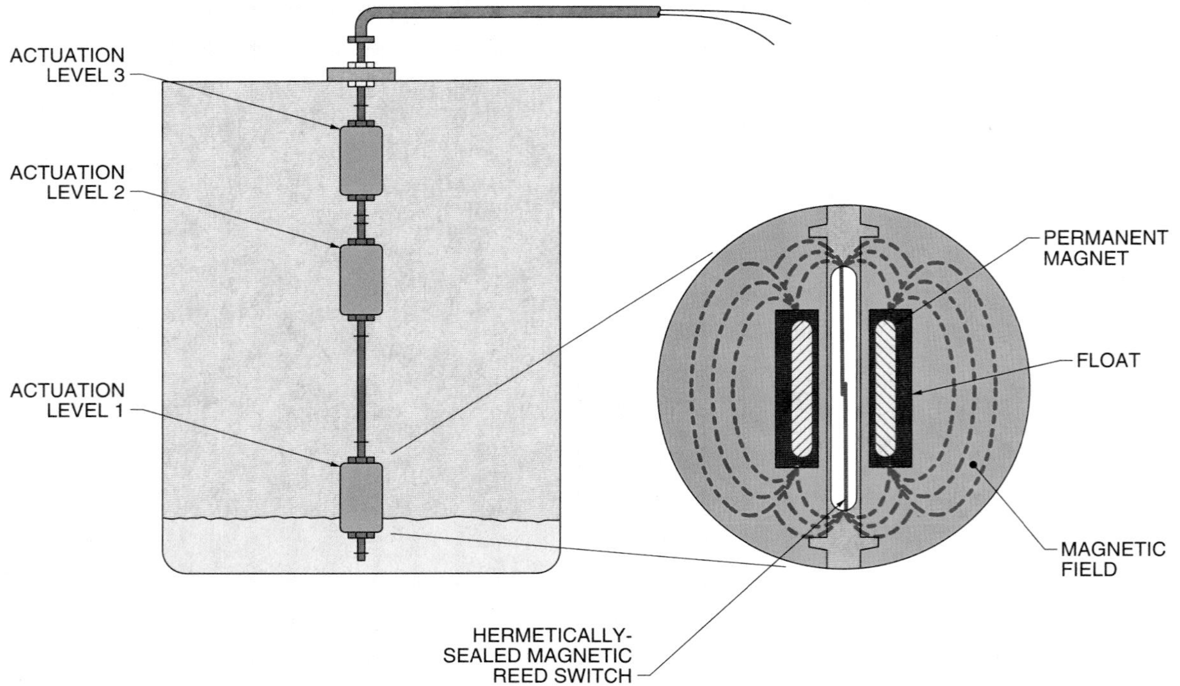

Figure 8-35. A magnetic level switch uses a magnetically-operated reed switch and a moving magnet to detect the level of a liquid.

Conductive Probe

A *conductive probe level switch* is a level switch that uses liquid to complete the electrical path between two conductive probes. The voltage that is applied to the probes is 24 V or less. The liquid must be a fair to good electrical conductor. All water-based solutions conduct electricity to some degree. The conductance (decreased resistance) of water increases as salts and acids are added to the water.

A fluid with an electrical resistance of less than 25 kΩ allows a sufficient amount of current to pass through it to actuate the relay inside the conductive probe level control. The electrical resistance is high and no current passes through the probes when the conductive liquid is no longer between the two probes. See Figure 8-36.

The number of probes used and their length depends on the application. Two probes of the same length may be used to detect a liquid at a given height. The relay inside the level control is activated when the liquid reaches the probes. The relay is no longer activated when the liquid is no longer in contact with the probes.

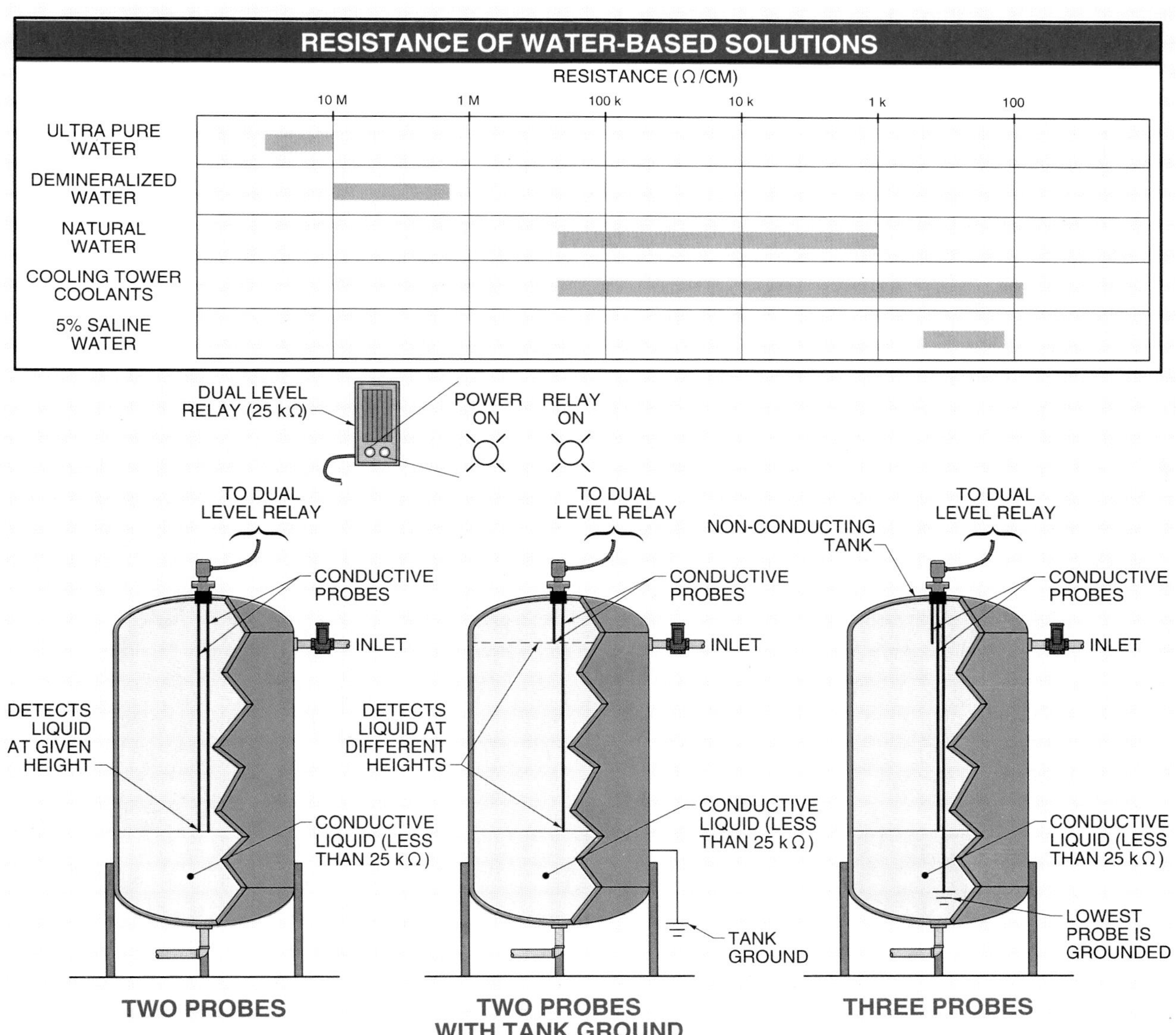

Figure 8-36. A conductive probe level switch uses liquid to complete the electrical path between two conductive probes.

Two probes of different lengths and a ground may be used to detect a liquid at different heights. The ground is connected to the conductive tank. The relay is activated when the liquid reaches the highest probe. The relay is not de-activated until the liquid no longer is in contact with the lowest probe. The probes may be any distance apart. Three probes of different lengths are used if the tank that holds the liquid is made of a non-conductive material. The lowest probe is connected to the ground wire of the level control.

Capacitive

A *capacitive level switch* is a level switch that detects the dielectric variation when the product is in contact (proximity) with the probe and when the product is not in contact with the probe. *Dielectric variation* is the range at which a material can sustain an electric field with a minimum dissipation of power. Capacitive level switches are used to detect solids or granules such as sand, sugar, grain, and chemicals in addition to some liquids. The capacitance of the sensor is changed when the product comes in proximity with the sensor. Materials with a dielectric constant of 1.2 or greater can be sensed. Some capacitive sensors are adjustable to allow only certain products to be sensed. Capacitive level switches are available that work well with hard-to-detect products such as plastic granules, shredded paper, copying machine toner, and fine powders. See Figure 8-37.

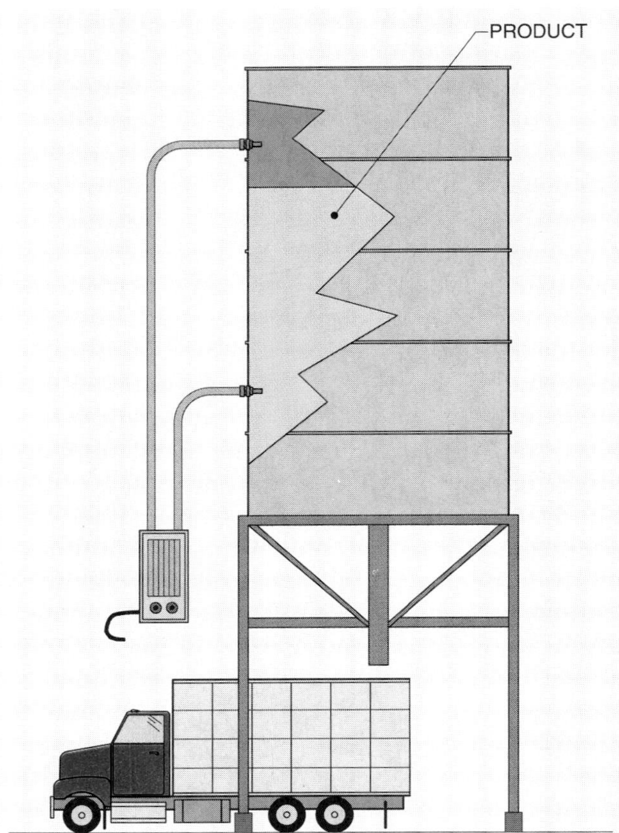

Figure 8-37. A capacitive level switch detects the dielectric variation when the product is in contact with the probe and when the product is not in contact with the probe.

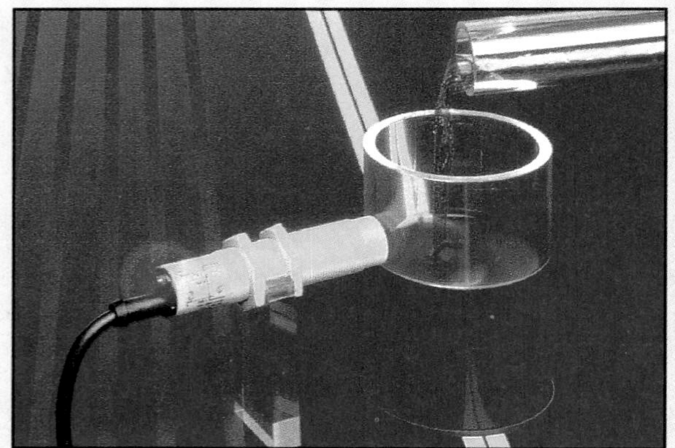

Carlo Gavazzi Inc. Electromatic Business Unit

Capacitive level switches are used in applications that require liquid detection through a container wall.

Optical

Optical level switches are level switches that use a photoelectric beam to sense the liquid. These switches are normally enclosed in a corrosion-resistant housing that makes them ideal for certain liquids. They are used to detect liquids such as oil, gas, beer, wine, milk, alcohol, and many acids. Because of the housing material, the sensor can be safely washed down with very hot (up to 212°) water or solvents.

The tip of the sensor forms a 90° angle that acts as a prism, reflecting the transmitted light beam to the receiver through the air. The beam is reflected into the liquid and a relay inside the level switch is activated if the sensor tip is immersed in a liquid having a refractive index different from that of air. See Figure 8-38.

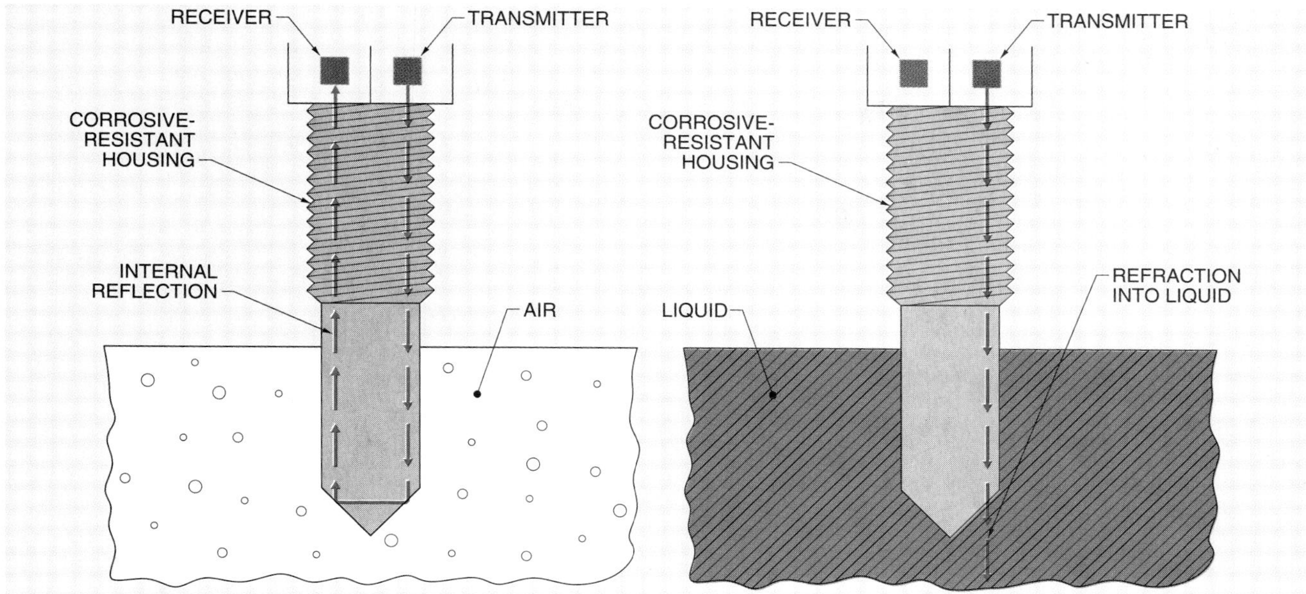

Figure 8-38. Optical level switches use a photoelectric beam to sense the liquid.

Charging and Discharging

Level switches detect and respond to the level of a material in a tank. The response is normally to charge or discharge the tank. Charging a tank is also known as pump control and discharging a tank is also known as sump control.

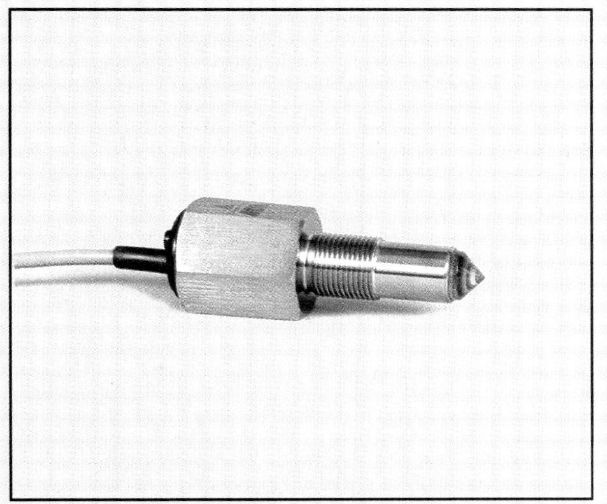

Carlo Gavazzi Inc. Electromatic Business Unit

Optical level switches are normally manufactured in corrosion-resistant housings which make them ideal for many hazardous location applications.

In a charging application, the level in a tank is maintained. As liquid is removed from the tank, the level switch signals the circuit to add liquid. Liquid may be added by opening a valve or starting a pump motor. The liquid is added until the level switch detects the correct height. Flow is stopped when the level switch detects the correct height. Liquid is added when the level switch is no longer in contact with the liquid. See Figure 8-39.

In a discharging application, the liquid in a tank is removed once it reaches a predetermined level. Liquid may be removed by a pump or through gravity when a valve is opened. The liquid is removed until the level switch detects the tank is empty.

One- or Two-Level Control

In level control applications, the distance between the high and low level must be considered. This distance may be small or large. In applications using a one-level switch, the distance is small. In applications using a two-level switch, the distance may be any length. See Figure 8-40. Although it may appear that a small distance maintained in a system is the best, this may not be true because the smaller the distance to be maintained, the greater the number of times the pump motor must cycle ON and OFF.

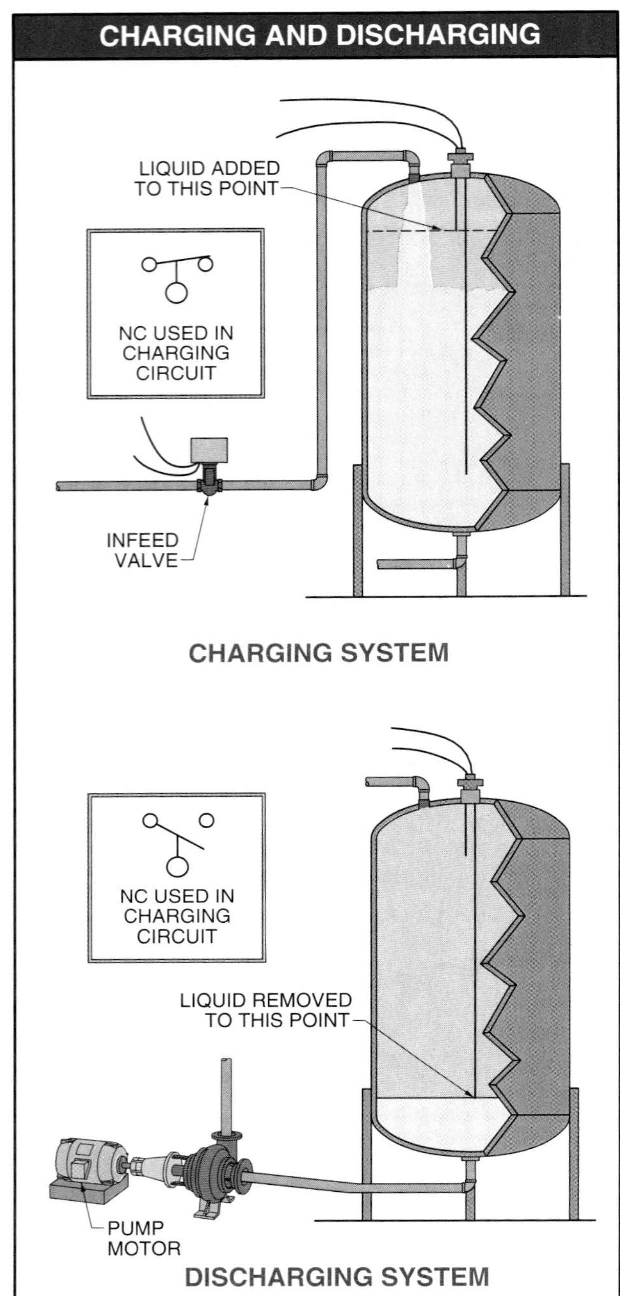

Figure 8-39. Level switches detect and respond to the level of a material in a tank.

Since motors draw much more current when starting than when running, excess heat is produced in a motor that must turn ON and OFF frequently. The faster the level in the tank drops, the faster the motor must cycle.

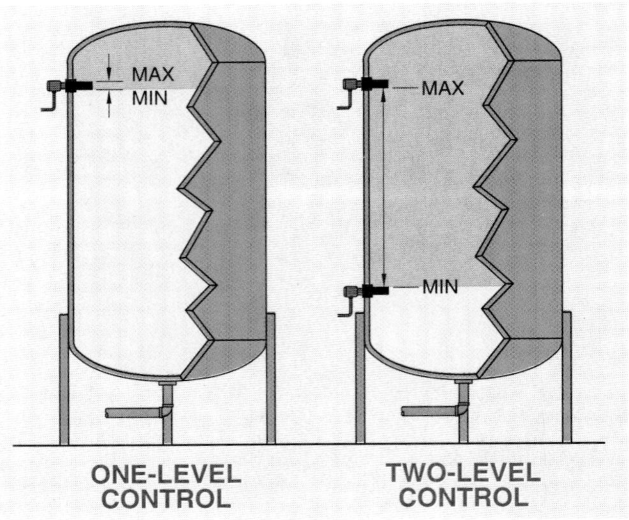

Figure 8-40. The distance controlled in one-level control is small. Any distance may be controlled in two-level control.

While it may appear that two-level registration is the best because the pump does not have to cycle as often, what is best for the motor may not be best for the total system. For example, if common house paint is to be maintained in a fill tank, problems develop if the length of time between the high level and low level is excessive. As the paint dries on the inside of the tank, it accumulates layer by layer. This causes skin to form which may clog or impede the pump or fill action if the skin falls into the product.

Therefore, product type must be taken into consideration when determining the distance and time between the high and low level. In general, one-level control is best when the liquid is emptied very slow from the tank. Two-level control is best when the liquid is emptied at a fast rate.

Temperature and Level Control Combinations

Temperature and level controls may be combined in some applications. See Figure 8-41. A contactor is used in the power circuit to control the heating element, and a temperature control relay is used in the control circuit. A magnetic motor starter is used to control the pump motor. A contactor can be used if a solenoid is used to open and close a valve that fills the tank. Circuit interlocking may be required to prevent the heating element from turning ON unless the level in the tank is at the maximum point.

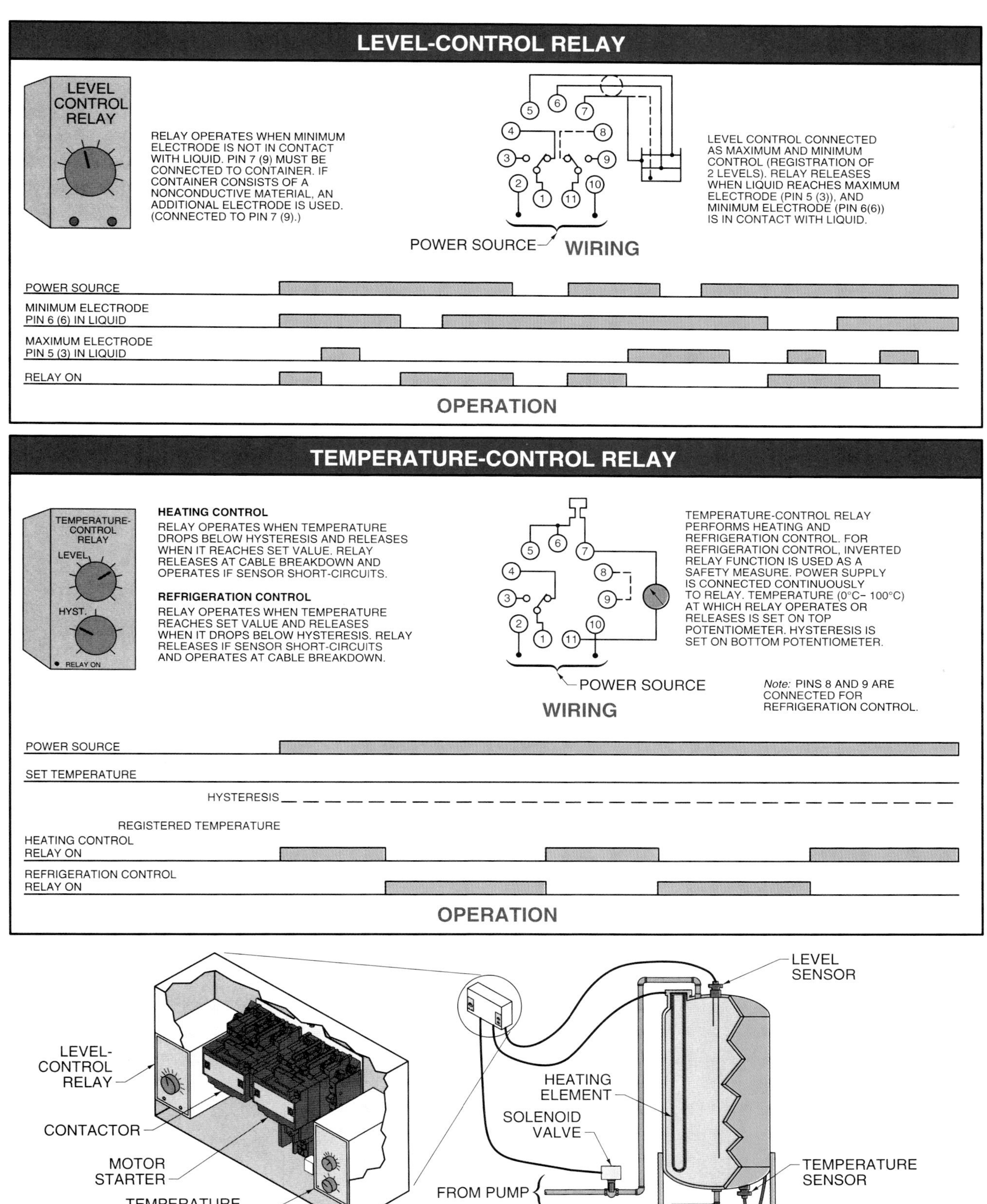

Figure 8-41. A temperature and level control may be combined to control the temperature and level of a fluid.

Solar Heating Control

Solar energy is plentiful and can provide substantial amounts of energy that can be used to replace other forms of more expensive or less available fuels. Although solar energy has been used in the past for many applications, such as supplying power in space, its main use is in space heating and hot water heating. Temperature control is required in each application.

In controlling a solar heating system, the temperature controllers must be able to measure the temperature inside the solar collector, at the water storage area, and circulate the heat accordingly. This circulation of heat must be accomplished based on the temperature setting and differential setting of the controller. To meet these requirements, the temperature controller must be able to take two separate signal inputs (collector and heated area), and make a decision that provides the correct heat circulation.

Solar Heating System Control Circuit. A solar heating control system needs two signal inputs (temperature sensor 1 and 2), a decision circuit, and an action circuit to start or stop the pump motor. Sensor 1 monitors the temperature at the solar collector. Sensor 2 monitors the temperature at the heating unit. The decision circuit compares temperatures at each sensor and starts or stops the pump based on demand. The decision circuit can be a logic module that has only the input and output connected to the module. The action circuit is a starter coil that energizes the pump motor based on demand.

Solar Heating System Temperature Control Circuit. In a solar heating system, the circulation pump is activated by the temperature differential between T1 and T2. See Figure 8-42. In this application, a water storage tank is used to hold hot water. A solar heat collector is used to absorb the sun's heat, and a pump motor is used to circulate the water.

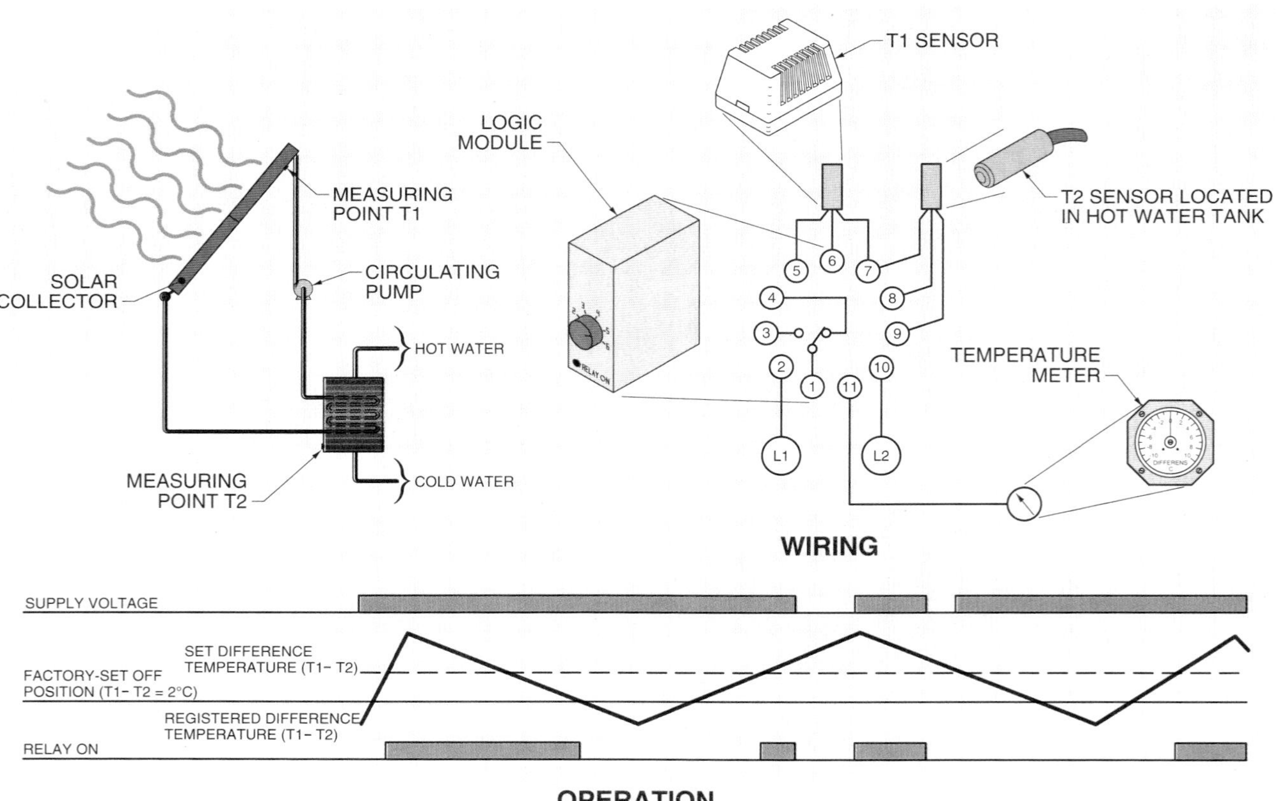

Figure 8-42. In a solar heating system, the circulation pump is activated by the temperature differential between T1 and T2.

This system does not include an auxiliary heating system, such as gas or electric, to supplement the solar system. A wiring diagram is used to properly install the control circuit of the solar heating system. See Figure 8-43.

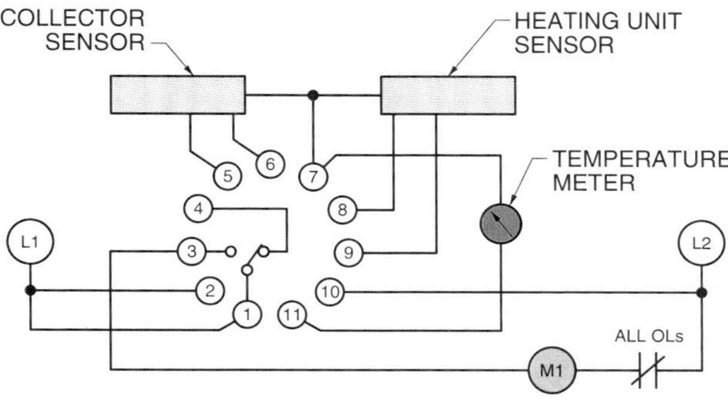

Figure 8-43. A wiring diagram is used to properly install the control circuit of the solar heating system.

The temperature relay is operated in conjunction with two semiconductor temperature sensors. These temperature sensors are available for use in liquids, gases, or may be constructed to measure the temperature of a metallic surface. These sensors are available in either a semiconductor (thermistor) or bulb type construction. The relay reacts to certain temperature differences as set on the adjustable knob.

Heidelberg Harris, Inc.

Air pollution control is required in any facility that contains highly sensitive manufacturing or production systems.

The relay operates when the temperature in the solar panel (measuring point T1) exceeds the temperature in the water tank used to accumulate the heat (measuring point T2) by a predetermined value which is set on the relay.

The smallest differential temperature (T1 – T2) at which the relay operates can be set to any desired value between 3°C to 10°C. The relay does not release until the temperature difference is reduced to 2°C (T1 – T2 = 2°C) as set by the factory. For example, if the relay is set for a 6°C temperature difference between T1 and T2, the pump turns ON at a 6°C temperature change between the two measuring points. The pump circulates the hot water from the solar collector to the storage tank. This circulation continues until there is a 2°C temperature difference between the two measuring points, at which point the relay turns OFF the pump.

A temperature meter may be added to the circuit to indicate the temperatures. A temperature meter with a neutral position in the middle of the scale must be used in cases where negative temperature differences may be displayed.

Air Pollution Control

The major emphasis on air pollution control today helped develop automatic air pollution control devices. The air pollution control devices detect and react to gases and smoke in the air. Such a system is similar to residential smoke detectors. Industrial applications often require the detection of other pollutants in addition to smoke. These applications may require the detection of gases or carbon in the air. Such detection may be used for safety or for measuring pollutants to meet recommended safety levels.

The air pollution control device must be able to measure the pollutants in the air and compare them to a set level. It must also take appropriate action if the level is exceeded. This action may be the sounding of an alarm or some corrective measure such as turning ON an exhaust fan. In meeting these requirements, the system must include a pollution detector (signal input), a logic module or control circuit that can react to the detector (decision), and an output (action). See Figure 8-44. The circuit uses an air pollution sensor and matching relay to detect flammable gases, smoke, and carbon in the air.

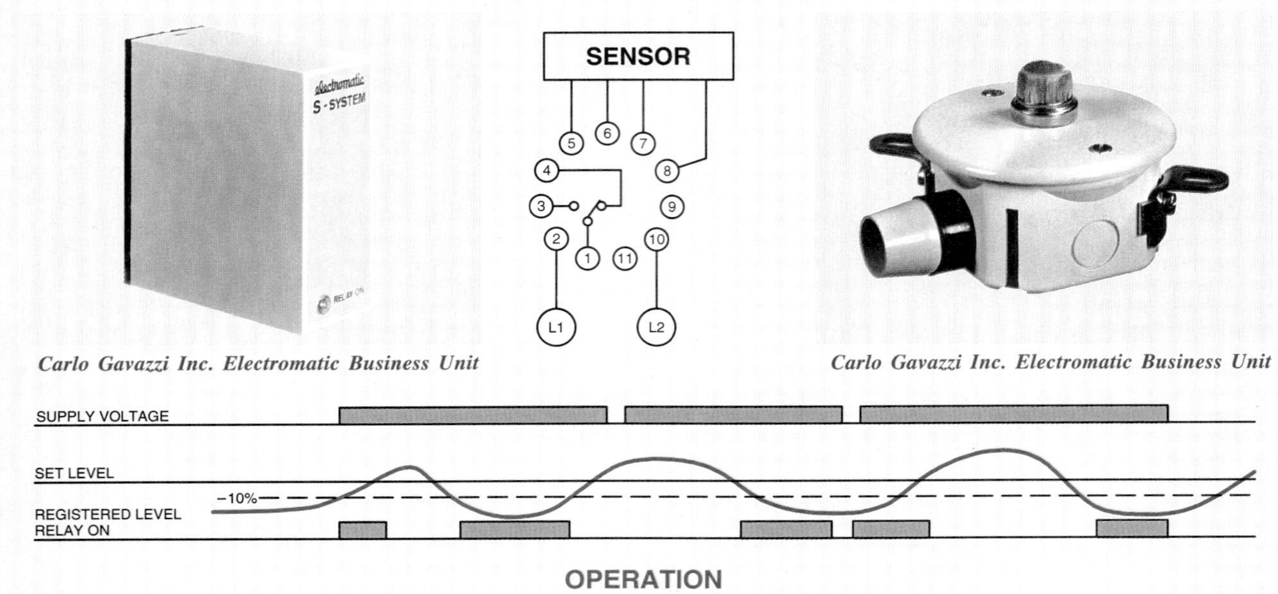

Carlo Gavazzi Inc. Electromatic Business Unit

Carlo Gavazzi Inc. Electromatic Business Unit

Figure 8-44. An air pollution detector uses a logic module and air pollution sensor to measure the pollutants in the air and compare them to a set level.

The relay is used with an air pollution detector that detects very small concentrations of all reducing or flammable gases such as hydrogen, carbon dioxide, methane, propane, butane, acetylene, and sulfur dioxide. Even very small concentrations, such as .02% [200 parts per million (ppm)], are detected. The detector also reacts to a smoke-filled carbonaceous atmosphere.

The relay immediately operates when the supply voltage is applied. The relay releases when the registered pollution level exceeds the set level. The relay does not operate again until the registered level is about 10% lower than the set level. The activation of the relay can be used to signal an alarm or record the time and duration of high levels of pollution.

WIND METERING

Windmills are used to harness the wind to generate power. A typical windmill consists of one or two generators depending on the size. A windmill with two generators commonly has a small generator with a synchronous speed of 1000 rpm and a large generator with a synchronous speed of 1500 rpm. The small generator is first connected to the system. The large generator is connected after a short time delay when the maximum power of the small generator is reached.

On a windmill, a control device (anemometer) is required to measure and react to wind velocity. On some windmill applications, a control device (wind vane) is also required to determine relative wind direction. See Figure 8-45.

Bergey Windpower Co., Inc.

The BWC Excel wind turbine system from Bergey Windpower Co., Inc. is a 10 kW wind turbine designed for use in areas with an average wind speed of 12 mph.

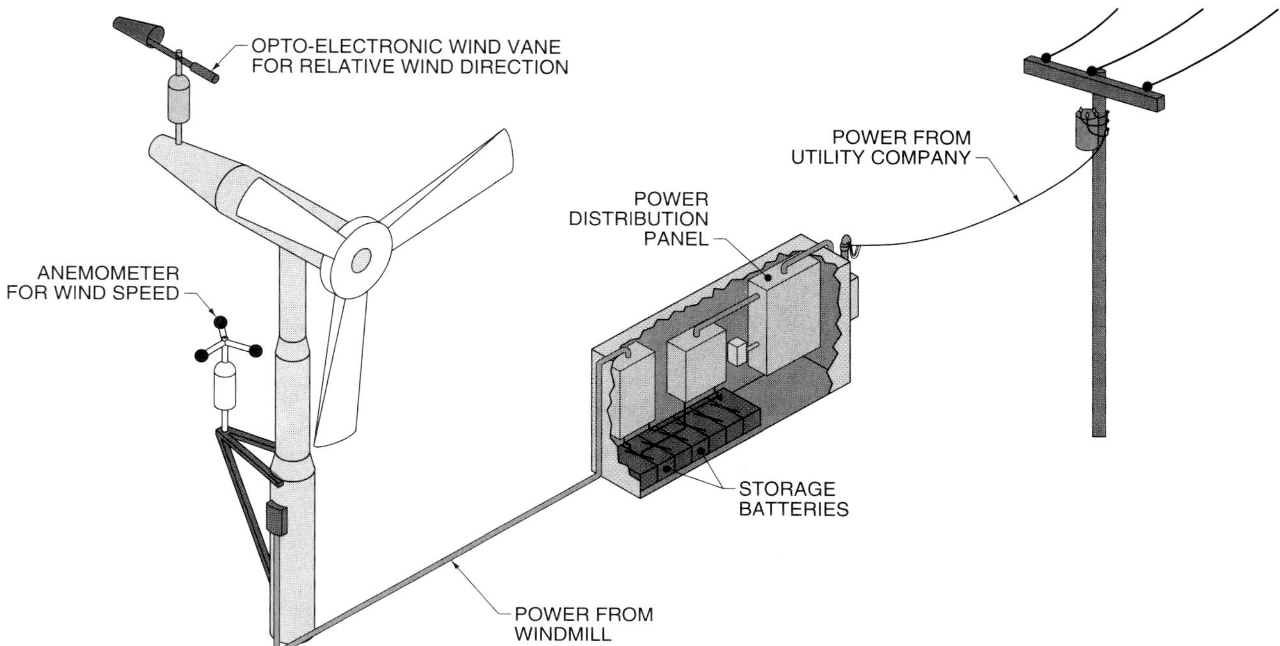

Figure 8-45. Windmills are used to harness the wind to generate power.

An anemometer is used to stop the windmill at too low and too high of wind velocity. Stopping the windmill at low wind velocities prevents the constant connection and disconnection of the small generator. Stopping the windmill at high wind velocities helps to protect the windmill against damage and wear from the high wind speeds. To stop the windmill, the anemometer and wind velocity relay control a mechanical brake inside the windmill. The specifications for the anemometer and wind velocity relay are given in manufacturer data sheets. See Figure 8-46.

A wind vane is used to control the yawing function of the windmill. *Yawing* is a side-to-side movement. This is required to turn the windmill into the direction of the wind for windmills that are not freely allowed to turn. To do this, the wind direction relay incorporates a relay with a neutral center position. In this position, the top of the windmill is kept still, while the two working positions of the contacts turn to either the right or left of the windmill. The specifications for wind vanes and wind direction relays are given in manufacturer's data sheets. See Figure 8-47.

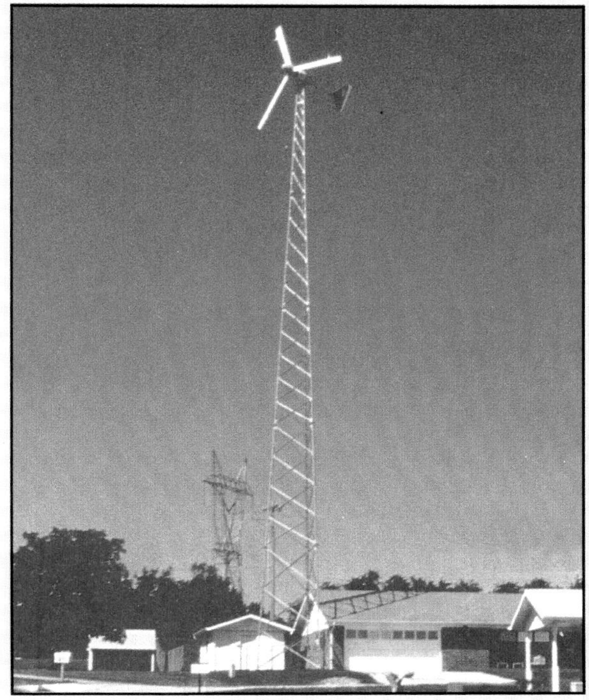

Bergey Windpower Co., Inc.

Wind turbines are used for supplying power for pumping, battery charging, and residential power demands.

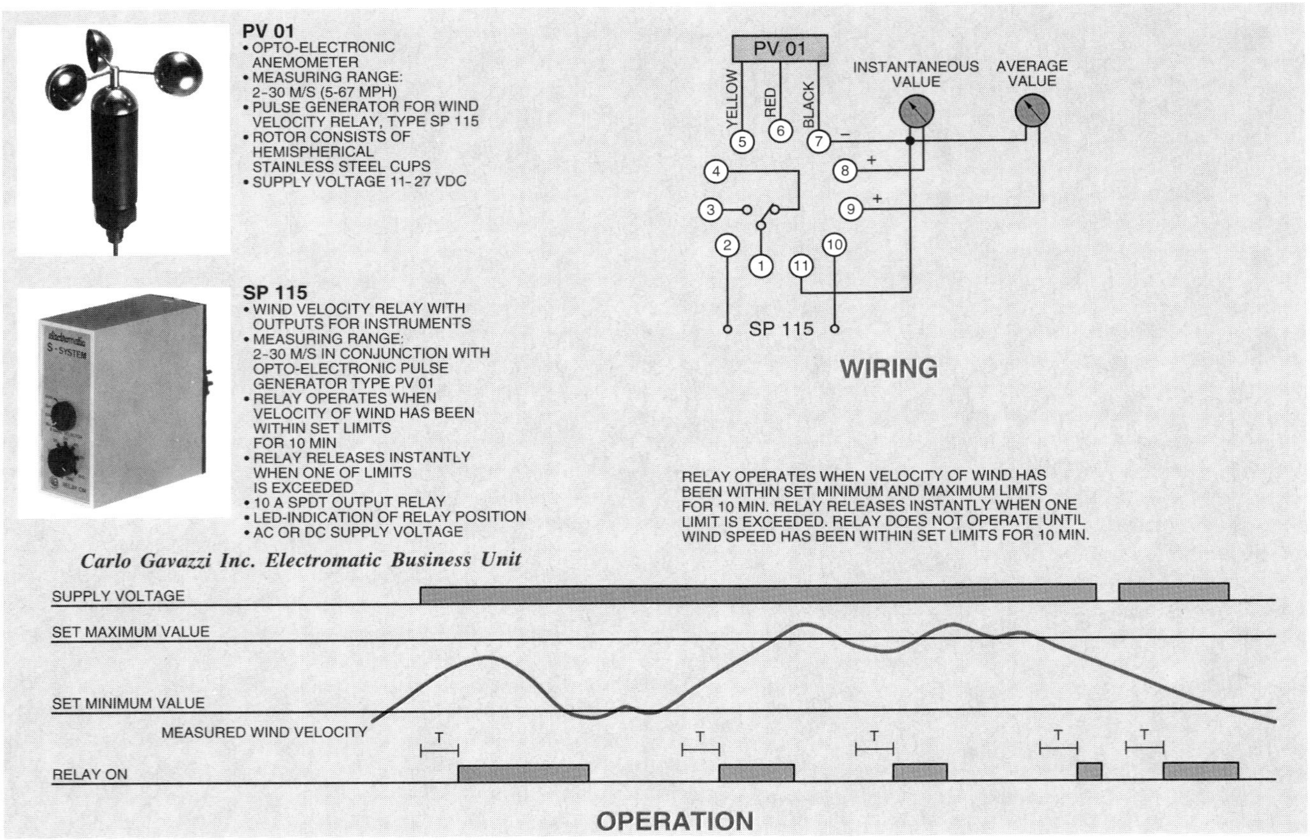

PV 01
- OPTO-ELECTRONIC ANEMOMETER
- MEASURING RANGE: 2-30 M/S (5-67 MPH)
- PULSE GENERATOR FOR WIND VELOCITY RELAY, TYPE SP 115
- ROTOR CONSISTS OF HEMISPHERICAL STAINLESS STEEL CUPS
- SUPPLY VOLTAGE 11-27 VDC

SP 115
- WIND VELOCITY RELAY WITH OUTPUTS FOR INSTRUMENTS
- MEASURING RANGE: 2-30 M/S IN CONJUNCTION WITH OPTO-ELECTRONIC PULSE GENERATOR TYPE PV 01
- RELAY OPERATES WHEN VELOCITY OF WIND HAS BEEN WITHIN SET LIMITS FOR 10 MIN
- RELAY RELEASES INSTANTLY WHEN ONE OF LIMITS IS EXCEEDED
- 10 A SPDT OUTPUT RELAY
- LED-INDICATION OF RELAY POSITION
- AC OR DC SUPPLY VOLTAGE

Carlo Gavazzi Inc. Electromatic Business Unit

PV 01
YELLOW
RED
BLACK
INSTANTANEOUS VALUE
AVERAGE VALUE
SP 115

WIRING

RELAY OPERATES WHEN VELOCITY OF WIND HAS BEEN WITHIN SET MINIMUM AND MAXIMUM LIMITS FOR 10 MIN. RELAY RELEASES INSTANTLY WHEN ONE LIMIT IS EXCEEDED. RELAY DOES NOT OPERATE UNTIL WIND SPEED HAS BEEN WITHIN SET LIMITS FOR 10 MIN.

SUPPLY VOLTAGE
SET MAXIMUM VALUE
SET MINIMUM VALUE
MEASURED WIND VELOCITY
RELAY ON

OPERATION

Figure 8-46. Anemometer and wind velocity relay specifications are given in manufacturer's data sheets.

In addition to wind speed and wind direction controls, other controls may be required depending on the application of the generated power. In a simple windmill application, the generated power can be used directly without synchronization with other power. An example of this is an application that connects the generated power to a set of heating elements.

In an application that connects the generated power to utility lines, synchronization of frequency and voltages must be built in. A rectifier circuit is required when connecting the generated power to a set of batteries.

AUTOMATED SYSTEMS

An automated system includes manual, mechanical, and automatic control devices. These control devices are interconnected to provide the required inputs to make the system function as designed. See Figure 8-48. This system could be used for paint, food products, beverages, or other products put in a container.

Honeywell's MICRO SWITCH Division

The SDS™ (Smart Distributed System) from Honeywell is a distributed machine control system that "empowers" sensors, actuators, and other devices to include error detection, error correction, and device and system diagnostics.

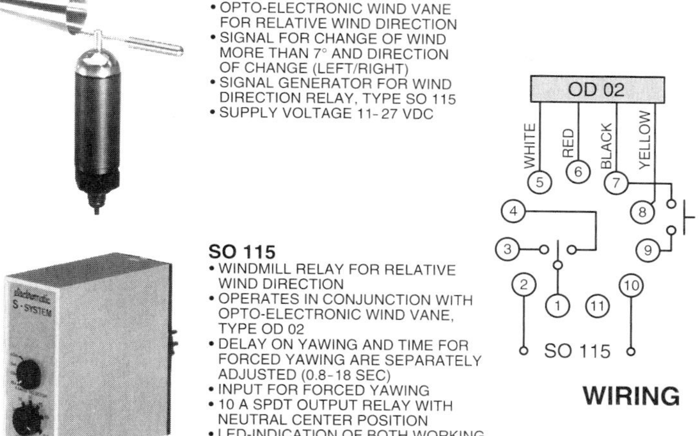

OD 02
• OPTO-ELECTRONIC WIND VANE FOR RELATIVE WIND DIRECTION
• SIGNAL FOR CHANGE OF WIND MORE THAN 7° AND DIRECTION OF CHANGE (LEFT/RIGHT)
• SIGNAL GENERATOR FOR WIND DIRECTION RELAY, TYPE SO 115
• SUPPLY VOLTAGE 11-27 VDC
SO 115
• WINDMILL RELAY FOR RELATIVE WIND DIRECTION
• OPERATES IN CONJUNCTION WITH OPTO-ELECTRONIC WIND VANE, TYPE OD 02
• DELAY ON YAWING AND TIME FOR FORCED YAWING ARE SEPARATELY ADJUSTED (0.8-18 SEC)
• INPUT FOR FORCED YAWING
• 10 A SPDT OUTPUT RELAY WITH NEUTRAL CENTER POSITION
• LED-INDICATION OF BOTH WORKING POSITIONS
• AC OR DC SUPPLY VOLTAGE
OD 02
WHITE
RED
BLACK
YELLOW
FORCED YAWING
SO 115
WIRING
SO 115 INCORPORATES A RELAY WITH NEUTRAL CENTER POSITION. IN THIS POSITION, TOP OF WINDMILL IS KEPT STILL WHILE TWO WORKING POSITIONS CAUSE A TURNING EITHER TO RIGHT OR LEFT OF WINDMILL.
TO AVOID INCESSANT TURNING OF TOP OF WINDMILL AT CHANGING DIRECTIONS OF WIND, AN ADJUSTABLE DELAY ON YAWING (UPPER KNOB) IS USED.
THE SO 115 ALSO HAS A BUILT-IN POTENTIOMETER (LOWER KNOB) FOR ADJUSTMENT OF YAWING TIME IN CASE OF ERROR IN CONDITIONS.
IF POSITION OF WINDMILL DEVIATES 180° FROM WIND DIRECTION ON STARTING, IT MAY BE NECESSARY TO FORCE START OF YAWING BY CONNECTING PIN 7 WITH 9 UNTIL WINDMILL STANDS APPROXIMATELY 45° TOWARDS WIND DIRECTION. AT A CHANGE OF WIND IN EXCESS OF 7°, SET DELAY PERIOD STARTS.
WHEN WIND VANE HAS HAD A CONSTANT DEFLECTION OF MORE THAN 7° FOR SET TIME, RELAY OPERATES, CAUSING TOP OF WINDMILL TO TURN TOWARDS NEW WIND DIRECTION.

Carlo Gavazzi Inc. Electromatic Business Unit

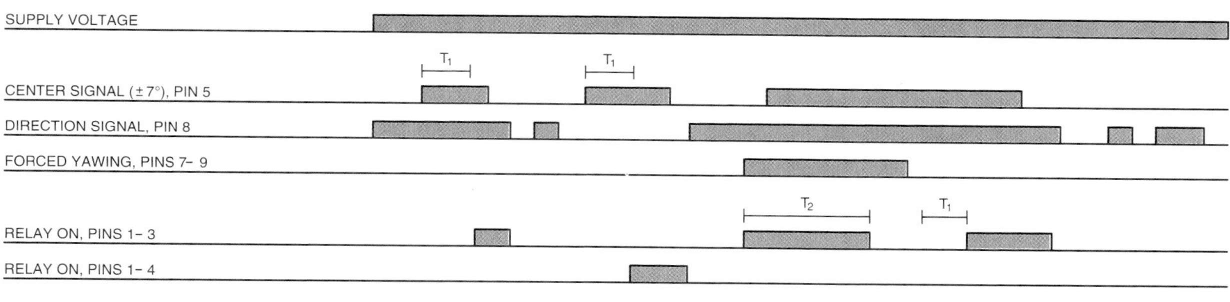

Figure 8-47. A wind vane and wind direction relay are used to control the yawing function of a windmill.

In this system, manual inputs such as pushbuttons and selector switches are required at the individual and main control stations. Automatic inputs such as pressure, temperature, flow, and level controls are required to control each step of the process from start to finish. Mechanical inputs are used to detect position as required.

All of these basic control devices provide a method for controlling the product or operation. Each control device must be selected and installed as if the entire operation depends on that one control input. In an automated system, the failure of any one control device could shut down the entire process.

Preventing Problems When Installing Control Devices

All electrical circuits must be controlled. For this reason, control devices are used in every type of control application. A control device must be properly protected and installed to ensure that the control device operates properly for a long time. Proper protection means that the switching contacts are operated within their electrical rating and are not subjected to destructive levels of current or voltage. Proper installation means that the control device is installed in such a manner as to ensure it operates as designed.

Protecting Switch Contacts

Control devices are used to switch ON, OFF, or redirect the flow of current in an electrical circuit. The control devices switch contacts that are rated for the amount of current they can safely switch. The switch rating is normally specified for switching a resistive load, such as small heating elements. Resistive loads are the least destructive loads to switch. However, most loads that are switched are inductive loads, such as solenoids and motor starter coils. Inductive loads are the most destructive loads to switch.

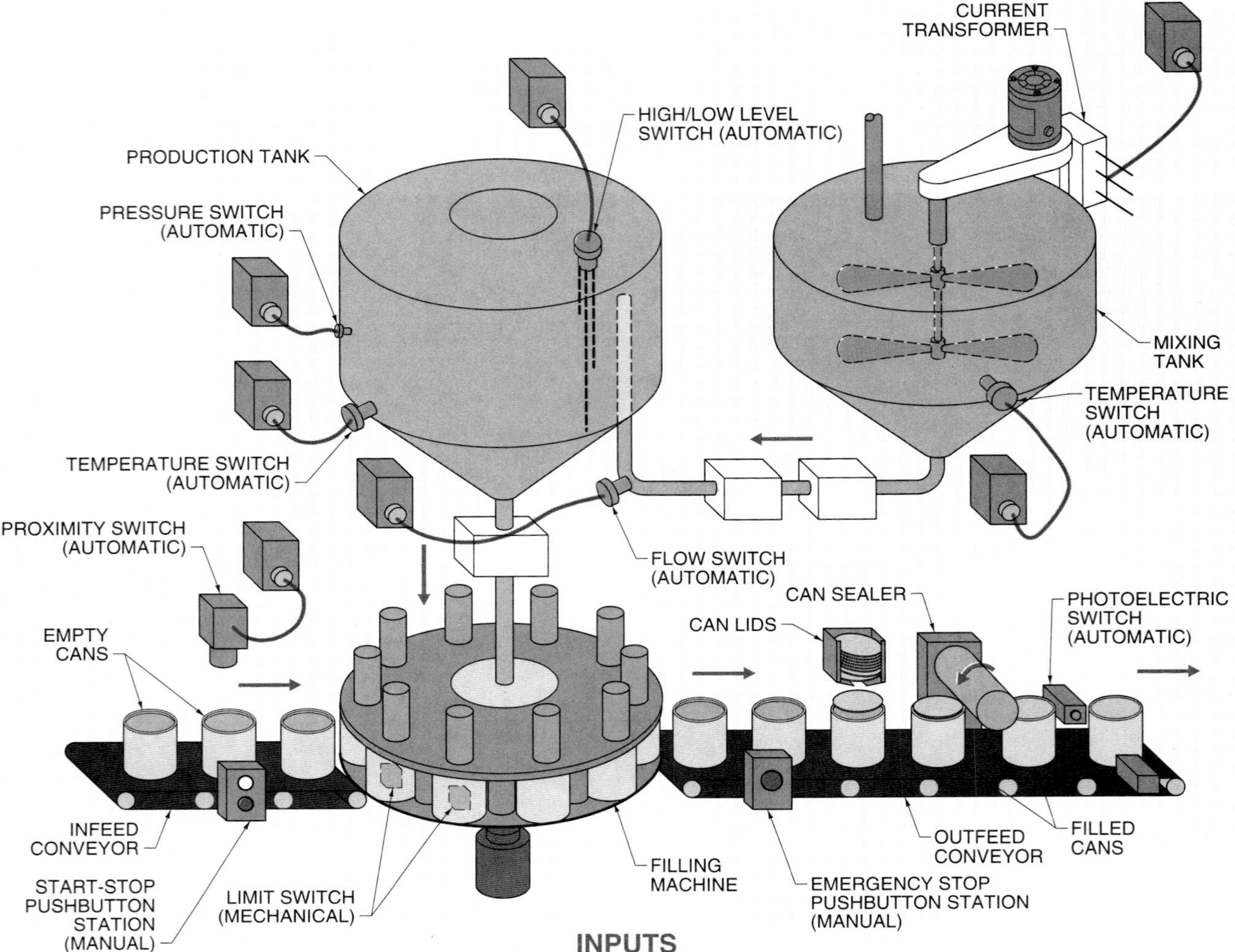

Figure 8-48. An automated system includes manual, mechanical, and automatic control devices that are interconnected to provide the required inputs to make the system function as designed.

A large induced voltage appears across the switch contacts when inductive loads are turned OFF. The induced voltage causes arcing at the switch contacts. Arcing may cause the contacts to burn, stick, or weld together. Contact protection should be added when frequently switching inductive loads to prevent or reduce arcing. See Figure 8-49.

A diode is added in parallel with the load to protect contacts that switch DC. The diode conducts only when the switch is open, providing a path for the induced voltage in the load to dissipate.

A resistor and capacitor are connected across the switch contacts to protect contacts that switch AC. The capacitor acts as a high impedance (resistor) load at 60 Hz, but becomes a short circuit at the high frequencies produced by the induced voltage of the load. This allows the induced voltage to dissipate across the resistor when the load is switched OFF.

Protecting Pressure Switches

A *pressure switch* is a switch that is designed to activate its contacts at a preset pressure. A pressure switch is rated according to its operating pressure range. A pressure switch may be damaged if its maximum pressure limit is exceeded.

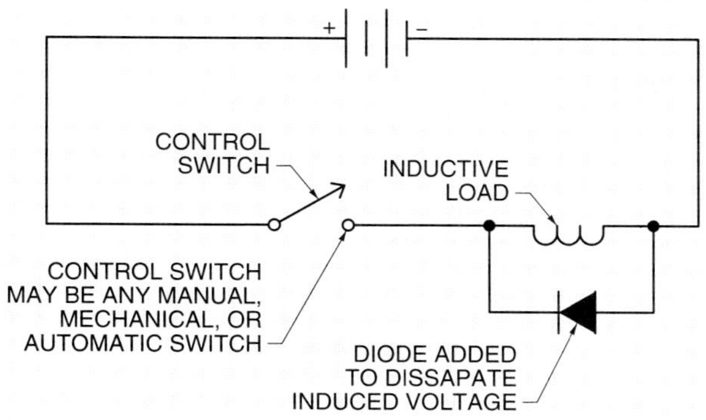

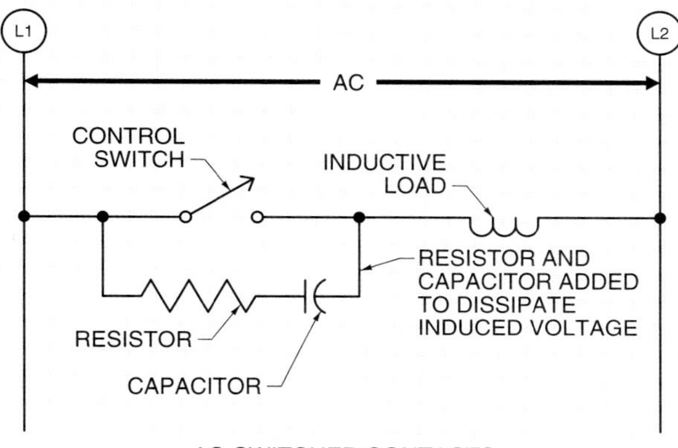

CONTACT PROTECTION

Figure 8-49. Contact protection may be added when switching large DC and AC inductive loads to prevent or reduce arcing at the switch contacts.

Protection for a pressure switch should be added in any system in which a higher pressure than the maximum limit is possible. A pressure-relief valve is installed to protect the pressure switch. A pressure-relief valve should be set just below the pressure switch's maximum limit. The valve opens when the system pressure increases to the setting of the relief valve.

Caution: The output of the relief valve must be connected to a proper drain (or return line) if the product under pressure is a gas or a fluid. See Figure 8-50.

Installing Flow Switches

A *flow switch* is a switch designed to detect the movement (flow) of a product through a pipe or duct. Most flow switches use a paddle to detect the movement of the product. The paddle is designed to detect the product movement with the least possible pressure drop across the switch. A flow switch must be installed correctly to ensure it does not interfere with the movement of the product. Most flow switches are designed to operate in the horizontal position. Allow a distance of at least three pipe diameters (ID) on each side of the flow switch when mounting a flow switch. See Figure 8-51.

For example, the minimum horizontal distance of straight pipe required on each side of a flow switch is $4\frac{1}{2}''$ ($1\frac{1}{2}'' \times 3 = 4\frac{1}{2}''$) when used in an application that moves a product through a $1\frac{1}{2}''$ diameter pipe.

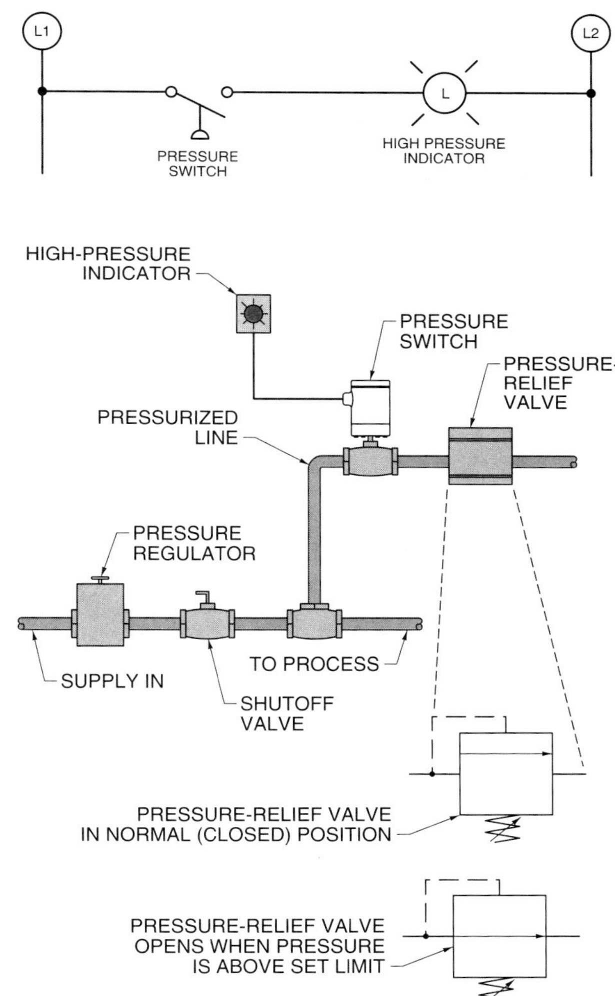

PRESSURE SWITCH PROTECTION

Figure 8-50. A pressure relief valve may be added to a circuit to protect a pressure switch from excessive pressure.

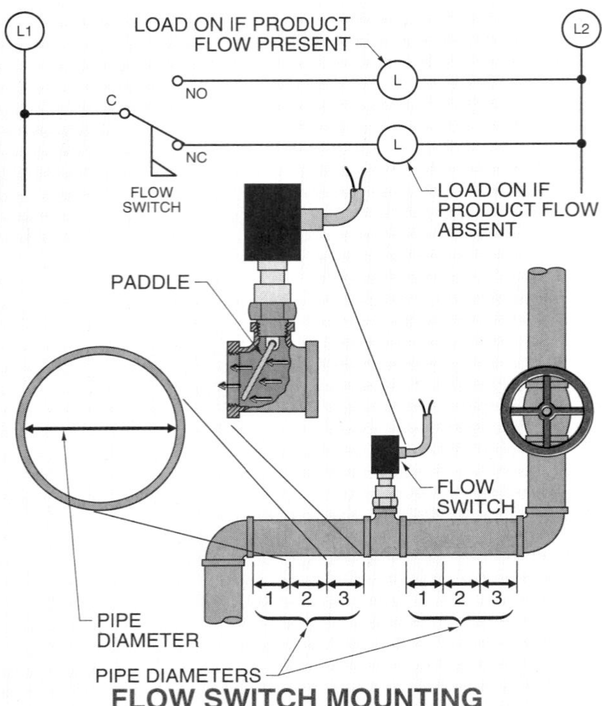

Figure 8-51. Allow a distance of at least three pipe diameters (ID) on each side of the flow switch when mounting a flow switch.

TROUBLESHOOTING CONTROL DEVICES

Control devices use mechanical or solid-state switches to control the flow of current. A *mechanical switch* is any switch that uses silver contacts to start and stop the flow of current in a circuit. Silver contacts can be used to switch either AC or DC loads. A solid-state switch has no moving parts (contacts). A *solid-state switch* is a switch that uses a triac, SCR, current sink (NPN) transistor, or current source (PNP) transistor to perform the switching function. The triac output is used for switching AC loads. The SCR output is used for switching high-power DC loads. The current sink and current source outputs are used for switching low-power DC loads.

Testing Electromechanical Switches

A suspected fault with an electromechanical switch is tested using a voltmeter. The voltmeter is used to test the voltage flowing into and out of the switch. See Figure 8-52.

To test a mechanical switch, apply the procedure:

1. Measure the voltage into the switch. Connect the voltmeter between the neutral and hot conductor feeding the switch. The voltmeter lead may be connected to ground instead of neutral if the neutral conductor is not available in the same box in which the switch is located. The problem is located upstream from the switch when there is no voltage present or the voltage is not at the correct level. The problem may be a blown fuse or open circuit. Voltage must be re-established to the switch before the switch may be tested.
2. Measure the voltage out of the switch. There should be a voltage reading when the switch contacts are closed. There should not be a voltage reading when the switch contacts are open. The switch has an open and must be replaced if there is no voltage reading in either switch position. The switch has a short and must be replaced if there is a voltage reading in both switch positions.

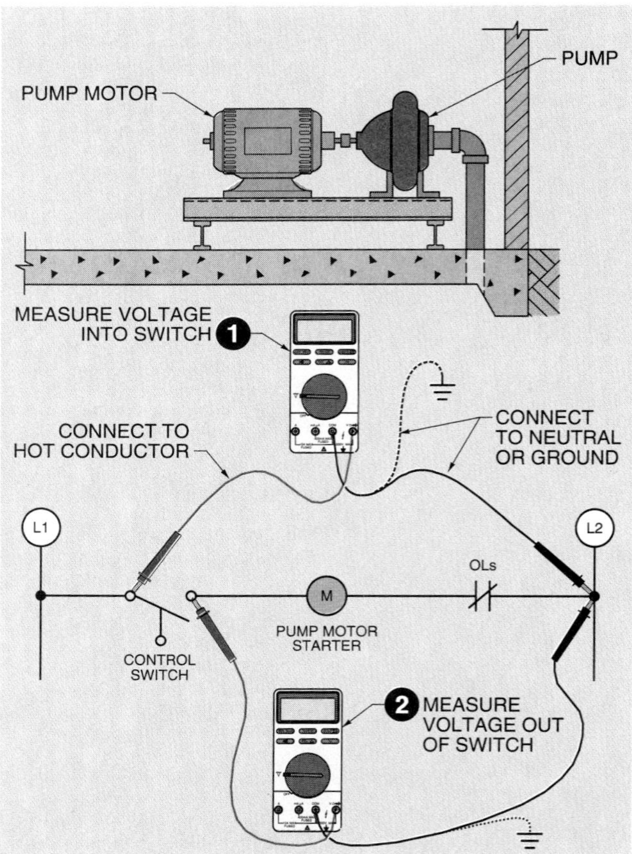

Figure 8-52. A voltmeter is used to test the operation of a switch.

Warning: Always ensure power is OFF before changing a control switch. Use a voltmeter to ensure the power is OFF.

Testing Solid-State Switches

A suspected fault in a two-wire solid-state switch may be tested using a voltmeter. The voltmeter is used to test the voltage into and out of the switch. See Figure 8-53.

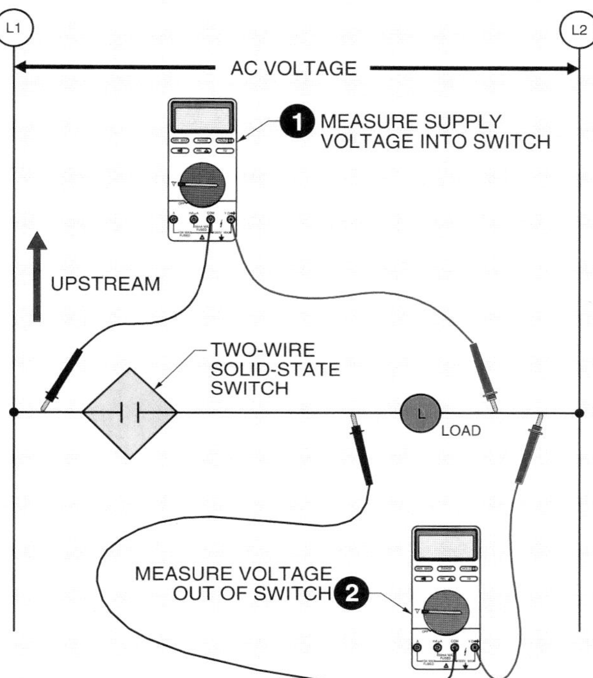

Figure 8-53. A voltmeter is used to measure the voltage into and out of a two-wire solid-state switch.

To test a two-wire solid-state switch, apply the procedure:

1. Measure the voltage into the switch. The problem is upstream from the switch when no voltage is present or the voltage is not at the correct level. The problem may be a blown fuse or open circuit. Voltage must be re-established before the switch may be tested.
2. Measure the voltage out of the switch. The voltage should equal the supply voltage minus the voltage drop (3 V to 8 V) when the switch is conducting (load ON). Replace the switch if the voltage output is not correct.

Warning: Always ensure power is OFF before changing a control switch. Use a voltmeter to ensure the power is OFF.

A suspected fault with a three-wire solid-state switch may be tested using a voltmeter. The voltmeter is used to test the voltage into and out of the switch. See Figure 8-54.

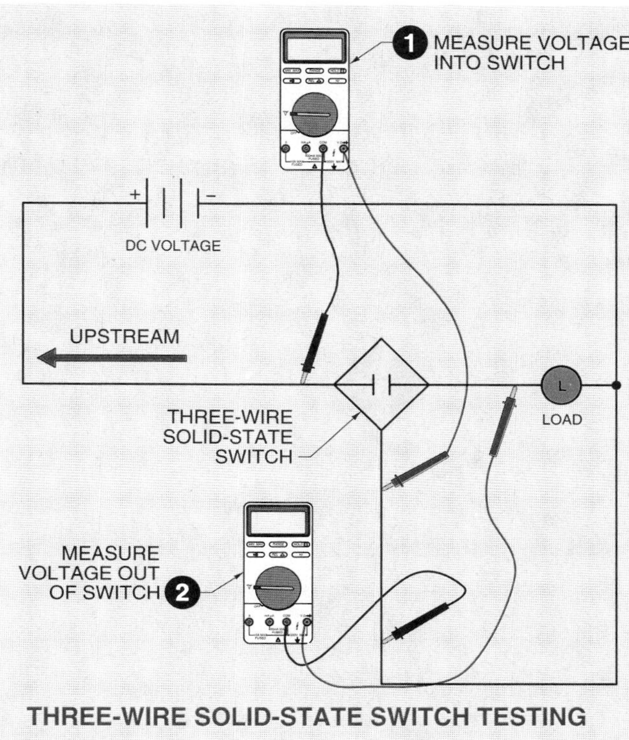

Figure 8-54. A voltmeter is used to measure the voltage into and out of a three-wire solid-state switch.

To test a three-wire solid-state switch, apply the procedure:

1. Measure the voltage into the switch. The problem is upstream from the switch when no voltage present or the voltage is not at the correct level. The problem may be a blown fuse or open circuit. Voltage must be re-established before the switch may be tested.
2. Measure the voltage out of the switch. The voltage should equal the supply voltage when the switch is conducting (load ON). Replace the switch if the voltage output is not correct.

Warning: Always ensure power is OFF before changing a control switch. Use a voltmeter to ensure the power is OFF.

1. What is the most common contact configuration used on pushbuttons?
2. What are the different operators that are available on pushbuttons?
3. What operator is best suited for an emergency stop pushbutton?
4. What NEMA enclosures are available for pushbutton stations?
5. What is the purpose of selector switches?
6. What is the purpose of a truth table?
7. What is a joystick?
8. How many positions can a joystick have?
9. What are the two methods used to illustrate the different positions of a joystick?
10. What types of contacts are normally included with limit switches?
11. What is an actuator?
12. What are the basic actuators available for use with limit switches?
13. What is a daylight switch?
14. What is a pressure switch?
15. What type of pressure switch contacts are used to signal an overpressure condition?
16. What is the dead band setting of a pressure switch?
17. What may happen if the pressure differential setting of a pressure switch is too small?
18. What are the uses of most pressure switches?
19. What is a temperature switch?
20. What are the different methods used to switch electrical contacts in response to a temperature change?
21. What is a flow switch?
22. Can gas detection be made to detect different groups of gas?
23. Can level switches only sense liquids?
24. What other two names are charging systems and discharging systems also known as?
25. How many single temperature inputs does a solar heating system need?
26. What is the purpose of an anemometer?
27. What is yawing?

9 Chapter

Reversing Motor Circuits

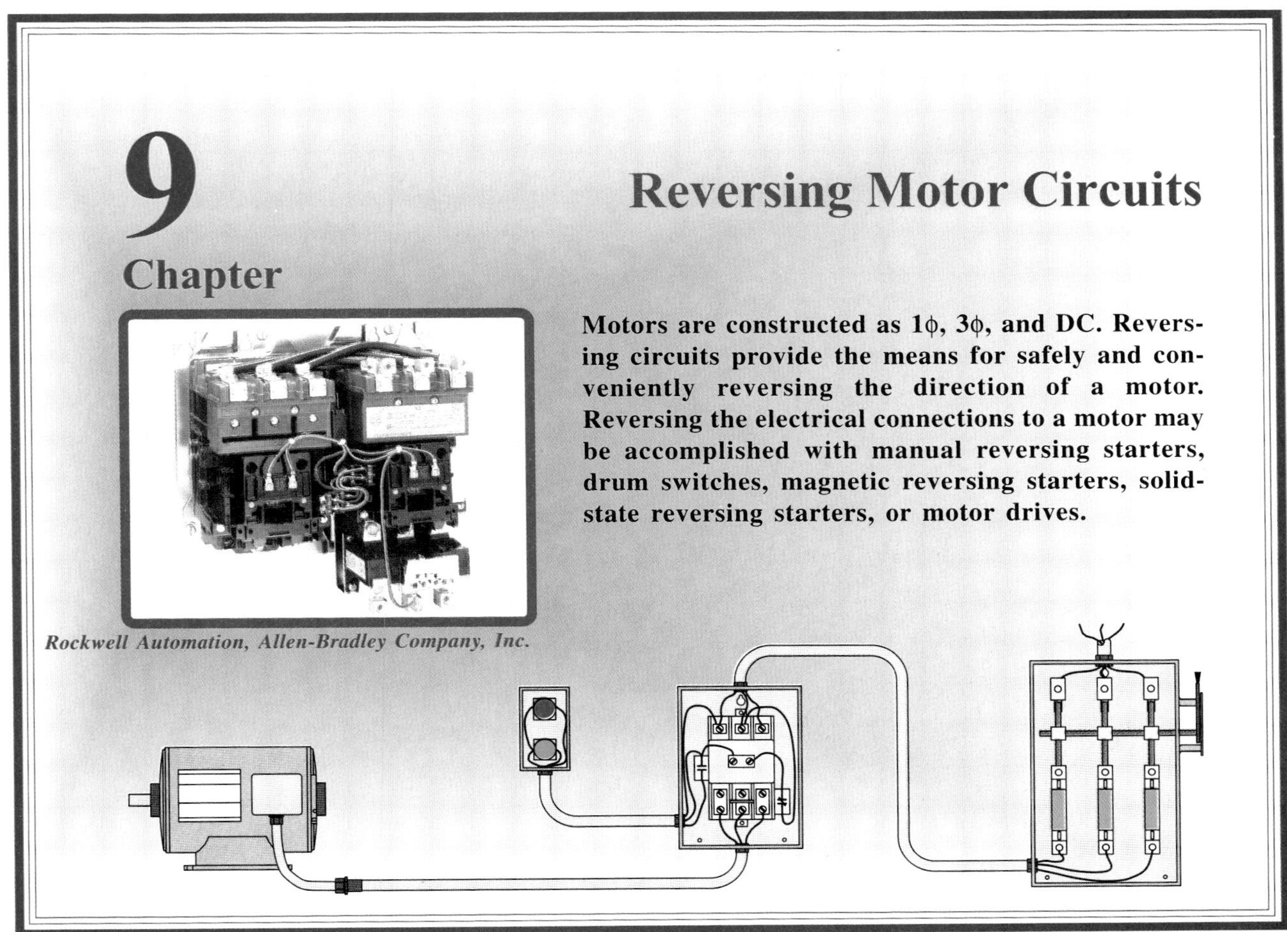

Rockwell Automation, Allen-Bradley Company, Inc.

Motors are constructed as 1ϕ, 3ϕ, and DC. Reversing circuits provide the means for safely and conveniently reversing the direction of a motor. Reversing the electrical connections to a motor may be accomplished with manual reversing starters, drum switches, magnetic reversing starters, solid-state reversing starters, or motor drives.

MOTORS

Almost 85% of the industrial machines in use today are driven by electric motors. Many factors must be considered when selecting, wiring, and maintaining electric motors.

Satisfying the electrical needs of a motor requires installing the unit according to the manufacturer's wiring diagrams. Additionally, motors shall be installed per Article 430 of the National Electrical Code® (NEC®).

Satisfying the mechanical needs of a motor requires that the unit selected meets the intended specifications of enclosures, bearings, frame size, and insulation. These specifications are usually established by the original manufacturer but must be understood when selecting any replacement part(s). The electrician must read the manufacturer's motor nameplate information to select the correct type of replacement motor.

THREE-PHASE MOTORS

Three-phase motors are the most common motor used in industrial applications. Three-phase motors are used in applications ranging from fractional horsepower to over 500 HP. They are used in most applications because they are simple in construction, require little maintenance, and cost less to operate than 1ϕ or DC motors.

The most common 3ϕ motor used in most applications is the induction motor. An *induction motor* is a motor that has no physical electrical connection to the rotor. Induction motors have no brushes that wear or require maintenance. Current in the rotor is induced by the rotating magnetic field of the stator.

The *stator* is the stationary part of an AC motor. The *rotor* is the rotating part of an AC motor. A rotating magnetic field is set up automatically in the stator when the motor is connected to 3ϕ power. The coils in the stator are connected to form three separate windings (phases). Each phase contains one-third of the total number of individual coils in the motor. These composite windings or phases are the A phase, B phase, and C phase. See Figure 9-1.

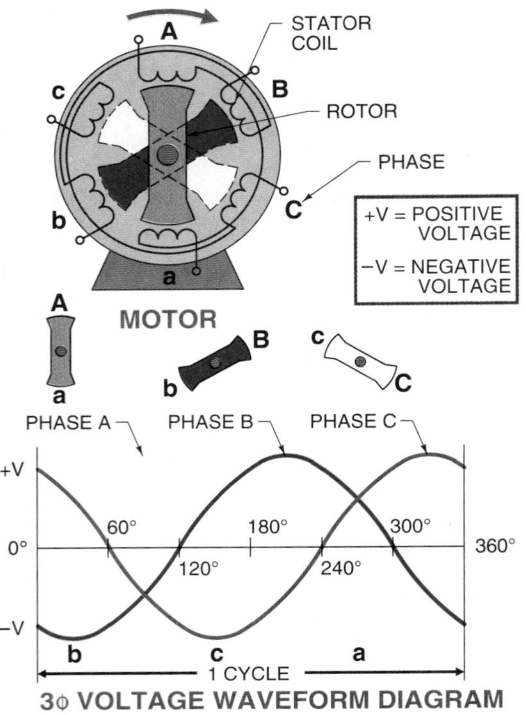

Figure 9-1. The coils in the stator are connected to form three separate windings (phases).

Leeson Electric Corporation

Electric motors are available in a wide range of sizes and types for any industrial application.

Each phase is placed in the motor so that it is 120° from the other phases. A rotating magnetic field is produced in the stator because each phase reaches its peak value 120° apart from the other phases. Three-phase motors are self-starting and do not require an additional starting method because of the rotating magnetic field in the motor.

Single-Voltage, 3ϕ Motor Construction

To develop a rotating magnetic field in the motor, the stator windings must be connected to the proper voltage level. This voltage level is determined by the manufacturer and stamped on the motor nameplate. Three-phase motors are either designed as single-voltage motors or as dual-voltage motors. A *single-voltage motor* is a motor that operates at only one voltage level. A *dual-voltage motor* is a motor that operates at more than one voltage level.

Single-voltage motors are less expensive to manufacture than dual-voltage motors, but are limited to locations having the same voltage as the motor. Common single-voltage, 3ϕ motor ratings are 230 V, 460 V, and 575 V. Other single-voltage, 3ϕ motor ratings are 200 V, 208 V, and 220 V.

Wye-Connected, 3ϕ Motors. All 3ϕ motors are wired so that the phases are connected in either a wye (Y) or delta (Δ) configuration. In a wye configuration, one end of each of the three phases is internally connected to the other phases. See Figure 9-2.

The remaining end of each phase is brought out externally and connected to the power line. The leads which are brought out externally are labeled terminals one (T1), two (T2), and three (T3). When connected, terminals T1, T2, and T3 are matched to the 3ϕ power lines labeled lines one (L1), two (L2), and three (L3). For the motor to operate properly, the 3ϕ supplying power to the motor must have the same voltage and frequency ratings as the motor.

Delta-Connected, 3ϕ Motors. In a delta configuration, each winding is wired end-to-end to form a completely closed loop circuit. See Figure 9-3. At each point where the phases are connected, leads are brought out externally to form terminals one (T1), two (T2), and three (T3). These terminals, like those of a wye-connected motor, are attached to power lines one (L1), two (L2), and three (L3). The 3ϕ lines supplying power to the delta motor must have the same voltage and frequency rating as the motor.

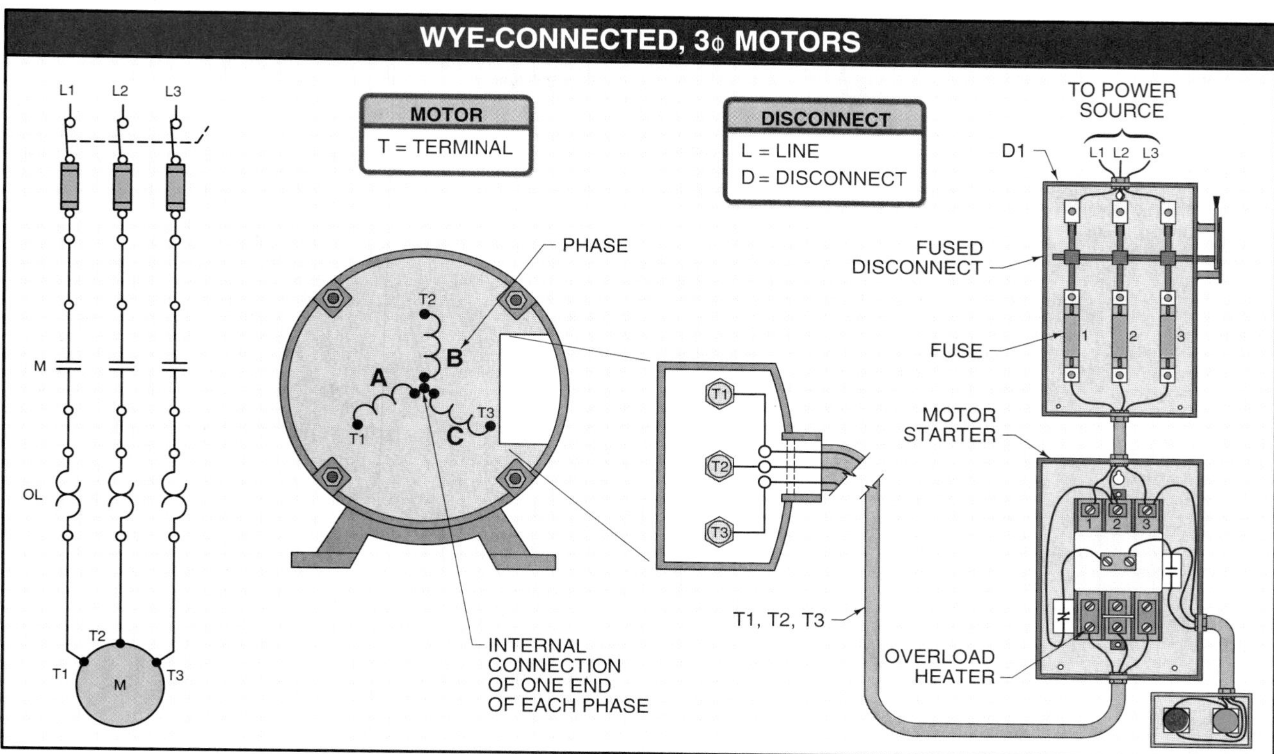

Figure 9-2. In a wye-connected, 3ϕ motor, one end of each phase is internally connected to the other phases.

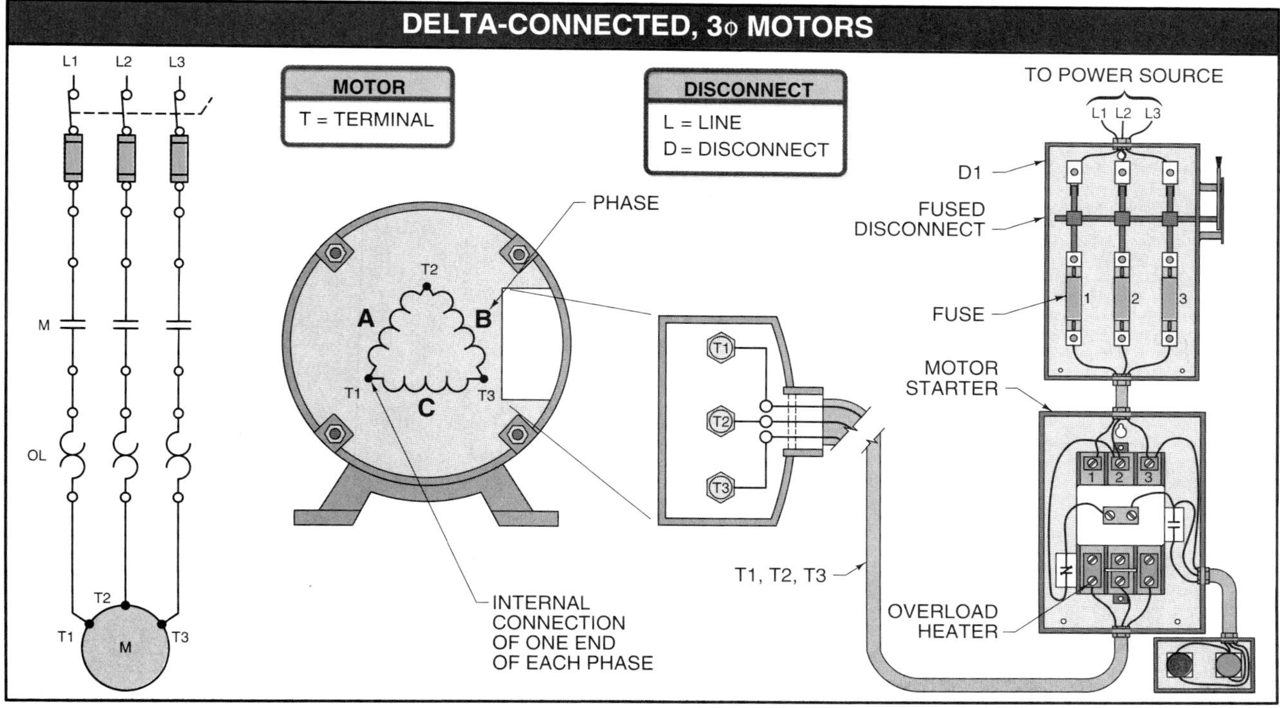

Figure 9-3. In a delta-connected, 3ϕ motor, each phase is wired end-to-end to form a completely closed loop.

Dual-Voltage, 3ϕ Motor Construction

Most 3ϕ motors are made so that they may be connected for either of two voltages. Making motors for two voltages enables the same motor to be used with two different power line voltages. The normal dual-voltage rating of industrial motors is 230/460 V. Always check the nameplate for proper voltage ratings.

The higher voltage is preferred when a choice between voltages is available. The motor uses the same amount of power and gives the same horsepower output for either high or low voltage, but as the voltage is doubled (230 V to 460 V), the current is cut in half. Using a reduced current enables the use of a smaller wire size which creates a savings on installation.

Wye-Connected, 3ϕ Motors. A wiring diagram is used to show the terminal numbering system for a dual-voltage, wye-connected, 3ϕ motor. See Figure 9-4. Nine leads are brought out of the motor. These leads are marked T1 through T9 and may be externally connected for either of the two voltages. The terminal connections for high and low voltage are shown below the wiring diagram. This information is normally provided on the motor nameplate.

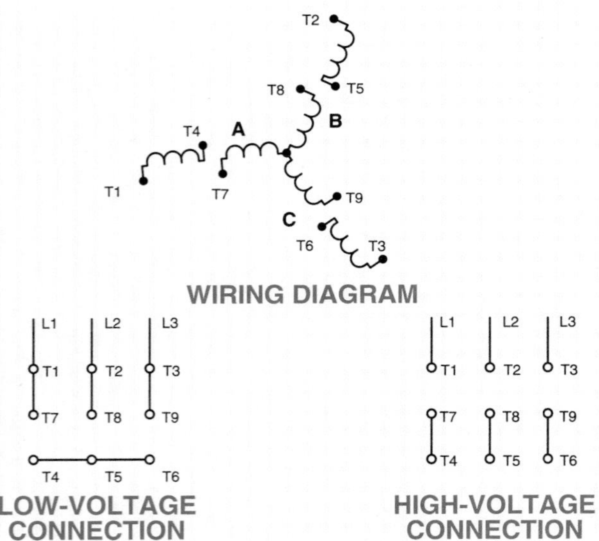

Figure 9-4. A wiring diagram shows the terminal numbering system of a dual-voltage, wye-connected, 3ϕ motor.

Leeson Electric Company

In a dual-voltage, 3ϕ motor, the nine leads connect the stator to the external electrical power source.

The nine leads are connected in either series (high voltage) or parallel (low voltage). See Figure 9-5. To connect a wye configuration for high voltage, connect L1 to T1, L2 to T2, L3 to T3, and tie T4 to T7, T5 to T8, and T6 to T9. This connects the individual coils in phases A, B, and C in series, each coil receiving 50% of the line-to-neutral point voltage. The neutral point equals the internal connecting point of all three phases.

To connect the wye motor for low voltage, connect L1 to T1 and T7, L2 to T2 and T8, L3 to T3 and T9, and tie T4, T5, and T6 together. This connects the individual coils in phases A, B, and C in parallel so that each coil receives 100% of the line-to-neutral point voltage.

SEW-EURODRIVE, Inc.

Variable frequency drives are used to control the speed of wye- or delta-connected, single- or dual-voltage, 3ϕ motors.

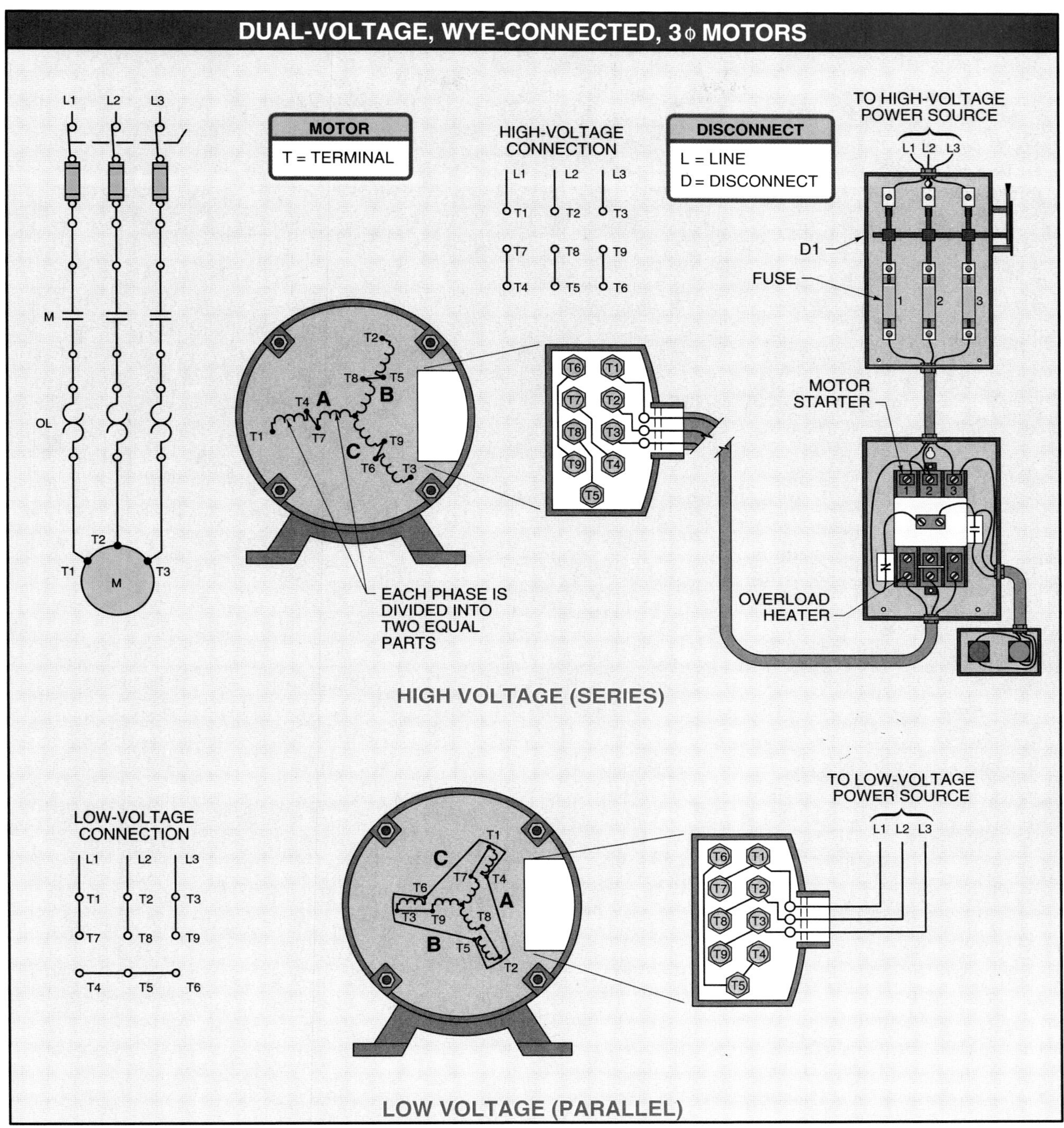

Figure 9-5. In a wye-connected, 3φ motor, each phase coil is divided into two equal parts.

Delta-Connected, 3φ Motors. A wiring diagram is used to show the terminal numbering system for a dual-voltage, delta-connected, 3φ motor. See Figure 9-6. The leads are marked T1 through T9 and a terminal connection chart is provided for wiring high- and low-voltage operations.

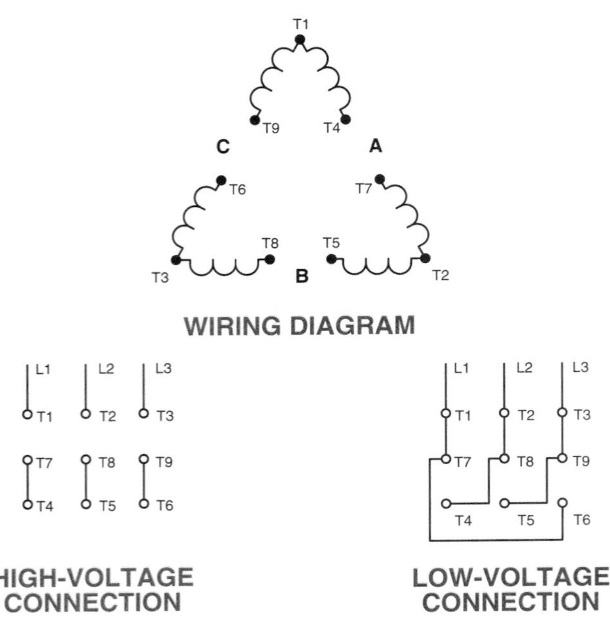

Figure 9-6. A wiring diagram shows the terminal numbering system of a dual-voltage, delta-connected, 3ϕ motor.

The nine leads are connected in either series or parallel for high or low voltage. See Figure 9-7. In the high-voltage configuration, the coils are wired in series. In the low-voltage configuration, the coils are wired in parallel to distribute the voltage according to the individual coil ratings.

Reversing Wye- and Delta-Connected, 3ϕ Motors

Reversing the direction of rotation of wye- and delta-connected, 3ϕ motors is accomplished by interchanging any two of the three main power lines to the motor. Although any two lines may be interchanged, the industry standard is to interchange L1 and L3. This standard is true for all 3ϕ motors including three, six, and nine lead wye- and delta-connected motors. Regardless of the type of 3ϕ motor, connect L1 to T1, L2 to T2, and L3 to T3 for forward rotation.

Connect L1 to T3, L2 to T2, and L3 to T1 for reverse rotation. If a 3ϕ motor has more than three leads coming out, these leads are connected according to the motor's wiring diagram.

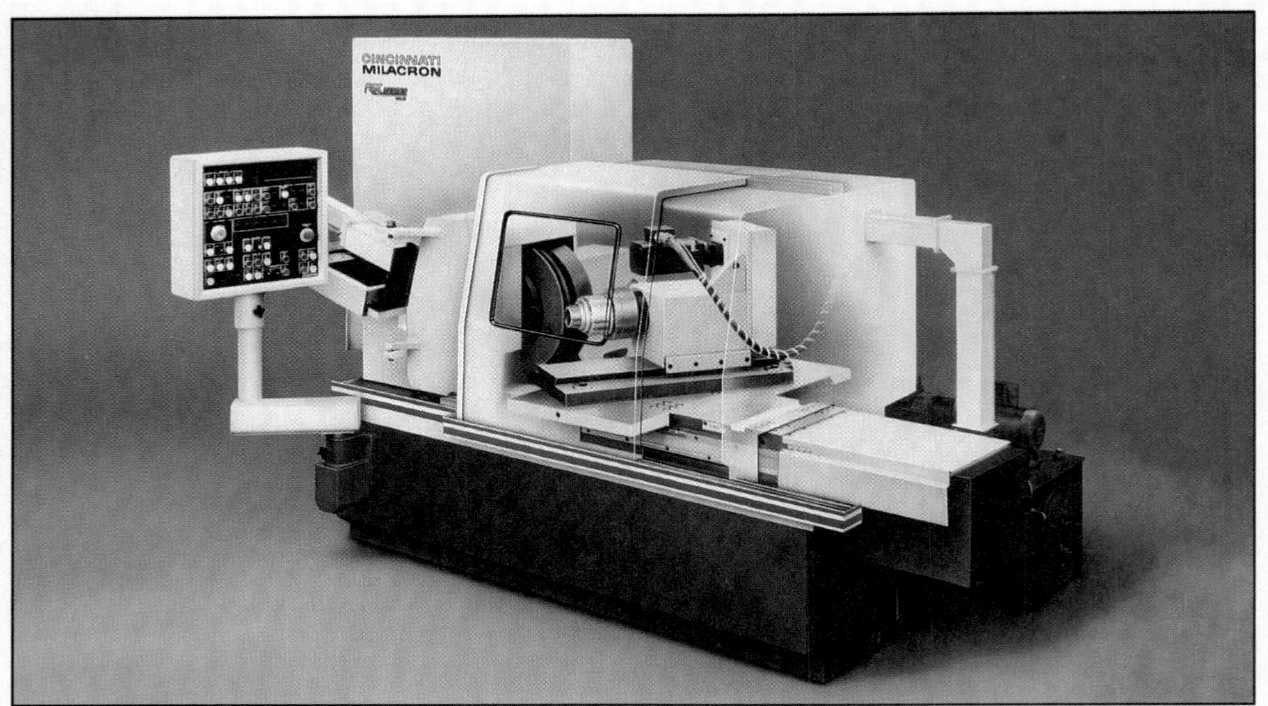

Cincinnati Milacron

Many machine tools contain reversible motors that are required for various machining and grinding operations.

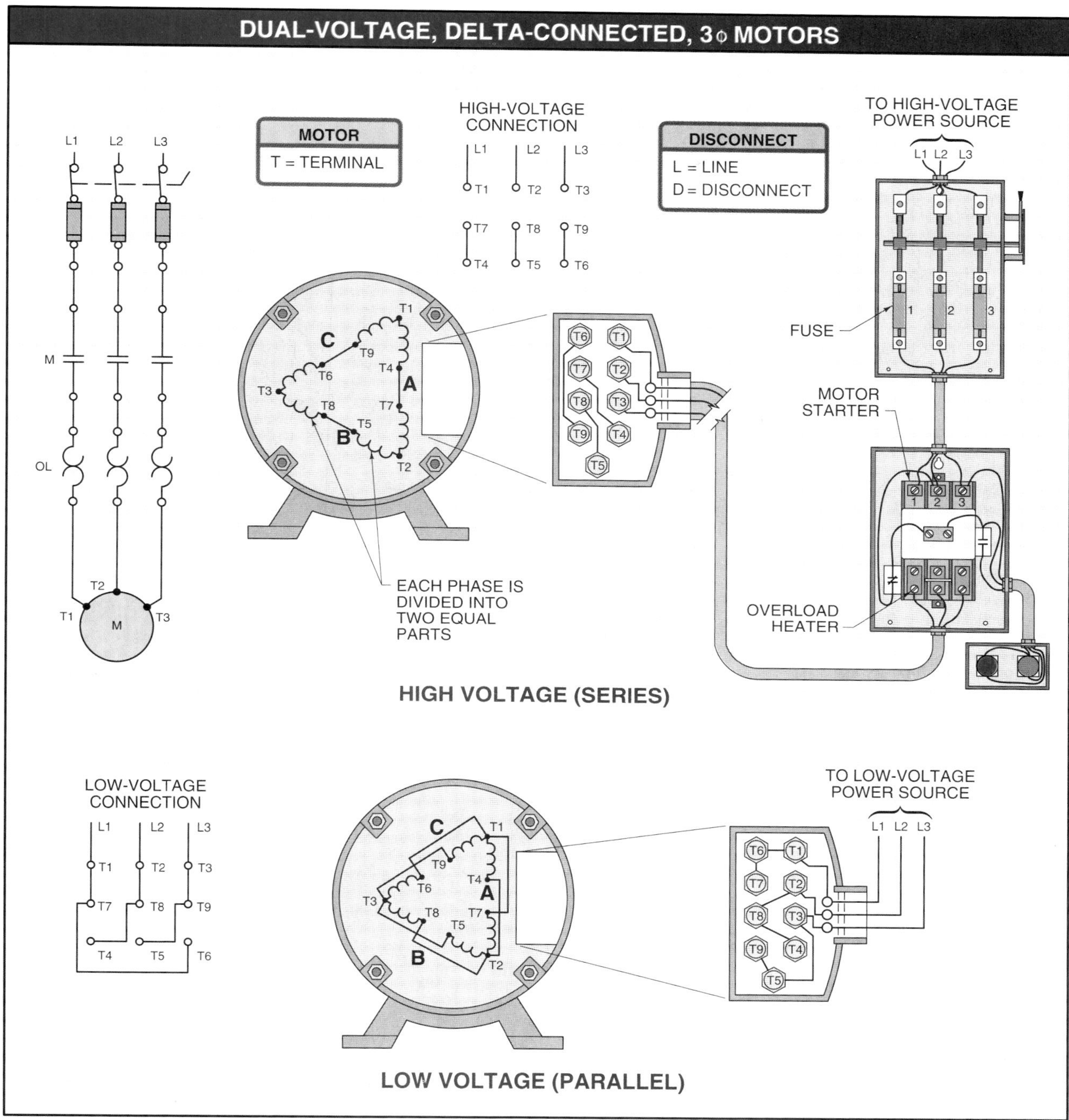

Figure 9-7. In a delta-connected, 3ϕ motor, each phase coil is divided into two equal parts.

Interchanging L1 and L3 is a standard for safety reasons. When first connecting a motor, the direction of rotation is not usually known until the motor is started. It is common practice to temporarily connect the motor to determine the direction of rotation before making permanent connections. Motor lead temporary connections are not taped. By always interchanging L1 and L3, L2 can be permanently connected to T2, creating an insulated barrier between L1 and L3.

SINGLE-PHASE MOTORS

Single-phase motors are most often found where a fractional horsepower motor drive is required or where no 3ϕ power is available. Single-phase large horsepower motors are not normally used because they are inefficient compared to 3ϕ motors and cost more to operate. With the exception of the universal motor, all 1ϕ motors must have an auxiliary means for developing starting torque. Single-phase motors include the split-phase, shaded-pole, capacitor, universal, repulsion, and synchronous motors.

Note: Always refer to the manufacturer's wiring diagram. Although there are many similarities in 1ϕ motors, minor differences among manufacturers have led to hundreds of different wiring diagrams.

Split-Phase Motors

A *split-phase motor* is a 1ϕ AC motor that includes a running winding (main winding) and a starting winding (auxiliary winding). A split-phase motor is an AC motor of fractional horsepower, usually $\frac{1}{20}$ HP to $\frac{1}{3}$ HP. Split-phase motors are commonly used to operate washing machines, oil burners, and small pumps and blowers.

A split-phase motor has a rotating part (rotor), a stationary part consisting of the running winding and starting winding (stator), and a centrifugal switch that is located inside the motor to disconnect the starting winding at about 60% to 80% of full-load speed. See Figure 9-8.

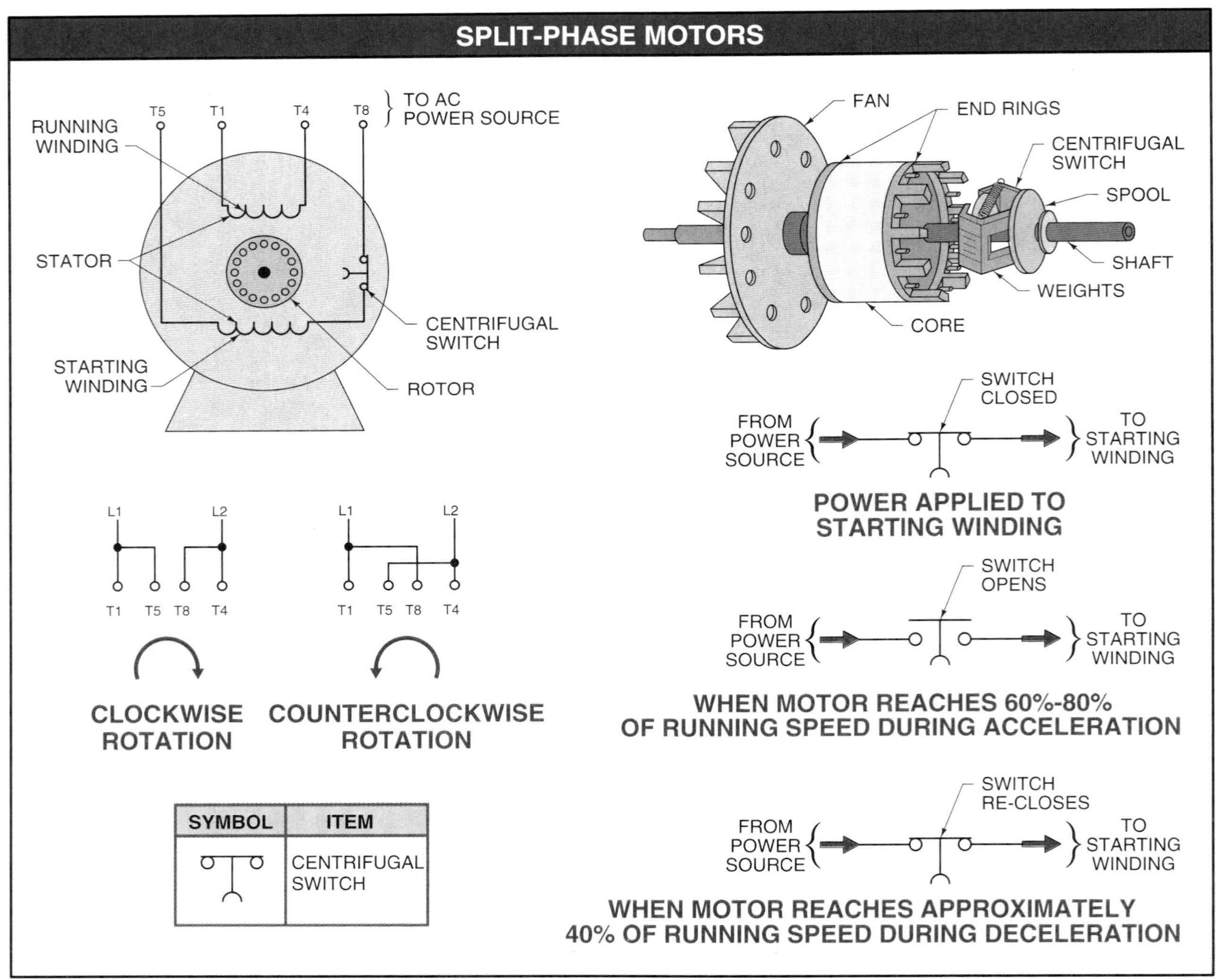

Figure 9-8. A split-phase motor is a 1ϕ, AC motor that includes a running winding and a starting winding.

When starting, both the running windings and the starting windings are in parallel. The running winding is normally made up of a heavy insulated copper wire and the starting winding is made of fine insulated copper wire. The centrifugal switch opens, disconnecting the starting winding from the circuit when the motor reaches approximately 75% of full speed. This allows the motor to operate on the running winding only. When the motor is turned OFF (power removed), the centrifugal switch recloses at approximately 40% of the motor speed.

Reversing Split-Phase Motors. Reversing the rotation of a split-phase motor is accomplished by interchanging the leads of the starting or running windings. The manufacturer's wiring diagram is used to determine the exact wires to interchange. An electrician can measure the resistance of the starting winding and running winding to determine which leads are connected to which windings if manufacturer's information is not available. The running winding is made of a heavier gauge wire than the starting winding, so the running winding shows a much lower resistance than the starting winding.

Dual-Voltage, Split-Phase Motors. In a dual-voltage, split-phase motor, the running winding is split into two sections. See Figure 9-9.

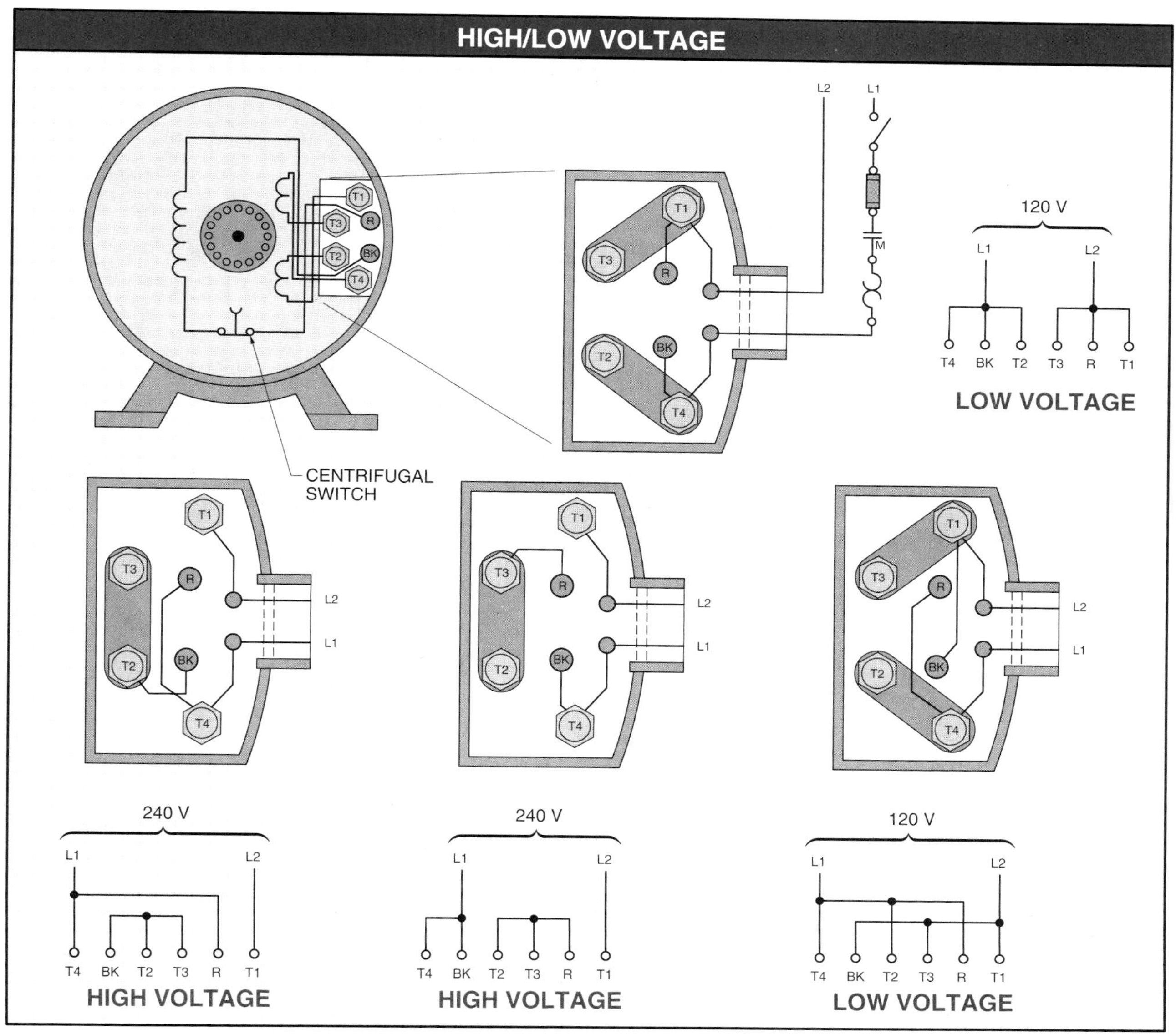

Figure 9-9. In a dual-voltage, split-phase motor, the running winding is split into two sections.

All the windings receive the same voltage when wired for low-voltage operation. Wiring for high voltage puts the running winding in series. The starting winding is parallel with only one running winding to provide proper voltage distribution. Always refer to nameplate or manufacturer's wiring diagrams.

Capacitor Motors

A *capacitor motor* is a 1φ, AC motor that includes a capacitor in addition to the running and starting windings. A capacitor motor is made in sizes ranging from 1/8 HP to 10 HP. It is used to operate refrigerators, compressors, washing machines, and air conditioners. The construction of a capacitor motor is similar to that of a split-phase motor, but a capacitor is connected in series with the starting winding. The addition of a capacitor in the starting winding gives the capacitor motor more torque than the split-phase motor. The three types of capacitor motors include capacitor-start, capacitor-run, and capacitor start-and-run motors.

Capacitor-Start Motors. A capacitor-start motor operates much the same as a split-phase motor in that it uses a centrifugal switch that operates at approximately 60% to 80% full speed. In this case, the starting winding and the capacitor are removed when the circuit opens. The capacitor used in the starting winding gives the capacitor-start motor a high starting torque. See Figure 9-10.

Heidelberg Harris, Inc.

Capacitor motors are used in automated processes where high-starting and/or running torque characteristics are required.

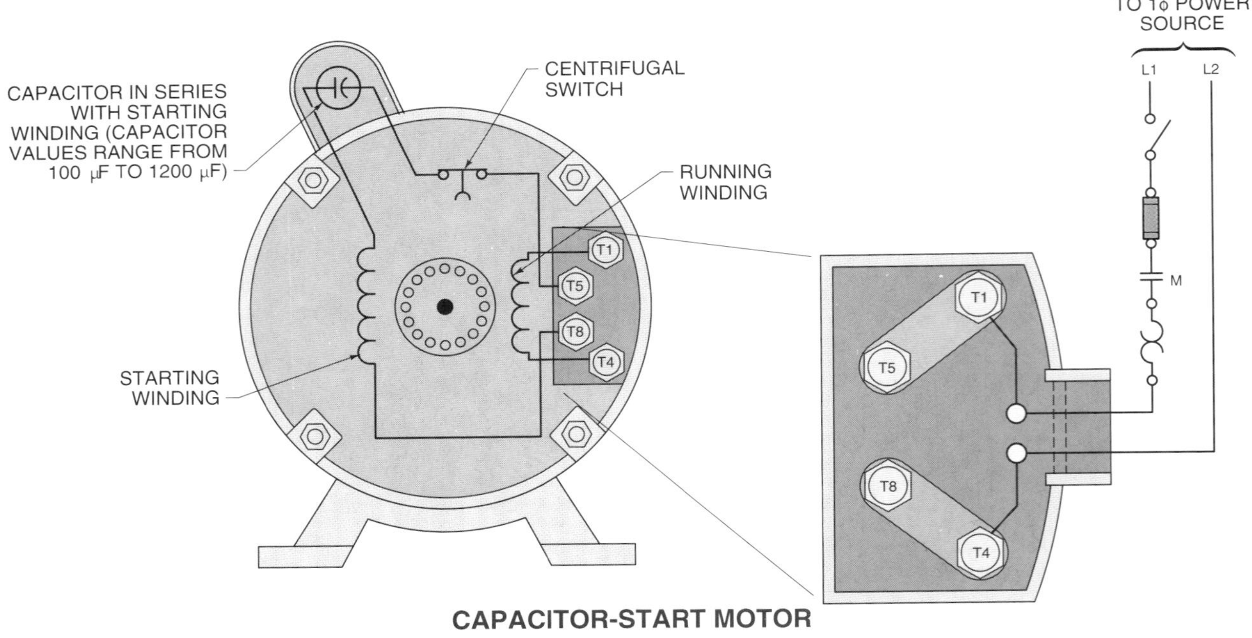

Figure 9-10. A capacitor-start motor has a capacitor in the starting winding which gives the capacitor-start motor a high starting torque.

Capacitor-Run Motors. A capacitor-run motor has the starting winding and capacitor connected in series at all times. A lower-value capacitor is used in a capacitor-run motor than in a capacitor-start motor because it remains in the circuit at full speed. This gives the capacitor-run motor medium starting torque and somewhat higher running torque than a capacitor-start motor. See Figure 9-11.

Capacitor Start-and-Run Motors. A capacitor start-and-run motor starts with a high- and low-value capacitor connected in parallel with each other, but in series with the starting winding to provide a very high starting torque. The centrifugal switch disconnects the high-value capacitor at about 75% full load speed, leaving the low-value capacitor in the circuit. See Figure 9-12.

The net result is that a capacitor start-and-run motor combines the advantage of a capacitor-start motor (high starting torque) with that of a capacitor-run motor (high running torque). In each case, the loads and the amount of torque (both starting and running) determine which capacitor motor to use.

Reversing Capacitor Motors. The direction of rotation of a capacitor motor can be changed by reversing the connections to the starting or running windings. Whenever possible, the manufacturer's wiring diagram should be referred to for the exact wires to interchange.

DC MOTORS

DC motors are usually found where the load requires an adjustable speed and simple torque control. Common applications for DC motors include printing presses, cranes, elevators, shuttle cars, and automobile starters.

A DC motor consists of a field circuit and an armature circuit. The field circuit consists of stationary windings, or permanent magnets, that provide a magnetic field around the armature. The armature circuit consists of laminated steel slots connected to the shaft and terminated at the commutator. When windings are used in the field circuit, power is applied to the windings through wires that are brought out from the windings. When permanent magnets are used for the field circuit, no power is applied to the windings, so no wires are brought out from the field circuit.

Power is applied to the armature circuit through the brushes that ride on the commutator as the armature rotates. The brushes are connected to power through wires that are brought out from the motor.

The four types of DC motors are the series, shunt, compound, and permanent-magnet motor. The main difference between the motors is in the way in which the field coil and armature coil circuits are wired. This relationship is either a series, parallel, or series-parallel connection.

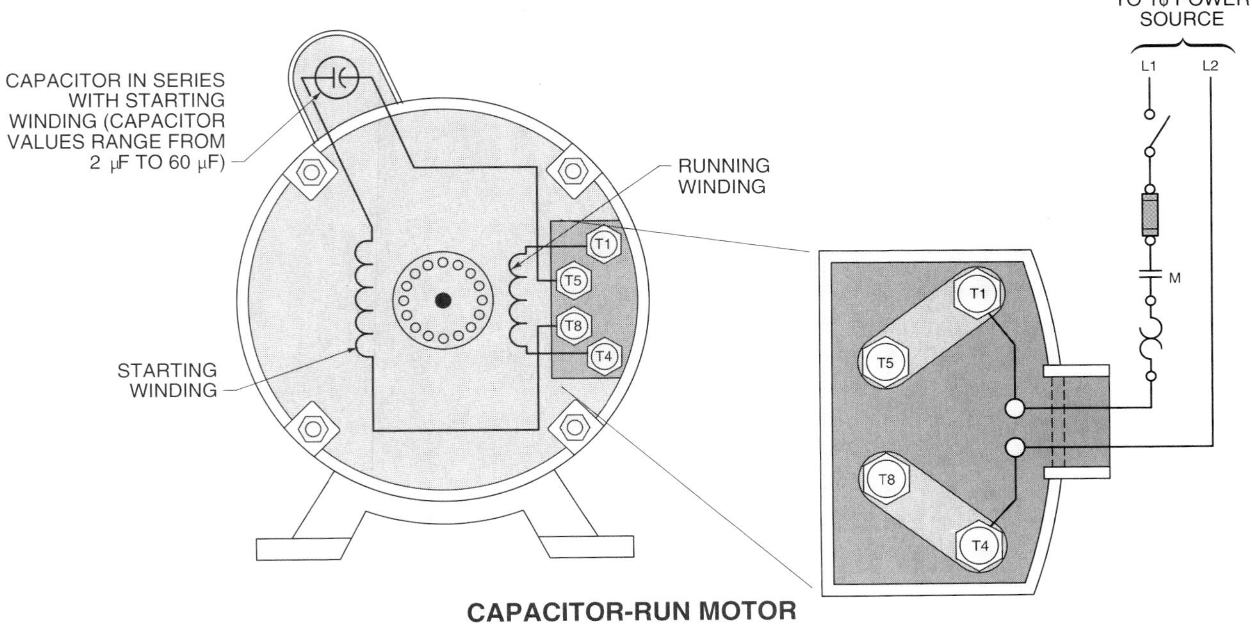

Figure 9-11. In a capacitor-run motor, the capacitor is not removed while the motor is running.

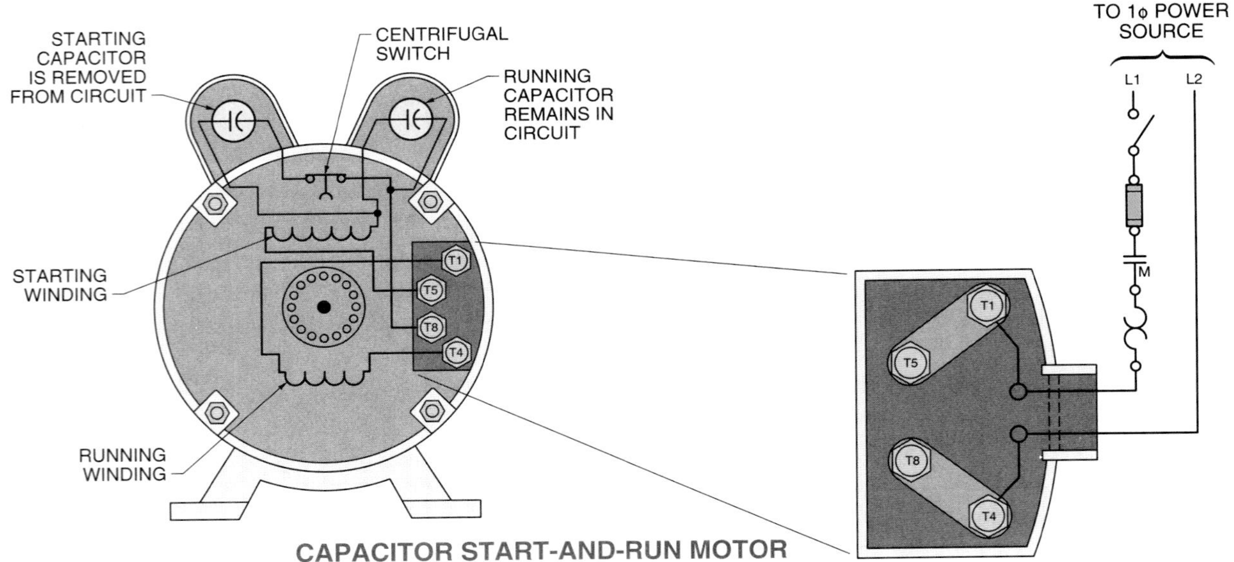

Figure 9-12. In a capacitor start-and-run motor, the starting capacitor is removed when the motor reaches speed, but the running capacitor remains in the circuit.

DC Series Motors

In a DC series motor, the series field coils are composed of a few turns of heavy gauge wire connected in series with the armature. The series coil wires are marked S1 and S2. The armature wires are marked A1 and A2. See Figure 9-13.

A DC series motor has a high starting torque and a variable speed. This allows the motor to start very heavy loads. The speed of a DC series motor increases as the load decreases. A DC series motor develops high starting torque because the same current that passes through the armature also passes through the field. If the armature calls for more current (developing more torque), this current also passes through the field, increasing the field strength.

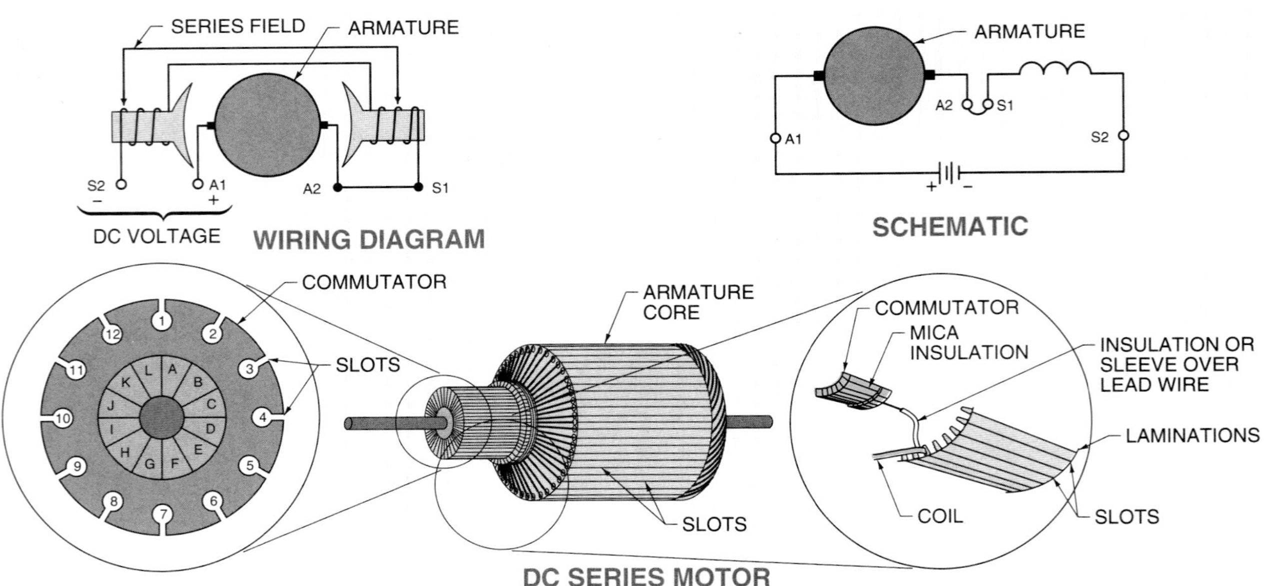

Figure 9-13. A DC series motor is a motor with the field connected in series with the armature.

The ability to draw more current is an advantage as long as a load is applied to the motor because the load tends to hold the motor speed under control. If the load is removed, the variable speed characteristic allows the motor to keep increasing speed after the load is moved. If left unchecked, the motor throws itself apart at uncontrolled speeds. For this reason, it is necessary to positively connect DC series motors to loads through couplings or gears that do not allow slip. Belt drives should not be used on DC series motors. In small DC series motors, brush friction, bearing friction, and winding loss may provide sufficient load to hold the speed to a safe level.

DC Shunt Motors

In a DC shunt motor, the armature and field circuits are wired in parallel, giving essentially constant field strength and motor speed. The windings extending from the shunt field are marked F1 and F2. See Figure 9-14. The shunt field can be either connected to the same power supply as the armature (self excited) or to another power supply (separately excited).

When DC shunt motors are separately excited, the motor can be speed controlled by varying the field current. Field speed control can be accomplished by inserting external resistances in series with the shunt field circuits. As resistance is increased in the field circuit, field current is reduced and the speed is increased. Conversely, as resistance is removed from the shunt field, field current is increased and the speed is decreased. By selecting the proper controller, the motor may be set for a specific speed control range. Standard DC shunt field voltages are 100 V, 150 V, 200 V, 240 V, and 300 VDC.

Baldor Electric Co.

The wiring diagram on a motor nameplate must always be checked because some DC motors include a dual-voltage field.

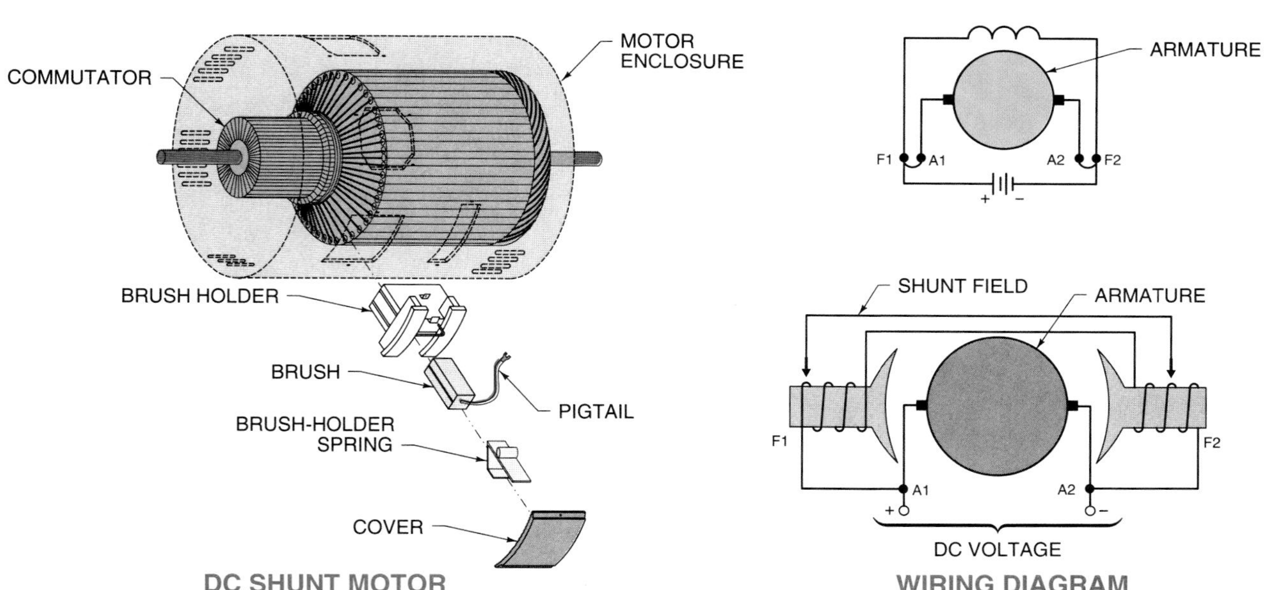

Figure 9-14. A DC shunt motor is a motor with the field connected in shunt (parallel) with the armature.

DC Compound Motors

In a DC compound motor, the field coil is a combination of the series and shunt fields. See Figure 9-15. The series field is connected in series with the armature and the shunt field is connected in parallel with the series field and armature. This arrangement combines the characteristics of both the series and shunt motor. DC compound motors have high starting torque and fairly good speed torque characteristics at its rated load. Only large bidirectional DC compound motors are normally built because of the complicated circuits needed to control compound motors.

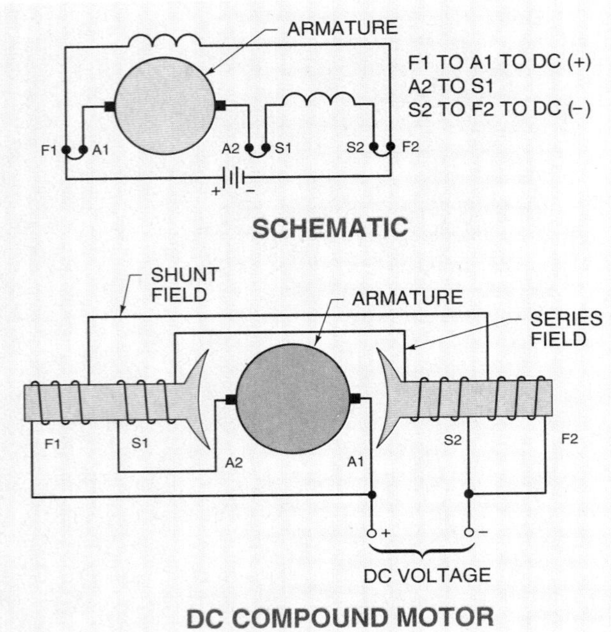

Figure 9-15. In a DC compound motor, the field coil is a combination of the series and shunt fields.

Some smaller DC motors may be slightly compounded to improve starting characteristics. A DC compound motor is used for drives needing fairly high starting torque and reasonably constant running speed.

DC Permanent-Magnet Motors

A DC permanent-magnet motor is a motor that uses magnets, not a coil of wire, for the field winding. A DC permanent-magnet motor has molded magnets mounted into a steel shell. The permanent magnets are the field coils. DC power is supplied only to the armature. See Figure 9-16.

A DC permanent-magnet motor is used in automobiles to control power seats, power windows, and windshield wipers. They produce relatively high torque at low speeds and provide some dynamic (self) braking when removed from power.

Reversing DC Motors

The direction of rotation of DC series, shunt, and compound motors may be reversed by reversing the direction of the current through the field without changing the direction of the current through the armature, or by reversing the direction of the current through the armature, but not both. The industrial standard is to reverse the current through the armature. The direction of rotation of a DC permanent-magnet motor is reversed by reversing the direction of the current through the armature only, since there are no field connections available.

In a DC compound motor, the series and shunt field relationship to the armature must be left unchanged. The shunt must be connected in parallel and the series field in series with the armature. Reversal is accomplished by reversing the armature connections only. If the motor has commutating pole windings, these windings are considered a part of the armature circuit and the current through them must be reversed when the current through the armature is reversed. Commutating windings (interpoles) are used to prevent sparking at the brushes in some DC motors.

Sprecher + Schuh

Overhead cranes driven by electric motors are used to move heavy metal beams.

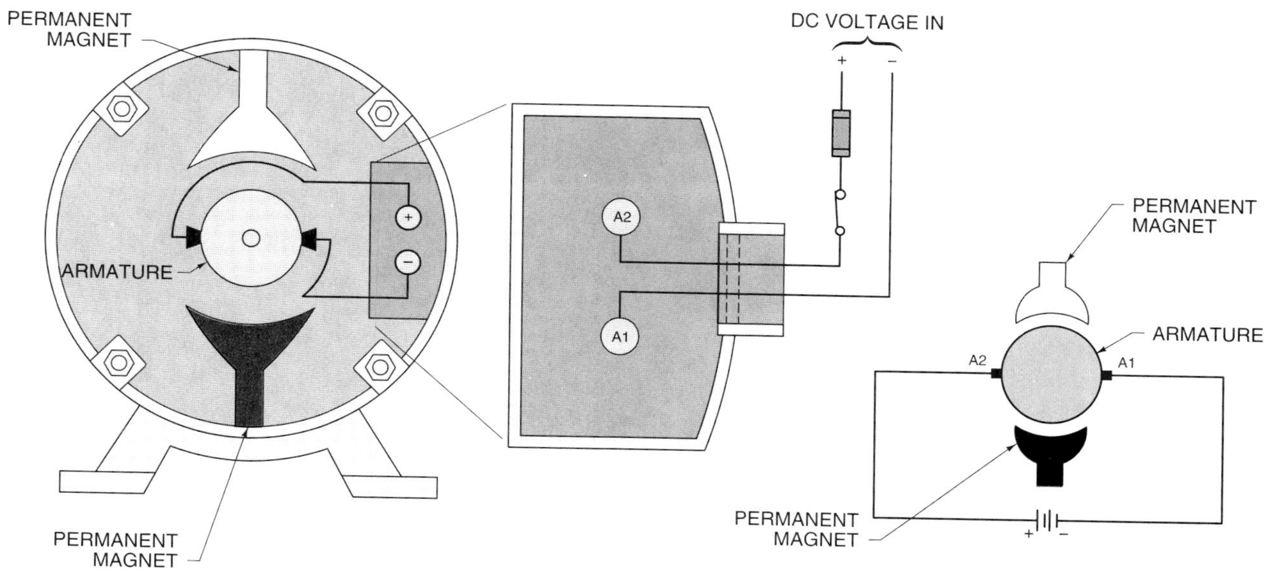

Figure 9-16. A DC permanent-magnet motor is a motor that uses magnets, not a coil of wire, for the field winding.

REVERSING MOTORS USING MANUAL STARTERS

Manual reversing starters used to change the direction of rotation in 3ϕ, 1ϕ, and DC motors are made by connecting two manual starters. See Figure 9-17.

A manual starter is used to run low horsepower motors such as those found on fans, small machines, pumps, and blowers in forward and reverse. The individual manual starters are marked start/stop instead of forward/stop or reverse/stop. See Figure 9-18. This is common when two manual starters are placed in the same enclosure to make up a manual reversing starter. The electrician must correctly label the unit once it is properly wired.

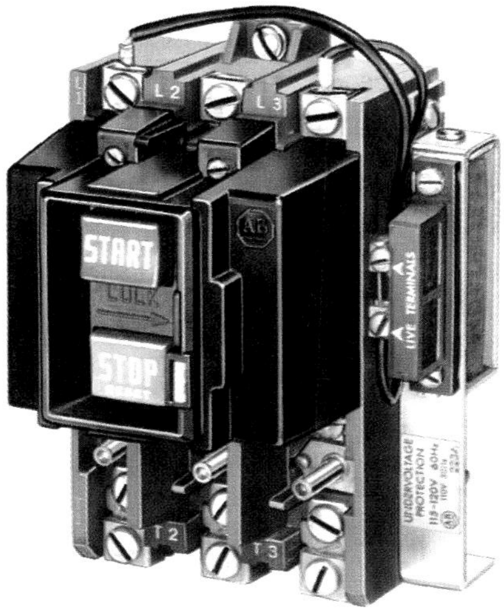

Rockwell Automation, Allen-Bradley Company, Inc.

Figure 9-17. Manual starters are used individually to control motors or in pairs to reverse motors.

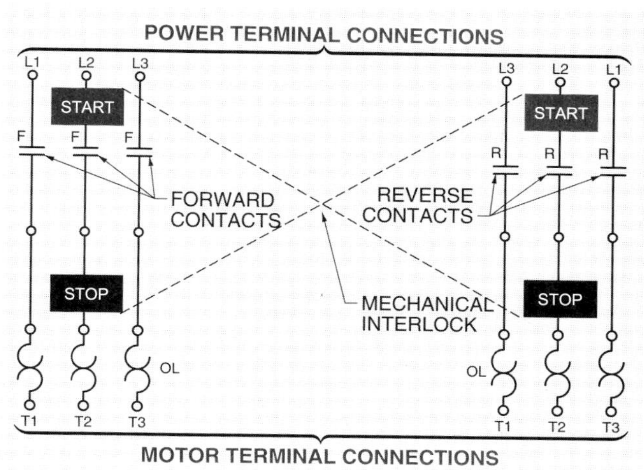

Figure 9-18. A manual pushbutton reversing motor starter consists of two manual starters mechanically interlocked.

Crossing dashed lines are used between the manual starter in the wiring diagram to indicate a mechanical interlock. Since the motor cannot run in both directions at the same time, some means must be included to prevent both starters from energizing at the same time. The manual reversing starter uses a mechanical interlock for separating the contacts on a reversing manual starter. These mechanical devices are inserted between the two starters as they are installed to ensure that both switching mechanisms cannot be energized at the same time. The electrician must ensure that the interlock is provided if the unit is not pre-assembled.

Reversing 3ϕ Motors Using Manual Starters

A wiring diagram illustrates the electrical connections necessary to properly reverse a 3ϕ motor using a manual reversing starter. See Figure 9-19. Only one set of overloads is required.

When using a manual reversing starter to reverse a 3ϕ motor, L1, L2, and L3 are connected directly in sequence to T1, T2, and T3 when the forward contacts close. See Figure 9-20.

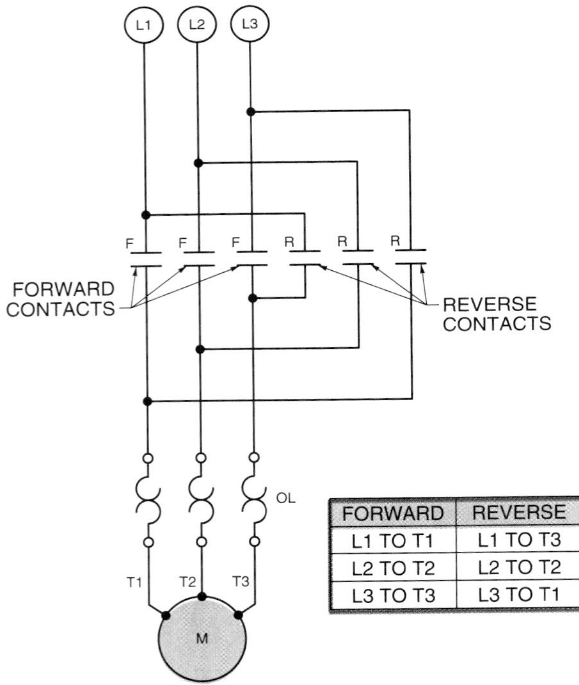

FORWARD	REVERSE
L1 TO T1	L1 TO T3
L2 TO T2	L2 TO T2
L3 TO T3	L3 TO T1

Figure 9-19. A wiring diagram illustrates the electrical connections necessary to properly reverse a 3ϕ motor using a manual reversing starter.

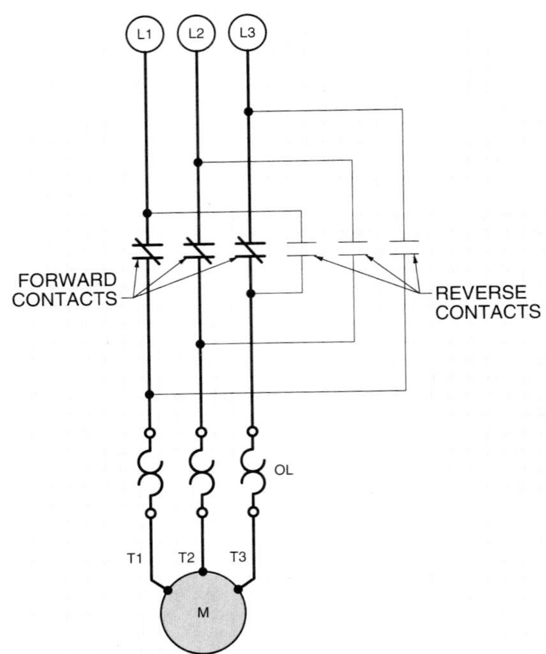

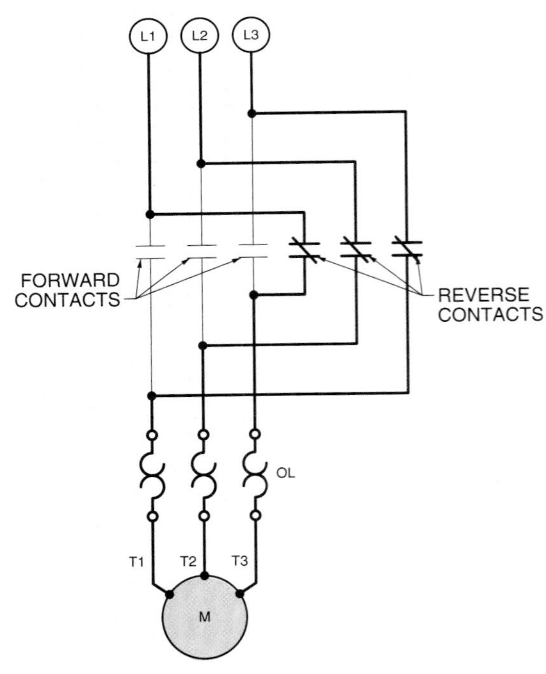

Figure 9-20. When using a manual reversing starter to reverse a 3ϕ motor, L1, L2, and L3 are connected directly in sequence to T1, T2, and T3 when the forward contacts close.

Line 1 is connected to T3, L2 is connected to T2, and L3 is connected to T1 when the reverse contacts close and the forward contacts open. The motor changes directions each time forward or reverse is depressed because it is necessary to interchange only two leads on a 3ϕ motor to reverse rotation.

Reversing 1ϕ Motors Using Manual Starters

A wiring diagram illustrates the wiring necessary to properly reverse a 1ϕ motor using a manual reversing starter. See Figure 9-21. *Note:* Always check the manufacturer's wiring diagrams for proper reversal of 1ϕ motors.

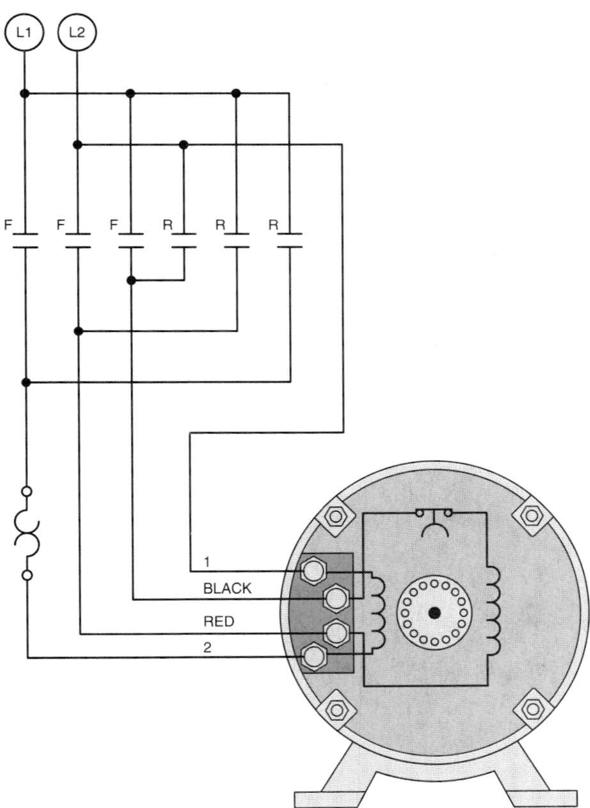

Figure 9-21. A wiring diagram illustrates the wiring necessary to properly reverse a 1ϕ motor using a manual reversing starter.

Line 1 is connected to the black lead of the starting winding and side 1 of the running winding, and L2 is connected to the red lead of the starting winding and side 2 of the running windings when the forward contacts close. See Figure 9-22.

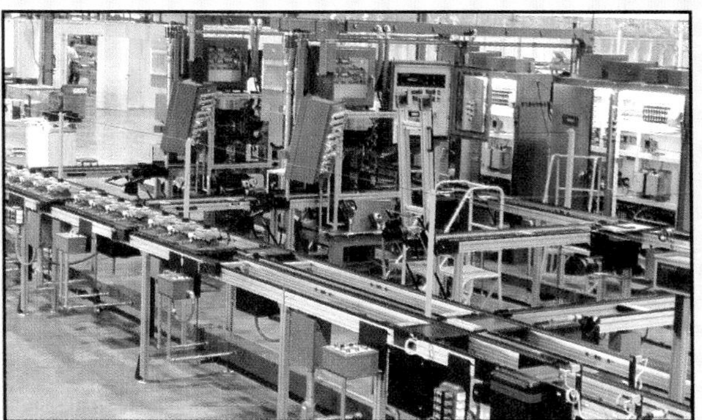

Advanced Assembly Automation Inc.

Most conveyor systems are designed to run in either direction to make production lines more flexible.

Line 1 is connected to the red lead of the starting winding and side 1 of the running winding and L2 is connected to the black lead of the starting winding and side 2 of the running winding when the reverse contacts close and the forward contacts open. The starting windings are interchanged while the running windings remain the same.

The motor changes direction each time forward or reverse is depressed because it is necessary to interchange only the starting windings on a 1ϕ motor to reverse rotation. *Note:* Always check the manufacturer's wiring diagrams when reversing 1ϕ motors to determine which leads are connected to the starting winding. The red and black wires are the ones normally used for reversal.

Rockwell Automation, Allen-Bradley Company, Inc.

Allen-Bradley's full-voltage reversing starters with the SMP-3™ overload relay enables auto/manual reset, communication capability, and phase loss, ground fault, and jam protection.

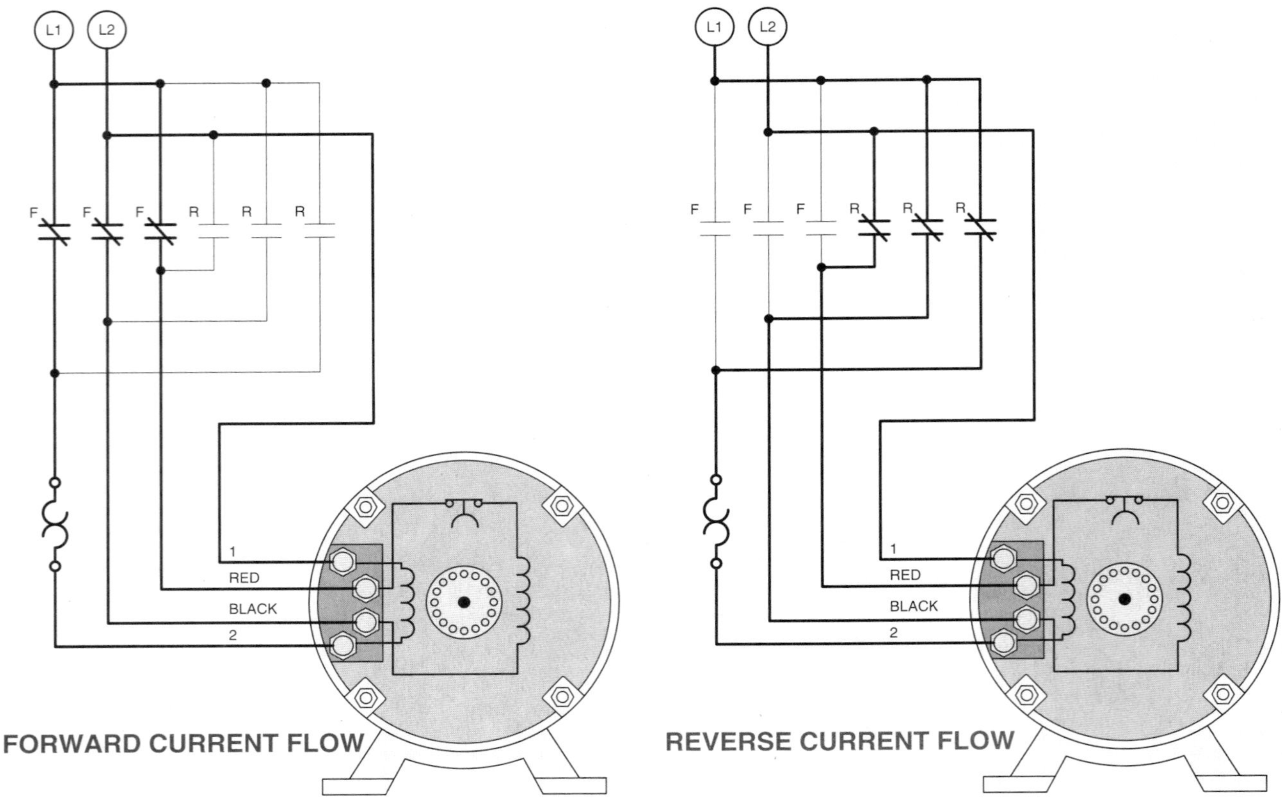

Figure 9-22. Line 1 is connected to the black lead of the starting winding and side 1 of the running winding, and L2 is connected to the red lead of the starting winding and side 2 of the running windings when the forward contacts close.

Reversing DC Motors Using Manual Starters

The direction of rotation of all DC motors is reversed by changing the direction of current flow through the armature. A manual starter can be used to reverse the direction of current flow through the armature of all DC motors. The motor is wired to the starter so that polarity of the applied DC voltage remains the same on the field for either direction, but the polarity on the armature is opposite for each direction.

A wiring diagram is used to properly wire a DC series motor for reversing. See Figure 9-23. A DC series motor is wired to the starter so that A2 is positive and A1 is negative when the forward contacts are closed, and A2 is negative and A1 is positive when the reverse contacts are closed. Regardless of whether the forward contacts or reverse contacts are closed, S2 is always positive and S1 is always negative. The motor reverses direction for each position of the starter because only the polarity of the armature reverses direction.

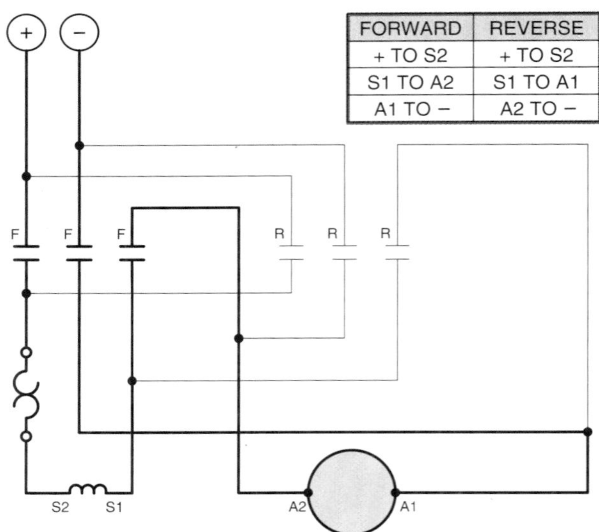

FORWARD	REVERSE
+ TO S2	+ TO S2
S1 TO A2	S1 TO A1
A1 TO –	A2 TO –

Figure 9-23. A DC series motor is wired to the starter so that A2 is positive and A1 is negative when the forward contacts are closed, and A2 is negative and A1 is positive when the reverse contacts are closed.

A wiring diagram is used to properly wire a DC shunt motor for reversing. See Figure 9-24. A DC shunt motor is wired to the starter so that A2 is positive and A1 is negative when the forward contacts are closed and A2 is negative and A1 is positive when the reverse contacts are closed. Regardless of whether the forward contacts or reverse contacts are closed, F2 is always positive and F1 is always negative. The motor reverses direction for each position of the starter because only the polarity of the armature reverses direction.

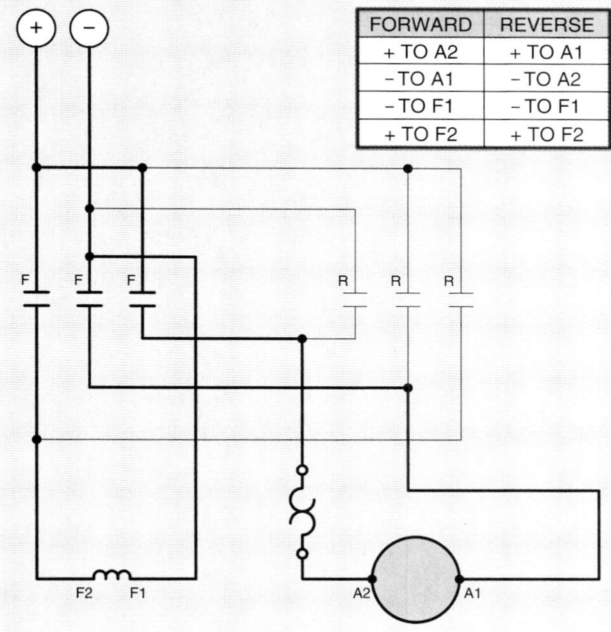

FORWARD	REVERSE
+ TO A2	+ TO A1
− TO A1	− TO A2
− TO F1	− TO F1
+ TO F2	+ TO F2

Figure 9-24. A wiring diagram is used to properly wire a DC shunt motor for reversing.

A wiring diagram is used for properly wiring a DC compound motor for reversing. See Figure 9-25. A DC compound motor is wired to the starter so that A2 is positive and A1 is negative when the forward contacts are closed, and A2 is negative and A1 is positive when the reverse contacts are closed. Regardless of whether the forward contacts or reverse contacts are closed, S2 and F2 are always positive and S1 and A1 are always negative. The motor reverses direction for each position of the starter because only the polarity of the armature reverses direction.

A wiring diagram is used for properly wiring a DC permanent-magnet motor for reversing. See Figure 9-26.

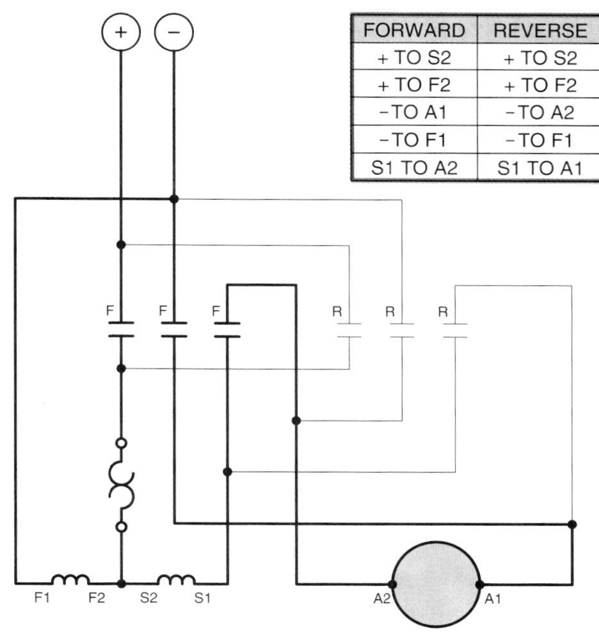

FORWARD	REVERSE
+ TO S2	+ TO S2
+ TO F2	+ TO F2
− TO A1	− TO A2
− TO F1	− TO F1
S1 TO A2	S1 TO A1

Figure 9-25. A DC compound motor is wired to the starter so that A2 is positive and A1 is negative when the forward contacts are closed, and A2 is negative and A1 is positive when the reverse contacts are closed.

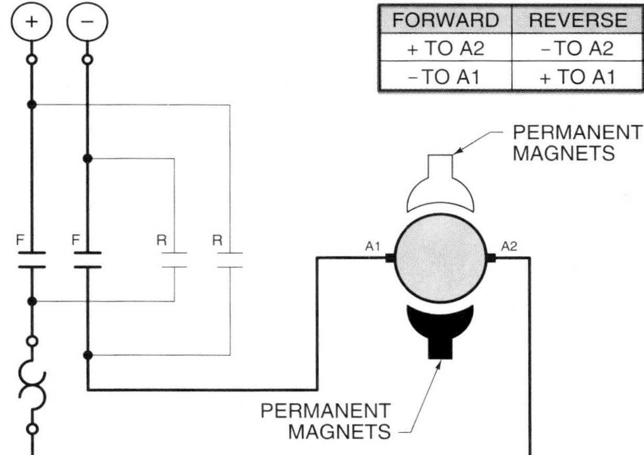

FORWARD	REVERSE
+ TO A2	− TO A2
− TO A1	+ TO A1

Figure 9-26. A wiring diagram is used for properly wiring a DC permanent-magnet motor for reversing.

The DC permanent-magnet motor is wired to the starter so that A2 is positive and A1 is negative when the forward contacts are closed and A2 is negative and A1 is positive when the reverse contacts are closed. A permanent-magnet field never reverses its direction of polarity regardless of the polarity to which the armature is connected.

REVERSING MOTORS USING MAGNETIC STARTERS

A magnetic reversing starter performs the same function as a manual reversing starter. See Figure 9-27. The only difference between manual and magnetic reversing starters is the addition of forward and reverse coils and the use of auxiliary contacts. The forward and reverse coils replace the pushbuttons of a manual starter and the auxiliary contacts provide additional electrical protection and circuit flexibility. The reversing circuit is the same for both manual and magnetic starters.

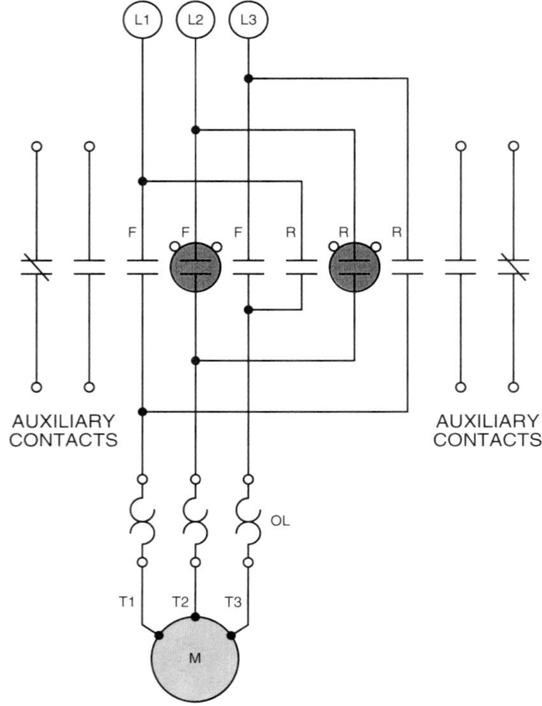

Figure 9-27. A magnetic reversing starter has forward and reverse coils which replace the pushbuttons of a manual starter and auxiliary contacts which provide additional electrical protection and circuit flexibility.

Mechanical Interlocking

A magnetic reversing starter may be controlled by a forward and reverse pushbutton. See Figure 9-28. *Note:* A line diagram does not show the power contacts. The power contacts are found in the wiring diagram. The broken lines running from the forward coil to the reverse coil indicate that the coils are mechanically interlocked like those of a manual reversing starter. This mechanical interlock is normally factory-installed by most manufacturers.

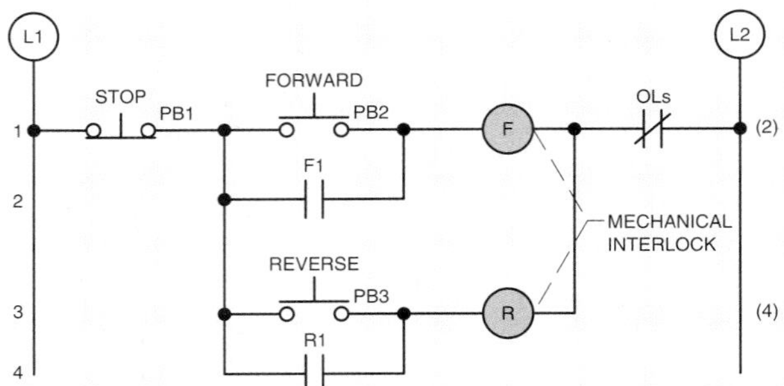

Figure 9-28. A magnetic reversing starter may be controlled by a forward and reverse pushbutton.

Depressing forward pushbutton PB2 completes the forward coil circuit from L1 to L2, energizing coil F. Coil F energizes auxiliary contacts F1, providing memory. Mechanical interlocking keeps the reversing circuit from closing. Depressing stop pushbutton PB1 opens the forward coil circuit, causing coil F to de-energize and contacts F1 to return to their NO position. Depressing reverse pushbutton PB3 completes the reverse coil circuit from L1 to L2, energizing coil R. Coil R energizes auxiliary contacts R1, providing memory.

Mechanical interlocking keeps the forward circuit from closing. Depressing stop pushbutton PB1 opens the reverse coil circuit, causing coil R to de-energize and contacts R1 to return to their NO position. Overload protection is provided both in forward and reverse by the same set of overloads.

Auxiliary Contact Interlocking

Although most magnetic reversing starters provide mechanical interlock protection, some circuits are provided with a secondary back-up or safety back-up system that uses auxiliary contacts to provide electrical interlocking. See Figure 9-29. One NO set and one NC set of contacts are activated when the forward coil circuit is energized. The NO contacts close, providing memory and the NC contacts open, providing electrical isolation in the reverse coil circuit. When the forward coil circuit is energized, the reverse coil circuit is automatically opened or isolated from the control voltage. Even if the reverse pushbutton is closed, no electrical path is available in the reverse circuit. For the reverse circuit to operate, the stop pushbutton must be pressed so that the forward circuit de-energizes and returns the contacts to their normal position. Depressing the reverse pushbutton provides the same electrical interlock for the reverse circuit when the forward contacts are in their normal position.

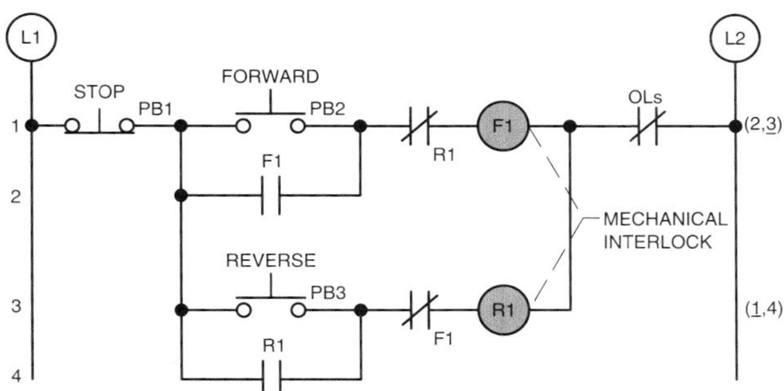

Figure 9-29. Although most magnetic reversing starters provide mechanical interlock protection, some circuits are provided with a secondary back-up system that uses auxiliary contacts to provide electrical interlocking.

Pushbutton Interlocking

Pushbutton interlocking may be used with either or both mechanical and auxiliary interlocking. Pushbutton interlocking uses both NO and NC contacts mechanically connected on each pushbutton. See Figure 9-30.

The NC contacts wired into the R coil circuit open, providing electrical isolation, when NO contacts on the forward pushbutton close to energize the F coil circuit. Conversely, the NC contacts wired into the F coil circuit open, providing electrical isolation, when the NO contacts on the reverse pushbutton close to energize the R coil circuit. Mechanical and auxiliary contact interlocking are also provided in the circuit.

Caution: In many cases motors, or the equipment they are powering, cannot withstand a rapid reversal of direction. Care must be exercised to determine what can be safely reversed under load. Also consider the braking that must be provided to slow the machine to a safe speed before reversal.

MAGNETIC REVERSING STARTER PRACTICAL APPLICATIONS

Many applications can be built around a basic magnetic reversing starter because magnetic reversing starters are controlled electrically. These circuits include starting and stopping motors in forward and reverse and controlling the motors with various control devices.

Starting and Stopping in Forward and Reverse with Indicator Lights

Operators are often required to know the direction of rotation of a motor at a given moment. An example is a motor controlling a crane which raises and lowers a load. See Figure 9-31. The line diagram is capable of indicating, through lights, the direction the motor is operating. By the electrician adding nameplates, these lights could indicate up and down directions of the hoist.

Atlas Technologies Inc.

Multiple forward and reversing circuits are required on large industrial machine operations.

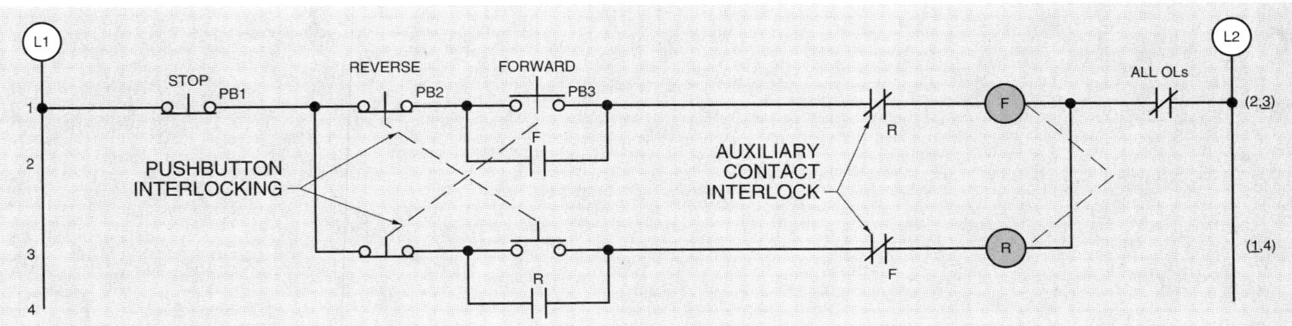

Figure 9-30. Pushbutton interlocking uses both NO and NC contacts mechanically connected on each pushbutton.

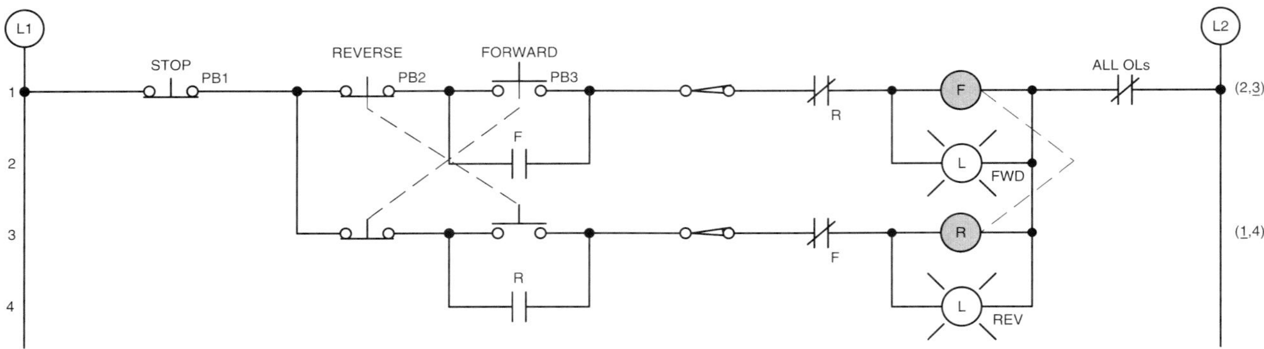

Figure 9-31. A start/stop/forward/reverse circuit with indicator lights enables an operator to know the direction of rotation of a motor at a given moment.

Pushing the momentary contact forward pushbutton causes the NO and NC contacts to move simultaneously. The NO contacts close, energizing coil F while this pushbutton is depressed. Coil F causes the memory contacts F to close and the NC electrical interlock to open, isolating the reversing circuit. The forward pilot light turns ON when holding contacts F are closed. For the period of time the pushbutton is depressed, the NC contacts of the forward pushbutton open and isolate the reversing coil R.

Pushing the momentary contact reverse pushbutton causes the NO and NC contacts to move simultaneously. The opening of the NC contacts de-energizes coil F. With coil F de-energized, the memory contacts F1 open and the electrical interlock F closes. The closing of the NO contacts energizes coil R. Coil R causes the holding contacts R1 to close and the NC electrical interlock to open, isolating the forward circuit. The reverse pilot light turns ON when the memory contacts R close. Pushing the stop pushbutton with the motor running in either direction stops the motor and causes the circuit to return to its normal position.

Overload protection for the circuit is provided by the heater coils. Operation of the overload contacts breaks the circuit opening the overload contacts. The motor cannot be restarted until the overloads are reset and the forward or reverse pushbutton is pressed.

This circuit provides protection against low voltage or a power failure. A loss of voltage de-energizes the circuit and hold-in contacts F or R open. This design prevents the motor from starting automatically after the power returns.

Starting and Stopping in Forward and Reverse with Limit Switches Controlling Reversing

Limit switches may be used to provide automatic control of reversing circuits. See Figure 9-32. This circuit uses limit switches and a control relay to automatically reverse the direction of a machine at predetermined points. This circuit could control the table of an automatic grinding machine where the operation must be periodically reversed.

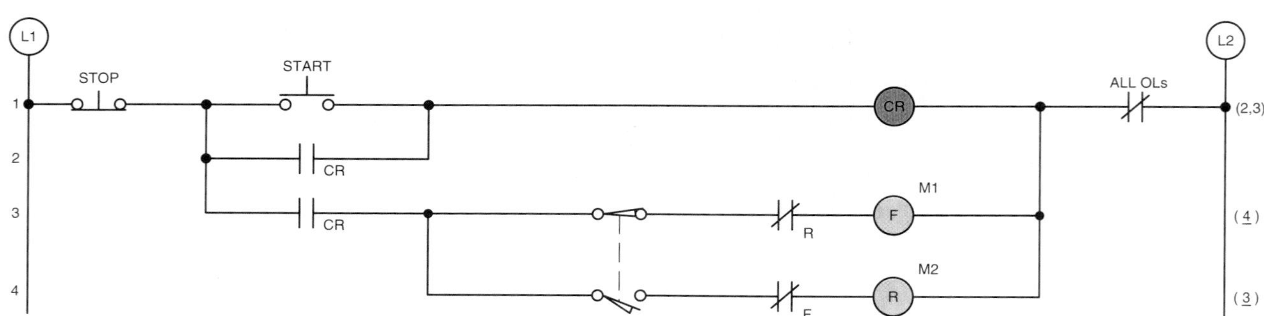

Figure 9-32. Limit switches may be used to provide automatic control of reversing circuits.

Pushing the start pushbutton causes control relay CR to become energized. The auxiliary CR contacts close when control relay CR is energized. One set of contacts form the holding contacts and the other contacts connect the limit switch circuit. The motor runs when the limit switch circuit is activated. The motor runs in the forward direction if the forward limit switch is closed. The motor runs in the opposite direction if the reverse limit switch is closed.

Overload protection for the circuit is provided by the heater coils. Operation of the overload contacts breaks the circuit. The motor cannot be restarted until the overloads are reset and the start pushbutton is pressed.

This circuit provides protection against low voltage or a power failure. A loss of voltage de-energizes the circuit and hold-in contacts CR open. This prevents the motor from starting automatically after the power returns.

Starting and Stopping in Forward and Reverse with Limit Switch Acting as Safety Stop at Certain Points in Either Direction

For safety reasons, it may be necessary to ensure that a load controlled by a reversing motor does not go beyond certain operating points in the system. A box should not go too far down a conveyor system or a hydraulic lift should not raise too high. Limit switches are incorporated to shut the operation down if it goes far enough to be unsafe. See Figure 9-33. The circuit provides overtravel protection through the use of limit switches.

Pushing the forward pushbutton activates coil F. Coil F pulls in the holding contacts F and opens the electrical interlock F, isolating the reversing circuit. The motor runs in the forward mode until either the stop pushbutton is pressed or the limit switch is activated. The circuit is broken and the holding contacts and electrical interlock return to their normal state if either control is activated.

Sprecher + Schuh

The troubleshooting of motor control circuits is done at the motor control center.

Pressing the reverse pushbutton activates coil R. Coil R pulls in the holding contacts R and opens the electrical interlock R, isolating the forward circuit. The motor runs in the reverse mode until either the stop pushbutton is pressed or the limit switch is activated. The circuit is broken and the holding contacts and electrical interlock return to their normal state if either control is activated. The circuit may still be reversed to clear a jam or undesirable situation if either limit switch is opened.

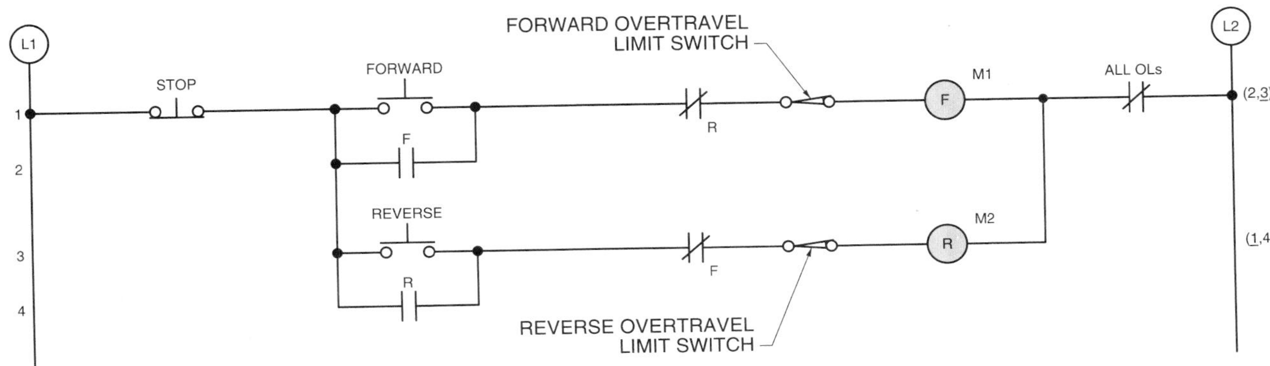

Figure 9-33. Limit switches may be used in a circuit to provide overtravel protection.

Overload protection for the circuit is provided by the heater coils. Operation of the overload contacts breaks the circuit. The motor cannot be restarted until the overloads are reset and the forward pushbutton is pressed.

This circuit provides protection against low voltage or a power failure in that a loss of voltage de-energizes the circuit and hold-in contacts F or R open. This prevents the motor from starting automatically after the power returns.

Selector Switch Used to Determine Direction of Motor Travel

A selector switch and a basic start/stop station can be used to reverse a motor. See Figure 9-34. The motor can be run in either direction, but the desired direction must be set by a selector switch before starting.

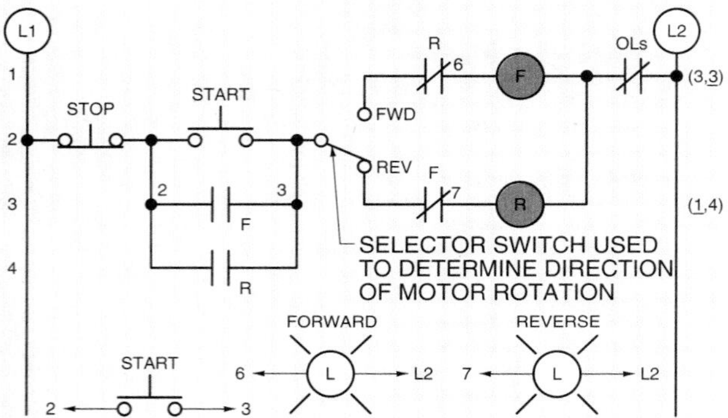

Figure 9-34. A selector switch and a basic start/stop station can be used to reverse a motor.

Pushing the start pushbutton with the selector switch in the forward position energizes coil F. Coil F closes the holding contacts F and opens the electrical interlock F, isolating the reversing circuit. Pushing the stop pushbutton de-energizes coil F, which releases the holding contacts and the electrical interlock. Pushing the start pushbutton with the selector switch in the reverse position energizes coil R. Coil R closes the holding contacts R and opens the electrical interlock R, isolating the forward circuit.

Overload protection for the circuit is provided by the heater coils. Operation of the overload contacts break the circuit. The motor cannot be restarted until the overloads are reset and the start pushbutton is pressed.

This circuit provides protection against low voltage or a power failure in that a loss of voltage de-energizes the circuit and hold-in contacts F or R open. This prevents the motor from starting automatically after the power returns.

This circuit also illustrates the proper connections for adding forward and reverse indicator lights. The forward indicator light is connected to wire 6 and L2. The reverse indicator light is connected to wire 7 and L2. Additional start pushbuttons are connected to wire 2 and 3. It is standard industrial practice to mark the NO memory contacts 2 and 3. It is also standard to mark the wire coming from the forward coil and leading to the NC reverse contact (used for interlocking) as wire 6. Likewise, the wire coming from the reverse coil and leading to the NC forward contact is marked 7. These numbers are usually printed on the magnetic starters to help in wiring the circuit.

Starting, Stopping, and Jogging in Forward and Reverse with Jogging Controlled Through a Selector Switch

In certain industrial operations, it may be necessary to reposition equipment a little at a time for small adjustments. A jogging circuit allows the operator to start a motor for short times without memory. See Figure 9-35. In the circuit, small adjustments may be made in forward and reverse motor rotation or in continuous operation, depending on the position of the selector switch.

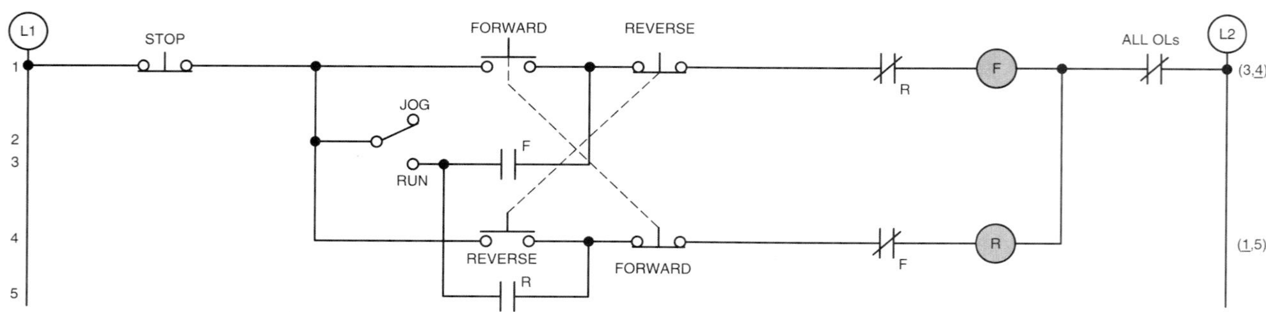

Figure 9-35. A jogging circuit allows the operator to start a motor for short times without memory.

Advanced Assembly Automation Inc.

Most complete automated systems include manual controls in addition to automatic controls.

Pressing the forward pushbutton with the selector switch in the run position activates coil F. Coil F pulls in the NO holding contacts F and opens the NC electrical interlock F, isolating the reverse circuit. The motor starts and continues to run. Pressing the reverse pushbutton with the selector switch in the run position activates Coil R. Coil R pulls in the NO holding contacts R and opens the NC electrical interlock R, isolating the forward circuit. The motor starts in the reverse direction and continues to run.

Pressing the stop pushbutton in either direction breaks the circuit and returns the circuit contacts to their normal positions. Pressing the forward pushbutton with the selector switch in the jog position activates Coil F and the motor only for the period of time that the forward pushbutton is depressed. In addition, the NC electrical interlock F opens and isolates the reverse circuit. Pressing the reverse pushbutton with the selector switch in the jog position activates Coil R and the motor only for the period of time that the reverse pushbutton is depressed. In addition, the NC electrical interlock R opens and isolates the forward circuit.

Overload protection is provided by the heater coils. Operation of the overload contacts breaks the circuit. The motor cannot be restarted until the overloads are reset and the start pushbutton is pressed.

This circuit provides protection against low voltage or a power failure in that a loss of voltage de-energizes the circuit and hold-in contacts F or R open. This prevents the motor from starting automatically after the power returns.

REVERSING MOTORS USING DRUM SWITCHES

A *drum switch* is a manual switch made up of moving contacts mounted on an insulated rotating shaft. See Figure 9-36. The moving contacts make and break contact with stationary contacts within the controller as the shaft is rotated.

Furnas Electric Co.

Figure 9-36. A drum switch is a manual switch with moving contacts on an insulated rotating shaft.

Drum switches are totally enclosed and an insulated handle provides the means for moving the contacts from point to point. Drum switches are available in several sizes and can have different numbers of poles and positions. Drum switches are usually used where the operator's eyes must remain on a particular operation such as a crane raising and lowering a load.

A drum switch may be purchased with maintained contacts or spring-return contacts. In either case, when the motor is not running in forward or reverse, the handle is in the center (OFF) position. To reverse a running motor, the handle must first be moved to the center position until the motor stops and then moved to the reverse position.

Drum switches are not motor starters because they do not contain protective overloads. Separate overload protection is normally provided by placing a non-reversing manual or magnetic starter in line before the drum switch. This provides the required overload protection and acts as a second disconnecting means. The drum switch is used only as a means for controlling the direction of a motor by switching the leads of the motor.

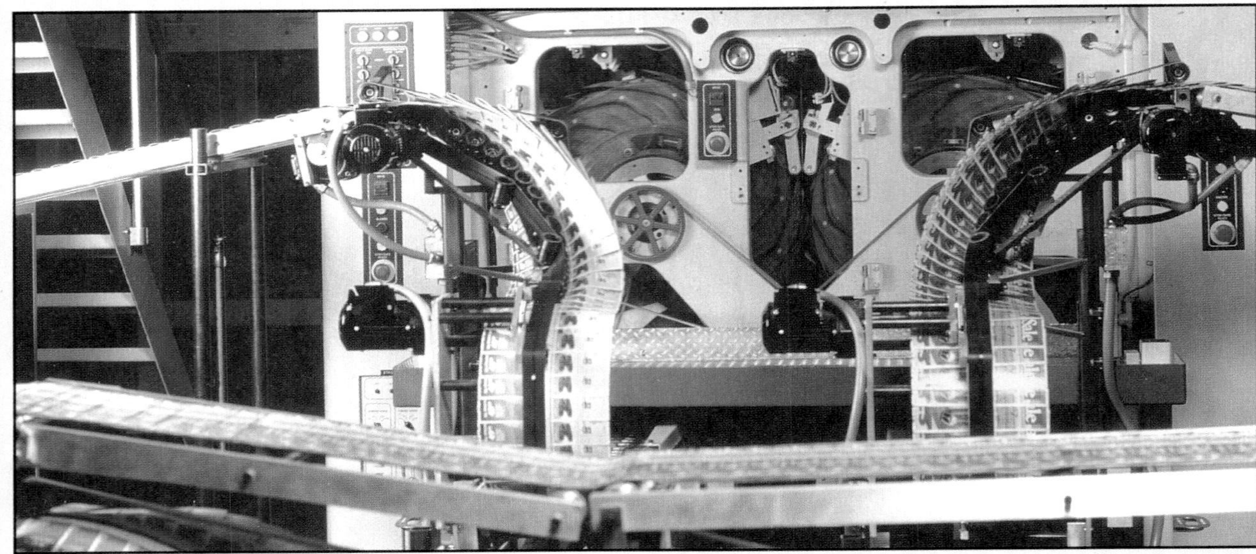

Heidelberg Harris, Inc.

Reversing circuits are used in complex systems to allow flexibility of product flow.

Reversing 3ϕ Motors Using Drum Switches

A 3ϕ motor may be connected to the contacts of a drum switch to forward or reverse the rotation of the motor. See Figure 9-37. Charts provide internal switching operation of the drum controller and the resulting wiring seen by the motor connections for forward and reverse. L1 and L3 are interchanged as the drum controller is moved from the forward to the reverse position. The motor changes directions each time the drum switch is moved to forward or reverse because only the two leads on a 3ϕ motor must be interchanged to reverse rotation.

Reversing 1ϕ Motors Using Drum Switches

A split-phase, 1ϕ motor may be connected to the contacts of a drum switch to forward or reverse the rotation of the motor. See Figure 9-38. Charts indicate the internal switching of the drum controller. Always consult the manufacturer's wiring diagrams to ensure proper wiring.

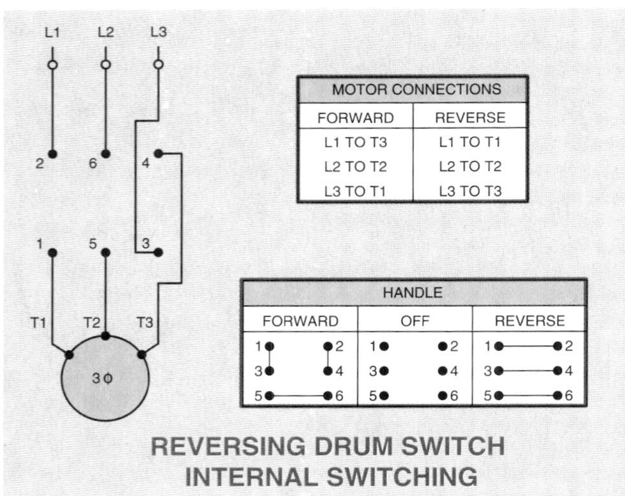

Figure 9-37. A 3ϕ motor may be connected to the contacts of a drum switch to forward or reverse the rotation of the motor.

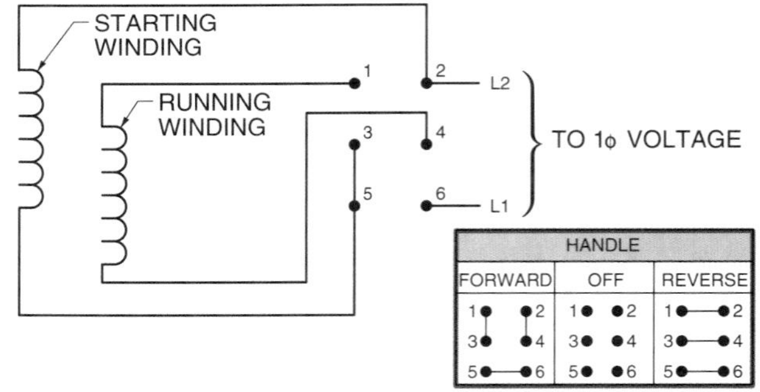

Figure 9-38. A split-phase, 1ϕ motor may be connected to the contacts of a drum switch to forward or reverse the rotation of the motor.

The motor changes direction each time the drum controller is moved to forward or reverse. This occurs because only the starting windings must be interchanged on a 1f motor to reverse rotation.

Reversing DC Motors Using Drum Switches

The direction of rotation of any DC shunt, compound, permanent-magnet or series motor may be reversed by reversing the direction of the current through the fields without changing the direction of the current through the armature, or by reversing the direction of the current through the armature without changing the direction of the currents through the fields. A drum switch may be connected to forward and reverse any DC shunt, compound, permanent-magnet, or series motor. See Figure 9-39. In each circuit, the current through the armature is changed. Some DC motors have commutating windings (interpoles) that are used to prevent sparking at the brushes in the motor. For this reason, reverse the armature circuit (armature and commutating windings) on all DC motors with commutating windings to reverse the motor.

TROUBLESHOOTING REVERSING CIRCUITS

The problem may be electrical or mechanical when a reversing motor circuit does not operate properly. The control circuit and power circuit are tested using a voltmeter to check for proper electrical operation. Troubleshooting starts inside the control cabinet when testing reversing control circuits or power circuits.

Troubleshooting Reversing Control Circuits

When troubleshooting reversing control circuits, a line diagram is used to illustrate circuit logic and a wiring diagram is used to locate the actual test points at which a meter is connected. See Figure 9-40. To troubleshoot a reversing control circuit, apply the procedure:

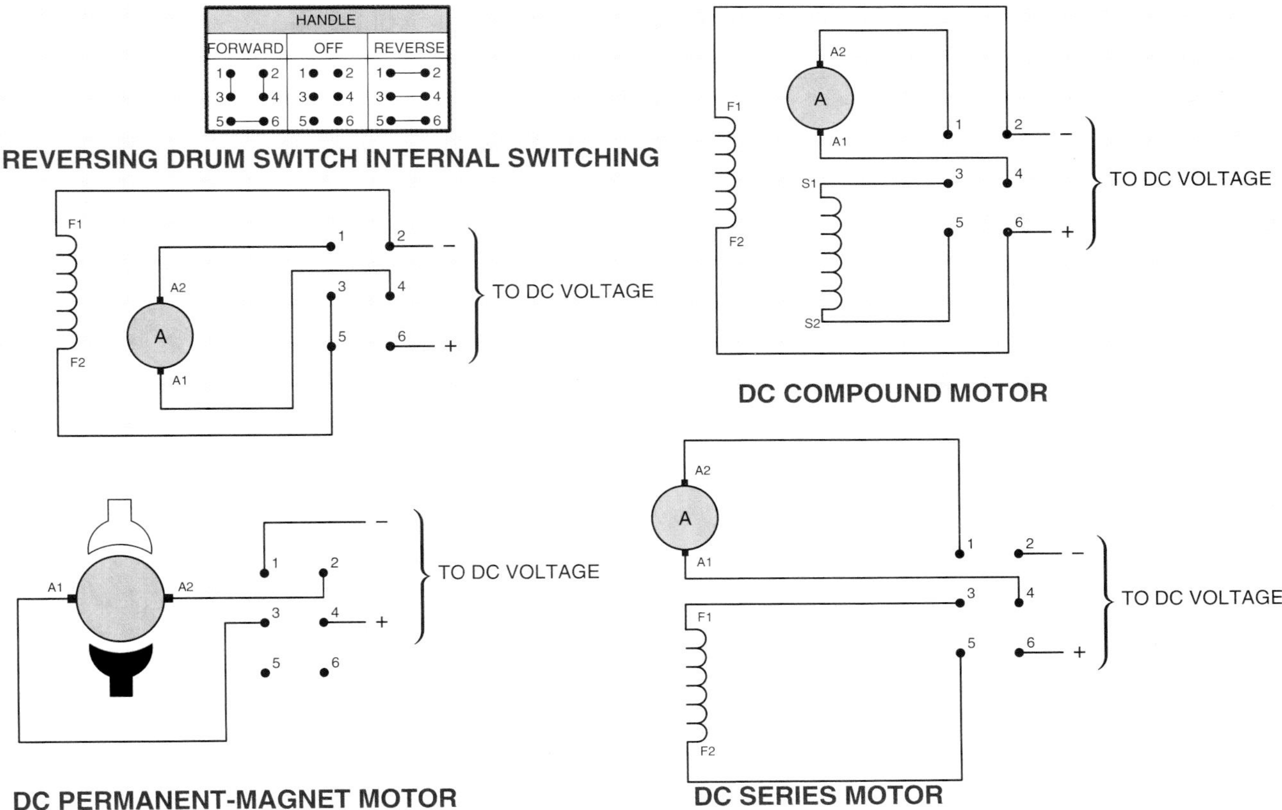

Figure 9-39. A drum switch may be connected to forward and reverse any DC shunt, compound, permanent-magnet, or series motor.

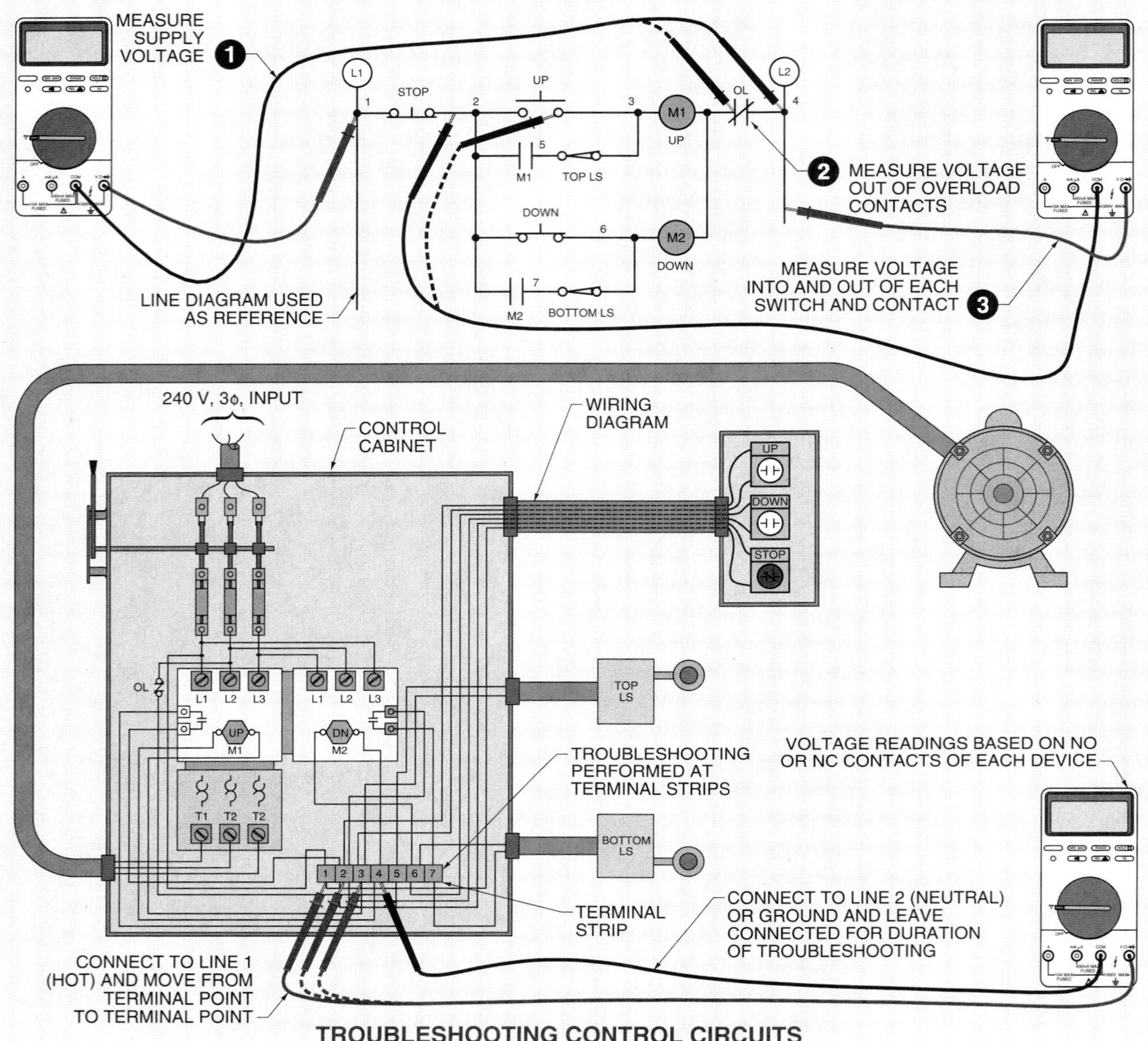

Figure 9-40. When troubleshooting reversing control circuits, a line diagram is used to show circuit logic and a wiring diagram is used to find the actual test points at which a meter is connected.

1. Measure the supply voltage of the control circuit by connecting a voltmeter between line 1 (hot conductor) and line 2 (neutral conductor). The voltage must be within 10% of the control circuit's rating. Test the power circuit if the voltage is not correct. The control circuit voltage rating is determined by the voltage rating of the loads used in the control circuit (motor starter coils, etc.).
2. Measure the voltage out of the overload contacts to ensure the contacts are closed. The contacts are tripped or are faulty if no voltage is present. Reset the overloads if tripped. Overloads are installed to protect the motor during operation. The control circuit does not operate when the overloads are tripped.
3. Measure the voltage into and out of the control switch or contacts. NC switches (stop pushbuttons, etc.) should have a voltage output before they are activated. NO switches (start pushbuttons, memory contacts, etc.) should have a voltage output only after they are activated.

Troubleshooting Reversing Power Circuits

A *power circuit* is the part of an electrical circuit that connects the loads to the main power lines. Troubleshooting reversing power circuits normally involves determining the point in the system where power is lost. See Figure 9-41. To troubleshoot a reversing power circuit, apply the procedure:

1. Measure the incoming voltage between each pair of power leads. Incoming voltage must be within 10% of the voltage rating of the motor. Measure the voltage at the main power panel feeding the control cabinet if no voltage is present or if the voltage is not at the correct level.
2. Measure the voltage out of each fuse or circuit breaker. The fuse or breaker is open if no voltage reading is obtained. Replace any blown fuse or tripped circuit breaker.
3. Measure the voltage out of the motor starter. The voltage should be present when either the forward power contacts or reverse power contacts are closed. The contacts can be closed manually at most motor starters if the power contacts cannot be closed by using the control circuit pushbuttons. Disconnect the incoming power and check the motor starter contacts for burning or wear if the voltage is not at the correct level.
4. Measure the voltage at the motor terminals. The voltage must be within 10% of the motor's rating and equal on each power line. There is a problem with the motor or mechanical connection if the voltage is correct and the motor does not operate.

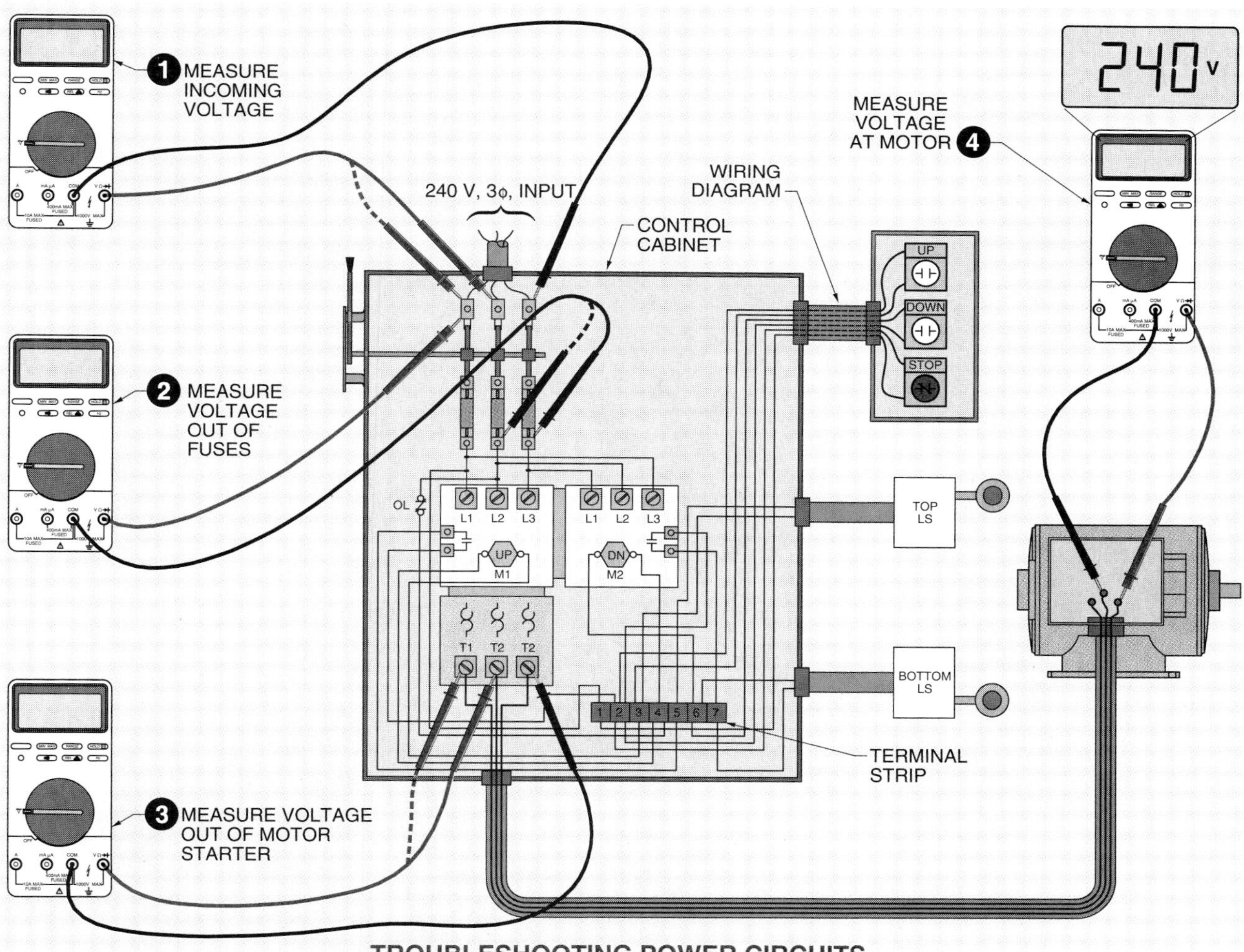

Figure 9-41. Troubleshooting reversing power circuits normally involves determining the point in the system where power is lost.

1. What is the name of the stationary part of an AC motor?
2. What is the name of the rotating part of an AC motor?
3. How are motor windings connected to form a wye-connected, 3ϕ motor?
4. How are motor windings connected to form a delta-connected, 3ϕ motor?
5. Why is the higher voltage of a dual-voltage, 3ϕ motor preferred over the lower voltage?
6. How are the nine leads (T1–T9) of a dual-voltage, wye-connected, 3ϕ motor connected to the power lines (L1–L3) for low voltage?
7. How are the nine leads (T1–T9) of a dual-voltage, wye-connected, 3ϕ motor connected to the power lines (L1–L3) for high voltage?
8. How are the nine leads (T1–T9) of a dual-voltage, delta-connected, 3ϕ motor connected to the power lines (L1–L3) for low voltage?
9. How are the nine leads (T1–T9) of a dual-voltage, delta-connected, 3ϕ motor connected to the power lines (L1–L3) for high voltage?
10. How is a 3ϕ motor reversed?
11. What is a split-phase motor?
12. What function does the centrifugal switch perform in a motor circuit?
13. How is a split-phase motor reversed?
14. What function does the capacitor perform in a capacitor motor?
15. Why are two capacitors used with some motors?
16. How is a capacitor-start motor reversed?
17. What are the four basic types of DC motors?
18. How are DC permanent-magnet motors reversed?
19. Why is mechanical interlock used on forward and reversing starter combinations?
20. What type of interlocking uses an NC contact on the starter to lock out the other starter?
21. What kind of contacts must a pushbutton have to use pushbutton interlocking?
22. Why are indicator lights often used with forward and reversing circuits?
23. Are NO or NC contacts used when a limit switch is used to stop a motor that is running in one direction?
24. Why is a drum switch not considered to be a motor starter even though it can start and stop a motor in either direction?
25. What is a power circuit?

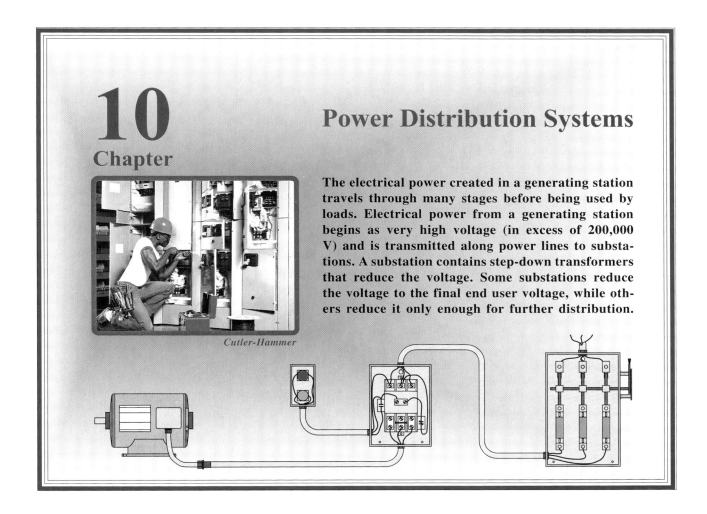

10
Chapter

Power Distribution Systems

The electrical power created in a generating station travels through many stages before being used by loads. Electrical power from a generating station begins as very high voltage (in excess of 200,000 V) and is transmitted along power lines to substations. A substation contains step-down transformers that reduce the voltage. Some substations reduce the voltage to the final end user voltage, while others reduce it only enough for further distribution.

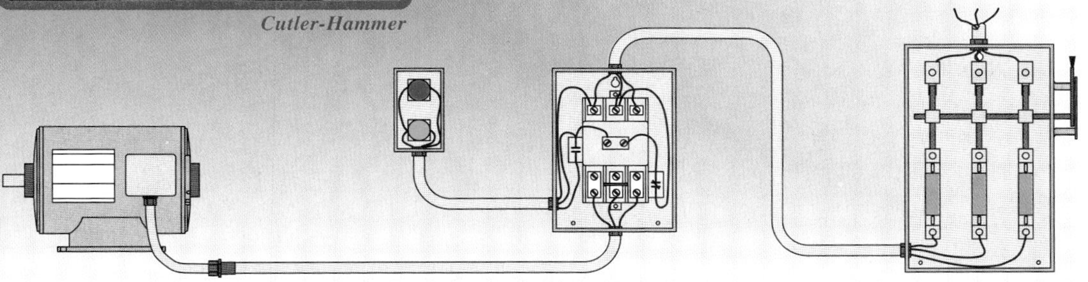

Cutler-Hammer

POWER DISTRIBUTION SYSTEMS

The distribution system from the generating source, to the plant, and within the plant must be in good working order and properly maintained. *Power distribution* is the process of delivering electrical power to where it is needed. Power control, protection, transformation, and regulation must take place before any power can be delivered. See Figure 10-1.

Electrical Energy Sources

Although several minor sources of electrical energy exist (static, chemical, thermal, solar, etc.), the major source of electrical power is an alternator. An *alternator* is an alternating current generator. An alternator provides an efficient means for the conversion of energy. Energy is converted from fuel to heat, heat to mechanical energy, and mechanical energy to electrical energy. The other equipment located in a power generating plant controls, protects, transports, and monitors the energy conversion process. See Figure 10-2.

Cutler-Hammer

Cutler-Hammer manufactures a wide variety of power distribution equipment to measure, step down, and distribute all incoming power.

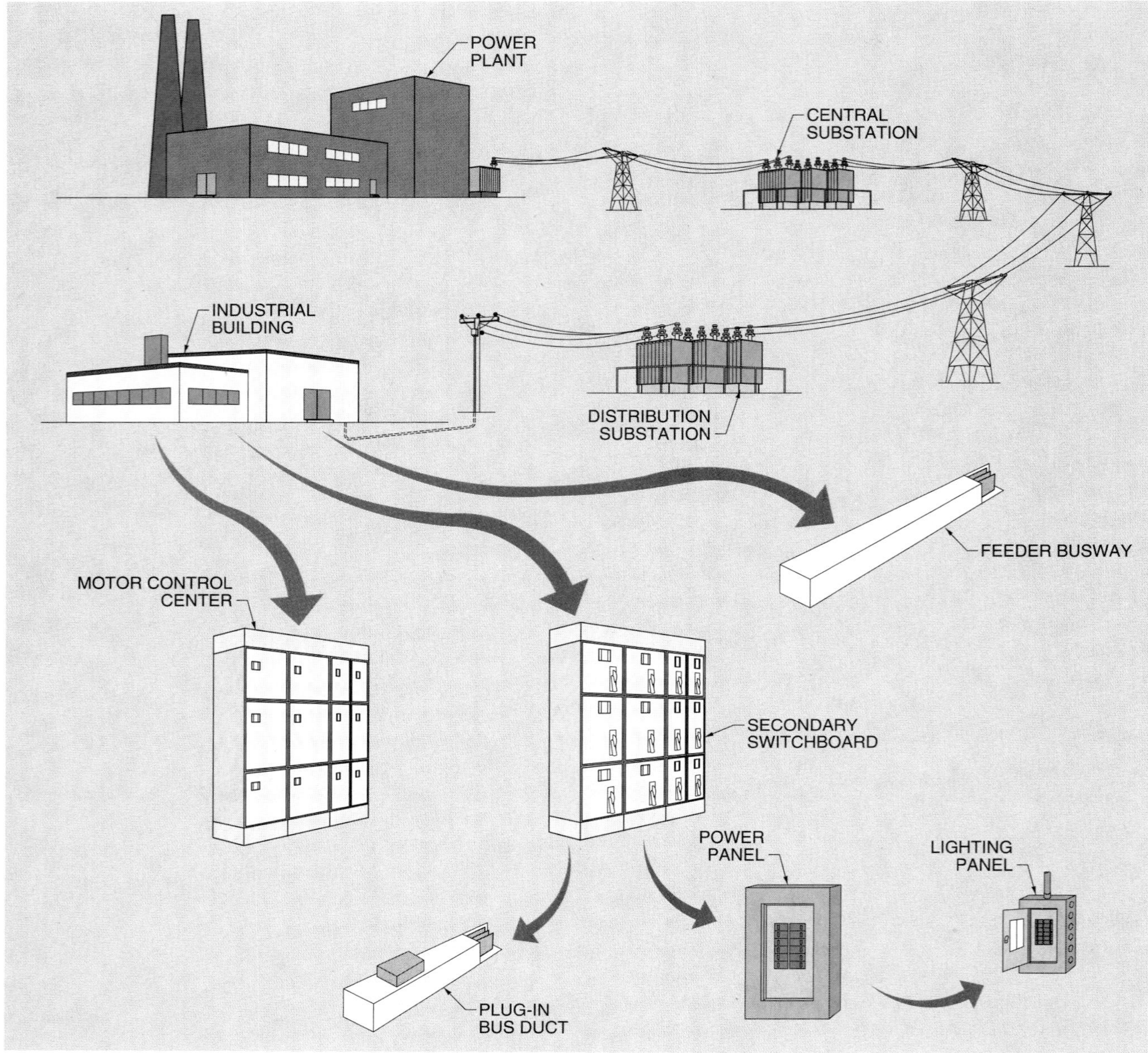

Figure 10-1. A distribution system has many steps through which power is passed to deliver the power required for an industrial user.

Today's generating plants produce alternating current (AC) because AC permits efficient transmission of electrical power between power stations and end users. Direct current (DC) is limited by the distance over which it can be economically transmitted. If direct current is required by an end user or equipment, alternating current power is rectified (changed) from AC to DC by a rectifier. A *rectifier* is a circuit that converts AC to DC.

AC Alternator Operation

Alternators operate on the theory of electromagnetic induction, which states that when conductors are moved through magnetic fields, voltages are induced into the conductors. The electron flow generator rule states that if a conductor is moved at a right angle through a magnetic field, a voltage is generated in the conductor. See Figure 10-3.

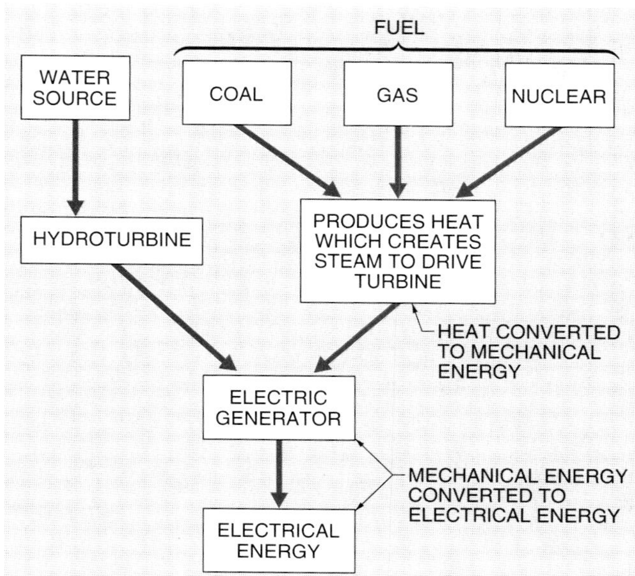

Figure 10-2. Energy is converted from fuel to heat, heat to mechanical energy, and mechanical energy to electrical energy.

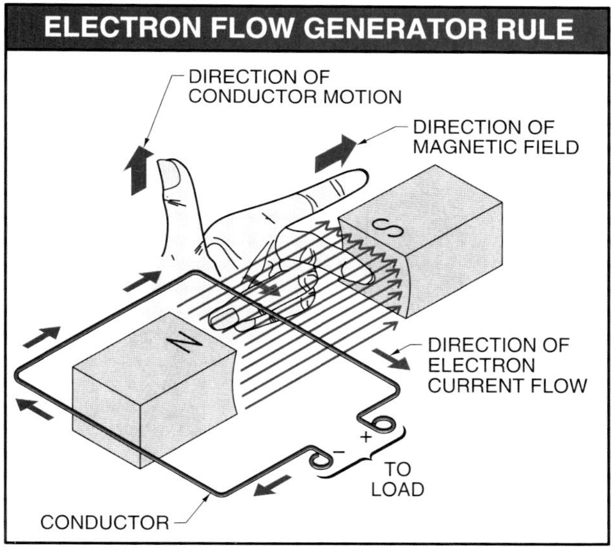

Figure 10-3. The electron flow generator rule states that if a conductor is moved at a right angle through a magnetic field, a voltage is generated in the conductor.

To determine the direction of current flow, the thumb indicates the direction of conductor motion and the index finger indicates the direction of the magnetic field. The middle finger indicates the direction of current flow. Reversing the direction of motion reverses the direction of current flow, providing the magnetic field remains constant.

An alternator is created by positioning a rotating wire coil between the poles of a permanent magnet. See Figure 10-4. In position 1, the rotor is just about to rotate in a clockwise direction. There is no current flow at this point because the rotor is not cutting any magnetic flux lines. As the rotor rotates from position 1 to position 2, the rotor begins to cut across the magnetic flux lines. The voltage in segments AB and CD increase as the rotor rotates. The maximum number of magnetic flux lines are cut when the rotor is in position 2. The induced voltage is greatest in this position.

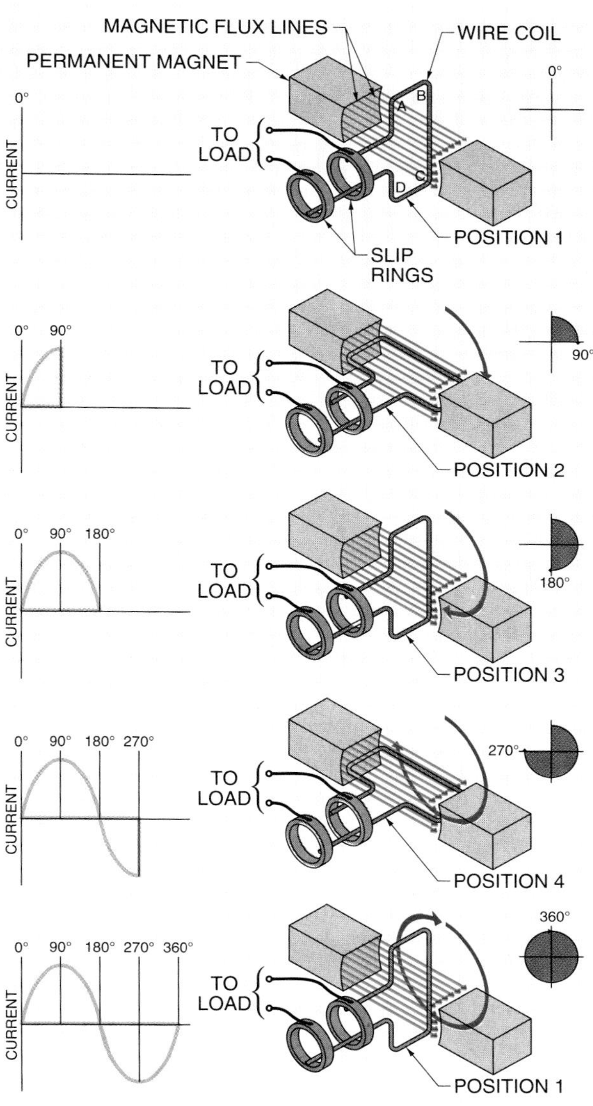

Figure 10-4. An alternator is created by positioning a rotating wire coil between the poles of a permanent magnet.

From position 2 to position 3, the voltage decreases to zero because the rotor is cutting less and less magnetic flux lines, and finally none at position 3. As the rotor continues to rotate from position 3 to position 4, the voltage increases, but in the opposite direction. The voltage reaches a maximum negative value at position 4, then returns to zero at position 1. A continuous sine wave of alternating voltage is produced if the rotor is allowed to rotate.

Slip rings are used on an alternator so that the external load may be easily attached to the rotor without interfering with its rotation. In an alternator with two magnetic poles, the rotor is rotated 3600 times per minute (60 revolutions per second) to produce the standard 60 Hz line voltage used throughout the U.S. In other countries, 50 Hz is common.

Most alternators contain two or three pairs of electromagnetic poles that are spaced in the alternator to allow the rotor to rotate at slower speeds and continue to generate the standard 60 Hz voltage. A 4-pole alternator can produce 360 electrical degrees by rotating only 180 mechanical degrees. See Figure 10-5. Additional pairs of poles may be added to further decrease the required speed of the rotor.

The power output of a 1ϕ alternator occurs in pulses. This is satisfactory for small power demands. For large power outputs, the physical size needed and the pulsating power become a problem. A 1ϕ alternator is not practical for producing large amounts of power. For this reason, three 1ϕ alternators are coupled to form one 3ϕ alternator. The 3ϕ alternator produces smoother power and provides more economical use of the magnetic field and space. For this reason, nearly all large amounts of power are generated and transmitted by 3ϕ alternators.

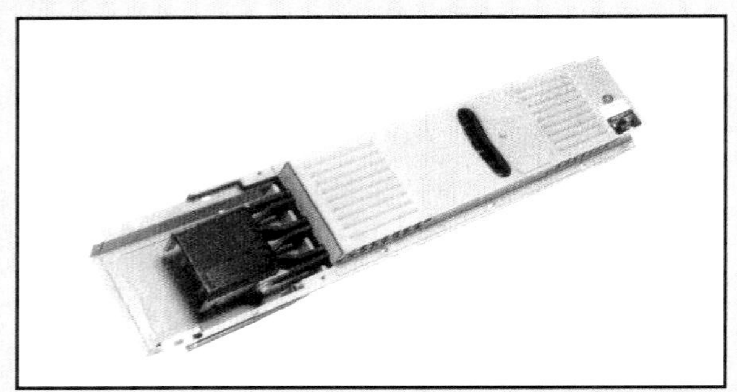

Park Detroit

Low-voltage busways from Park Detroit are manufactured in accordance with NEMA standards and are available as ventilated or totally enclosed non-ventilated with either copper or aluminum bus bars.

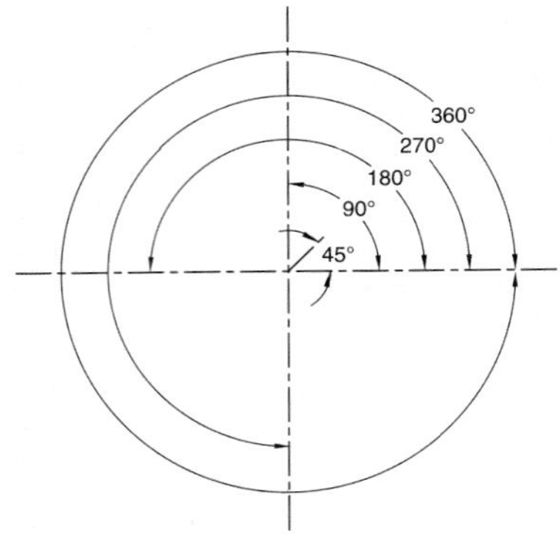

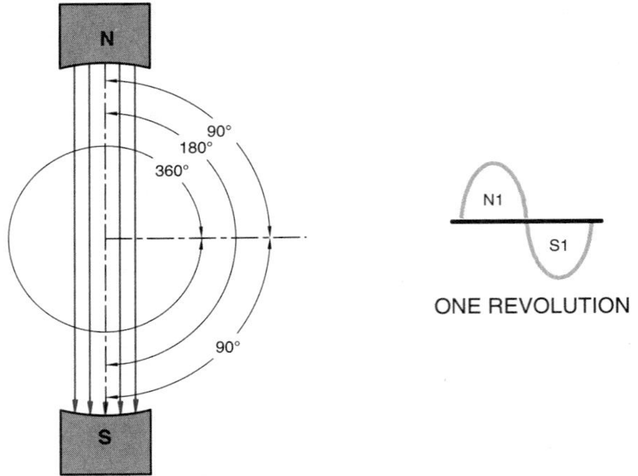

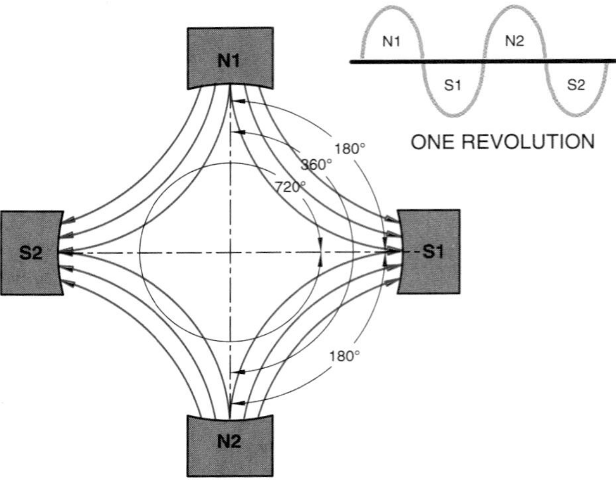

Figure 10-5. A 4-pole alternator can produce 360 electrical degrees by rotating only 180 mechanical degrees.

In 3ϕ power generation, three separate coils are spaced 120 electrical degrees apart. See Figure 10-6. By having three separate coils equally spaced around the armature by 120 electrical degrees, each of the coils' generated voltages are also spaced 120 electrical degrees apart. These individually-generated phases are phase 1, phase 2, and phase 3, or phase A, phase B, and phase C.

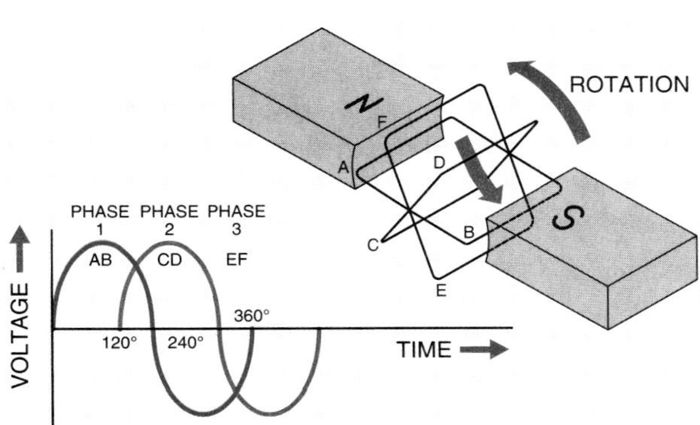

Figure 10-6. In 3ϕ power generation, three separate coils are spaced 120 electrical degrees apart.

Alternator Phase Connections

In a 3ϕ alternator, three individual phases are present. Six wires extend from the alternator because each phase coil has a beginning and end. The three coils with six wires may be connected internally or externally in wye (star) or delta connections.

Wye (Y) Connections. Three lights may be connected to each separate phase of a wye-connected alternator. See Figure 10-7. Each light illuminates from the generated 1ϕ power delivered from each phase. The A2, B2, and C2 wires return to the alternator together. This circuit can be simplified by using only one wire and connecting it to the A2, B2, and C2 phase ends. This common wire is the neutral wire.

The three ends can be safely connected at the neutral point because no voltage difference exists between them. As phase A is maximum, phases B and C are opposite to A. If the equal opposing values of B and C are added vectorially, the opposing force of B and C combined is exactly equal to A. See Figure 10-8. For example, if three people are pulling with the same amount of force on ropes tied together at a single point, the resulting forces cancel each other and the resultant force is zero in the center (neutral point).

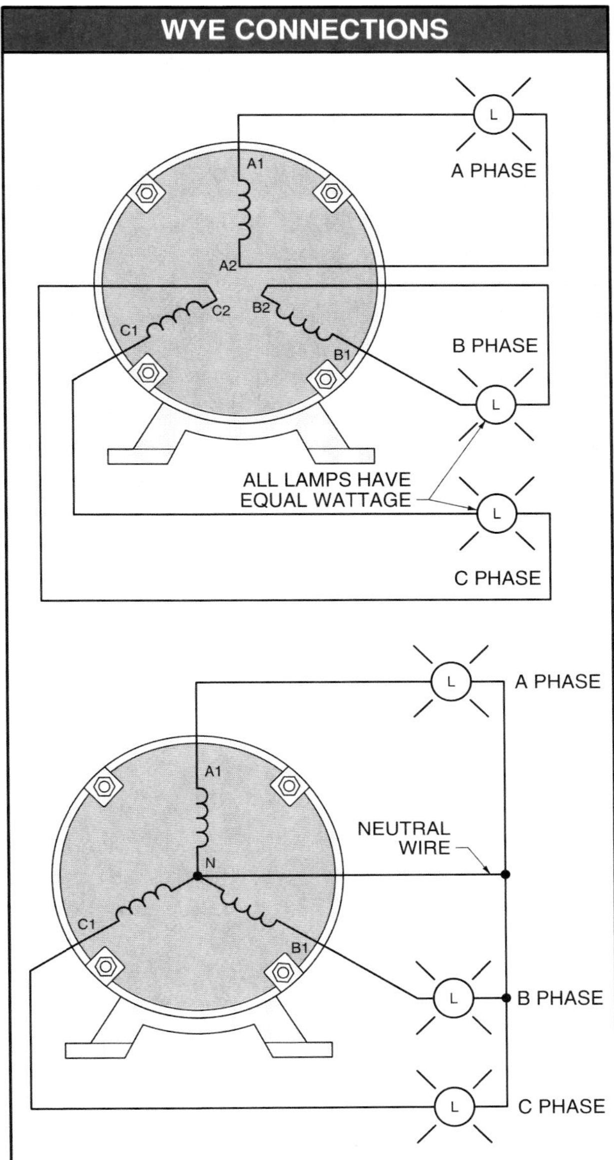

Figure 10-7. A common neutral wire can safely connect the internal leads of a wye-connected alternator to form a common return for the lighting loads.

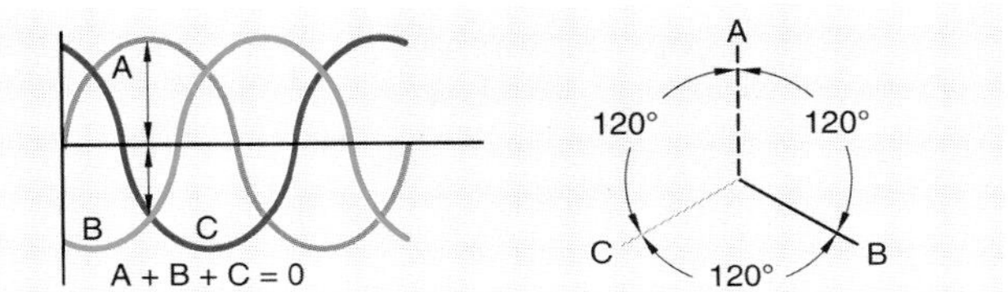

Figure 10-8. The 3ϕ voltages of a wye-connected alternator effectively cancel each other at the neutral point, allowing the three leads of the alternator to be connected.

The net effect is a large voltage (pressure difference) between the A1, B1, and C1 coil ends, but no pressure difference between the A2, B2, and C2 coil ends. In a 3ϕ wye-connected lighting circuit, the 3ϕ circuit is balanced because the loads are all equal in power consumption. In a balanced circuit, there is no current flow in the neutral wire because the sum of all the currents is zero.

All large power distribution systems are designed as 3ϕ systems with the loads balanced across the phases as closely as possible. The only current that flows in the neutral wire is the unbalanced current. This is normally kept to a minimum because most systems can be kept fairly balanced. It is customary to connect the neutral wire to a ground such as the earth. See Figure 10-9. The voltages available from a wye-connected system are phase-to-neutral, phase-to-phase, or phase-to-phase-to-phase.

In a 3ϕ wye-connected system, the phase-to-neutral voltage is equal to the voltage generated in each coil. For example, if an alternator produces 120 V from A1 to A2, the equivalent 120 V is present from B1 to B2 and C1 to C2. Thus, in a 3ϕ wye-connected system, the output voltage of each coil appears between each phase and the neutral.

In a 3ϕ wye-connected system, the voltage values must be added vectorially because the coils are set 120 electrical degrees apart. In such an arrangement, the phase-to-phase voltage is obtained by multiplying the phase-to-neutral voltage by 1.73.

Extra care must be taken when working around high-voltage transformers and power lines because of the exposed electrical terminals.

Similarly, on large wye-connected systems, a phase-to-neutral voltage of 2400 V creates 4160 V line-to-line, and 7200 V phase-to-neutral creates 12,470 V line-to-line. One of the benefits wye-connected systems bring to the utility company is that even though its alternators are rated at 2400 V or 7200 V per coil, they can transmit at a higher phase-to-phase voltage with a reduction in losses and can provide better voltage regulation. This is because the higher the transmitted voltage, the less voltage losses. This is especially important in long rural lines.

In a wye-connected system, the neutral connection point is grounded and a fourth wire is carried along the system and grounded at every distribution transformer location. This solidly-grounded system is regarded as the safest of all distribution systems.

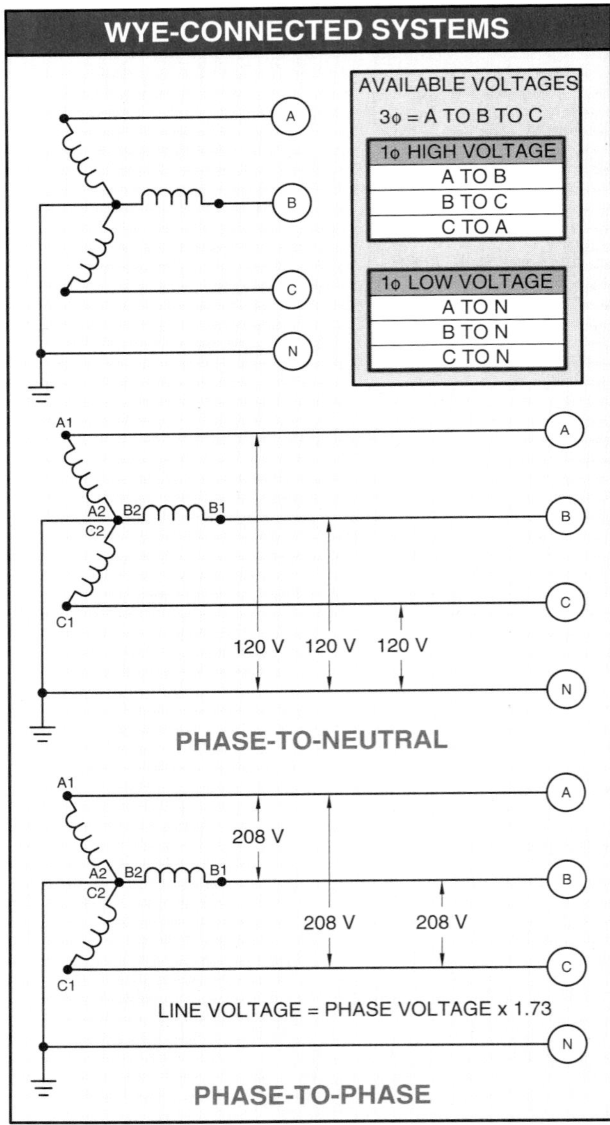

Figure 10-9. In wye-connected systems, the neutral wire is connected to ground and has various available voltages.

In a 3ϕ wye-connected system, the current in the line is the same as the current in the coil (phase) windings. This is because the current in a series circuit is the same throughout all parts of the circuit.

Delta (Δ) Connections. The alternator coil windings of a 3ϕ system can also be connected as a delta connection. A *delta connection* has each coil end connected end-to-end to form a closed loop. See Figure 10-10. As in a wye-connected system, the coil windings are spaced 120 electrical degrees apart.

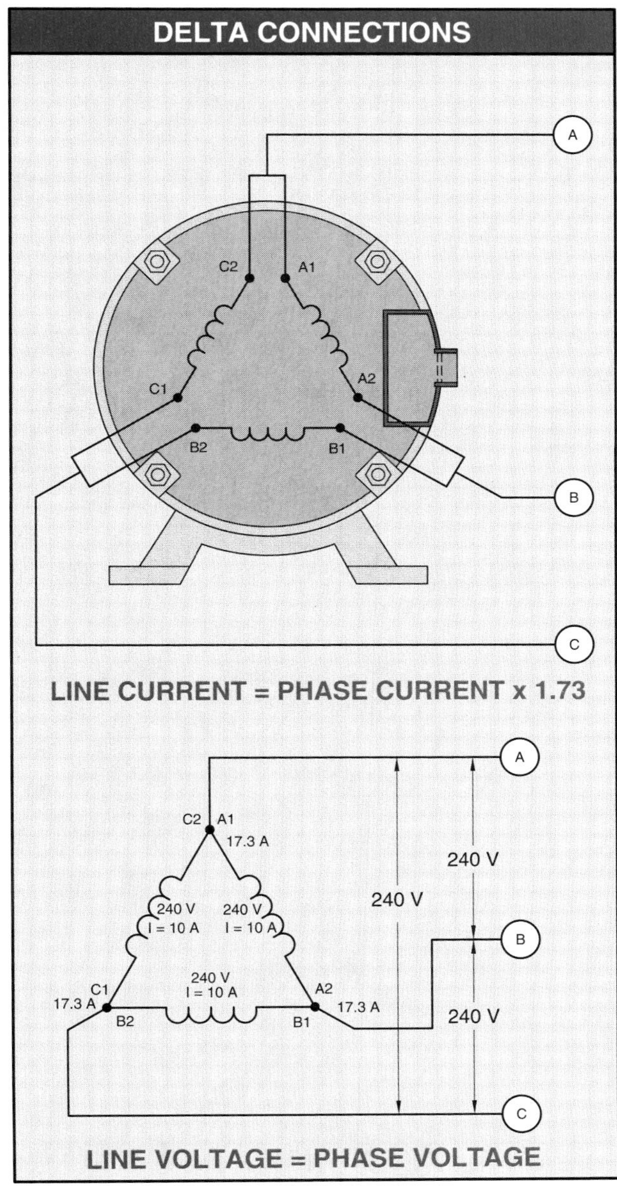

Figure 10-10. A delta connection has each coil end connected end-to-end to form a closed loop.

In a delta-connected system, the voltage measured across any two lines is equal to the voltage generated in the coil winding. This is because the voltage is measured directly across the coil winding. For example, if the generated coil voltage is equal to 240 V, the voltage between any two lines equals 240 V.

Following any line in a delta-connected system back to the connection point shows that the current supplied to that line is supplied by two coils. Phase A can be traced back to connection point A1, C2. However, as in a wye-connected system, the coils are 120 electrical degrees apart. Therefore, the line current is the vector sum of the two coil currents. In a balanced system, the phase currents are equal. In a balanced 3ϕ delta-connected system, the line current is equal to 1.73 times the current in one of the coils. For example, if each coil current is equal to 10 A, the line current is equal to 17.3 A ($1.73 \times 10 = 17.3$ A).

In a delta-connected system, only three wires appear in the system. None of the three lines are normally connected to ground. However, when a delta-connected system is not grounded, it is possible for one phase to accidentally become grounded without anyone being aware of this. The problem is not apparent until another phase also grounds. For this reason, some plants deliberately ground one corner of the delta-connected system so that inadvertent faults on the other two phases cause a fuse or breaker to trip. Although it is not common, some plants also may make a ground in a delta-connected system by grounding the midpoint of one of the phases.

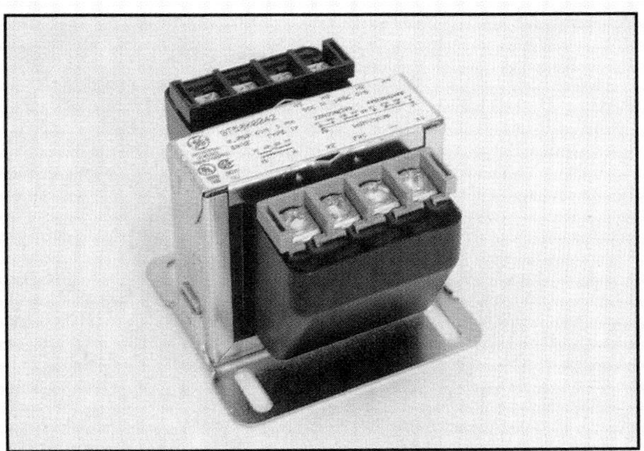

General Electric Company

Type IP core and coil transformers from General Electric are designed for panelboard, industrial control, machine tool, and general purpose applications.

A delta-connected system delivers different voltage possibilities. See Figure 10-11. Three-phase power (240 V) is available between A, B, and C. Single-phase power (120 V) is available from B to N and from C to N. Single-phase power (240 V) is available from A to B, B to C, and C to A. Also available is approximately 195 V, 1ϕ power from A to N. This voltage is the high (bastard) voltage, and should be avoided because it is an unreliable source of voltage that could damage equipment.

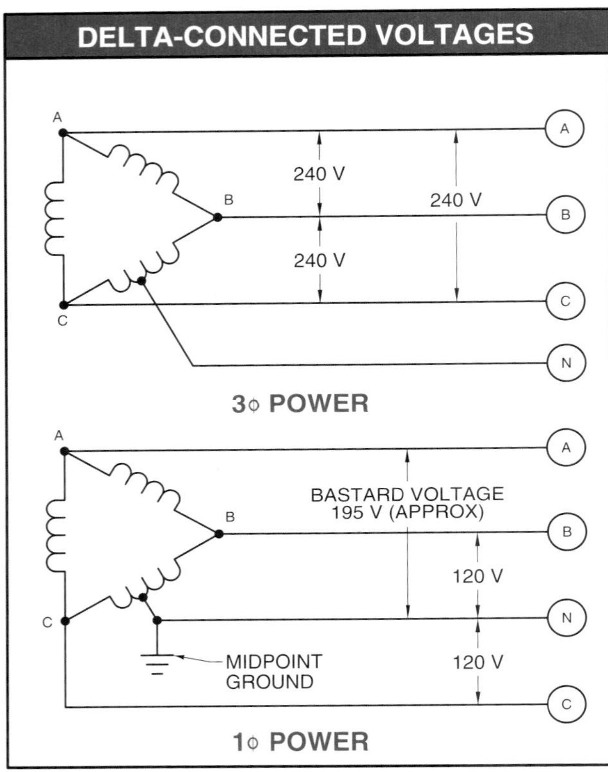

Figure 10-11. A delta-connected system delivers different voltage possibilities.

An open-delta connection makes 3ϕ power available anywhere along the distribution line with only two transformers rather than the usual three. Although this system delivers only 57.7% of the nominal full-load capability of a full bank of three transformers, it has the advantage of lower initial cost if the extra power is not presently needed. Another advantage is that in an emergency, this system allows reduced 1ϕ and 3ϕ power at a location where one transformer has burned out.

Transformers

A *transformer* is an electrical interface designed to change AC from one voltage level to another voltage level. Transformers are used in a distribution system to increase or decrease the voltage and current safely and efficiently. Transformers are used to increase generated voltage to a high level for transmission across the country and then decrease it to a low level for use by electrical loads. This allows power companies to distribute large amounts of power at a reasonable cost.

A transformer has a primary winding and a secondary winding wound around an iron core. See Figure 10-12. The *primary winding* is the coil that draws power from the source. The *secondary winding* is the coil that delivers the energy at a transformed or changed voltage to the load.

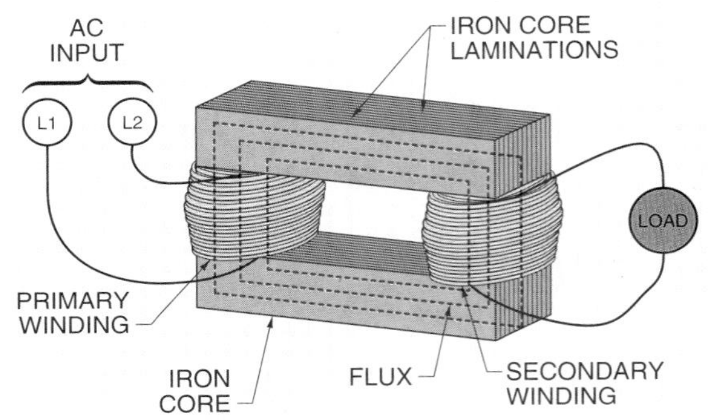

Figure 10-12. A transformer has a primary winding and a secondary winding wound around an iron core.

A transformer core provides a controlled path for the magnetic flux generated in the transformer by the current flowing through the windings. The core is constructed of many layers (laminations) of thin sheet steel. The core is laminated to help reduce heating, which creates power losses. Because the two circuits are not electrically connected, the core transfers electrical power into the secondary winding through magnetic action.

Transformer Operation. A transformer transfers AC energy from one circuit to another. The energy transfer is made magnetically through the iron core. A magnetic field builds up around a wire when AC is passed through the wire. The magnetic field builds up and collapses each half cycle because the wire is carrying AC. See Figure 10-13.

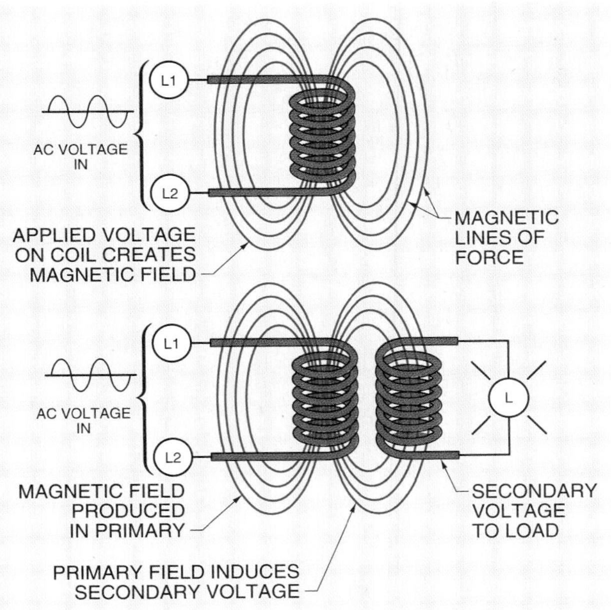

Figure 10-13. In a transformer, magnetic lines of force created by one coil induce a voltage into a second coil.

The primary magnetic field induces a secondary voltage in any wire that is within this magnetic field. The induced voltage is also alternating because the magnetic field reverses direction every half-cycle in the primary. In other words, the alternating voltage connected to the primary coil produces an alternating magnetic field in the iron core. This magnetic field cuts through the secondary coil and induces an alternating voltage on the transformer's secondary.

Fluke Corporation

Fluke meters are used to troubleshoot low voltage and other problems which are determined by measuring the voltage and current in and out of transformer banks.

The primary coil of the transformer supplies the magnetic field for the iron core. The secondary coil supplies the load with an induced voltage proportional to the number of conductors cut by the magnetic field of the core. Depending on the number of lines cut, the transformer is either a step-up or step-down transformer. See Figure 10-14.

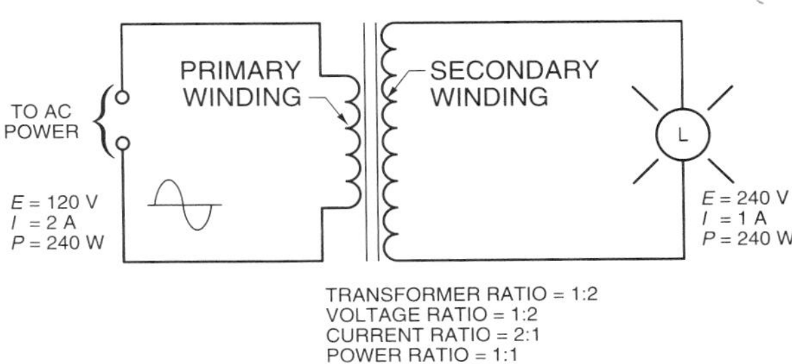

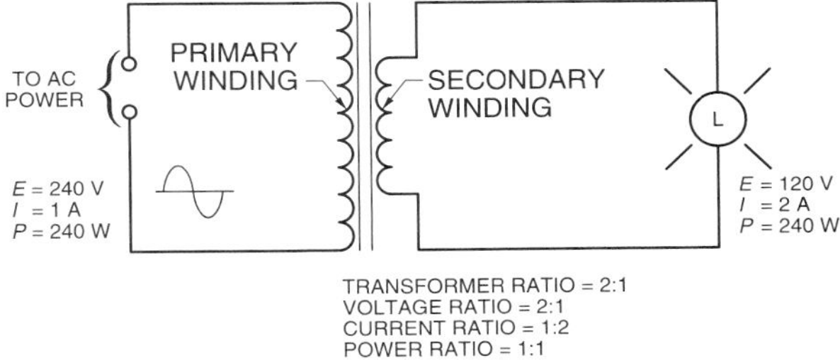

Figure 10-14. Voltage and current change from the primary to secondary winding in step-up and step-down transformers.

If only half as many turns are on the secondary, only half the voltage is induced on the secondary. The ratio of primary to secondary is 2:1, making it a step-down transformer. If twice as many turns are on the secondary, twice the voltage is induced on the secondary. The ratio of primary to secondary is 1:2, making it a step-up transformer.

In a step-up transformer, a ratio of 1:2 doubles the voltage. This may seem like a gain or a multiplication of voltage without any sacrifice. However, this is not true. The amount of power transferred in a transformer is equal on both the primary and the secondary when ignoring small losses.

Because power is equal to voltage times current ($P = E \times I$) and power is always equal on both sides of a transformer, the voltage cannot change without changing the current. For example, when voltage is stepped down from 240 V to 120 V in a 2:1 ratio, the current increases from 1 A to 2 A, keeping the power equal on each side of the transformer. By contrast, when the voltage is stepped up from 120 V to 240 V in a 1:2 ratio, the current is reduced from 2 A to 1 A to maintain power balance. In other words, voltage and current may be changed for particular reasons, but power is merely transferred from one point to another.

One advantage of increasing voltage and reducing current is that power may be transmitted through smaller gauge wire, thus reducing the cost of power lines. For this reason, the generated voltages are stepped up very high for distribution across large distances, and then stepped back down to meet the consumer needs. Although both the voltage and current can be stepped up or down, the terms step-up or step-down, when used with transformers, always applies to the voltage.

Transformers Connected for Wye and Delta Distribution Systems. Large amounts of power are generated using a 3ϕ system. The generated voltage is stepped up and down many times before it reaches the loads in dwellings or plants. The transformation can be accomplished by using wye- or delta-connected transformers or a combination of both wye and delta transformers with differing voltage ratio transformers.

Three 1ϕ transformers may be connected for a wye-to-wye, step-down transformer bank. See Figure 10-15. The line voltage is equal to 1.73 times the coil voltage and the line current and coil current are equal in a balanced wye system.

The wye-connected secondary provides three different types of service: a 3ϕ, 208 V service for 3ϕ motor loads, a 1ϕ, 208 V service for 1ϕ, 208 V motor loads, and a 1ϕ, 120 V service for lighting loads. This type of system is commonly used in schools, commercial stores, and offices.

On the primary side of the transformer bank, the grounded neutral wire is connected to the common points of all three high-voltage primary coil windings. These coil windings are marked H1 and H2 on each transformer to indicate the high-voltage side of the transformer. The low-voltage side is marked with X1 and X2. The voltage from the neutral to any phase of the three power lines is 2400 V. The voltage across the three power lines is 4152 V (1.73 × 2400 V = 4152 V).

ABB Power T&D Company Inc.

ABB Power T&D Company Inc. manufactures a variety of 3ϕ transformers which can be connected as wye and/or delta distribution systems.

As on the primary side, the grounded neutral is connected to the common points of all three low-voltage secondary coil windings. This allows for a 120 V output on each of the secondary coils. The voltage across the three secondary power lines is 208 V (1.73 × 120 V = 208 V).

To help maintain a balanced transformer bank, the loads should be connected to evenly distribute them among the three transformers. This naturally occurs when connecting a 3ϕ motor because a 3ϕ motor draws the same amount of current from each line. Care should be taken to balance the 1ϕ loads. Three 1ϕ transformers of the same power rating are used in most wye-to-wye systems. The capacity of transformers is rated in kilovolt-amperes (kVA). The total kVA capacity of a transformer bank is found by adding the individual kVA ratings of each transformer in the bank. For example, if each transformer is rated at 50 kVA, the total capacity of the bank is 150 kVA (50 kVA + 50 kVA + 50 kVA = 150 kVA).

Three 1ϕ transformers may be connected for a delta-to-delta, step-down transformer bank. See Figure 10-16. The line voltage is equal to the coil voltage and the line current is equal to 1.73 times the coil current in a balanced delta system.

The delta-connected secondary, with one coil centered tapped, provides three different types of service: a 3ϕ, 240 V service for 3ϕ motor loads, a 1ϕ, 120 V service for lighting loads, and a 1ϕ, 240 V service for 1ϕ, 240 V motor loads. The high phase of line C to N is not used.

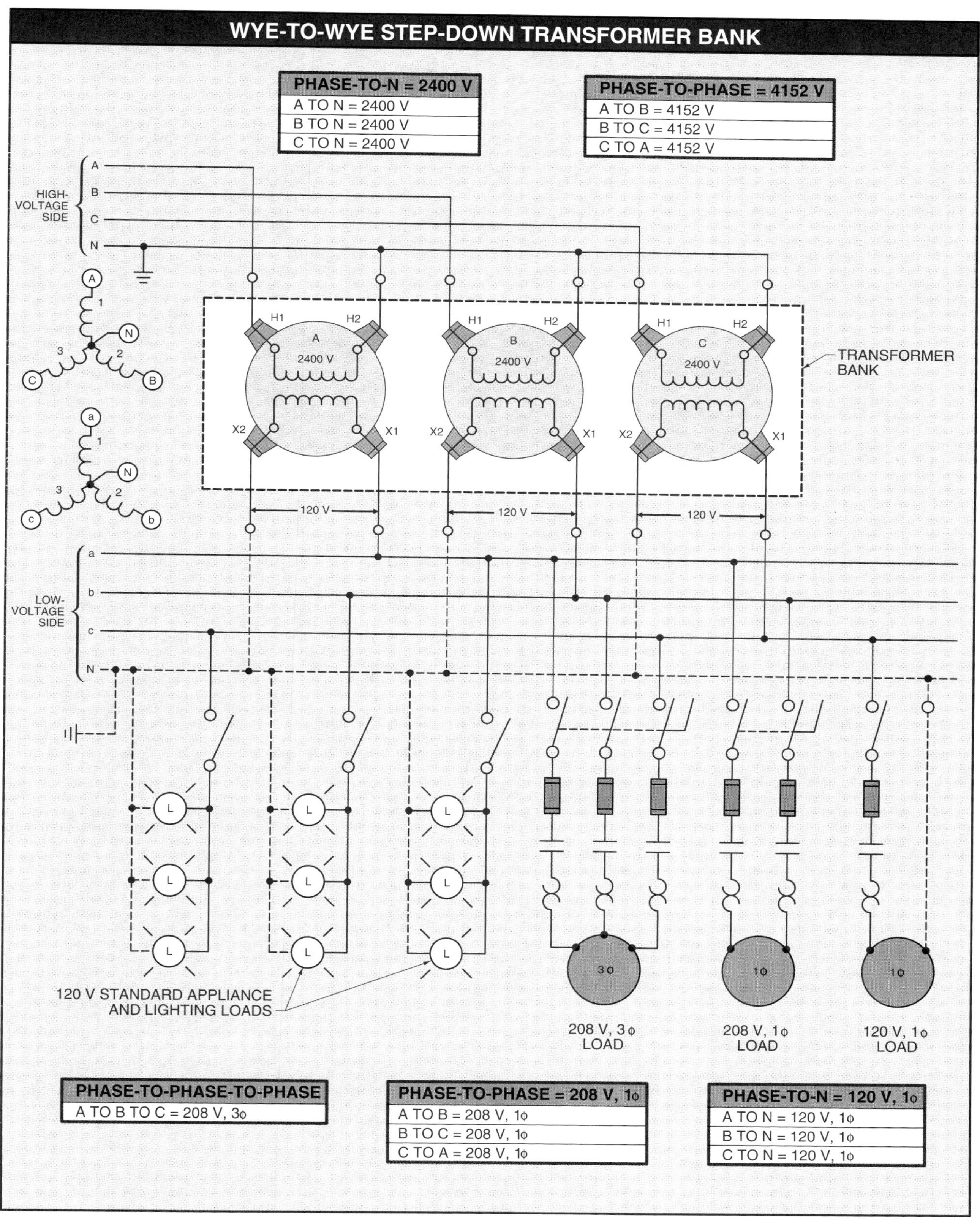

Figure 10-15. Three 1ϕ transformers may be connected for a wye-to-wye, step-down transformer bank.

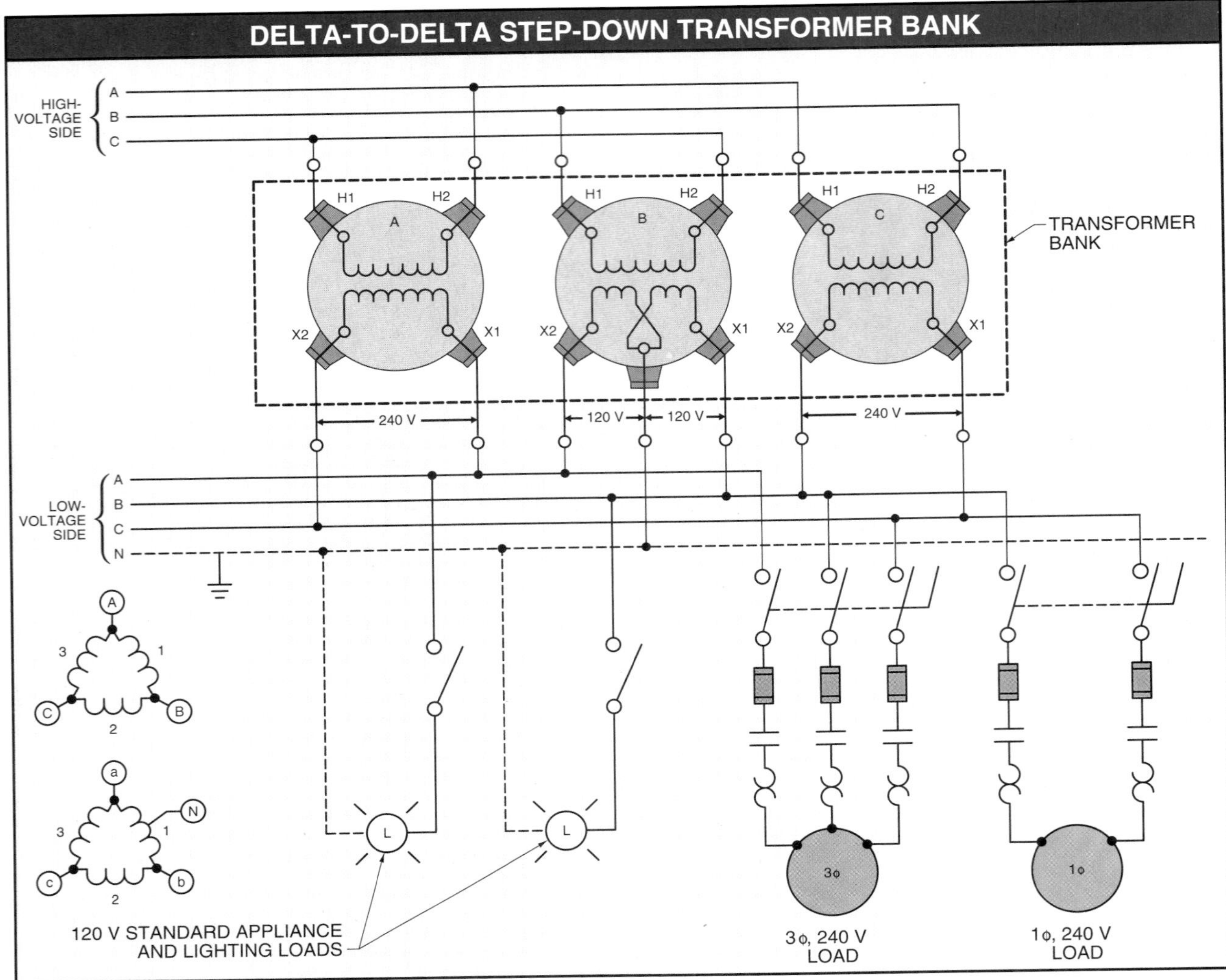

Figure 10-16. Three 1φ transformers may be connected for a delta-to-delta, step-down transformer bank.

As with the wye-connected transformer bank, a closed delta bank delivers a total kVA output equal to the sum of the individual transformer ratings. One advantage of a delta system is that if one transformer is damaged or removed from service, the other two can be connected in an open-delta connection. This type of connection enables power to be maintained at a reduced level. The level is reduced to 57.7% of a full delta-connected transformer bank.

Substations

Substations serve as a source of voltage transformation and control along the distribution system. Their function includes:

- Receiving voltage generated and increasing it to a level appropriate for transmission
- Receiving the transmitted voltage and reducing it to a level appropriate for customer use
- Providing a safe point in the distribution system for disconnecting the power in the event of problems
- Providing a place to adjust and regulate the outgoing voltage
- Providing a convenient place to take measurements and check the operation of the distribution system
- Providing a switching point where different connections may be made between various transmission lines

Substations have three main sections, which include the primary switchgear, transformer, and secondary switchgear sections. See Figure 10-17. Depending on the function of the substation (stepping up or down voltage), the primary or secondary switchgear section may be the high-voltage or low-voltage section. In step-up substations, the primary switchgear section is the low-voltage section and the secondary switchgear section is the high-voltage section. In step-down substations, the primary switchgear section is the high-voltage section and the secondary switchgear section is the low-voltage section. The substation sections normally include breakers, junction boxes, and interrupter switches.

Substations may be entirely enclosed in a building or totally in the open, as in the case of outdoor substations located along a distribution system. The location for a substation is generally selected so that the station is near as possible to the area to be served.

Substations can be built to order or purchased from factory-built, metal-enclosed units. The purchased units are unit substations. A unit substation offers standardization and flexibility for future changes when quick replacements are needed.

A transformer's function in a substation is the same as that of any transformer. Transformers are broadly classified as wet or dry.

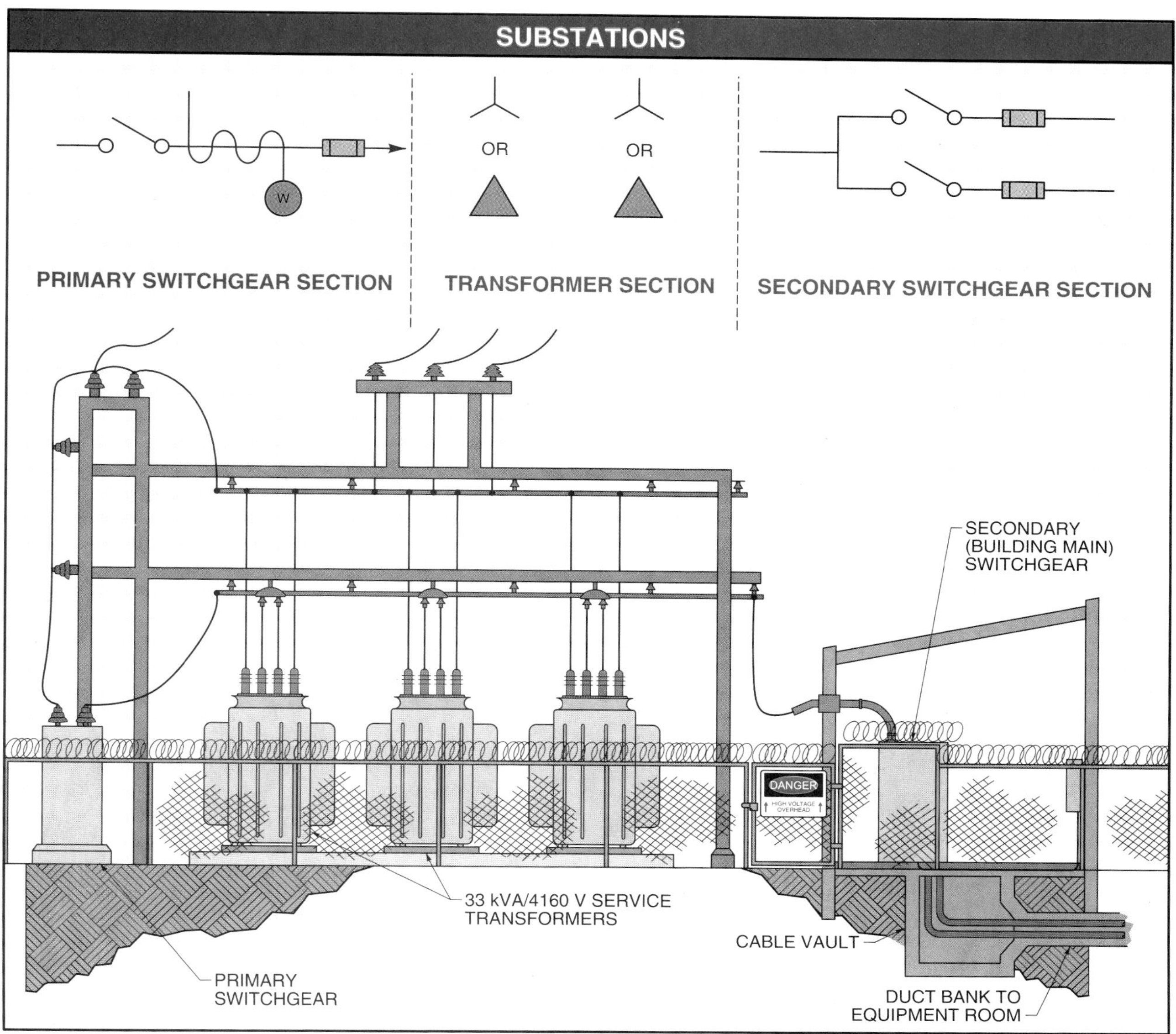

Figure 10-17. The three main sections of a substation are the primary switchgear, transformer, and secondary switchgear sections.

In wet types, oil or some other liquid serves as a heat-transfer medium and insulation. Dry types use air or inert gas in place of the liquid. Fans may be used for forced-air cooling on transformers to provide additional power for peak demand periods where the surrounding air is hot and the transformers are not able to handle their full-rated load without exceeding their recommended temperature. The normal loads are handled by natural circulation.

Most transformers used in substations include a voltage regulator. The regulator on a transformer has taps that allow for a variable output. These taps are needed because transmission lines rarely deliver the transformer input voltage for which they are rated because of transmission line losses, line loading, and other factors. See Figure 10-18.

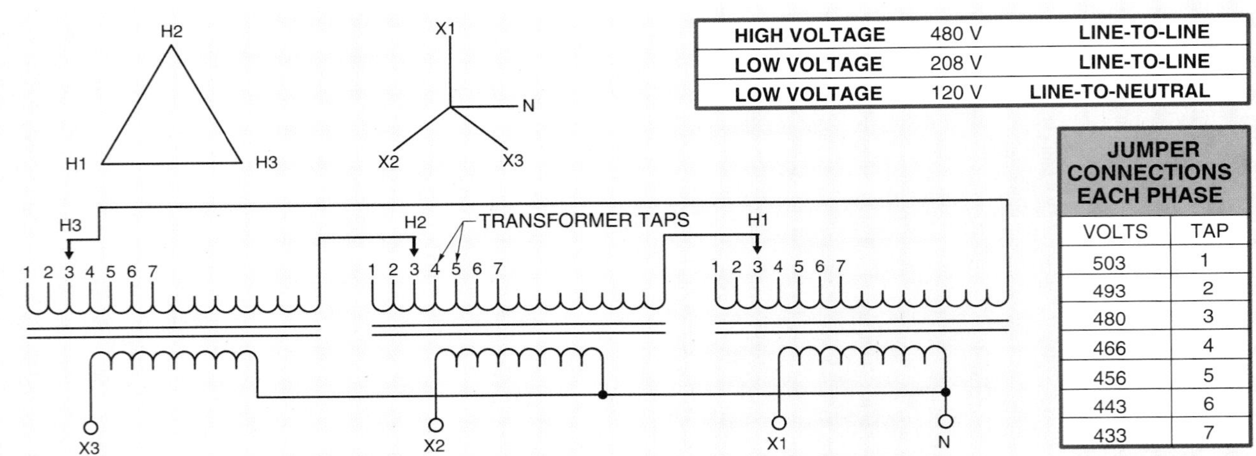

HIGH VOLTAGE	480 V	LINE-TO-LINE
LOW VOLTAGE	208 V	LINE-TO-LINE
LOW VOLTAGE	120 V	LINE-TO-NEUTRAL

JUMPER CONNECTIONS EACH PHASE	
VOLTS	TAP
503	1
493	2
480	3
466	4
456	5
443	6
433	7

Figure 10-18. Taps are built into a transformer to compensate for voltage differences.

The taps are normally provided at $2\frac{1}{2}\%$ increments for adjusting above and below the rated voltage. The adjustment may be manual or automatic. In an automatic regulator, a control circuit automatically changes the tap setting on the transformer's windings. This allows the outgoing voltage to be kept nearly constant even though the incoming primary voltage or load demands may vary.

Switchboards

Electrical power is delivered to industrial, commercial, and residential buildings through a distribution and transmission system. Once the power is delivered to a building, it is up to the building electrician to further distribute the power to where it is required within the building.

A switchboard is the link between the power delivered to a building and the start of the local distribution system throughout the building. The switchboard is the last point on the power distribution system as far as the power company is concerned and the beginning of the distribution system as far as the building electrician is concerned.

A *switchboard* is the piece of equipment in which a large block of electric power is delivered from a substation and broken down into smaller blocks for distribution throughout a building. See Figure 10-19. Switchboards are rated as to the maximum voltage and current they can handle. For example, a switchboard may have a 600 V rating and a bus rating up to 5000 A.

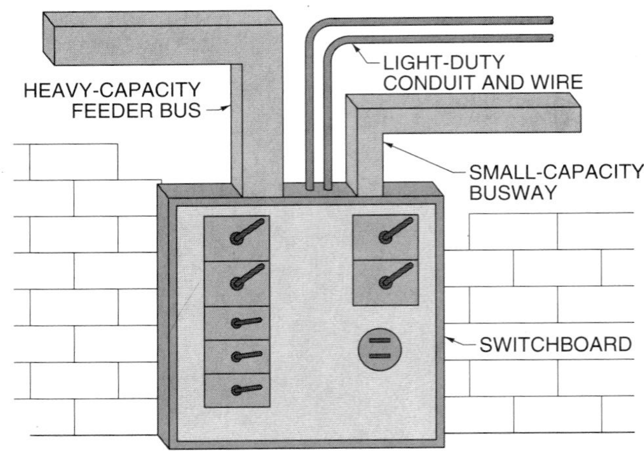

Figure 10-19. A switchboard is used to divide incoming power into smaller branch circuits.

In addition to dividing the incoming power, a switchboard may contain all the equipment needed for controlling, monitoring, protecting, and recording the functions of the substation. Switchboards are designed for use in three categories: service-entrance, distribution, and service-entrance/distribution.

A service-entrance switchboard has space and mounting provisions required by the local power company for metering equipment, as well as overcurrent protection and disconnect means for the service conductors. Provision for grounding the service neutral conductor when a ground is needed is also provided. A distribution switchboard contains the protective devices and feeder circuits required to distribute the power throughout a building. A distribution switchboard may contain either circuit breakers or fused switches. See Figure 10-20.

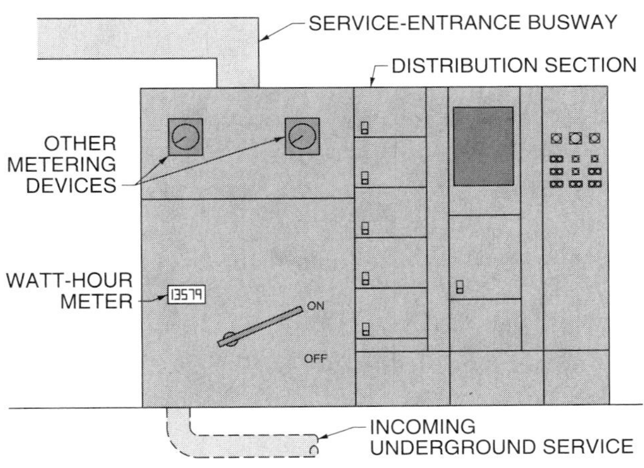

Figure 10-20. A distribution switchboard contains the protective devices and feeder circuits required to distribute the power throughout a building.

A distribution switchboard has the space and mounting provisions required by the local power company for metering their equipment and incoming power. To meter the incoming power, the switchboard must have a watt-hour meter to measure power usage. Metering is always located on the incoming line side of the disconnect. The compartment cover is sealed to prevent tapping power ahead of the power company's metering equipment.

Other meters and indicator lights, such as ammeters and voltmeters, may also be built into the meter compartment. In most cases, these are not requirements but options, depending on the application and the plant's requirements. A voltmeter is used to indicate to the maintenance person the various incoming and outgoing voltages. An ammeter is used to indicate the various currents throughout the system. A wattmeter is used to indicate the power used throughout the system. Each of these instruments can be of the indicating type, recording type, or both. A recording instrument is used to keep track of the various values over a period of time.

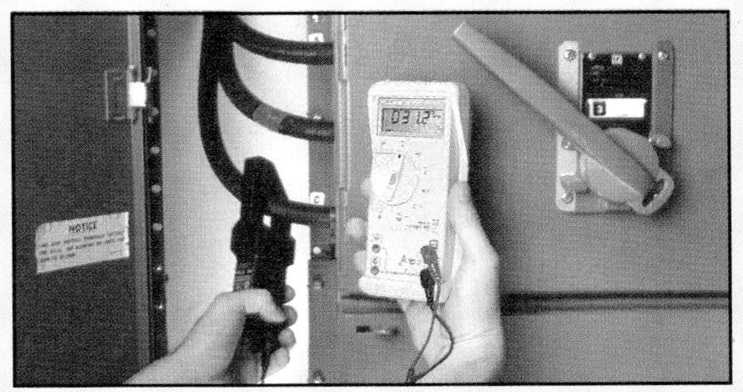

Fluke Corporation

Clamp-on attachments are available from Fluke Corporation for use with their meters to measure high circuit currents.

In addition to measuring the voltage, current, and power of a system, a distribution switchboard also controls the power. Control is achieved through the use of switches and overcurrent and overvoltage relays that are used to disconnect the power. These devices protect the distribution system in the event of a fault.

Switchboards that have more than six switches or circuit breakers must include a main switch to protect or disconnect all circuits. Switchboards with more than one, but not more than six switches or circuit breakers do not require a main switch. A switchboard with six switches or circuit breakers does not require a main disconnect. In a switchboard with more than six switches or breakers, the service-entrance section of the switchboard may have any number of feeder circuits added to the rated capacity of the main. A switchboard with a main section can easily contain more than one distribution section. This depends on the number of feeder circuits required in addition to the blank spaces needed for future expansion.

In addition to distributing the power throughout the building, the distribution section of a switchboard may contain the provisions for motor starters and other control devices. See Figure 10-21. The addition of starters and controls to the switchboard allows for motors to be connected to the switchboard. This combination can be used when the motors to be controlled are located near the switchboard. This combination allows for high-current loads such as motors to be connected to the source of power without further power distribution.

Figure 10-21. A switchboard may contain the provisions for motor starters and other control devices.

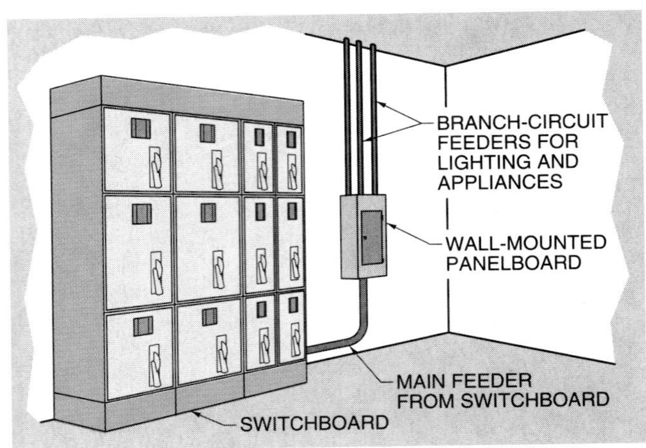

Figure 10-22. A panelboard is a wall-mounted distribution cabinet containing overcurrent and short-circuit protection devices.

Panelboards and Branch Circuits

A *panelboard* is a wall-mounted distribution cabinet containing a group of overcurrent and short-circuit protection devices for lighting, appliance, or power distribution branch circuits. The wall-mounted feature distinguishes the panelboard from a switchboard, which normally stands on the floor. See Figure 10-22.

A panelboard is normally supplied from a switchboard and further divides the power distribution system into smaller parts. Panelboards are the part of the distribution system that provides the last centrally-located protection for the final power run to the load and its control circuitry. Panelboards are classified according to their use in the distribution system.

A panelboard provides the required circuit control and overcurrent protection for all circuits and power-consuming loads connected to the distribution system. See Figure 10-23. The panelboards are located throughout a plant or building, providing the necessary protection for the branch circuits feeding the loads.

A *branch circuit* is the portion of a distribution system between the final overcurrent protection device and the outlet or load connected to it. The basic requirements for panelboards and overcurrent protection devices are given in Article 240 of the National Electrical Code® and must be used for individual applications. In addition, check local power company, city, and county regulations.

Overcurrent protection devices used for protecting branch circuits include fuses or circuit breakers. Overcurrent protection devices must provide for proper overload and short-circuit protection. The size (in amperes) of the overcurrent protection device is based on the rating of the panelboard and load. The overcurrent protection device must protect the load and be within the rating of the panelboard. If the overcurrent protection device exceeds the ampacity of the bus bars in the panelboard, the panelboard is undersized for the load(s) that are to be connected.

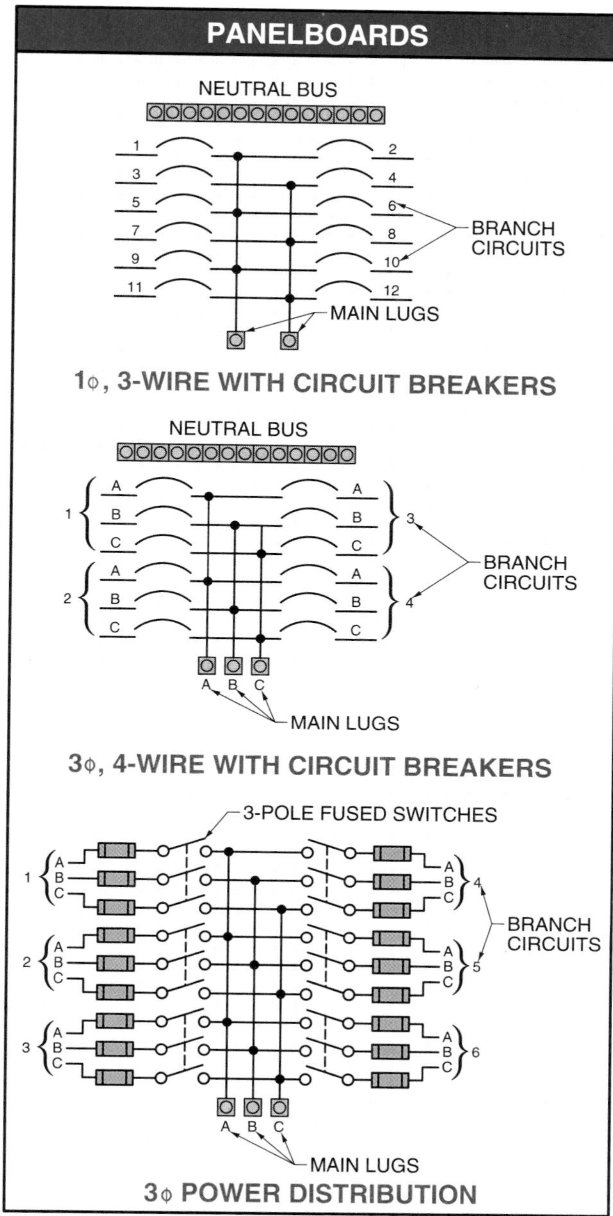

Figure 10-23. A panelboard provides the required circuit control and overcurrent protection for all circuits and power-consuming loads connected to the distribution system.

A panelboard may be compared to a load center found in most residential dwellings. The load center in residential dwellings contains the fuses or breakers which control the individual branch circuits throughout the dwelling.

Although panelboards and load centers perform about the same function, they are separated by certain distinct features when used in commercial and industrial applications. Panelboard features not shared by load centers include:

- A box made of Underwriters Laboratories Inc.® (UL®) approved corrosion-resistant galvanized steel
- A minimum of a 4″ wiring gutter on all sides
- Combination catch and lock in addition to hinges
- Bus bars listed to 1200 A (load center's main bus bars are generally 200 A maximum)
- Greater enclosure depth to accommodate 2½″ or greater conduit
- Main and branch terminal lugs

Motor Control Centers

In a power distribution system, many different kinds of loads are connected to the system. The loads vary considerably from application to application, as does their degree of control. For example, a light may be connected to a system requiring only a switch for control (along with proper protection). However, other loads, such as motors, may require complicated and lengthy control and protection circuits. The more complicated a control circuit becomes, the more difficult it is to wire into the system.

The most common loads requiring simple and complex control are electric motors. The need to simplify and consolidate motor control circuits is required because an electric motor is the backbone of almost all production and industrial applications. To do this, a control center takes the incoming power, control circuitry, required overload and overcurrent protection, and any transformation of power, combining them into one convenient motor control center.

A motor control center combines individual control units into standard modular structures joined on formed sills. A motor control center is normally supplied from a panelboard or switchboard. A motor control center is different from a switchboard containing motor panels in that the motor control center is a modular structure designed specifically for plug-in type control units and motor control. See Figure 10-24.

A motor control center receives the incoming power and delivers it to the control circuit and motor loads. The center provides space for the control and load wiring in addition to providing required control components. The control inputs into the center are the control devices such as pushbuttons, liquid level and limit switches, and other devices that provide a signal. The output of the control center is the wire connecting the motors. All other control devices are located in the motor control center. These control devices include relays, control transformers, motor starters, overload and overcurrent protection, timers, counters, and any other required control devices.

General Electric Company

Figure 10-24. A motor control center combines the incoming power, control circuitry, overload and overcurrent protection, and any transformation of power, into one convenient location.

One advantage of a motor control center is that it provides one convenient place for installing and troubleshooting control circuits. This is especially useful in applications that require individual control circuits to be related to other control circuits. An example includes assembly lines in which one machine feeds the next.

A second advantage is that individual units can be easily removed, replaced, added to, and interlocked at one central location. Manufacturers of motor control centers produce factory preassembled units to meet all the standard motor functions, such as start/stop, reversing, reduced-voltage starting, and speed control. This leaves only the connecting of the control devices (limit switches, etc.) and the motors to the center.

Common preassembled motor control unit panels are available from the factory, along with their schematic diagram. See Figure 10-25. The only required wiring by the electrician is the connection to control inputs, terminal blocks, and the motor.

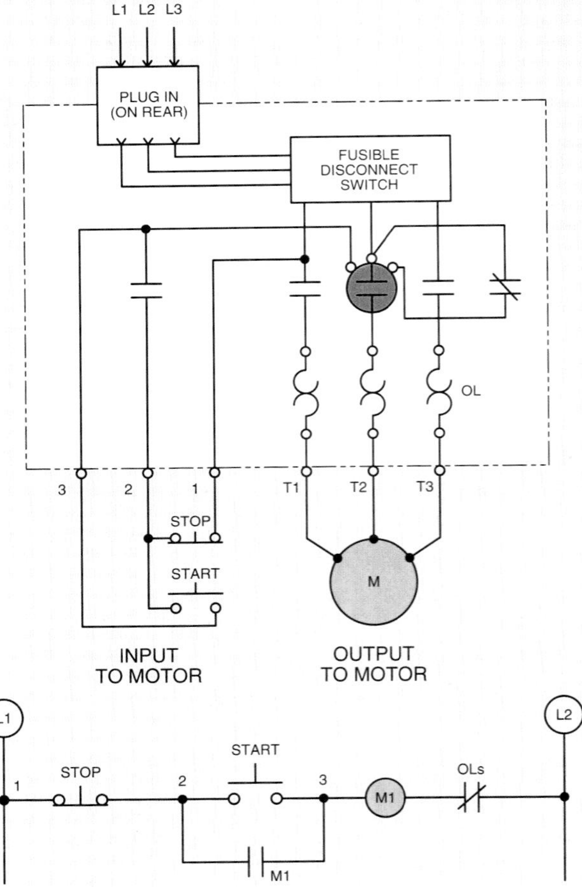

Figure 10-25. Common preassembled motor control unit panels are available from the factory, along with their schematic diagram.

The motor is connected to T1, T2, and T3. The control inputs are connected to the terminal blocks marked 1, 2, and 3. If a 2-wire control, like a liquid level switch, is connected to the circuit, it is connected to terminals 1 and 3 only. Also provided on each unit are predrilled holes to allow for easy additions to the circuit. These holes match the manufacturer's standard devices, and most manufacturers provide templates for easy layout and circuit designs.

Feeders and Busways

The electrical distribution system in a plant must transport the electrical power from the source of supply to the loads. In today's industry, that may consist of distribution over large areas with many different electrical requirements. See Figure 10-26. In many cases, the distribution system must be changed from time to time where shifting of production machinery is common. A *busway* is a metal-enclosed distribution system of bus bars available in prefabricated sections. Prefabricated fittings, tees, elbows, and crosses simplify the connecting and reconnecting of the distribution system. By bolting sections together, the electrical power is available at many locations and throughout the system.

A busway does not have exposed conductors. This is because the power in a plant distribution system is at a high level. To offer protection from the high voltage, the conductors of a busway are supported with insulating blocks and covered with an enclosure to prevent accidental contact. A typical busway distribution system provides for fast connection and disconnection of machinery. Plants can be retooled or reengineered without major changes in the distribution system.

The most common length of busways is 10′. Shorter lengths are used as needed. Prefabricated elbows, tees, and crosses make it possible for the electrical power to run up, down, around corners, and to be tapped off from the distribution system. This allows the distribution system to have maximum flexibility with simple and easy connections when working on installations.

The two basic types of busways are feeder and plug-in busways. See Figure 10-27. Feeder busways deliver the power from the source to a load-consuming device. Plug-in busways serve the same function, but also allow load-consuming devices to be conveniently added along the bus structure. A plug-in power module is used on a plug-in busway system.

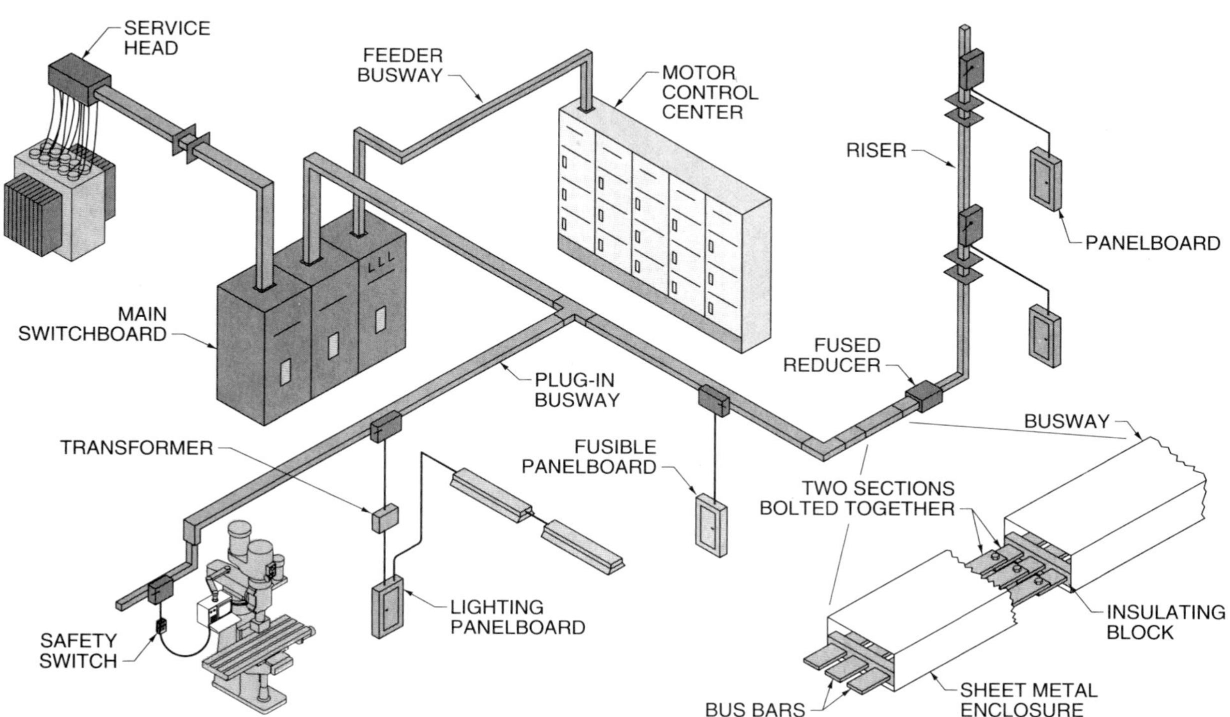

Figure 10-26. The electrical distribution system in a plant must transport the electrical power from the source of supply to the loads.

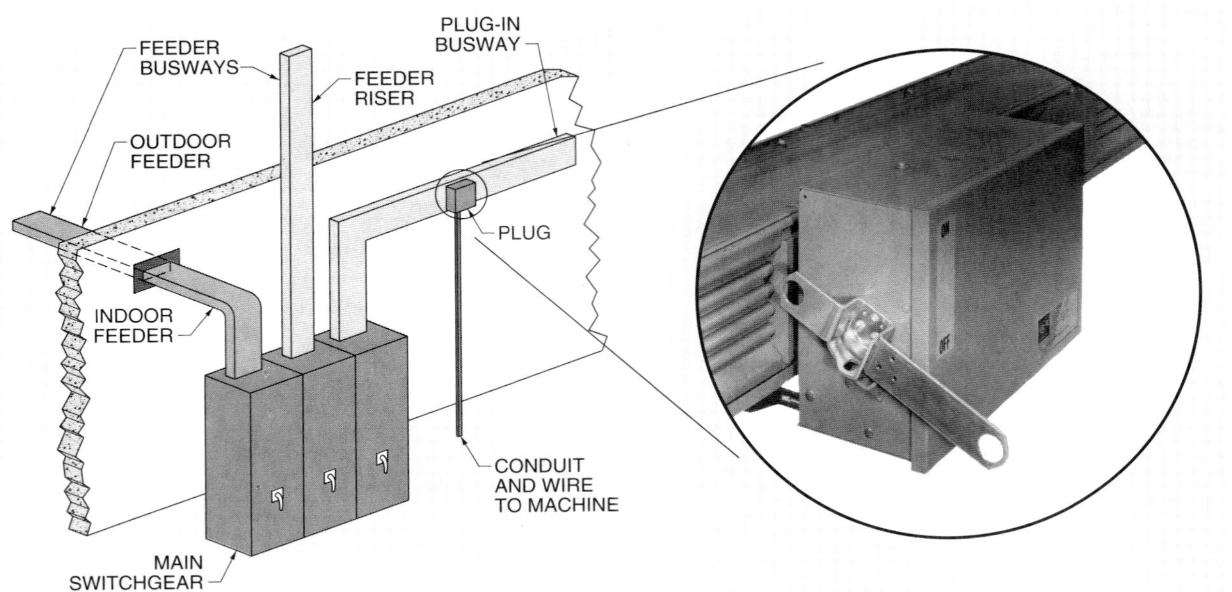

Figure 10-27. The two basic types of busways are feeder and plug-in busways.

The three general types of plug-in power panels used with busways are fusible switches, circuit breakers, and specialty plugs (duplex receptacles with circuit breakers, twistlock receptacles, etc.). It is from these fusible switches and circuit breaker plug-in panels that the conduit and wire is run to a machine or load. Generally, cords may be used only for portable equipment.

The loads connected to the power distribution system are often portable or unknown at the time of installation. For this reason, the power distribution system must often terminate in such a manner as to provide for a quick connection of a load at some future time. To do this, an electrician installs receptacles throughout the building or plant to serve the loads as required. With these receptacles, different loads can be connected easily.

Because the distribution system's wiring and protection devices determine the size of the load that can be connected to it, a method is required for distinguishing the rating in voltage and current of each termination. This is especially true in industrial applications which require a variety of different currents, voltages, and phases.

The National Electrical Manufacturers Association (NEMA) has established a set of standard plug and receptacle configurations that clearly indicate the type of termination. See Appendix. The standard configurations enable the identification of the voltage and current rating of any receptacle or plug simply by looking at the configuration.

Grounding

Equipment grounding is required throughout the entire distribution system. This means connecting to ground all noncurrent-carrying metal parts including conduit, raceways, transformer cases, and switch gear enclosures. The objective of grounding is to limit the voltage between all metal parts and the earth to a safe level.

Grounding is accomplished by connecting the noncurrent-carrying metal to a ground bus with an approved grounding conductor and fitting. A grounding bus is a network that ties solidly to grounding electrodes. A *grounding electrode* is a conductor embedded in the earth to provide a good ground.

The ground bus should surround the transmission station or building. This bus must be connected to the grounding electrodes in several spots. The size of the ground bus is based on the amount of current that flows through the grounding system and the length of time it flows.

In addition to grounding all noncurrent-carrying metal, lightning arresters may be needed. A *lightning arrester* is a device which protects the transformers and other electrical equipment from voltage surges caused by lightning. A lightning arrester provides a path over which the surge can pass to ground before it has a chance to damage electrical equipment.

Gould Inc.

The various classes of fuses available from Gould Inc. are used to protect load centers, panelboards, motors, motor controllers, and transformers, as well as other applications.

Troubleshooting Fuses

A *fuse* is an overcurrent protection device (OCPD) with a fusible link that melts and opens the circuit on an overcurrent condition. Fuses are connected in series with a circuit to protect a circuit from overcurrents or shorts. Fuses may be one-time or renewable. One-time fuses are fuses that cannot be reused after they have opened. One-time fuses are the most common. *Renewable fuses* are OCPDs designed so that the fusible link can be replaced. A multimeter or voltmeter is used to test fuses. See Figure 10-28.

To troubleshoot fuses, apply the procedure:

1. Turn the handle of the safety switch or combination starter OFF.
2. Open the door of the safety switch or combination starter. The operating handle must be capable of opening the switch. If it is not, replace the switch.

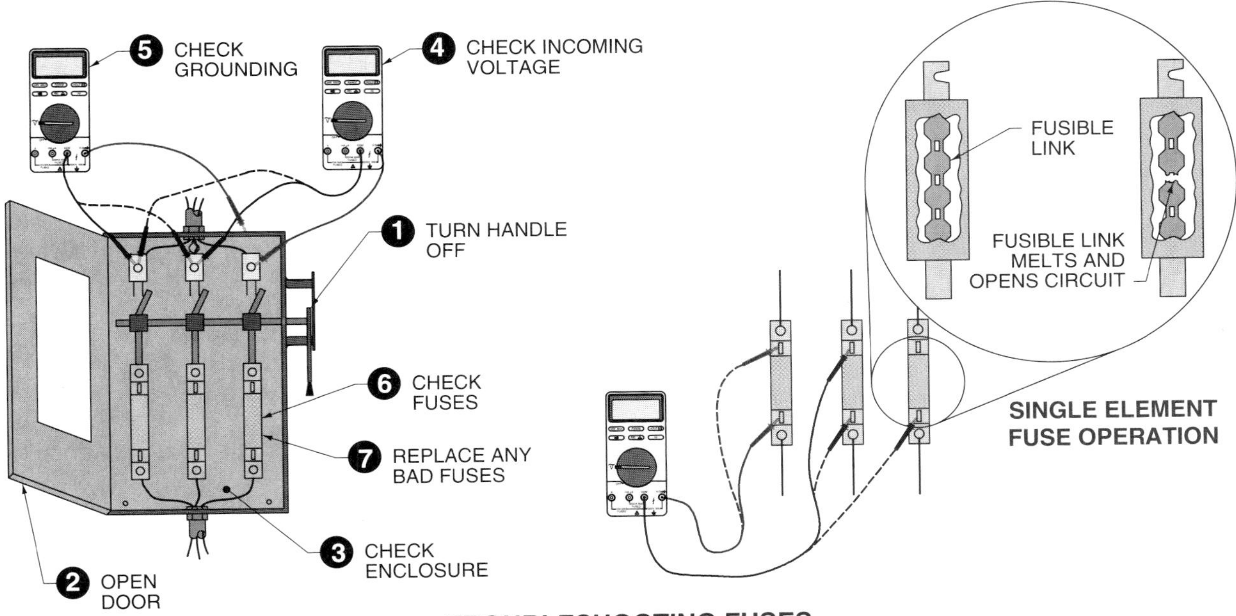

Figure 10-28. Fuses are connected in series with a circuit to protect a circuit from overcurrents or shorts.

3. Check the enclosure and interior parts for deformation, displacement of parts, and burning. Such damage may indicate a short, fire, or lightning strike. Deformation requires replacement of the part or complete device. Any indication of arcing damage or overheating, such as discoloration or melting of insulation, requires replacement of the damaged part(s).
4. Check the voltage between each pair of power leads. Incoming voltage should be within 10% of the voltage rating of the motor. A secondary problem exists if voltage is not within 10%. This secondary problem may be the reason the fuses have blown.
5. Test the enclosure for grounding if voltage is present and at the correct level. To test for grounding, connect one side of a voltmeter to an unpainted metal part of the enclosure and touch the other side to each of the incoming power leads. A voltage difference is indicated if the enclosure is properly grounded. The line-to-ground voltage probably does not equal the line-to-line voltage reading taken in Step 4.
6. Turn the handle of the safety switch or combination starter ON to test the fuses. One side of a voltmeter is connected to one side of an incoming power line at the top of one fuse. The other side of the voltmeter is connected to the bottom of each of the remaining fuses. A voltage reading indicates the fuse is good. If no voltage reading is obtained, the fuse is open and no voltage passes through. The fuse must be replaced (not at this time). Repeat this procedure for each fuse. When testing the last fuse, the voltmeter is moved to a second incoming power line.
7. Turn the handle of the safety switch or combination starter OFF to replace the fuses. Use a fuse puller to remove bad fuses. Replace all bad fuses with the correct type and size replacement. Close the door on the safety switch or combination starter and turn the circuit ON.

Troubleshooting Circuit Breakers

A *circuit breaker (CB)* is a reusable OCPD that opens a circuit automatically at predetermined overcurrent. CBs are connected in series with the circuit. They protect a circuit from overcurrents or short circuits. CBs are thermally- or magnetically-operated and are reset after an overload. A multimeter is used to test CBs. CBs perform the same function as fuses and are tested the same way. See Figure 10-29.

To troubleshoot CBs, apply the procedure:

1. Turn the handle of the safety switch or combination starter OFF.
2. Open the door of the safety switch or combination starter. The operating handle must be capable of opening the switch. Replace the operating handle if it does not open the switch.
3. Check the enclosure and interior parts for deformation, displacement of parts, and burning.
4. Check the voltage between each pair of power leads. Incoming voltage should be within 10% of the voltage rating of the motor.
5. Test the enclosure for grounding if voltage is present and at the correct level.
6. Examine the CB. It is in one of three positions, ON, TRIPPED, or OFF.
7. If no evidence of damage is present, reset the CB by moving the handle to OFF and then to ON. CBs must be cooled before they are reset. CBs are designed so they cannot be held in the ON position if an overload or short is present. Check the voltage of the reset CB if resetting the CB does not restore power. Replace all faulty CBs. Never try to service a faulty CB.

Fluke Corporation

The Model 867 graphical multimeter from Fluke Corporation is used to check the voltage at the power panel, which is a common point to start checking the system when troubleshooting a circuit.

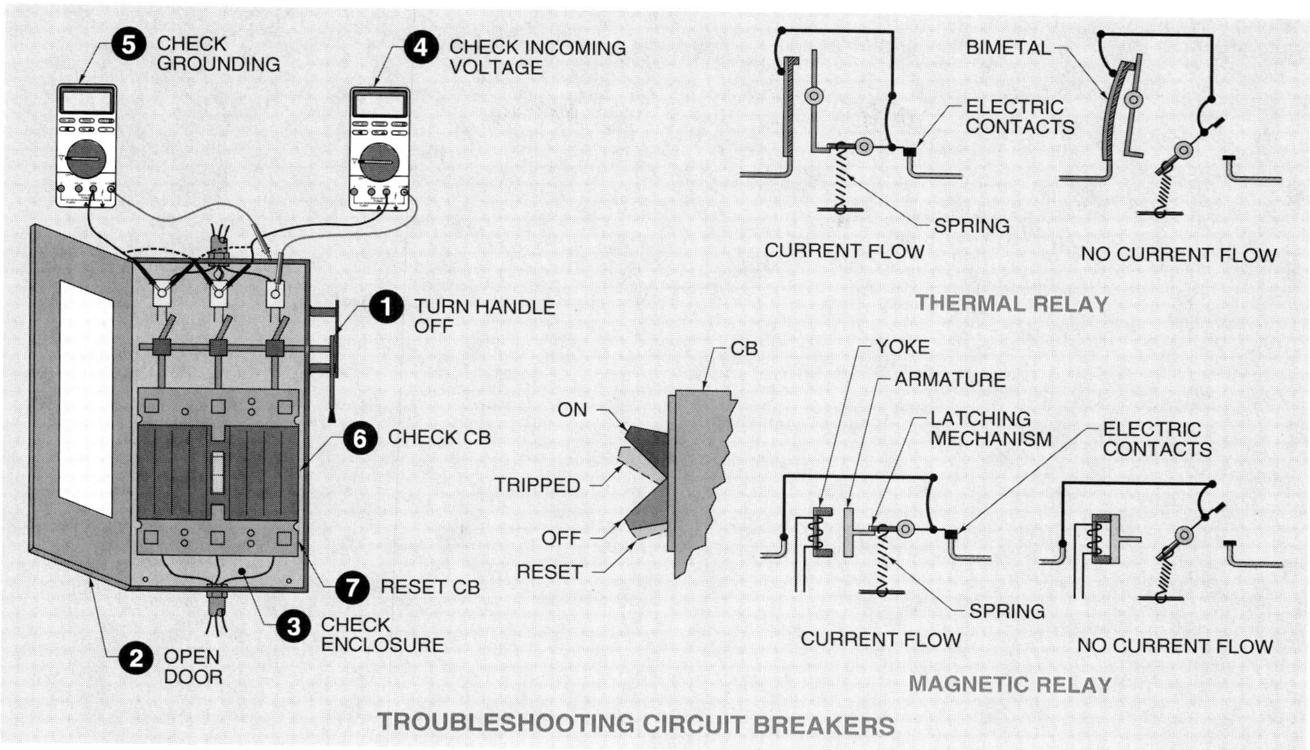

Figure 10-29. Circuit breakers (CBs) perform the same function as fuses and are tested the same way.

General Electric Company

Circuit breakers are available from General Electric Company in a variety of sizes and configurations for use in various applications.

1. What is the major source of electrical power?
2. How is electricity generated in an alternator?
3. Why are slip rings used on an AC alternator?
4. What is the effect of adding pairs of poles to an alternator?
5. Why is 3ϕ power generated instead of 1ϕ power whenever possible?
6. What are the voltages available from a wye-connected system with a common neutral?
7. What is the phase-to-phase voltage if the phase-to-neutral voltage is 208 V in a wye-connected system?
8. What are the voltages available from a 240 V delta-connected system without a neutral wire?
9. What is a transformer?
10. What are the names of the two windings of a transformer?
11. What is the function of the transformer core?
12. Is a transformer with a ratio of 4:1 a step-up or step-down transformer?
13. What happens to the current on the output of a transformer if the voltage is doubled on the output of the transformer?
14. How is the high side of a transformer identified?
15. How is the low side of a transformer identified?
16. How is a transformer rated for power output?
17. Why is it important to balance a transformer bank?
18. What are the three main parts of a substation?
19. What are the functions of the taps on a transformer?
20. What is the last point on a power distribution system as far as the power company is concerned?
21. What is the difference between a service-entrance switchboard and a distribution switchboard?
22. What is a panelboard?
23. What is a branch circuit?
24. What is the function of a motor control center in a power distribution system?

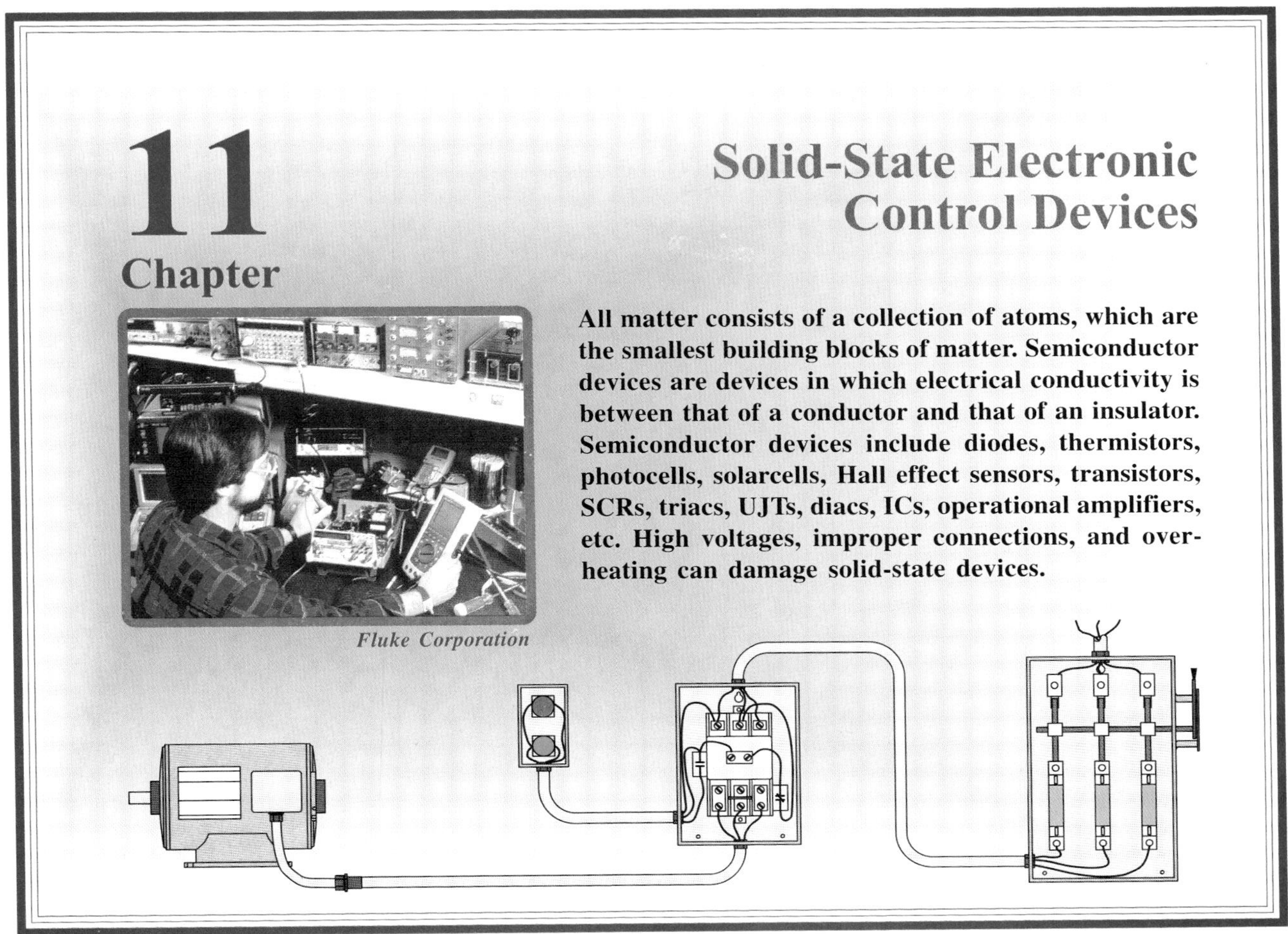

11 Chapter

Solid-State Electronic Control Devices

All matter consists of a collection of atoms, which are the smallest building blocks of matter. Semiconductor devices are devices in which electrical conductivity is between that of a conductor and that of an insulator. Semiconductor devices include diodes, thermistors, photocells, solarcells, Hall effect sensors, transistors, SCRs, triacs, UJTs, diacs, ICs, operational amplifiers, etc. High voltages, improper connections, and overheating can damage solid-state devices.

Fluke Corporation

SEMICONDUCTOR THEORY

All matter consists of an organized collection of atoms. *Atoms* are the smallest building blocks of matter that cannot be divided into smaller units without changing their basic character. The three fundamental particles contained in atoms are protons, neutrons, and electrons. Protons and neutrons make up the nucleus, and electrons whirl about the nucleus in orbits or shells.

The *nucleus*, which contains protons and neutrons, is the heavy, dense center of the atom and has a positive electrical charge. *Protons* are particles with a positive electrical charge of one unit. *Neutrons* are particles with no electrical charge. The nucleus is surrounded by one or more electrons. *Electrons* are negatively charged particles whirling around the nucleus at great speeds in shells. Each shell can hold a specific number of electrons. The innermost shell can hold two electrons. The second shell can hold eight electrons. The third shell can hold 18 electrons, etc. The shells are filled starting with the inner shell and working outward, so that when the inner shells are filled with as many electrons as they can hold, the next shell is started. Electrons and protons have equal amounts of opposite charges. There are as many electrons as there are protons in an atom, which leaves the atom electrically neutral.

Valence Electrons

Most elements do not have a completed outer shell with the maximum allowable number of electrons. *Valence electrons* are electrons in the outermost shell of an atom. Valence electrons determine the conductive or insulative value of a material. Conductors normally have only one or two valence electrons in their outer shell. See Figure 11-1. Insulators normally have several electrons in their outer shell which is either almost or completely filled with electrons. Semiconductor materials fall between the low resistance offered by a conductor and the high resistance offered by an insulator. Semiconductors are made from materials that have four valence electrons.

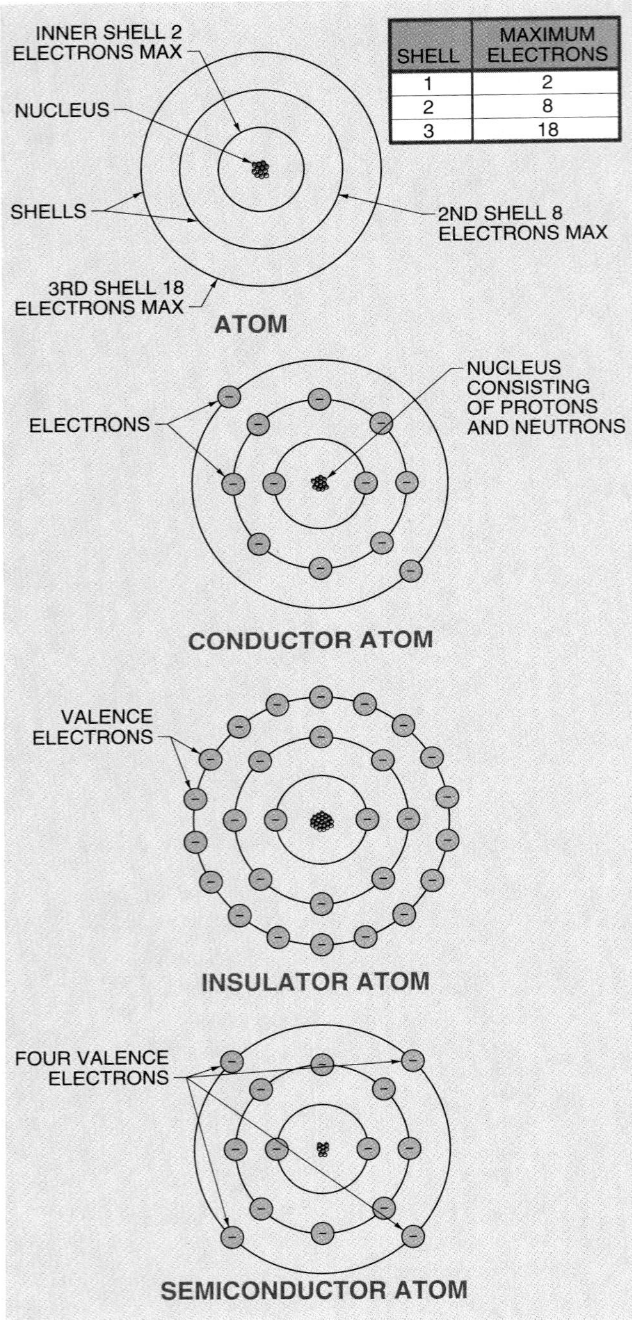

Figure 11-1. Valence electrons determine the amount of conductivity or insulating characteristics of a given material.

Doping

The basic material used in most semiconductor devices is either germanium or silicon. In their natural state, germanium and silicon are pure crystals. These pure crystals do not have enough free electrons to support a significant current flow. To prepare these crystals for use as a semiconductor device, their structure must be altered to permit significant current flow.

Doping is the process by which the crystal structure is altered. In doping, some of the atoms in the crystal are replaced with atoms from other elements. The addition of new atoms in the crystal creates N-type material and P-type material.

N-Type Material. *N-type material* is material created by doping a region of a crystal with atoms from an element that has more electrons in its outer shell than the crystal. Adding these atoms to the crystal results in more free electrons. Free electrons (carriers) support current flow. Current flows from negative to positive through the crystal when voltage is applied to N-type material. The material is N-type material because electrons have a negative charge. See Figure 11-2.

Elements commonly used for creating N-type material are arsenic, bismuth, and antimony. The quantity of doping material used ranges from a few parts per billion to a few parts per million. By controlling these small quantities of impurities in a crystal, the manufacturer controls the operating characteristics of the semiconductor.

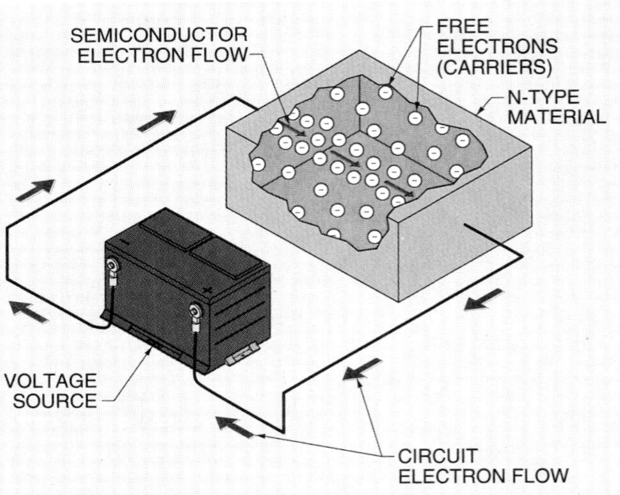

Figure 11-2. Current flows from negative to positive and is assisted by free electrons when voltage is applied to N-type material.

P-Type Material. *P-type material* is material with empty spaces (holes) in its crystalline structure. To create P-type material, a crystal is doped with atoms from an element that has fewer electrons in its outer shell than the crystal. *Holes* are the missing electrons in the crystal structure. The holes are represented as positive charges.

In P-type material, the holes act as carriers. The holes are filled with free electrons when voltage is applied, and the free electrons move from negative potential to positive potential through the crystal. See Figure 11-3. Movement of the electrons from one hole to the next makes the holes appear to move in the opposite direction. Hole flow is equal to and opposite of electron flow. Typical elements used for doping a crystal to create P-type material are gallium, boron, and indium.

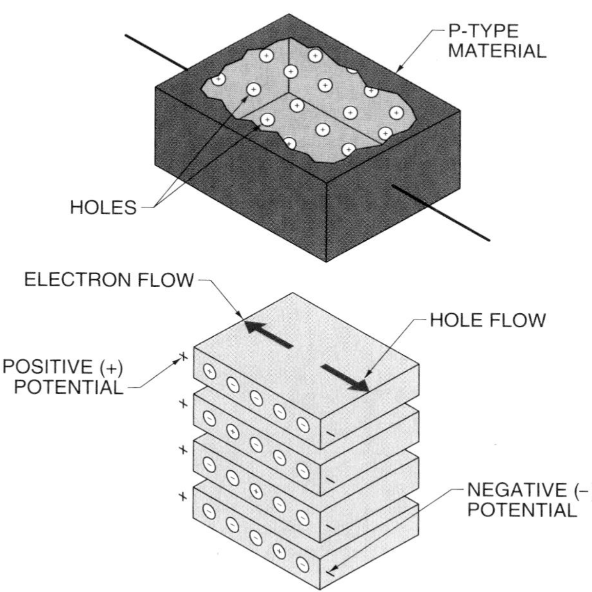

Figure 11-3. When voltage is applied to P-type material, the holes are filled with free electrons that move from the negative potential to the positive potential through the crystal.

SEMICONDUCTOR DEVICES

Semiconductor devices are devices in which electrical conductivity is between that of a conductor (high conductivity) and that of an insulator (low conductivity). The electrical conductivity of a semiconductor device is nearly as great as the conductivity of a metal at high temperatures and is nearly absent at low temperatures.

Semiconductor devices are often mounted on a PC board. A *PC board* is an insulating material such as fiberglass or phenolic with conducting paths laminated to one or both sides of the board. PC boards provide electrical paths of sufficient size to ensure a reliable electronic circuit. See Figure 11-4. *Pads* are small round conductors to which component leads are soldered. *Traces (foils)* are conducting paths used to connect components on a PC board. They are used to interconnect two or more pads. A *bus* is a large trace extending around the edge to provide conduction from several sources.

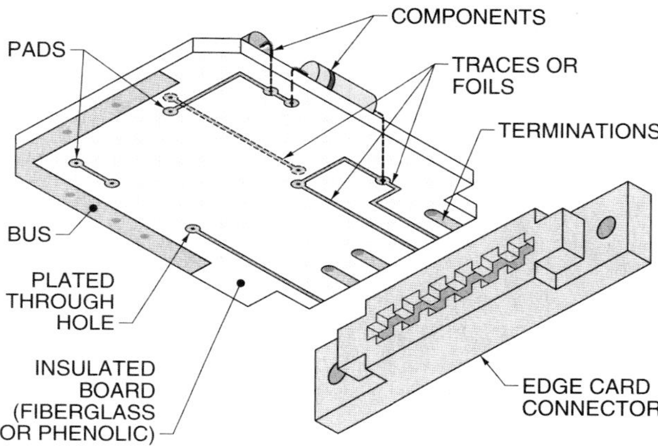

Figure 11-4. A PC board is constructed of an insulating material such as a fiberglass or phenolic with conducting paths laminated on one or both sides of the board.

An *edge card* is a PC board with multiple terminations (terminal contacts) on one end. Most edge cards have terminations made from copper which is the same material as the traces. In some instances, the terminations are gold plated, allowing for the lowest possible contact resistance. An *edge card connector* allows the edge card to be connected to the system's circuitry with the least amount of hardware.

Semiconductor control devices are normally mounted on one side of a PC board. See Figure 11-5. In some cases where space is a premium, components may be mounted on both sides of the PC board. Component leads extend through the board and are connected to the pads, traces, and bus with solder. PC boards may have markings next to each component to help identify the component in relation to the schematic.

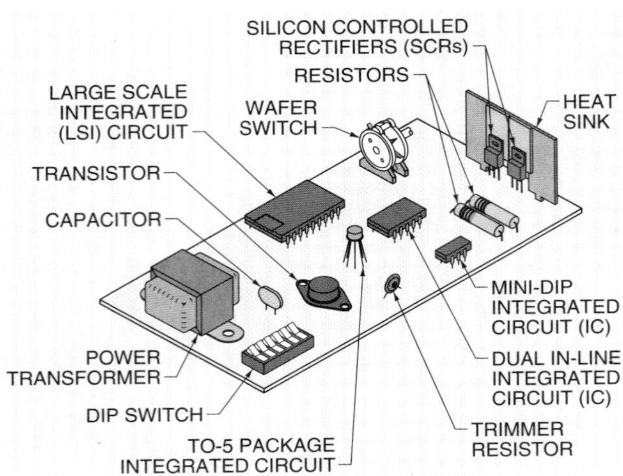

Figure 11-5. Semiconductor control devices are normally mounted on one side of a PC board.

Siemens Corporation

The SIPLACE 80 S from Siemens Corporation is a high-speed surface-mount system for PC board manufacturing.

Diodes

Diodes are electronic components that allow current to pass through them in only one direction. This is made possible by the doping process, which creates N-type material and P-type material on the same component. The P-type and N-type materials exchange carriers at the junction of the two materials, creating a thin depletion region. See Figure 11-6.

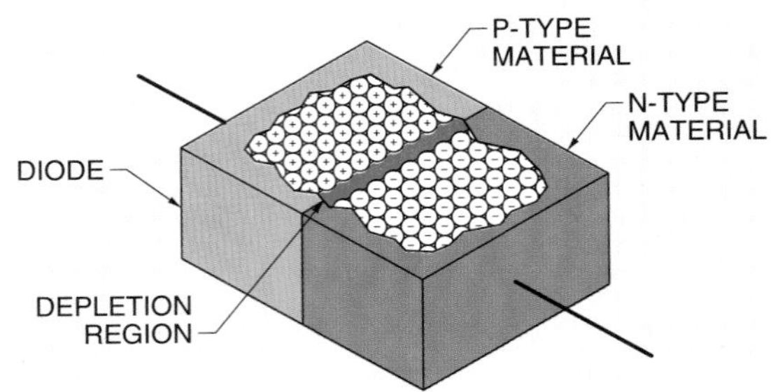

Figure 11-6. P-type and N-type materials exchange carriers at the junction of the two materials, creating a thin depletion region.

The thin depletion region responds rapidly to voltage changes. The operating characteristics of a specific diode can be determined through the use of its operating characteristic curve. See Figure 11-7.

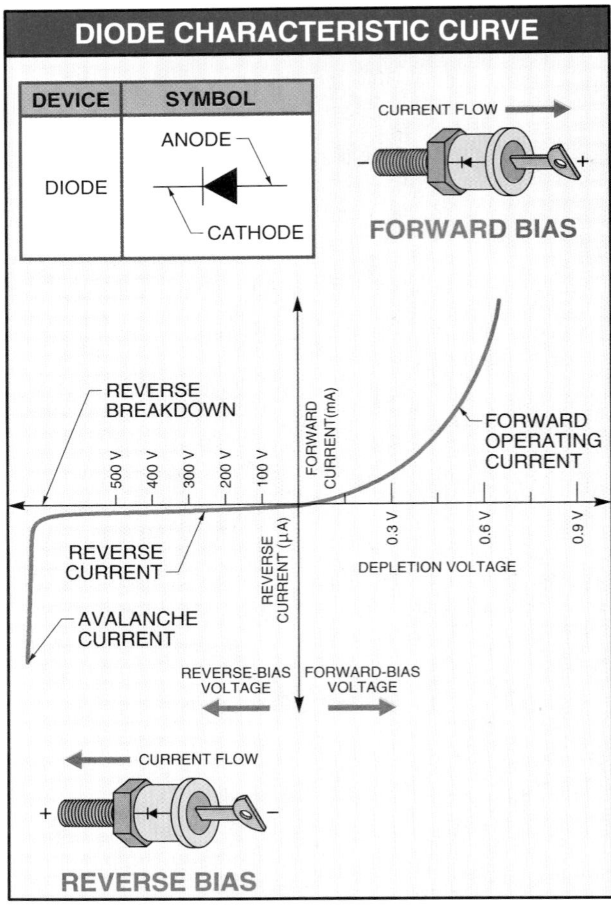

Figure 11-7. A diode characteristic curve indicates the response of a diode when subjected to different forward- and reverse-bias voltages.

Fluke Corporation

A Model 87 true rms multimeter from Fluke Corporation is used to check the rectifiers in automobiles that change the generated AC from the alternator into DC.

When voltage is applied to the diode, the action occurring in the depletion region either blocks current flow or passes current. *Forward-bias voltage* is the application of the proper polarity to a diode. Forward bias results in forward current. *Reverse-bias voltage* is the application of the opposite polarity to a diode. Reverse bias results in a reverse current which should be very small (normally 1 mA).

Peak inverse voltage (PIV) is the maximum reverse bias voltage that a diode can withstand. The PIV ratings for most diodes used in industry range from a few volts to several thousand volts. The diode breaks down and passes current freely if the reverse bias applied to the diode exceeds its PIV rating. *Avalanche current* is current passed when a diode breaks down. Avalanche current can destroy diodes. Diodes with the correct voltage rating must be used to avoid avalanche current.

Rectification of Alternating Current

Alternating current (AC) power is more efficiently and economically generated and transmitted than direct current (DC) power. AC must be changed into DC because machinery and other loads often need DC to operate. *Rectification* is the changing of AC into DC.

Single-Phase Rectifiers. A half-wave rectifier is used to convert AC to pulsating DC. See Figure 11-8. A *half-wave rectifier* is a circuit containing a diode which permits only the positive half-cycles of the AC sine wave to pass. Half-wave rectification is accomplished because current is allowed to flow only when the anode terminal is positive with respect to the cathode. Current is not allowed to flow through the rectifier when the cathode is positive with respect to the anode.

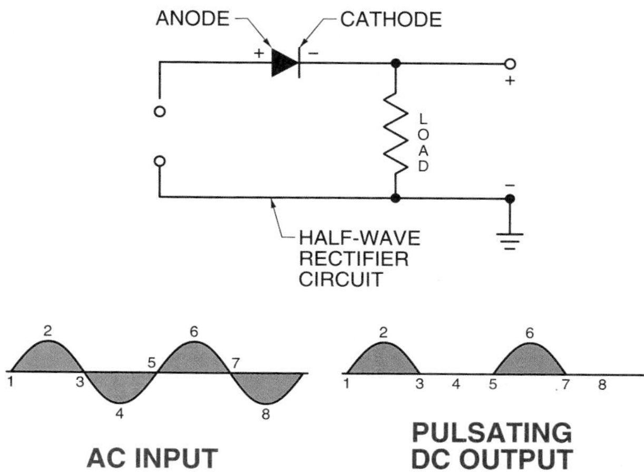

Figure 11-8. A half-wave rectifier converts AC to pulsating DC.

The output voltage of a half-wave rectifier is considered pulsating DC with half of the AC sine wave cut off. The rectifier passes either the positive or negative half-cycle of the input AC sine wave, depending on the way the diode is connected into the circuit. Half-wave rectification is inefficient for most applications because one-half of the input sine wave is not used.

A full-wave rectifier circuit uses both halves of the input AC sine wave. Full-wave rectification may be obtained from a 1ϕ AC source by using two diodes with a center-tapped transformer or by using a bridge rectifier circuit.

In the circuit using two diodes and a center-tapped transformer, when voltage is induced in the secondary from point A to B, point A is positive with respect to point N. Current flows from A to N, through the load, and through diode 1 (D1). See Figure 11-9. D1 conducts current and diode 2 (D2) blocks current because A is positive with respect to N.

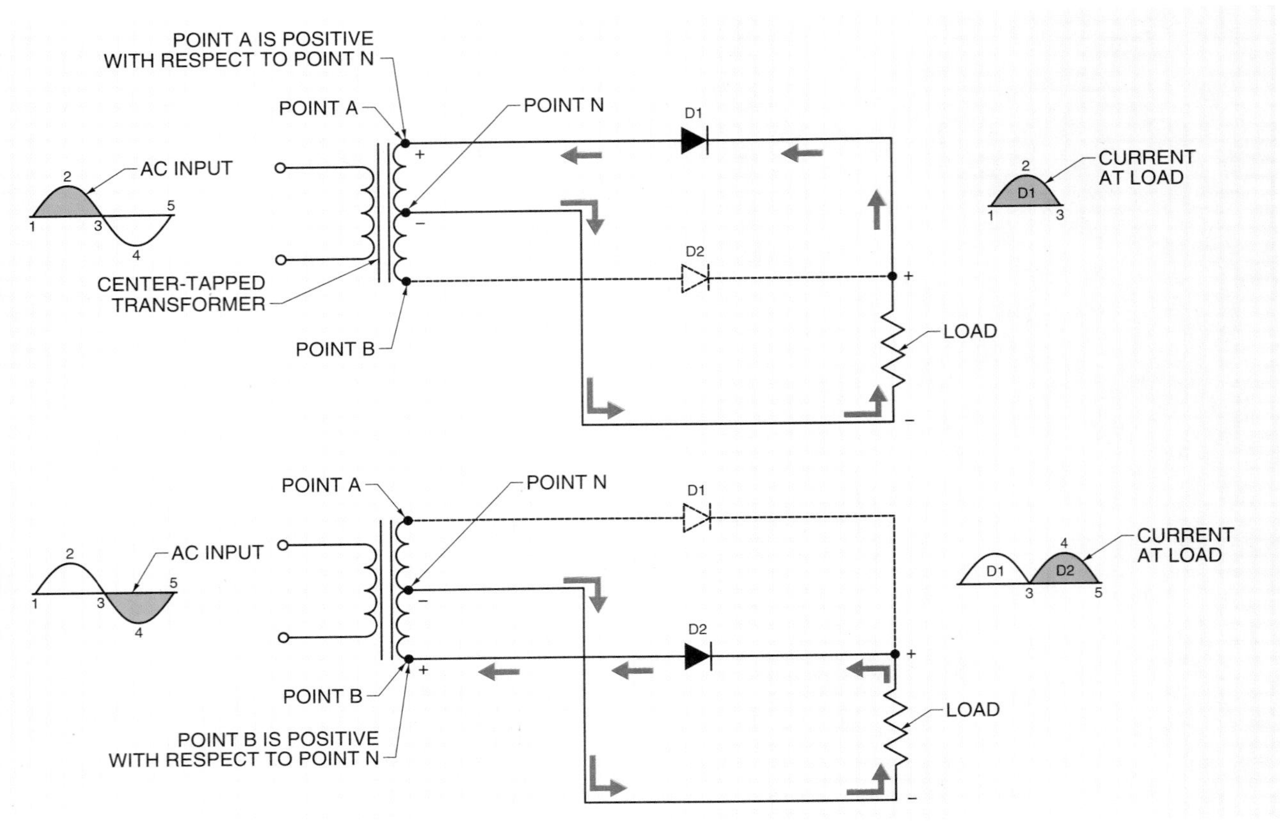

Figure 11-9. Full-wave rectification may be obtained from a 1ϕ AC source by using two diodes with a center-tapped transformer.

Fluke Corporation

Fluke multimeters are used to bench test electronic components mounted on printed circuit boards.

When the voltage across the secondary reverses during the negative half-cycle of the AC sine wave, point B is positive with respect to point N. Current then flows from B to N, through the load, and through D2. D2 conducts and D1 blocks current because B is positive with respect to N. This is repeated every cycle of the AC sine wave, producing a full-wave DC output.

A bridge rectifier circuit produces the same full-wave DC output. See Figure 11-10. A bridge rectifier circuit requires four diodes and eliminates the need for a transformer. A bridge rectifier circuit is more efficient than the center-tapped circuit because each diode blocks only half as much reverse voltage for the same output voltage.

In this circuit, when the AC supply voltage is positive at point A and negative at point B, current flows from point B, through D2, the load, D1, and to point A. When the AC supply voltage is positive at point B and negative at point A, current flows from point A, through diode 4 (D4), the load, diode 3 (D3), and to point B.

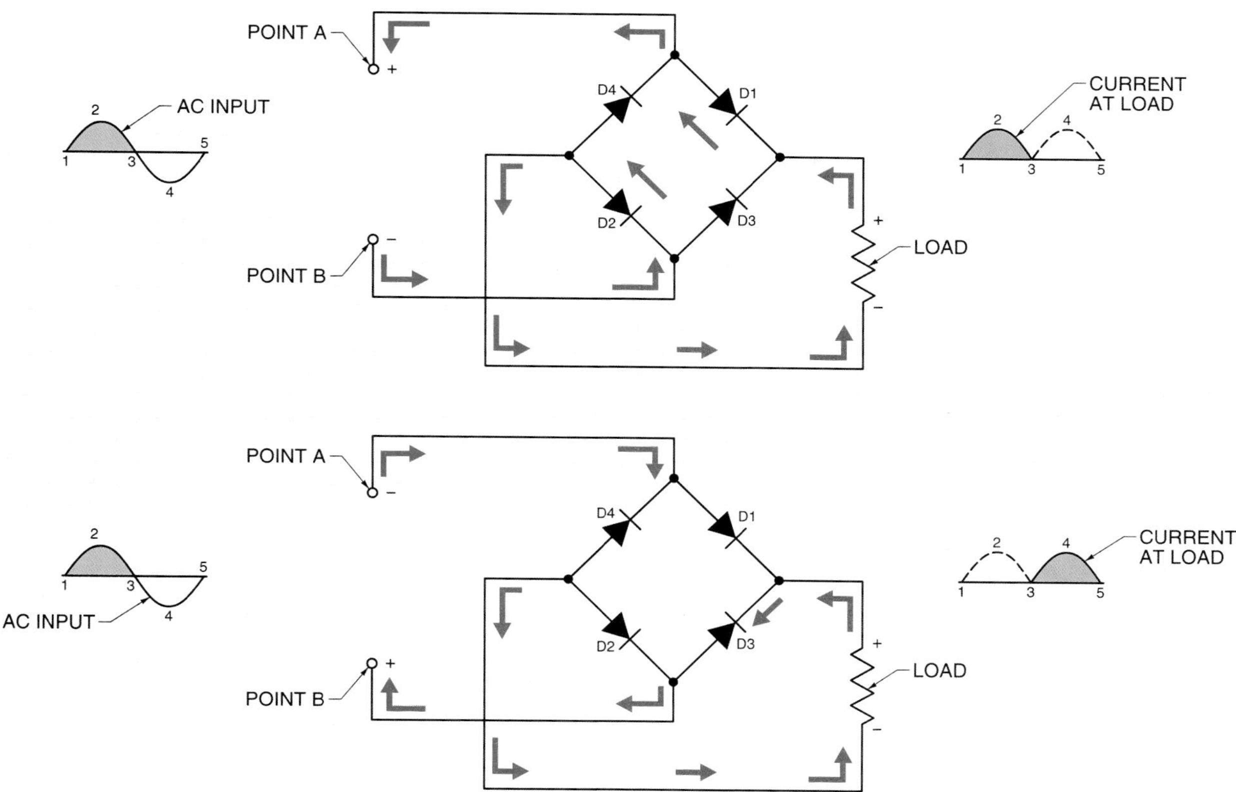

Figure 11-10. A bridge rectifier circuit is more efficient than a center-tapped circuit because each diode blocks only half as much reverse voltage for the same output voltage.

The output of a full-wave rectifier is pulsating DC and must be filtered or smoothed out before it can be used in most electronic equipment. This filtering is done by a filter circuit connected to the output of the rectifier circuit. This filter circuit normally consists of one or more capacitors, inductors, or resistors connected in different combinations. The choice of a filter circuit is determined by the load (how much ripple it can take), cost, and available space.

Filtered DC eliminates pulsations and provides DC at a constant intensity. See Figure 11-11. This is accomplished because the pulsating voltage no longer drops to zero at the end of each pulsation. This results in the average voltage delivered by the rectifier circuit being higher. The purpose of a filter is to smooth and increase the DC voltage output of the circuit.

Three-Phase Rectifiers. A DC output can also be supplied from a 3ϕ power source. The advantage of using 3ϕ power is that it is possible to obtain a smooth DC output without the use of a filter circuit. This is possible in a 3ϕ circuit because when any one phase becomes negative, at least one of the other phases becomes positive. The result is a relatively smooth output without any filtering.

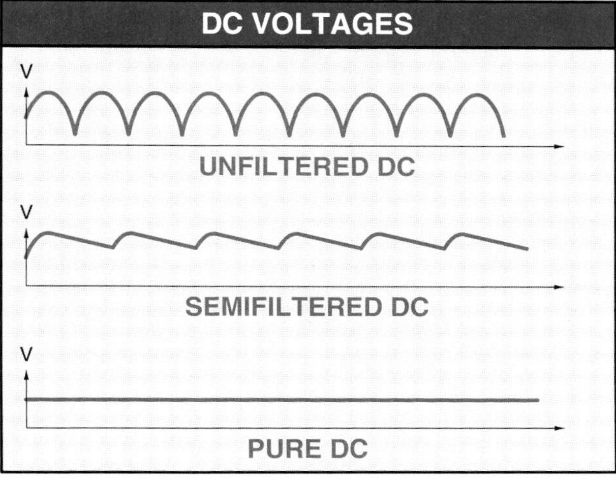

Figure 11-11. Filtered DC eliminates pulsations and provides DC at a constant intensity.

A 3ϕ rectifier circuit uses three diodes connected to a wye circuit with a neutral tap. See Figure 11-12. Each diode conducts in succession while the remaining two are blocking. The output voltage never goes below a certain voltage level. This circuit delivers the same smooth DC output as a filtered 1ϕ bridge rectifier circuit.

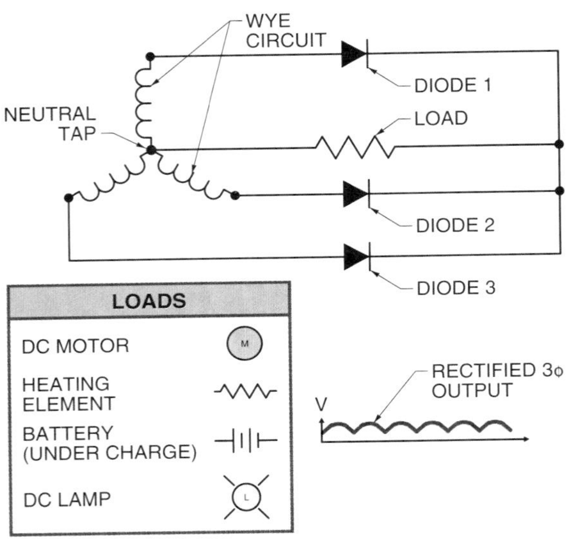

Figure 11-12. A 3ϕ rectifier circuit uses three diodes connected to a wye circuit with a neutral tap.

Zener Diodes

A *zener diode* is a silicon PN junction that differs from a rectifier diode in that it operates in the reverse breakdown region. A *PN junction* is the area on a semiconductor material between the P-type and N-type material. A zener diode acts as a voltage regulator either by itself or in conjunction with other semiconductor devices. A zener diode symbol differs from a standard diode symbol in that the normally vertical cathode line is bent slightly at each end. Standard diodes normally conduct in forward bias and can be destroyed if the reverse voltage or bias is exceeded. A zener diode is often referred to as an avalanche diode because the zener diode normally operates in reverse breakdown.

Zener Diode Operation. The forward breakover voltage and current characteristics are similar to a standard diode when a source voltage is applied to a zener diode in the forward direction. See Figure 11-13. When a source voltage is applied to the zener diode in the reverse direction, the current remains low until the reverse voltage reaches reverse breakdown (zener breakdown). The zener diode conducts heavily (avalanches) at zener breakdown. Reverse current flow through a zener diode must be limited by a resistor or other device to prevent diode destruction. The maximum current that may flow through a zener diode is determined by diode size. Like the forward voltage drop of a standard diode, the reverse voltage drop or zener voltage of a zener diode remains essentially constant despite large current fluctuations.

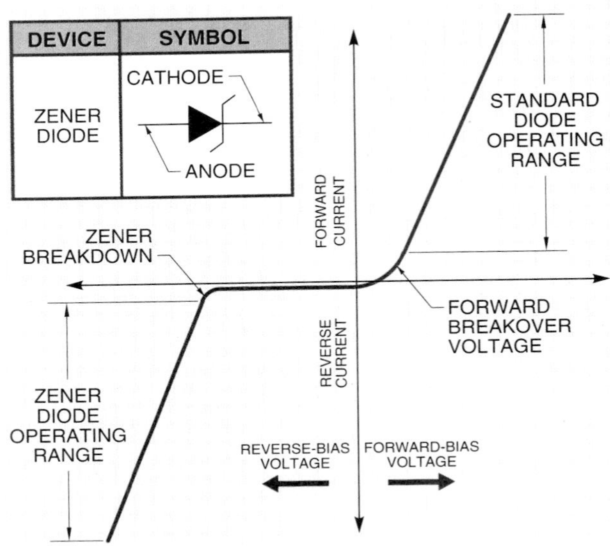

Figure 11-13. A zener diode is often referred to as an avalanche diode because the zener diode normally operates in reverse breakdown.

A zener diode is capable of being a constant voltage source because of the resistance changes that take place within the PN junction. The resistance of the PN junction remains high and should produce leakage current in the microampere range when a source of voltage is applied to the zener diode in the reverse direction. However, as the reverse voltage is increased, the PN junction reaches a critical voltage and the zener diode avalanches. As the avalanche voltage is reached, the normally high resistance of the PN junction drops to a low value and the current increases rapidly. The current is normally limited by a circuit resistor or load resistance.

Thermistors

A *thermistor* is a temperature-sensitive resistor that changes its electrical resistance with a change in temperature. See Figure 11-14. The resistance of a negative temperature coefficient thermistor decreases and current flow increases as heat is placed under the thermistor. The resistance increases to its original state (resistance value) when the heat is removed.

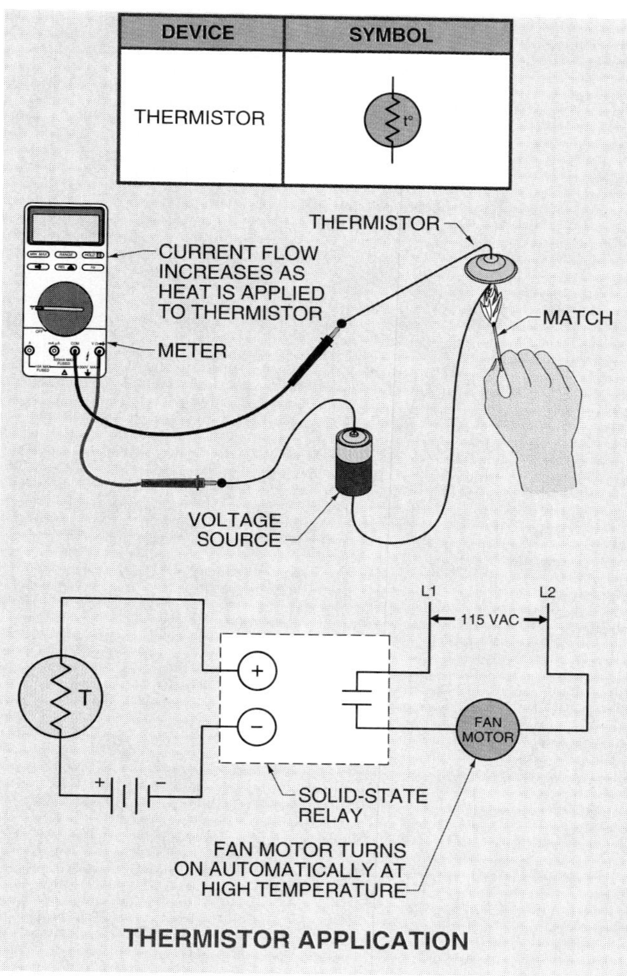

Figure 11-14. A thermistor is a thermally-sensitive resistor whose resistance changes with a change in temperature.

The operation of a thermistor is based on the electron-hole theory. As the temperature of the semiconductor increases, the generation of electron-hole pairs increases due to thermal agitation. Increased electron-hole pairs cause a drop in resistance.

Thermistors are popular because of their small size. They can be mounted in places that are inaccessible to other temperature-sensing devices. Thermistors may be directly heated or indirectly heated.

Controlling a fan motor is a typical example of using a thermistor. As the thermistor is heated, its resistance decreases and more current flows through the circuit. When enough current flows through the circuit, the solid-state relay turns ON. The solid-state relay is used to switch ON a fan motor at high temperatures. Such a circuit can be used to automatically reduce heat in attics or to circulate warm air.

Photoconductive Cells

A *photoconductive cell (photocell)* is a device which conducts current when energized by light. Current increases with the intensity of light because resistance decreases. A photocell is, in effect, a variable resistor. See Figure 11-15. A photocell is formed with a thin layer of semiconductor material such as cadmium sulfide or cadmium selenide deposited on a suitable insulator. Leads are attached to the semiconductor material and the entire assembly is hermetically sealed in glass. The transparency of the glass allows light to reach the semiconductor material. For maximum current-carrying capacity, the photocell is manufactured with a short conduction path having a large cross-sectional area.

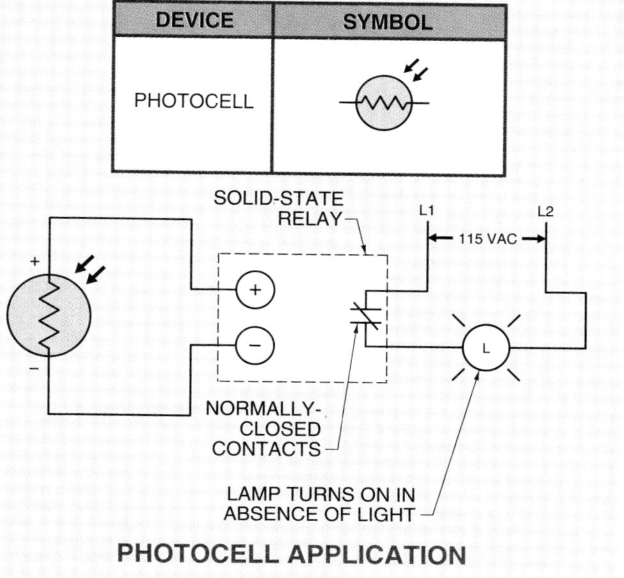

Figure 11-15. A photocell is a device which conducts current when energized by light.

Ruud Lighting, Inc.

Outdoor and landscape lighting systems from Ruud Lighting, Inc. are available with photocells to control the operation of the lighting systems.

Controlling an outdoor lamp is a typical example of using a photocell. When the photocell is dark, its resistance is high and the solid-state relay is OFF. If a solid-state relay with normally closed contacts is used to control the lamp, the lamp is energized whenever the photocell is dark.

Photovoltaic Cells

A *photovoltaic cell (solarcell)* is a device that converts solar energy to electrical energy. A solarcell is sensitive to light and produces a voltage without an external source. Several different solarcells are available. The device is equivalent to a single-cell voltage source like those found in batteries.

The use of solarcells as a remote power source is becoming more popular. Many manufacturers are designing solarcells into their products on individual and multi-cell applications. For example, most handheld calculators are powered by solarcells and require no batteries.

Solarcell Operation. A solarcell generates energy by using a PN junction to convert light energy into electrical energy. See Figure 11-16. It produces a potential difference between a pair of terminals only when exposed to light.

At the junction of N-type material and P-type material, some recombination of the electrons and holes occurs, but the junction itself acts as a barrier between the two charges. The electrical field at the junction maintains the negative charges on the N-type material side and the positive charges in the holes on the P-type material side.

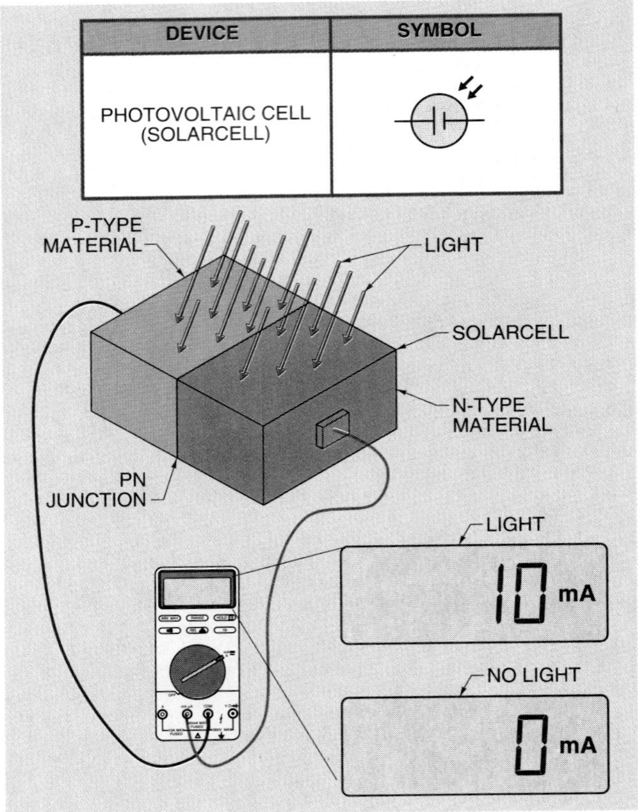

Figure 11-16. A solarcell generates energy by using a PN junction to convert light energy into electrical energy.

Current flows with light acting as a generator if the load is connected across the PN junction. Electron-hole pairs formed by light energy recombine and return to the normal condition prior to the application of light when current flows through the load. Consequently, there is no loss or addition of electrons to the silicon during the process of converting light energy to electrical energy. A solarcell should have no limit to its life span, provided it is not damaged.

Photoconductive Diodes

A *photoconductive diode (photodiode)* is a diode which is switched ON and OFF by a light. A photodiode is internally similar to a regular semiconductor diode. The primary difference is the addition of a lens in the housing for focusing light on the PN junction. See Figure 11-17.

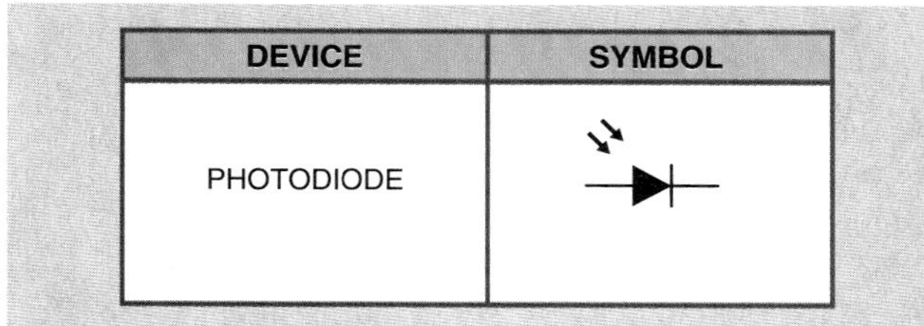

Figure 11-17. A photodiode is a diode which is switched ON and OFF by a light.

Photodiode Operation. The conductive properties change when light strikes the surface of the PN junction in a photodiode. Without light, the resistance of the photodiode is high. The resistance reduces proportionately when the photodiode is exposed to light.

Hall Effect Sensors

A *Hall effect sensor* is a sensor that detects the proximity of a magnetic field. The output of a Hall generator depends on the presence of a magnetic field and the current flow in the Hall generator. See Figure 11-18. A constant control current passes through a thin strip of semiconductor material (Hall generator).

When a permanent magnet is brought near, a small voltage (Hall voltage) appears at the contacts that are placed across the narrow dimension of the strip. As the magnet is removed, the Hall voltage reduces to zero. Thus, the Hall voltage depends on the presence of a magnetic field and on the current flowing through the Hall generator. The output of the Hall generator is zero if the current or the magnetic field is removed. In most Hall effect sensors, the control current is held constant and the flux density is changed by movement of a permanent magnet.

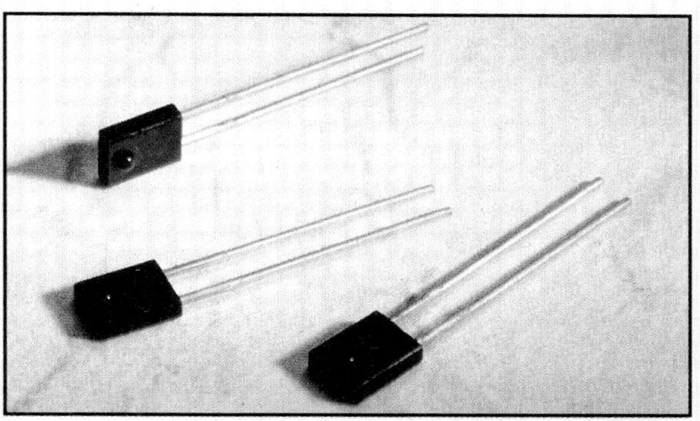

Honeywell's MICRO SWITCH Division

The new SDP8276-001 PN sidelooker photodiode from Honeywell's MICRO SWITCH Division provides a low-power-consumption detection solution when interfaced with CMOS amplifier battery-powered systems.

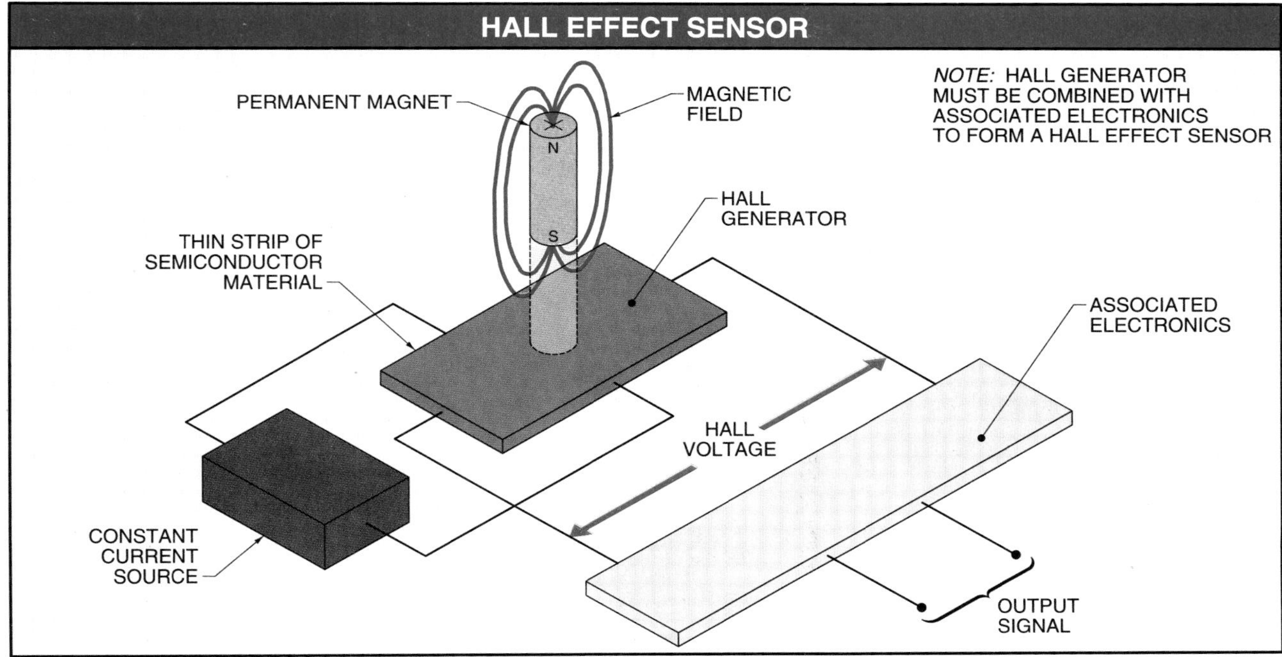

Figure 11-18. A Hall effect sensor is a sensor which produces a voltage depending on the strength of the magnetic field applied to the sensor.

Note: The Hall generator must be combined with an association of electronic circuits to form a Hall effect sensor. Because all the circuitry is normally on an IC, the Hall effect sensor can be considered a single device with a voltage output.

Solid-State Pressure Sensors

A *solid-state pressure sensor* is a transducer that changes resistance with a corresponding change in pressure. See Figure 11-19. A pressure sensor is designed to activate or deactivate when its resistance reaches a predetermined value. A pressure sensor is used for high- or low-pressure control, depending on the switching circuit design. It is suited for a wide variety of pressure measurements on compressors, pumps, and other similar equipment.

A pressure sensor can detect low pressure, high pressure, or it can trigger a relief valve. A pressure sensor is also used to measure compression in various types of engines because it is extremely rugged.

DEVICE	SYMBOL
PRESSURE SENSOR	

PRESSURE-SENSING ELEMENT

Figure 11-19. A solid-state pressure sensor is a transducer that changes resistance with a corresponding change in pressure.

Light Emitting Diodes

A *light emitting diode (LED)* is a diode which produces light when current flows through it. As the electrons move across the depletion region, they give up extra kinetic energy. The extra energy is converted into light. An electron must acquire additional energy to get through the depletion region. This additional energy comes from the positive field of the anode. The electron does not get through the depletion region and no light is emitted if the positive field is not strong. See Figure 11-20.

DEVICE	SYMBOL
LIGHT EMITTING DIODE	

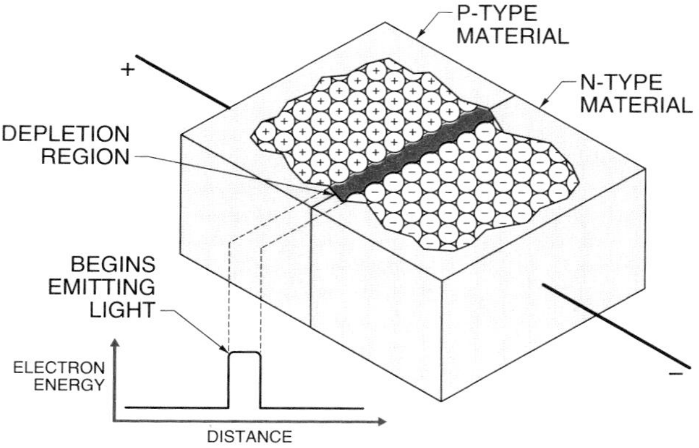

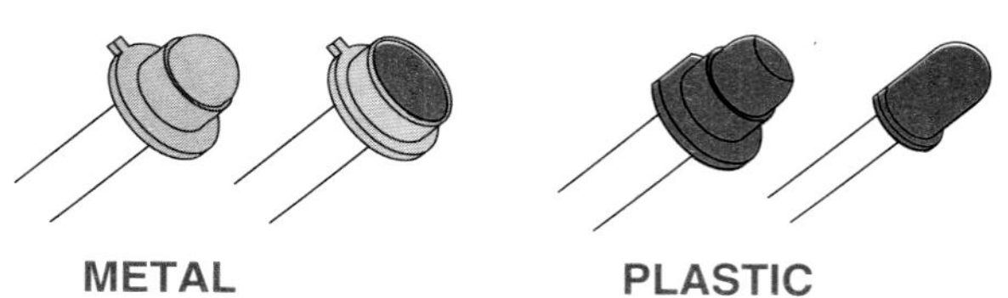

Figure 11-20. A light emitting diode (LED) is a diode which produces light when current flows through it.

For a standard silicon diode, a minimum of 0.6 V must be present before the diode conducts. For a germanium diode, 0.3 V must be present before the diode conducts. Most LED manufacturers make a larger depletion region that requires 1.5 V for the electrons to get across the depletion region.

LED Construction. Manufacturers of LEDs normally use a combination of gallium and arsenic with silicon or germanium to construct semiconductors. By adding and adjusting other impurities to the base semiconductor, different wavelengths of light can be produced. LEDs are capable of producing infrared light. *Infrared light* is light that is not visible to the human eye. LEDs may emit a visible red or green light. Colored plastic lenses are available if different colors are desired.

Like standard semiconductor diodes, there must be a method for determining which end of an LED is the anode and which end is the cathode. The cathode lead is identified by the flat side of the device or it may have a notch cut into the ridge.

A colored plastic lens focuses the light produced at the junction of the LED. Without the lens, the small amount of light produced at the junction is diffused and becomes virtually unusable as a light source. The size and shape of the LED package determines how it is positioned for proper viewing.

The schematic symbol for an LED is exactly like that of a photodiode, but the arrows point away from the diode. The LED is forward biased and a current-limiting resistor is normally present to protect the LED from excessive current.

Boeing Commercial Airplane Group

Many aircraft systems contain electrical, electronic, and fluid power circuits controlled by solid-state electronic components.

Transistors

A *transistor* is a three-terminal device that controls current through the device depending on the amount of voltage applied to the base. Transistors may be PNP or NPN transistors. A PNP transistor is formed by sandwiching a thin layer of N-type material between two layers of P-type material. An NPN transistor is formed by sandwiching a thin layer of P-type material between two layers of N-type material. See Figure 11-21. Transistor terminals are the emitter (E), base (B), and collector (C). The emitter, base, and collector are located in the same place for both symbols. The only difference is the direction in which the emitter arrow points. In both cases, the arrow points from the P-type material toward the N-type material. Transistors are bipolar devices. A *bipolar device* is a device in which both holes and electrons are used as internal carriers for maintaining current flow.

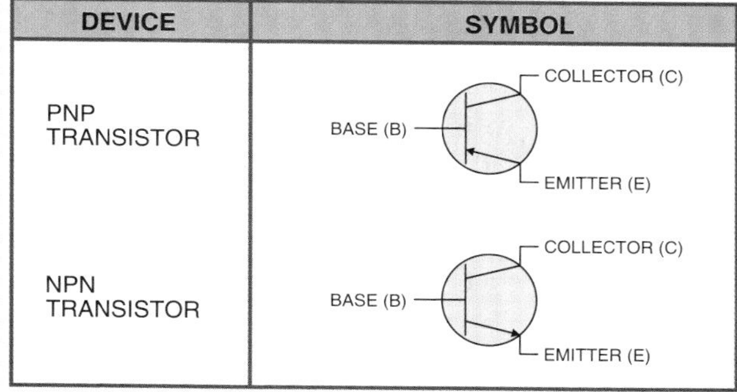

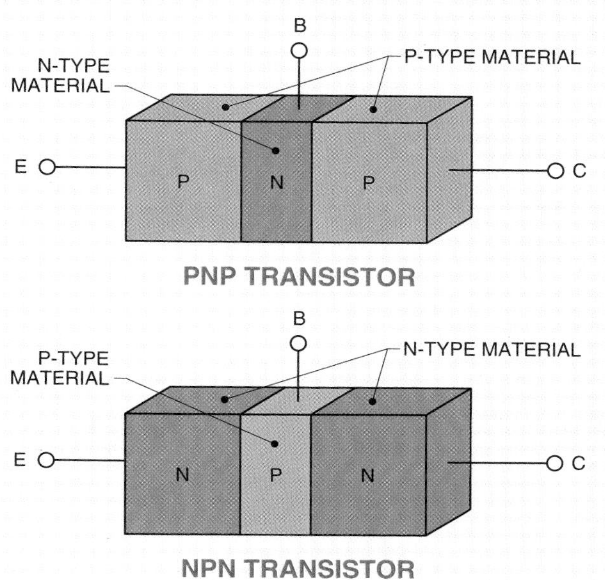

Figure 11-21. A PNP transistor is formed by sandwiching a thin layer of N-type material between two layers of P-type material. An NPN transistor is formed by sandwiching a thin layer of P-type material between two layers of N-type material.

Transistor Terminal Arrangements. Transistors are manufactured with two or three leads extending from their case. A transistor's outline (TO) number is used as a reference when a specific-shaped transistor must be used. See Figure 11-22. TO numbers are determined by individual manufacturers. *Note:* The bottom view of transistor TO-3 shows only two leads (terminals). Frequently, transistors use the metal case as the collector-pin lead.

Spacing can also be used to identify transistor leads. Normally, the emitter and base leads are close together and the collector lead is farther away. The base lead is normally in the middle. A transistor with an index pin must be viewed from the bottom. An *index pin* is a metal extension from the transistor case. The leads are identified in a clockwise direction from the index pin. For example, the loads on TO-5 are identified as E, B, and C. The emitter is closest to the index pin. Refer to a transistor manual or to manufacturer specification sheets for detailed information on transistor construction and identification.

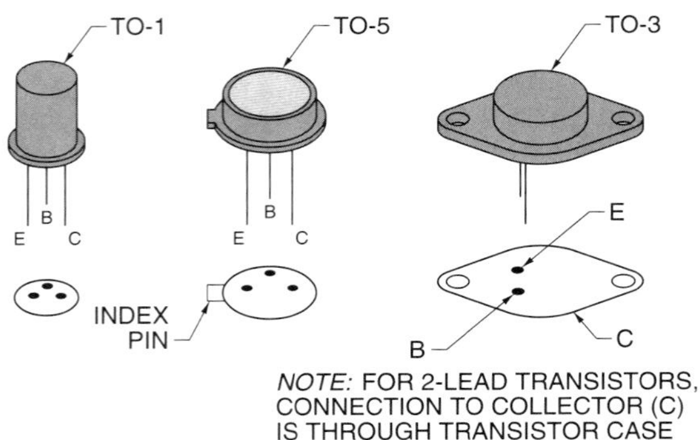

Figure 11-22. Transistors are manufactured with two or three leads extending from their case.

Fluke Corporation

The proper test equipment must be used when bench testing solid-state components, such as transistors, to prevent damage to the component.

Biasing Transistor Junctions. In any transistor circuit, the base/emitter junction must always be forward biased and the base/collector junction must always be reverse biased. See Figure 11-23. The external voltage (bias voltage) is connected so that the positive terminal connects to the P-type material (base) and the negative terminal connects to the N-type material (emitter). This arrangement forward biases the base/emitter junction. Current flows from the emitter to the base. The action that takes place is the same as the action that occurs for a forward-biased semiconductor diode.

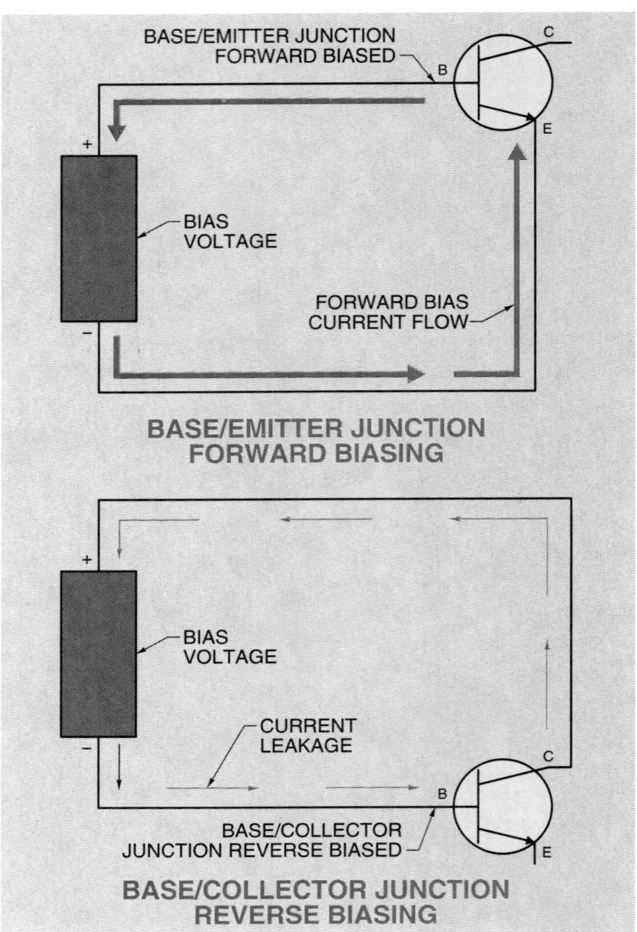

Figure 11-23. In any transistor circuit, the base/emitter junction must always be forward biased and the base/collector junction must always be reverse biased.

In any transistor circuit, the base/collector junction must always be reverse biased. The external voltage is connected so that the negative terminal connects to the P-type material (base) and the positive terminal connects to the N-type material (collector). This arrangement reverse biases the base collector junction. Only a very small current (leakage current) flows in the external circuit. The action that takes place is the same as the action that occurs for a semiconductor diode with reverse bias applied.

Transistor Current Flow. Individual PN junctions can be used in combination with two bias arrangements. See Figure 11-24. The base/emitter junction is forward biased while the base/collector junction is reverse biased. This circuit arrangement results in an entirely different current path than the path that occurs when the individual circuits are biased separately.

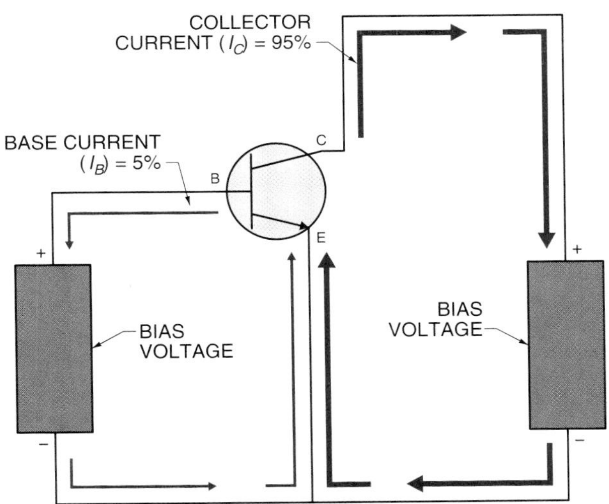

Figure 11-24. An entirely different current path is created when both junctions are biased simultaneously than when each junction is biased separately.

The forward bias of the base/emitter circuit causes the emitter to inject electrons into the depletion region between the emitter and the base. Because the base is less than .001″ thick for most transistors, the more positive potential of the collector pulls the electrons through the thin base. As a result, the greater percentage (95%) of the available free electrons from the emitter passes directly through the base (I_C) into the N-type material, which is the collector of the transistor.

Control of Base Current. The base current (I_B) is a critical factor in determining the amount of current flow in a transistor because the forward biased junction has a very low resistance and could be destroyed by heavy current flow. Therefore, the base current must be limited and controlled.

Transistors as DC Switches. Transistors were mainly developed to replace mechanical switches. Transistors have no moving parts and can switch ON and OFF quickly. Mechanical switches have two conditions: open and closed or ON and OFF. The switch has a very high resistance when open and a very low resistance when closed.

A transistor can be made to operate like a switch. For example, it can be used to turn a pilot light ON or OFF. See Figure 11-25. In this circuit, the resistance between the collector (C) and the emitter (E) is determined by the current flow between the base (B) and emitter (E). When no current flows between B and E, the collector/emitter resistance is high, like that of an open switch. The pilot light does not glow because there is no current flow.

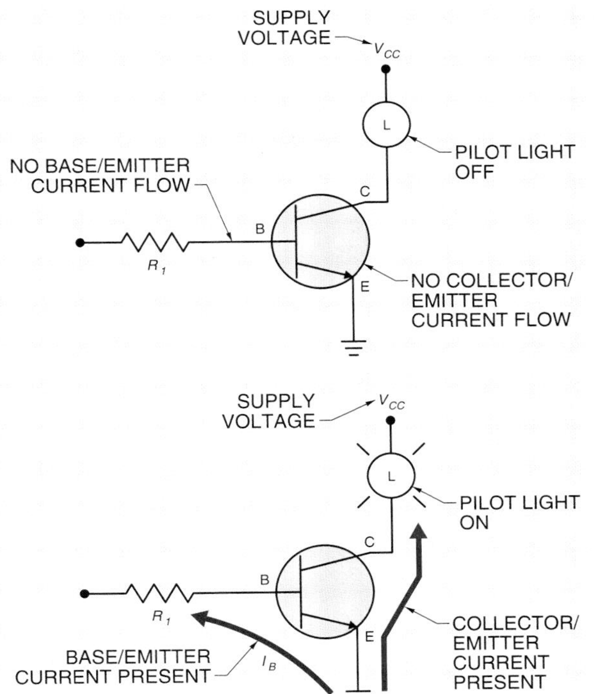

Figure 11-25. A transistor can be made to operate like a switch.

If a small current flows between B and E, the collector/emitter resistance is reduced to a very low value, like that of a closed switch. The pilot light is switched ON. A transistor switched ON is normally operating in the saturation region. The *saturation region* is the maximum current that can flow in the transistor circuit. At saturation, the collector resistance is considered zero and the current is limited only by the resistance of the load.

When the circuit reaches saturation, the resistance of the pilot light is the only current-limiting device in the circuit. When the transistor is switched OFF, it is operating in the cutoff region. The *cutoff region* is the point at which the transistor is turned OFF and no current flows. At cutoff, all the voltage is across the open switch (transistor) and the collector-emitter voltage is equal to the supply voltage V_{CC}.

Transistors as AC Amplifiers. Transistors may be used as AC amplification devices as well as DC switching devices. *Amplification* is the process of taking a small signal and increasing its size. In control systems, transistor AC amplifiers are used to increase small signal currents and voltages so they can do useful work. Amplification is accomplished by using a small signal to control the energy output from a large source, such as a power supply.

Amplifier Gain. The primary objective of an amplifier is to produce gain. *Gain* is a ratio of the amplitude of the output signal to the amplitude of the input signal.

Gain is a ratio of output to input and has no unit of measure, such as volts or amps, attached to it. Gain is used to describe current gain, voltage gain, and power gain. In each case, the output is compared to the input.

A single amplifier may not provide enough gain to increase the amplitude for the output signal needed. In such a case, two or more amplifiers can be used to obtain the gain required. *Cascaded amplifiers* are two or more amplifiers connected to obtain the required gain. For many amplifiers, gain is in the hundreds and even thousands.

Transistor Amplifiers. The three basic transistor amplifiers are the common-emitter, common-base, and common-collector. See Figure 11-26. Each amplifier is named after the transistor connection that is common to both the input and the load. For example, the input of a common-emitter circuit is across the base and emitter, while the load is across the collector and emitter. Thus, the emitter is common to the input and load.

Classes of Operation. The four main classes of operation for an amplifier are designated by the letters A, B, AB, and C. In each case, the letter is a reference to the level of an amplifier operation in relation to the cutoff condition. The *cutoff condition* is the point at which all collector current is stopped by the absence of base current.

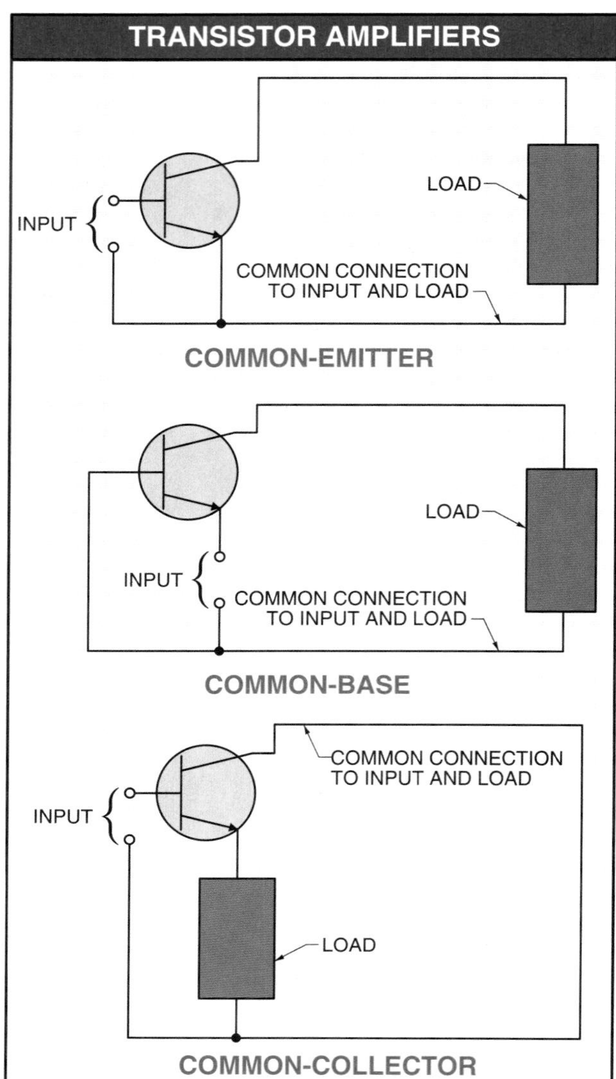

Figure 11-26. The three basic transistor amplifiers are the common-emitter, common-base, and common-collector.

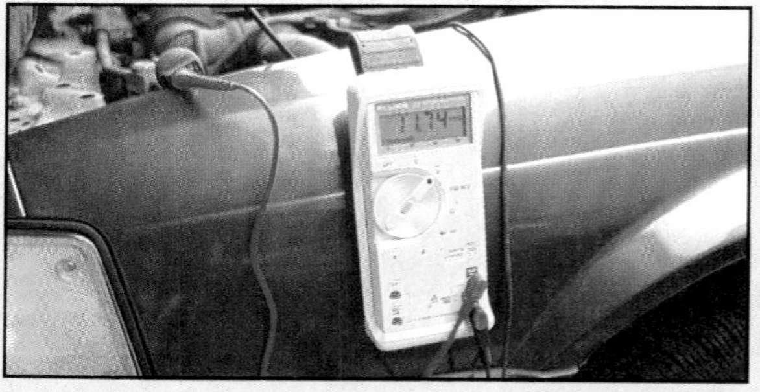

Fluke Corporation

Fluke multimeters are used to check the many solid-state components in today's automobiles, such as diodes in the alternator, and SCRs and transistors in electronic ignition systems.

Silicon Controlled Rectifiers (SCRs)

A *silicon controlled rectifier (SCR)* is a solid-state rectifier with the ability to rapidly switch heavy currents. It uses three electrodes for normal operation. See Figure 11-27. The three electrodes are the anode, cathode, and gate. The anode and cathode of the SCR are similar to the anode and cathode of an ordinary semiconductor diode.

The gate serves as the control point for the SCR. The SCR differs from an ordinary semiconductor diode in that it does not pass significant current, even when forward biased, unless the anode voltage equals or exceeds the forward breakover voltage. *Forward breakover voltage* is the voltage required to switch the SCR into a conductive state. The SCR switches ON and becomes highly conductive when forward breakover voltage is reached. The SCR is unique because the gate current is used to reduce the level of breakover voltage necessary for the SCR to conduct.

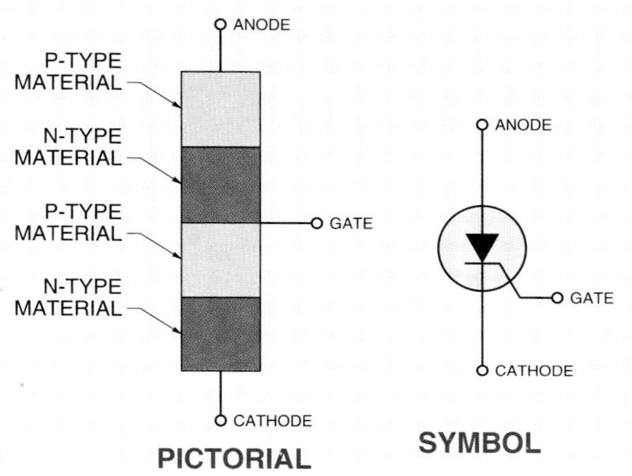

Figure 11-27. An SCR is a four-layer (PNPN) semiconductor device that does not pass significant current, even when forward biased, unless the anode voltage equals or exceeds the forward breakover voltage.

Low-current SCRs can operate with an anode current of less than 1 mA. High-current SCRs can handle load currents in the hundreds of amperes. The size of an SCR increases with an increase in its current rating.

SCR Characteristic Curves. The voltage-current characteristic curve of an SCR shows that the SCR operates much like a regular diode in reverse bias. See Figure 11-28. With reverse bias, there is a small current until avalanche is reached. After avalanche is reached, the current increases dramatically. This current can cause damage if thermal runaway begins.

When the SCR is forward biased, there is also a small forward leakage current (forward blocking current). This current stays relatively constant until the forward breakover voltage is reached. At that point, the current increases rapidly and is often referred to as the forward avalanche region. In the forward avalanche region, the resistance of the SCR is very low.

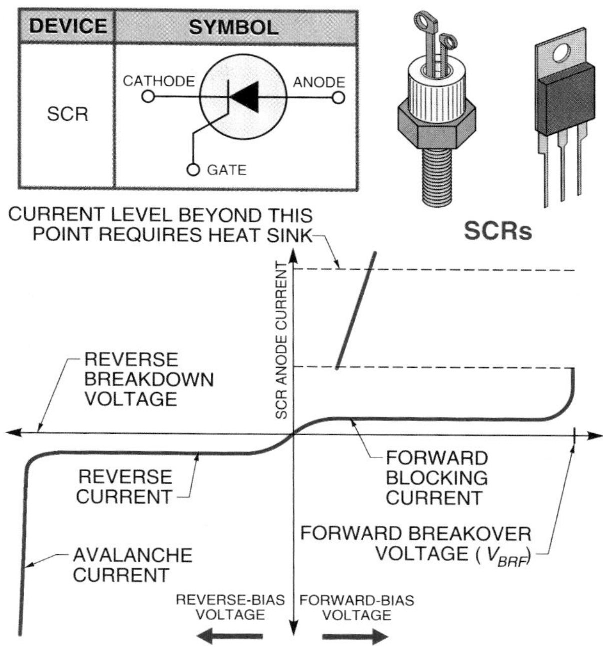

Figure 11-28. The voltage-current characteristic curve of an SCR shows that the SCR operates much like a regular diode in reverse bias.

The SCR acts much like a closed switch and the current is limited only by the external load resistance. A short in the load circuit of an SCR can destroy the SCR if overload protection is not adequate.

Operating States of SCRs. An SCR operates much like a mechanical switch. An SCR is either ON or OFF. The SCR is ON (fires) when the applied voltage is above the forward breakover voltage (V_{BRF}). The SCR remains ON as long as the current stays above the holding current. *Holding current* is the minimum current necessary for an SCR to continue conducting. An SCR returns to its OFF state when voltage across the SCR drops to a value too low to maintain the holding current.

Gate Control of Forward Breakover Voltage. The value of forward breakover voltage can be reduced when the gate is forward biased and current begins to flow in the gate/cathode junction. Increasing values of forward bias can be used to reduce the amount of forward breakover voltage (V_{BRF}) necessary to get the SCR to conduct.

Once the SCR has been turned ON by the gate current, the gate current loses control of the SCR forward current. Even if the gate current is completely removed, the SCR remains ON until the anode voltage has been removed. The SCR also remains ON until the anode voltage has been significantly reduced to a level where the current is not large enough to maintain the proper level of holding current.

Process Control Using SCRs. SCRs may be used in circuits to provide heat control. For example, an SCR can bring a chemical mixture stored in a vat to a specific temperature and maintain that temperature. See Figure 11-29. With the proper circuitry, the temperature of the mixture can be precisely controlled. Using a bridge circuit, the temperature can be maintained within 1°F over a temperature range of 20°F to 150°F.

In this circuit, transformer T1 has two secondary windings, W1 and W2. W1 furnishes voltage through the SCR to relay coil K1. W2 furnishes AC voltage to the gate circuit of the SCR. Primary control over this circuit is accomplished through the use of the bridge circuit. The bridge circuit is formed by thermistor R1, fixed resistors R2 and R3, and potentiometer R4. Resistor R5 is a current-limiting resistor used to protect the bridge circuit. The fuse is used to protect the primary of the transformer.

Siemens Corporation

The SIPLACE 80 S high-speed surface mount system manufactured by Siemens includes solid-state components used to precisely control the heat produced when assembling printed circuit boards.

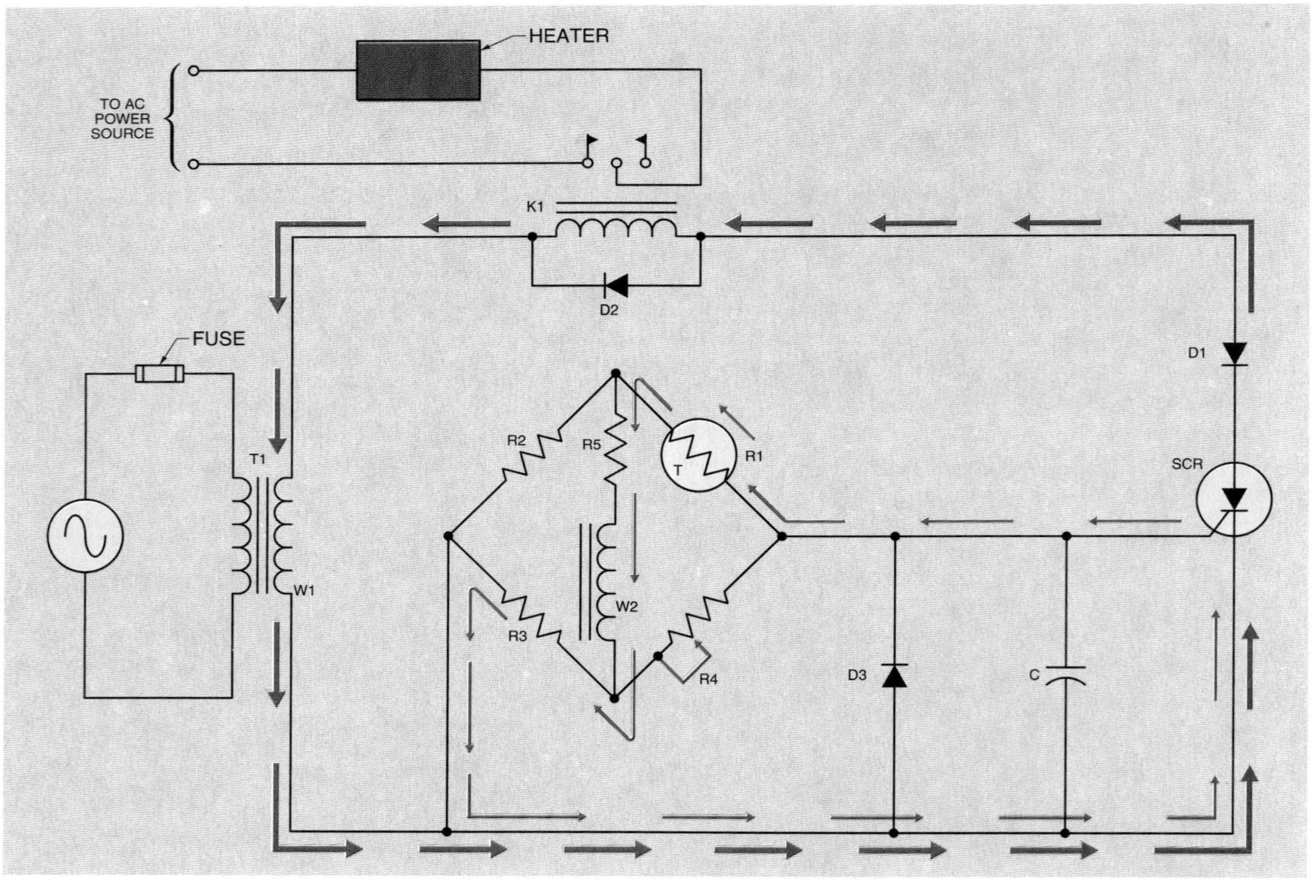

Figure 11-29. SCRs may be used in circuits to provide heat control.

The bridge is balanced when the resistance of R1 equals the resistance setting on R4. None of the AC voltage introduced into the bridge by winding W2 is applied to the gate of the SCR. The relay coil K1 remains de-energized and its normally closed contacts apply power to the heating elements.

The resistance of thermistor R1 decreases if the temperature increases above a preset level. The bridge becomes unbalanced such that a current flows to the gate of the SCR while the anode of the SCR is still positive. This turns ON the SCR and energizes the relay coil K1, thereby switching power from the load through the relay contact. R1 unbalances the bridge in the opposite direction if the temperature falls below the preset temperature setting. A negative signal is applied to the gate of the SCR when the anode of the SCR is positive. The negative signal stops the SCR from conducting and allows current to continue to flow to the heating elements.

Triacs

A *triac* is a three-electrode AC semiconductor switch. It is triggered into conduction in both directions by a gate signal in a manner similar to the action of an SCR. Triacs were developed to provide a means for producing improved controls for AC power. Triacs are available in a variety of packaging arrangements. Triacs can handle a wide range of amperages and voltages. Triacs normally have relatively low-current capabilities compared to SCRs. Triacs are normally limited to less than 50 A and cannot replace SCRs in high-current applications.

Triac Construction. The terminals of a triac are the gate, main terminal 1 (MT1), and main terminal 2 (MT2). There is no designation of anode and cathode. Current may flow in either direction through MT1 and MT2. MT2 is the case- or metal-mounting tab to which the heat sink can be attached. A triac can be considered two NPN switches sandwiched together on a single N-type material wafer.

Triac Operation. A triac blocks current in either direction between MT1 and MT2. A triac can be triggered into conduction in either direction by a momentary pulse in either direction supplied to the gate. A triac operates much like a pair of SCRs connected in a reverse parallel arrangement. The triac conducts if the appropriate signal is applied to the gate.

A triac characteristic curve shows the characteristics of a triac when triggered into conduction. See Figure 11-30. The triac remains OFF until the gate is triggered. The trigger circuit pulses the gate and turns ON the triac, allowing current to flow. The trigger circuit can be designed to produce a pulse that varies at any point in the positive or negative half-cycle. Therefore, the average current supplied to the load may vary.

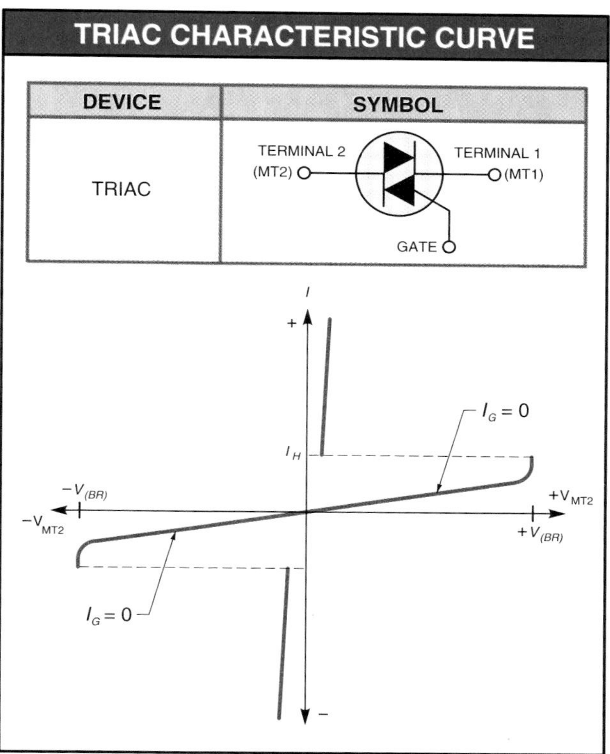

Figure 11-30. A triac characteristic curve shows the characteristics of a triac when triggered into conduction.

One advantage of the triac is that virtually no power is wasted by being converted to heat. Heat is generated when current is impeded, not when current is switched OFF. A triac is either fully ON or fully OFF. It never partially limits current. Another important feature of the triac is the absence of a reverse breakdown condition of high voltage and high current, such as those found in diodes and SCRs. The triac turns ON if the voltage across the triac goes too high. The triac can conduct a reasonably high current when turned ON.

Unijunction Transistors (UJTs)

A *unijunction transistor (UJT)* is a transistor consisting of N-type material with a region of P-type material doped within the N-type material. The N-type material functions as the base and has two leads, base 1 (B1) and base 2 (B2). The lead extending from the P-type material is the emitter (E). See Figure 11-31.

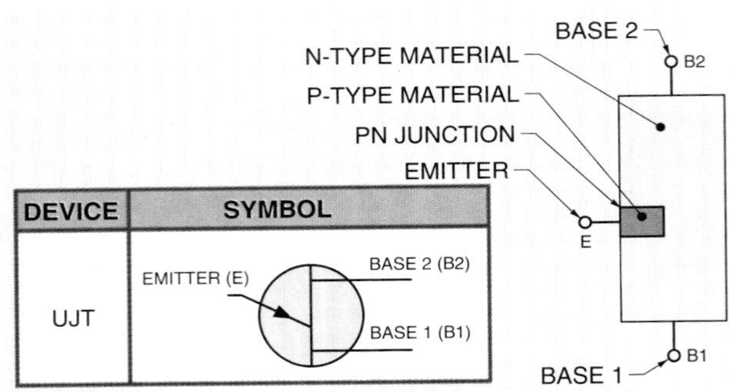

Figure 11-31. A UJT consists of N-type material with a region of P-type material doped within the N-type material.

A UJT is used primarily as a triggering device because it serves as a step-up device between low-level signals and SCRs and triacs. Outputs from photocells, thermistors, and other transducers can be used to trigger UJTs, which fire SCRs and triacs. UJTs are also used in oscillators, timers, and voltage/current-sensing applications.

UJT Biasing. In normal operation, B1 is negative, and a positive voltage is applied to B2. The internal resistance between B1 and B2 divides at the emitter (E), with approximately 60% of the resistance between E and B1. The remaining 40% of resistance is between E and B2. The net result is an internal voltage split. This split provides a positive voltage at the N-type material of the emitter junction, creating an emitter junction that is reverse biased. As long as the emitter voltage remains less than the internal voltage, the emitter junction remains reverse biased, even at a very high resistance.

The junction of a UJT is forward biased when the emitter voltage is greater than the internal value. This rapidly drops the resistance between E and B1 to a very low value. A UJT characteristic curve shows the dramatic change in voltage due to this resistance change. See Figure 11-32.

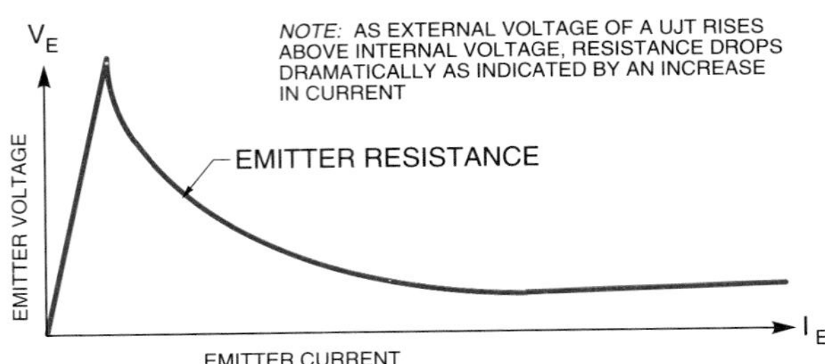

Figure 11-32. A UJT characteristic curve shows the dramatic change in voltage due to the resistance change when the device is forward biased.

Diacs

A *diac* is a three-layer bidirectional device used primarily as a triggering device. Unlike a transistor, the two junctions are heavily and equally doped. Each junction is almost identical to the other.

A diac acts much like two zener diodes that are series connected in opposite directions. The diac is used primarily as a triggering device. It accomplishes this through the use of its negative resistance characteristic. *Negative resistance characteristic* is the characteristic that current decreases with an increase of applied voltage. The diac has negative resistance because it does not conduct current until the voltage across it reaches breakover voltage. See Figure 11-33.

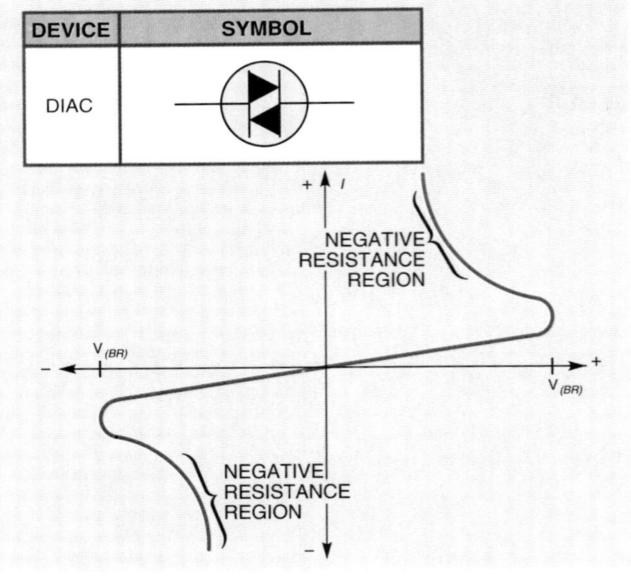

Figure 11-33. A diac rapidly switches from a high-resistance state to a low-resistance state when a positive or negative voltage reaches the breakover voltage.

A diac rapidly switches from a high-resistance state to a low-resistance state when a positive or negative voltage reaches the breakover voltage. Because the diac is a bidirectional device, it is ideal for controlling triacs, which are also bidirectional.

Integrated Circuits

An *integrated circuit (IC)* is a circuit composed of thousands of semiconductors providing a complete circuit function in one small semiconductor package. ICs are popular because they provide a complete circuit function in one package. ICs are often referred to as chips, which are actually a part of the IC. See Figure 11-34. Although many processes have been developed to create these devices, the end result has always been a totally-enclosed system with specific inputs and specific outputs.

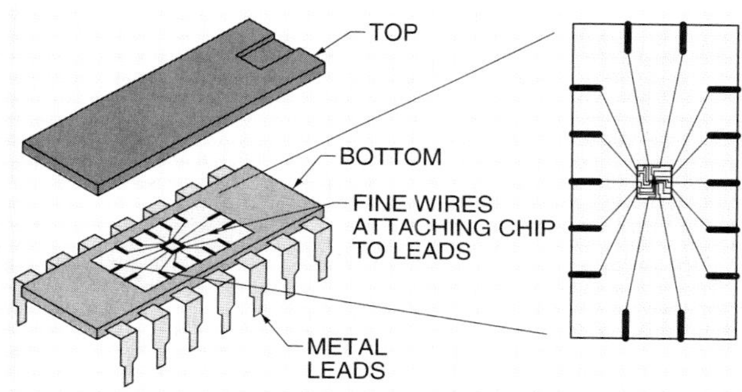

Figure 11-34. ICs are thousands of semiconductors providing a complete circuit function in one small semiconductor package.

Because of the nature of ICs, a technician must approach ICs in an entirely different manner from individual solid-state components. An IC is a system within a system. The entire system of an IC and what it does must first be understood. Data books and manufacturer's specification sheets can normally provide this information. The inputs and outputs of the system must be studied by using meters and an oscilloscope when data books and manufacturer's specification sheets are not available.

Troubleshooting ICs also requires knowledge of how the system functions and what the input and output should be. ICs must be replaced if they are defective because ICs cannot be repaired.

IC Packages. IC shapes and sizes range from standard transistor shapes, such as TO-5 packages, to the latest in large-scale integration (LSI). Metal-oxide substrate (MOS) is a type of LSI. ICs are also designed with flatpack construction for applications where space is a premium. See Figure 11-35.

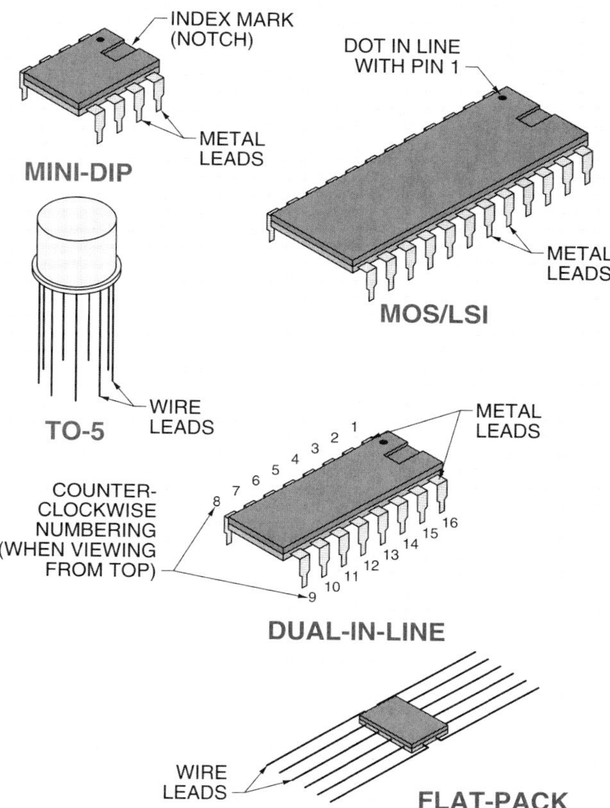

Figure 11-35. IC shapes and sizes range from standard transistor shapes, such as the TO-5 package, to the latest in large-scale integration (LSI).

The dual-in-line package (DIP) with 14, 16, or 24 pins is the most widely used configuration. The mini-DIP is a smaller dual-in-line package with 8 pins. A modified TO-5 is available with 8, 10, or 12 pins. The housings for ICs may be metal, plastic, or ceramic. Ceramic is used in applications where high temperatures may be a factor.

Pin Numbering System. All manufacturers use a standardized pin numbering system for their devices. Consult manufacturer's data sheets when unsure about pin numbering patterns.

Dual-in-line packages and flatpacks have index marks and notches at the top for reference. Before removing an IC, note where the index mark is in relation to the board or socket to aid in installation of the new unit. The numbering of the pins is always the same. The notch is at the top of the chip. To the left of the notch is a dot that is in line with pin 1. The pins are numbered counterclockwise around the chip when viewed from the top.

Operational Amplifiers (Op-Amps)

An operational amplifier (op-amp) is one of the most widely used ICs. An *op-amp* is a very high gain, directly-coupled amplifier that uses external feedback to control response characteristics. An example of this feedback control is gain. The gain of an op-amp can be controlled externally by connecting feedback resistors between the output and input. A number of different amplifier applications can be achieved by selecting different feedback components and combinations. With the right component combinations, gains of 500,000 to 1,000,000 are common.

The schematic symbol for an op-amp may be shown in two ways. See Figure 11-36. In each case, the two inputs of the op-amp are the inverting (–) and the non-inverting (+). The two inputs are normally drawn with the inverting input at the top. The exception to the inverting input being at the top is when it complicates the schematic. In either case, the two inputs should be clearly identified by polarity symbols on the schematic symbol.

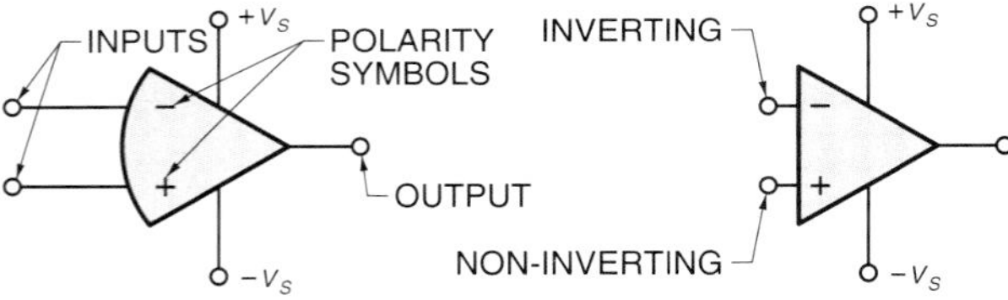

Figure 11-36. The schematic symbol for an op-amp which has two inputs should be clearly identified by polarity symbols.

Internal Op-Amp Operation. An op-amp consists of a high-impedance differential amplifier, a high-gain stage, and a low-output impedance power-output stage. The high-impedance differential amplifier provides the wide bandwidth and the high impedance. The high-gain stage boosts the signal. The power-output stage isolates the gain stage from the load and provides for the power output.

The operation of the differential amplifier is unique. Current to the emitter-coupled transistors Q1 and Q2 are supplied by the source Q3. The characteristics of Q1 and Q2, along with their biasing resistors R1, R2, and R3, are closely matched to make them as equal as possible. See Figure 11-37.

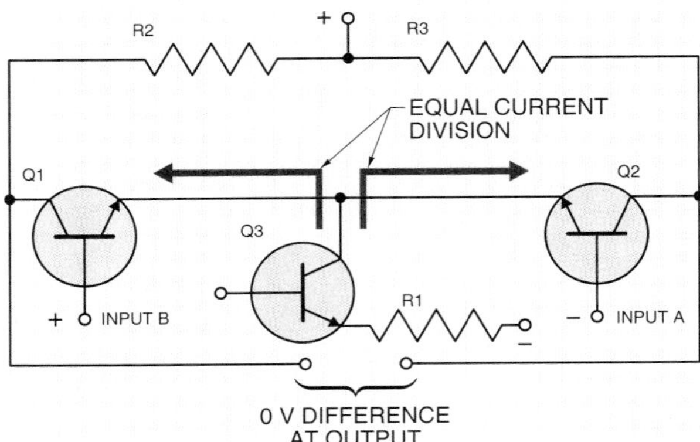

Figure 11-37. An op-amp is a very high gain, directly-coupled amplifier that uses external feedback to control response characteristics.

As long as the two input voltages, A and B, are either zero or equal in amplitude and polarity, the amplifier is balanced because the collector currents are equal. Zero voltage difference exists between the two collectors when balanced.

The sum of the emitter currents is always equal to the current supplied by Q3. Thus, if the input to one transistor causes it to draw more current, the current in the other decreases and the voltage difference between the two collectors changes in a differential manner. The differential swing, or output signal, is greater than the simple variation that can be obtained from only one transistor. Each transistor amplifies in the opposite direction so that the total output signal is twice that of one transistor. This swing is amplified through the high-gain stage and matched to the load through the power-output stage. By changing op-amps to different configurations, they can be made into oscillators, pulse generators, and level detectors.

555 Timer

A *555 timer* is an integrated circuit designed to output timing pulses for control of certain types of circuits. A 555 timer consists of a voltage-divider network (R1, R2, and R3), two comparators (Comp 1 and

Comp 2), two control transistors (Q1 and Q2), a power output amplifier, and a flip-flop. See Figure 11-38. A *flip-flop* is an electronic circuit having two stable states or conditions normally designated "set" and "reset". It has two outputs (high and low). When one is high, the other is low and vice versa.

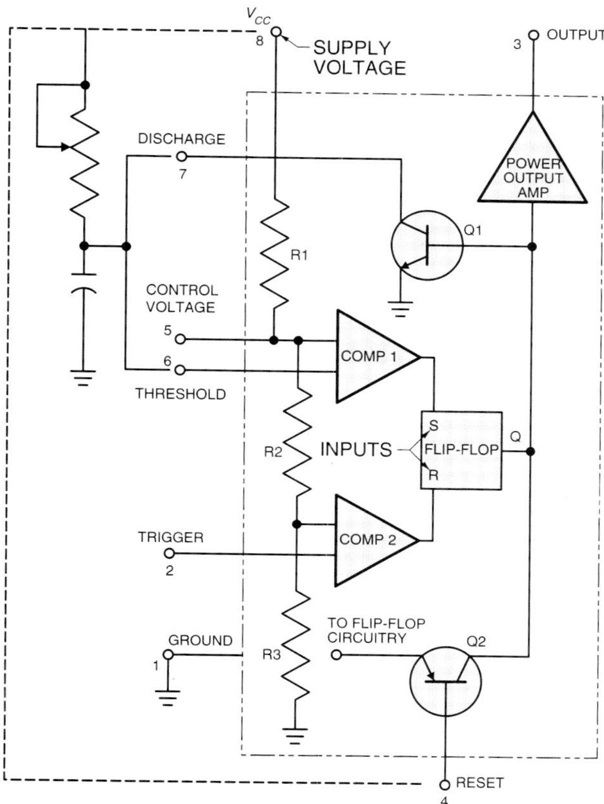

Figure 11-38. A 555 timer is an integrated circuit designed to output timing pulses for control of certain types of circuits.

The comparators compare the input voltages to internal reference voltages that are created by the voltage divider, which consists of resistors R1, R2, and R3. Because the resistors are of equal value, the reference voltage provided by two resistors is two-thirds of the supply voltage (V_{CC}). The other resistor provides one-third of V_{CC}. The value of V may change (9 V, 12 V, 15 V, etc.) from chip to chip. However, the $\frac{2}{3}$:$\frac{1}{3}$ ratio always remains the same.

The comparator goes into saturation and produces a signal that triggers the flip-flop when the input voltage to either one of the comparators is higher than the reference voltage. In this IC circuit, the flip-flop has two inputs: S and R.

Note: The two comparators feed signals into the flip-flop. Comparator 1 is the threshold comparator, and comparator 2 is the trigger comparator. Comparator 1 is connected to the S input of the flip-flop, and comparator 2 is connected to the R input of the flip-flop.

The output of the flip-flop is high whenever the voltage at S is positive and the voltage at R is zero. The output of the flip-flop is low whenever the voltage at S is zero and the voltage at R is positive. The output from the flip-flop at point Q is applied to transistors Q1 and Q2 and to the output amplifier simultaneously. Q1 turns ON such that pin 7 (discharge pin) is grounded through the emitter-collector circuit if the signal is high. Q1 is then in a position to turn ON pin 7 to ground through the emitter-collector circuit. *Note:* Pin 7 is the discharge pin because it is connected to the timing capacitor. When Q1 conducts, pin 7 is grounded, and the capacitor can be discharged.

The flip-flop signal is also applied to Q2. A signal to pin 4 can be used to reset the flip-flop. Pin 4 can be activated when a low-level voltage signal is applied. Once applied, this signal overrides the output signal from the flip-flop. The reset pin (pin 4) forces the output of the flip-flop to be low, no matter the state of the other inputs.

The flip-flop signal is also applied to the power output amplifier. The power output amplifier boosts the signal and the 555 timer delivers up to 200 mA of current when operated at 15 V. The output can be used to drive other transistor circuits and even a small audio speaker. The output of the power output amplifier is always an inverted signal compared to the input. The output is low if the input to the power output amplifier is high. The output is high if the input is low.

Digital ICs

Electronic signals may be analog or digital. Analog signals (voltage and current) vary smoothly or continuously. Digital signals are a series of pulses that change levels between the OFF or ON state.

The analog and digital processes can be seen in a simple comparison between a light dimmer and light switch. A light dimmer varies the intensity of light from fully OFF to fully ON. This is an example of an analog process. A standard light switch has only two positions. It is either fully OFF or fully ON. This is an example of a digital process. Electronic circuits that process these quickly-changing pulses are digital or logic circuits. The four most common gates used in digital electronics are the AND, OR, NAND, and NOR gates.

AND Gates. An *AND gate* is a device with an output that is high only when both of its inputs are high. The quad AND gate is one type of IC chip. See Figure 11-39. The manufacturer places four AND gates in one package. By using the numbering system on the chip, any one or all four AND gates may be used. In this case, voltage is applied to the circuit at pins 14 and 7.

Fluke Corporation

The Fluke 867 graphical multimeter may be used to troubleshoot solid-state electrical control devices.

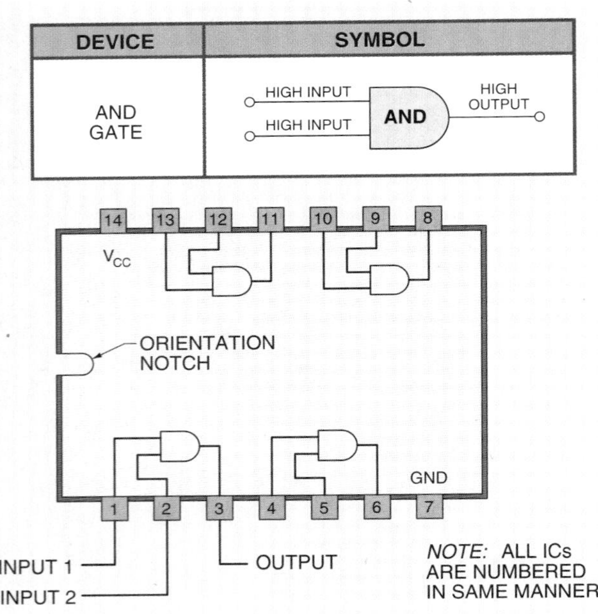

Figure 11-39. An AND gate is a device with an output that is high only when both of its inputs are high.

To connect to the AND gate, pins 1, 2, and 3 of the quad AND gate chip could be used. Pins 1 and 2 are the input and pin 3 is the output. An application of an AND gate is in an elevator control circuit. See Figure 11-40. The elevator cannot move unless the inner and outer doors are closed. Once both doors are closed, the output of the AND gate could be fed to an op-amp, which fires a triac that starts the elevator motor.

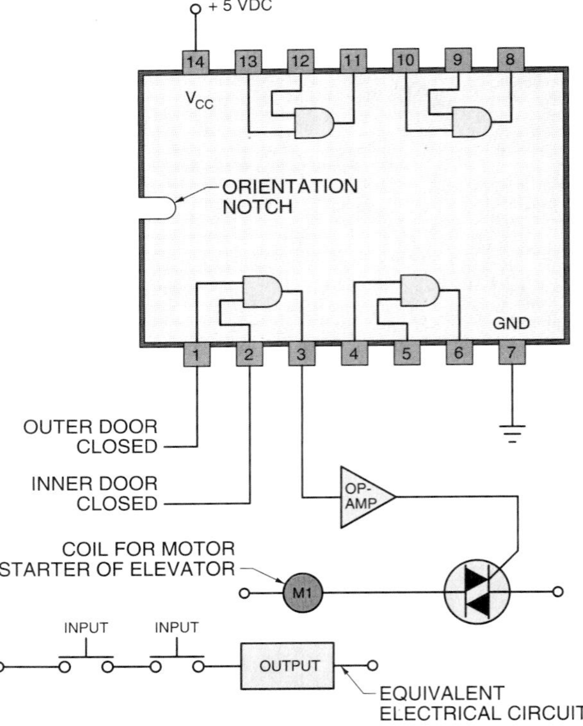

Figure 11-40. An AND gate may be used in an elevator control circuit.

OR Gates. An *OR gate* is a device with an output that is high when either or both inputs are high. See Figure 11-41. A practical application of an OR gate is in a burglar alarm circuit. A signal is sent to the burglar alarm circuit if the front door or the back door is opened. The electrical equivalent of an OR gate is two pushbuttons in parallel.

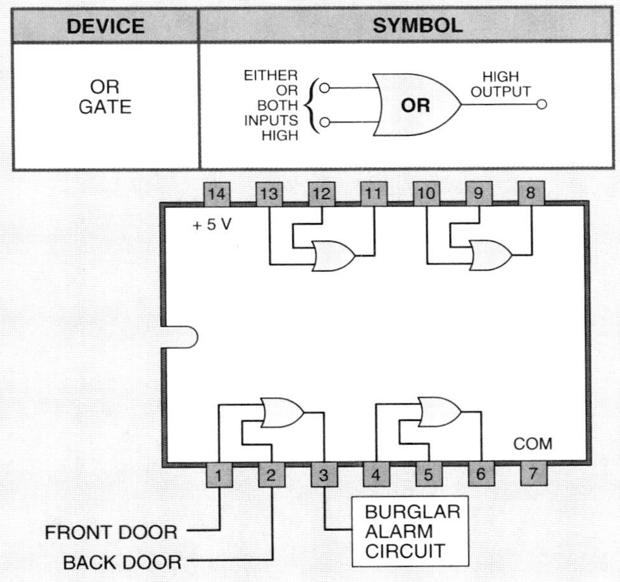

Figure 11-41. An OR gate is a device with an output that is high when either or both inputs are high.

NAND Gates. A NAND (NOT-AND) gate is an inverted AND function. A *NAND gate* is a device that provides a low output when both inputs are high. The NAND gate is represented by the AND symbol followed by a small circle indicating an inversion of the output. See Figure 11-42.

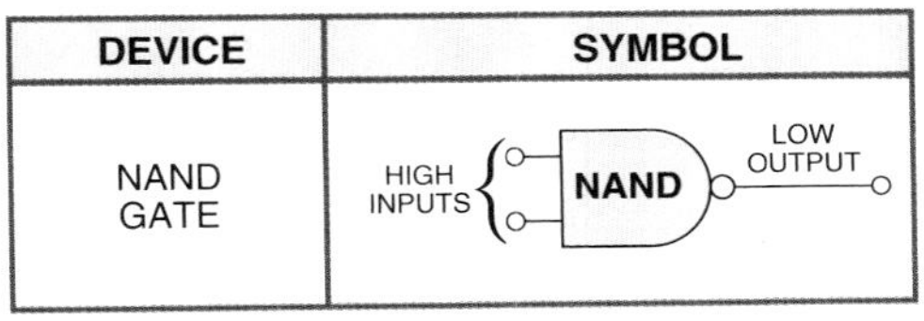

DEVICE	SYMBOL
NAND GATE	HIGH INPUTS — NAND — LOW OUTPUT

Figure 11-42. A NAND gate is a device that provides

A NAND gate is a universal building block of digital logic. It is normally used in conjunction with other elements to implement more complex logic functions. NAND gates are also available in quad IC packaging.

NOR Gates. A NOR (NOT-OR) gate is the same as an inverted OR function. A *NOR gate* is a device that provides a low output when either or both inputs are high. A NOR gate is represented by the OR gate symbol followed by a small circle indicating an inversion of the output. See Figure 11-43. The NOR gate is a universal building block of digital logic. It is normally used in conjunction with other elements to implement more complex logic functions. NOR gates are also available in quad IC packaging.

DEVICE	SYMBOL
NOR GATE	SMALL CIRCLE INDICATES AN INVERSION OF OUTPUT; EITHER OR BOTH INPUTS HIGH — NOR — LOW OUTPUT

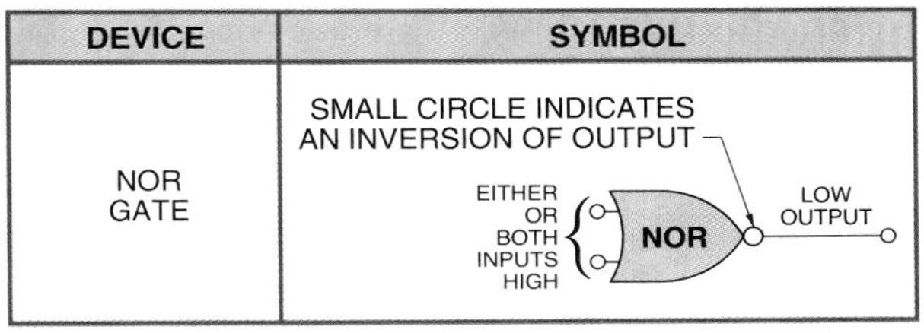

Figure 11-43. A NOR gate is a device that provides a low output when either or both inputs are high.

FIBER OPTICS

Fiber optics is a technology that uses a thin flexible glass or plastic optical fiber to transmit light. Fiber optics is most commonly used as a transmission link. As a link, it connects two electronic circuits consisting of a transmitter and a receiver. See Figure 11-44.

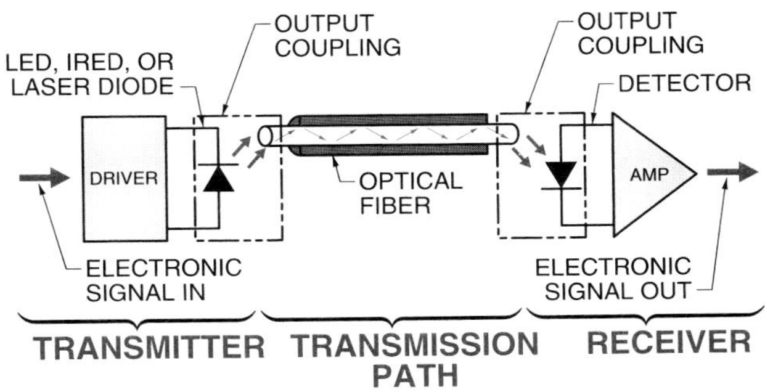

Figure 11-44. Fiber optics uses a thin flexible glass or plastic optical fiber to transmit light.

The central part of the transmitter is its source. The source consists of a light emitting diode (LED), infrared emitting diode (IRED), or laser diode, which changes electrical signals into light signals. The receiver normally contains a photodiode that converts light back into electrical signals. The receiver output circuit also amplifies the signal and produces the desired results, such as voice transmission or video signals. Advantages of fiber optics include large bandwidth, low loss (attenuation), electromagnetic interference (EMI) immunity, small size, light weight, and security.

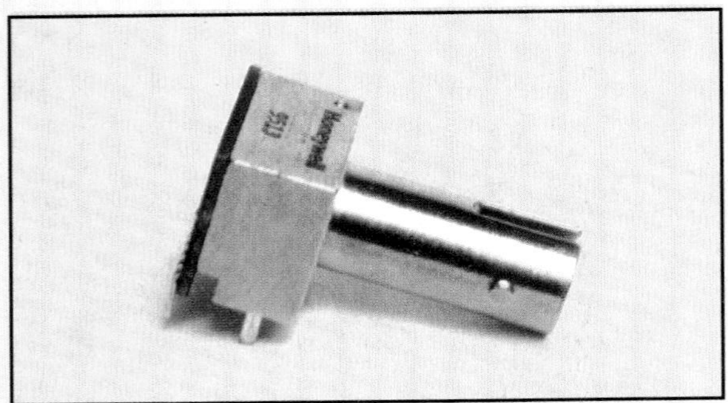

Honeywell's MICRO SWITCH Division

The HFM1220 Series fiber-optic receivers from Honeywell's MICRO SWITCH Division feature differential data and signal quality detect outputs, adjustable signal quality detect levels, and 500 Ω output drive capability.

Optical Fibers

Optical fibers consist of a core, cladding, and protective jacket. See Figure 11-45. The *core* is the actual path for light. The core is normally made of glass but may occasionally be constructed of plastic. *Cladding* is the first layer of protection for the glass or plastic core of the optical fiber cable. A glass or plastic cladding layer is bonded to the core. The cladding is enclosed in a jacket for additional protection.

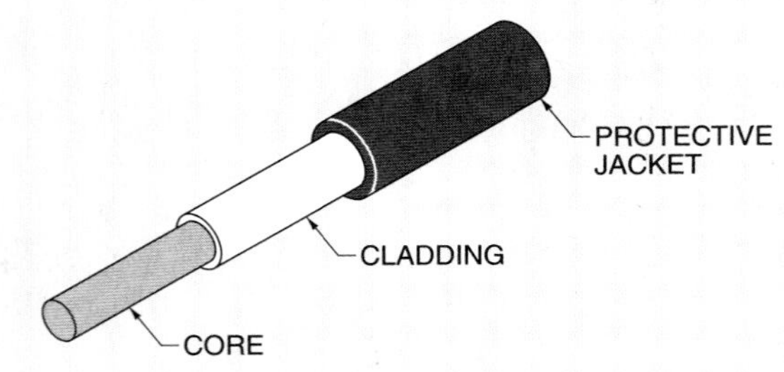

Figure 11-45. Optical fibers consist of a core, cladding, and protective jacket.

Light Source

The light source feeding the cable must be properly matched to the light-activated device for a fiber-optic cable to operate effectively. The source must also be of sufficient intensity to drive the light-activated device.

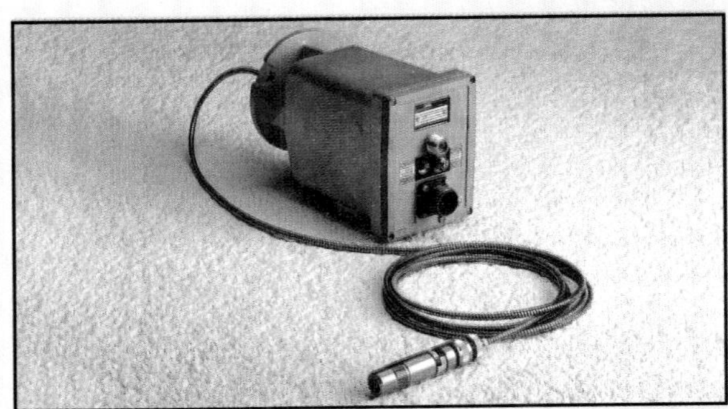

Ircon, Inc.

The fiber-optic SR Series infrared thermometer from Ircon, Inc. is designed to operate at temperature ranges between 1300°F and 6500°F.

Laser Diodes

A *laser diode* is a diode similar to an LED but has an optical cavity, which is required for lasing production (emitting coherent light). The optical cavity is formed by coating opposite sides of a chip to create two highly reflective surfaces. See Figure 11-46.

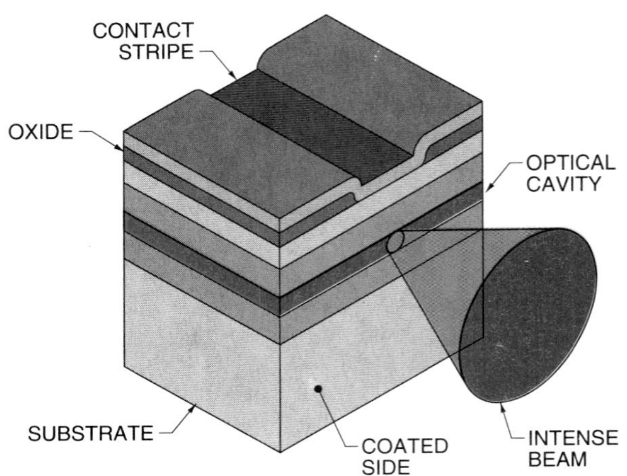

Figure 11-46. A laser diode is similar to an LED but has an optical cavity, which is required for lasing production (emitting coherent light).

Fiber Couplings

The ideal interconnection of one fiber to another is an interconnection that has two fibers that are optically and physically identical. These two fibers are held together by a connector or splice that squarely aligns them on their center axes. The joining of the fibers is so nearly perfect that the interface between them has no influence on light propagation. A perfect connection is limited by variations in fibers and the high tolerances required in the connector or splice. These two factors affect cost and ease of use.

Fiber-Coupling Hardware

Splices and fiber interconnections are often more of a negative factor than poor quality materials because of alignment problems that can arise. The elimination of alignment problems can be accomplished through proper installation of fiber splices, connectors, and couplers. See Figure 11-47.

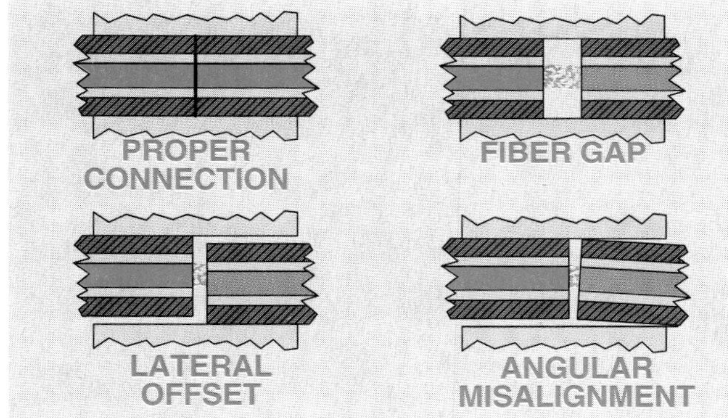

Figure 11-47. Improper connection of fiber-optic cable can result in improper transmission.

Honeywell's MICRO SWITCH Division

Surface-mount emitters and detectors from Honeywell's MICRO SWITCH Division are photodiode or phototransistor detectors used for optical encoders for motion control, computer peripherals, smoke detectors, and medical equipment.

LIGHT-ACTIVATED DEVICES

Once light rays have passed through the optical fiber, they must be detected and converted back into electrical signals. The detection and conversion is accomplished with light-activated devices, such as PIN photodiodes, phototransistors, light-activated SCRs, phototriacs, and optocouplers.

PIN Photodiodes

A *PIN photodiode* is a diode with a large intrinsic region sandwiched between P-type and N-type regions. PIN stands for P-type material, insulator, and N-type material. The operation of a PIN photodiode is based on the principle that light radiation, when exposed to a PN junction, momentarily disturbs the structure of the PN junction. The disturbance is due to a hole created when a high-energy photon strikes the PN junction and causes an electron to be ejected from the junction. Thus, light creates electron-hole pairs, which act as current carriers. PIN photodiodes are used in gas detectors, spectrometers, and gas analyzers. See Figure 11-48.

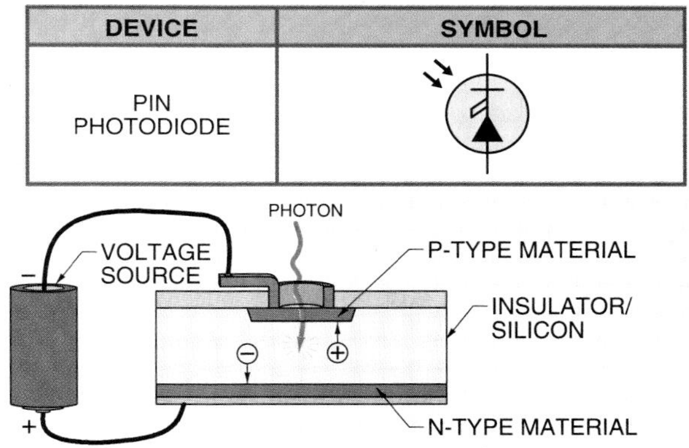

Figure 11-48. A PIN photodiode is a diode with a large intrinsic region sandwiched between P-type and N-type regions.

Phototransistors

A *phototransistor* is a device that combines the effect of a photodiode and the switching capability of a transistor. See Figure 11-49. A phototransistor, when connected in a circuit, is placed in series with the bias voltage so that it is forward biased.

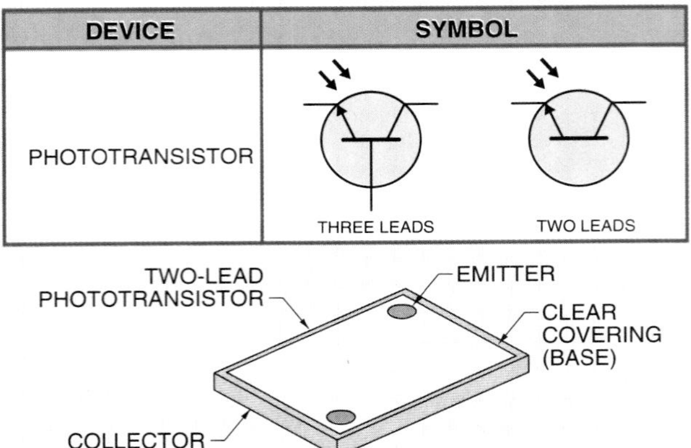

Figure 11-49. A phototransistor is a device that combines the effect of a photodiode and the switching capability of a transistor.

With a two-lead phototransistor, the base lead is replaced by a clear covering. This covering allows light to fall on the base region. Light falling on the base region causes current to flow between the emitter and collector. The collector-base junction is enlarged and works as a reverse-biased photodiode controlling the phototransistor. The phototransistor conducts more or less current, depending on the light intensity. If light intensity increases, resistance decreases and more emitter-to-base current is created. Although the base current is relatively small, the amplifying capability of the small base current is used to control the large emitter-to-collector current. The collector current depends on the light intensity and the DC current gain of the phototransistor. In darkness, the phototransistor is switched OFF with the remaining leakage current (collector dark current).

Light-Activated SCRs (LASCRs)

A *light-activated SCR (LASCR)* is an SCR that is activated by light. The symbol of an LASCR is identical to the symbol of a regular SCR. The only difference is that arrows are added in the LASCR symbol to indicate a light-sensitive device. See Figure 11-50.

DEVICE	SYMBOL
LIGHT-ACTIVATED SCRs	ANODE GATE CATHODE

Figure 11-50. An LASCR is an SCR that is activated by light.

Like the photodiode, current is of a very low level in an LASCR. Even the largest LASCRs are limited to a maximum of a few amps. When larger current requirements are necessary, the LASCR can be used as a trigger circuit for a conventional SCR.

The primary advantage of an LASCR over an SCR is its ability to provide isolation. Because the LASCR is triggered by light, the LASCR provides complete isolation between the input signal and the output load current.

Phototriacs

A *phototriac* is a triac that is activated by light. The gate of the phototriac is light sensitive. It triggers the triac at a specified light intensity. See Figure 11-51. In darkness, the triac is not triggered. The remaining leakage current is referred to as peak blocking current. A phototriac is bilateral and is designed to switch AC signals.

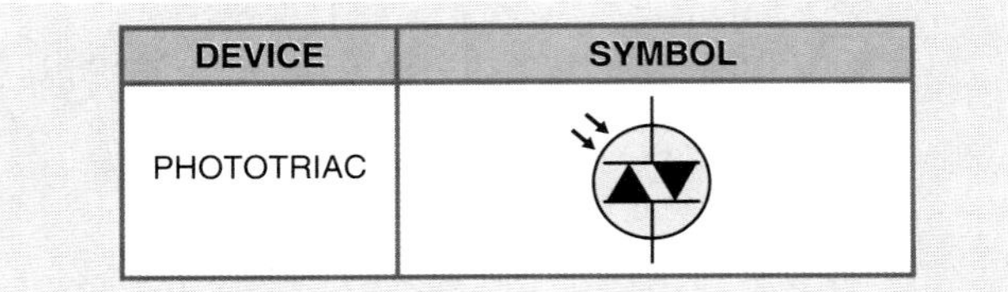

Figure 11-51. A phototriac is bilateral and is designed to switch AC signals.

Optocouplers

An *optocoupler* is a device that consists of an IRED as the input stage and a silicon NPN phototransistor as the output stage. An optocoupler is normally constructed as a dual in-line plastic package. See Figure 11-52.

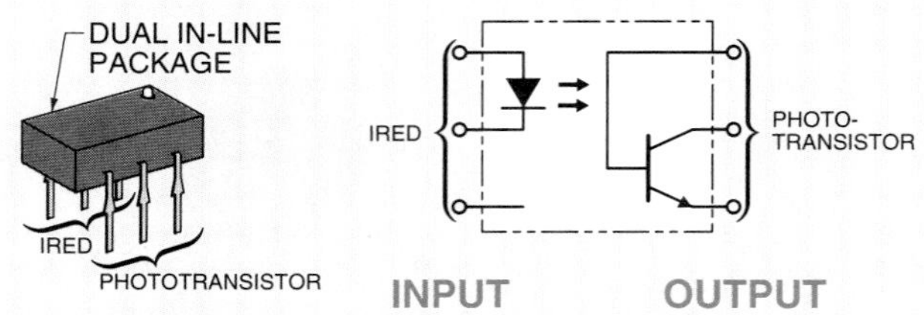

Figure 11-52. An optocoupler consists of an IRED as the input stage and a silicon NPN phototransistor as the output stage.

An optocoupler uses a glass dielectric sandwich to separate input from output. The coupling medium between the IRED and sensor is the infrared transmitting glass. This provides one-way transfer of electrical signals from the IRED to the photodetector (phototransistor) without electrical connection between the circuitry containing the devices.

Photons emitted from the IRED (emitter) have wavelengths of about 900 nm (nanometers). The detector (transistor) responds effectively to photons with this same wavelength. Input and output devices are always spectrally matched for maximum transfer characteristics. The signal cannot go back in the opposite direction because the emitters and detectors cannot reverse their operating functions.

TROUBLESHOOTING SOLID-STATE DEVICES

High voltages, improper connections, and overheating can damage solid-state devices. An electrician or technician may be responsible for determining the condition of solid-state devices.

Multimeter Diode Test

Testing a diode using an ohmmeter may not indicate whether a diode is good or bad. Testing a diode that is connected in a circuit with an ohmmeter may give false readings because other components may be connected in parallel with the diode under test. The best way to test a diode is to measure the voltage drop across the diode when it is forward biased.

A good diode has a voltage drop across it when it is forward biased and conducting current. The voltage drop is between .5 V and .8 V for the most commonly used silicon diodes. Some diodes are made of germanium and have a voltage drop between .2 V and .3 V.

A multimeter in the diode test position is used to test the voltage drop across a diode. In this position, the meter produces a small voltage between the test leads. The meter displays the voltage drop when the leads are connected across a diode. See Figure 11-53.

To test a diode using the diode test position on a multimeter, apply the procedure:

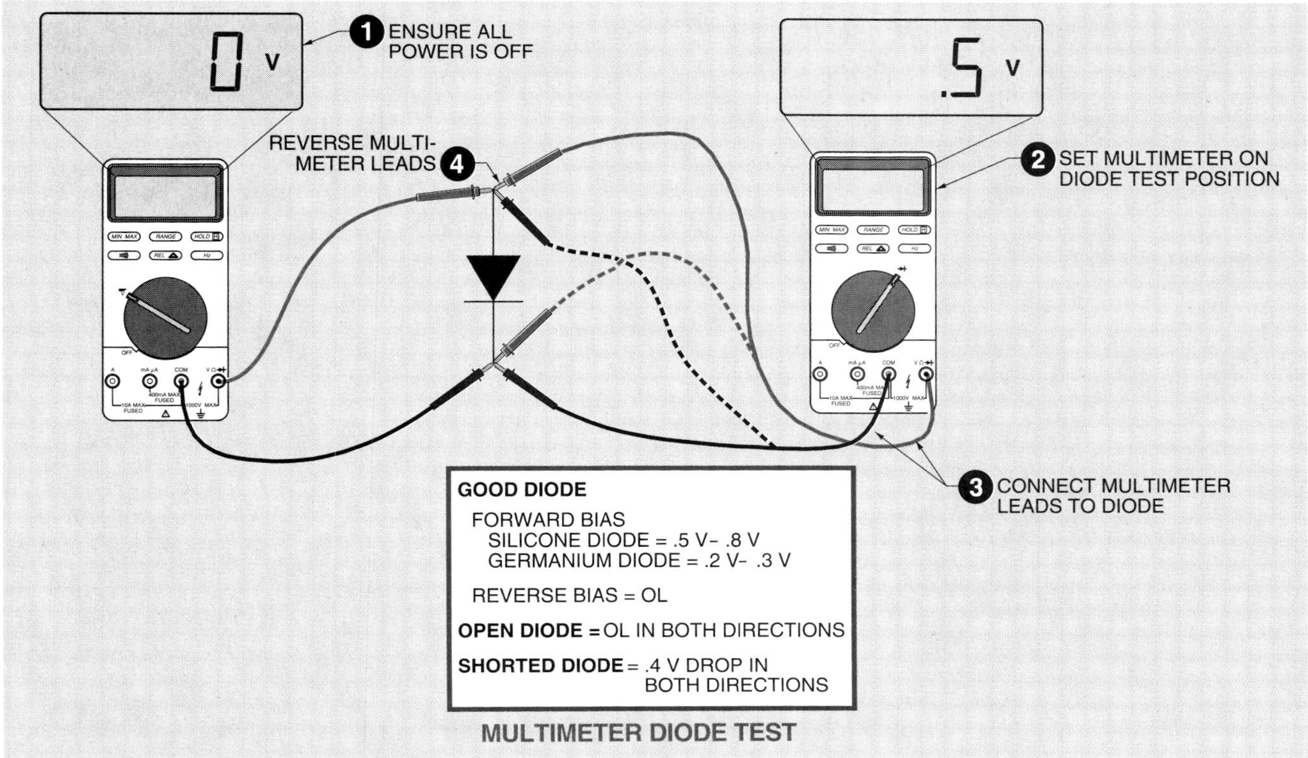

Figure 11-53. A good diode has a voltage drop across it when it is forward biased and conducting current.

1. Ensure that all power in the circuit is OFF. Test for voltage using a voltmeter to ensure power is OFF.
2. Set the meter on the diode test position.
3. Connect the meter leads to the diode. Record the meter reading.
4. Reverse the meter leads. Record the meter reading.

The meter displays a voltage drop between .5 V and .8 V (silicon diode) or .2 V and .3 V (germanium diode) when a good diode is forward biased. The meter displays an OL when a good diode is reverse biased. The OL reading indicates that the diode is acting like an open switch. An open (bad) diode does not allow current to flow through it in either direction. The meter displays an OL reading in both directions when the diode is open. A shorted diode gives the same voltage drop reading in both directions. This reading is normally about .4 V.

Zener Diode Test

A zener diode either provides voltage regulation or it fails. The zener diode must be replaced to return the circuit to proper operation if it fails. Occasionally, a zener diode may appear to fail only in certain situations. To check for intermittent failures, a zener diode must be tested while in operation. An oscilloscope is used for testing the characteristics of a zener diode in an operating situation. An oscilloscope displays the dynamic operating characteristics of the zener diode. See Figure 11-54.

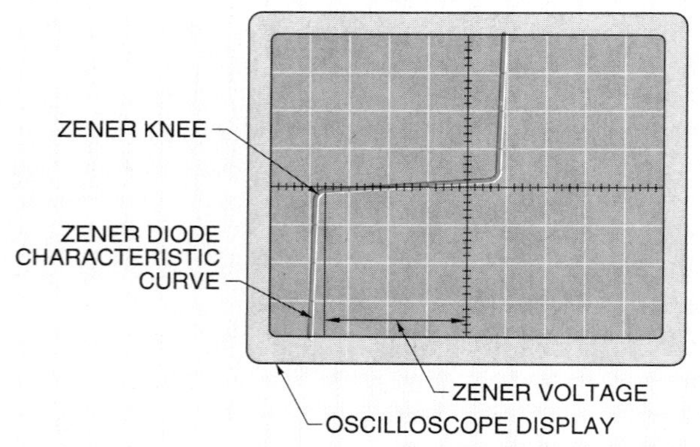

Figure 11-54. An oscilloscope test display indicates if a zener diode is good.

Testing Thermistors

A thermistor must be connected properly to an electronic circuit. Loose or corroded connections create a high resistance in series with the thermistor resistance. The control circuit may sense the additional resistance as a false temperature reading.

The hot and cold resistance of a thermistor can be checked with a multimeter. See Figure 11-55. To test the hot and cold resistance of a thermistor, apply the procedure:

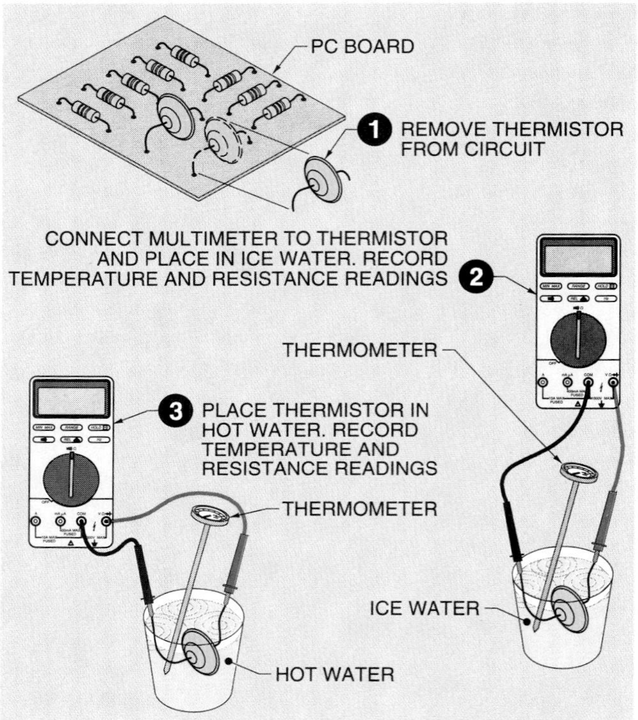

Figure 11-55. The hot and cold resistance of a thermistor can be checked with a multimeter.

1. Remove the thermistor from the circuit.
2. Connect the multimeter leads to the thermistor leads and place the thermistor and a thermometer in ice water. Record the temperature and resistance readings.
3. Place the thermistor and thermometer in hot water (not boiling). Record the temperature and resistance readings.

Compare the hot and cold readings with the manufacturer's specification sheet or with a similar thermistor that is known to be good.

Testing Solid-State Pressure Sensors

Solid-state pressure sensors are tested by checking the resistance of the device at low and high pressure and then comparing the value to manufacturer's specification sheets. See Figure 11-56.

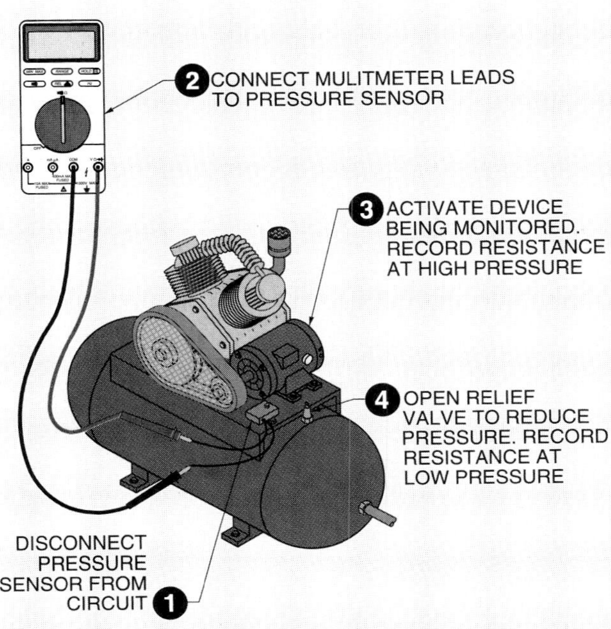

Figure 11-56. Solid-state pressure sensors are tested by checking the resistance of the device at low and high pressure and then comparing the value to manufacturer specification sheets.

To test the condition of a solid-state pressure sensor, apply the procedure:

1. Disconnect the pressure sensor from the circuit.
2. Connect the multimeter leads to the pressure sensor.
3. Activate the device being monitored (compressor, air tank, etc.) until pressure builds up. Record the resistance of the pressure sensor at the high-pressure setting.
4. Open the relief or exhaust valve and reduce the pressure on the sensor. Record the resistance of the pressure sensor at the low-pressure setting.

Compare the high and low resistance readings with manufacturer specification sheets. Use a replacement pressure sensor that is known to be good when manufacturer specification sheets are not available.

Testing Photocells

Humidity and contamination are the primary causes of photocell failure. See Figure 11-57. The use of quality components that are hermetically sealed is essential for long life and proper operation. Some plastic units are less rugged and more susceptible to temperature changes than glass units.

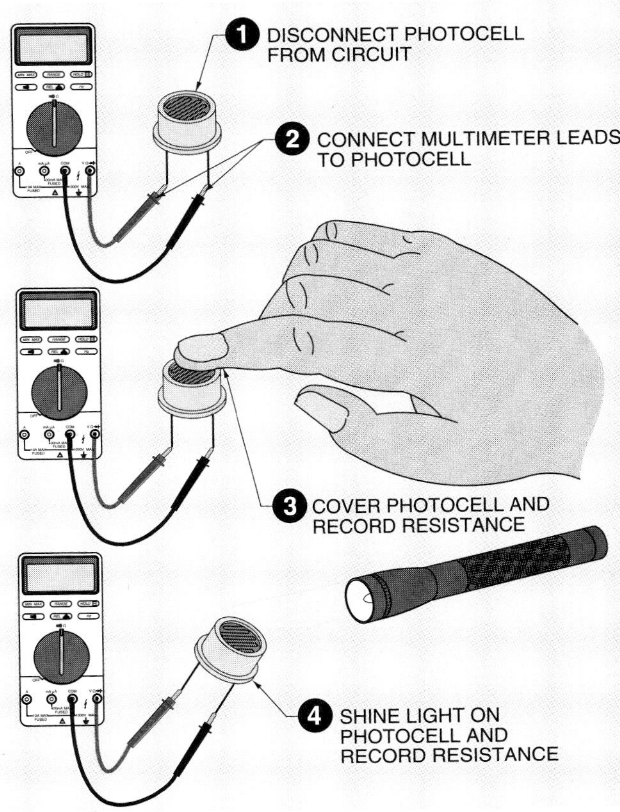

Figure 11-57. Humidity and contamination are the primary causes of photocell failure.

To test the resistance of a photocell, apply the procedure:

1. Disconnect photocell from the circuit.
2. Connect the multimeter leads to the photocell.
3. Cover the photocell and record dark resistance.
4. Shine a light on the photocell and record light resistance.

Compare the resistance readings with manufacturer's specification sheets. Use a similar photocell that is known to be good when specification sheets are not available. All connections should be tight and corrosion free.

Transistor Testing

A transistor becomes defective from excessive current or temperature. A transistor normally fails due to an open or shorted junction. The two junctions of a transistor may be tested with an ohmmeter. See Figure 11-58.

To test an NPN transistor for an open or shorted junction, apply the procedure:

1. Connect a multimeter to the emitter and base of the transistor. Measure the resistance.
2. Reverse the meter leads and measure the resistance. The emitter/base junction is good when the resistance is high in one direction and low in the opposite direction.

Note: The ratio of high to low resistance should be greater than 100:1. Typical resistance values are 1 kΩ (with the positive lead of the meter on the base) and 100 kΩ (with the positive lead of the meter on the emitter). The junction is shorted when both readings are low. The junction is open when both readings are high.

3. Connect the meter to the collector and base of the transistor. Measure the resistance.
4. Reverse the meter leads and measure the resistance. The collector/base junction is good when the resistance is high in one direction and low in the opposite direction.

Note: The ratio of high to low resistance should be greater than 100:1. Typical resistance values are 1 kΩ (with the positive lead of the meter on the base) and 100 kΩ (with the positive lead of the meter on the collector).

5. Connect the meter to the collector and emitter of the transistor. Measure the resistance.

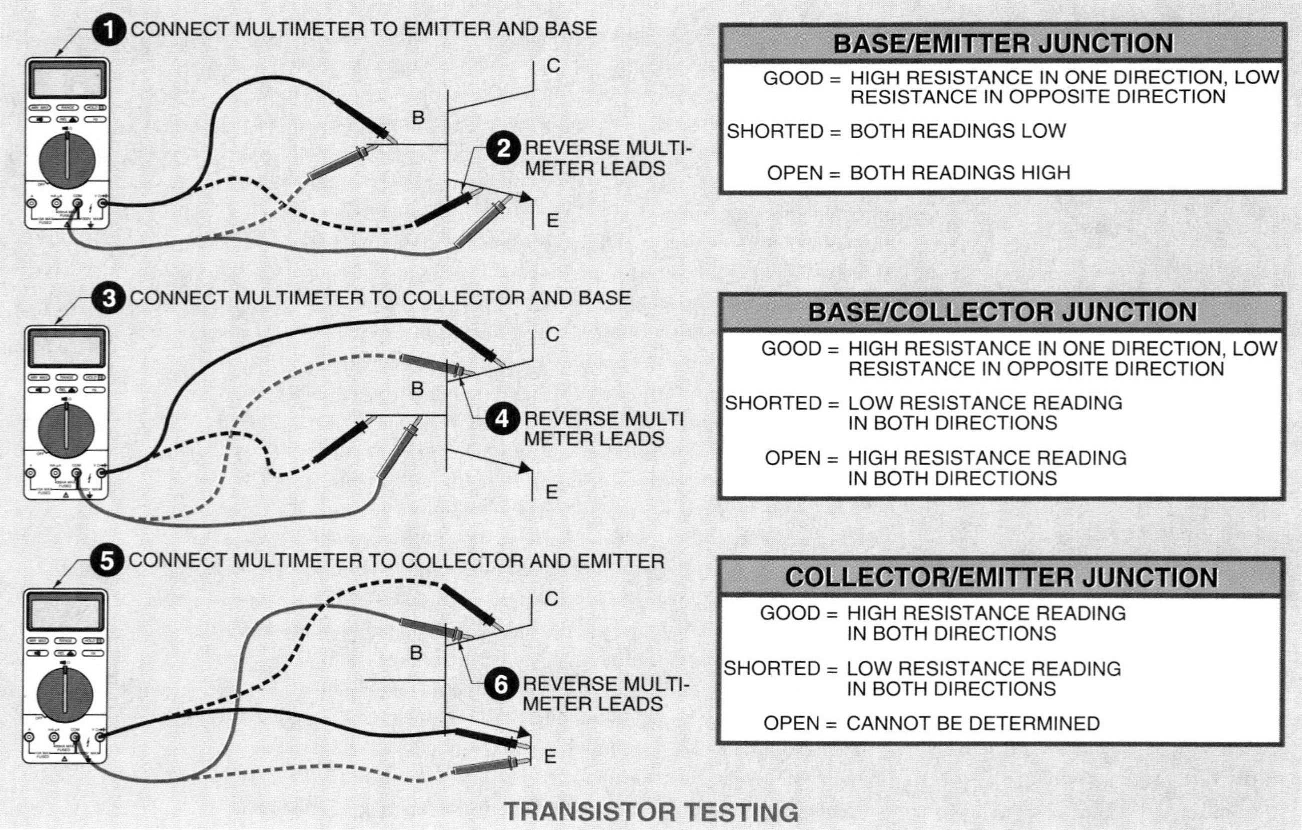

Figure 11-58. A transistor normally fails due to an open or shorted junction.

6. Reverse the meter leads and measure the resistance. The collector/emitter junction is good when the resistance reading is high in both directions.

The same test used for an NPN transistor is used for testing a PNP transistor. The difference is that the meter test leads must be reversed to obtain the same results.

SCR Testing

An oscilloscope is needed to properly test an SCR under operating conditions. A rough test using a test circuit can be made with a multimeter. See Figure 11-59. To test an SCR using a multimeter, apply the procedure:

1. Set the multimeter on the Ω scale.
2. Connect the negative lead of the multimeter to the cathode.
3. Connect the positive lead of the multimeter to the anode. The multimeter should read infinity.
4. Short circuit the gate to the anode using a jumper wire. The multimeter should read almost 0 Ω. Remove the jumper wire. The low-resistance reading should remain.
5. Reverse the multimeter leads so that the positive lead is on the cathode and the negative lead is on the anode. The multimeter should read almost infinity.
6. Short circuit the gate to the anode with a jumper wire. Resistance on the meter should remain high.

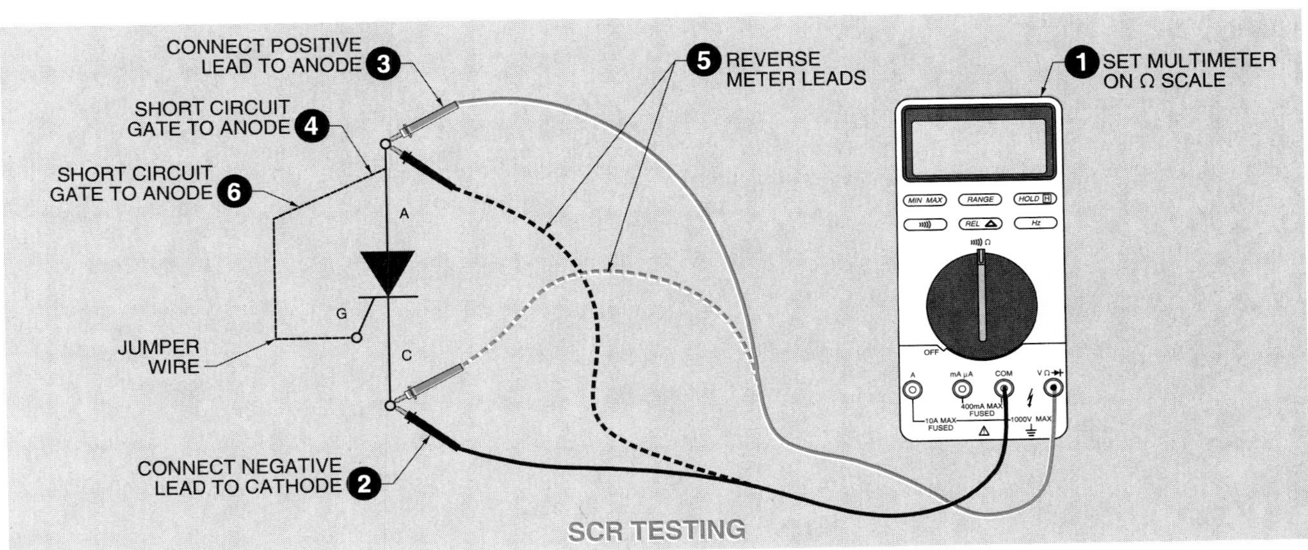

Figure 11-59. A rough test using a test circuit can be made on an SCR with a multimeter.

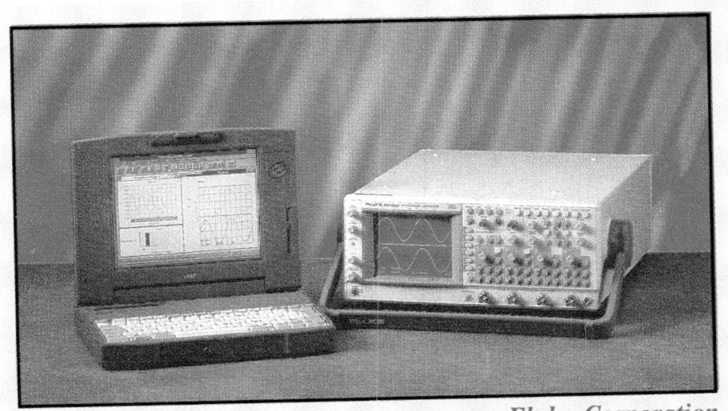
Fluke Corporation

Test equipment that displays voltage and current patterns, such as Fluke Corporation's PM3394A Autoranging Combiscope™ is used to test solid-state electronic circuits and components.

Diac Testing

A multimeter may be used to test a diac for a short circuit. See Figure 11-60. To text a diac for a short circuit, apply the procedure:

1. Set the multimeter on the Ω scale.
2. Connect the multimeter leads to the leads of the diac and record the resistance reading.
3. Reverse the multimeter leads and record the resistance reading.

Both resistance readings should show high resistance because the diac is essentially two zener diodes connected in series. Testing a diac in this manner only shows that the component is shunted.

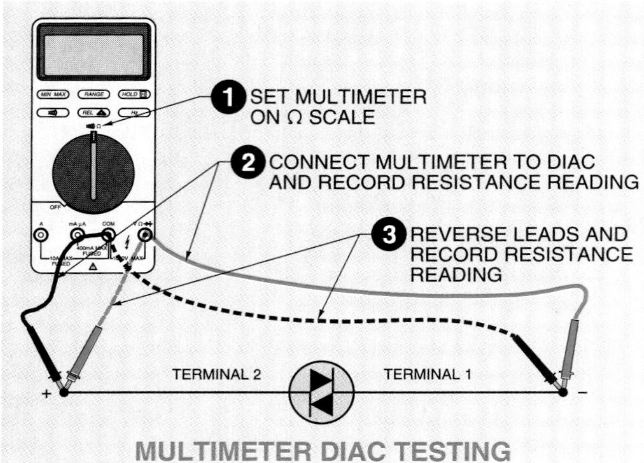

Figure 11-60. A multimeter may be used to test a diac for a short circuit.

A diac should be tested using an oscilloscope if the diac is suspected of being open. See Figure 11-61. To test a diac using an oscilloscope, apply the procedure:

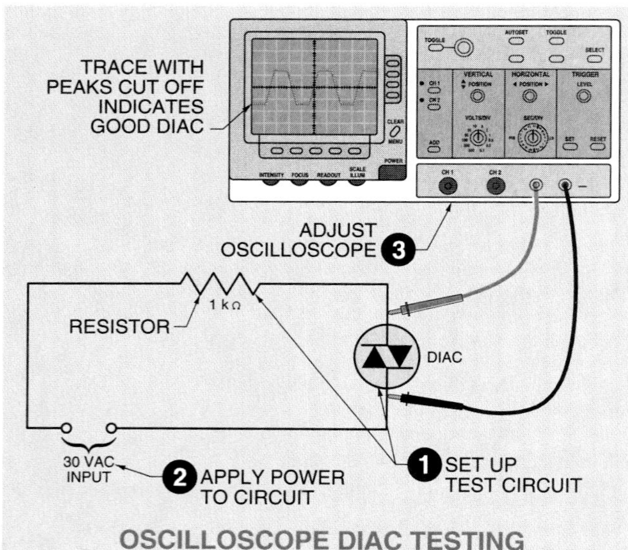

Figure 11-61. A diac should be tested using an oscilloscope if the diac is suspected of being open.

1. Set up the test circuit.
2. Apply power to the circuit.
3. Adjust the oscilloscope.

A trace of an AC sine wave with the peaks cut off indicates that the diac is good.

Fluke Corporation

Portable test equipment, such as a Fluke Model 96B Scopemeter™, is used when troubleshooting solid-state components in the field.

Triac Testing

Triacs should be tested under operating conditions using an oscilloscope. A multimeter may be used to make a rough test with the triac out of the circuit. See Figure 11-62.

To test a triac using a multimeter, apply the procedure:

1. Set the multimeter on the Ω scale.
2. Connect the negative multimeter lead to main terminal 1.
3. Connect the positive multimeter lead to main terminal 2. The multimeter should read infinity.
4. Short circuit the gate to main terminal 2 using a jumper wire. The multimeter should read almost 0 Ω. The zero reading should remain when the lead is removed.
5. Reverse the multimeter leads so that the positive lead is on main terminal 1 and the negative lead is on main terminal 2. The multimeter should read infinity.
6. Short circuit the gate of the triac to main terminal 2 using a jumper wire. The multimeter should read almost 0 Ω. The zero reading should remain after the lead is removed.

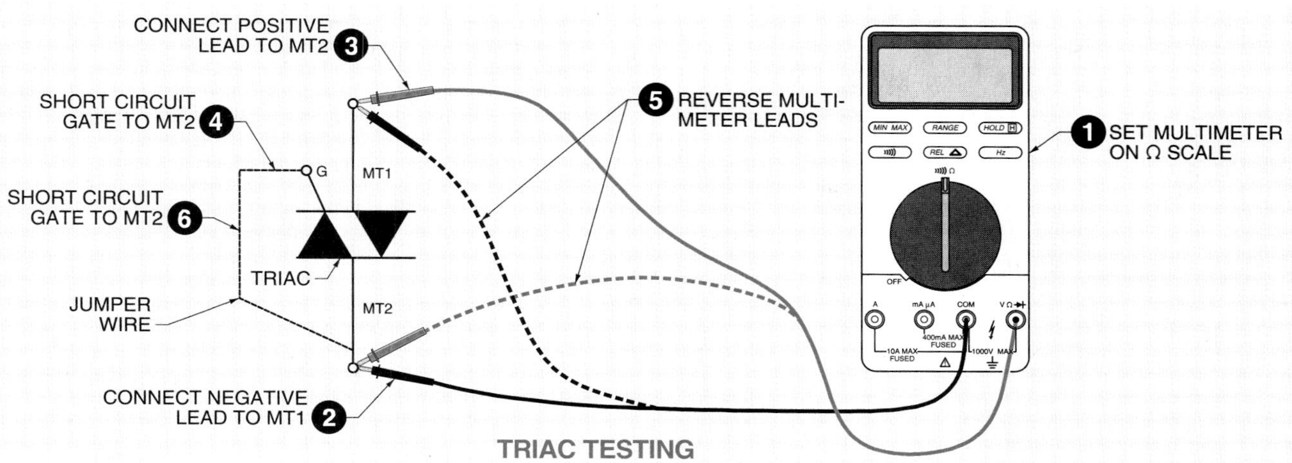

Figure 11-62. A multimeter may be used to make a rough test with the triac out of the circuit.

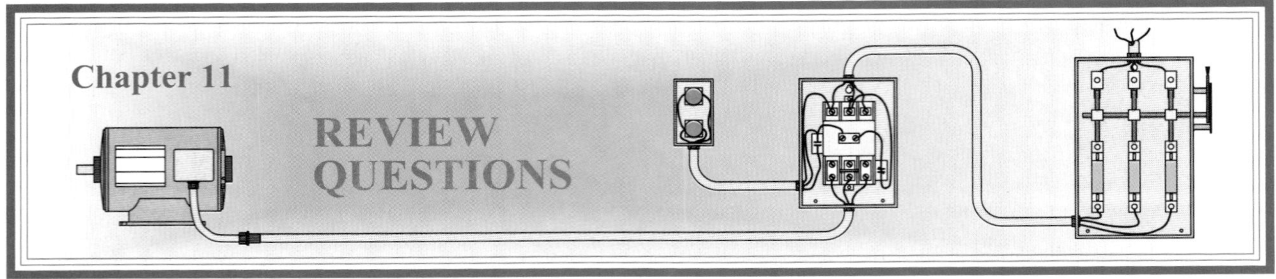

1. What determines whether an element is an insulator, conductor, or semiconductor?
2. How is the crystal structure in most semiconductors altered?
3. What is a PC board?
4. What are forward bias and reverse bias in semiconductors?
5. What is a half-wave rectifier?
6. How does a zener diode operate in a circuit?
7. What is a thermistor?
8. How does a photoconductive cell differ from a photovoltaic cell?
9. What is a Hall effect sensor?
10. What is a solid-state pressure sensor?
11. What is an LED?
12. What is a transistor?
13. How is the shape of a transistor identified?
14. What is rectification?
15. What is gain?
16. What are the three types of transistor amplifiers?
17. What is forward breakover voltage?
18. What is a triac?
19. How does a triac function?
20. How does a diac function?
21. What is an IC?
22. How are pins on an IC identified?
23. What is an op amp?
24. What are the two basic types of electronic signals?
25. What are the four most common gates in digital electronics?

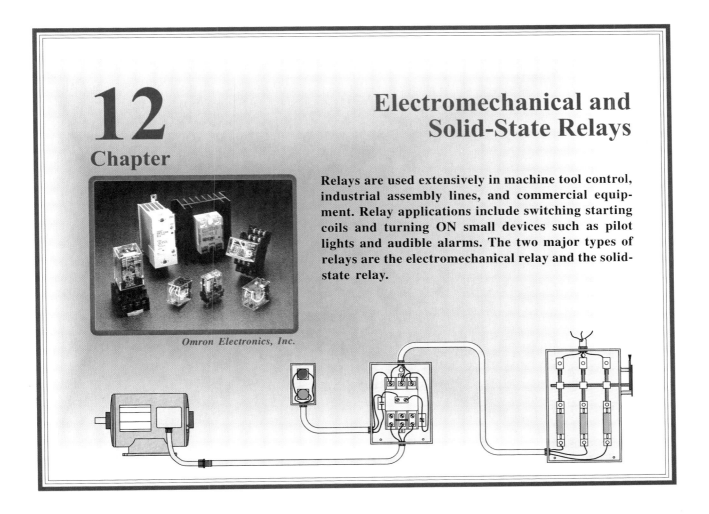

12 Chapter

Electromechanical and Solid-State Relays

Omron Electronics, Inc.

Relays are used extensively in machine tool control, industrial assembly lines, and commercial equipment. Relay applications include switching starting coils and turning ON small devices such as pilot lights and audible alarms. The two major types of relays are the electromechanical relay and the solid-state relay.

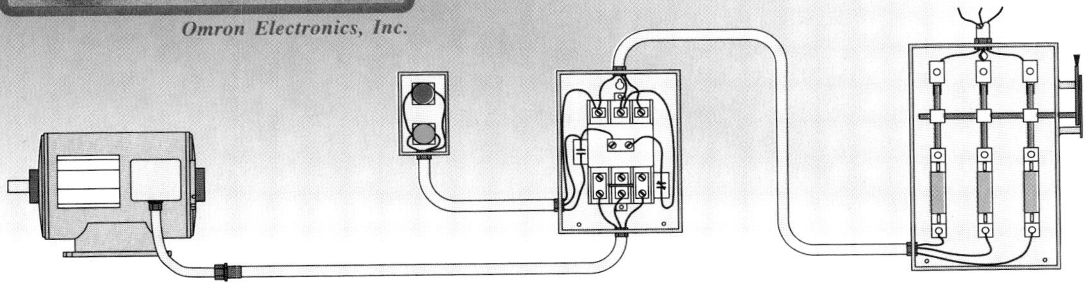

RELAYS

A *relay* is a device that controls one electrical circuit by opening and closing contacts in another circuit. Depending on design, relays normally do not control power-consuming devices (except for small loads which draw less than 15 A). Relays are used extensively in machine tool control, industrial assembly lines, and commercial equipment. Relays are used to switch starting coils in contactors and motor starters, heating elements, pilot lights, audible alarms, and some small motors (less than ⅛ HP).

A small voltage applied to a relay results in a large voltage being switched. See Figure 12-1. For example, applying 24 V to the relay coils may operate a set of contacts that is controlling a 230/460 V circuit. The relay acts as an amplifier of the voltage or current in the control circuit because relay coils require a low current or voltage to switch, but can energize large currents or voltages.

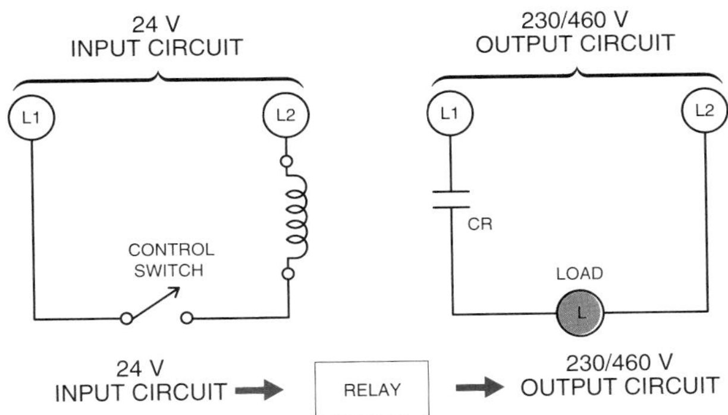

Figure 12-1. Relays may be compared to amplifiers in that small voltage input results in large voltage output.

Another example of a relay providing an amplifying effect is when a single input to the relay results in several other circuits being energized. See Figure 12-2. An input may be considered amplified because certain mechanical relays provide eight or more sets of contacts controlled from any one input.

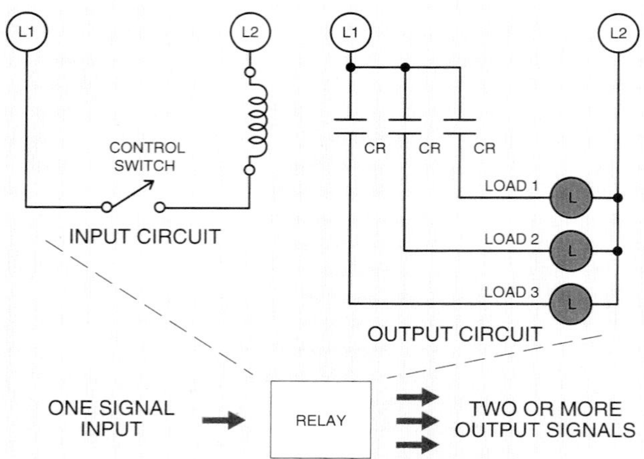

Figure 12-2. Relays may be compared to amplifiers in that a single input may result in multiple outputs.

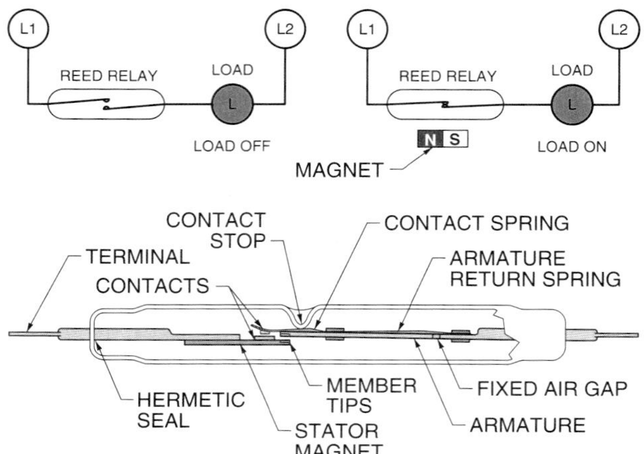

Figure 12-3. A reed relay is a fast-operating, single-pole, single-throw switch that is activated by a magnetic field.

The two major types of relays are the electromechanical relay and the solid-state relay. An *electromechanical relay (EMR)* is a switching device that has sets of contacts which are closed by a magnetic effect. A *solid-state relay (SSR)* is a switching device that has no contacts and switches entirely by electronic means. A *hybrid relay* is a combination of electromechanical and solid-state technology used to overcome unique problems which cannot be resolved by one or the other devices. Hybrid relays are generally considered EMRs.

ELECTROMECHANICAL RELAYS (EMRs)

EMRs which are common to commercial and industrial applications may be reed, general purpose, or machine control relays. The major difference between the types of EMRs is their intended use in the circuit, cost, and the life expectancy of the device.

Reed Relays

A *reed relay* is a fast-operating, single-pole, single-throw switch with normally open (NO) contacts hermetically sealed in a glass envelope. See Figure 12-3. During the sealing operation, dry nitrogen is forced into the tube, creating a clean inner atmosphere for the contacts. Because the contacts are sealed, they are unaffected by dust, humidity, and fumes. The life expectancy of reed relay contacts is quite high.

A reed relay includes a very low current-rated contact (less than .25 mA) that is activated by the presence of a magnetic field. Reed relays may be activated in a variety of ways, which allows them to be used in circuit applications where other relay types are inappropriate.

Reed relays are designed to be actuated by an external movable permanent magnet or DC electromagnet. When a magnetic field is brought close to the two reeds, the ferromagnetic (easily magnetized) ends assume opposite magnetic polarity. The attracting force of the opposing poles overcomes the stiffness of the reed drawing the contacts together if the magnetic field is strong enough. Removing the magnetizing force allows the contacts to spring open. AC electromagnets are not suitable for reed relays because the reed relay switches so fast that it would energize and de-energize on alternate half-cycles of a standard 60 Hz line.

Reed Contacts. To obtain a low and consistent contact resistance, the overlapping ends of the contacts may be plated with gold, rhodium, silver alloy, or other low-resistance metals. Contact resistance is often under 0.1 Ω when closed. Reed contacts have an open contact resistance of several million ohms.

Most reed contacts are capable of direct switching of industrial solenoids, contactors, and motor starters. Reed relay contact ratings indicate the maximum current, voltage, and volts/amperes that may be switched by the relay. Under no circumstances should these values be exceeded.

Reed Relay Actuation

A permanent magnet is the most common actuator for a reed relay. Permanent-magnet actuation can be arranged in several ways depending on the switching requirement. The most commonly used arrangements are proximity motion, rotary motion, shielding, and biasing.

Proximity Motion. The proximity motion arrangement uses the presence of a magnetic field that is brought within a specific proximity (close distance) to the reed relay to close the contacts. The distance for activating any given relay depends on the sensitivity of the relay and the strength of the magnet. A more sensitive relay or stronger magnet needs less distance for actuation. Methods of proximity motion operation are the pivoted motion, perpendicular motion, parallel motion, and front-to-back motion. See Figure 12-4. In each method, either the magnet or relay is moved. In some applications, both the magnet and relay are in motion. The contacts operate quickly with snap action and little wear. The application and switching requirements determine the best method.

Rotary Motion. The rotary motion arrangement involves revolving the magnet or relay which results in relay contact operation every 180° or two operations every 360°. See Figure 12-5. The contacts are closed when the magnet and relay are parallel. The contacts are opened when the magnet and relay are perpendicular. Although the magnetic poles reverse every 180°, they induce the magnetic field of opposite poles on the relay and close the contacts.

Furnas Electric Co.

The MT/46 machine tool relay from Furnas Electric Co. has convertible contact cartridges and a common mechanical tie between the contact cartridges and relay armature.

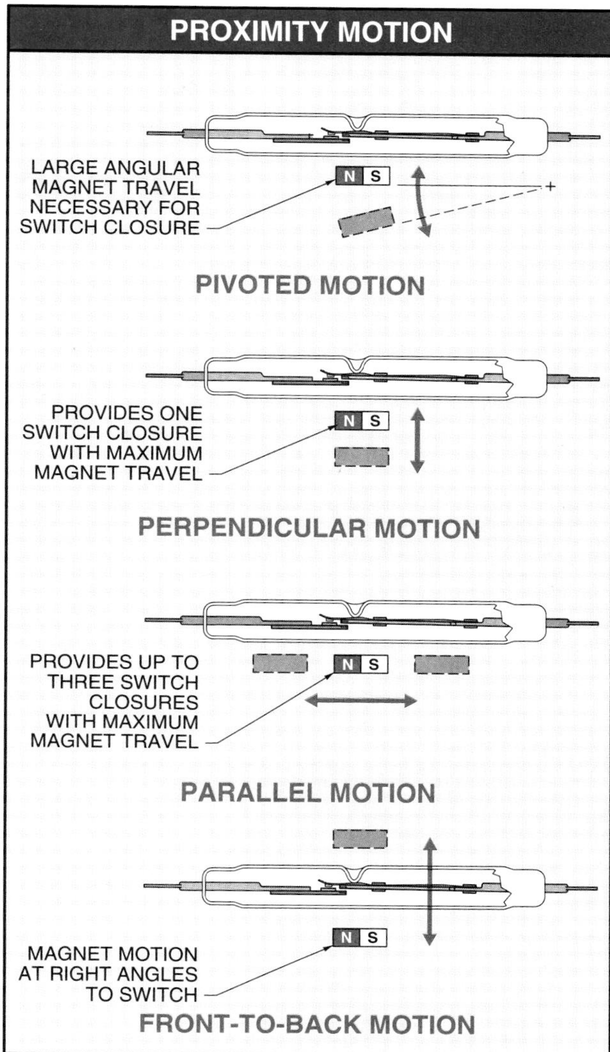

Figure 12-4. The proximity motion arrangement uses the presence of a magnetic field brought within a specific proximity to the reed relay to close the contacts.

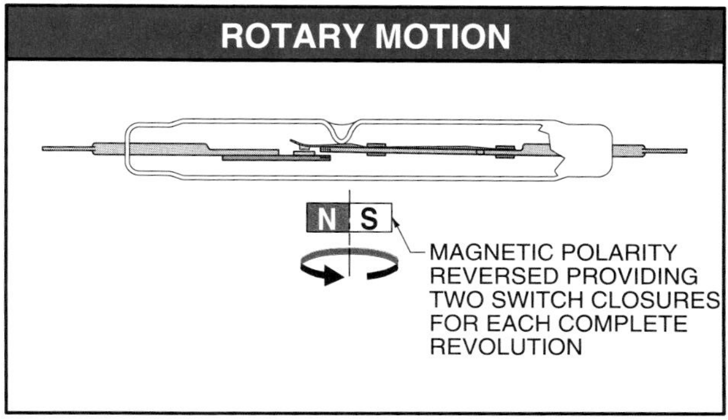

Figure 12-5. A reed relay may be activated by rotary motion.

Shielding. The shielding arrangement involves permanently fixing the magnet and relay so that the relay's contacts are held closed. See Figure 12-6. The contacts are open as ferromagnetic (iron based) material is passed between the magnet and relay. The ferromagnetic material acts like a short circuit or shunt for the magnetic field and eliminates the magnetic field holding the contacts. As the shield is removed, the contacts are closed. It makes no difference at what angle the shield is passed between the magnet and relay. This method may be used to signal that a protective shield, such as a cover on a high voltage box, has been removed.

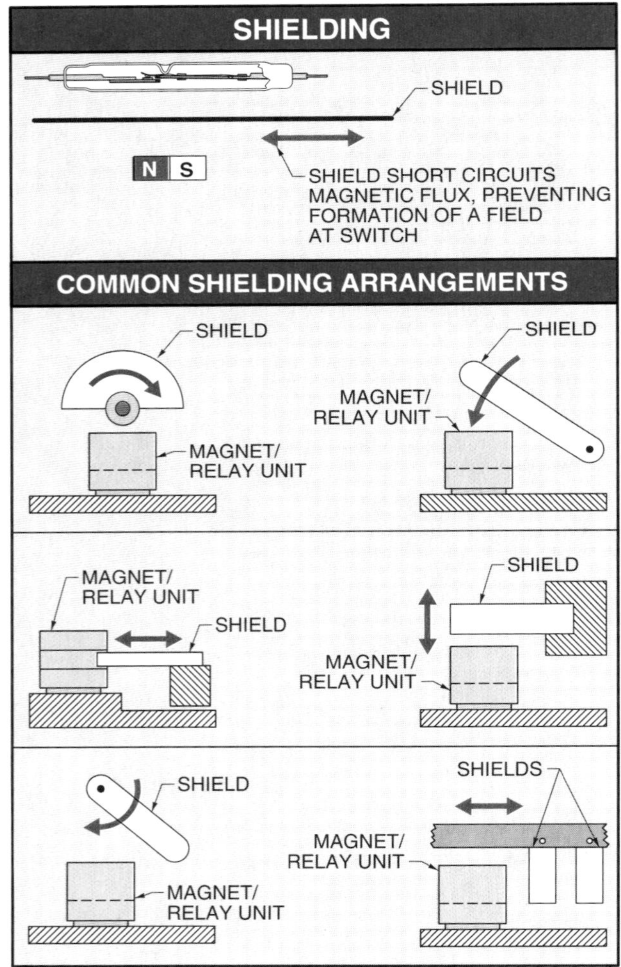

Figure 12-6. The shielding arrangement involves permanently fixing the magnet and relay so that the relay's contacts are held closed. The contacts are opened when a ferromagnetic material is passed between the magnet and relay.

Biasing. In the biasing arrangement, a bias magnet holds the switch closed until an actuating magnet cancels the magnetic field of the bias magnet and opens the switch. See Figure 12-7. The actuating magnet approaches the bias magnet with opposite polarity which cancels the magnetic field of the bias magnet and opens the relay's contacts.

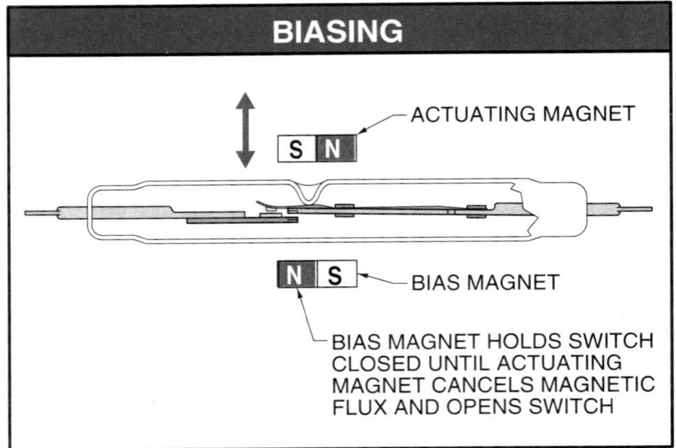

Figure 12-7. In the biasing arrangement, a bias magnet holds the switch closed until an actuating magnet cancels the magnetic field of the bias magnet and opens the switch.

The relay's contacts open only if the magnetic fields of the two magnets cancel each other. The correct strength of the magnetic field and distance is required. This application may be used for detecting magnetic polarity or other similar applications.

General Purpose Relays

General purpose relays are EMRs that include several sets (normally two, three, or four) of nonreplaceable NO and NC contacts (normally rated at 5 A to 15 A) that are activated by a coil. A general purpose relay is a good relay for applications that can use a throw away, plug-in relay to simplify troubleshooting and reduce costs. Special attention must be given to the contact current rating when using general purpose relays because the contact rating for switching DC is less than the contact rating for switching AC. For example, a 15 A AC rated contact normally is only rated for 8 A to 10 A DC.

Several different styles of general purpose relays are available. These relays are designed for commercial and industrial applications where economy and fast replacement are high priorities. Most general purpose relays have a plug-in feature that makes for quick replacement and simple troubleshooting. See Figure 12-8.

Omron Electronics, Inc.

Figure 12-8. General purpose relays are EMRs that include several sets of nonreplaceable NO and NC contacts that are activated by a coil.

A *general purpose relay* is a mechanical switch operated by a magnetic coil. See Figure 12-9. General purpose relays are available in AC and DC designs. These relays are available with coils that can open or close contacts ranging from millivolts to several hundred volts. Relays with 6 V, 12 V, 24 V, 48 V, 115 V, and 230 V are the most common. General purpose relays are available that require as little as 4 mA at 5 VDC, or 22 mA at 12 VDC, making them IC compatible to TTL and CMOS logic gates. These relays are available in a wide range of switching configurations.

Contacts. A *contact* is the conducting part of a switch that operates with another conducting part of the switch to make or break a circuit. Relay contacts switch electrical circuits. The most common contacts are the single-pole, double-throw (SPDT), double-pole, double-throw (DPDT), and the three-pole, double-throw (3PDT) contacts. Relay contacts are described by their number of poles, throws, and breaks. See Figure 12-10.

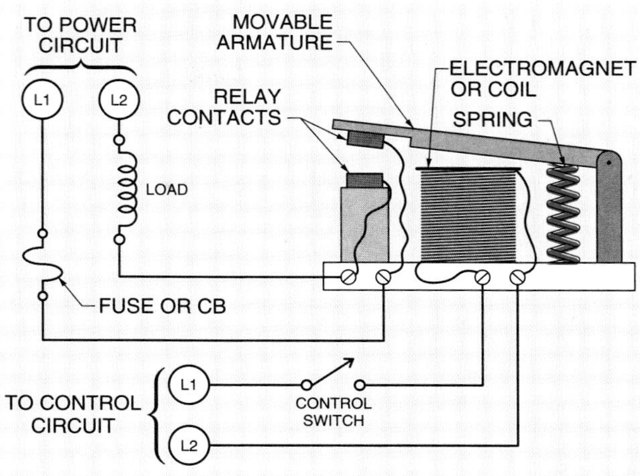

Figure 12-9. A general purpose relay is a mechanical switch operated by a magnetic coil.

A *break* is the number of separate places on a contact that open or close an electrical circuit. For example, a single-break contact breaks an electrical circuit in one place. A double-break (DB) contact breaks the electrical circuit in two places. All contacts are single break or double break. Single-break (SB) contacts are normally used when switching low-power devices such as indicating lights. Double-break contacts are used when switching high-power devices such as solenoids.

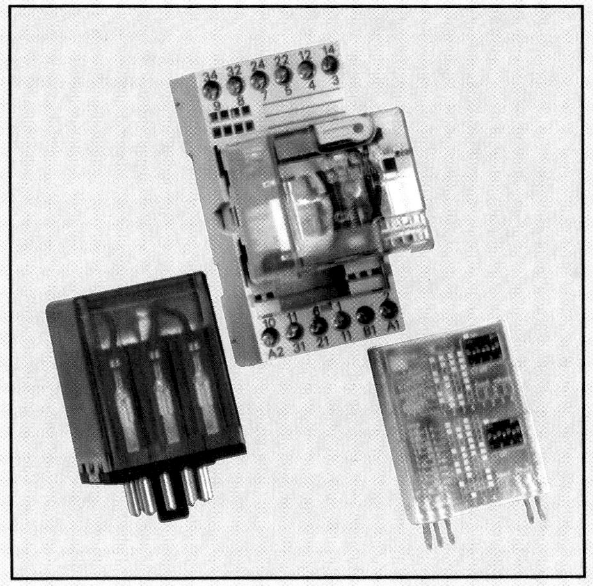

Sprecher + Schuh

Sprecher + Schuh's CH1 control relays are plug-in units that are available in a standard 8 pin or 11 pin profile depending on the number of contacts required for the application.

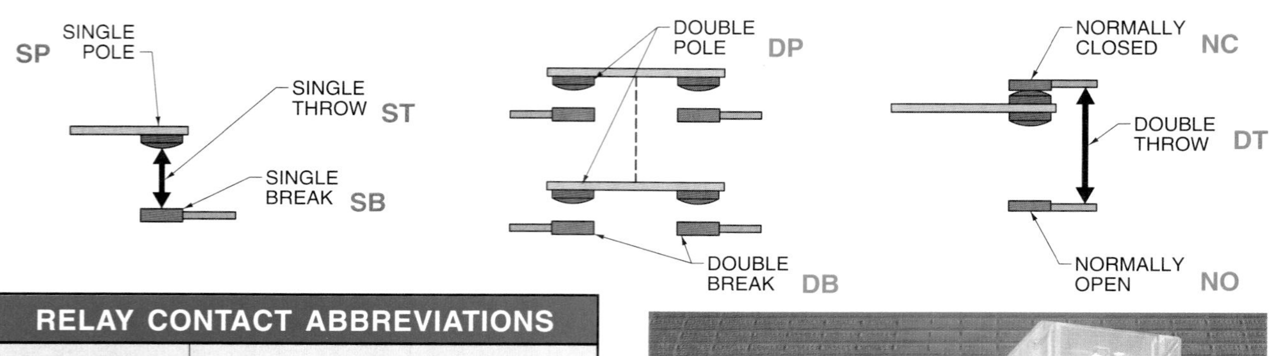

RELAY CONTACT ABBREVIATIONS	
Abbreviation	**Meaning**
SP	Single pole
DP	Double pole
3P	Three pole
ST	Single throw
DT	Double throw
NO	Normally open
NC	Normally closed
SB	Single break
DB	Double break

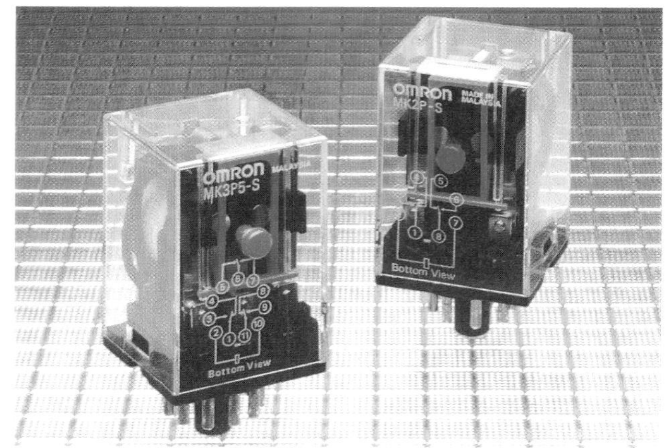

Omron Electronics, Inc.

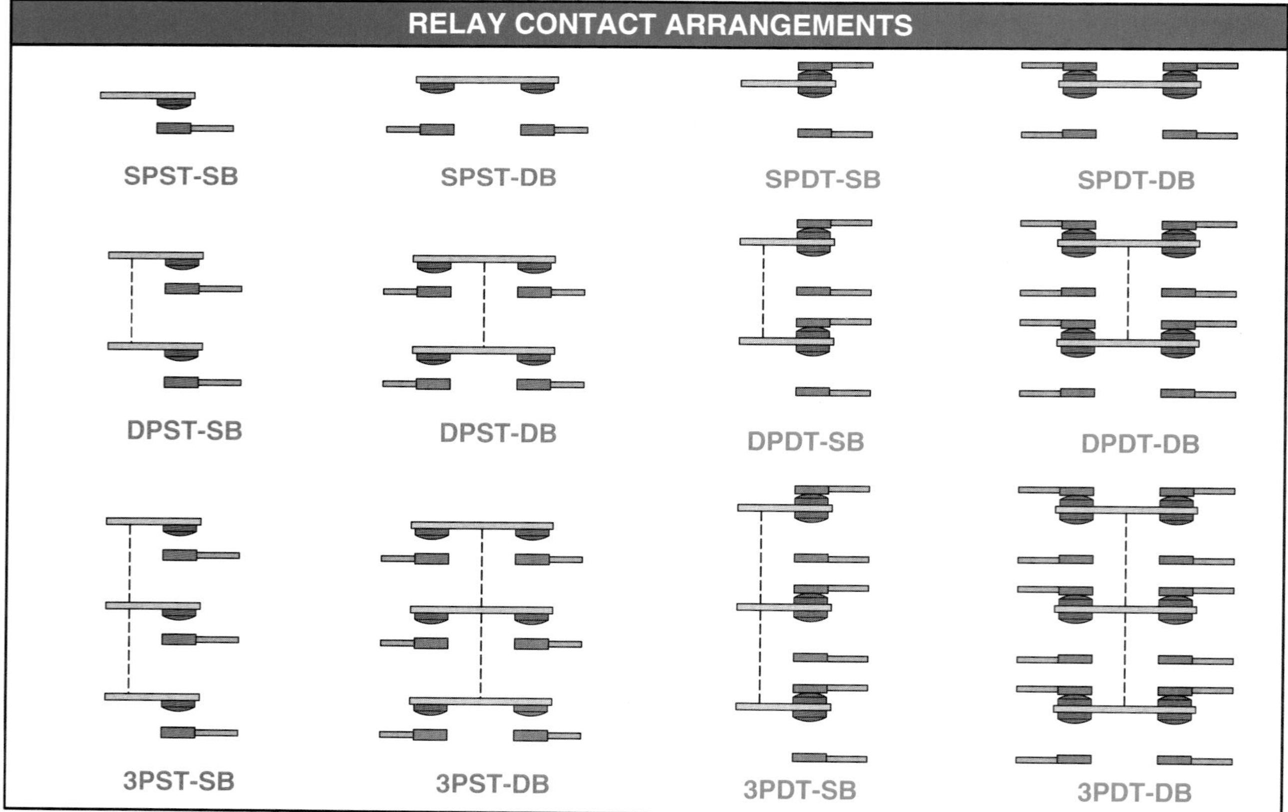

Figure 12-10. Relay contacts are described by their number of poles, throws, and breaks.

A *pole* is the number of completely isolated circuits that a relay can switch. A single-pole contact can carry current through only one circuit at a time. A double-pole contact can carry current through two circuits simultaneously. In a double-pole contact, the two circuits are mechanically connected to open or close simultaneously and are electrically insulated from each other.

The mechanical connection is represented by a dashed line connecting the poles. Relays are available with 1 to 12 poles. A *throw* is the number of closed contact positions per pole. A single-throw contact can control only one circuit. A double-throw contact can control two circuits.

Relay manufacturers use a common code to simplify the identification of relays. See Figure 12-11. This code uses a form letter to indicate the type of relay. For example, Form A has one contact that is NO, and closes (makes) when the coil is energized. Form B has one contact that is NC, and breaks (opens) when the coil is energized. Form C has one pole that first breaks one contact and then makes a second contact when the coil is energized.

RELAY FORM INDENTIFICATION

Design	Sequence	Symbol	Form
SPST-NO	MAKE (1)		A
SPST-NC	MAKE (1)		B
SPDT	BREAK (1) MAKE (2)		C
SPDT	MAKE (1) BEFORE BREAK (2)		D
SPDT (B-M-B)	BREAK (1) MAKE (2) BEFORE BREAK (3)		E
SPDT-NO	CENTER OFF		K
SPST-NO (DM)	DOUBLE MAKE (1)		X
SPST-NC (DB)	DOUBLE BREAK (1)		Y
SPDT-NC-NO (DB-DM)	DOUBLE BREAK (1) DOUBLE MAKE (2)		Z

Figure 12-11. Relay manufacturers use a common code (form letter) to simplify the identification of relays.

In some electrical applications, the exact order in which each contact operates (makes or breaks) must be known so the circuit can be designed to reduce arcing. Arcing occurs at any electrical contact that has current flowing through it when the contact is opened.

Machine Control Relays

A *machine control relay* is an EMR that includes several sets (usually 2 to 8) of NO and NC replaceable contacts (typically rated at 10 A to 20 A) that are activated by a coil. See Figure 12-12. Machine control relays are the backbone of electromechanical control circuitry and are expected to have long life and minimum problems. Machine control relays are used extensively in machine tools for direct switching of solenoids, contactors, and starters. Machine control relays provide easy access for contact maintenance and may provide additional features like time-delay, latching, and convertible contacts for maximum circuit flexibility. Convertible contacts are mechanical contacts that can be placed in either a NO or NC position. Machine control relays are also known as heavy-duty or industrial control relays.

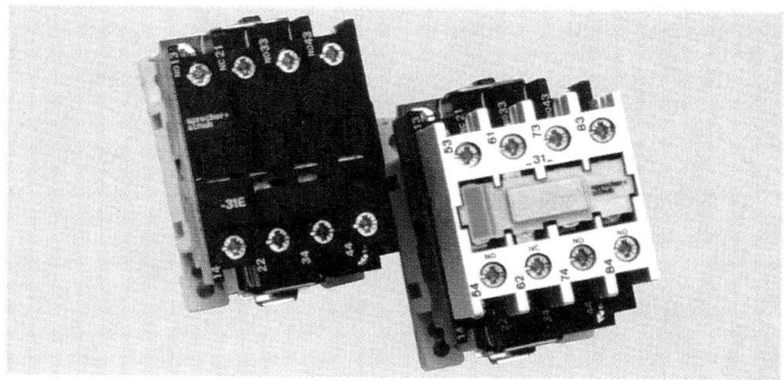

Sprecher + Schuh

Figure 12-12. Machine control relays are used extensively in machine tools for direct switching of solenoids, contactors, and starters.

The popularity of machine control relays stems from their good quality and reliability, along with their extreme flexibility. In a machine control relay, each contact is a separate removable unit that may be installed to obtain any combination of NO and NC switching. These contacts are also convertible from NO to NC and from NC to NO. See Figure 12-13. The unit may be used as either an NO or NC contact by changing the terminal screws and rotating the unit 180°. Relays of 1 to 12 contact poles are readily assembled from stock parts.

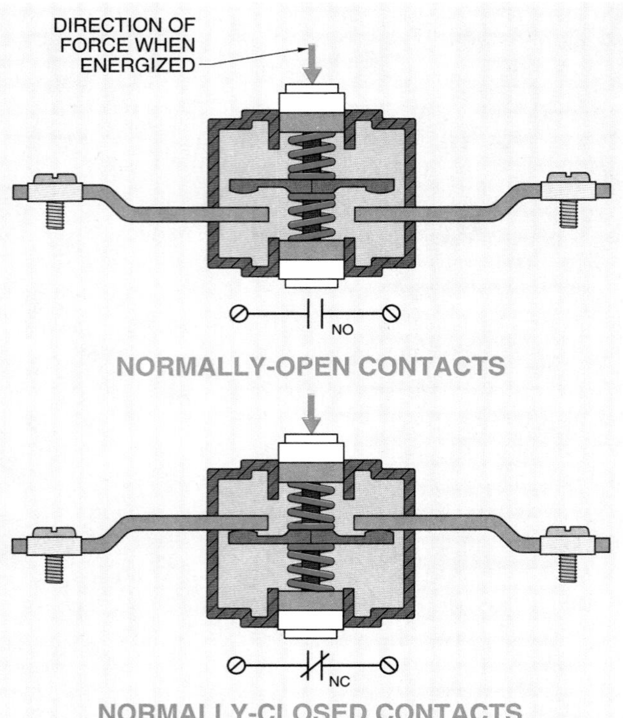

Figure 12-13. Machine control relay contacts are separate, removable units that are installed to obtain any combination of NO and NC switching.

The control coils for machine control relays are easily changed from one control voltage to another and are available in AC or DC standard ratings. Machine control relays have a large number of accessories that may be added to the relay unit. These include indicating lights, transient suppression, latching controls, and time controls.

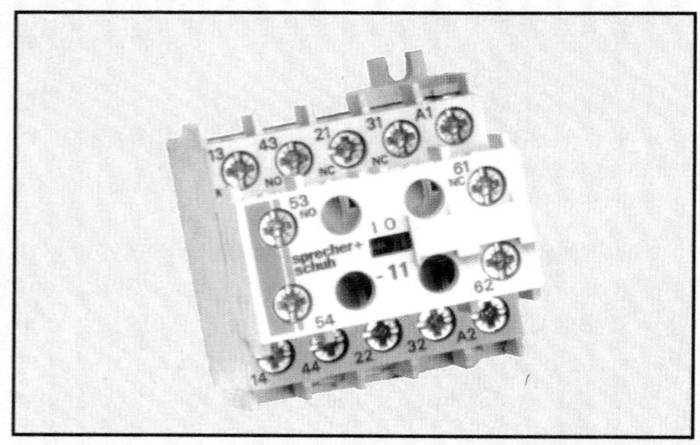

Sprecher + Schuh

CS4C industrial control relays from Sprecher + Schuh are available with auxiliary components that allow the basic 4-pole relays to be converted into 6- or 8-pole relays, 4-, 6-, or 8-pole relays with electronic time delay, and multiple 4-, 6-, or 8-pole relays with mechanical interlocks.

EMR Life

Electromechanical relay life expectancy is rated in contact life and mechanical life. *Contact life* is the number of times a relay's contacts switch the load controlled by the relay before malfunctioning. Typical contact life ratings are 100,000 to 500,000 operations. *Mechanical life* is the number of times a relay's mechanical parts operate before malfunctioning. Typical mechanical life ratings are 1,000,000 to 10,000,000 operations.

Relay contact life expectancy is lower than mechanical life expectancy because the life of a contact depends on the application. The contact rating of a relay is based on the contact's full-rated power. Contact life is increased when contacts switch loads less than their full-rated power. Contact life is reduced when contacts switch loads that develop destructive arcs. *Arcing* is the discharge of an electric current across a gap, such as when an electric switch is opened. Arcing causes contact burning and temperature rise. See Figure 12-14.

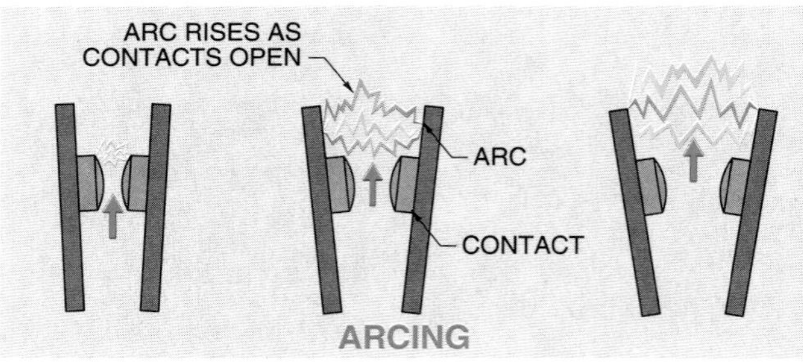

Figure 12-14. Arcing is the discharge of an electric current across a gap, such as when an electric switch is opened.

Arcing is minimized by an arc suppressor and by using the correct contact material for the application. An *arc suppressor* is a device that dissipates the energy present across opening contacts. Arc suppression is used in applications that switch arc producing loads such as solenoids, coils, motors, and other inductive loads.

Arc suppression is also accomplished by using a contact protection circuit. A *contact protection circuit* is a circuit that protects contacts by providing a nondestructive path for generated voltage as a switch is opened. A contact protection circuit may contain a diode, a resistance/capacitance (RC) circuit, or a varistor.

A diode is used as contact protection in DC circuits. The diode does not conduct electricity when the load is energized. The diode conducts electricity and shorts the generated voltage when the switch is opened. By shorting the generated voltage, the voltage is dissipated across the diode and not the relay contacts. See Figure 12-15.

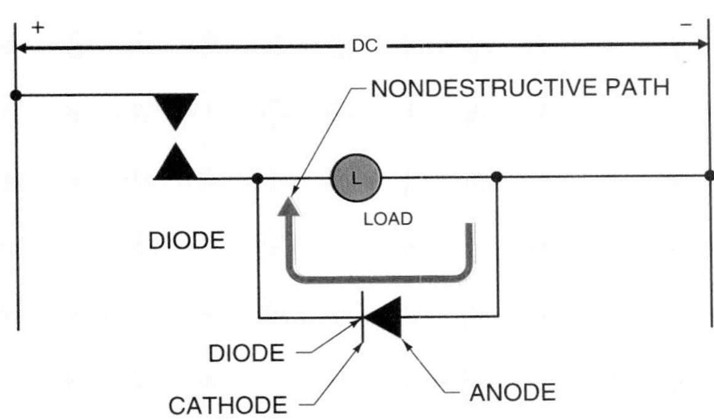

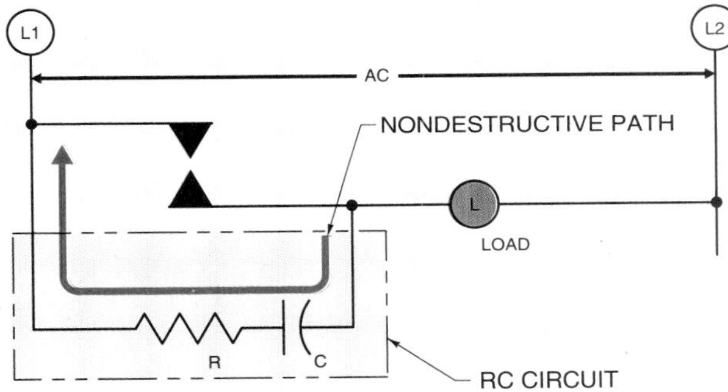

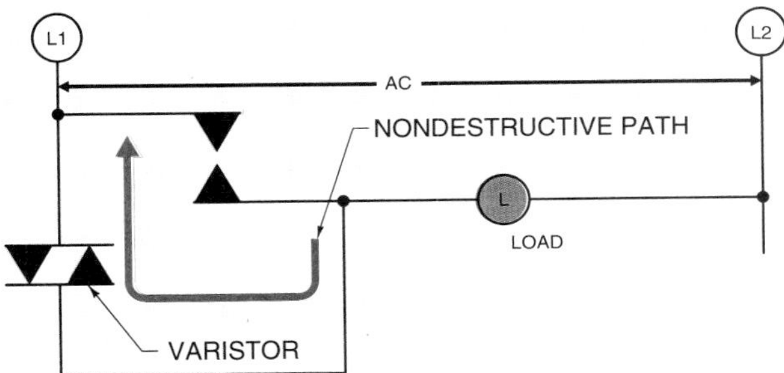

CONTACT PROTECTION CIRCUITS

Figure 12-15. A contact protection circuit is a circuit that protects contacts by providing a nondestructive path for generated voltage as a switch is opened.

An RC circuit and varistor are used as contact protection in AC circuits. The capacitor in an RC circuit is a high impedance to the 60 Hz line power and a short circuit to generated high frequencies. The short circuit dissipates generated voltage. A *varistor* is a resistor whose resistance is inversely proportional to the voltage applied to it. The varistor becomes a low impedance circuit when its rated voltage is exceeded. The low impedance circuit dissipates generated voltage when a switch is opened.

Contact Material

Relay contacts are available in fine silver, silver-cadmium, gold-flashed silver, and tungsten. Fine silver has the highest electrical and thermal properties of all metals. Fine silver sticks, welds, and is subject to sulfidation when used for many applications. *Sulfidation* is the formation of film on the contact surface. Sulfidation increases the resistance of the contacts. Silver is alloyed with other metals to reduce sulfidation.

Silver is alloyed with cadmium to produce a silver-cadmium contact. Silver-cadmium contacts have good electrical characteristics and low resistance, which helps the contact resist arcing but not sulfidation. Silver or silver alloy contacts are used in circuits that switch several amperes at more than 12 V, which burns off the sulfidation.

Sulfidation can damage silver contacts when used in intermittent applications. Gold-flashed silver contacts are used in intermittent applications to minimize sulfidation and provide a good electrical connection. Gold contacts are not used in high-current applications because the gold burns off quickly. Gold-flashed silver contacts are good for switching loads of 1A or less.

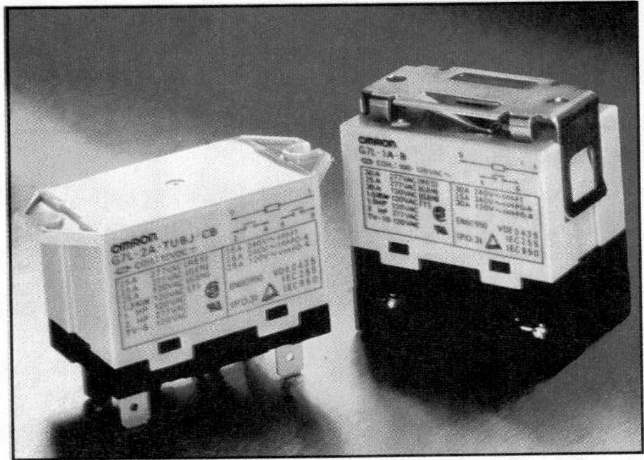

Omron Electronics, Inc.

G7L general-purpose relays from Omron Electronics, Inc. allow for the switching of 25 A and 30 A loads and produce no contact chattering for momentary voltage drops up to 50% of rated voltage.

Tungsten contacts are used in high-voltage applications because tungsten has a high melting temperature and is less affected by arcing. Tungsten contacts are used when high repetitive switching is required.

Contact Failure

In most applications, a relay fails due to contact failure. In some low-current applications, the relay contacts may look clean but may have a thin film of sulfidation, oxidation, or contaminates on the contact surface. This film increases the resistance to the flow of current through the contact. Normal contact wiping or arcing usually removes the film. In low-power circuits this action may not take place. In most applications, contacts are oversized for maximum life. Low-power circuit contacts are not oversized to the extent that they switch just a small fraction of their rated value.

Contacts are often subject to high-current surges. High-current surges reduce contact life by accelerating sulfidation and contact burning. For example, a 100 W incandescent lamp has a current rating of about 1 A. The life of the contacts is reduced if a relay with 5 A contacts is used to switch the lamp because the lamp's filament has a low resistance when cold. When first turned ON, the lamp draws 12 A or more. The 5 A relay switches the lamp, but does not switch it for the rated life of the relay. Contacts are oversized in applications that have high-current surges.

SOLID-STATE RELAYS (SSRs)

The industrial control market has moved to solid-state electronics. Due to their declining cost, high reliability, and immense capability, solid-state devices are replacing many devices that operated on mechanical and electromechanical principles. The selection of either a solid-state or electromechanical relay is based on the electrical, mechanical, and financial characteristic of each device and the required application.

SSR Switching Methods

The SSR used in an application depends on the load to be controlled. The different SSRs are designed to properly control certain loads. The four basic SSRs are the zero switching (ZS), instant ON (IO), peak switching (PS), and analog switching (AS).

Zero Switching. A *zero switching relay* is an SSR that turns ON the load when the control voltage is applied and the voltage at the load crosses zero (or within a few volts of zero). The relay turns OFF the load when the control voltage is removed and the current in the load crosses zero. See Figure 12-16.

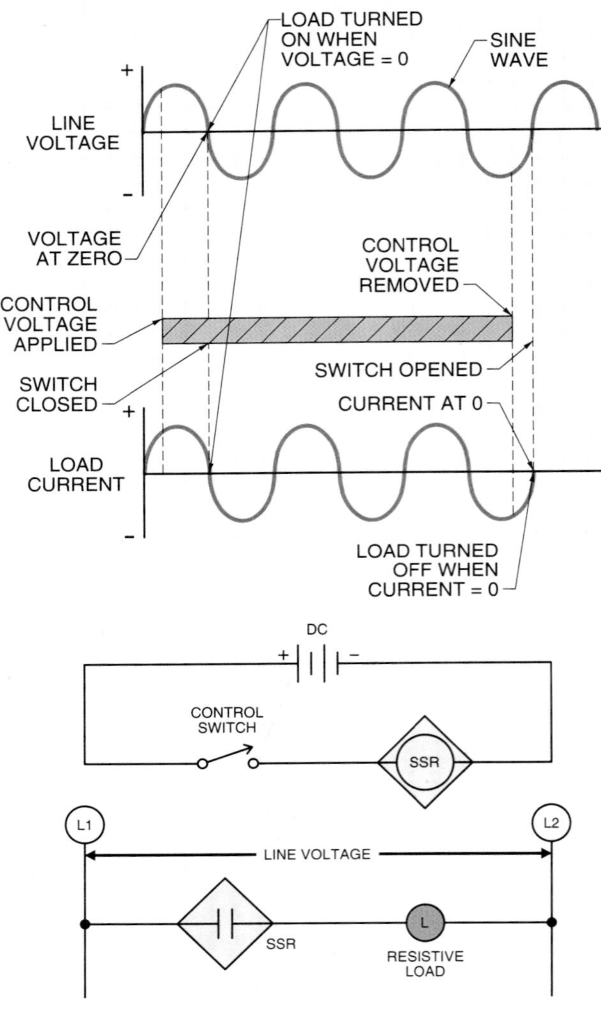

Figure 12-16. A zero switching relay turns ON the load when the control voltage is applied and the voltage at the load crosses zero.

The zero switching relay is the most widely-used relay. Zero switching relays are designed to control resistive loads. Zero switching relays control the temperature of heating elements, soldering irons, extruders for forming plastic, incubators, and ovens. Zero switching relays control the switching of incandescent lamps, tungsten lamps, flashing lamps, and programmable controller interfacing.

Instant ON. An *instant ON switching relay* is an SSR that turns ON the load immediately when the control voltage is present. This allows the load to be turned ON at any point on the AC sine wave.

The relay turns OFF when the control voltage is removed and the current in the load crosses zero. Instant ON switching is exactly like electromechanical switching because both switching methods turn ON the load at any point on the AC sine wave. See Figure 12-17.

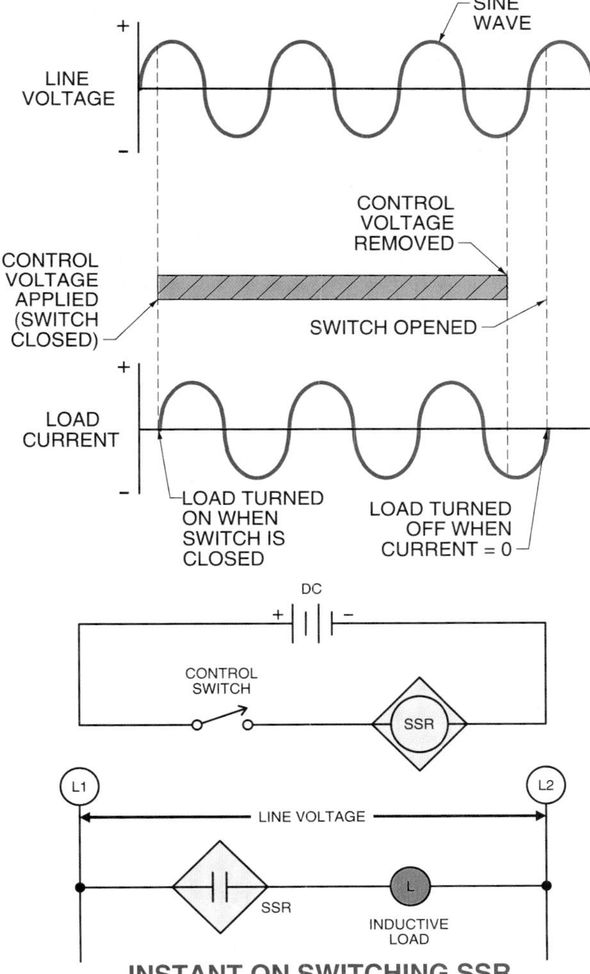

Figure 12-17. An instant ON switching relay turns ON the load immediately when the control voltage is present.

Carlo Gavazzi Inc. Electromatic Business Unit

Solid-state relays from Carlo Gavazzi Inc. use direct copper bonding technology and superior housing design to optimize heat dissipation.

Instant ON relays are designed to control inductive loads. In inductive loads, voltage and current are not in phase, and turn ON at a different point other than the zero voltage point is preferred. Instant ON relays control the switching of contactors, magnetic valves and starters, valve positioning, magnetic brakes, small motors (used for position control), 1ϕ motors, small 3ϕ motors, lighting systems (fluorescent and HID), programmable controller interfaces, and phase control (by pulsing the input).

Peak Switching. A *peak* switching relay is an SSR that turns ON the load when the control voltage is present, and the voltage at the load is at its peak. The relay turns OFF when the control voltage is removed and the current in the load crosses zero. Peak switching is preferred when the voltage and the current are about 90° out of phase because switching at peak voltage is switching at close to zero current. See Figure 12-18.

Peak switching relays control transformers and other heavy inductive loads and limit the current in the first half-period of the AC sine wave. Peak switching relays control the switching of transformers, large motors, DC loads, high inductive lamps, magnetic valves, and small DC motors.

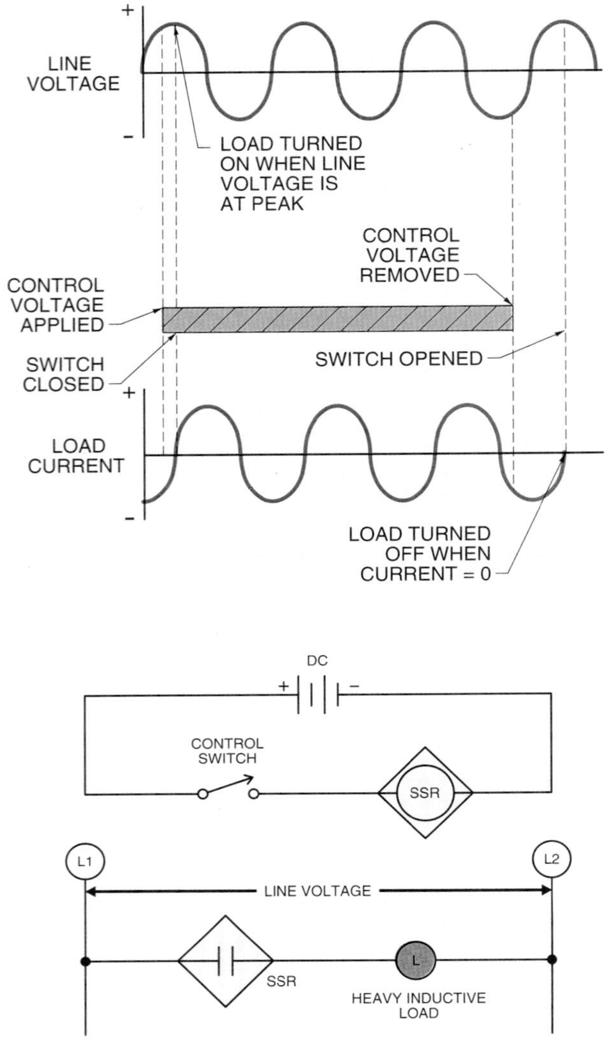

Figure 12-18. A peak switching relay turns ON the load when the control voltage is present, and the voltage at the load is at peak.

Analog Switching. An *analog switching relay* is an SSR that has an infinite number of possible output voltages within the relay's rated range. An analog switching relay has a built-in synchronizing circuit that controls the amount of output voltage as a function of the input voltage. This allows for a ramp-up function of the load. In a ramp-up function, the voltage at the load starts at a low level and is increased over a period of time. The relay turns OFF when the control voltage is removed and the current in the load crosses zero. See Figure 12-19.

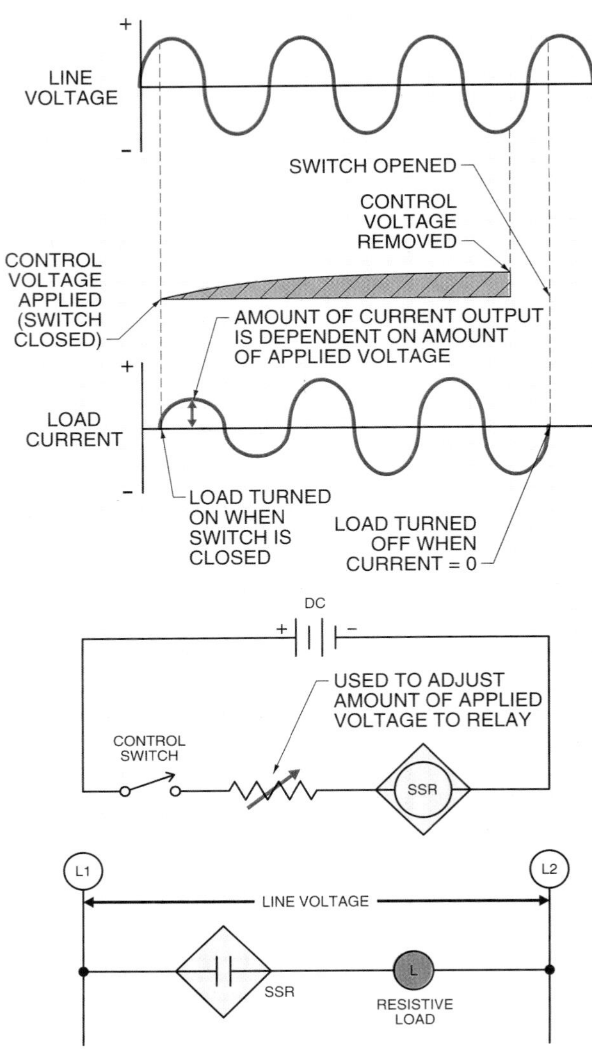

Figure 12-19. An analog switching relay has an infinite number of possible output voltages within the relay's rated range.

A typical analog switching relay has an input control voltage of 0 VDC to 5 VDC that corresponds respectively to no switching and full switching on the output load. For any voltage between 0 VDC and 5 VDC, the output is a percentage of the available output voltage. However, the output is normally nonlinear when compared to the input, and the manufacturer's data must be checked.

Analog switching relays are designed for closed-loop applications. A closed-loop application is a temperature control with feedback from a temperature sensor to the controller.

In a closed-loop system, the amount of output is directly proportional to the amount of input signal. For example, if there is a small temperature difference between the actual temperature and the set temperature, the load (heating element) is given low power. However, if there is a large temperature difference between the actual temperature and the set temperature, the load (heating element) is given high power. This relay may also be used for starting high-power incandescent lamps to reduce the inrush current.

SSR Circuits

An SSR circuit consists of an input circuit, a control circuit, and an output (load switching) circuit. These circuits may be used in any combination providing many different solid-state switching applications. See Figure 12-20.

Input Circuits. The *input circuit* of an SSR is the part of the relay to which the control component is connected. The input circuit performs the same function as the coil of an EMR. The input circuit is activated by applying a voltage to the input of the relay that is higher than the specified pickup voltage of the relay. The input circuit is deactivated when a voltage less than the specified minimum dropout voltage of the relay is applied. Some SSRs have a fixed input voltage rating, such as 12 VDC. Most SSRs have an input voltage range, such as 3 VDC to 32 VDC. The voltage range allows a single SSR to be used with most electronic circuits.

The input voltage of an SSR may be controlled (switched) through mechanical contacts, transistors, digital gates, etc. Most SSRs may be switched directly by low-power devices, which include integrated circuits, without adding external buffers or current-limiting devices. Variable input devices, such as thermistors, may also be used to switch the input voltage of an SSR.

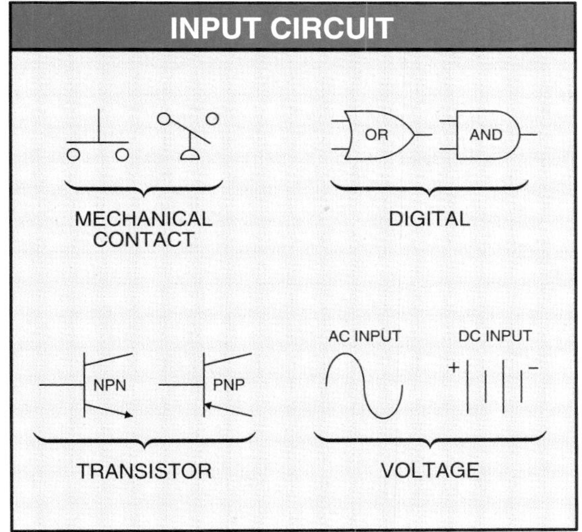

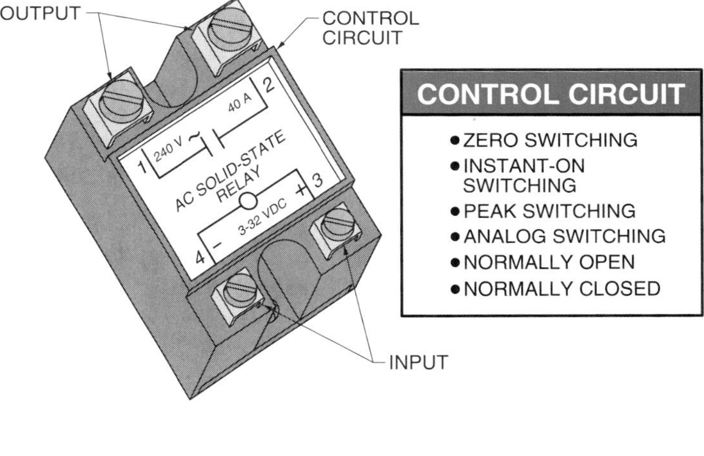

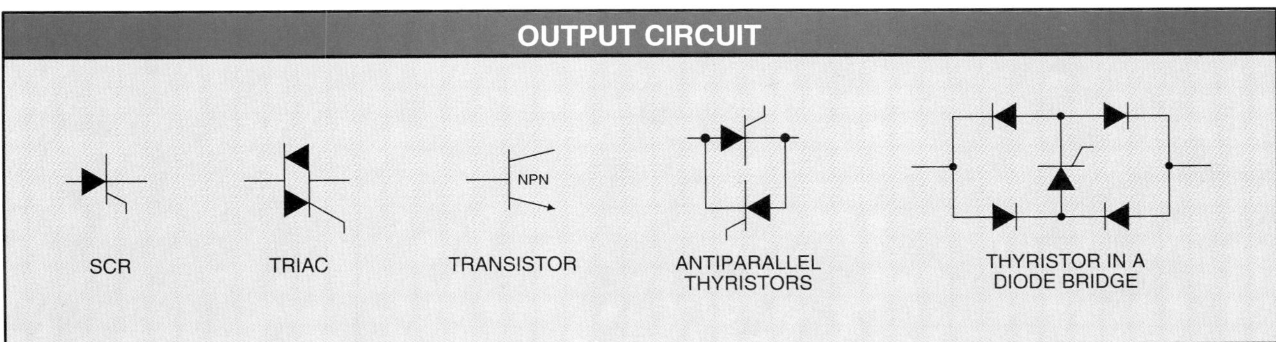

Figure 12-20. An SSR circuit consists of an input circuit, a control circuit, and an output (load switching) circuit.

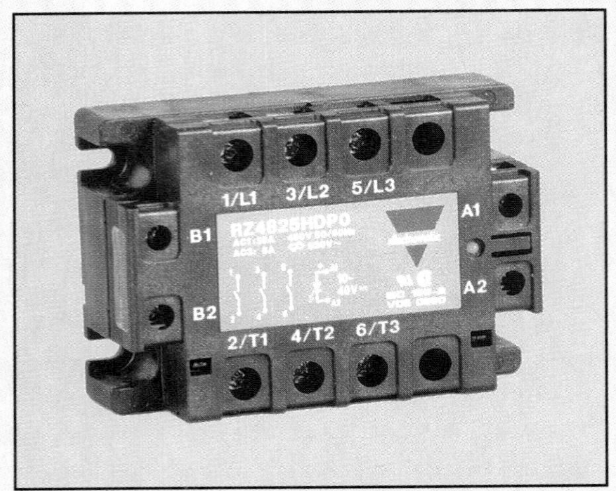

Carlo Gavazzi Inc. Electromatic Business Unit

Three-phase SSRs from Carlo Gavazzi Inc. have power switching elements for switching 3ϕ resistive loads, such as heating elements, and inductive loads, such as motors and transformers.

Control Circuits. The *control circuit* of an SSR is the part of the relay that determines when the output component is energized or de-energized. The control circuit functions as the coupling between the input and output circuits. This coupling is accomplished by an electronic circuit inside the SSR. In an EMR, the coupling is accomplished by the magnetic field produced by the coil.

The circuit is switched or not switched depending on whether the relay is a zero switching, instant ON, peak switching, or an analog switching relay when the control circuit receives the input voltage.

Each relay is designed to turn ON the load-switching circuit at a predetermined voltage point. For example, a zero switching relay allows the load to be turned ON only after the voltage across the load is at or near zero. The zero switching function provides a number of benefits, such as the elimination of high inrush currents on the load.

Output (Load Switching) Circuits. The *output (load switching) circuit* of an SSR is the load switched by the SSR. The output circuit performs the same function as the mechanical contacts of an electromechanical relay. However, unlike the multiple contact outputs of EMRs, SSRs normally have only one output contact.

Most SSRs use a thyristor as the output-switching component. Thyristors change from the OFF state (contacts open) to the ON state (contacts closed) very quickly when their gate switches ON. This fast switching action allows for high-speed switching of loads. The output switching device used depends on the type of load to be controlled. Different outputs are required when switching DC circuits than are required when switching AC circuits. Common outputs used in SSRs include:

- SCRs–Used to switch high-current DC loads.
- Triacs–Used to switch low-current AC loads.
- Transistors–Used to switch low-current DC loads.
- Antiparallel Thyristors–Used to switch high-current AC loads. They are able to dissipate more heat than a triac.
- Thyristors in Diode Bridges–Used to switch low-current AC loads.

SSR Circuit Capabilities

An SSR can be used to control most of the same circuits that an EMR is used to control. Because the SSR differs from the EMR in function, the control circuits for SSRs differ from those of the EMR. This difference is in how the relay is connected into the circuit. An SSR performs the same circuit requirements as an EMR only with a slightly different control circuit.

Two-Wire Control. An SSR may be used to control a load using a momentary control such as a pushbutton. See Figure 12-21.

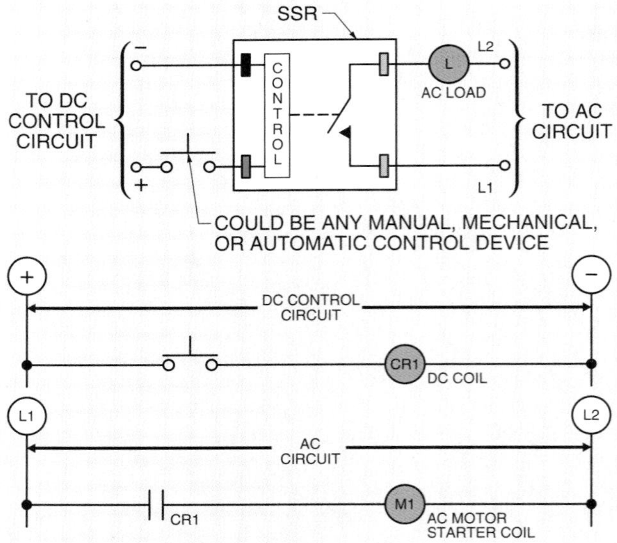

Figure 12-21. An SSR may be used to control a load using a momentary control such as a pushbutton.

In this circuit, the pushbutton signals the SSR which turns ON the load. To keep the load turned ON, the pushbutton must be held down. The load is turned OFF when the pushbutton is released. This circuit is identical in operation to the standard two-wire control circuit used with EMRs, magnetic motor starters, and contactors. For this reason, the pushbutton could be changed to any manual, mechanical, or automatic control device for simple ON/OFF operation. The same circuit may be used for liquid level control if the pushbutton is replaced with a float switch.

Three-Wire Memory Control. An SSR may be used with a silicon controlled rectifier (SCR) for latching the load ON. See Figure 12-22. This circuit is identical in operation to the standard three-wire memory control circuit. An SCR is used to add memory after the start pushbutton is pressed. An SCR acts as a current-operated OFF-to-ON switch. The SCR does not allow the DC control current to pass through until a current is applied to its gate. There must be a flow of a definite minimum current to turn the SCR ON. This is accomplished by pressing the start pushbutton. Once the gate of the SCR has voltage applied, the SCR is latched in the ON condition and allows the DC control voltage to pass through even after the start pushbutton is released. Resistor R1 is used as a current-limiting resistor for the gate, and is determined by gate current and supply voltage.

The circuit must be opened to stop the anode-to-cathode flow of DC current to the SCR. This is accomplished by pressing the stop pushbutton. Additional start pushbuttons are added in parallel with the start pushbutton. Additional stops may be added to the circuit by placing them in series with the stop pushbutton. The additional start/stops may be any manual, mechanical, or automatic control.

Equivalent NC Contacts. An SSR may be used to simulate an equivalent NC contact condition. See Figure 12-23.

Fluke Corporation

The proper heat sink for a solid-state relay is selected based on the load current to be switched and the ambient temperature of the area where the relay is mounted.

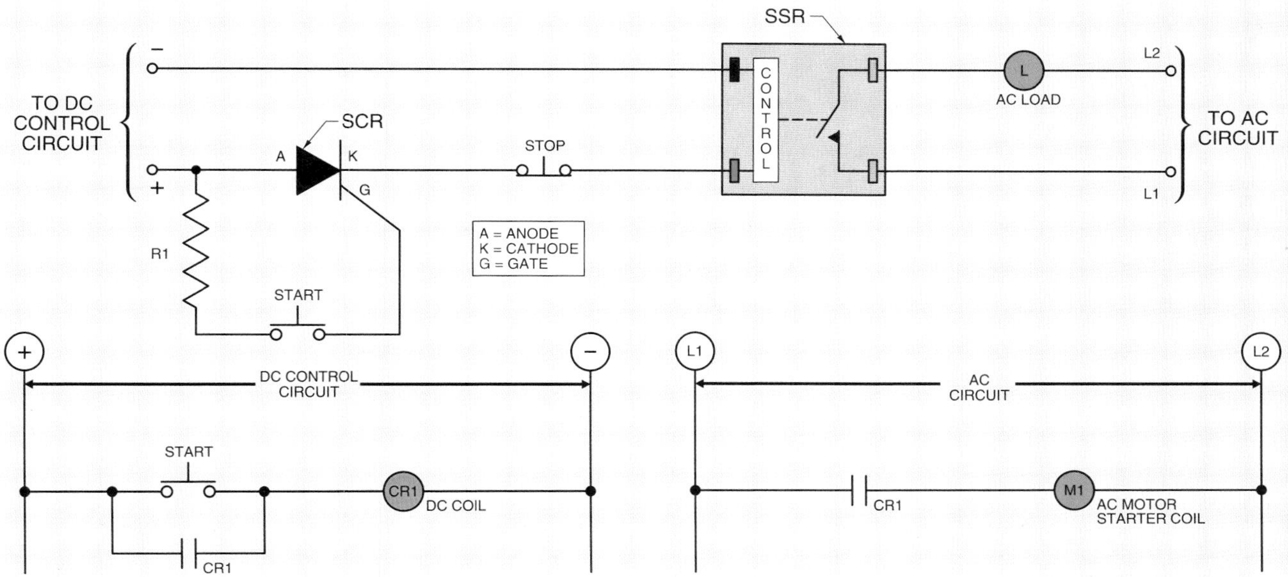

Figure 12-22. An SSR may be used with an SCR to latch a load ON.

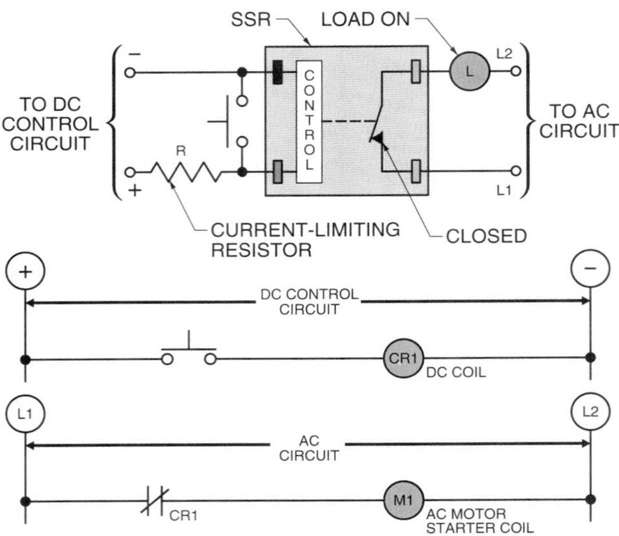

Figure 12-23. An SSR may be used to simulate an equivalent NC contact condition.

An NC contact must be electrically made because most SSRs have the equivalence of an NO contact. This is accomplished by allowing the DC control voltage to be connected to the SSR through a current-limiting resistor (R). The load is held in the ON condition because the control voltage is present on the SSR. The pushbutton is pressed to turn OFF the load. This allows the DC control voltage to take the path of least resistance and electrically remove the control voltage from the relay. This also turns OFF the load until the pushbutton is released.

Transistor Control. SSRs are also capable of being controlled by electronic control signals from integrated circuits and transistors. See Figure 12-24. In this circuit, the SSR is controlled through an NPN transistor which receives its signal from IC logic gates, etc. The two resistors (R1 and R2) are used as current-limiting resistors.

Carlo Gavazzi Inc. Electromatic Business Unit

Three-phase relays manufactured by Carlo Gavazzi Inc. may be used for direct and delta switching of motor loads and direct switching of motor loads with shunting by an electromechanical contactor.

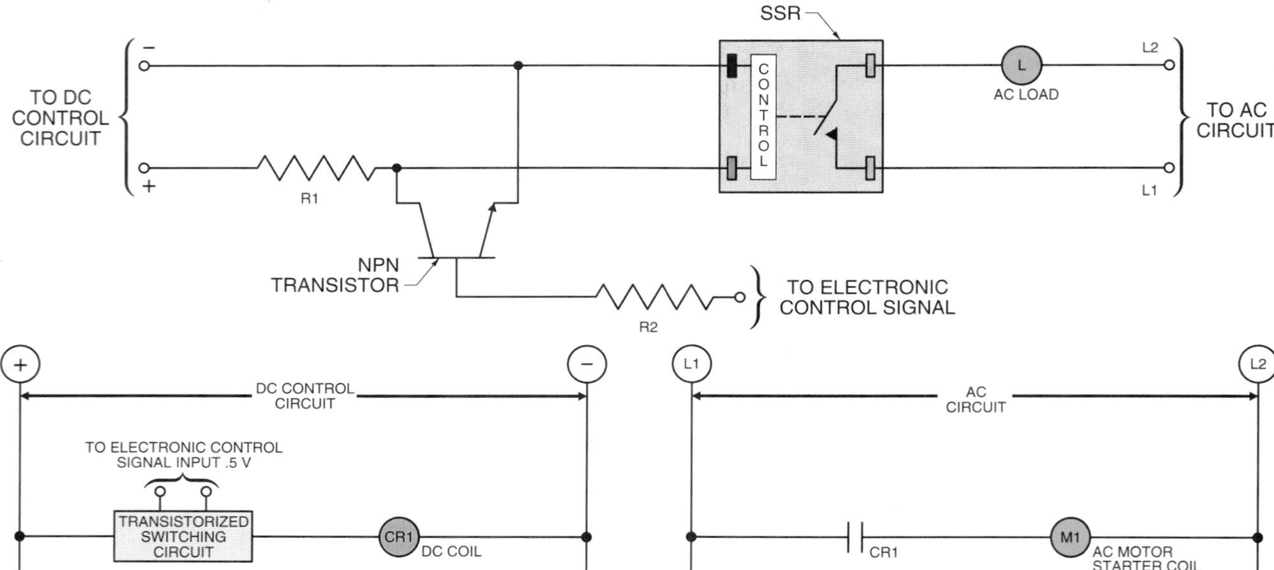

Figure 12-24. SSRs may be controlled by electronic control signals from integrated circuits and transistors.

Series and Parallel Control of SSRs. SSRs can be connected in series or parallel to obtain multi-contacts that are controlled by one input device. Multi-contact SSRs may also be used. Three SSR control inputs may be connected in parallel so that when the switch is closed all three are actuated. See Figure 12-25. This controls the 3ϕ circuit.

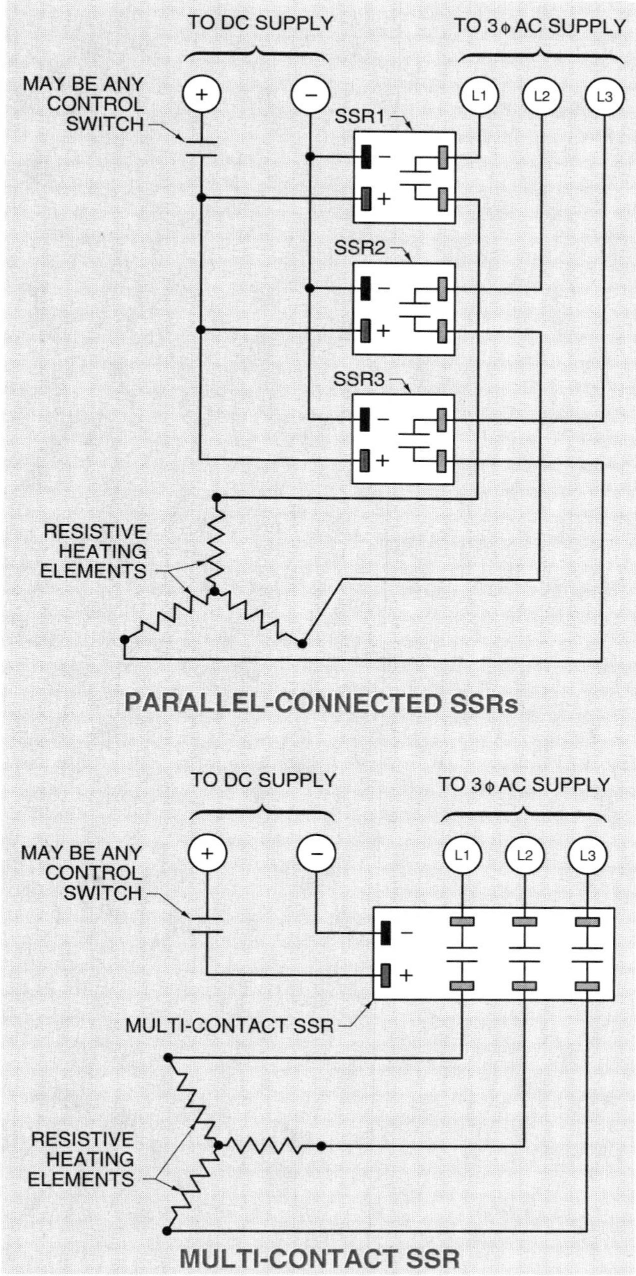

Figure 12-25. Three SSRs may be connected in parallel to control a 3ϕ circuit, or a multi-contact SSR may be used.

In this application, the DC control voltage across each SSR is equal to the DC supply voltage because they are connected in parallel. When a multi-contact SSR is used, there is only one input that controls all output switches.

SSRs can be connected in series to control a 3ϕ circuit. See Figure 12-26. The DC supply voltage is divided across the three SSRs when the switch is closed. For this reason, the DC supply voltage must be at least three times greater than the minimum operating voltage of each relay.

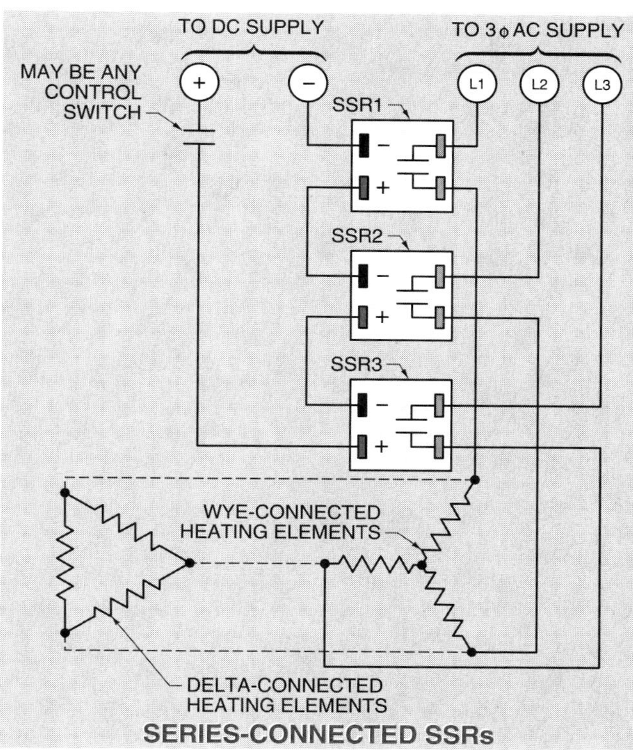

Figure 12-26. Three SSRs may be connected in series to control a 3ϕ circuit.

SSR Temperature Problems

Temperature rise is the largest problem in applications using an SSR. As temperature increases, the failure rate of SSRs increases. As temperature increases, the number of operations of an SSR decreases. The higher the heat in an SSR, the more problems occur. See Figure 12-27.

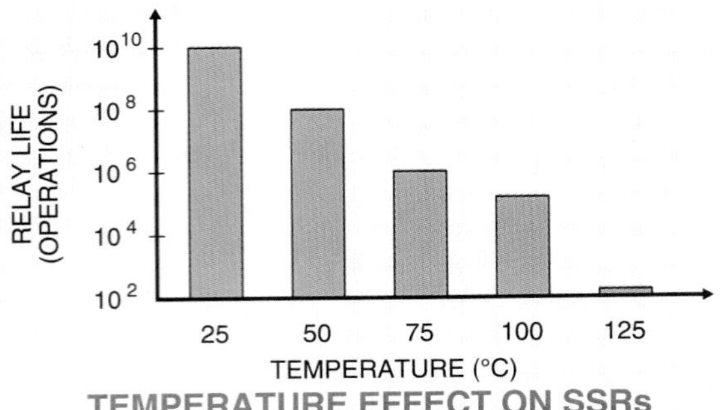

Figure 12-27. As temperature increases, the number of operations of an SSR decreases.

The failure rate of most SSRs doubles for every 10°C temperature rise above an ambient temperature of 40°C. An ambient temperature of 40°C is considered standard by most manufacturers.

Solid-state relay manufacturers specify the maximum relay temperature permitted. The relay must be properly cooled to ensure that the temperature does not exceed the specified maximum safe value. Proper cooling is accomplished by installing the SSR to the correct heat sink. A heat sink is chosen based on the maximum amount of load current controlled.

Heat Sinks

The performance of an SSR is affected by ambient temperature. The ambient temperature of a relay is a combination of the temperature of the relay location and the type of enclosure used. The temperature inside an enclosure may be much higher than the ambient temperature of an enclosure that does not allow good air flow.

The temperature inside an enclosure increases if the enclosure is located next to a heat source or in the sun. The electronic circuit and SSR also produce heat. Forced cooling is required in some applications.

Selecting Heat Sinks

A low resistance to heat flow is required to remove the heat produced by an SSR. The opposition to heat flow is thermal resistance. *Thermal resistance (R_{TH})* is the ability of a device to impede the flow of heat. Thermal resistance is a function of the surface area of a heat sink and the conduction coefficient of the heat sink material. Thermal resistance is expressed in degrees Celsius per Watt (°C/W). See Figure 12-28.

HEAT SINK SELECTIONS

Type	H × W × L (mm)	R_{TH} (°C/W)
01	15 × 79 × 100	2.5
02	15 × 100 × 100	2.0
03	25 × 97 × 100	1.5
04	37 × 120 × 100	.9
05	40 × 60 × 150	.5
06	40 × 200 × 150	.4

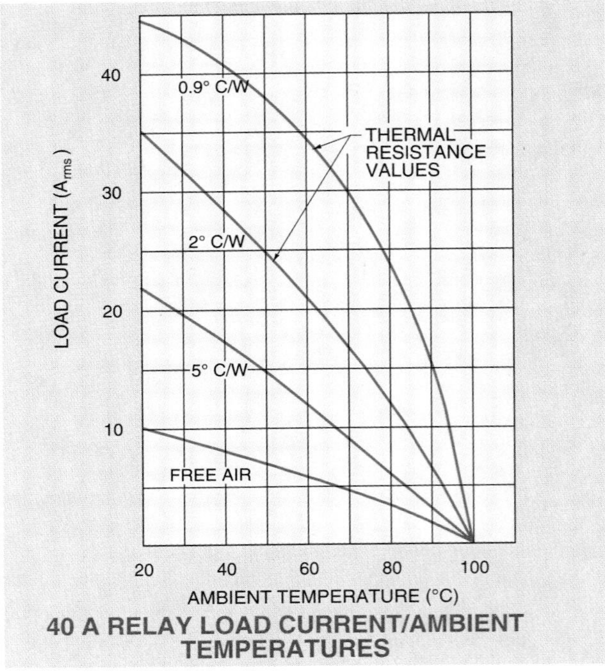

Figure 12-28. Thermal resistance (R_{TH}) is the ability of a device to impede the flow of heat.

Heat sink manufacturers list the thermal resistance of heat sinks. The lower the thermal resistance number, the easier the heat sink dissipates heat. The larger the thermal resistance number, the less effectively the heat sink dissipates heat. The thermal resistance value of a heat sink is used with an SSR Load Current/Ambient Temperature chart to determine the size of the heat sink required.

A relay can control a large amount of current when a heat sink with a low thermal resistance number is used. A relay can control the least amount of current when no heat sink (free air mounting) is used. To maximize heat conduction through a relay and into a heat sink:

- Use heat sinks made of a material that has a high thermal conductivity. Silver has the highest thermal conductivity rating. Copper has the highest practical thermal conductivity rating. Aluminum has a good thermal conductivity rating, and is the most cost effective and widely used heat sink.
- Keep the thermal path as short as possible.
- Use the largest cross-sectional surface area in the smallest space.
- Always use thermal grease or pads between the relay housing and the heat sink to eliminate air gaps and aid in thermal conductivity.

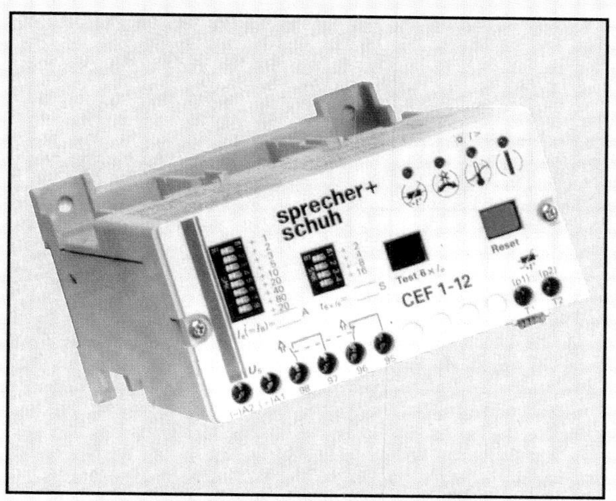

Sprecher + Schuh

Sprecher + Schuh's CEF1 electronic motor protection relay uses solid-state technology to provide accurate thermal overload protection, thermistor overtemperature protection, and overcurrent warning.

Mounting Heat Sinks

A heat sink must be correctly mounted to assure proper heat transfer. To properly mount a heat sink:

- Choose a smooth mounting surface. The surfaces between a heat sink and a solid-state device should be as flat and smooth as possible. Ensure that the mounting bolts and screws are securely tightened.
- Locate heat producing devices so that the temperature is spread over a large area. This helps prevent increased temperature areas.
- Use heat sinks with fins to maintain as large a surface area as possible.
- Ensure that the heat from one heat sink does not add to the other.

Relay Current Problems

The overcurrent passing through an SSR must be kept below the maximum load current rating of the relay. An overload protection fuse is used to prevent overcurrents from damaging an SSR.

An overload protection fuse opens the circuit when the current is increased to a higher value than the nominal load current. The fuse should be an ultra-fast fuse used for the protection of semiconductors. See Figure 12-29.

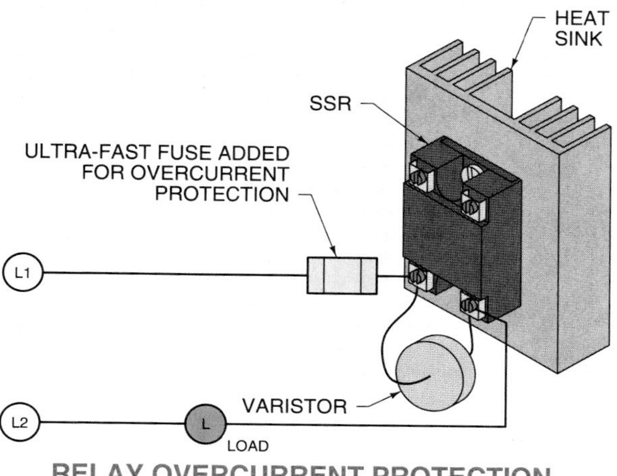

Figure 12-29. An overload protection fuse opens the circuit when the current is increased to a higher value than the nominal load current.

Relay Voltage Problems

Most AC power lines contain voltage spikes superimposed on the voltage sine wave. Voltage spikes are produced by switching motors, solenoids, transformers, motor starters, contactors, and other inductive loads. Large spikes are also produced by lightning striking the power distribution system.

The output element of a relay can exceed its breakdown voltage and turn ON for part of a half period if overvoltage protection is not provided. This short turn ON can cause problems in the circuit.

Varistors are added to the relay output terminals to prevent an overvoltage problem. A varistor should be rated 10% higher than the line voltage of the output circuit. The varistor bypasses the transient current. See Figure 12-30.

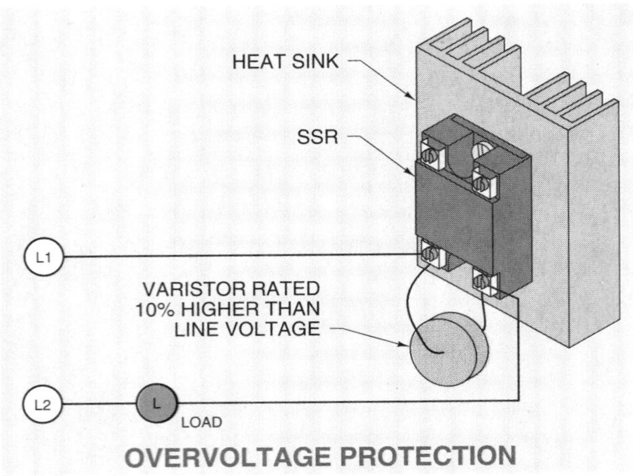

Figure 12-30. Varistors are added to relay output terminals to prevent an overvoltage problem.

Voltage Drop

A voltage drop in the switching component is unavoidable in an SSR. The voltage drop produces heat. The larger the current passing through the relay, the greater amount of heat produced. The generated heat affects relay operation and can destroy the relay if not removed. See Figure 12-31.

The voltage drop in an SSR is usually 1 V to 1.6 V, depending on the load current. For small loads (less than 1 A), the heat produced is safely dissipated through the relay's case. High-current loads require a heat sink to dissipate the extra heat. See Figure 12-32.

For example, if the load current in a circuit is 1 A and the SSR switching device has a 2 V drop, the power generated is 2 W. The 2 W of power generates heat that can be dissipated through the relay's case.

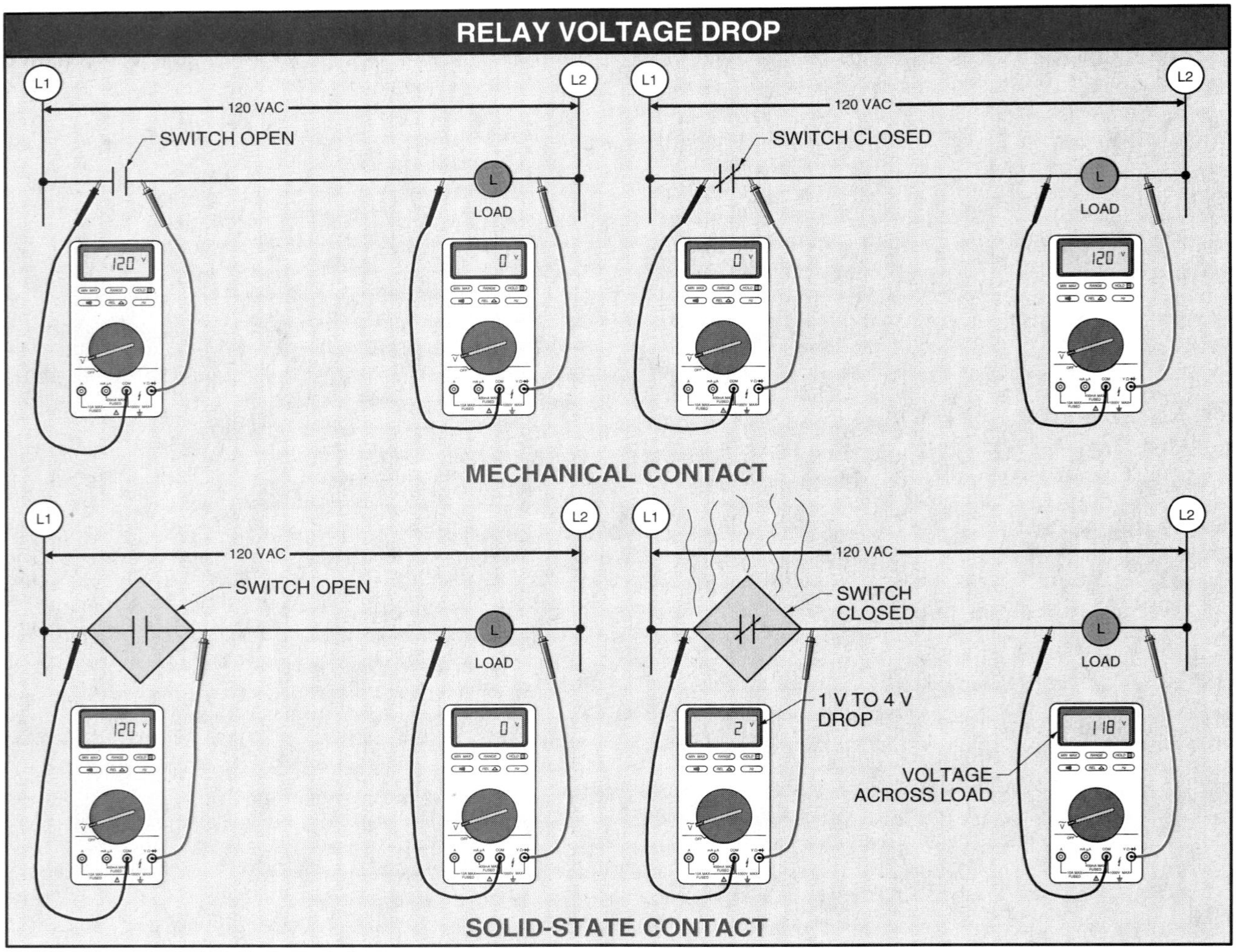

RELAY VOLTAGE DROP

L1 — 120 VAC — L2
SWITCH OPEN
LOAD
SWITCH CLOSED
MECHANICAL CONTACT
1 V TO 4 V DROP
VOLTAGE ACROSS LOAD
SOLID-STATE CONTACT

Figure 12-31. The voltage drop in the switching component of an SSR produces heat which can destroy the relay if not removed.

SSR VOLTAGE DROP		
Load Current (in A)	Voltage Drop (in V)	Power at Switch (in W)
1	2	2
2	2	4
5	2	10
10	2	20
20	2	40
50	2	100

Figure 12-32. For small loads (less than 1 A), the heat produced in an SSR is safely dissipated through the relay's case.

If the load current in a circuit is 20 A and the SSR switching device has a 2 V drop, the power generated in the device is 40 W. The 40 W of power generates heat that requires a heat sink to safely dissipate the heat.

Electromechanical and Solid-State Relay Comparison

EMRs and SSRs are designed to provide a common switching function. An EMR provides switching through the use of electromagnetic devices and sets of contacts. An SSR depends on electronic devices such as SCRs and triacs to switch without contacts. In addition, the physical features and operating characteristics are different in EMRs and SSRs. See Figure 12-33.

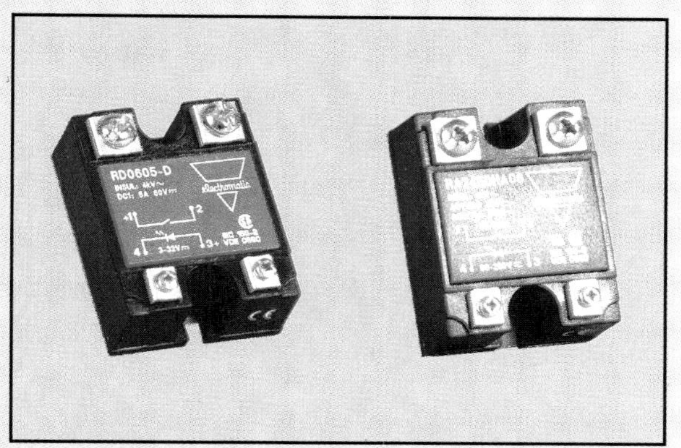

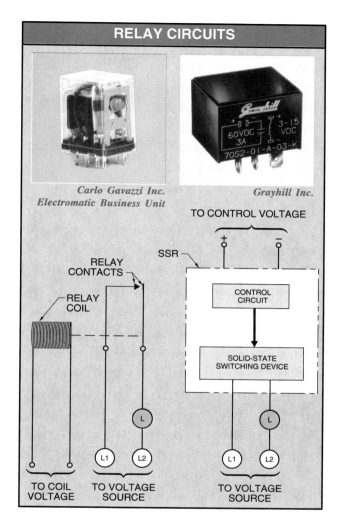

Carlo Gavazzi Inc. Electromatic Business Unit

Carlo Gavazzi Inc. manufactures industrial 1ϕ AC and DC solid-state relays that have current ratings of 10 A, 25 A, and 50 A AC and 1 A and 5 A DC.

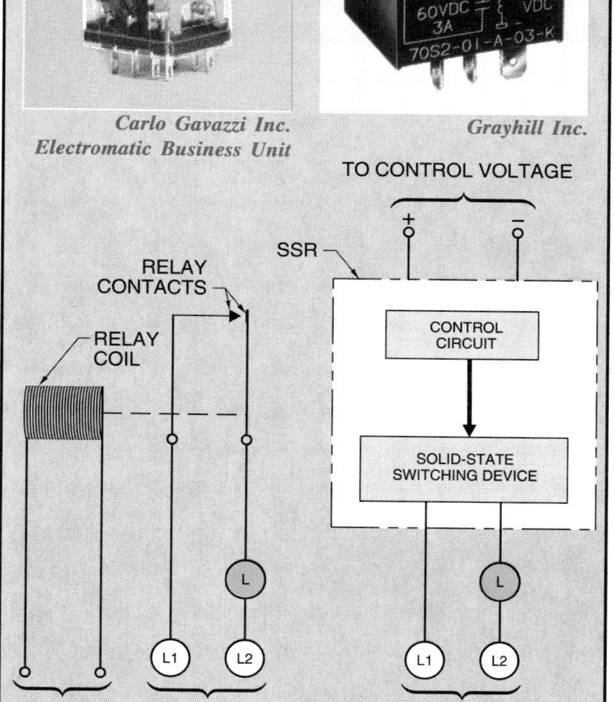

Figure 12-33. An EMR provides switching using electromagnetic devices. An SSR depends on SCRs and triacs to switch without contacts.

An equivalent terminology chart is used as an aid in the comparison of EMRs and SSRs. Because the basic operating principles and physical structures of the devices are so different, it is difficult to find a direct comparison of the two. Differences arise almost immediately both in the terminology used to describe the devices and in their overall ability to perform certain functions. See Figure 12-34.

Advantages and Limitations

EMRs and SSRs are used in many applications. The relay used depends on the application's electrical requirements, cost requirements, and life expectancy.

EMR/SSR EQUIVALENT TERMINOLOGY CHART			
EMRs		SSRs	
Term	Definition	Term	Definition
Coil Voltage	Minimum voltage necessary to energize or operate relay. Also referred to as pick-up voltage	Control Voltage	Minimum voltage required to gate or activate control circuit of SSR. Generally, a maximum value is also specified
Coil Current	Amount of current necessary to energize or operate relay	Control Current	Minimum current required to turn ON solid-state control circuit. Generally, a maximum value is also specified
Holding Current	Minimum current required to keep a relay energized	Control Current	
Drop-Out Voltage	Maximum voltage at which the relay is no longer energized	Control Voltage	
Pull-in Time	Amount of time required to operate (open or close) relay contacts after coil voltage is applied	Turn-ON Time	Elapsed time between application of control voltage and application of voltage to load circuit
Drop-Out Time	Amount of time required for the relay contacts to return to their normal de-energized position after coil voltage is removed	Turn-OFF Time	Elapsed time between removal of control voltage and removal of voltage from load circuit
Contact Voltage Rating	Maximum voltage rating that contacts of relay are capable of safely switching	Load Voltage	Maximum output voltage handling capability of an SSR
Contact Current Rating	Maximum current rating that contacts of relay are capable of safely switching	Load Current	Maximum output current handling capability of an SSR
Surge Current	Maximum peak current which contacts on a relay can withstand for short periods of time without damage	Surge Current	Maximum peak current which an SSR can withstand for short periods of time without damage
Contact Voltage Drop	Voltage drop across relay contacts when relay is operating (usually low)	Switch-ON Voltage Drop	Voltage drop across SSR when operating
Insulation Resistance	Amount of resistance measured across relay contacts in open position	Switch-OFF Resistance	Amount of resistance measured across an SSR when turned OFF
No equivalent or comparison		OFF-State Leakage Current	Amount of leakage current through SSR when turned OFF but still connected to load voltage
No equivalent or comparison		Zero Current Turn-OFF	Turn-OFF at zero crossing of load current that flows through an SSR. A thyristor turns OFF only when current falls below minimum holding current. If input control is removed when current is a higher value, turn-OFF is delayed until next zero current crossing
No equivalent or comparison		Zero Voltage Turn-ON	Initial turn-ON occurs at a point near zero crossing of AC line voltage. If input control is applied when line voltage is at a higher value, initial turn-ON is delayed until next zero crossing

Figure 12-34. An equivalent terminology chart is used as an aid in the comparison of EMRs and SSRs.

Although SSRs are replacing EMRs in many applications, EMRs are still very common. EMRs offer many advantages that make them very cost-effective. However, they do have disadvantages that limit their use in some applications. SSRs provide many advantages such as small size, fast switching, long life, and the ability to handle complex switching requirements. SSRs have some limitations that restrict their use in some applications. The advantages and limitations include:

EMR Advantages

- Normally have multipole, multithrow contact arrangements
- Contacts can switch AC or DC
- Low initial cost
- Very low contact voltage drop, thus no heat sink is required
- Very resistant to voltage transients
- No OFF-state leakage current through open contacts

EMR Limitations

- Contacts wear, thus have a limited life
- Short contact life when used for rapid switching applications or high current loads
- Generates electromagnetic noise and interference on the power lines
- Poor performance when switching high inrush currents

SSR Advantages

- Very long life when properly applied
- No contacts to wear
- No contact arcing to generate electromagnetic interference
- Resistant to shock and vibration because they have no moving parts
- Logic compatible to programmable controllers, digital circuits, and computers
- Very fast switching capability
- Different switching modes (zero switching, instant-ON, etc.)

SSR Limitations

- Normally only one contact available per relay
- Heat sink required due to voltage drop across switch
- Can switch only AC or DC
- OFF-state leakage current when switch is open
- Normally limited to switching only a narrow frequency range such as 40 Hz to 70 Hz

Input Signals. Applying a voltage to the input coil of an electromagnetic device creates an electromagnet which is capable of pulling in an armature with a set of contacts attached to control a load circuit. It takes more voltage and current to pull in the coil than to hold it in due to the initial air gap between the magnetic coil and the armature. The specifications used to describe the energizing and de-energizing process of an electromagnetic device are coil voltage, coil current, holding current, and drop-out voltage.

An SSR has no coil or contacts and requires only minimum values of voltage and current to turn it ON and turn it OFF. The two specifications needed to describe the input signal for an SSR are control voltage and control current.

The electronic nature of an SSR and its input circuit allows easy compatibility with digitally controlled logic circuits. Many SSRs are available with minimum control voltages of 3 V and control currents as low as 1 mA, making them ideal for a variety of current state-of-the-art logic circuits.

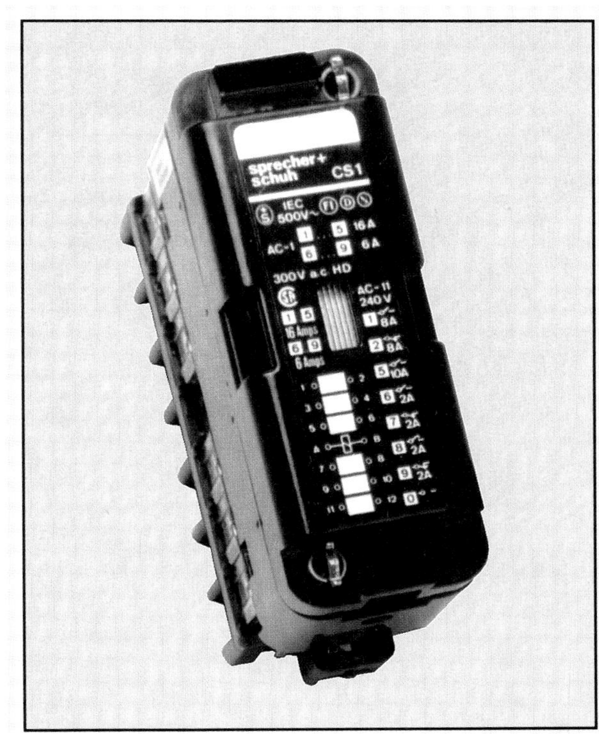

Sprecher + Schuh

The CS1 industrial control relay from Sprecher + Schuh allows 2 A, 10 A, or 16 A contact elements to be rearranged into one plug-in relay of up to six poles.

Response Time. One of the significant advantages of an SSR over an EMR is its response time (ability to turn ON and turn OFF). An EMR may be able to respond hundreds of times per minute. An SSR is capable of switching thousands of times per minute with no chattering or bounce.

Omron Electronics, Inc.

Electromechanical relays can be used without heat sinks, but solid-state relays require heat sinks to dissipate the heat produced by the relay.

DC switching times for an SSR are in the microsecond range, while AC switching time, with the use of zero-voltage turn-ON, is less than 9 ms. The reason for this advantage is that the SSR may be turned ON and turned OFF electronically much more rapidly than a relay may be electromagnetically pulled in and dropped out.

The higher speed of SSRs has become increasingly more important as industry demands higher productivity from processing equipment. The more rapidly the equipment can process or cycle its output, the greater the productivity of the machine.

Voltage and Current Ratings. Electromechanical relays and SSRs have certain limitations which determine how much voltage and current each device can safely handle. The values vary from device to device and from manufacturer to manufacturer. Data sheets are used to determine if a given device can safely switch a given load. The advantages of SSRs are that they have a capacity for arcless switching, have no moving parts to wear out, and are totally enclosed, thus being able to be operated in potentially explosive environments without special enclosures.

The advantage of EMRs is the possibility for replacement of contacts when the device receives an excessive surge current. In an EMR, the contacts may be replaced. In an SSR, the complete device must be replaced.

Voltage Drop. When a set of contacts on an EMR closes, the contact resistance is normally low unless the contacts are pitted or corroded. The SSR, however, being constructed of semiconductor materials, opens and closes a circuit by increasing or decreasing its ability to conduct. Even at full conduction, the device presents some residual resistance which can create a voltage drop of up to approximately 1.5 V in the load circuit. This voltage drop is usually considered insignificant because it is small in relation to the load voltage, and in most cases presents no problems. This unique feature may have to be taken into consideration when load voltages are small. A method of removing the heat produced at the switching device must be used when load currents are high.

Omron Electronics, Inc.

Electromechanical relays are normally available with several NO and NC contacts to control several different circuits.

Insulation and Leakage. The air gap between a set of open contacts provides an almost infinite resistance through which no current flows. SSRs, because of their unique construction, provide a very high but measurable resistance when turned OFF. SSRs have a switched OFF resistance not found on EMRs.

It is possible for small amounts of current (OFF-state leakage) to pass through an SSR because some conductance is still possible through an SSR even though it is turned OFF. OFF-state leakage current is not found on EMRs.

Off-state leakage current is the amount of current that leaks through an SSR when the switch is turned OFF, normally about 2 mA to 10 mA. The rating of OFF-state leakage current in an SSR is usually determined at 200 VDC across the output and should not usually exceed more than 200 mA at this voltage.

This leakage current normally presents no problem unless the load device is affected by low values of leakage current. For example, small neon indicator lights and some programmable controllers cannot be switched OFF, and remain ON because of the leakage current.

TROUBLESHOOTING RELAYS

When troubleshooting EMRs, the input and output of the relay are checked to determine if the circuit on the input side of the relay is the problem, if the circuit on the output side of the relay is the problem, or if the relay itself is the problem. The relay coil and contacts are checked to determine if the relay is the problem. The correct voltage must be applied to the relay's coil before it energizes. The relay contacts are checked by energizing and de-energizing the coil. The contacts should have little to no voltage drop across them when closed. The contacts should have nearly full voltage across them when open.

SSRs require periodic inspection. Dirt, burning, or cracking should not be present on an SSR. Printed circuit (PC) boards should be properly seated. Ensure that the board locking tabs are in place if used. Consider adding locking tabs if a PC board without locking tabs loosens. Check to ensure that any cooling provisions are working and free of obstructions.

Troubleshooting EMRs

Check for contact sticking or binding if the relay is not functioning properly. Tighten any loose parts. Replace any broken, bent, or badly worn parts. Check all contacts for signs of excessive wear and dirt buildup. Contacts are not harmed by discoloration or slight pitting. Vacuum or wipe contacts with a soft cloth to remove dirt. Never use a contact cleaner on relay contacts. Contacts require replacement when the silver surface has become badly worn. Replace all contacts when severe contact wear is evident on any contact. Replacing all contacts prevents uneven and unequal contact closing. Never file a contact.

Relay coils should be free of cracks and burn marks. Replace the coil if there is any evidence of overheating, cracking, melting, or burning. Check the coil terminals for the correct voltage level. Overvoltage or undervoltage conditions of more than 10% should be corrected. Use only replacement parts recommended by the manufacturer when replacing parts of a relay. Using non-approved parts can void the manufacturer's warranty and may transfer product liability from the manufacturer. Relays are tested by manual operation and by using a multimeter.

Manual Relay Operation

Most relays can be manually operated. Manually operating a relay determines whether the circuit that the relay is controlling (output side) is working correctly. A relay is manually operated by pressing down at a designated area on the relay. This closes the relay contacts. Electromechanical relays may include a push-to-test button. See Figure 12-35.

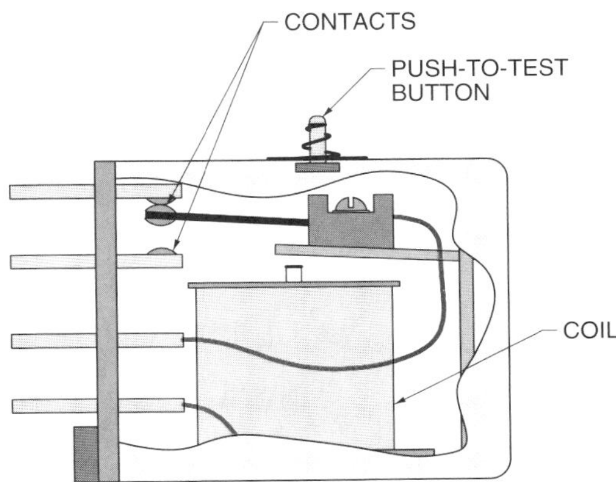

MANUAL RELAY OPERATION

Figure 12-35. Manually operating a relay determines whether the circuit that the relay is controlling (output side) is working correctly.

When manually operating relay contacts, the circuit controlling the coil is bypassed. Troubleshoot from the relay through the control circuit when the load controlled by the relay operates manually. Troubleshoot the circuit that the relay is controlling if the load controlled by the relay does not operate when the relay is manually operated.

Multimeter Test

A multimeter is also used to test an electromechanical relay. A multimeter is connected across the input and output side of a relay. Troubleshoot from the input of the relay through the control circuit when no voltage is present at the input side of the relay. The relay is the problem if the relay is not delivering the correct voltage.

Troubleshoot from the output of the relay through the power circuit when the relay is delivering the correct voltage. The supply voltage measured across an open contact indicates that the multimeter is completing the circuit across the contact. The contacts are not closing and the relay is defective if the voltage measured across the contact remains at full voltage when the coil is energized and de-energized. The contacts are welded closed and the relay is defective if the voltage measured across the contacts remains zero (or very low) when the coil is energized and de-energized. See Figure 12-36.

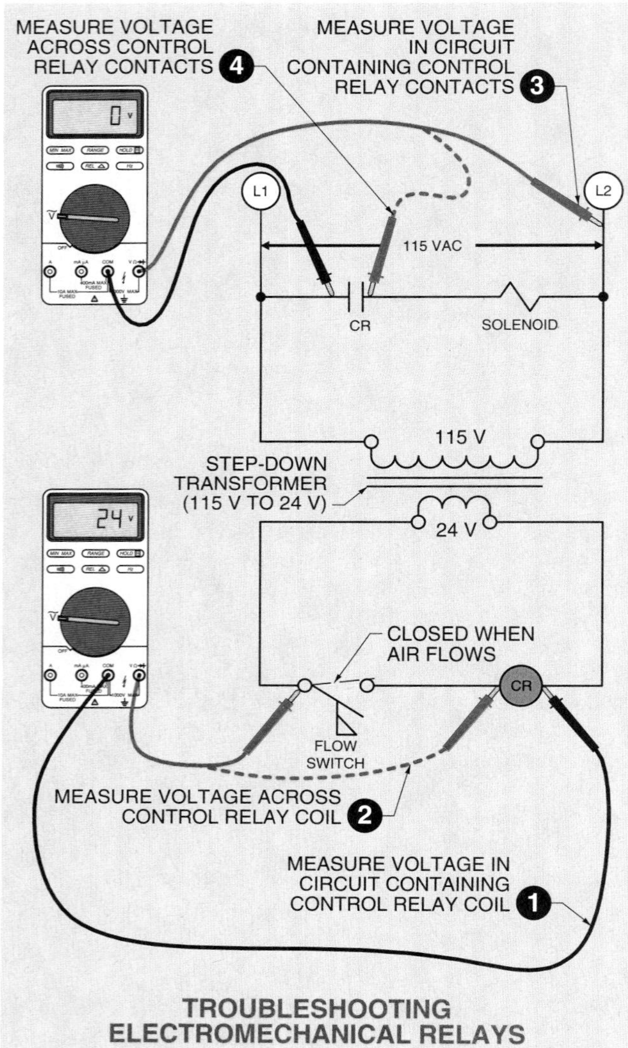

Figure 12-36. A multimeter is connected across the input and output side to test an EMR.

To troubleshoot an electromechanical relay, apply the procedure:

1. Measure the voltage in the circuit containing the control relay coil. The voltage should be within 10% of the voltage rating of the coil. The relay coil cannot energize if the voltage is not present. The coil may not energize properly if the voltage is not at the correct level. Troubleshoot the power supply when the voltage level is incorrect.
2. Measure the voltage across the control relay coil. The voltage across the coil should be within 10% of the coil's rating. Troubleshoot the switch controlling power to the coil when the voltage level is incorrect.
3. Measure the voltage in the circuit containing the control relay contacts. The voltage should be within 10% of the rating of the load. Troubleshoot the power supply if the voltage level is incorrect.
4. Measure the voltage across the control relay contacts. The voltage across the contacts should be less than 1 V when the contacts are closed and nearly equal to the supply voltage when open. The contacts have too much resistance and are in need of service if the voltage is more than 1 V when the contacts are closed. Troubleshoot the load when the voltage is correct at the contacts and the circuit does not work.

Troubleshooting SSRs

Troubleshooting an SSR is accomplished by either the exact replacement method or the circuit analysis method. The *exact replacement method* is a method of SSR replacement which a bad relay is replaced with a relay of the same type and size. The exact replacement method involves making a quick check of the relay's input and output voltages. The relay is assumed to be the problem and is replaced when there is only an input voltage being switched.

The *circuit analysis method* is a method of SSR replacement in which a logical sequence is used to determine the reason for the failure. Steps are taken to prevent the problem from recurring once the reason for a failure is known. The circuit analysis method of troubleshooting is based on three improper relay operations, which are:

- The relay fails to turn OFF the load
- The relay fails to turn ON the load
- Erratic relay operation

Relay Fails to Turn OFF Load

A relay may not turn OFF the load to which it is connected when a relay fails. This condition occurs either when the load is drawing more current than the relay can withstand, the relay's heat sink is too small, or transient voltages are causing a breakover of the relay's output.

Transient voltages are temporary, unwanted voltages in an electrical circuit. Overcurrent permanently shorts the relay's switching device if the load draws more current than the rating of the relay. High temperature causes thermal runaway of the relay's switching device if the heat sink does not remove the heat. Replace the relay with one of a higher voltage rating and/or add a transient suppression device to the circuit if the power lines are likely to have transients (usually from inductive loads connected on the same line). See Figure 12-37.

To troubleshoot an SSR that fails to turn OFF a load, apply the procedure:

1. Disconnect the input leads from the SSR. See Step 3 if the relay load turns OFF. The relay is the problem if the load remains ON and the relay is normally open.
2. Measure the voltage of the circuit that the relay is controlling. The line voltage should not be higher than the rated voltage of the relay. Replace the relay with a relay that has a higher voltage rating if the line voltage is higher than the relay's rating. Check to ensure that the relay is rated for the type of line voltage (AC or DC) being used.
3. Measure the current drawn by the load. The current draw must not exceed the relay's rating. For most applications, the current draw should not be more than 75% of the relay's maximum rating.
4. Reconnect the input leads and measure the input voltage to the relay at the time when the control circuit should turn the relay OFF. The control circuit is the problem and needs to be checked if the control voltage is present. The relay is the problem if the control voltage is removed and the load remains ON. Before changing the relay, ensure that the control voltage is not higher than the relay's rated limit when the control circuit delivers the supply voltage. Ensure that the control voltage is not higher than the relay's rated drop-out voltage when the control circuit removes the supply voltage. This condition may occur in some control circuits using solid-state switching.

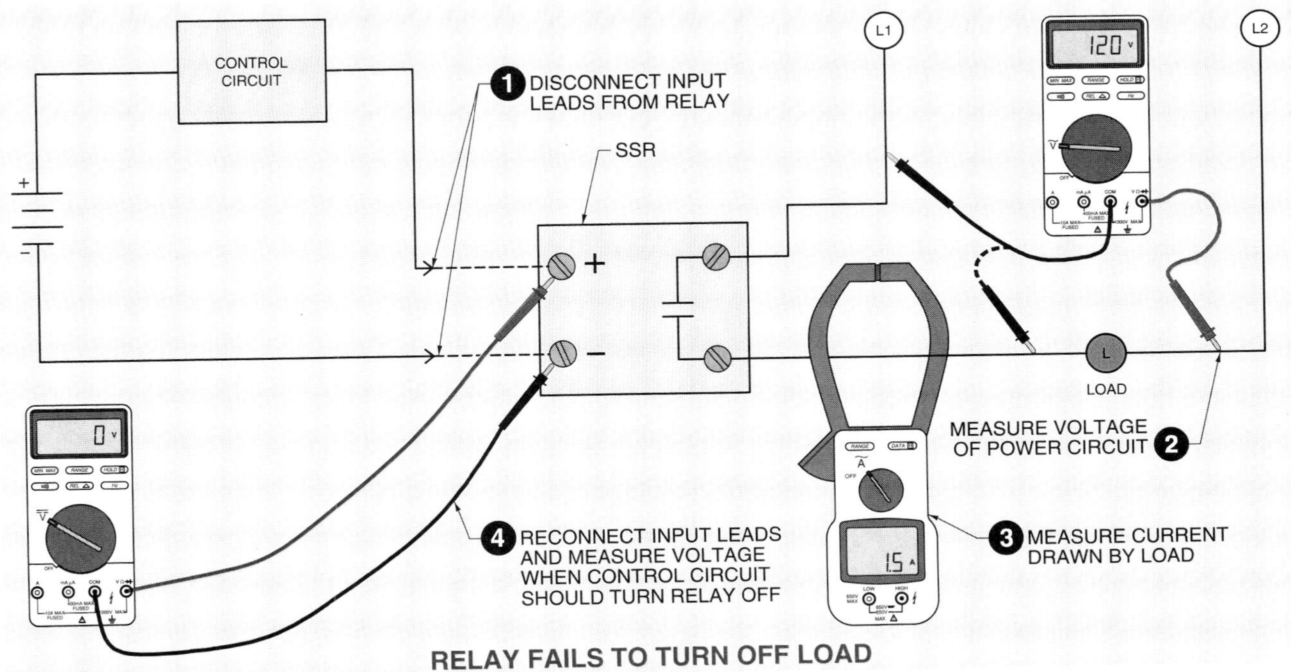

Figure 12-37. A relay may not turn OFF the load to which it is connected when the relay fails.

Relay Fails to Turn ON Load

A relay may fail to turn ON the load to which it is connected when the relay fails. This condition occurs when the relay's switching device receives a very high voltage spike or the relay's input is connected to a higher-than-rated voltage. A high voltage spike blows open the relay's switching device, preventing the load from turning ON. Excessive voltage on the relay's input side destroys the relay's electronic circuit.

Replace the relay with one that has a higher voltage and current rating and/or add a transient-voltage-suppression device to the circuit if the power lines are likely to have high voltage spikes. See Figure 12-38.

To troubleshoot an SSR that fails to turn ON a load, apply the procedure:

1. Measure the input voltage when the relay should be ON. Troubleshoot the circuit ahead of the relay's input if the voltage is less than the relay's rated pick-up voltage. The circuit ahead of the relay is the problem if the voltage is greater than the relay's rated pick-up voltage. The higher voltage may have destroyed the relay. The relay may be a secondary problem caused by the primary problem of excessive applied voltage. Correct the high-voltage problem before replacing the relay. The relay or output circuit is the problem if the input voltage is within the pick-up limits of the relay.
2. Measure the voltage at the output of the relay. The relay is probably the problem if the relay is not switching the voltage. See Step 3. The problem is in the output circuit if the relay is switching the voltage. Check for an open circuit in the load.
3. Insert an ammeter in series with the input leads of the relay. Measure the current when the relay should be ON. The relay input is open if no current is flowing. Replace the relay. The relay is bad if the current flow is within the relay's rating. Replace the relay. The control circuit is the problem if current is flowing but is less than that required to operate the relay.

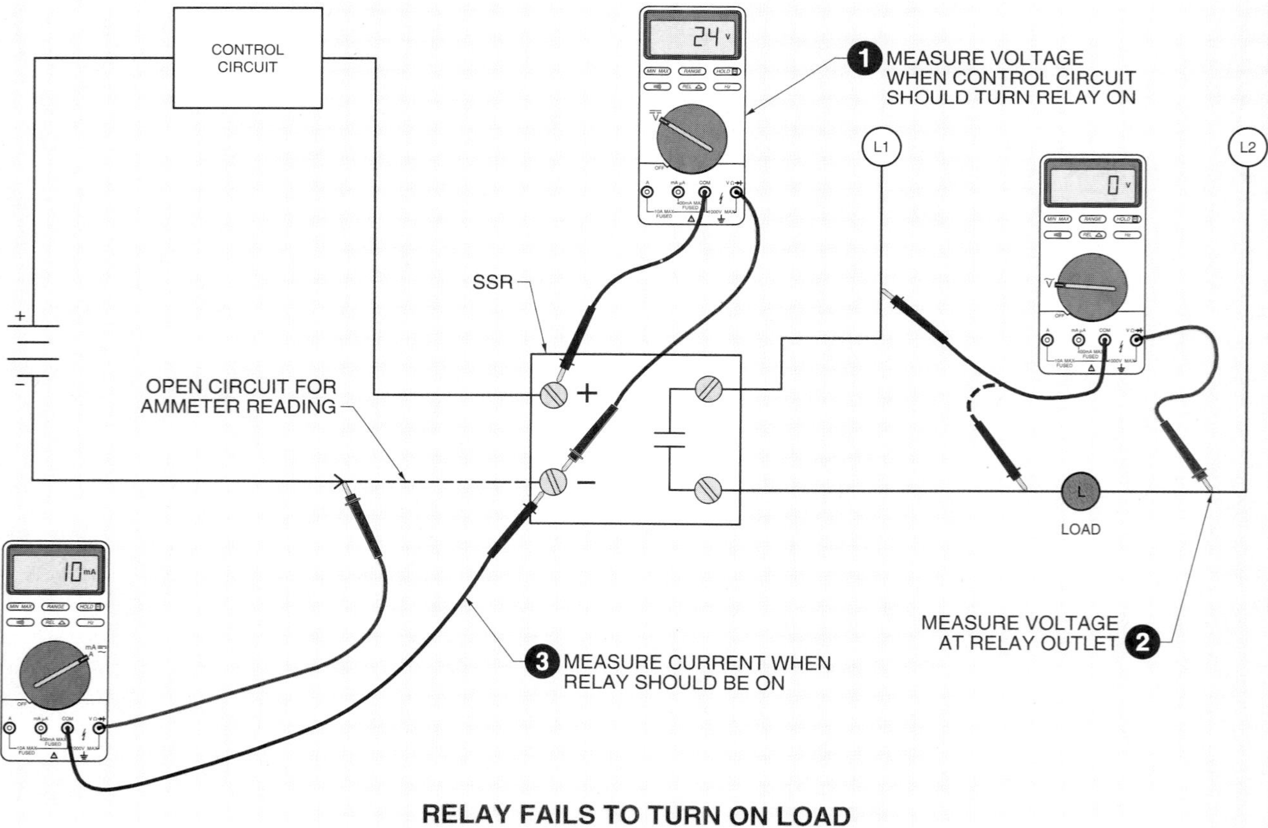

Figure 12-38. A relay may fail to turn ON the load to which it is connected when the relay fails.

Erratic Relay Operation

Erratic relay operation is the proper operation of a relay at times, and the improper operation of the relay at other times. Erratic relay operation is caused by mechanical problems (loose connections), electrical problems (incorrect voltage), or environmental problems (high temperature). See Figure 12-39.

To troubleshoot erratic relay operation, apply the procedure:

1. Check all wiring and connections for proper wiring and tightness. Loose connections cause many erratic problems. No sign of burning should be present at any terminal. Burning at a terminal usually indicates a loose connection.
2. Ensure that the input control wires are not next to the output line or load wires. The noise carried on the output side may cause unwanted input signals.
3. The relay may be half-waving if the load is a chattering AC motor or solenoid. *Half-waving* occurs when a relay fails to turn OFF because the current and voltage in the circuit reach zero at different times. Half-waving is caused by the phase shift inherent in inductive loads. The phase shift makes it difficult for some solid-state relays to turn OFF. Connecting an RC or another snubber circuit across the output load should allow the relay to turn OFF. An *RC circuit* is a circuit in which resistance (R) and capacitance (C) are used to help filter the power in a circuit.

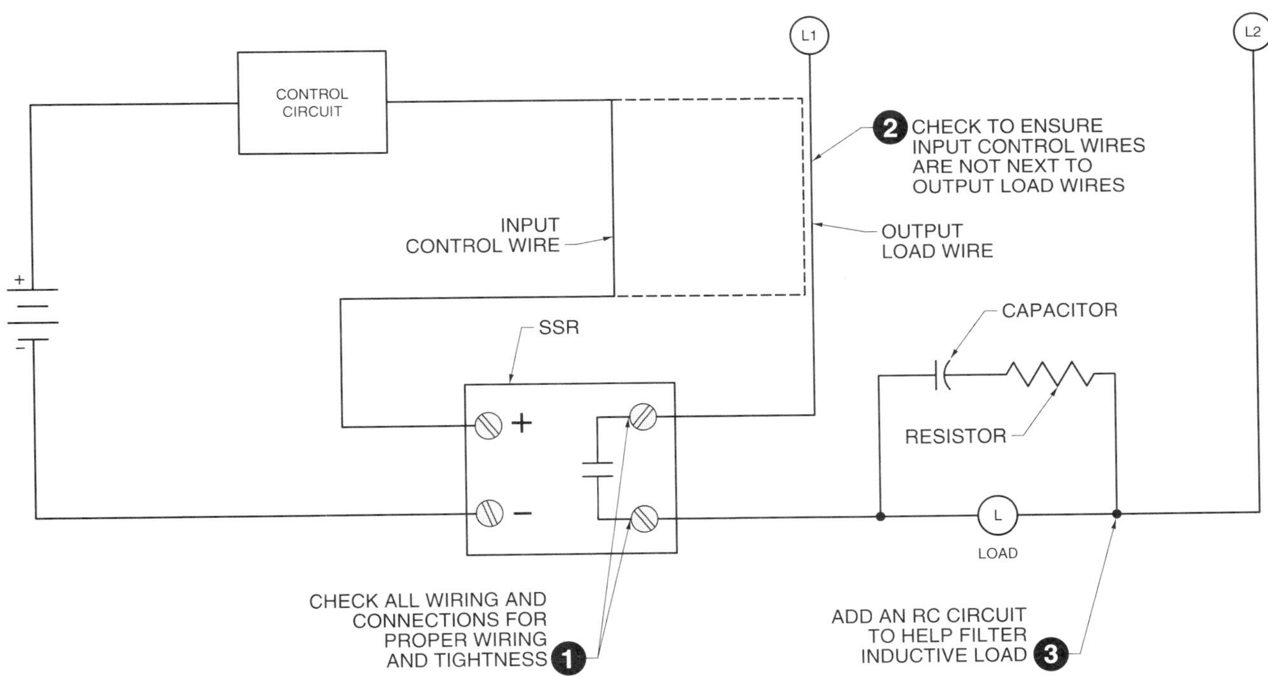

Figure 12-39. Erratic relay operation is the proper operation of a relay at times, and the improper operation of the relay at other times.

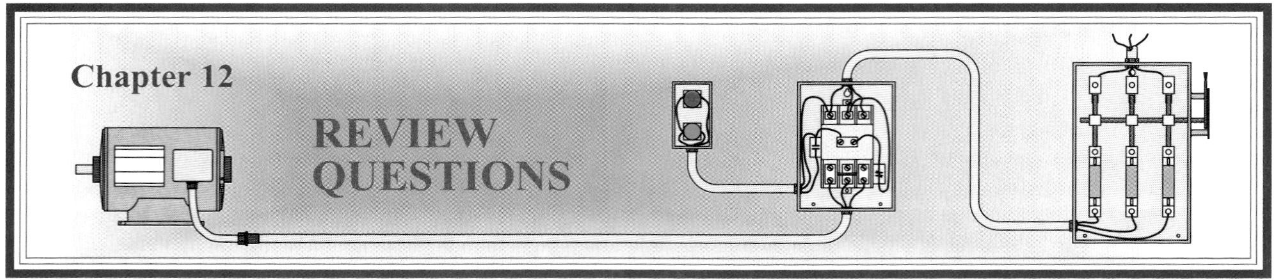

1. What are the two major types of relays?
2. What type of relay is activated by the presence of a magnetic field?
3. Do reed relays normally have more than one set of contacts?
4. Do reed relays normally have NC contacts?
5. Do general purpose relays normally have more than one set of contacts?
6. Do general purpose relays normally have NC contacts?
7. What does SPDT stand for?
8. What does 3PDT stand for?
9. How many breaks can contacts have?
10. What is the name of a contact which can carry current through two circuits simultaneously?
11. How many circuits can single-throw contacts control?
12. What are convertible contacts?
13. Does an electromechanical relay normally have a higher electrical life rating or a higher mechanical life rating than a solid-state relay?
14. What is sulfidation?
15. What type of SSR is designed to control resistive loads and is the most widely used relay?
16. What type of SSR turns ON a load immediately when the control voltage is present?
17. What is a ramp up function?
18. What output is used in SSRs to switch high-current DC loads?
19. What output is used in SSRs to switch low-current AC loads?
20. What output is used in SSRs to switch low-current DC loads?
21. What happens to the failure rate of SSRs as the temperature rises?
22. What is thermal resistance?
23. What is OFF-state leakage current?
24. What are transient voltages?
25. What is half-waving?

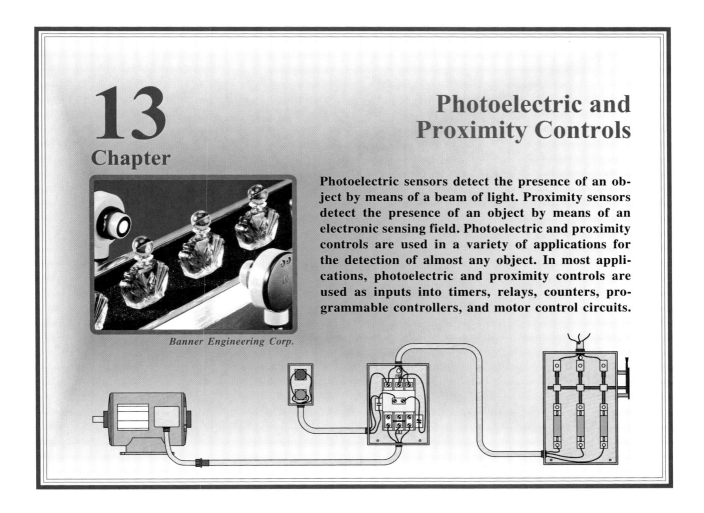

13 Chapter

Photoelectric and Proximity Controls

Photoelectric sensors detect the presence of an object by means of a beam of light. Proximity sensors detect the presence of an object by means of an electronic sensing field. Photoelectric and proximity controls are used in a variety of applications for the detection of almost any object. In most applications, photoelectric and proximity controls are used as inputs into timers, relays, counters, programmable controllers, and motor control circuits.

Banner Engineering Corp.

PHOTOELECTRIC SWITCHES

A *photoelectric switch* is a solid-state sensor that can detect the presence of an object without touching the object. A photoelectric switch detects the presence of an object by means of a beam of light. Photoelectric switches can detect most materials, and have a longer sensing distance than proximity sensors. Depending on the model, photoelectric switches can detect objects from several millimeters to over 100′ away.

The maximum sensing distance of any switch is determined by size, shape, color, and character of the surface of the object to be detected. Many switches include an adjustable sensing distance, making it possible to exclude detection of the background of the object. Photoelectric switches are used in applications in which the object to be detected is excessively light, heavy, hot, or untouchable. See Figure 13-1.

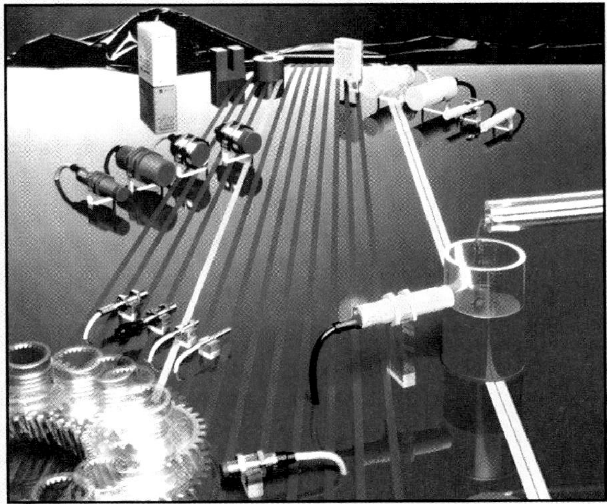

Carlo Gavazzi Inc. Electromatic Business Unit

Photoelectric and proximity switches from Carlo Gavazzi Inc. are available in a variety of types to meet the many different industry requirements.

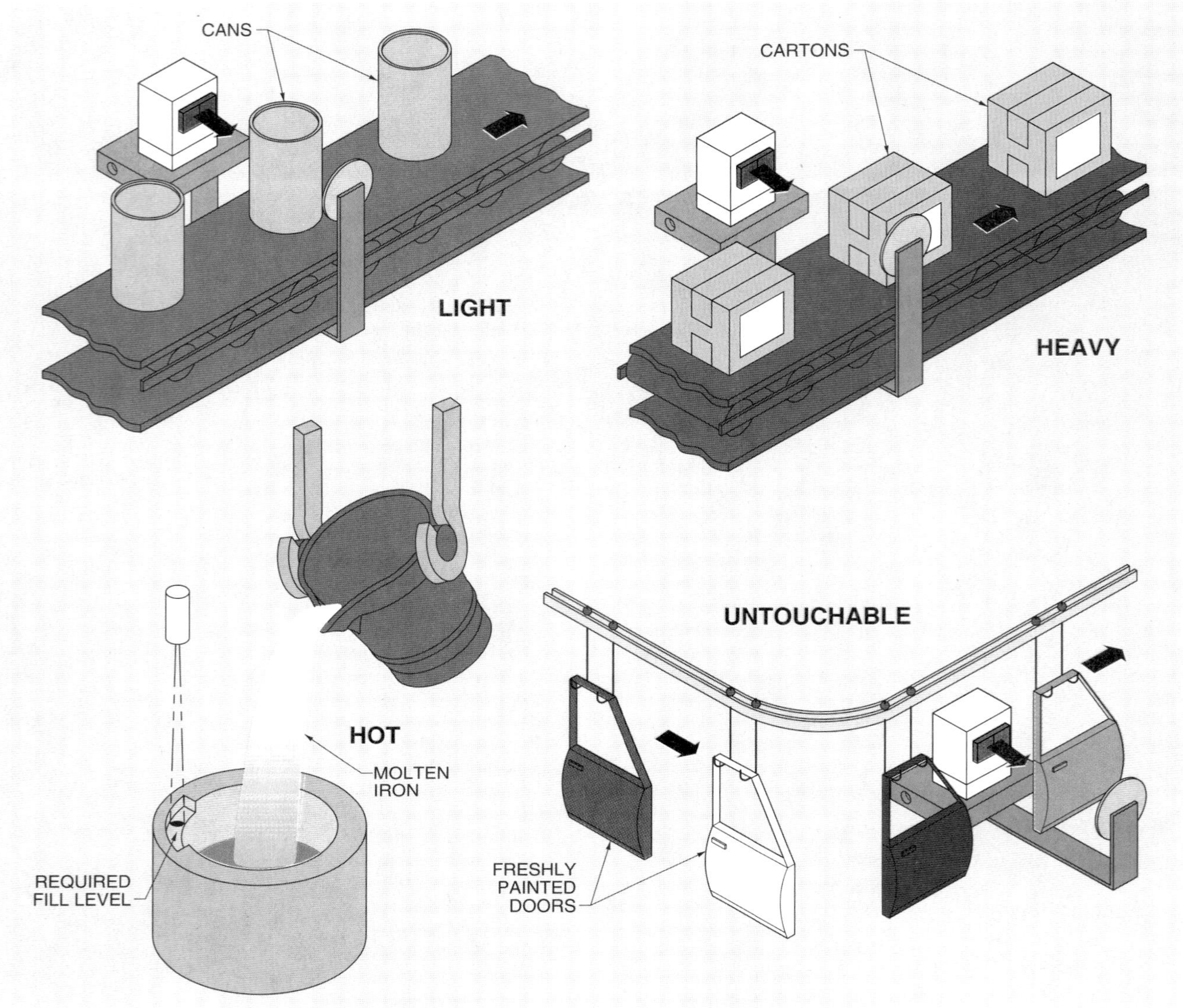

Figure 13-1. Photoelectric switches are used to detect objects without touching the object.

Scanning Techniques

Photoelectric switches are comprised of two separate major components: a light source (phototransmitter), and a photosensor (photoreceiver). The light source emits a beam of light and the photosensor detects the beam of light. The light source and photosensor may be housed in the same enclosure or in separate enclosures. *Scanning* is the process of using the light source and photosensor together to allow the light source and the photosensor to measure a change in light intensity when a target is present in, or absent from, the transmitted light beam.

When the photosensor detects the target, it sends a signal to the control circuit. The control circuit processes the signal and activates a solid-state output switch (thyristor or transistor). The output switch energizes or de-energizes a solenoid, relay, magnetic motor starter, or other load.

The phototransmitter and photoreceiver may be set up in several different methods. The best method depends on the particular application. Factors that determine the best scanning technique include:

- Scanning distance. An application may require the target to be a few millimeters or several feet away.

- Size of target. The target may be as small as a needle or as large as a truck.
- Reflectance level of target. All targets reflect the transmitted light. Lighter targets reflect more light than darker targets.
- Target positioning. Targets may enter the detection area in the same position or in different positions.
- Differences in color and reflective properties between the background and the target. The transmitted light beam is reflected by the background as well as the target.
- Changes in the ambient light intensity. The photosensor may be affected by the amount of natural (or artificial) light at the location.
- Condition of the surrounding air. The transmitted light beam is affected by the quality (amount of impurities) in the air in which the transmitted light beam must travel through. Impurities reduce the range of the transmitted light beam.

Several common scanning methods are used with photoelectric switches. These include the direct, retroreflective, polarized, specular, diffuse, and convergent beam scanning methods.

Direct Scan. *Direct scan (opposed scan)* is a method of scanning in which the transmitter and receiver are placed opposite each other so that the light beam from the transmitter shines directly at the receiver. The target must pass directly between the transmitter and receiver. The target size should be at least 50% of the diameter of the receiver lens to block enough light. For very small targets, a special converging lens or aperture may be used. See Figure 13-2.

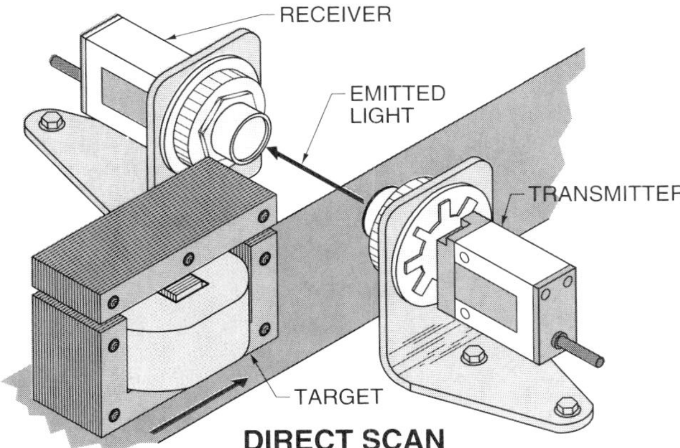

Figure 13-2. Direct scan is a method of scanning in which the target is detected as it passes between the transmitter and receiver.

The direct scan method should generally be the first choice for scanning targets that block most of the light beam. Because the light beam travels in only one direction, direct scan provides long-range sensing and works well in areas of heavy dust, dirt, mist, etc. Direct scan may be used at distances of over 100′.

Retroreflective Scan. *Retroreflective scan (retro scan)* is a method of scanning in which the transmitter and receiver are housed in the same enclosure and the transmitted light beam is reflected back to the receiver from a reflector. The light beam is directed at a reflector which returns the light beam to the receiver when no target is present. When a target blocks the light beam, the output switch is activated. The reflector may be a common bicycle disc reflector or reflective tape. See Figure 13-3.

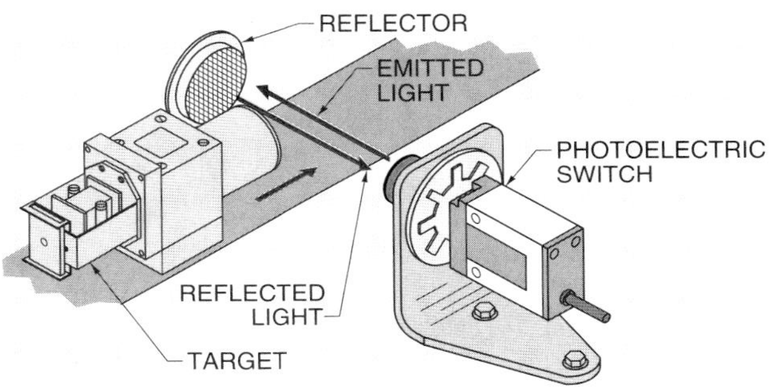

Figure 13-3. Retroreflective scan is a method of scanning in which the target is detected as it passes between the photoelectric switch and reflector.

Alignment is not critical with retroreflective scan. This makes it a good choice for high vibration applications. Retroreflective scan is used in applications in which sensing is possible from only one side at distances up to about 40′. Retroreflective scan does not work well in applications that require the detection of translucent or transparent materials.

Polarized Scan. *Polarized scan* is a method of scanning in which the receiver responds only to the depolarized reflected light from corner cube reflectors or polarized sensitive reflective tape. The light source (emitter) and photoreceiver are placed on the same side of the object to be detected. A special lens filters the emitter's beam of light so that it is projected in one plane only. See Figure 13-4. The receiver ignores the light reflected from most varieties of shrink-wrap materials, shiny luggage, aluminum cans, or common reflective objects.

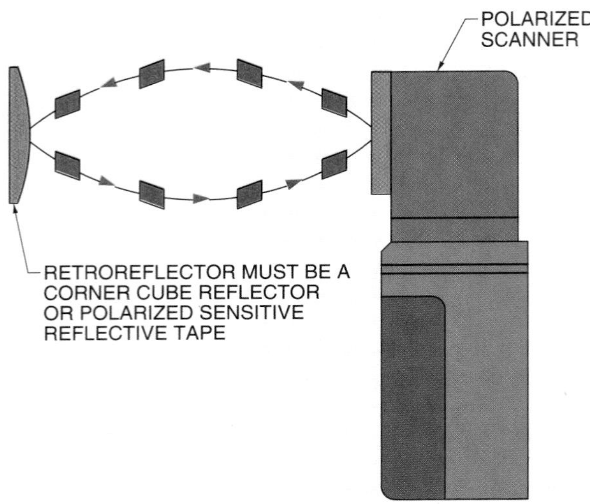

Figure 13-4. Polarized scan uses a special lens which filters the emitter's beam of light so that it is projected in one plane only.

Specular Scan. *Specular scan* is a method of scanning in which the transmitter and receiver are placed at equal angles from a highly reflective surface. With reflective surfaces, the angle at which light strikes the reflecting surface equals the angle at which it reflects from the surface. This is similar to billiards in which a ball leaves the cushion at an angle equal to the angle it struck the cushion. See Figure 13-5.

Honeywell's MICRO SWITCH Division

The CCS Series adaptive clear-container sensors from Honeywell's MICRO SWITCH Division contain patented technology which allows microprocessor-based electronics to sense and adjust to environmental changes.

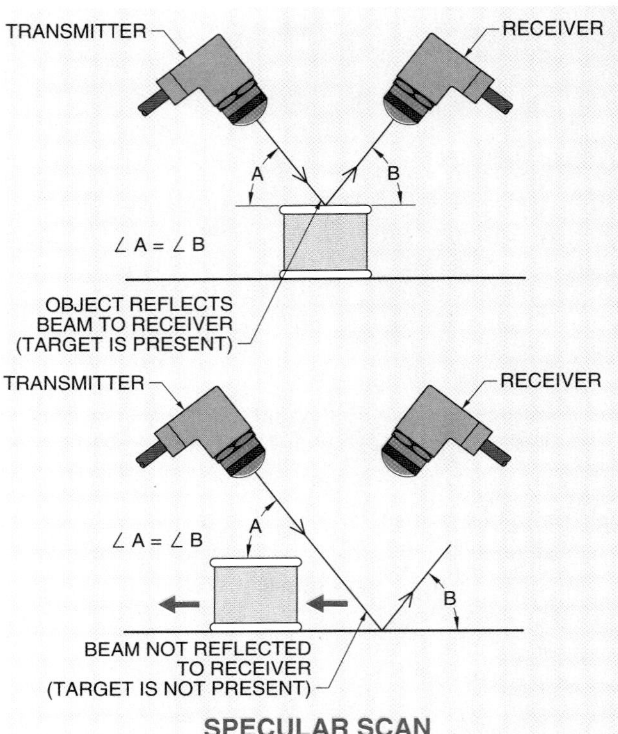

Figure 13-5. Specular scan is a method of scanning in which the transmitter and receiver are placed at equal angles from a highly reflective surface.

Specular scan distinguishes between shiny and nonshiny (matte) surfaces. For example, specular scan may detect a break when printing a newspaper. The newspaper may be moved over a stainless steel plate in which a photoelectric switch is positioned. When a break occurs, the photoelectric switch detects the break and stops the press.

Diffuse Scan. *Diffuse scan (proximity scan)* is a method of scanning in which the transmitter and receiver are housed in the same enclosure and a small percentage of the transmitted light beam is reflected back to the receiver from the target. In diffuse scan, the transmitter and the receiver are placed in the same enclosure so the receiver picks up some of the diffused (scattered) light. The target detected may be large or small. See Figure 13-6.

Diffuse scan is used in color-mark detection to detect the amount of light reflected from a printed surface. Color marks (registration or index marks) are used for registering a specific location on a product. For example, registration marks are used in packaging applications to determine the cutoff point and to identify the point for adding printed material.

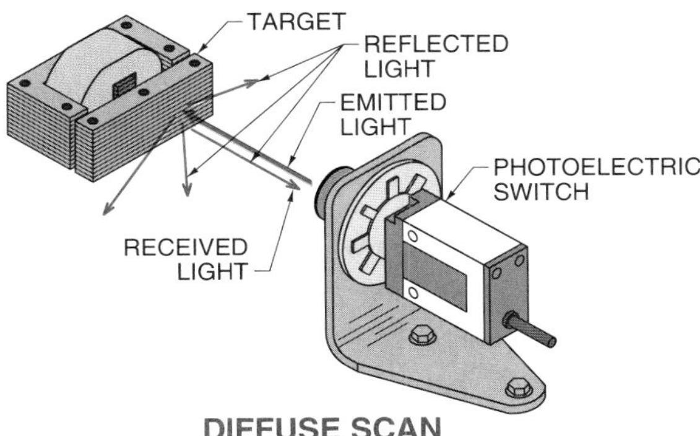

Figure 13-6. Diffuse scan is a method of scanning in which the target is detected when some of the emitted, reflected light is received.

The color of the registration mark is selected to provide enough contrast so that the diffuse scanner can detect the difference between the registration mark and the background material. Black marks against a white background provide the best contrast. However, to provide a better selection of sensors, manufacturers offer transmitters with infrared, visible red, green, or white light sources. By using different colors of transmitted light, many different color registration marks may be used with different color backgrounds.

Convergent Beam. *Convergent beam scan* is a method of scanning which simultaneously focuses and converges a light beam to a fixed focal point in front of the photoreceiver. See Figure 13-7.

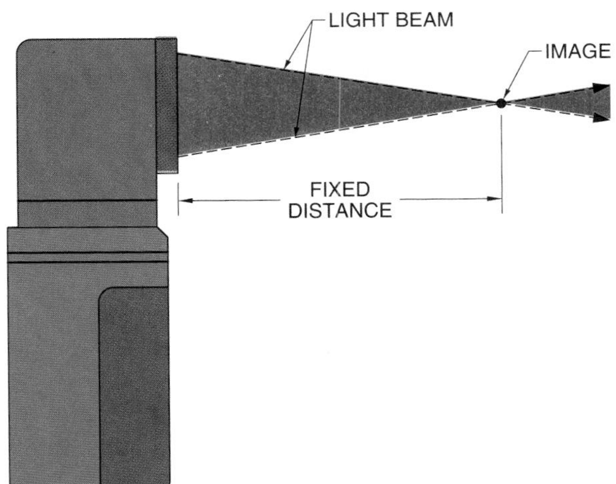

Figure 13-7. Convergent beam scan is a method of scanning which simultaneously focuses and converges a light beam to a fixed focal point in front of the photoreceiver.

Convergent beam scanning is used to detect products that are inches away from another reflective surface. It is the first choice for edge-guiding or positioning clear or translucent materials. The well-defined beam makes convergent beam scanning a good choice for position sensing of opaque materials.

The convergent beam scanner's optical system can only sense light reflected back from an object in its focal point. The scanner is blind a short distance before and beyond the focal point. Operation is possible when highly reflective backgrounds are present. Convergent beam scanning is used for detecting the presence or absence of small objects while ignoring nearby background surfaces.

Parts can be sensed on a conveyor from above while ignoring the conveyor belt. Parts may also be sensed from the side without detecting guides or rails directly in back of the object. Convergent beam scanning can detect the presence of fine wire, resistor leads, needles, bottle caps, pencils, stack height of material, and fill level of clear liquids. It is also capable of sensing black code marks against a contrasting background.

Fiber Optics

Fiber optics is a technology that uses a thin flexible glass or plastic optical fiber to transmit light. Optical fiber cables are used with photoelectric switches to conduct transmitted light into and out of the sensing area. The optical fiber cables are used as light pipes.

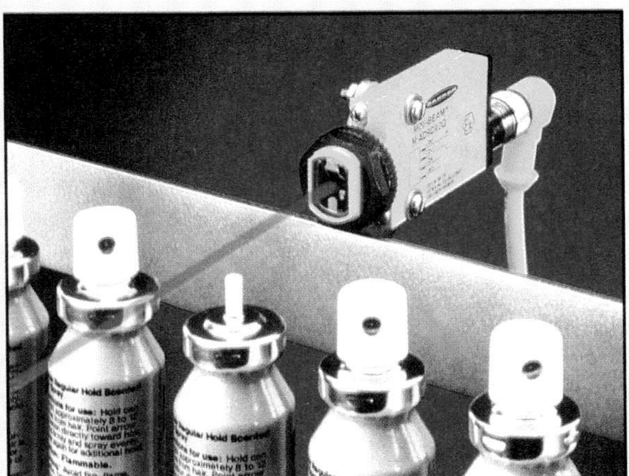

Banner Engineering Corp.

Banner NAMUR sensors are available in sensing modes including retroreflective, diffuse, convergent beam, and glass fiber optics.

The control beam is transmitted through an optical fiber cable and returned through a separate cable either combined in the same cable assembly (bifurcated) or within a separate cable assembly to the receiver. See Figure 13-8. Retroreflective and diffuse scan use a bifurcated cable, and direct scan uses two separate cables (emitter and receiver). Scan distances commonly vary from 0.4″ to 54″ depending on the scanning technique. An optical lens accessory that attaches to some cable ends significantly increases scan distances.

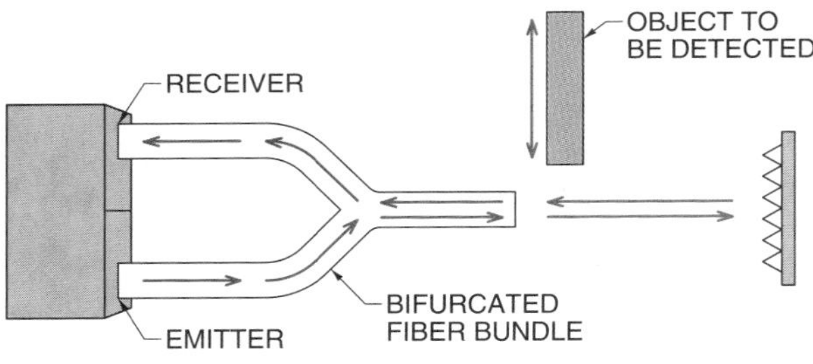

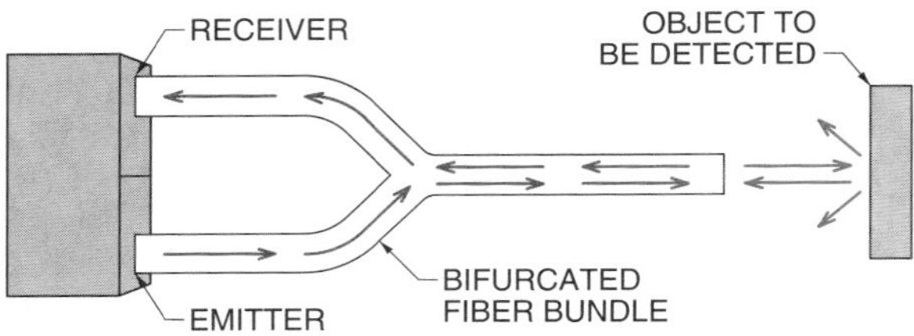

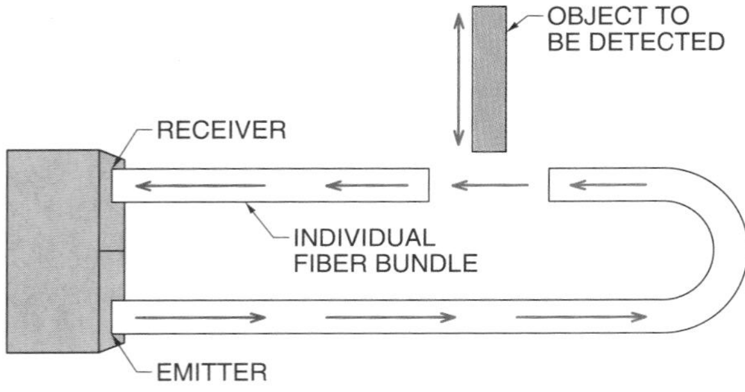

Figure 13-8. Fiber optics use transparent fibers of glass or plastic to conduct and guide light energy.

Fiber-optic controllers are available in different sizes and configurations. See Figure 13-9. Combining the optical fiber cables with photoelectric controls enables use in limited mounting spaces, small parts detection, and detection in applications having high temperature, high vibration, or high electrical noise levels.

The light-emitting and receiving components in fiber optics are located remotely at the control's housing, and only optical fiber cables are exposed to the severe environmental or hazardous areas.

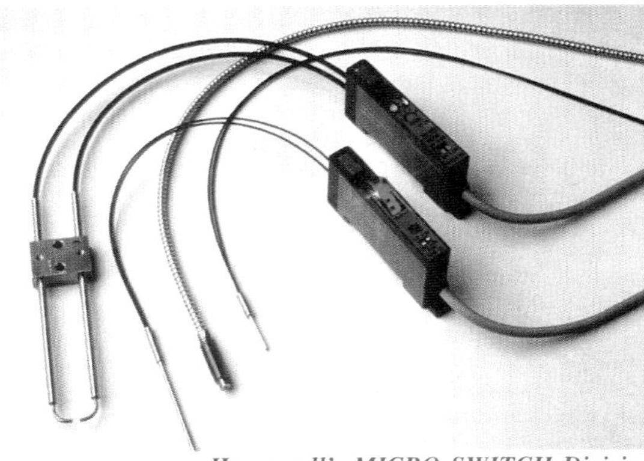

Honeywell's MICRO SWITCH Division

Figure 13-9. Fiber-optic controllers are available in different sizes and configurations.

Selection of Scanning Methods

The scanning method used for an application depends on the environment of the scanning area. In many applications, several methods of scanning work, but normally one method is the best. See Figure 13-10.

Honeywell's MICRO SWITCH Division

The HDMP Series harsh-duty modular photoelectric sensors from Honeywell's MICRO SWITCH Division can withstand washdown spray pressures up to 1200 psi and water or cleaning solution temperatures up to 140°F.

SCANNING METHODS			
Methods	**Features**		
	Configuration	**Advantages**	**Disadvantages**
Direct	Transmitter on one side sends signal to receiver on other side. Object to be detected passes between the two	Reliable performance in contaminated areas; long range scanning; most well-defined effective beam of all scanning techniques	Wiring and alignment required for both transmitter and receiver; high installation cost
Retroreflective	Transmitter and receiver are housed in one package and are placed on same side of object to be detected. Signal from transmitter is reflected to receiver by retroreflector	Ease of installation in that wiring on only one side is required; alignment need not be exact; more tolerant to vibration	Sensitive to contamination since light source must travel to retroreflector and back; hard to detect transparent or translucent materials; for no good small part detection
Polarized	Transmitter and receiver housed in one package and placed on side of object to be detected. Special lens is used to filter light beam to project it in one plane only	Only depolarized light from transmitter is detected, thus ignoring other unwanted light sources	Detection distance and plane of detection limited
Specular	Transmitter sends signal to receiver by reflecting signal of object to be detected. Transmitter and receiver are not housed in same package and receiver must be positioned precisely to receive reflected light	Good for detecting shiny versus dull surfaces; depth of field can be changed by changing transmitter/receiver angle	Wiring required for both transmitter and receiver; alignment is important
Diffused	Transmitter and receiver are housed in one package. Object being detected reflects signal back to detector. No retroreflector is used	Ease of installation in that wiring on only one side is required since detected object returns signal; alignment is not critical; best scanning technique for transparent or translucent materials	Limited range since object is used to reflect transmitted light; performance changes from one type of object to be detected to another
Convergent Beam	Transmitter and receiver are housed in one package and are placed on side of object to be detected. Light beam is focused to a fixed point in front of controller	Detection point is fixed so that objects before or beyond focal point are not detected	Detection point is very small, thus not allowing for much variation in distance that may be caused by such factors as vibration

Figure 13-10. The scanning method used for an application depends on the environment of the scanning area.

Modulated and Unmodulated Light

Although some older photoelectric transmitters use white light, modern photoelectric light sources produce infrared light. Photoelectric controls use either infrared modulated light or unmodulated light. See Figure 13-11.

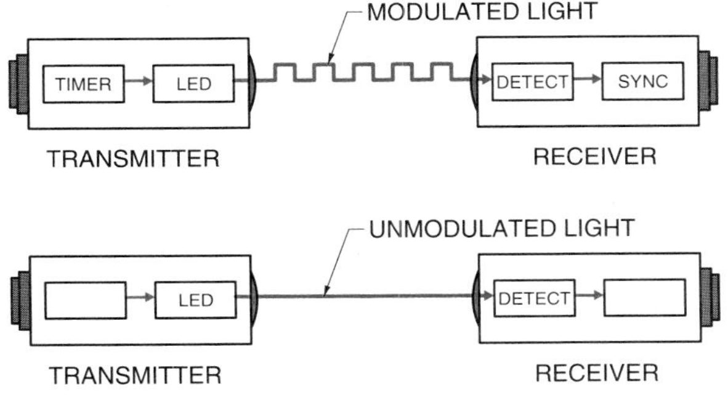

Figure 13-11. Photoelectric controls use either infrared modulated or unmodulated light.

In a modulated light source, the light source is turned ON and OFF at a very high frequency, normally several kilohertz (kHz). The control responds to this modulated frequency rather than just the intensity of the light. Because the receiver circuitry is tuned to the phototransmitter modulating frequency, the control does not respond to ambient light. This feature also helps to reject other forms of light (noise). In an unmodulated light source, the light beam is constantly ON and is not turned ON and OFF. Most manufacturers offer both types of photoelectric controls.

A modulated light source is generally considered first when using a photoelectric control. Unmodulated light is considered when the scanning range is very short and when dirt, dust, and bright ambient light conditions are not a problem. Unmodulated light sources are also used for high-speed counting because the beam is continually transmitting and responds quickly.

Response Time

The *response time* of a photoelectric control is the number of pulses (objects) per second the controller can detect. This is important when the object to be detected moves past the beam at a very high speed or when the object to be detected is not much bigger than the effective beam of the controller. This information is listed in the specification sheet of the photoelectric control. For example, a photoelectric control may have an activating frequency of 10 pulses per second. This means the photoelectric control, on average, can detect an object passing by it every 1⁄10 (.1) sec. See Figure 13-12.

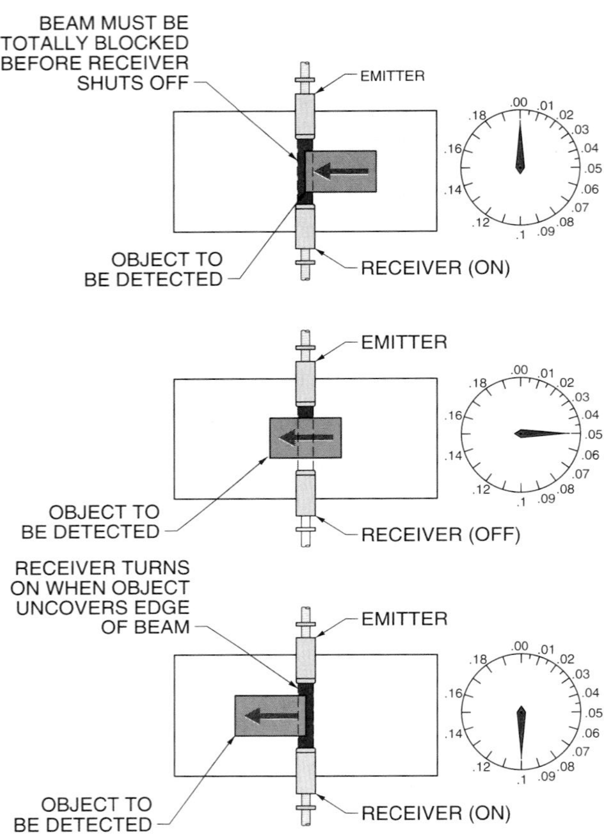

Figure 13-12. Response time of a photoelectric control is the number of pulses per second the controller can detect.

The beam must be totally blocked before the receiver shuts OFF. The receiver turns ON when the object uncovers the edge of the beam. This has the effect of shortening the size of the object to be detected as seen by the photoelectric control. The length of time that an object breaks the beam is found by applying the formula:

$$t = \frac{w - D}{s}$$

where

t = time object takes to break beam (in sec)

w = width of object moving through beam (in in.)

D = effective beam diameter (in in.)

s = speed of object (in in./sec)

Example: Calculating Object Beam Break Time

What is the length of time it takes an object that is 2¼″ wide to pass a ¼″ diameter beam when the object is moving 2″ per second?

$$t = \frac{w - D}{s}$$

$$t = \frac{2.25 - .25}{2}$$

$$t = \frac{2}{2}$$

t = **1 sec**

A 10 pulse-per-second rated photoelectric control may be used because the object takes 1 sec to pass the photoelectric control. If the speed of the object is increased to 10″ per second, the length of time it takes the object to move past the photoelectric control is .2 sec ($2.25 - {}^{.25}/_{10}$ = .2 sec). A 10 pulse-per-second rated photoelectric control may also be used because the object takes .2 sec to pass the photoelectric control.

Sensitivity Adjustment

Many photoelectric switches have a sensitivity adjustment that determines the operating point or the level of light intensity which triggers the output. This adjustment allows the sensitivity to be set between a minimum and maximum range. The adjustment is made after the unit has been installed and the maximum and minimum settings for the application are experimentally determined. The sensitivity adjustment is normally set half-way between these two points. Low sensitivity setting may be desirable, especially when there is bright ambient light, electrical noise interference, or when detecting translucent objects. Reducing the sensitivity may avoid false triggering of the control in these conditions.

Light-Operated/Dark-Operated

A *light-operated photoelectric control* energizes the output switch when the target is missing (removed from the beam). Basic photoelectric controls are designed to be dark-operated. A *dark-operated photoelectric control* energizes the output switch when a target is present (breaks the beam). Some photoelectric controls include an optional feature that allows the control to be set in a light-operated mode. See Figure 13-13.

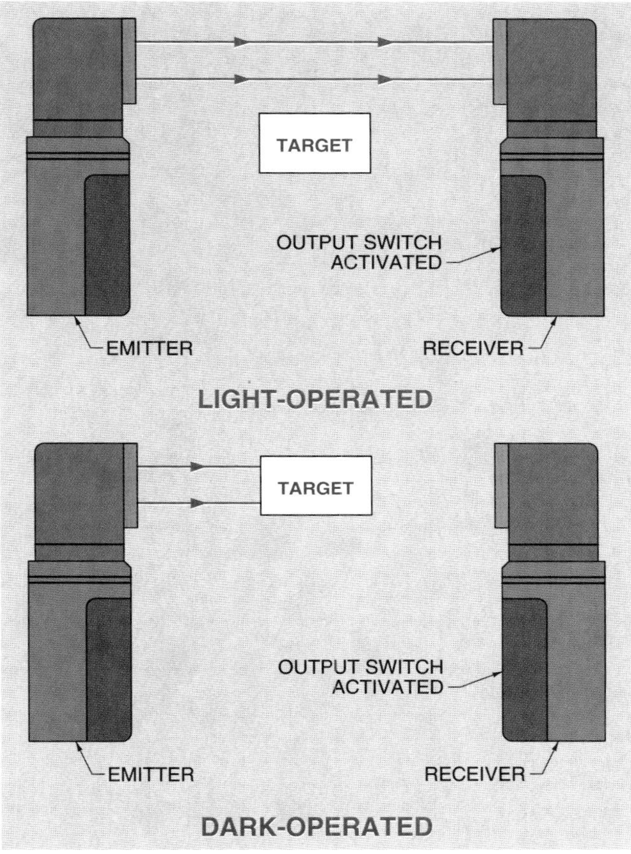

Figure 13-13. A photoelectric control can have a light-operated or dark-operated mode.

The setting of a dark-operated or light-operated mode is usually accomplished by a selector switch located on the photoelectric control. This switch is typically marked with a LO/DO position (light-operated/dark-operated), or an INV (invert) position. In a security system, the dark-operated mode is used to activate the switch contacts that sound an alarm when a person walks into the beam. The light-operated mode is used to activate the switch contacts that sound an alarm when a person removes an object (toolbox, painting, etc.) from the light beam.

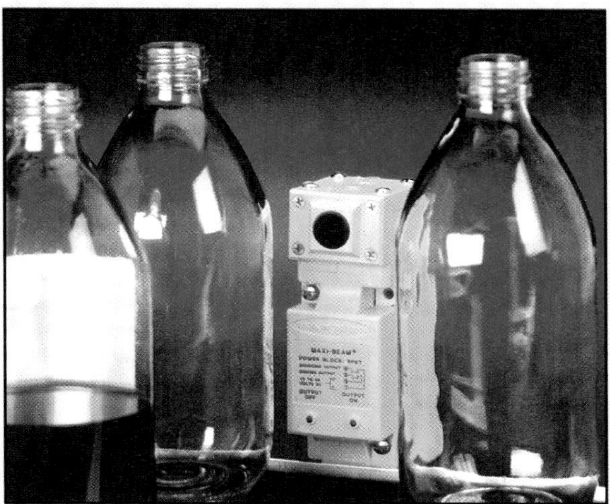

Banner Engineering Corp.

Banner's new MAXI-BEAM® diffuse sensors sense at a precise distance which reduces false triggering due to background objects.

PHOTOELECTRIC APPLICATIONS

Photoelectric controls are used in a variety of applications for the detection of almost any object. They are used to detect objects smaller than a needle and larger than a truck. Along a production line, photoelectric controls are used for counting, positioning, safety, and sorting. In security systems, they are used to detect the presence or removal of an object. They are also used to detect the presence of vehicles at toll gates, parking areas, and truck docks. In most applications, photoelectric controls are used as inputs into timers, relays, counters, programmable controllers, and motor control circuits.

Height and Distance Monitoring

Photoelectric controls may be used to monitor a truck loading bay for clearance and distance. See Figure 13-14. In this application, a truck in a loading bay is monitored for necessary clearance and distance. The dimensions vary depending on particular needs.

Any truck 14′ – 0″ or higher must be unloaded at another bay. In the control circuit, photoelectric control 1 (photo 1) turns ON an alarm in line 2 if the truck is too high. Photoelectric control 2 (photo 2) starts a recycle timer (TR) that flashes a yellow light ON and OFF at a distance of 2′ from the dock. Photoelectric control 3 (photo 3) starts a recycle timer that turns ON a red light at a distance of 6″ from the dock.

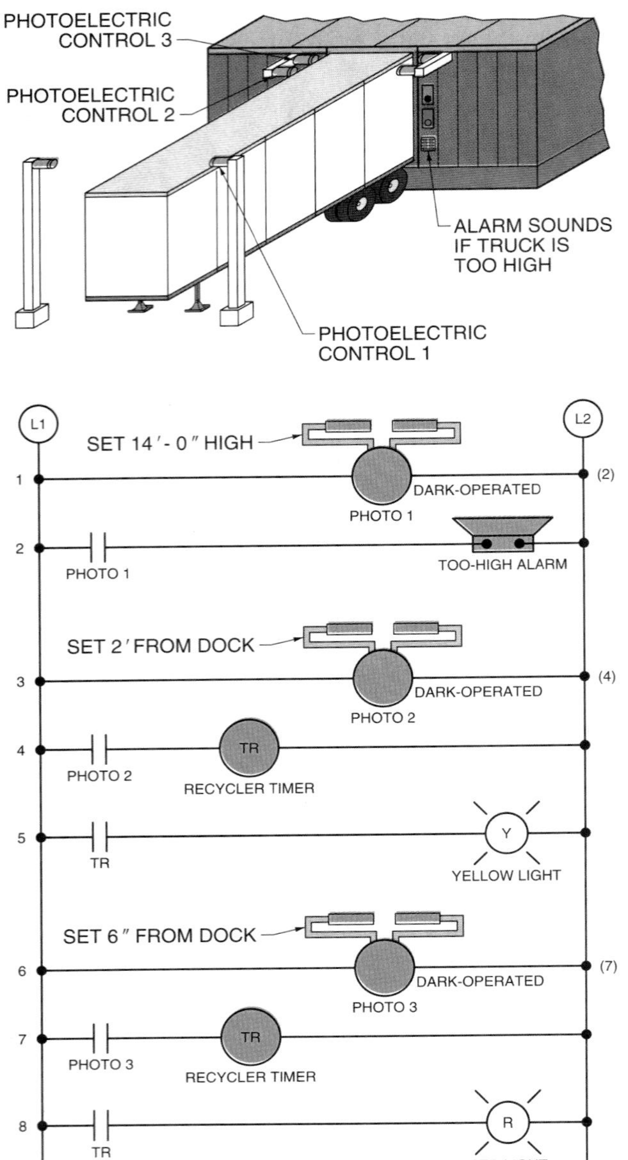

Figure 13-14. Photoelectric controls may be used to monitor a truck loading bay for clearance and distance.

Direct scan using modulated controls is the best scan method for this application. The photoelectric control is connected for a dark operation of the controller, allowing for operation only when the beam is blocked by the truck.

Product Monitoring

Photoelectric controls may be used to detect a backup of a product on a conveyor line. See Figure 13-15.

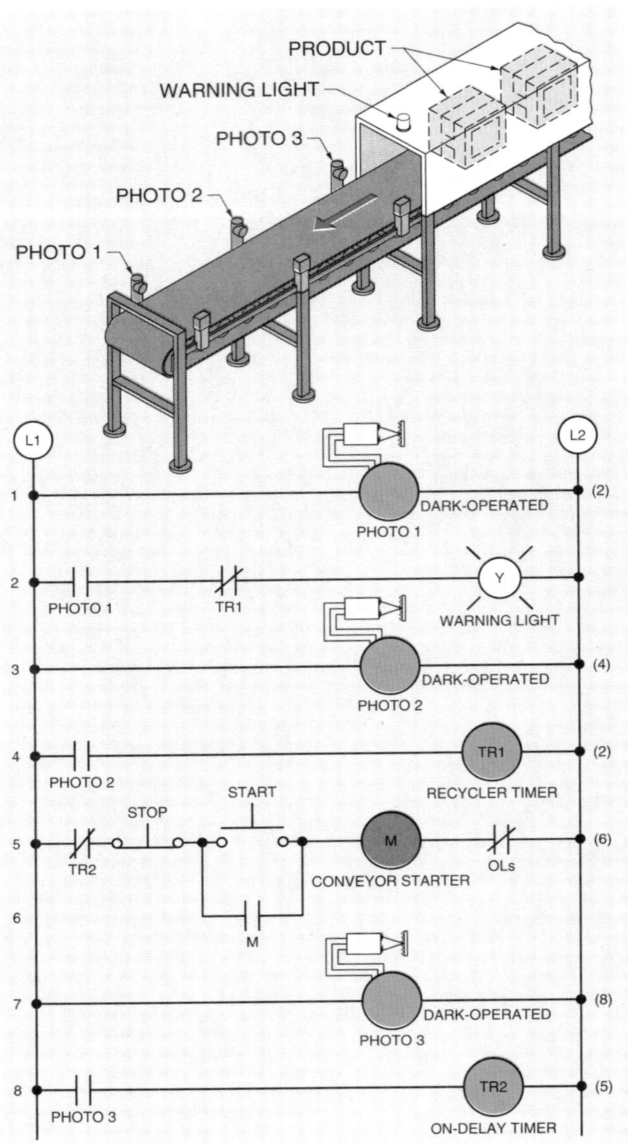

Figure 13-15. Photoelectric controls may be used to detect a backup of a product on a conveyor line.

In this application, three photoelectric controls are used to turn ON a warning light and turn OFF the conveyor motor if required.

Photoelectric control 1 (photo 1) turns ON a warning light, indicating product is at the end of the conveyor line. At this time an operator may remove the product or wait until more products are on the line. This allows for best use of the worker's time. If the product backs up to photoelectric control 2 (photo 2), a recycle timer is activated which flashes the warning light.

At this time, the operator should unload the conveyor. If the conveyor is not unloaded, and product backs up to photoelectric control 3 (photo 3), an ON-delay timer is activated which, after a few seconds, stops the conveyor motor thus preventing a problem.

In this application, retroreflective scan is used for ease of installation and because of vibration of the conveyor line. All upstream conveyors and machines must also be turned OFF to prevent a jam.

PROXIMITY SWITCHES

A *proximity switch* is a solid-state sensor that detects the presence of an object by means of an electronic sensing field. A proximity sensor does not come into physical contact with the object. Proximity sensors can detect the presence or absence of almost any solid or liquid. They are extremely versatile, safe, reliable, and may be used in applications where other limit switches and mechanical level switches cannot.

Proximity sensors can detect very small objects such as microchips, and very large objects, such as automobile bodies. All proximity sensors have encapsulated solid-state circuits which may be used in high-vibration areas, wet locations, and fast-switching applications. To meet as many application requirements as possible, proximity sensors are available in an assortment of different sizes and shapes. See Figure 13-16.

Banner Engineering Corp.

Figure 13-16. Proximity switches are solid-state sensors that detect the presence of an object by means of an electronic sensing field.

The two basic proximity sensors are the inductive proximity sensor and capacitive proximity sensor. An *inductive proximity sensor* is a sensor that detects conductive substances only. Inductive proximity sensors detect only metallic targets. A *capacitive proximity sensor* is a sensor that detects either conductive or nonconductive substances. Capacitive proximity sensors detect solid, fluid, or granulated targets whether conductive (metallic) or nonconductive. The proximity sensor used depends on the type and material of the target.

Inductive Proximity Sensors

Inductive proximity sensors operate on the Eddy Current Killed Oscillator (ECKO) principle. The ECKO principle states that an oscillator produces an alternating magnetic field that varies in strength depending on whether or not a metallic target is present. The generated alternating field operates at a radio frequency. See Figure 13-17.

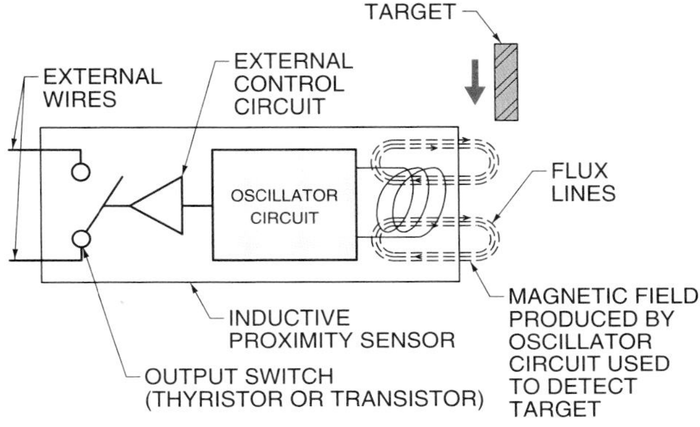

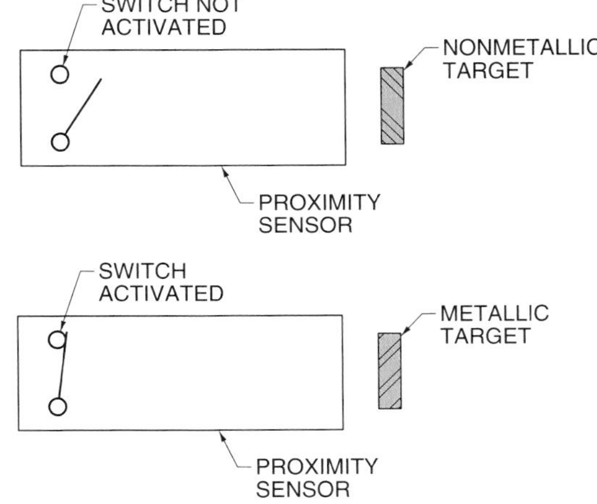

Figure 13-17. Inductive proximity sensors use a magnetic field to detect the presence of a target.

When a metallic target is in front of an inductive proximity sensor, the RF field causes eddy currents to be set up on the surface of the target material.

These eddy currents upset the AC inductance of the sensor's oscillator circuit causing the oscillations to be reduced. When the oscillations are reduced to a certain level, the sensor triggers indicating the presence of a metallic object. Inductive proximity sensors detect ferrous materials (containing iron, nickel, or cobalt) more readily than nonferrous materials (all other metals, such as aluminum, brass, etc.).

Nominal sensing distances range from .5 mm to about 40 mm. Sensitivity varies depending on the size of the object and the type of metal. Iron may be sensed at about 40 mm. Aluminum may be sensed at approximately 20 mm. Applications of inductive proximity sensors include positioning of tools and parts, metal detection, drill bit breakage, and solid-state replacement of mechanical limit switches.

Capacitive Proximity Sensors

A capacitive proximity sensor is a sensor that measures a change in capacitance, which is caused by the approach of an object into the electrical field of a capacitor. A capacitive sensor detects all materials which are good conductors in addition to insulators that have a relatively high dielectric constant. Capacitive sensors can detect materials such as plastics, glass, water, moist wood, etc. A *dielectric* is a nonconductor of direct electric current. See Figure 13-18.

Namco Controls Corporation

The new inductive proximity sensors developed by Namco Controls feature a "hard coat" non-stick material which prevents adherence of weld splatter and weld field immune electronics.

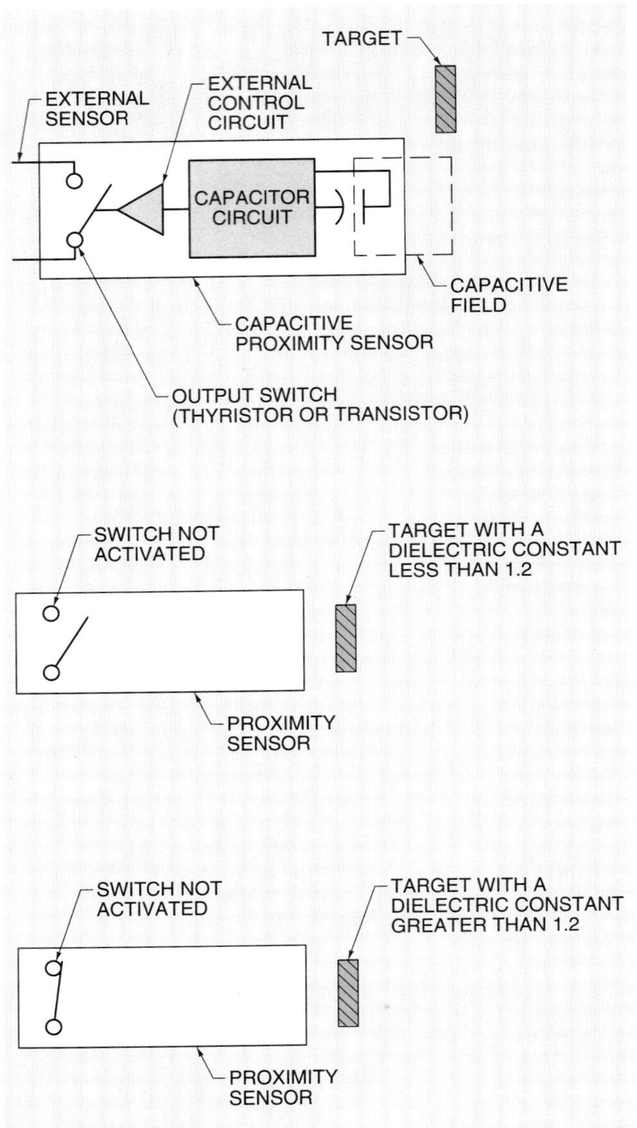

Figure 13-18. Capacitive proximity sensors use a capacitive field to detect the presence of a target.

Two small plates that form a capacitor are located directly behind the front of the sensor. When an object approaches the sensor, the dielectric constant of the capacitor changes, thus changing the oscillator frequency which activates the sensor's output. Nominal sensing distances range from 3 mm to about 15 mm. The maximum sensing distance depends on the physical and electrical characteristics (dielectric) of the object to be detected. The larger the dielectric constant, the easier it is for a capacitive sensor to detect the material. Generally, any material with a dielectric constant greater than 1.2 may be detected. See Figure 13-19.

DIELECTRIC CONSTANT	
Material	**Number**
Acetone	19.5
Acrylic Resin	2.7 – 4.5
Air	1.000264
Ammonia	15 – 24
Aniline	6.9
Aqueous Solutions	50 – 80
Benzene	2.3
Carbon Dioxide	1.000985
Carbon Tetrachloride	2.2
Cement Powder	4
Cereal	3 – 5
Chlorine Liquid	2.0
Ebonite	2.7 – 2.9
Epoxy Resin	2.5 – 6
Ethanol	24
Ethylene Glycol	38.7
Fired Ash	1.5 – 1.7
Flour	2.5 – 3.0
Freon R22 & 502 (liquid)	6.11
Gasoline	2.2
Glass	3.7 – 10
Glycerine	47
Marble	8.5
Melamine Resin	4.7 – 10.2
Mica	5.7 – 6.7
Nylon	4 – 5
Paraffin	1.9 – 2.5
Paper	1.6 – 2.6
Perapex	3.5
Petroleum	2.0 – 2.2
Phenol Resin	4 – 12
Polyacetal	3.6 – 3.7
Polyester Resin	2.8 – 8.1
Polypropylene	2.0 – 2.2
Polyvinyl Chloride Resin	2.8 – 3.1
Porcelain	5 – 7
Powdered Milk	3.5 – 4
Pressboard	2 – 5
Rubber	2.5 – 35
Salt	6
Sand	3 – 5
Shellac	2.5 – 4.7
Shell Lime	1.2
Silicon Varnish	2.8 – 3.3
Soybean Oil	2.9 – 3.5
Styrene Resin	2.3 – 3.4
Sugar	3.0
Sulfur	3.4
Tetrafluoroethylene Resin	2.0
Toluene	2.3
Turpentine	2.2
Urea Resin	5 – 8
Vaseline	2.2 – 2.9
Water	80
Wood, Dry	2 – 6
Wood, Wet	10 – 30

Figure 13-19. Capacitive sensors work on the dielectric of the material to be sensed.

Hall Effect Sensors

A *Hall effect sensor* is a sensor that detects the proximity of a magnetic field. The Hall effect principle was discovered in 1879 by Edward H. Hall at Johns Hopkins University. Hall found that when a magnet was placed in a position where its field was perpendicular to one face of a thin rectangle of gold through which current was flowing, a difference of potential appeared at the opposite edges. He found this voltage was proportional to the current flowing through the conductor and the magnetic induction was perpendicular to the conductor.

Today, semiconductors are used for the Hall element. Hall voltages obtained with semiconductors are much higher than those obtained with gold and are also less expensive.

Theory of Operation. A *Hall generator* is a thin strip of semiconductor material through which a constant control current is passed. See Figure 13-20. When a magnet is brought near, with its field directed at right angles to the face of the semiconductor, a small voltage (Hall voltage) appears at the contacts placed across the narrow dimension of the Hall generator. As the magnet is removed, the Hall voltage reduces to zero. The Hall voltage is dependent on the presence of the magnetic field and on the current flowing in the Hall generator. The output of the Hall generator is zero if the current or the magnetic field is removed. In most Hall effect devices, the control current is held constant and the magnetic induction is changed by movement of a permanent magnet.

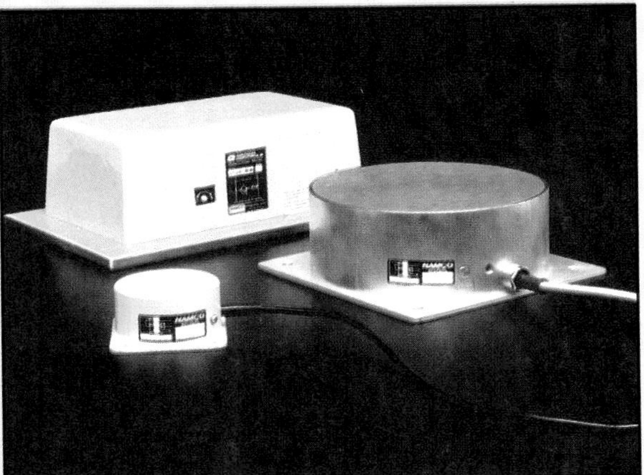

Namco Controls Corporation

The Namco ER500 Series proximity sensors can withstand extreme shock and vibration and have 70 mm to 250 mm sensing ranges.

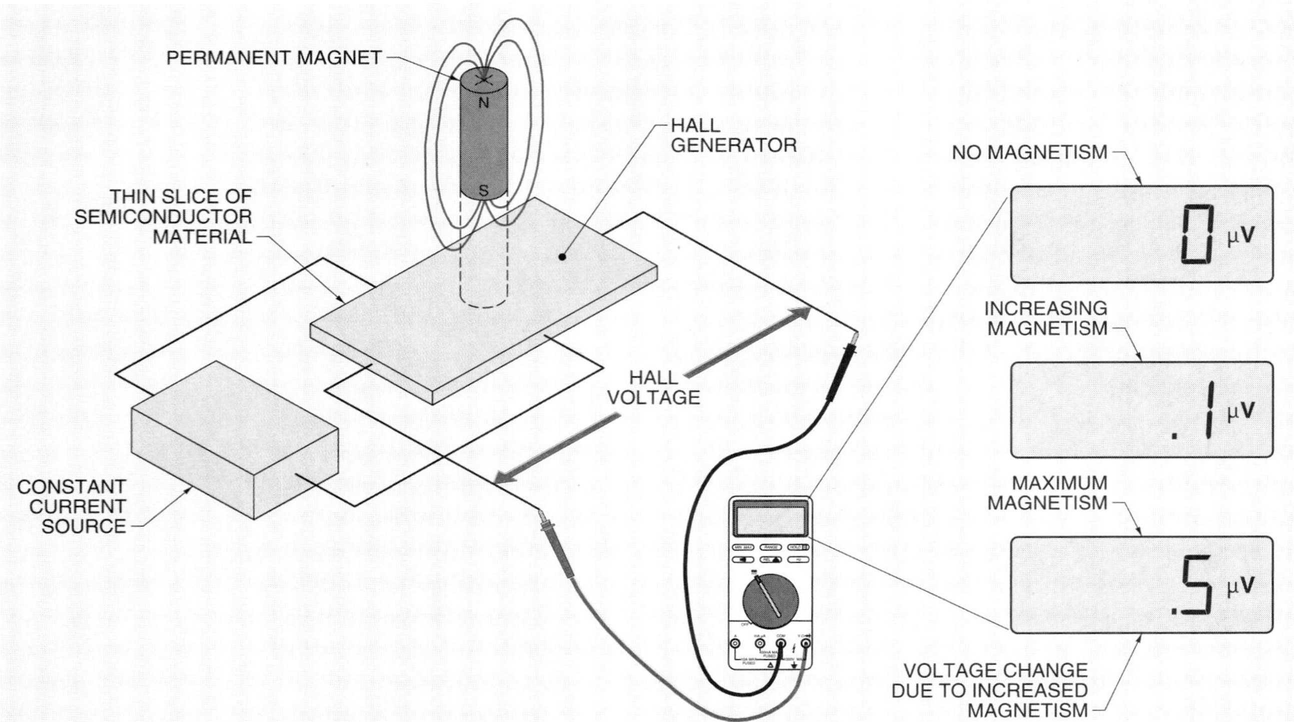

Figure 13-20. A Hall generator is a thin strip of semiconductor material through which a constant control current is passed.

Sensor Packaging

To meet many different application requirements, Hall effect sensors are packaged in a number of different types of configurations. Typical configurations include cylinder, proximity, vane, and plunger. The cylinder and proximity types are used to detect the presence of the magnet. The vane type includes the sensor on one side and the magnet on the other, and is used to detect an object passing through the opening. The plunger type includes a magnet that is moved by an external force acting against the lever. See Figure 13-21.

Hall Effect Sensor Actuation

Hall effect sensors may be activated by head-on, slide-by, pendulum, rotary, vane, ferrous proximity shunt, and electromagnetic actuation. The actuation method depends on the application.

Head-On Actuation. *Head-on actuation* is an active method of sensor activation in which a magnet is oriented perpendicularly to the surface of the sensor and is usually centered over the point of maximum sensitivity. See Figure 13-22.

Honeywell's MICRO SWITCH Division

Figure 13-21. Hall effect sensors are available in a variety of packages for different applications.

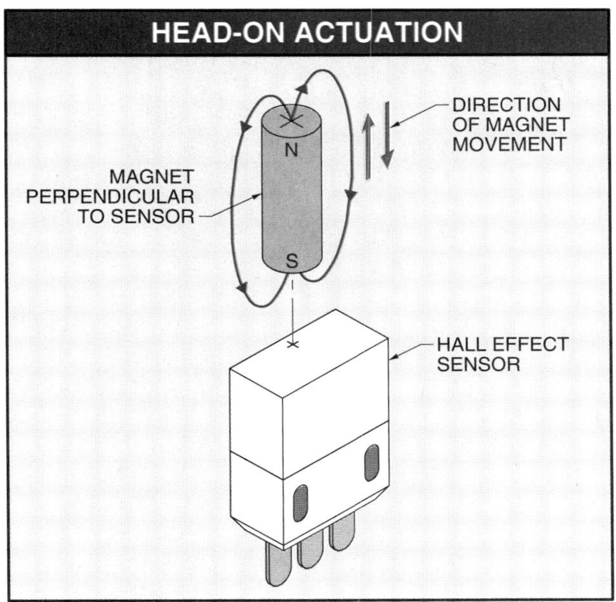

Figure 13-22. In head-on actuation, a magnet is oriented perpendicularly to the surface of the sensor and is usually centered over the point of maximum sensitivity.

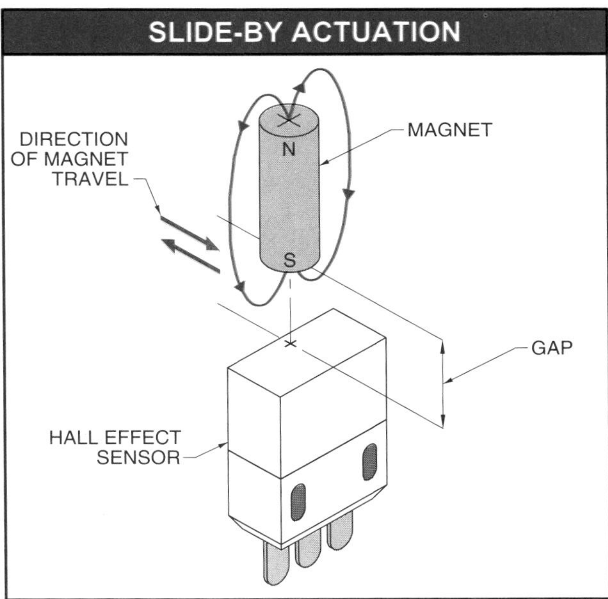

Figure 13-23. In slide-by actuation, a magnet is moved across the face of the Hall effect sensor at a constant distance.

The direction of movement is directly toward and away from the Hall effect sensor. The actuator and Hall effect sensor are positioned so the S pole of the magnet approaches the sensitive face of the sensor.

Slide-By. *Slide-by actuation* is an active method of sensor activation in which a magnet is moved across the face of the Hall effect sensor at a constant distance (gap). See Figure 13-23. The primary advantage of the slide-by actuation over the head-on actuation is that less actuator travel is needed to produce a signal large enough to cycle the device between operate and release.

Honeywell's MICRO SWITCH Division

The 40 FY Series Hall effect sensors from Honeywell's MICRO SWITCH Division are designed for door "open" detection, guard-in-place detection, and gate/access door positioning.

Pendulum Actuation. *Pendulum actuation* is a method of sensor activation which is a combination of the head-on and the slide-by actuation methods. See Figure 13-24.

The two methods of pendulum actuation are single-pole and multiple-pole. Single or multiple signals are generated by one actuator.

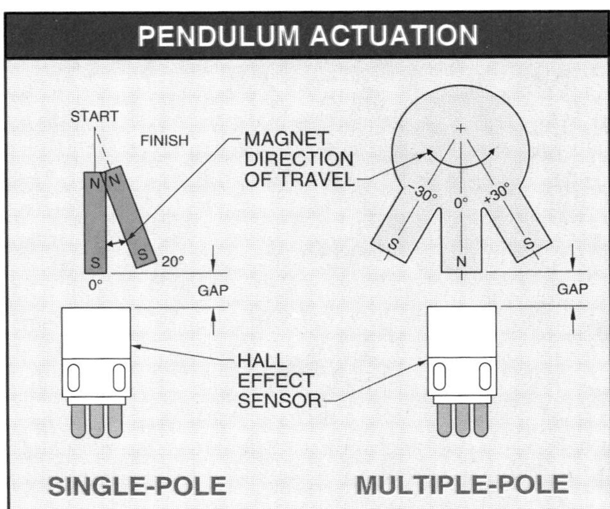

Figure 13-24. Pendulum actuation is a combination of the head-on and the slide-by actuation methods.

Rotary Actuation. *Rotary actuation* is an active method of sensor activation in which a multipolar ring magnet or collection of magnets is used to produce an alternating magnetic pattern. See Figure 13-25. The induction pattern (+ and –) is seen by the sensor located in the threaded tubular housing. Ring magnets with up to 60 or more poles are available for multiple output signals.

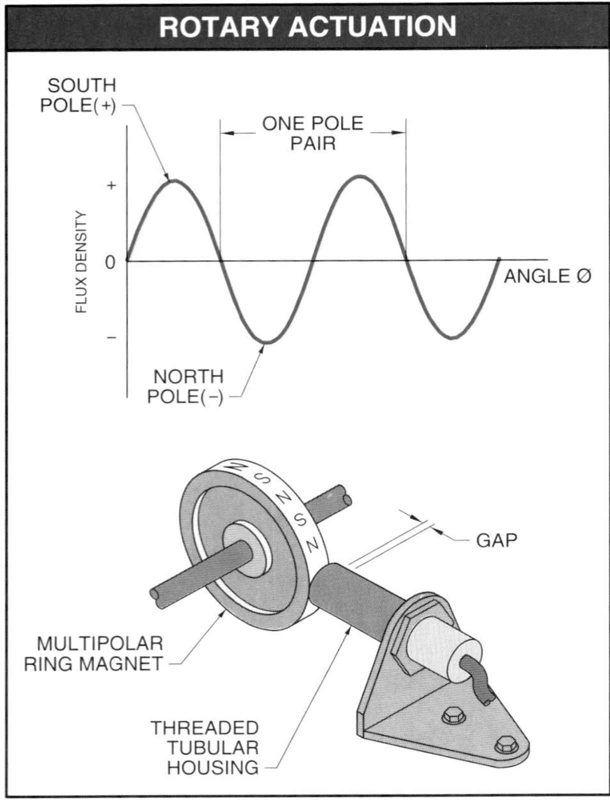

Figure 13-25. Rotary actuation uses a multipolar ring magnet or collection of magnets to produce an alternating magnetic pattern.

Vane Actuation. *Vane actuation* is a passive method of sensor activation in which an iron vane shunts or redirects the magnetic field in the air gap away from the Hall effect sensor. See Figure 13-26. When the iron vane is moved through the air gap between the Hall effect sensor and magnet, the sensor is turned ON and OFF sequentially at any speed due to the shunting effect. The same effect is achieved with a rotary-operated vane.

Ferrous Proximity Shunt Actuation. *Ferrous proximity shunt actuation* is a passive method of sensor activation in which the magnetic induction around the Hall effect sensor is shunted with a gear tooth. See Figure 13-27.

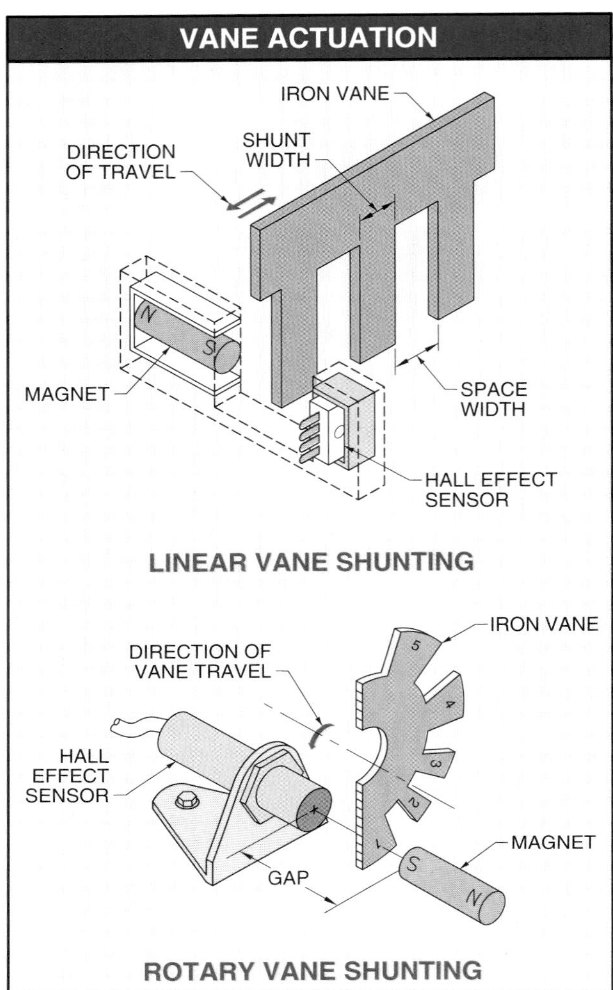

Figure 13-26. In vane actuation, an iron vane shunts or redirects the magnetic field in the air gap away from the Hall effect sensor.

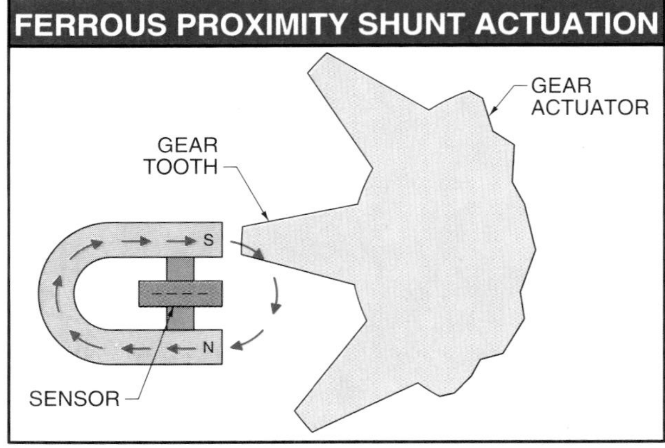

Figure 13-27. In ferrous proximity shunt actuation, the magnetic induction around the Hall effect sensor is shunted with a gear tooth.

The gear tooth causes the magnetic induction to be shunted from the sensor when a tooth is present and passes through the sensor when the tooth is absent. This variable magnetic field concentration causes the Hall effect sensor to be activated, producing an output signal.

Electromagnetic Actuation. *Electromagnetic actuation* is a passive method of sensor activation in which a magnetic field produced by a coil of wire is used to activate a Hall effect sensor. See Figure 13-28. The more current and the more turns of wire, the stronger the magnetic field. The coil serves the same purpose as a bar magnet. Attention must be given to thermal considerations because electromagnets dissipate heat. Care must be exercised to see that the current and temperature limits are not exceeded.

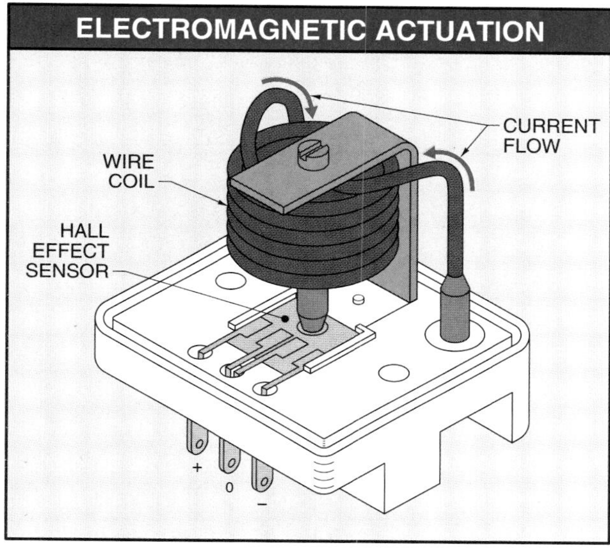

Figure 13-28. In electromagnetic actuation, a magnetic field produced by a coil of wire is used to activate a Hall effect sensor.

Hall Effect Sensor Applications

Hall effect sensors are used in a wide range of applications requiring the detection of the presence (proximity) of most objects. They are used in slow-moving and fast-moving applications to detect movement. They are used to replace mechanical limit switches in applications that require the detection of an object's position. They are also used to provide a solid-state output in applications that normally use a mechanical output, such as a standard pushbutton on a level switch.

Conveyor Belt. A Hall effect sensor may be used for monitoring a remote conveyor operation. A cylindrical Hall effect sensor is mounted to the frame of the conveyor. See Figure 13-29. A magnet mounted on the tail pulley revolves past the sensor to cause an intermittent visual or audible signal at a remote location to assure that the conveyor is running. Any shutdown of the conveyor interferes with the normal signal and alerts the operators of trouble. Maintenance is minimal because the sensor makes no physical contact and has no levers or linkages to break.

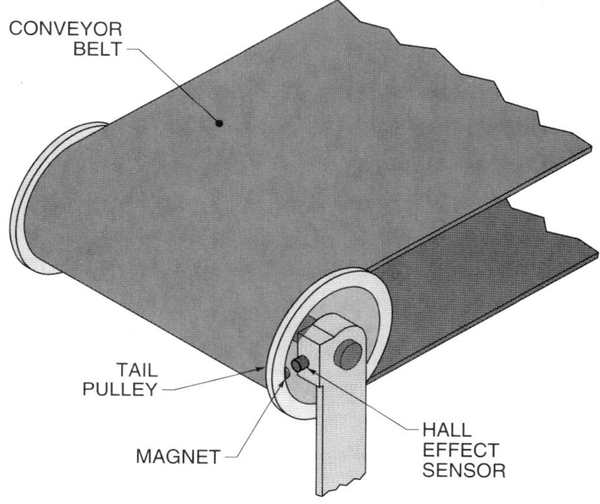

Figure 13-29. A Hall effect sensor may be used for monitoring a remote conveyor operation.

Current Sensor. A Hall effect current sensor may be used to develop a fast-acting, automatically resetting circuit. See Figure 13-30.

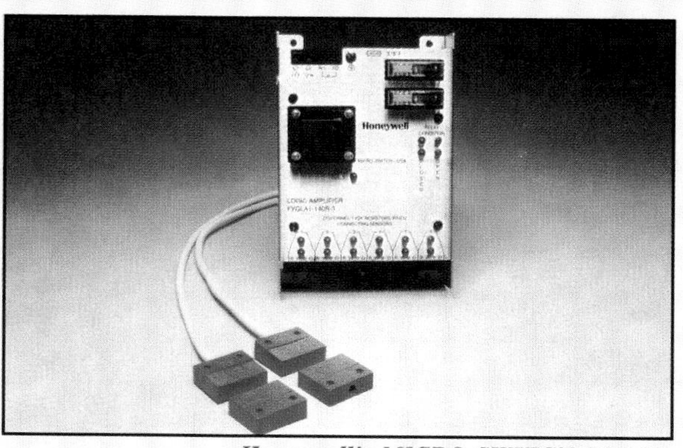

Honeywell's MICRO SWITCH Division

The new Hall effect door-interrupt system from Honeywell's MICRO SWITCH Division conforms to the technical requirements of EN60730-2-1, as applicable to an electronic incorporated (for Class I equipment) sensing control with Type 2 and 2B action for continuous operation in normal pollution situations.

Honeywell's MICRO SWITCH Division

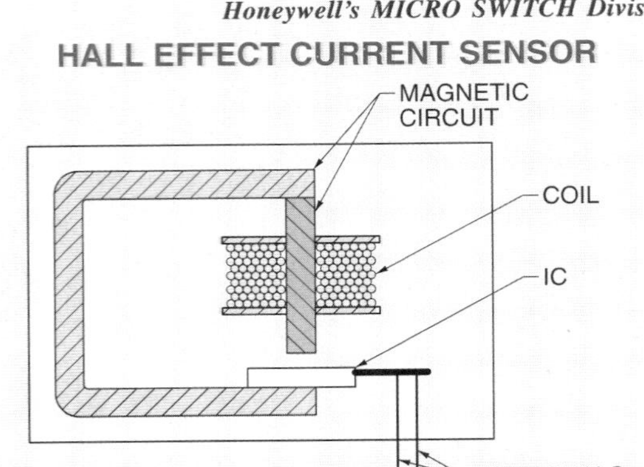

Figure 13-30. A Hall effect current sensor may be used to develop a fast-acting, automatically resetting circuit.

An overload signal changes state from low to high, or vice versa, when the current exceeds the circuit design trip point. This signal is used to trigger a warning alarm or to control the current directly by electronic means. Hall effect current sensors are built in sizes ranging from the diameter of a dime to about half the size of a deck of playing cards. They indicate overload currents from a few milliamperes to several amperes.

Different reset characteristics can be achieved by tailoring the electromagnetic design to the sensor's requirements. In addition, use of a linear Hall effect sensor in place of a digital Hall effect sensor can provide an output signal proportional to the input current, but electrically isolated from it.

Speed Sensing. A Hall effect sensor may be used with a ring magnet in speed-sensing applications. See Figure 13-31. The sensor's output provides an electrical waveform in which repetition frequency varies directly with shaft speed, but in which amplitude does not.

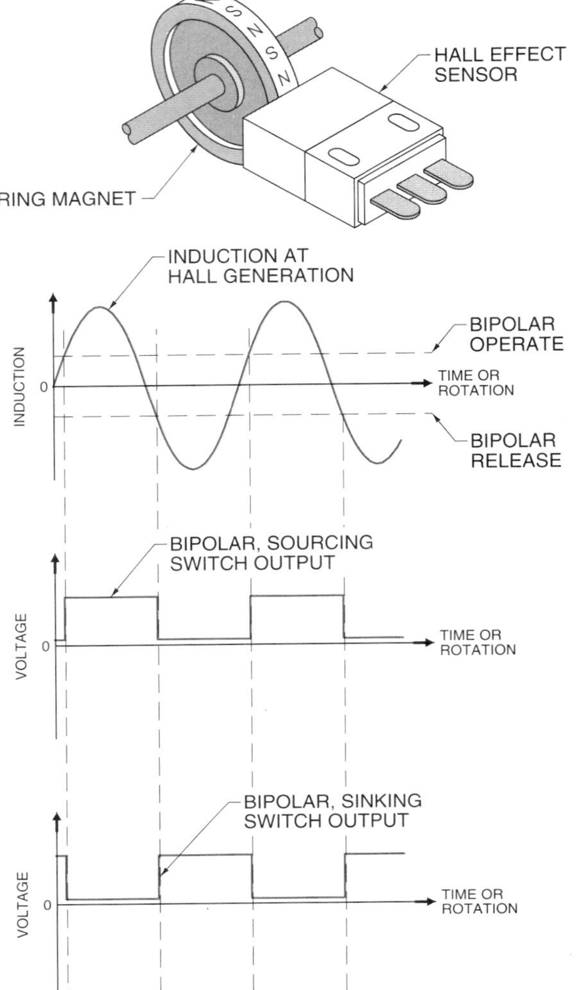

Figure 13-31. A Hall effect sensor may be used with a ring magnet in speed-sensing applications.

The use of a ring magnet with a Hall effect sensor provides a valuable alternative to coil pickup or variable reluctance speed sensing methods. Techniques which depend on inducing a voltage in a coil, either by passing magnets near it or by changing the air gap reluctance of a fixed magnet, have the disadvantage that both the frequency and amplitude of the voltage waveform change with speed. Hall sensors have no minimum speed of operation.

Instrumentation. A Hall effect sensor and ring magnet combination may be used in an instrumentation application. See Figure 13-32. The Hall effect sensor is actuated by a ring magnet which initiates resistance measurements of electrical circuits being life tested.

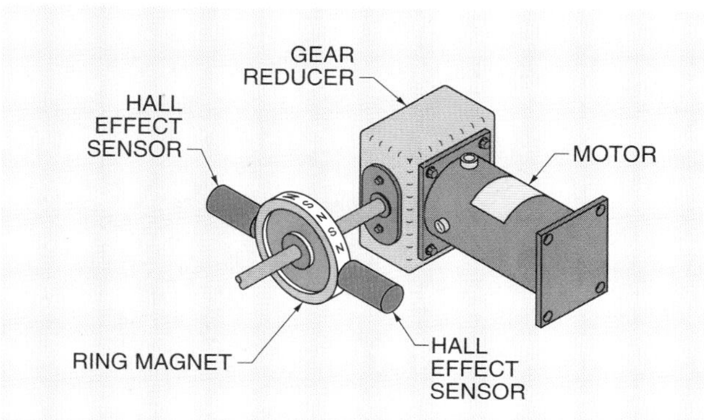

Figure 13-32. A Hall effect sensor and ring magnet combination may be used in an instrumentation application.

A second Hall effect sensor actuated 180° after the first sensor checks that the circuits are open. The no-touch actuation and long life are ideal in instruments and apparatus designs.

Beverage Gun. Hall effect sensors are used in beverage gun applications because of their small size, sealed construction, and reliability. See Figure 13-33. The Hall effect sensor's small size allows seven sensors to be installed in a hand-held device. The beverage gun cannot be contaminated by syrups, etc. It is completely submersible in water for easy cleaning and requires less maintenance.

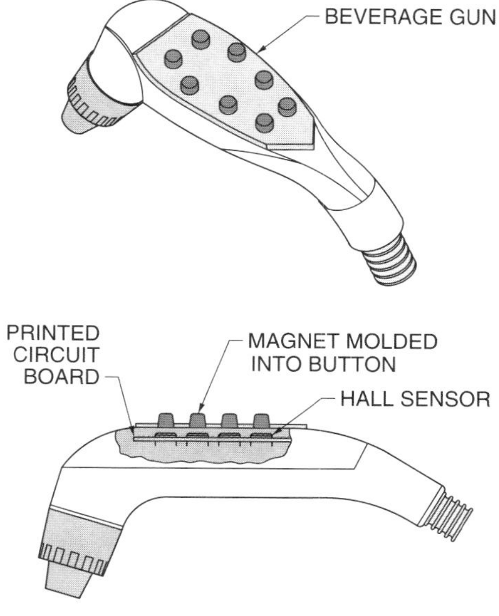

Figure 13-33. Hall effect sensors are used in beverage gun applications because of their small size, sealed construction, and reliability.

Sequencing Operations. Sequencing operations can be controlled by activating Hall effect sensors through the use of metal discs clamped to a common shaft. See Figure 13-34. The metal discs are rotated in the gaps of Hall effect vane sensors. A disc rotating in tandem with others can be used to create a binary code which is used to establish a sequence of operation. Programs can be altered by replacing the discs with others having a different air-to-ferrous cam ratio.

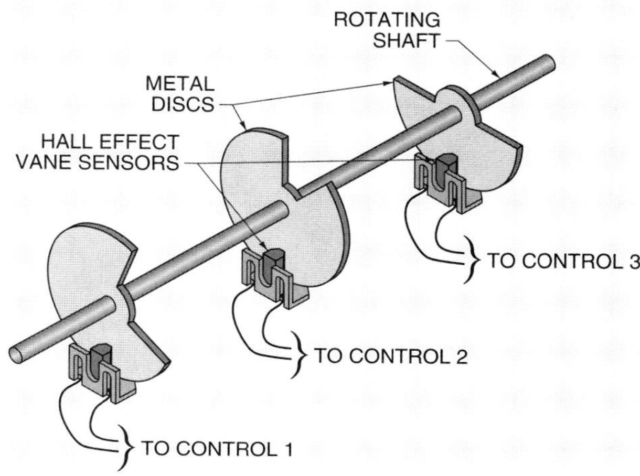

Figure 13-34. Sequencing operations can be controlled by activating Hall effect sensors through the use of metal discs clamped to a common shaft.

Length Measurement. Length measurement can be accomplished by mounting a disc with two notches on the extension of a motor drive shaft. See Figure 13-35. A Hall effect vane sensor is mounted so that the disc passes through the gap. Each notch represents a fixed length of material and can be used to measure tape, fabric, wire, rope, thread, aluminum foil, plastic bags, etc.

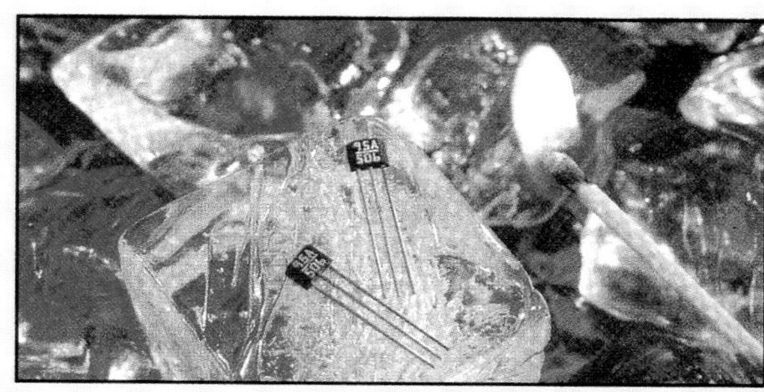

Honeywell's MICRO SWITCH Division

The new SS490 Series minature ratiometric linear sensor from Honeywell's MICRO SWITCH Division uses a Hall effect integrated circuit chip that provides increased temperature stability and sensitivity.

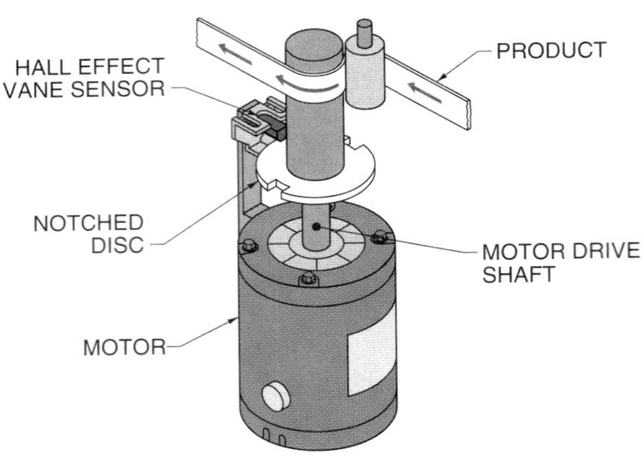

Figure 13-35. Length measurement can be accomplished by mounting a disc with two notches on the extension of a motor drive shaft.

Shaft Encoding. A cylindrical Hall effect sensor can be used in shaft encoding applications. See Figure 13-36. A ring magnet is mounted on the motor shaft. Each pair of N and S poles activates the Hall effect sensor. Each pulse represents angular movement.

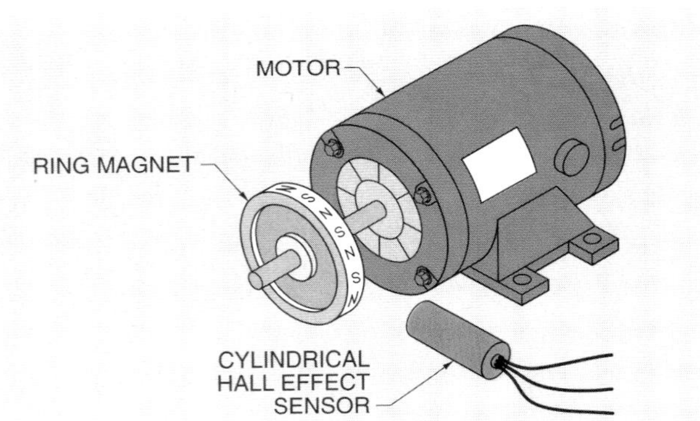

Figure 13-36. A cylindrical Hall effect sensor can be used in shaft encoding applications.

Level/Degree of Tilt. Hall effect sensors may be installed in the base of a machine to indicate the level or degree of tilt. See Figure 13-37. Magnets are installed above them in a pendulum fashion. The machine is level as long as the magnet remains directly over the sensor. A change in state of output (when a magnet swings away from a sensor) is indication that the machine is not level. The sensor/magnet combination may also be installed in such a manner as to indicate the degree of tilt.

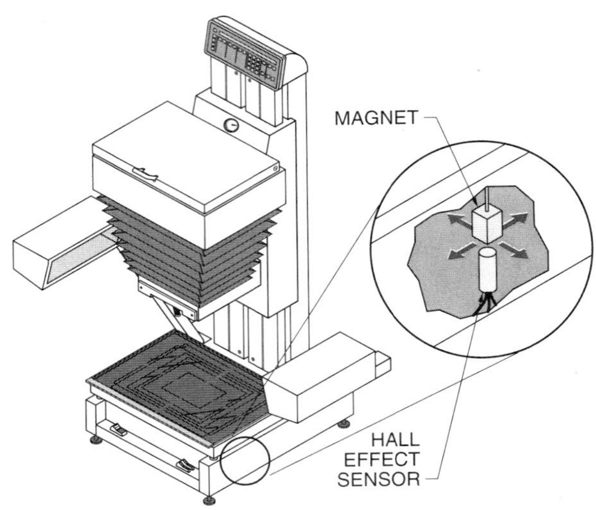

Figure 13-37. Hall effect sensors may be installed in the base of a machine to indicate the level or degree of tilt.

Joystick. A Hall effect sensor may be used in a joystick application. See Figure 13-38. The Hall effect sensors inside the joystick housing are actuated by a magnet on the joystick. The proximity of the magnet to the sensor controls activation of different outputs used to control cranes, operators, motor control circuits, wheelchairs, etc. Use of an analog device also achieves degree of movement measurements such as speed.

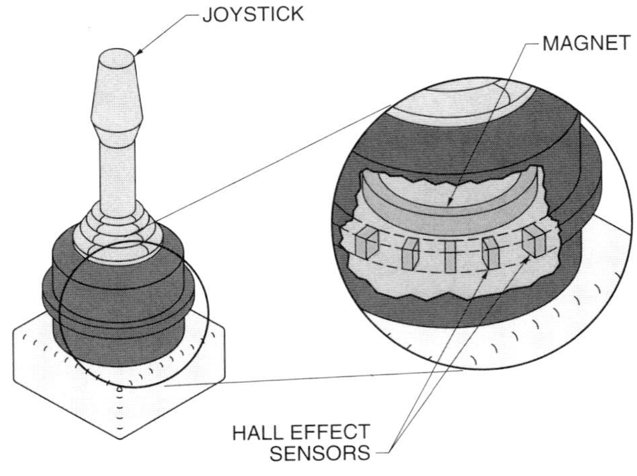

Figure 13-38. A Hall effect sensor may be used in a joystick application.

Paper Detector. A Hall effect plunger sensor can be used in printers to detect paper flow. See Figure 13-39. Hall effect sensors have no contacts to become gummy or corroded, extremely long life, low maintenance costs, and may be directly interfaced with logic circuitry.

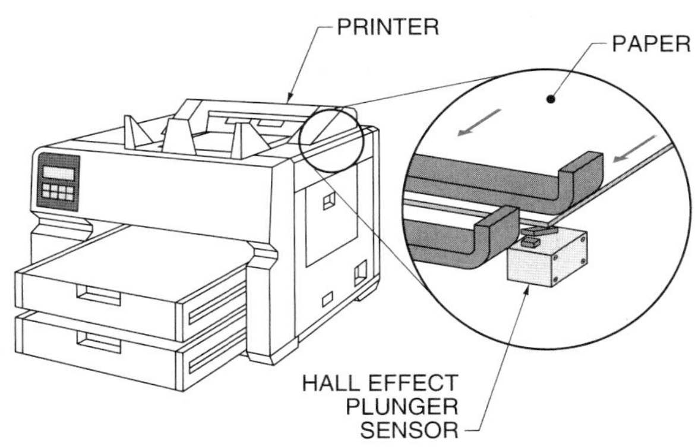

Figure 13-39. A Hall effect plunger sensor can be used in printers to detect paper flow.

PHOTOELECTRIC AND PROXIMITY OUTPUTS

Photoelectric and proximity sensors use solid-state outputs to control the flow of electric current. The solid-state output of a photoelectric or proximity sensor may be a thyristor, NPN transistor, or PNP transistor. The thyristor output is used for switching AC circuits. The NPN and PNP outputs are used for switching DC circuits. The output selected depends on specific application needs. Considerations that affect the solid-state output include:

- Voltage type to be switched (AC or DC).
- Amount of current to be switched. Most proximity sensors can only switch a maximum of a few hundred milliamperes. An interface is needed if higher current switching is required. The solid-state relay is the most common interface used with photoelectric and proximity sensors.
- Electrical requirements of the device to which the output of the proximity sensor is to be connected. Compatibility to a controller such as a programmable controller may require one type of solid-state output or the other to be used as an input into that controller.
- Required polarity of switched DC output. NPN outputs deliver a negative output and PNP outputs deliver a positive output.
- Consideration must always be given to such electrical characteristics as load current, operating current, and minimum holding current because all solid-state outputs are never truly open or closed.

AC Photoelectric and Proximity Sensors

An AC photoelectric and proximity sensor is a switch used to switch alternating current circuits. An AC sensor is connected in series with the load that it controls. The sensor is connected between Line 1 and the load to be controlled. See Figure 13-40.

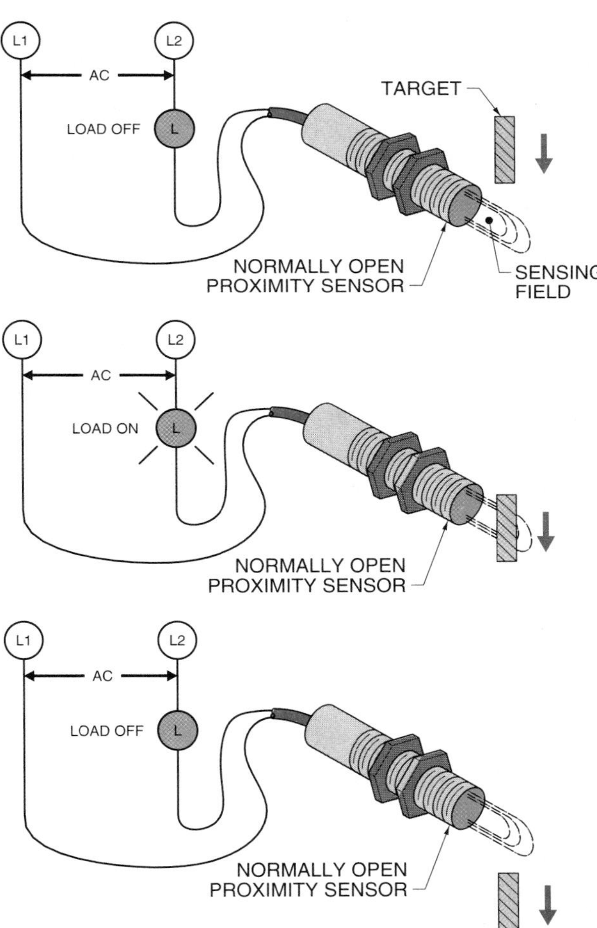

Figure 13-40. AC photoelectric and proximity sensors are connected in series with the load.

Because AC sensors are connected in series with the load, special precautions must be taken. The three main factors considered when AC sensors are connected include:

- Load current
- Operating (residual) current
- Minimum holding current

Load Current. *Load current* is the amount of current drawn by the load when energized. Since a solid-state sensor is wired in series with the load, the current drawn by the load must pass through the solid-state sensor. For example, if a load draws 5 A, the sensor must be able to safely switch 5 A. Five amperes burns out most solid-state proximity sensors because they are normally rated for a maximum of less than 0.5 A. An electromechanical or solid-state relay must be used as an interface to control the load if a solid-state sensor must switch a load above its rated maximum current. A solid-state relay is the preferred choice of interface with a sensor. See Figure 13-41.

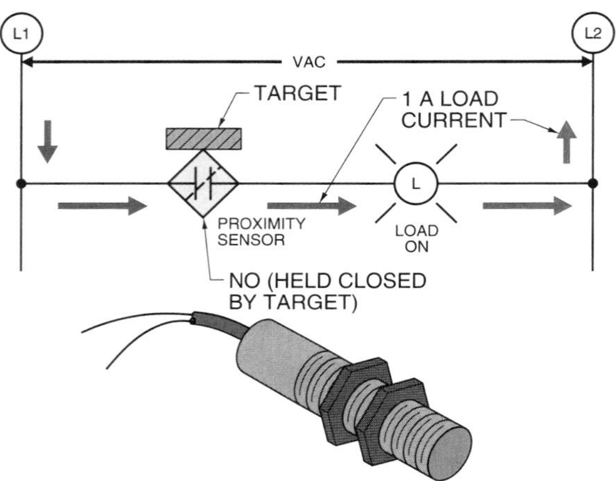

Figure 13-41. Load current is the amount of current drawn by the load when energized, and flows through photoelectric and proximity sensors.

Honeywell's MICRO SWITCH Division

Hazardous location stainless steel proximity sensors from Honeywell's MICRO SWITCH Division carry a UL® 698 explosion-proof rating for Division 1, Class I, Groups A, B, C, D, Class II, Groups E, F, G, and Class III hazardous locations.

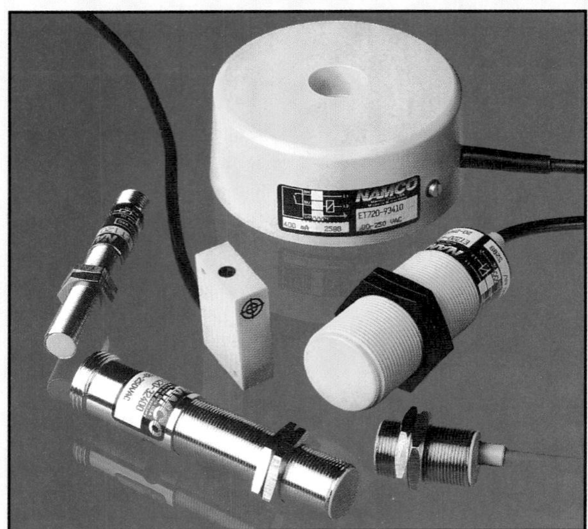

Namco Controls Corporation

The Namco ER500 Series proximity sensors are designed for special-duty applications that require extended range and the ability to withstand extreme shock and vibration.

Operating Current. *Operating current* (residual or leakage current) is the amount of current a sensor draws from the power lines to develop a field that can detect the target. When a sensor is in the OFF condition (target not detected), a small amount of current passes through both the sensor and the load. This operating current is required for the solid-state detection circuitry housed within the sensor. Operating currents are normally in the range of 1.5 mA to 7 mA for most sensors. See Figure 13-42.

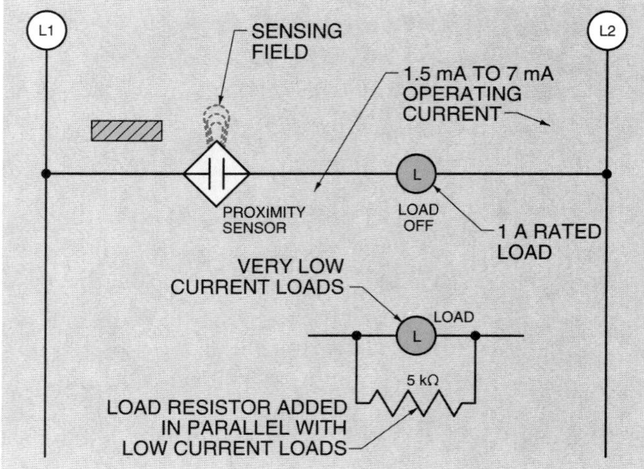

Figure 13-42. Operating current is the amount of current a sensor draws from the power lines to develop a field that can detect the target.

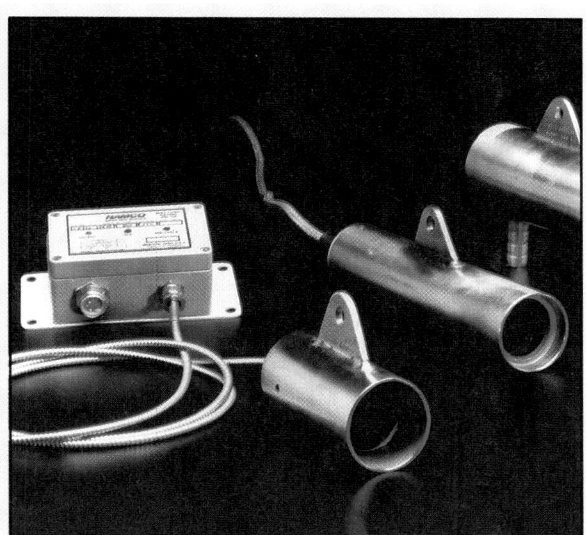

Namco Controls Corporation

The Namco ER900 Series sensors respond to infrared energy radiated by heated sufaces and are available in three preset sensing ranges to detect items up to 350°C, 450°C, and 800°C.

The small operating current normally does not have a negative effect on low-impedance loads or circuits such as mechanical relays, solenoids, and magnetic motor starters. However, the operating current may be enough to activate high impedance loads, such as programmable controllers, electronic timers, and other solid-state devices. The load is activated regardless of whether a target is present or not. This problem may be corrected by placing a resistor in parallel with the load. The resistance value should be selected to ensure the effective load impedance (load plus resistor) is reduced to a level that prevents false triggering due to the operating current, and also ensures the minimum current required to operate the load is provided. This resistance value is normally in the range of 4.5 kΩ to 7.5 kΩ. A general rule is to use a 5 kΩ, 5 W resistor for most conditions.

Minimum Holding Current. *Minimum holding current* is the minimum amount of current required to keep a sensor operating. When the sensor has been triggered and is in the ON condition (target detected), the current drawn by the load must be sufficient to keep the sensor operating. Minimum holding currents range from 3 mA to 20 mA for most solid-state sensors. The amount of current a load draws is important for the proper operation of a sensor. Excessive current (operating current) burns up the sensor. Low current (below minimum holding current) prevents proper operation of the sensor. See Figure 13-43.

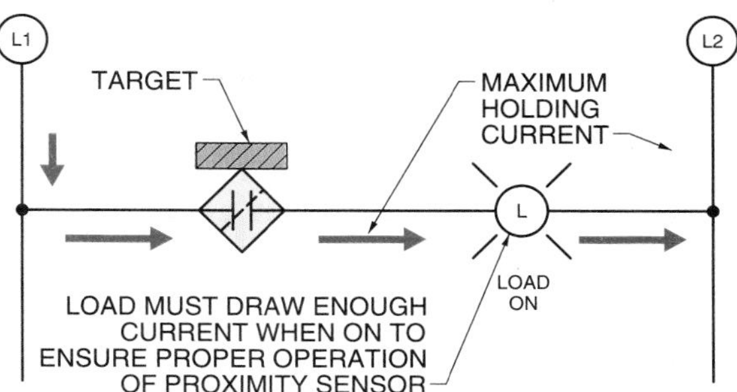

Figure 13-43. Minimum holding current is the minimum amount of current required to keep a sensor operating.

Series/Parallel Connections. All AC, 2-wire photoelectric and proximity sensors may be connected in series or in parallel to provide AND and OR control logic. When connected in series (AND logic), all sensors must be activated to energize the load. When connected in parallel (OR logic), any one sensor that is activated energizes the load. See Figure 13-44.

As a general rule, a maximum of three sensors may be connected in series to provide AND logic. Factors that limit the number of AC, two-wire sensors that may be wired in series to provide AND logic include:

- AC supply voltage. Generally, the higher the supply voltage, the higher the number of sensors that may be wired in series.
- Voltage drop across the sensor. This varies for different sensors. The lower the voltage drop, the higher the number of sensors that may be connected in series.

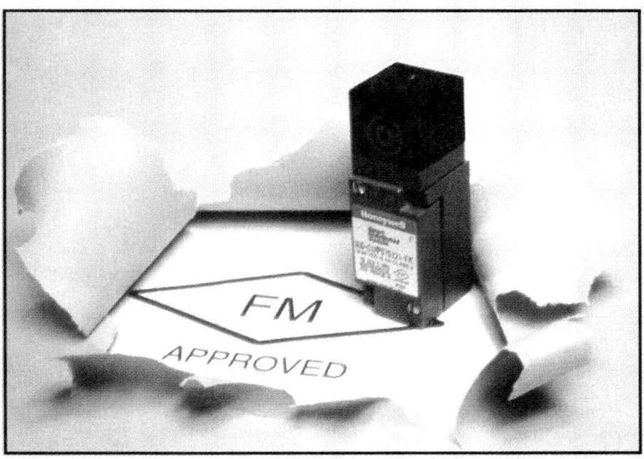

Honeywell's MICRO SWITCH Division

The smart distributed system harsh-duty limit switch-style proximity sensor from Honeywell's MICRO SWITCH Division is factory mutual approved for use in harsh environments.

- Minimum operating load voltage. This varies depending upon the load that is controlled. For every proximity sensor added in series with the load, less supply voltage is available across the load.

As a general rule, a maximum of three sensors may be connected in parallel to provide OR logic. Factors that limit the number of AC, two-wire sensors that may be wired in parallel to provide OR logic include:

- Photoelectric and proximity switch operating current. The total operating current flowing through the load is equal to the sum of each sensor's operating current. The total operating current must be less than the minimum current required to energize the load.
- Amount of current the load draws when energized. The total amount of current the load draws must be less than the maximum current rating of the lowest rated sensor. For example, if three sensors rated at 125 mA, 250 mA, and 275 mA are connected in parallel, the maximum rating of the load cannot exceed 125 mA.

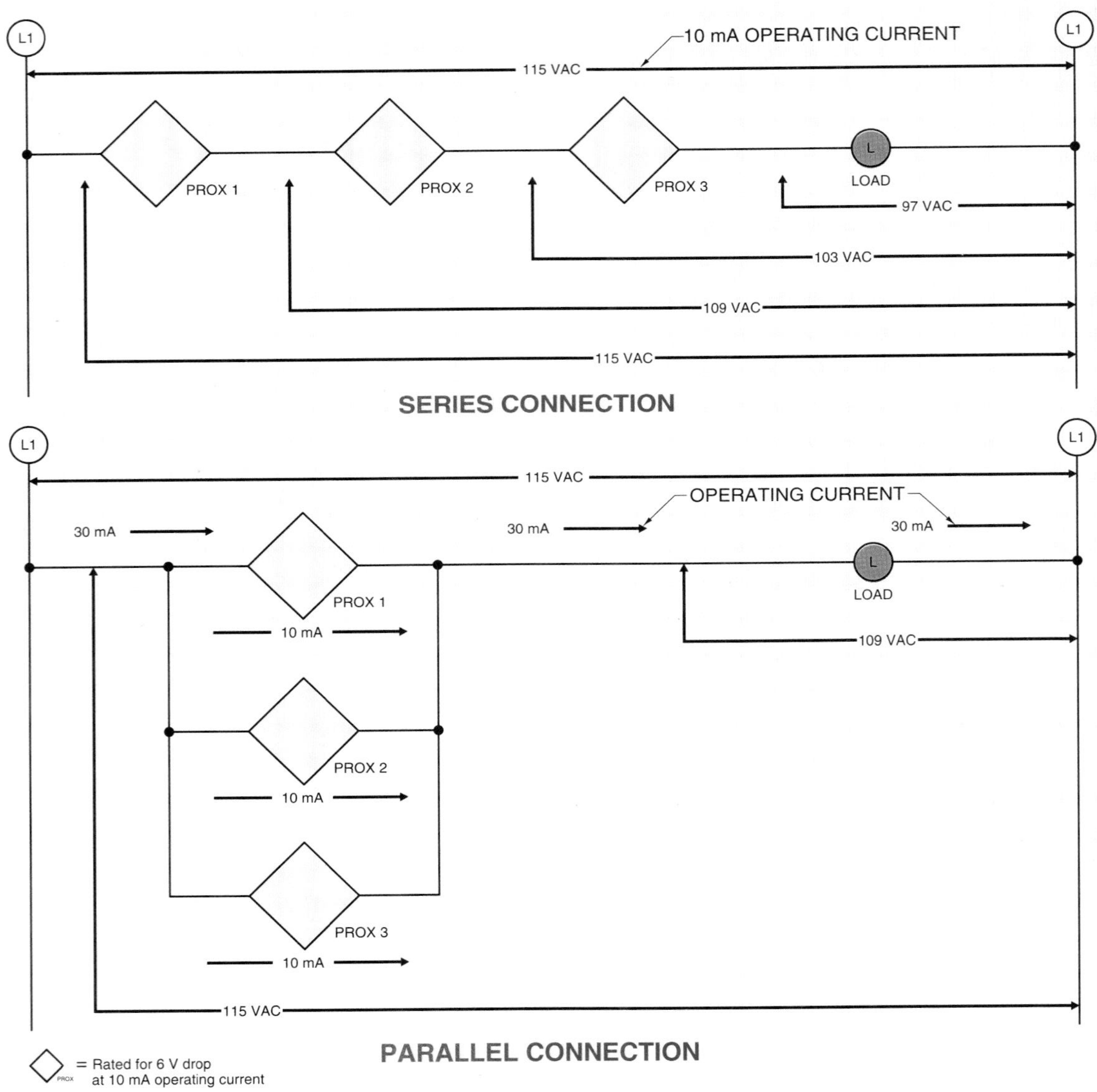

Figure 13-44. AC photoelectric and proximity sensors may be connected in series or parallel.

DC Photoelectric and Proximity Sensors

Photoelectric and proximity sensors that switch DC circuits normally use transistors as the switching element. The sensors use NPN transistors or PNP transistors. For most applications, the exact transistor used does not matter, as long as the switch is properly connected into the circuit. However, NPN transistor sensors are far more common than PNP transistor sensors.

NPN Transistor Switching. When using an NPN transistor, the load is connected between the positive terminal of the supply voltage and the output terminal (collector) of the sensor. When the sensor detects a target, current flows through the transistor and the load is energized. See Figure 13-45.

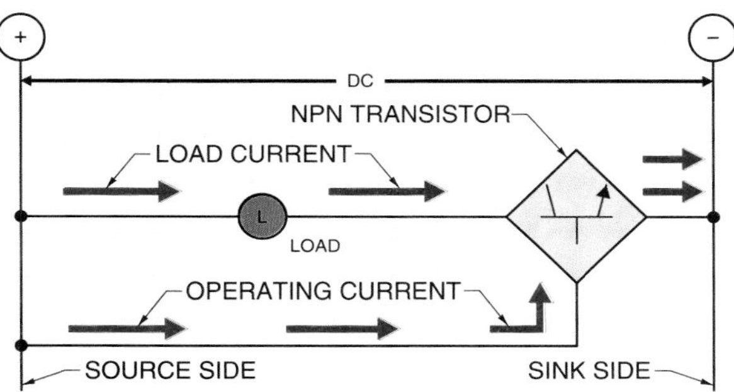

Figure 13-45. When an NPN (current sink) transistor is used, the load is connected between the positive terminal of the supply voltage and the output terminal of the sensor.

Output devices that use an NPN transistor as the switching element are current sink devices. This is because the negative terminal of a DC system is the sink because conventional current flows into it. A current sinking switch "sinks" the current from the load.

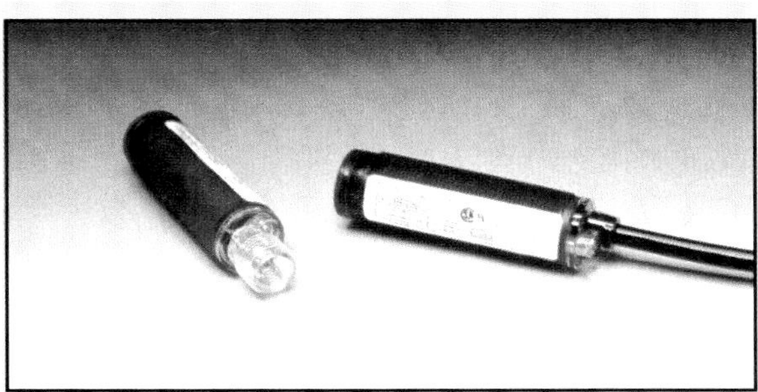

Honeywell's MICRO SWITCH Division

The new CP18 Series polarized photoelectric control from Honeywell's MICRO SWITCH Division was developed for applications that require the reliable sensing of highly reflective materials.

PNP Transistor Switching. When using a PNP transistor, the load is connected between the negative terminal of the supply voltage and the output terminal (collector) of the sensor. When the sensor (current source) detects a target, current flows through the transistor and the load is energized. See Figure 13-46.

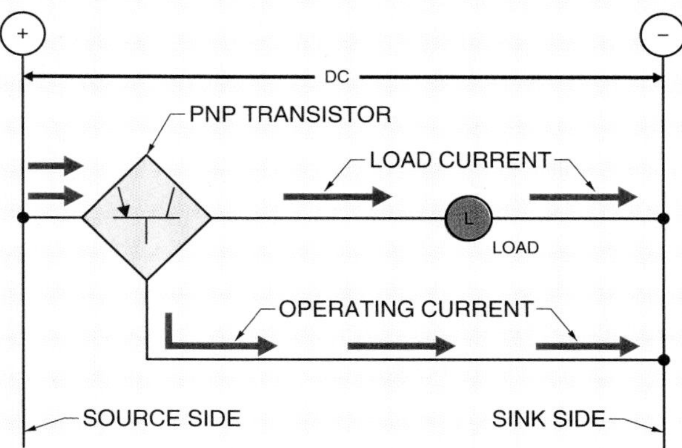

Figure 13-46. When a PNP (current source) transistor is used, the load is connected between the negative terminal of the supply voltage and the output terminal of the sensor.

PROXIMITY SENSOR INSTALLATION

Proximity sensors have a sensing head that produces a radiated sensing field. The sensing field detects the target of the sensor. The sensing field must be kept clear of interference for proper operation. *Interference* is any object other than the object to be detected that is sensed by the sensor. Interference may come from objects close to the sensor or from other sensors. General clearances are required for most proximity sensors.

Flush-Mounted Inductive and Capacitive Proximity Sensors

A distance equal to or greater than twice the diameter of the sensors is required between sensors when flush mounting inductive and capacitive proximity sensors. The diameter of the largest sensor is used for installation when two sensors of different diameters are used. For example, at least 16 mm is required between sensors if two 8 mm inductive proximity sensors are flush mounted. See Figure 13-47.

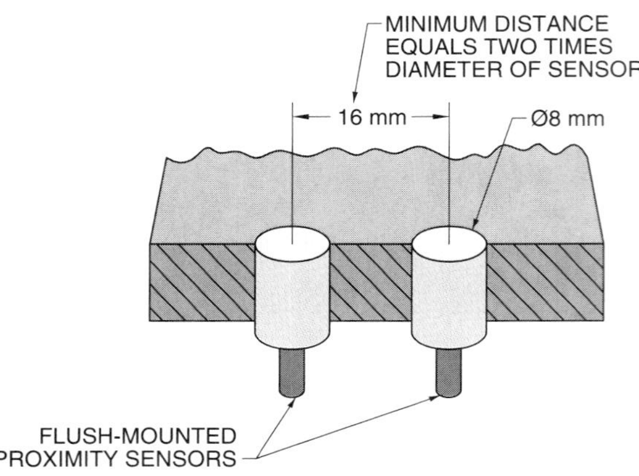

Figure 13-47. A distance equal to or greater than twice the diameter of the sensors is required between sensors when flush mounting inductive and capacitive proximity sensors.

Nonflush-Mounted Inductive and Capacitive Proximity Sensors

A distance of three times the diameter of the sensor is required within or next to a material that may be detected when using nonflush-mounting inductive and capacitive proximity sensors. For example, at least 48 mm is required between sensors if two 16 mm capacitive proximity sensors are nonflush-mounted. See Figure 13-48.

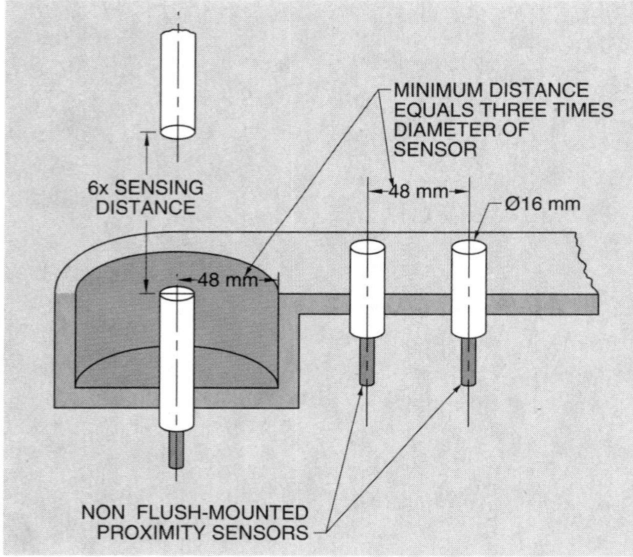

Figure 13-48. A distance of three times the diameter of the sensor is required within or next to a material that may be detected when using nonflush-mounting inductive and capacitive proximity sensors.

Three times the diameter of the largest sensor is required when inductive and capacitive proximity sensors are installed next to each other. Spacing is measured from center to center of the sensors.

Six times the rated sensing distance is required for proper operation when inductive and capacitive proximity sensors are mounted opposite each other. Six times the rated sensing distance is required because the sensing field causes false readings on the other.

Mounting Photoelectric Sensors

A photoelectric sensor transmits a light beam. The light beam detects the presence (or absence) of an object. Only part of the light beam is effective when detecting the object. The effective light beam is the area of light that travels directly from the transmitter to the receiver. The object is not detected if the object does not completely block the effective light beam.

The receiver is positioned to receive as much light as possible from the transmitter when mounting photoelectric sensors. Greater operating distances are allowed and more power is available for the system to see through dirt in the air and on the transmitter and receiver lenses because more light is available at the receiver. The transmitter is mounted on the clean side of the detection zone because light scattered by dirt on the receiver lens affects the system less than light scattered by dirt on the transmitter lens. See Figure 13-49.

Omron Electronics, Inc.

Omron's E35-AD short-range diffuse reflective photoelectric sensors have a strong, short-range beam that can detect dark or dull objects while ignoring background objects.

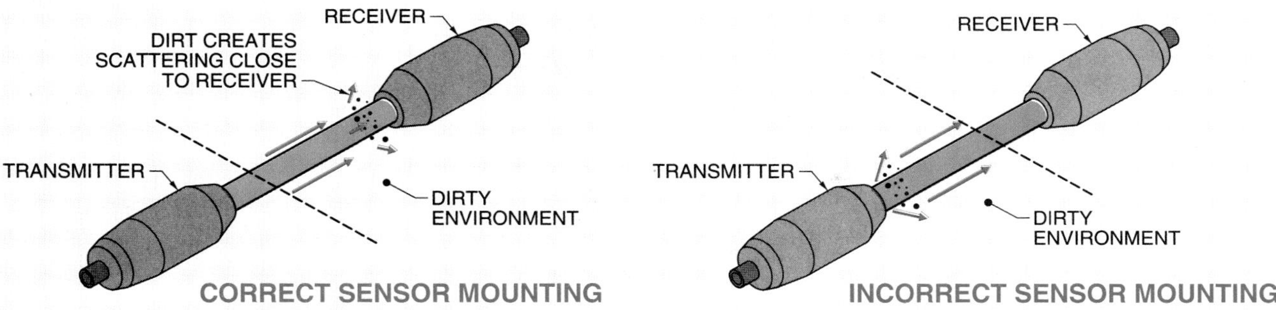

Figure 13-49. The transmitter is mounted on the clean side of the detection zone because light scattered by dirt on the receiver lens affects the system less than light scattered by dirt on the transmitter lens.

TROUBLESHOOTING PHOTOELECTRIC AND PROXIMITY SWITCHES

Photoelectric and proximity switches typically have solid-state output switches. A solid-state switch has no moving parts (contacts). A solid-state switch uses a triac, SCR, current sink (NPN) transistor, or current source (PNP) transistor output to perform the switching function. The triac output is used for switching AC loads. The SCR output is used for switching high-power DC loads. The current sink and current source outputs are used for switching low-power DC loads. Solid-state switches include normally-open, normally-closed, or combination switching outputs. See Figure 13-50.

OUTPUT SWITCHING DEVICES

Device	Use
TRIAC	SWITCH AC LOADS
SCR	SWITCH HIGH-POWER DC LOADS
NPN TRANSISTOR, PNP TRANSISTOR	SWITCH LOW-POWER DC LOADS

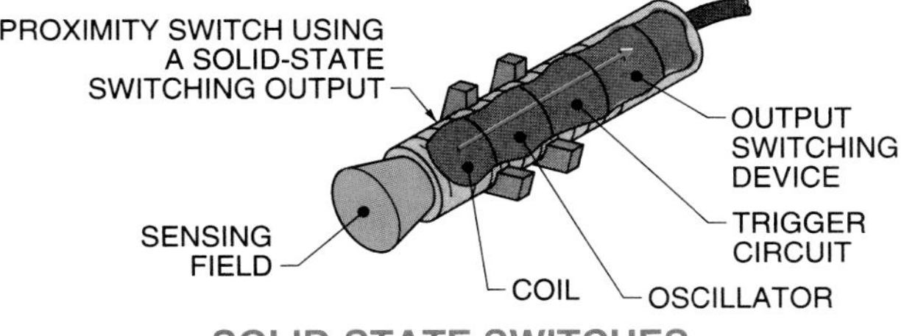

SOLID-STATE SWITCHES

Figure 13-50. A solid-state switch uses a triac, SCR, current sink transistor, or current source transistor output to perform the switching function.

Two-Wire Solid-State Switches

A two-wire solid-state switch has two connecting terminals or wires (exclusive of ground). A two-wire switch is connected in series with the controlled load. A two-wire solid-state switch is also referred to as a load-powered switch because it draws operating current through the load. The operating current flows through the load when the switch is not conducting (load OFF). This operating current is inadequate to energize most loads. Operating current is also referred to as residual current or leakage current by some manufacturers. Operating current may be measured with an ammeter when the load is OFF. See Figure 13-51.

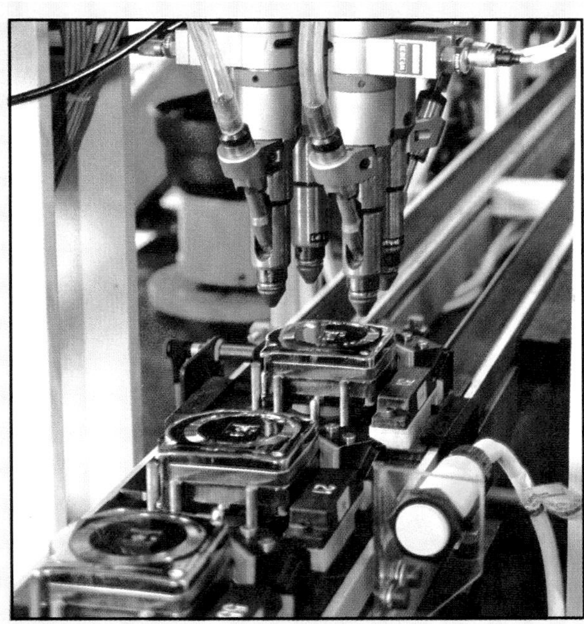

Advanced Assembly Automation Inc.

Proximity sensors are used in manufacturing and assembly systems to control the positioning of products.

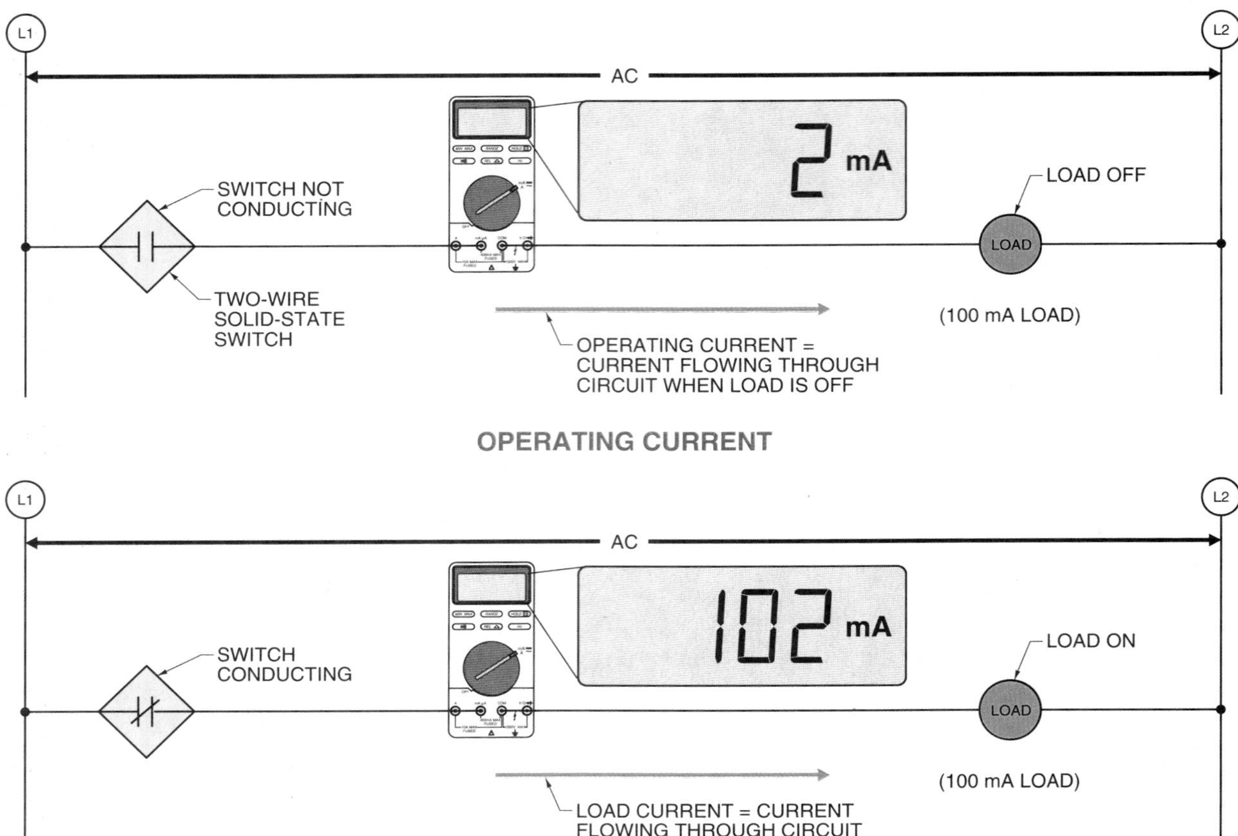

Figure 13-51. A two-wire solid-state switch is also referred to as a load-powered switch because it draws operating current through the load.

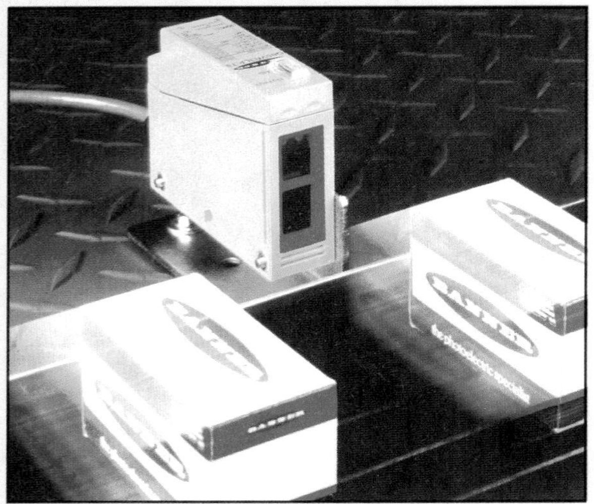

Banner Engineering Corp.

The new Conveyor-Beam™ Sensors from Banner are designed to operate on either 24-240 VAC or 12-240 VDC and draw a maximum of only 2 W with a sensing range of 13.1′.

The current in a circuit is a combination of the operating current and load current when a switch is conducting (load ON). A solid-state switching device must be rated high enough to carry the current of the load. Load current is measured with an ammeter when the load is ON.

The current draw of a load must be sufficient to keep the solid-state switch operating when the switch is conducting (load ON). Minimum holding current is the minimum current that ensures proper operation of a solid-state switch. Minimum holding current values range from 2 mA to 20 mA.

Operating current and minimum holding current values are normally not a problem when a solid-state switch controls a low-impedance load, such as a motor starter, a relay, or a solenoid. Operating current and minimum holding current values may be a problem when a solid-state switch controls a high-impedance load, such as a PLC or other solid-state device.

The operating current may be high enough to affect the load when the switch is not conducting. For example, a programmable controller may see the operating current as an input signal.

A load resistor must be added to the circuit to correct this problem. A load resistor is connected in parallel with the load. The load resistor acts as an additional load which increases the total current in the circuit. Load resistors range in value from 4.5 kΩ to 7 kΩ. A 5 kΩ, 5 W resistor is used in most applications. See Figure 13-52.

Two-wire solid-state switches connected in series affect the operation of the load because of the voltage drop across the switches. A two-wire switch drops about 3 V to 8 V. The total voltage drop across the switches equals the sum of the voltage drop across each switch. No more than three solid-state switches should be connected in series. See Figure 13-53.

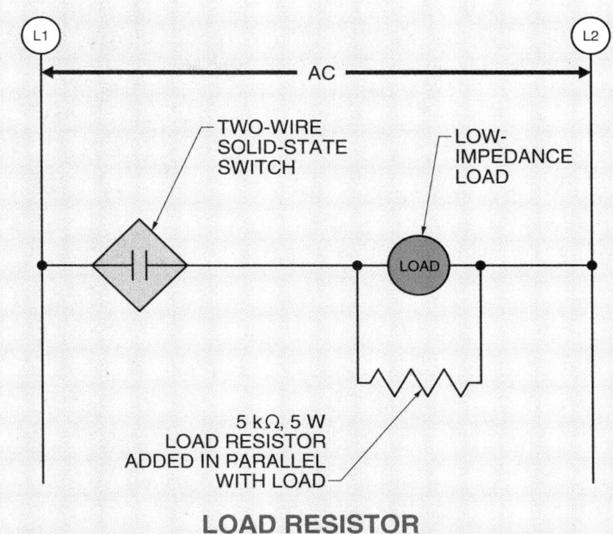

Figure 13-52. A load resistor acts as an additional load which increases the total current in the circuit.

TWO-WIRE SOLID-STATE SWITCHES

L1
L2
120 VAC
SWITCH 1
SWITCH 2
SWITCH 3
LOAD
6 V DROP
6 V DROP
6 V DROP
102 V ACROSS LOAD
SERIES-CONNECTED

L1
L2
SWITCH 1
2 mA OPERATING CURRENT
SWITCH 2
2 mA OPERATING CURRENT
SWITCH 3
2 mA OPERATING CURRENT
LOAD
6 mA OPERATING CURRENT
PARALLEL-CONNECTED

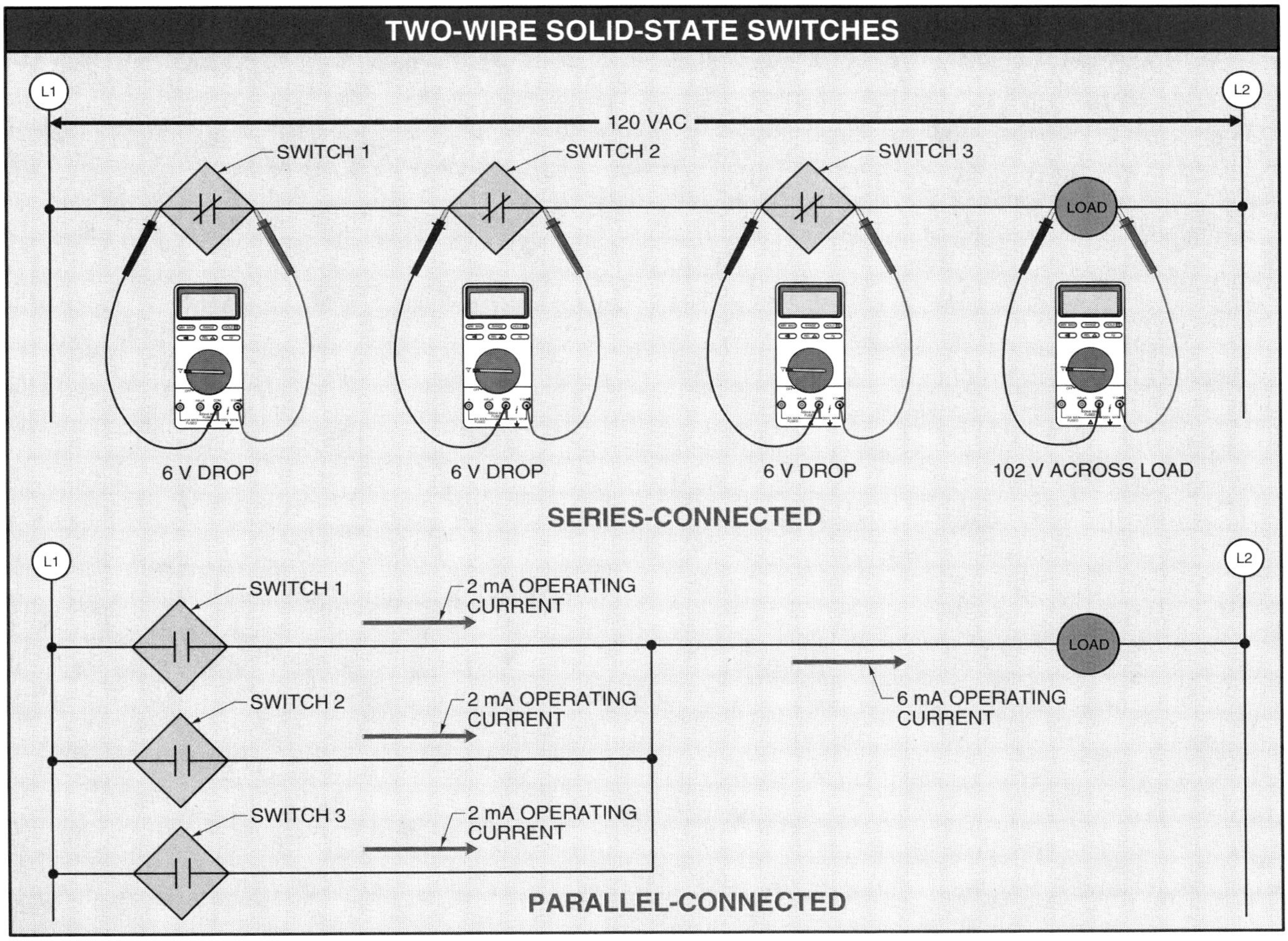

Figure 13-53. A two-wire solid-state switch has two connecting terminals or wires (exclusive of ground).

Two-wire solid-state switches connected in parallel affect the operation of the load because each switch has its operating current flowing through the load. The load may turn ON if the current through the load becomes excessive. The total operating current equals the sum of the operating current of each switch. No more than three solid-state switches should be connected in parallel.

A suspected fault in a two-wire solid-state switch may be tested using a voltmeter. The voltmeter is used to test the voltage into and out of the switch. See Figure 13-54.

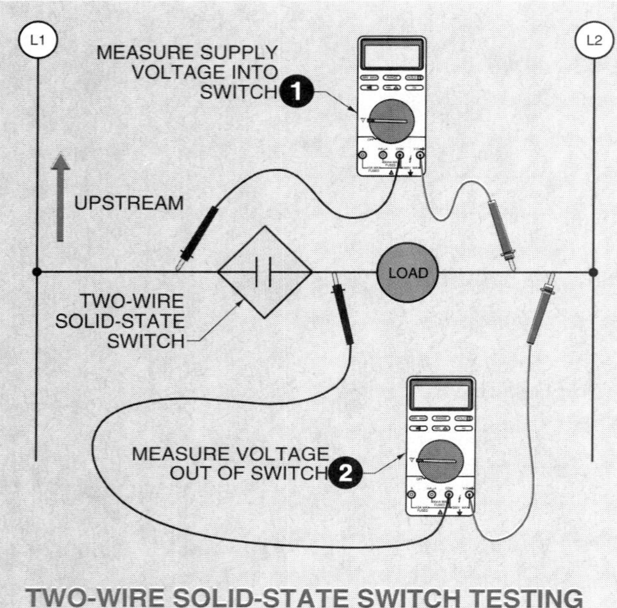

Figure 13-54. A suspected fault in a two-wire solid-state switch may be tested using a voltmeter.

To test a two-wire solid-state switch, apply the procedure:

1. Measure the voltage into the switch. The problem is located upstream from the switch when there is no voltage present or the voltage is not at the correct level. The problem may be a blown fuse or open circuit. Voltage must be re-established to the switch before the switch may be tested.
2. Measure the voltage out of the switch. The voltage should equal the supply voltage minus the rated voltage drop (3 V to 8 V) of the switch when the switch is conducting (load ON). Replace the switch if the voltage output is not correct.

Warning: Always ensure power is OFF before changing a control switch. Use a voltmeter to ensure the power is OFF.

Three-Wire Solid-State Switches

Three-wire solid-state switches have three connecting terminals or wires (exclusive of ground). A three-wire solid-state switch draws its operating current directly from the power lines. The operating current does not flow through the switch. See Figure 13-55.

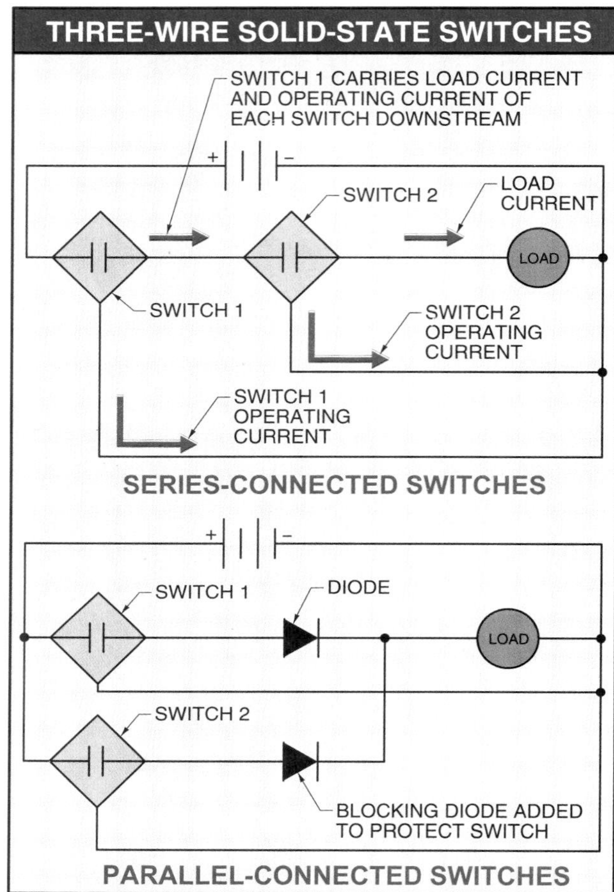

Figure 13-55. A three-wire solid-state switch draws its operating current directly from the power lines.

Three-wire solid-state switches connected in series affect the operation of the load because each switch downstream from the last switch must carry the load current and the operating current of each switch. An ammeter may be used to measure operating and load current values. The measured values must not exceed the manufacturer's maximum rating.

Three-wire solid-state switches connected in parallel affect the operation of the load because the non-conducting switch may be damaged due to reverse polarity. A blocking diode should be added to each switch output to prevent reverse polarity on the switch.

A suspected fault with a three-wire solid-state switch may be tested using a voltmeter. The voltmeter is used to test the voltage into and out of the switch. See Figure 13-56.

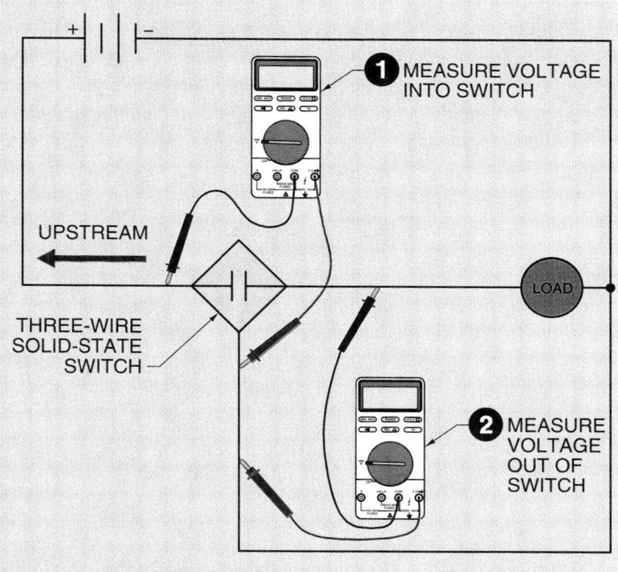

Figure 13-56. A suspected fault with a three-wire solid-state switch is tested by checking the voltage into and out of the switch.

To test a three-wire solid-state switch, apply the procedure:

1. Measure the voltage into the switch. The problem is located upstream from the switch when there is no voltage present or the voltage is not at the correct level. The problem may be a blown fuse or open circuit. Voltage must be re-established to the switch before the switch may be tested.
2. Measure the voltage out of the switch. The voltage should equal the supply voltage when the switch is conducting (load ON). Replace the switch if the voltage output is not correct.

Warning: Always ensure power is OFF before changing a control switch. Use a voltmeter to ensure the power is OFF.

Protecting Switch Contacts

Switches are rated for the amount of current they may switch. The switch rating is usually specified for switching a resistive load. Resistive loads are the least destructive loads to switch. Most loads that are switched are inductive loads, such as solenoids, relays, and motors. Inductive loads are the most destructive loads to switch.

A large induced voltage appears across the switch contacts when inductive loads are turned OFF. The induced voltage causes arcing at the switch contacts. Arcing may cause the contacts to burn, stick, or weld together. Contact protection should be added when frequently switching inductive loads to prevent or reduce arcing. See Figure 13-57.

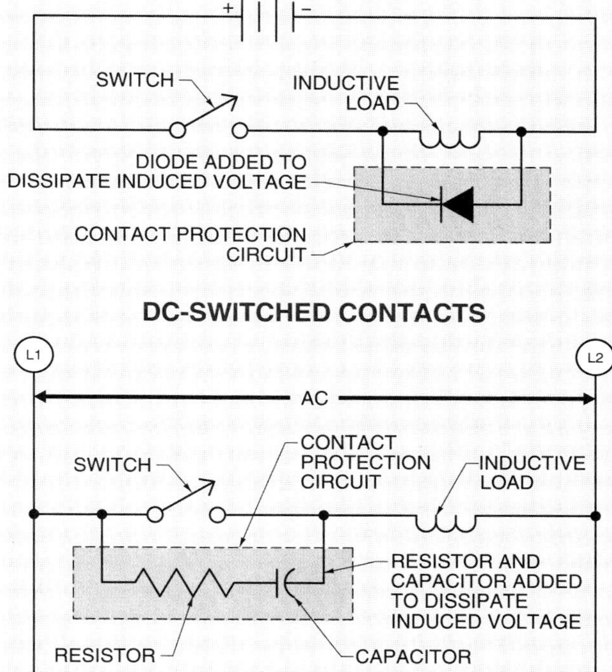

Figure 13-57. Contact protection should be added when frequently switching inductive loads to prevent or reduce arcing.

A diode is added in parallel with the load to protect contacts that switch DC. The diode does not conduct when the load is ON. The diode conducts when the switch is open, providing a path for the induced voltage in the load to dissipate.

A resistor and capacitor are connected across the switch contacts to protect contacts that switch AC. The capacitor acts as a high impedance (resistor) load at 60 Hz, but becomes a short circuit at the high frequencies produced by the induced voltage of the load. This allows the induced voltage to dissipate across the resistor when the load is switched OFF.

1. What is direct scan?
2. When may an aperture have to be used?
3. What is retroreflective scan?
4. What scan method works best in a high-vibration area?
5. What is specular scan?
6. What is diffuse scan?
7. What is convergent beam scan?
8. What light source (modulated or unmodulated) should generally be the first choice for a given application?
9. In what type of applications is the rated response time of a photoelectric control most important?
10. When does a dark-operated photoelectric control energize the output switch?
11. When does a light-operated photoelectric control energize the output switch?
12. What are the two basic types of proximity sensors?
13. What does a Hall effect sensor detect?
14. What type of solid-state output is used to switch AC circuits?
15. What type of solid-state output is used to switch DC circuits?
16. What is load current?
17. What is another name for operating current?
18. What is minimum holding current?
19. What is the maximum number of 2-wire AC sensors that should generally be connected in series?
20. What type of transistor (NPN or PNP) is used most frequently as an output?

14 Chapter

Programmable Controllers

Programmable controllers are used to automatically control many industrial processes. Programmable controllers are ruggedly constructed and allow for easy circuit modification. Programmable controllers include the power supply, input/output section, processor section, and programming section. Solid-state controls are easily interfaced with programmable controllers. Multiplexing permits the transmittal of more than one signal over a single transmission system.

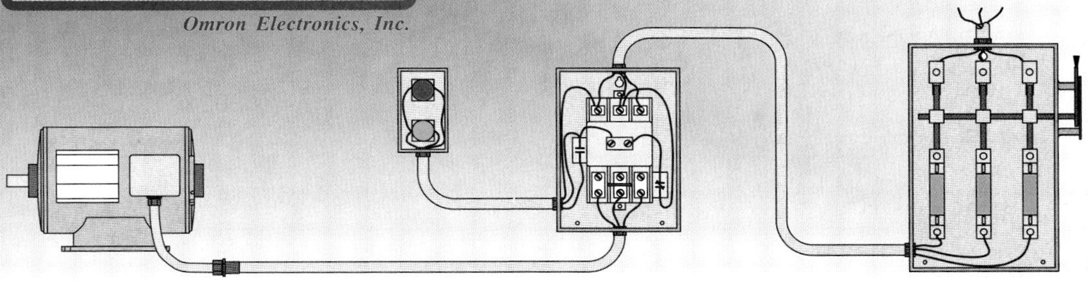

Omron Electronics, Inc.

PROGRAMMABLE CONTROLLERS

A *programmable controller (PLC)* is a solid-state control device that is programmed and reprogrammed to automatically control an industrial process or machine. PLCs are capable of many industrial functions and applications and are widely used in automated industrial applications.

The automotive industry was the first to recognize the advantages of PLCs. Annual model changes required constant modifications of production equipment controlled by relay circuitry. In some cases, entire control panels had to be scrapped and new ones designed and built with new components. This resulted in increased production costs.

The automotive industry was looking for equipment that could reduce changeover costs required by model changes. In addition, the equipment had to operate in a harsh factory environment of dirty air, vibration, electrical noise, and a wide range of temperatures and humidity.

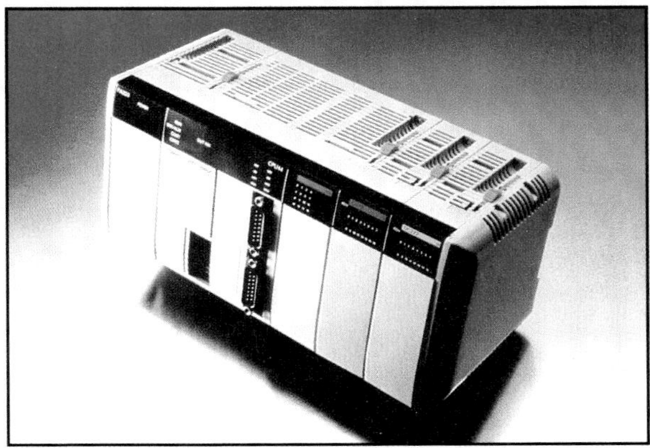

Omron Electronics, Inc.

Omron's modular CQM1 Series PLC offers 16 built-in DC inputs, programmable interrupts, a built-in RS-232C port, and a wide range of discrete I/O modules to control freestanding machines or to automate an in-line process.

To meet this need, a ruggedly constructed computer-like control was developed. The PLC could easily accommodate constant circuit change using a keyboard to introduce new operation instructions. In 1968, the first PLC was delivered to General Motors (GM) in Detroit by Modicon (now Gould).

The first PLCs were large and costly. Their initial use was in large systems with the equivalent of 100 or more relays. Today, PLCs are available in all sizes from micro, which is cost-effective to the equivalent of as few as 10 relays, to large units with the equivalent of thousands of inputs and outputs.

PLCs are popular because they can be programmed and reprogrammed using ladder (line) diagrams that plant personnel understand. Required machine operation is programmed and read as a line diagram showing open and closed contacts. This is the same approach used to describe relay logic circuits. This allowed the use of a computer-like device without learning a computer language. Advantages of using a PLC include:

- Reduction in hard wiring and wiring cost
- Reduced space requirements due to a small size as compared to using standard relays, timers, counters, and other control components
- Flexible control because all operations are programmable
- High reliability using solid-state components
- Storage of large programs and data due to microprocessor-based memory
- Improved on-line monitoring and troubleshooting by monitoring and diagnosing its own failures as well as the machines and processes it controls
- Elimination of the need to stop a controlled process to change set parameters
- Provides for analog, digital, and voltage inputs as well as discrete inputs, such as pushbuttons and limit switches
- Modular design allows components to be added, substituted, and rearranged as requirements change
- Programming languages used are familiar and follow industrial standards, such as line diagrams

Although the first PLCs were designed to replace relays, today's PLCs are used to achieve factory automation and interfacing with robots, numerical control (NC) equipment, CAD/CAM systems, and general-purpose computers. PLCs are used in almost all segments of industry where automation is required. See Figure 14-1.

Giddings & Lewis, Inc.

Figure 14-1. PLCs are used to achieve factory automation and interfacing with robots, numerical control equipment, CAD/CAM systems, and general-purpose computers.

PLC Usage

Some electrical components, like pushbuttons and fuses, are used in most types of residential, commercial, and industrial electrical systems. Other electrical components, like PLCs, are used primarily in only one type of electrical system. PLCs are commonly used in industrial electrical systems that are designed to manufacture a product. Industrial electrical systems designed to produce products are commonly divided into discrete parts manufacturing and process manufacturing. In discrete parts and process manufacturing, PLCs have become the standard component used to control the operation from start to finish.

Discrete Parts Manufacturing. The discrete parts manufacturing market area represents durable goods such as automobiles, washers, refrigerators, and tractors. Discrete parts manufacturing is done primarily by stand-alone machines that bend, drill, punch, grind, and shear metals. All of these machines can be automated with PLCs. See Figure 14-2.

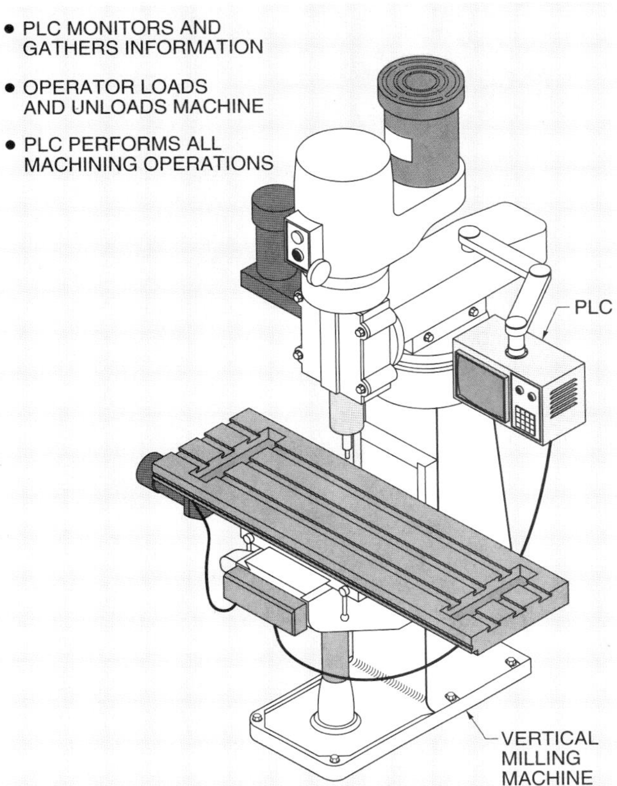

Figure 14-2. A PLC can be used to control all electrical functions on a machine used in discrete parts manufacturing.

The PLC allows each machine to have its own unique capability using standard hardware. The PLC allows easy modification of the controls when the functional requirements of the machine change. Modular replacement of PLCs reduces downtime of the machine.

PLCs allow flexibility in the machine function because every system is unique. PLC use helps reduce start-up and de-bug time, and allows manufacturers to incorporate additional user requirements for changes in machine operations after start-up.

Today, the PLC has become the standard for machine builders. Increased capabilities in a reduced size allow today's PLCs to control one machine or link up to many machines in any network configuration.

In addition to allowing each machine to have its own unique capabilities, a PLC can also be used to interface and control the operation of all (or parts) of the machines along a production line. PLCs can be used to control the speed of a production line, divert production to other lines when there is a problem, make product changes, and maintain documentation, such as inventory and losses. See Figure 14-3.

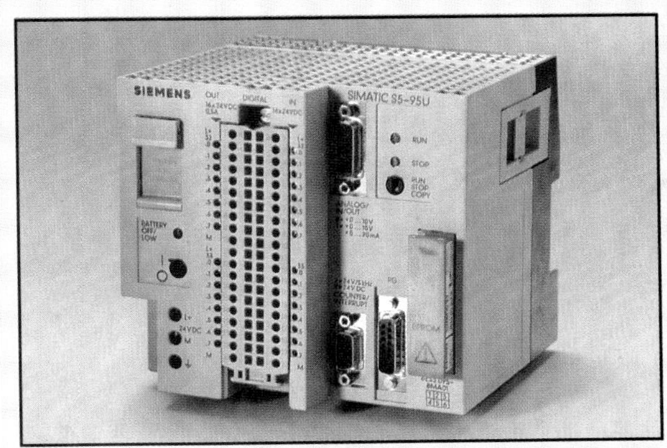

Siemens Corporation

The SIMATIC® S5-95U mini controller from Siemens is designed for jobs that require fast counting or a real time clock such as packaging and materials-handling operations.

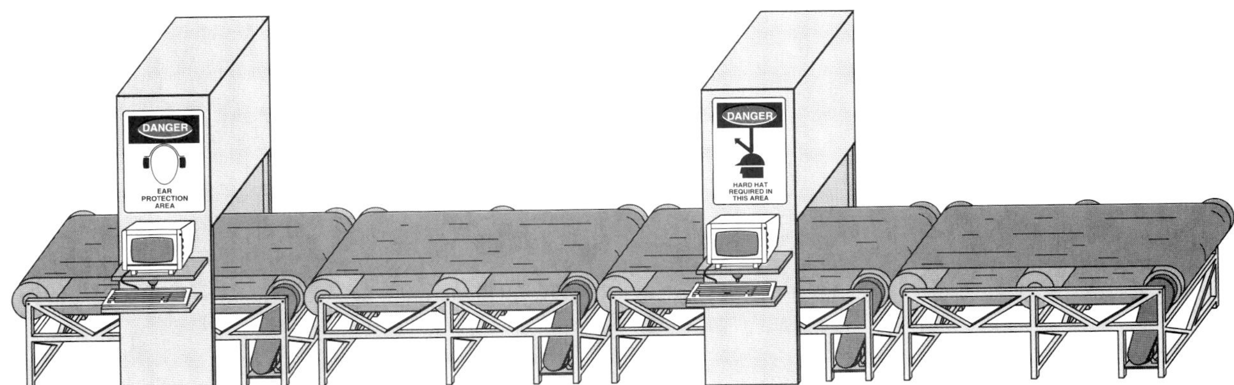

Figure 14-3. PLCs can be used to control the speed of a production line, divert production to other lines when there is a problem, make product changes, and maintain documentation, such as inventory and losses.

Process Manufacturing. The process manufacturing market area produces consumables such as food, gas, paint, pharmaceuticals, and chemicals. Most of these processes require systems to blend, cook, dry, separate, or mix ingredients. See Figure 14-4.

Automation is required for opening and closing valves and controlling motors in the proper sequence and at the correct time. A PLC allows for easy modifications to the system if the time, temperature, or flow requirements of the products change.

Today's PLCs control process manufacturing, such as the conveying of the product, palletizing, storage, treatment, alarms, interlocks, and preventive maintenance functions for the system. The PLC can also generate reports that are used to determine production efficiency.

PLC manufacturers offer a variety of PLCs from micro to very large units. See Figure 14-5. A micro or small PLC is the best choice for machines and processes that have limited capability and little potential for future expansion.

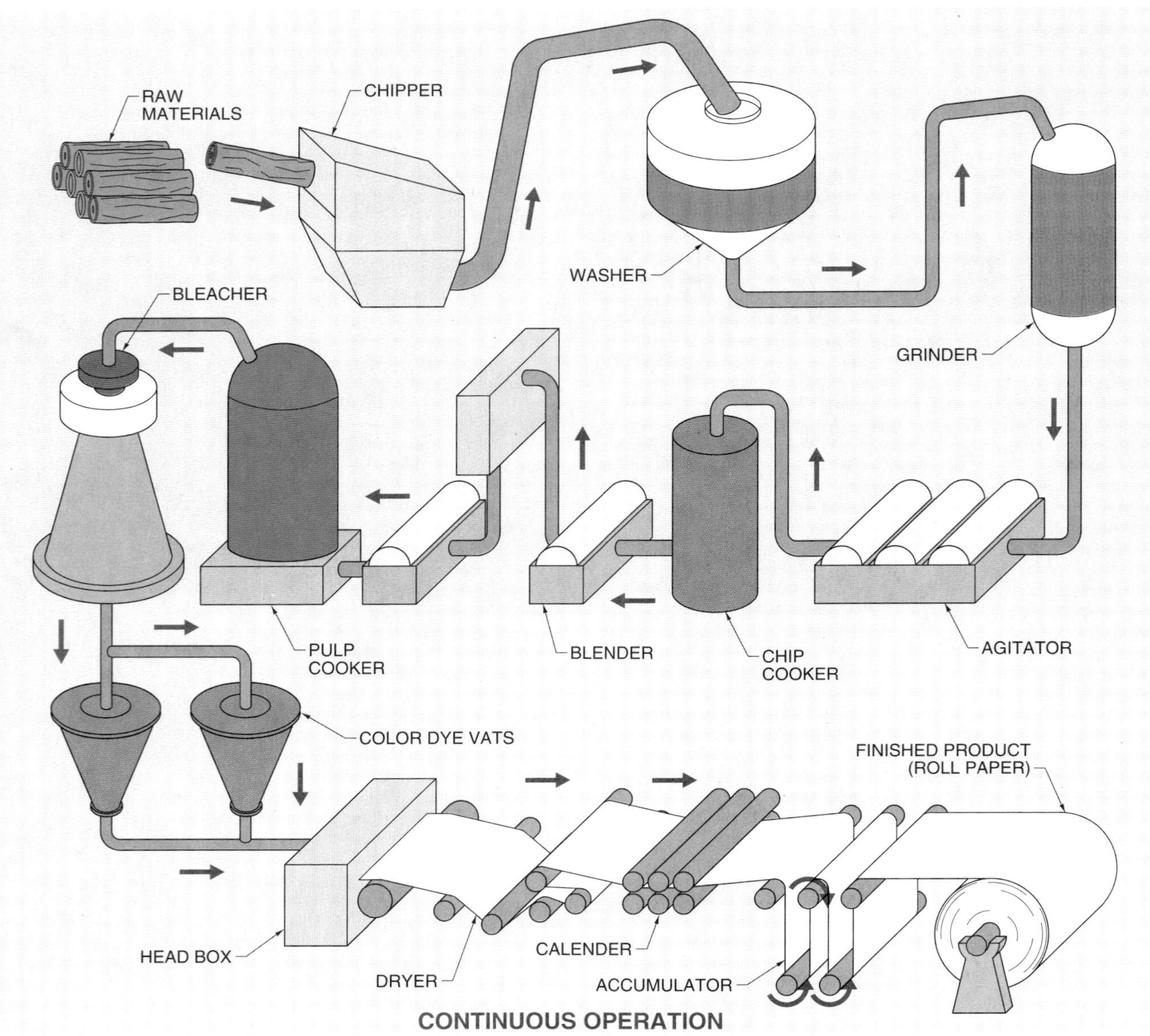

Figure 14-4. The process manufacturing market area produces consumables such as food, gas, paint, pharmaceuticals, and chemicals which require systems to blend, cook, dry, separate, or mix ingredients.

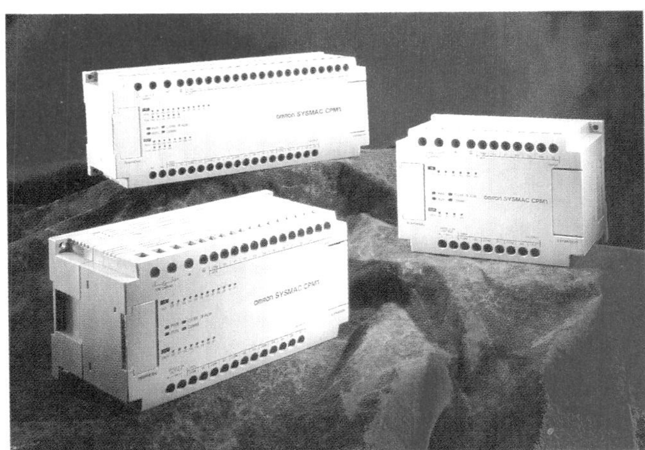

Omron Electronics, Inc.

Figure 14-5. PLC manufacturers offer a variety of PLCs for machines and processes that have limited capability and little potential for future expansion and for processes that have complex control requirements.

PLCs AND MICROCOMPUTERS

PLCs have grown in popularity in applications which were once performed exclusively by microcomputers. Microcomputers feature fast number manipulation and powerful text-handling capabilities. PLCs offer several advantages for industrial control applications that microcomputers do not.

The first difference between a PLC and a microcomputer is that a PLC is designed to communicate directly with inputs from the machine and process and control outputs. The PLC recognizes these inputs and outputs (I/Os) as part of its internally programmed system. Inputs include limit switches, pushbuttons, temperature controls, photoelectric controls, analog signals, American National Standard Code for Information Interchange (ASCII), serial data, and other inputs. The outputs include voltage or current levels that drive end devices such as solenoids, motor starters, relays, and lights. Other outputs are analog devices, digital binary coded decimal (BCD) displays, ASCII compatible devices, and other PLCs and computers.

The second difference between PLCs and microcomputers is the ease in programming. A PLC uses simple programming techniques that are easily learned and understood. Simple ladder (line) diagram programming does not require knowledge of computer languages. A PLC can be programmed and reprogrammed on-line while a process is running. Hardware modifications are not required.

The third difference between PLCs and microcomputers is that a PLC is designed specifically for use in an industrial environment. See Figure 14-6. Variations in levels of noise, vibration, temperature, and humidity do not adversely affect PLC operations. A microcomputer cannot withstand the typical industrial environment.

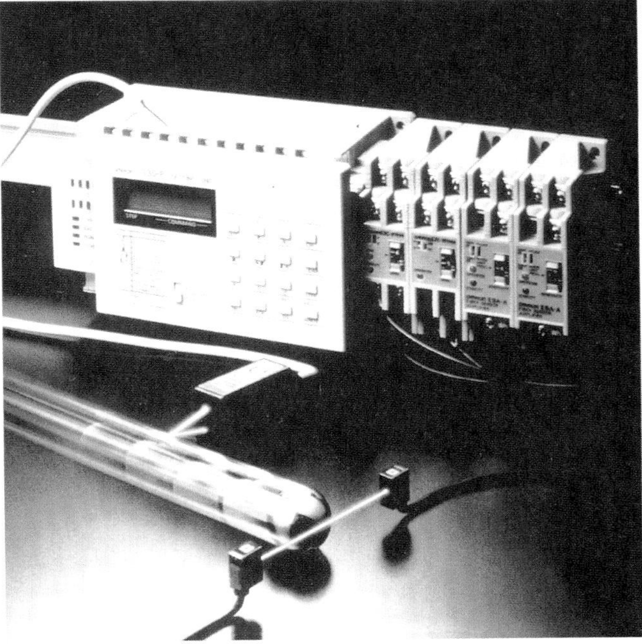

Omron Electronics, Inc.

Figure 14-6. PLCs are designed to withstand fluctuations in noise, vibration, temperature, and humidity in the industrial environment.

PLC PARTS

All PLCs have four basic parts which include the power supply, input/output section, processor section, and programming section. See Figure 14-7.

The programs used in manufacturing parts, equipment, etc. and processing goods and other consumables are stored and retrieved from memory as required. Sections of the PLC are interconnected and work together to allow the PLC to accept inputs from a variety of sensors, make a logical decision as programmed, and control outputs such as motor starters, solenoids, valves, and drives.

Power Supply

The power supply provides necessary voltage levels required for the internal operations of the PLC. In addition, it may provide power for the input/output modules. The power supply can be a separate unit or built into the processor section. It takes the incoming voltage (normally 120 VAC or 240 VAC) and changes the voltage as required (normally 5 VDC to 32 VDC).

The power supply must provide constant output voltage free of transient voltage spikes and other electrical noise. The power supply also charges an internal battery in the PLC to prevent memory loss when external power is removed. Memory retention time varies from hours up to 10 years.

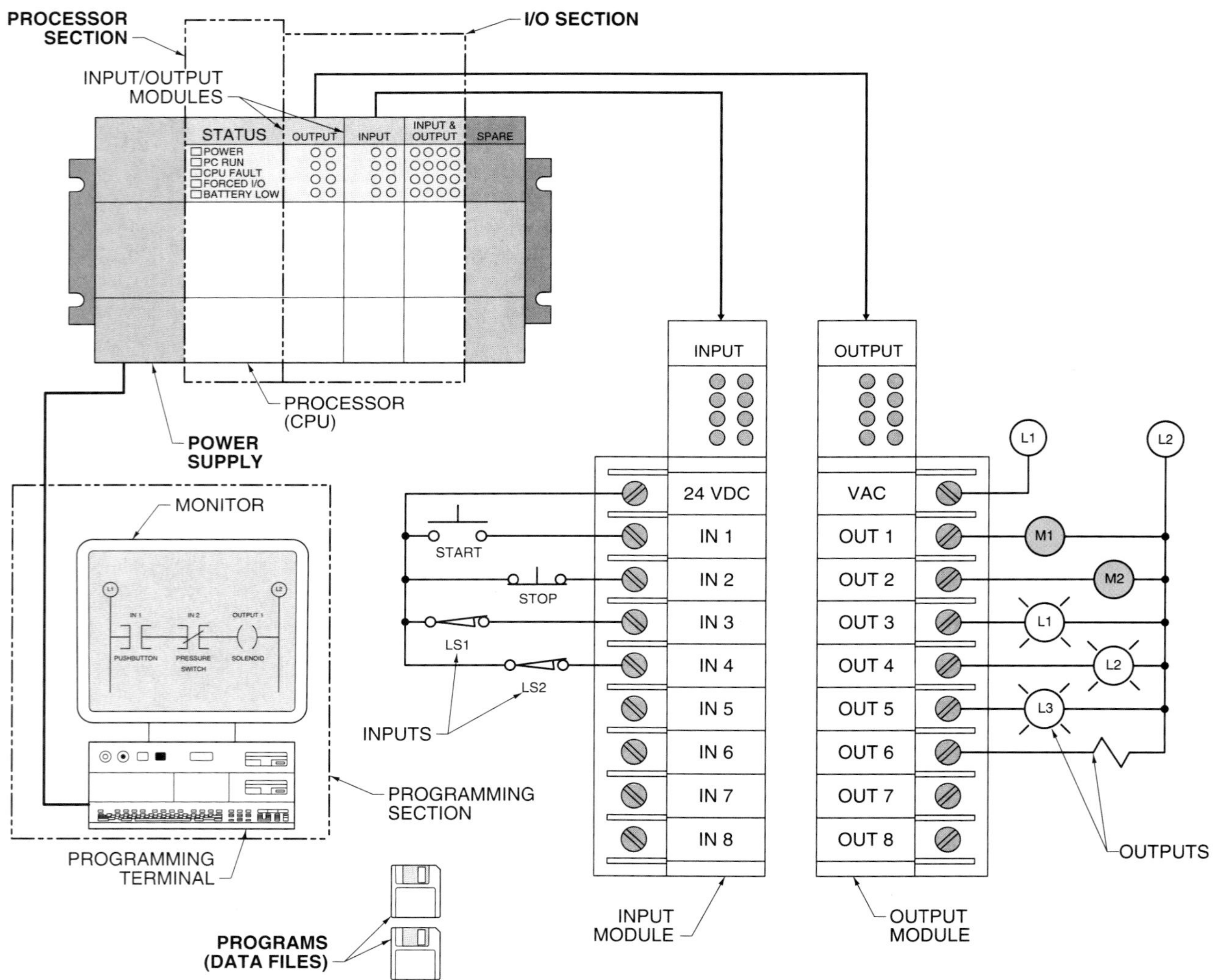

Figure 14-7. The four basic parts of a PLC include the power supply, input/output section, processor section, and programming section.

Input/Output Section

The input/output section functions as the eyes, ears, and hands of the PLC. The input section is designed to receive information from pushbuttons, temperature switches, pressure switches, photoelectric and proximity switches, and other sensors. The output section is designed to deliver the output voltage required to control alarms, lights, solenoids, starters, and other loads.

The input section receives incoming signals (normally at a high-voltage level) and converts them to low-power digital signals that are sent to the processor section. The processor then registers and compares the incoming signals to the program.

The output section receives low-power digital signals from the processor and converts them into high-power signals. These high-power signals can drive industrial loads that can light, move, grip, rotate, extend, release, heat, and perform other functions.

The input/output section can be either located on the PLC (onboard) or be part of expansion modules. Onboard inputs and outputs are a permanent part of the PLC package. Expansion modules are removable units that include inputs, outputs, or combinations of inputs and outputs.

Onboard inputs and outputs usually include a fixed number of inputs and outputs that define the limits of the PLC. For example, a small PLC may include up to 16 inputs and 8 outputs. This means that up to 16 inputs and 8 outputs may be connected to the PLC. PLCs that use expansion modules allow for the total number of input and/or outputs to be changed, by changing or adding modules. Onboard PLCs are normally used for individual machines and small systems. Expansion PLCs are normally used for large systems or small systems that require flexible changes.

Discrete I/Os. Discrete I/Os are the most common inputs and outputs. Discrete I/Os use bits, with each bit representing a signal that is separate and distinct, such as ON/OFF, open/closed, or energized/de-energized. The processor reads this as the presence or absence of power.

Examples of discrete inputs are pushbuttons, selector switches, joysticks, relay contacts, starter contacts, temperature switches, pressure switches, level switches, flow switches, limit switches, photoelectric switches, and proximity switches. Discrete outputs include lights, relays, solenoids, starters, alarms, valves, heating elements, and motors.

Data I/Os. In many applications, more complex information is required than the simple discrete I/O is capable of producing. For example, measuring temperature may be required as an input into the PLC and numerical data may be required as an output. Data I/Os are inputs and outputs that produce or receive a variable signal. They may be analog, which allows for monitoring and control of analog voltages and currents, or they may be digital, such as BCD (binary coded decimal) inputs and outputs.

When an analog signal (such as voltage or current) is input into an analog input card, the signal is converted from analog to digital by an analog to digital (A/D) converter. The converted value, which is proportional to the analog signal, is sent to the processor section. After the processor has processed the information according to the program, the processor outputs the information to a digital to analog (D/A) converter. The converted signal can provide an analog voltage or current output that can be used or displayed on an instrument in a variety of processes and applications.

Examples of data inputs are potentiometers, rheostats, encoders, bar code readers, and temperature, level, pressure, humidity, and wind speed transducers. Examples of data outputs are analog meters, digital meters, stepping motor (signals), variable voltage outputs, and variable current outputs.

I/O Capacity. The size of a PLC is based on the controller's I/O and capacity. Common I/O capacities of different size PLCs include:

- Mini/Micro – 32 or less I/Os, but may have up to 64
- Small – 64 to 128 I/Os, but may have up to 256
- Medium – 256 to 512 I/Os, but may have up to 1023
- Large – 1024 to 2048 I/Os, but may have many thousands more on very large units

The inputs and outputs may be directly connected to the PLC or may be in a remote location. I/Os in a remote location from the processor section can be hard wired to the controller, multiplexed over a pair of wires, or sent by a fiber-optic cable. In any case, the remote I/O is still under the control of the central processor section. Common PLCs may have 16, 32, 64, 128, or 256 remote I/Os.

Fiber-optic communication modules route signals to and from I/Os to the processor section. Fiber-optic communication modules are unaffected by noise interference and are commonly used for process applications in the food industry, petrochemicals, and hazardous locations.

Processor Section

The processor section is the brain of the PLC. The *processor section* is the section of a PLC that organizes all control activity by receiving inputs, performing logical decisions according to the program, and controlling the outputs. See Figure 14-8.

Rockwell Automation, Allen-Bradley Company, Inc.

Figure 14-8. The processor section organizes all control activity by receiving inputs, performing logical decisions, and controlling the outputs.

The processor section evaluates all input signals and levels. This data is compared to the memory in the PLC, which contains the logic of how the inputs are interconnected in the circuit. The interconnections are programmed into the processor by the programming section. The processor section controls the outputs based on the input conditions and program. The processor continuously examines the status of the inputs and outputs and updates them according to the program. See Figure 14-9.

Scan is the process of evaluating the input/output status, executing the program, and updating the system. *Scan time* is the time it takes a PLC to make a sweep of the program. Scan time is normally given as the time per 1 kilobyte of memory and normally is listed in milliseconds (ms). Scanning is a continuous and sequential process of checking the status of inputs, evaluating the logic, and updating the outputs.

The processor section of a PLC has different modes. The different modes allow the PLC to be taken on-line (system running) or off-line (system on stand-by). Processor modes include the program, run, and test modes. The program mode is used for developing the logic of the control circuit. In the program mode, the circuit is monitored and the program is edited, changed, saved, and transferred.

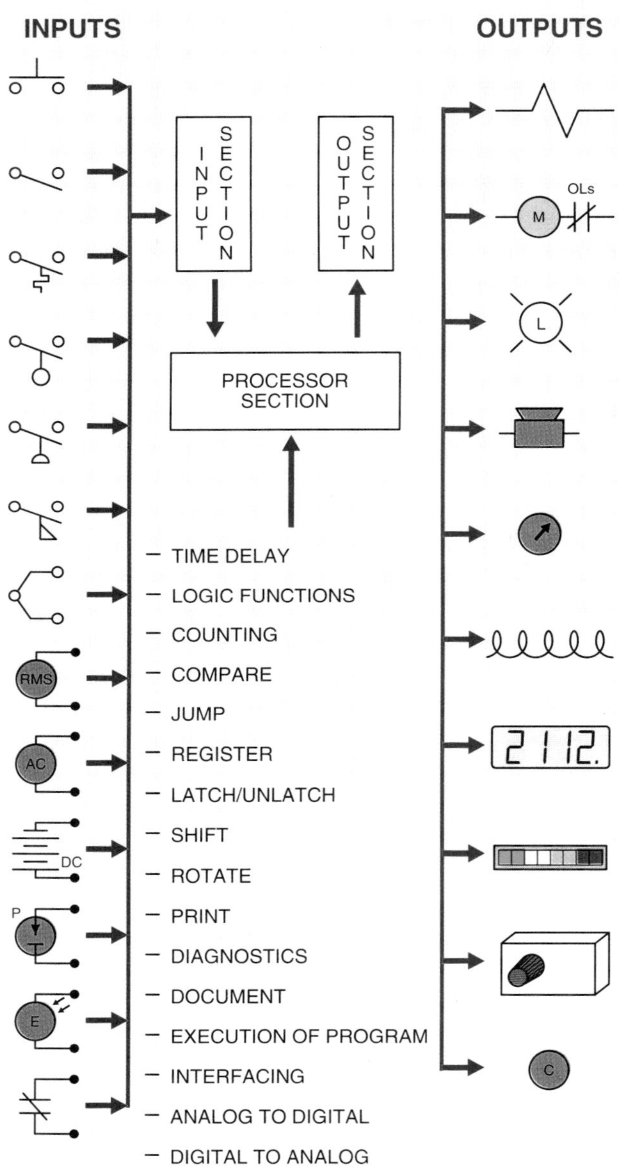

Figure 14-9. The processor section organizes all control activity by receiving inputs, performing logical decisions as programmed, and controlling the outputs.

The run mode is used to execute the program. In the run mode, the circuit may be monitored and the inputs and outputs forced. Program changes cannot be made in the run mode. The test mode is used to check the program without energizing output circuits or devices. In the test mode, the circuit is monitored and inputs and outputs are forced (without actually energizing the load connected to the output). See Figure 14-10.

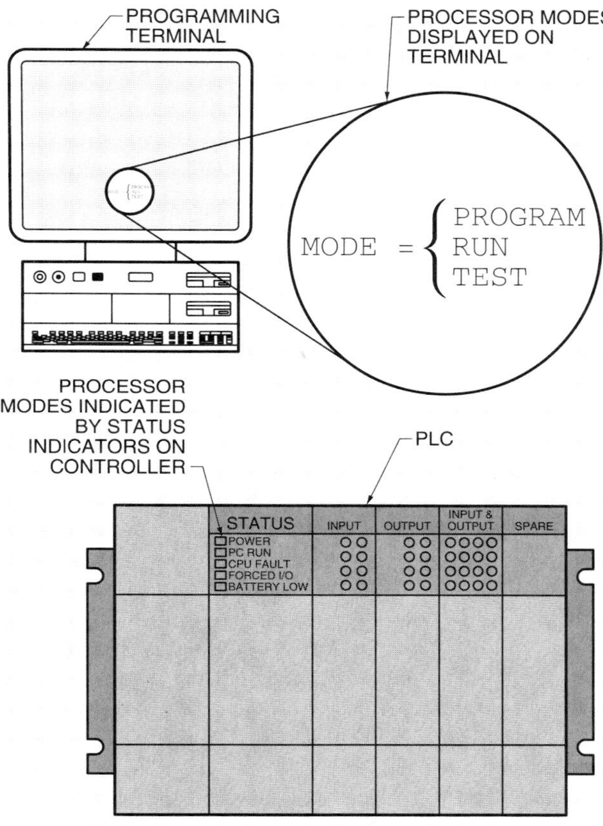

Figure 14-10. The processor section of a PLC has different modes which allow the PLC to be taken on-line (system running) or off-line (system on standby).

Warning: A PLC is switched from the program mode to the run mode by placing the controller in the run mode. The machine or process is started when the controller is placed in the run mode. Extreme care must be taken to ensure that no damage to personnel or equipment occurs when switching the controller to the run mode. Only qualified personnel should change processor modes, and key-operated switches should always be used in any dangerous application.

Programming Section

The *programming section* of a PLC is the section that allows input into the PLC through a keyboard. The processor must be given exact, step-by-step directions. This includes communicating to the processor such things as load, set, reset, clear, enter in, move, and start timing. Programming a PLC involves the programming device that allows access to the processor and the programming language that allows the operator to communicate with the processor section.

Programming Devices. Programming devices vary in size, capability, and function. Programming devices are available as simple, small, text display units or complex color CRTs with monitoring and graphics capabilities. See Figure 14-11.

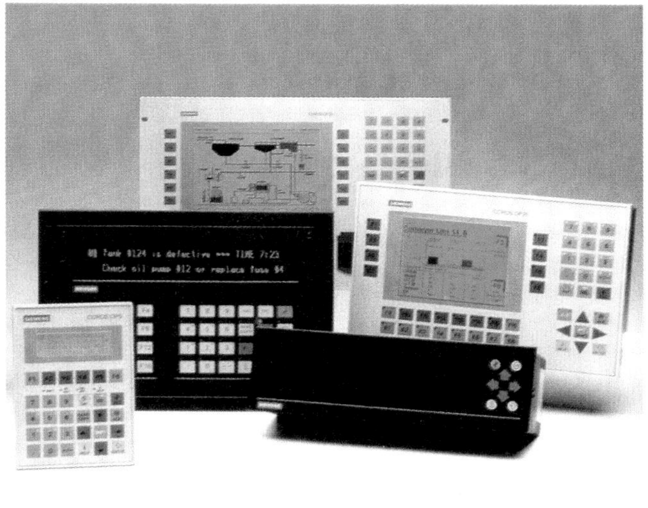

Siemens Corporation

Figure 14-11. Programming devices are available as simple, small, text display units or complex color CRTs with monitoring and graphics capabilities.

A programming device may be connected permanently to the PLC or connected only while the program is being entered. Once a program is entered, the programming device is no longer needed, except to make changes in the program or for monitoring functions. Some PLCs are designed to use an existing personal computer (PC) for programming. *Off-line programming* is the use of a personal computer to program a PLC. This permits the computer to be used for other purposes when not being used with the PLC.

PLC Language. The first PLCs used line diagrams as a language for inputting information. Line diagrams are still commonly used as a language for PLCs throughout the world. Other languages used include Boolean, Functional Blocks, and English Statement. Line diagrams and Boolean are basic PLC languages. Functional Blocks and English Statement are higher-level languages required to execute more powerful operations, such as data manipulations, diagnostics, and report generation.

The line diagram is drawn in a series of rungs. Each rung contains one or more inputs and the output (or outputs) controlled by the inputs. The rung relates to the machine or process controls, and the programming instructions relate the desired logic to the processor.

Basic logic functions are used to enter the circuit's logical operation into the processor section. See Figure 14-12. The program is entered into the controller through the keyboard.

BASIC LOGIC FUNCTIONS

FUNCTION	DIGITAL CIRCUIT	PROGRAMMABLE CIRCUIT	CIRCUIT DESCRIPTION
AND	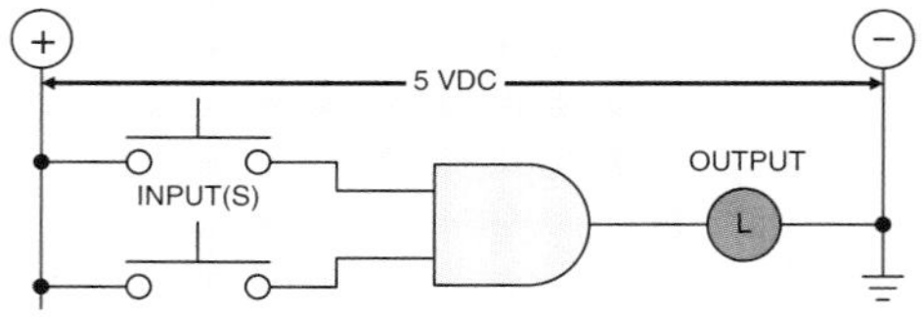 + − 5 VDC INPUT(S) OUTPUT L	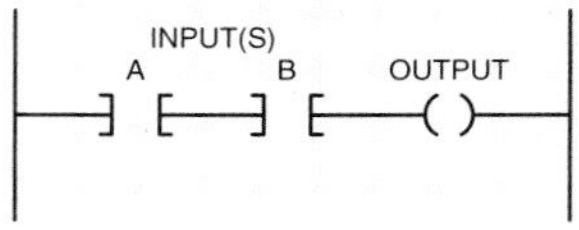 INPUT(S) A B OUTPUT	**ENERGIZED** Output energized if all inputs are activated **DE-ENERGIZED** Output de-energized if any one input is deactivated
OR	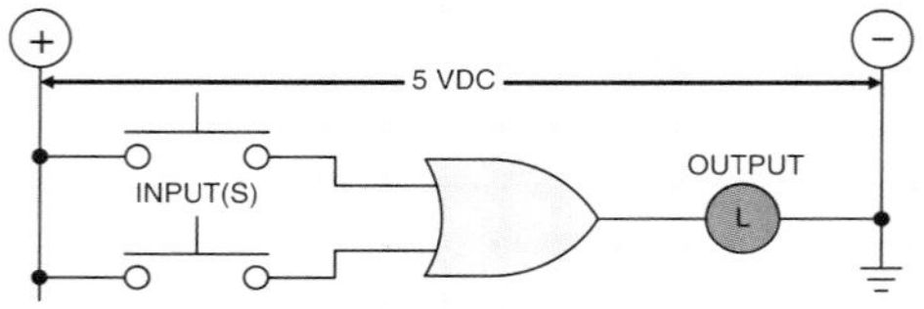 + − 5 VDC INPUT(S) OUTPUT L	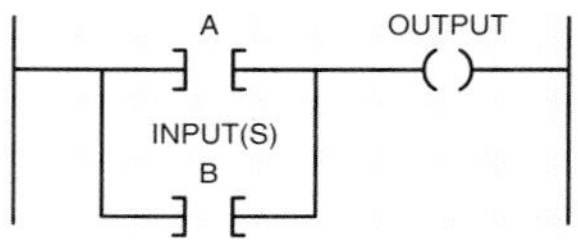 A OUTPUT INPUT(S) B	**ENERGIZED** Output energized if one or more inputs are activated **DE-ENERGIZED** Output de-energized if both inputs are deactivated
NOT	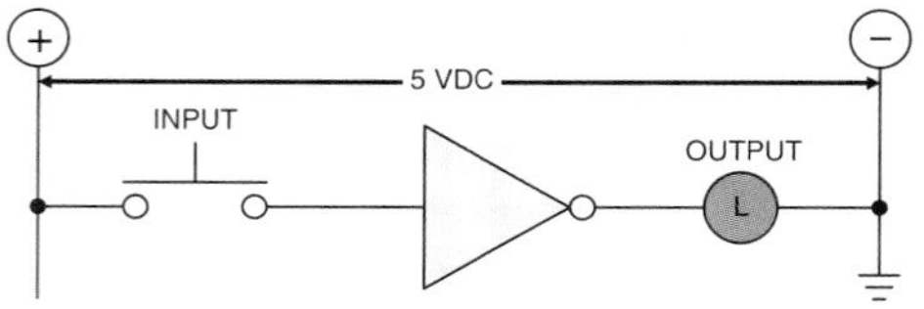 + − 5 VDC INPUT OUTPUT L	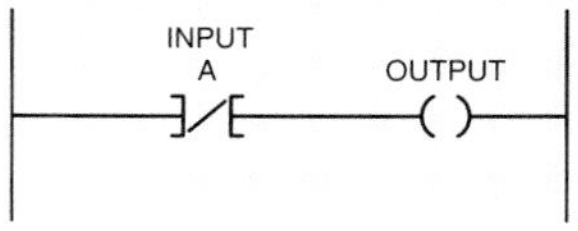 INPUT A OUTPUT	**ENERGIZED** Output energized if input is not activated **DE-ENERGIZED** Output de-energized if input is activated
NOR	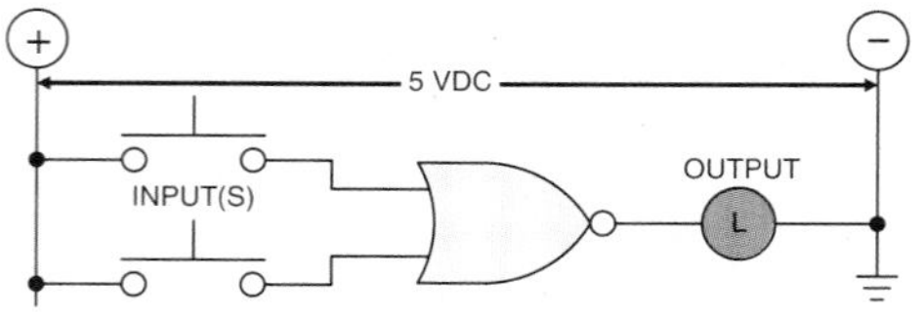 + − 5 VDC INPUT(S) OUTPUT L	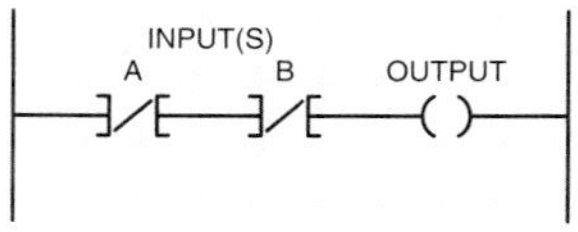 INPUT(S) A B OUTPUT	**ENERGIZED** Output energized if no inputs are activated **DE-ENERGIZED** Output de-energized if one or more inputs are activated
NAND	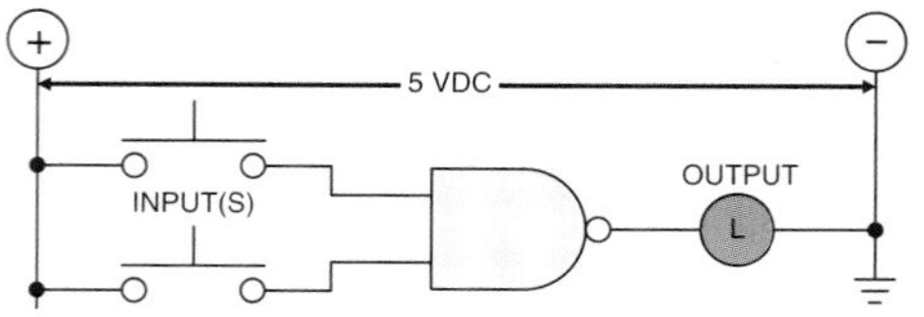 + − 5 VDC INPUT(S) OUTPUT L	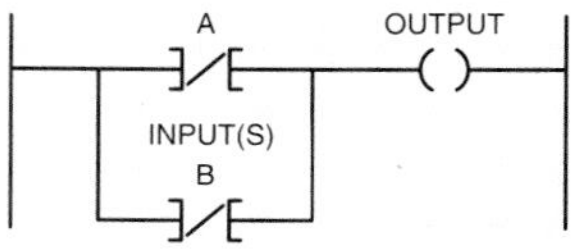 A OUTPUT INPUT(S) B	**ENERGIZED** Output energized unless all inputs are activated **DE-ENERGIZED** Output de-energized if all inputs are activated

Figure 14-12. Basic logic functions are used to enter the circuit's logical operation into the processor section.

Programming a PLC follows a logical process. Inputs and outputs are entered into the controller in the same manner as if connecting them by hard wiring. The difference in programming is that although a circuit is the same, each manufacturer has a different method of entering that circuit. There are more similarities than differences from manufacturer to manufacturer.

Developing Typical Programs. Several steps must be taken before a program can be entered into a PLC. The first step is to develop the logic required of the circuit into a line diagram. See Figure 14-13. In this circuit, pressing any one of the three start pushbuttons energizes the motor starter. Once the motor starter is energized, the start pushbutton may be released. The motor starter remains energized because the M1 contact closes and provides a parallel path for current flow around the start pushbuttons. Pressing one of the stop pushbuttons stops the flow of current through the motor starter and de-energizes it.

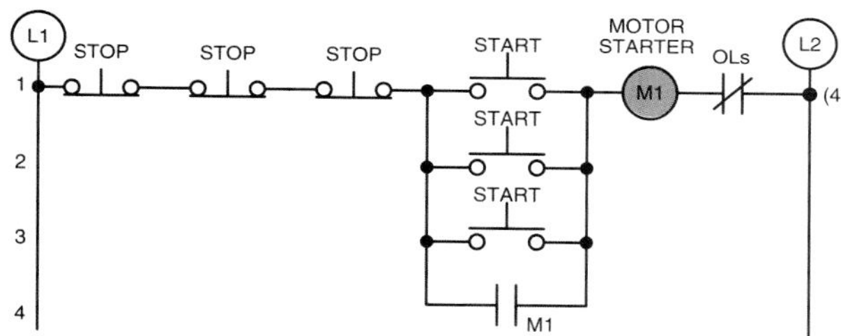

Figure 14-13. Circuit logic must be developed into a line diagram to enter the circuit into a PLC.

The line diagram shows the logic of the circuit, but not the actual location of each component. A wiring diagram shows the location of the components in an electrical circuit. See Figure 14-14. The wiring diagram of the three start/stop pushbutton stations shows the location of each pushbutton.

The phantom line around each start/stop pushbutton station indicates that the two pushbuttons are located in the same enclosure. Each pushbutton in the wiring diagram is connected in the exact manner as in the line diagram. Any additions (or changes) to this hard-wired control circuit requires that the circuit be rewired.

The second step is to take the line diagram and convert it into a programming diagram. A *programming diagram* is a line diagram that better matches the PLC's language. Like a standard (hard-wired) line diagram, a PLC line diagram shows the flow of current through the control circuit. The PLC line diagram does not use distinct symbols for each input/output. Instead, there are two basic symbols for inputs and one basic symbol for outputs. One of the input symbols represents normally open (NO) inputs and the other represents normally closed (NC) inputs. See Figure 14-15.

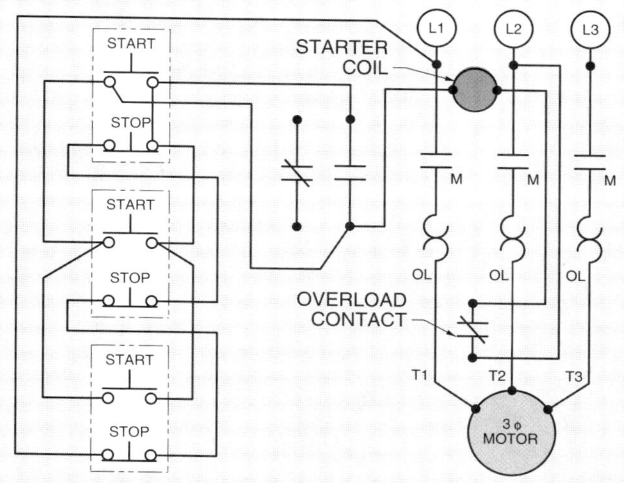

Figure 14-14. A wiring diagram shows the location of the components in an electrical circuit.

In this circuit, pressing any one of the three start pushbuttons energizes the motor starter. Once the motor starter is energized, the start pushbutton may be released and remains energized. This is because the contacts of output 1 close and provide a parallel path for current flow around the start pushbuttons. The motor starter de-energizes when the current flow to output 1 is de-energized by pressing the stop pushbutton.

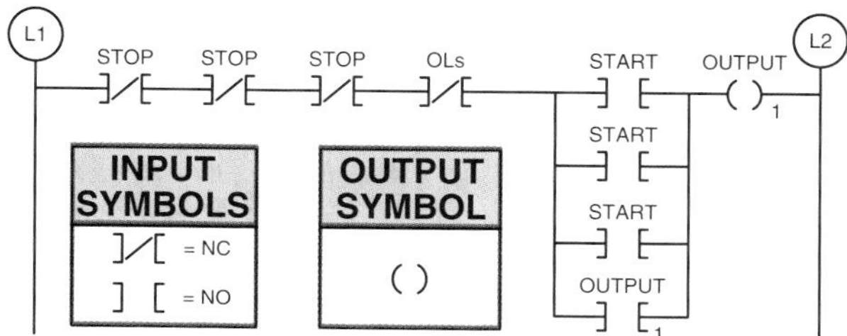

Figure 14-15. A programming diagram is a line diagram that better matches the PLC's language.

The PLC wiring diagram is much different than a hard-wired wiring diagram. See Figure 14-16. In a PLC wiring diagram, each input is wired to a designated input terminal, and each output to a designated output terminal. The way the inputs and outputs are connected does not determine the logic of the circuit's operation. The circuit's logic is controlled by the way the circuit is programmed into the PLC. Any changes to the circuit are made by changing the program, not the wiring of the inputs and outputs.

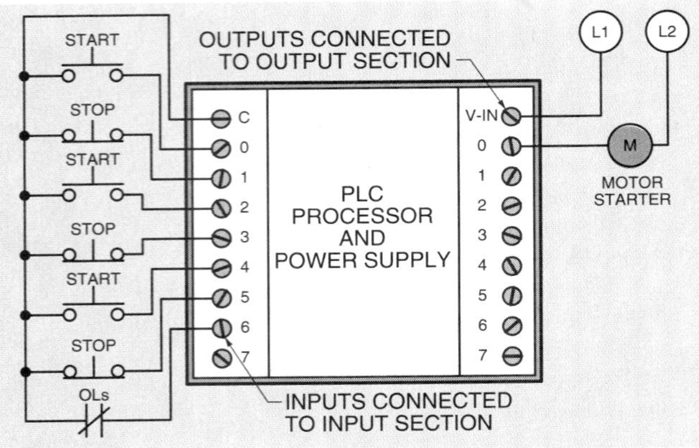

Figure 14-16. In a PLC wiring diagram, each input is wired to a designated input terminal, and each output is wired to a designated output terminal.

The third step is to enter the desired logic of the circuit into the controller. Every manufacturer has a slightly different set of steps and functions to enter the program into the PLC. The program is entered in the program mode and then saved for future use or downloaded to the PLC.

Storing and Documentation. Once a program has been developed it may be necessary to store the program outside of the controller or document the program by printing it out. See Figure 14-17. This allows for a means of storing and retrieving control programs, which makes for fast changes in a process or operation. Storage of a program is commonly achieved using standard diskettes. For example, one disk (or file) may have the program for filling 8 oz bottles and a second disk (or file) may have the program for filling 16 oz bottles.

When a change from one size bottle to another is required, the PLC is loaded with the correct disk (or file) to start the line for all the proper control settings. Even if the PLC is not likely to ever have its program changed, the program should be stored on a disk. This ensures the safety of the program in the event of a problem.

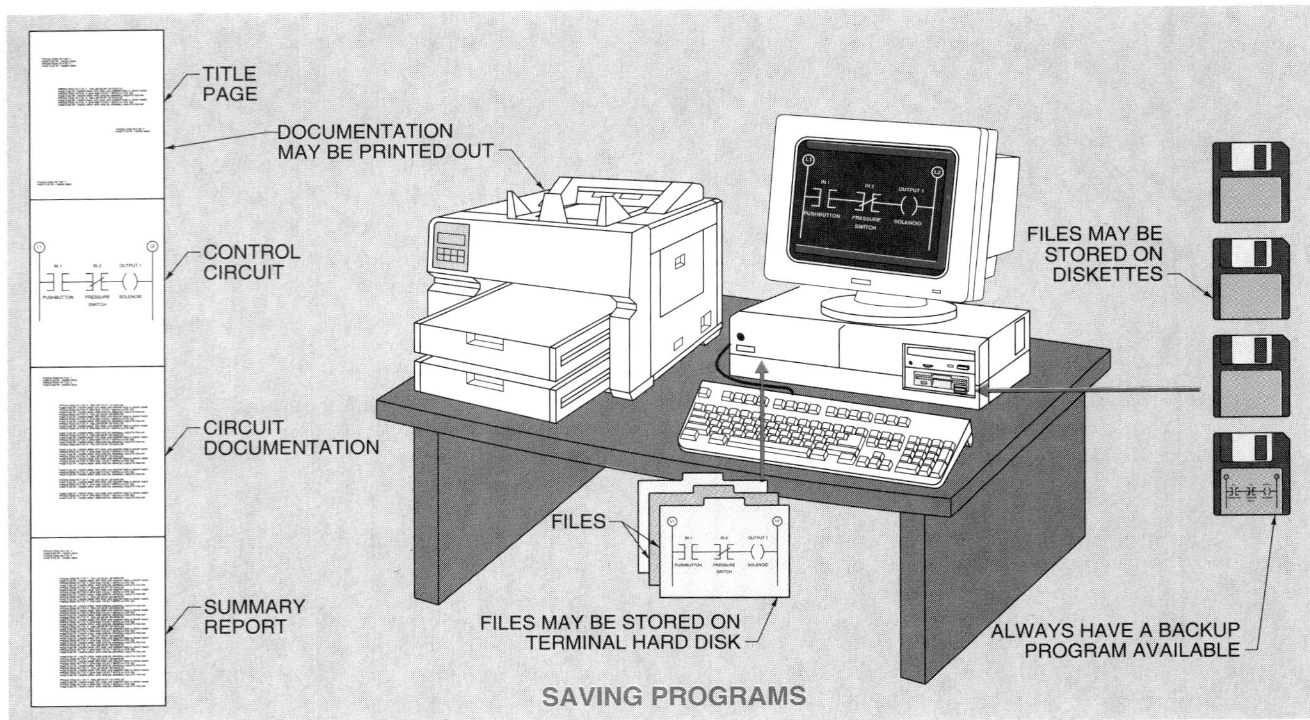

Figure 14-17. Storage of a program is commonly achieved using standard diskettes.

Once a program has been entered into the PLC, a copy of the program and other circuit documentation can be made by connecting the controller to a printer. The printout can be used as a hard copy of the program for documentation and future reference.

I/O Status Indicators

An I/O status indicator shows the status of the input and output devices. The status indicators on the input module are energized when an electrical signal is received at an input terminal. This occurs when an input contact is closed or a signal is present. The status indicators on the output module are energized when an output device is energized. Each input device and output device has its own status indicator. See Figure 14-18.

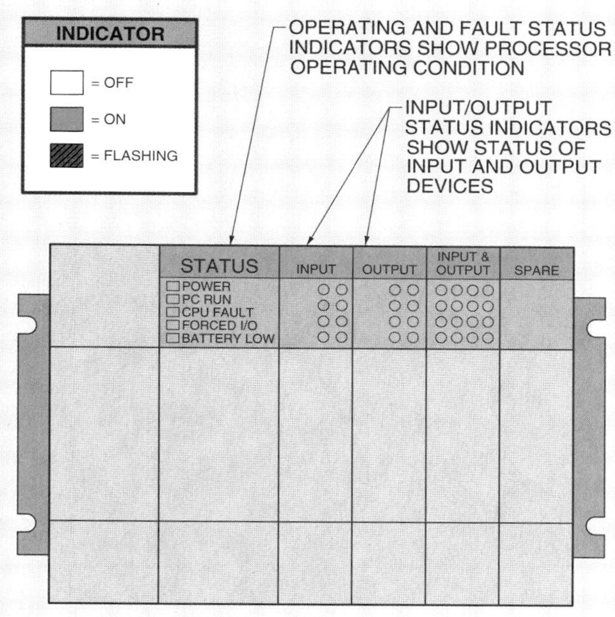

Figure 14-18. An I/O status indicator shows the status of the input and output devices.

Operating and Fault Status Indicators

The processor of a PLC is programmed to look for potential problems (self-diagnostics). The processor performs error checks as part of its normal operation and sends status information to the appropriate status indicators. Typical diagnostic checks include monitoring the input power, the processor's operating mode, CPU faults, forced inputs or outputs, and a low-battery condition.

Operating and fault status indicators include the power status, PC run, CPU fault, forced I/O, and battery low indicators. The power status indicator turns ON to indicate that the processor is energized and power is being applied. This status indicator should normally be ON. The PC run indicator turns ON when the processor is in the run mode. Care must be taken when the run indicator is ON because the controller activates the loads as programmed. This indicator is OFF when the processor is placed in the program mode. The CPU fault indicator turns ON when the processor has detected an error in the controller. The processor automatically turns OFF all loads and stops operation when this indicator is ON. An error message is normally displayed on the monitor screen indicating the problem.

The forced I/O indicator turns ON when one or more input or output device has been forced ON or OFF. All force commands must be removed from the program before normal operation is resumed. A battery is used to provide back-up power for the processor memory in case of an external power failure. The battery-low indicator turns ON when the battery should be replaced or if the battery is not charging.

The condition of the status indicators must be checked when troubleshooting a problem in a system using a PLC. Potential problems may be determined based on the condition of the status indicators.

Force and Disable

The force command opens or closes an input device or turns ON or OFF an output device. The force command is designed for use when troubleshooting the system. Forcing an input or output device allows checking the circuit using software. See Figure 14-19.

Omron Electronics, Inc.

PLCs include status indicators that are used when troubleshooting to verify circuit conditions.

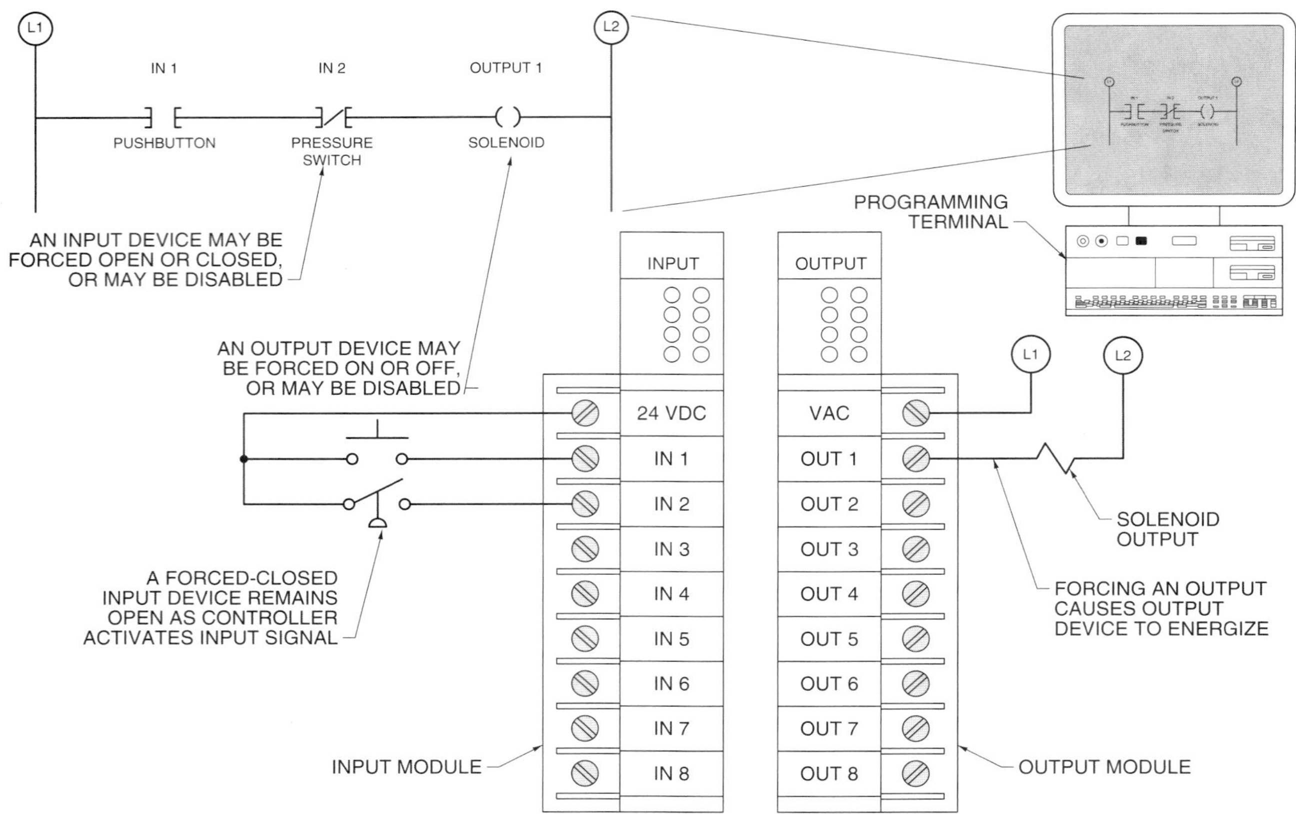

Figure 14-19. Forcing an input or output device allows checking the circuit using software.

An input device may be forced to test the circuit operation. Forcing an input device may also be used when service is required on a defective input device. The defective input device may be forced ON until the device may be serviced if the input device is not critical to production. The force command is removed after the device is fixed.

An output device turns ON regardless of the programmed circuit's logic when the force ON command is used. The output device remains ON until the force OFF command is used. Care must be taken when using the force command because it overrides all safety features designed for the program.

The disable command prevents an output device from operating. The disable command is the opposite of the force command. The disable command is used to prevent one or all of the output devices from operating. Ensure that all force and disable commands are removed before returning a system to normal operation.

INTERFACING SOLID-STATE CONTROLS

PLCs can have many types of inputs including pushbuttons, level switches, temperature controls, and photoelectric controls. Inputs, such as pushbuttons and temperature controls, are normally easy to input. However, more complex solid-state control inputs such as proximity and photoelectric inputs require special consideration because of their function.

Solid-state proximity and photoelectric controls are used in many automated systems. See Figure 14-20. These controls normally have a solid-state output and are ideal for inputting to PLCs. Photoelectric controls can be input into PLCs for detection, inspection, monitoring, counting, and documentation. Available outputs include two- and three-wire types with thyristor and transistor outputs that can be connected individually or in series/parallel combinations.

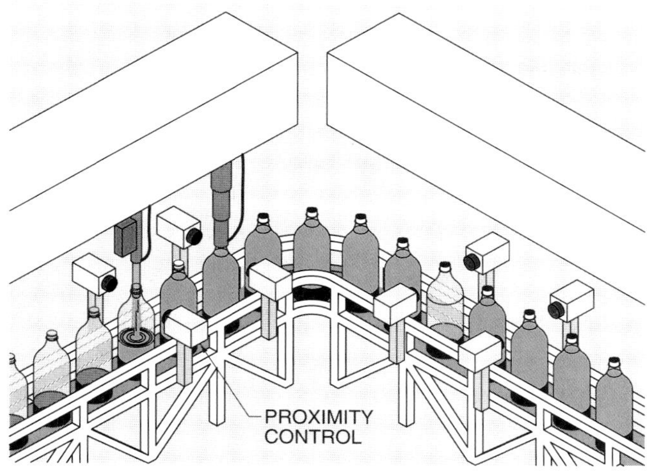

Figure 14-20. Solid-state proximity controls normally have a solid-state output and are ideal for inputting to PLCs.

Two-Wire Thyristor Output Sensors

Two-wire thyristor output sensors are available in a supply voltage range of 20 VAC to 270 VAC at about 180 mA to 500 mA range in either NO or NC versions. Two-wire thyristor output sensors have only two wires and are wired in series with the load like a mechanical switch. See Figure 14-21. The power to operate these sensors is received through the load when the load is not being operated. As with any thyristor output device, some consideration must be given to off-state leakage current and minimum load current. Unlike a mechanical switch, there is current consumed by the proximity sensor in the inactivated mode. The current is small enough that most industrial loads are not affected. This leakage current may be enough to activate the load on some high-impedance loads and PLCs.

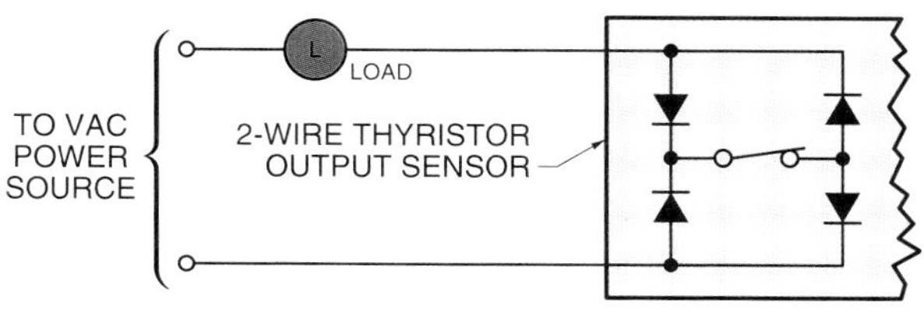

Figure 14-21. Two-wire thyristor output sensors have only two wires and are wired in series with the load like a mechanical switch.

This problem can be corrected by placing a load resistor across the load. See Figure 14-22. The resistor value should be chosen to ensure that minimum load current is exceeded and the effective load impedance is reduced. This prevents off-state leakage current turn ON. This resistance value is normally in the range of 4.5 kΩ to 7.5 kΩ. A general rule is to use a 5 kΩ, 5 W resistor for most applications.

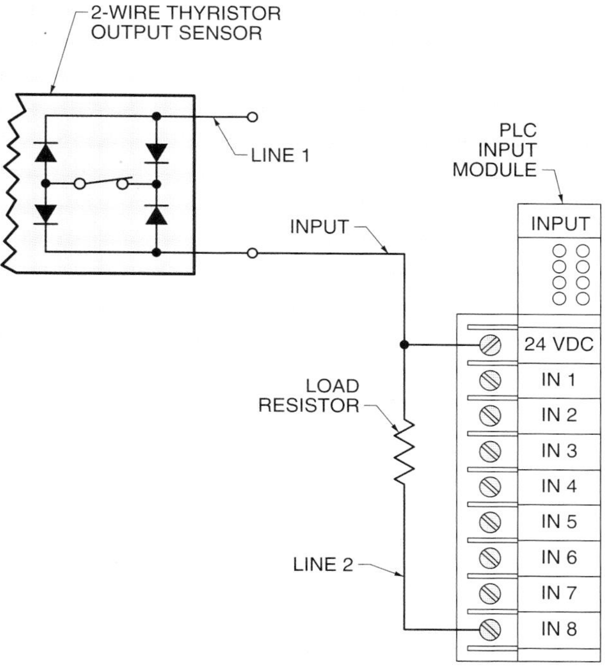

Figure 14-22. A load resistor may be required when connecting a sensor to the PLC to prevent leakage current of the sensor from inputting into the controller.

Electrical Noise Suppression

Electrical noise is unwanted signals that are present on a power line. Electrical noise enters through input devices, output devices, and power supply lines. Unwanted noise pickup may be reduced by placing the controller away from noise-generating equipment such as motors, motor starters, welders, and drives.

Noise suppression should be included in every PLC installation because it is impossible to eliminate noise in an industrial environment. Certain sensitive input devices (analog, digital, and thermocouple) require a shielded cable to reduce electrical noise.

A shielded cable uses an outer conductive jacket (shield) to surround the two inner signal-carrying conductors. The shield blocks electromagnetic interference. The shield must be properly grounded to be effective. Proper grounding includes grounding the shield at only one point. A shield grounded at two points tends to conduct current between the two grounds. See Figure 14-23.

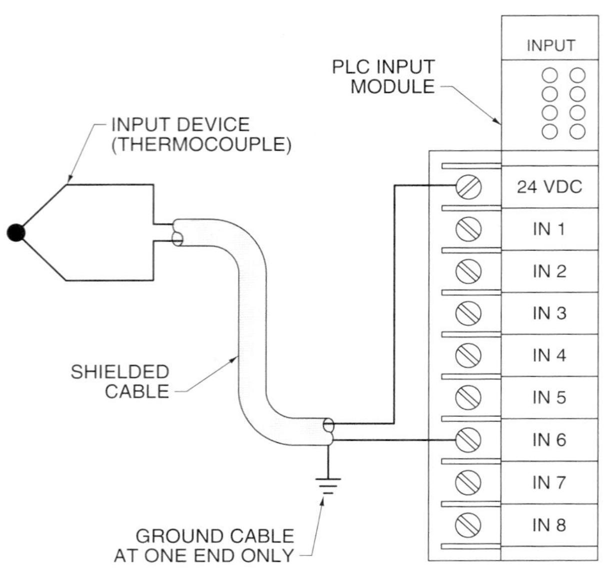

Figure 14-23. A shielded cable uses an outer conductive jacket (shield) which blocks magnetic interference from the two inner signal-carrying conductors.

A high-voltage spike is produced when inductive loads such as motors, solenoids, and coils are turned OFF. These spikes may cause problems in a PLC. High-voltage spikes should be suppressed to prevent problems. A snubber circuit is used to suppress a voltage spike. Typical snubber circuits use an RC (resistor/capacitor), MOV (metal oxide varistor), or a diode depending on the load. See Figure 14-24.

Three-Wire Transistor Output Sensors

Three-wire transistor output sensors are available in a supply range of 10 VDC to 40 VDC at about 200 mA. These sensors are easily interfaced with other electronic circuitry and PLCs. Output sensor types consist of either an open collector NPN or PNP transistor. Both NO or NC versions are available.

These sensors receive their power to operate through two of the leads (positive and negative respectively) from the power source. The third lead is used to switch power to the load either using the same source of power as the proximity switch or an independent source of power. See Figure 14-25. When an independent source is used, one lead of that source is common with one lead of the source used to power the sensor. The voltage level must be within the specifications of the sensor used when using an independent power source for the load.

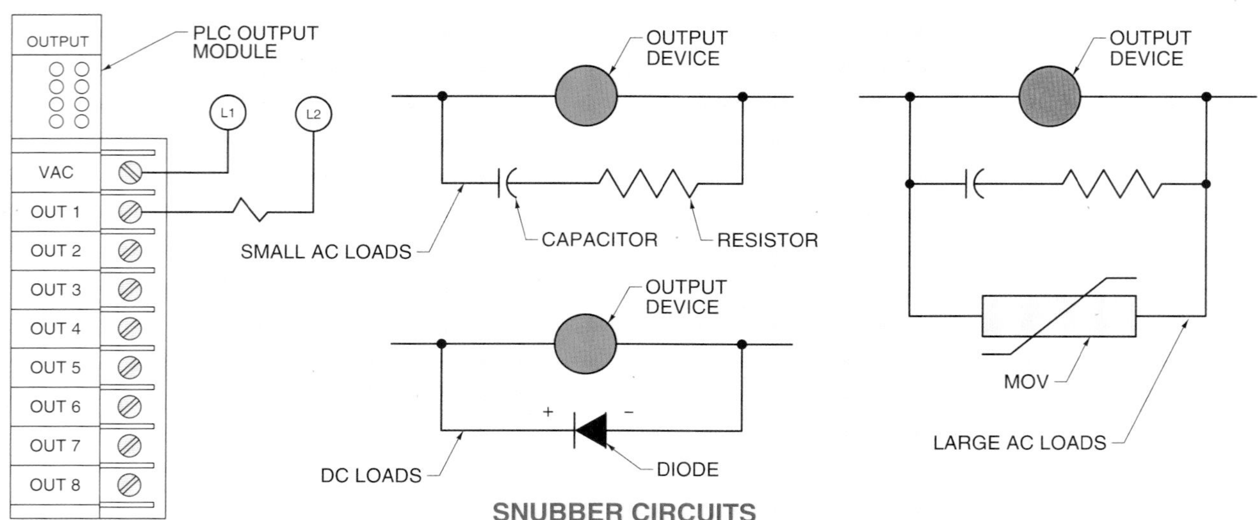

Figure 14-24. Snubber circuits are used to suppress voltage spikes in PLCs.

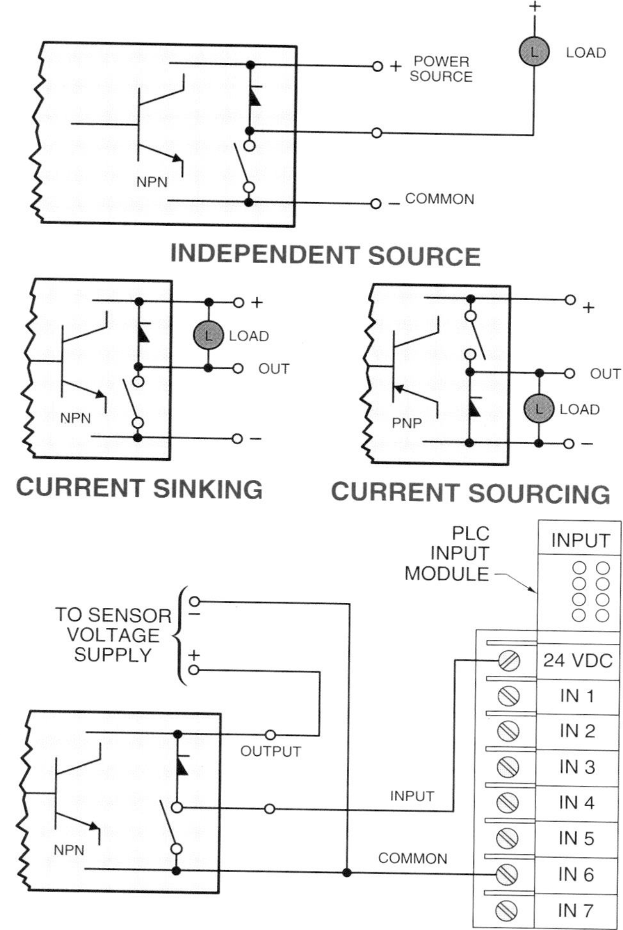

Figure 14-25. Three-wire transistor output sensors use either NPN or PNP transistors to control the load.

PLC APPLICATIONS

PLCs are useful in increasing production and improving overall plant efficiency. PLCs can control individual machines and link the machines together into a system. The flexibility provided by a PLC has resulted in many applications in manufacturing and process control.

Process control has gone through many changes in the past few years. In the past, process control was mostly accomplished by manual control. Flow, temperature, level, pressure, and other control functions were monitored and controlled at each stage by production workers.

Today, using PLCs, an entire process can automatically be monitored and controlled with little or no workers involved. Process applications in which PLCs are used include:

- Grain operations involving storage, handling, and bagging
- Syrup refinery involving product storage tanks, pumping, filtration, clarification, evaporators, and all fluid distribution systems
- Fats and oils processing involving filtration units, cookers, separators, and all charging and discharging functions
- Dairy plant operations involving all process control from raw milk delivered to finished dairy products
- Oil and gas production and refinement from the well pumps in the fields to finished product delivered to the customer
- Bakery applications from raw material to finished product
- Beer and wine processing, including the required quality control and documentation procedures

Welding

In manufacturing of discrete parts, welding is a major part of the system. PLCs may be used to control and automate industrial welding processes. See Figure 14-26. In this application, the PLC can control the length of the weld and the power required to produce the correct weld. The PLC is programmed to allow the weld to occur only if all inputs and conditions are correct. These include:

Fanuc Robotics North America

Figure 14-26. PLCs may be used to control and automate industrial welding processes.

- All the parts are present and in the correct position
- The correct weld cycle speed and power setting
- The correct rate of speed on the line for the given application
- All interlocks and safety features are functioning properly

In addition, the PLC can be used to determine if parts are running low and be set to automatically turn ON and OFF the line as required. Documentation of production efficiency can be generated for quality control and inventory requirements.

A PLC may be used to control and interlock many welders. Welders at one station may require more power than is available if all the welders are ON simultaneously. In this case, a large power draw can cause poor-quality welds. The requirements in a system using many welders are to limit the amount of power being consumed at any one point in time. This is accomplished by time-sharing the power feed to each welder.

A PLC may be programmed for a maximum power draw. The controller can determine if power is available when a welder requires power. The weld takes place if the correct power level is available. If not, the controller remembers the request and when power is available, it permits the welder to proceed with the weld cycle. The PLC can also be programmed to determine which welder has priority.

Machine Control

Controls must be synchronized when machines are linked together to form an automated system. See Figure 14-27. In this application, each machine may be controlled by a PLC, with another controller synchronizing the operation. This is likely if the machines are purchased from different manufacturers. In this case, each machine may include a PLC to control all the functions on that machine only. If the machines are purchased from one manufacturer or designed in-plant, it is possible to use one large PLC to control each machine and synchronize the process.

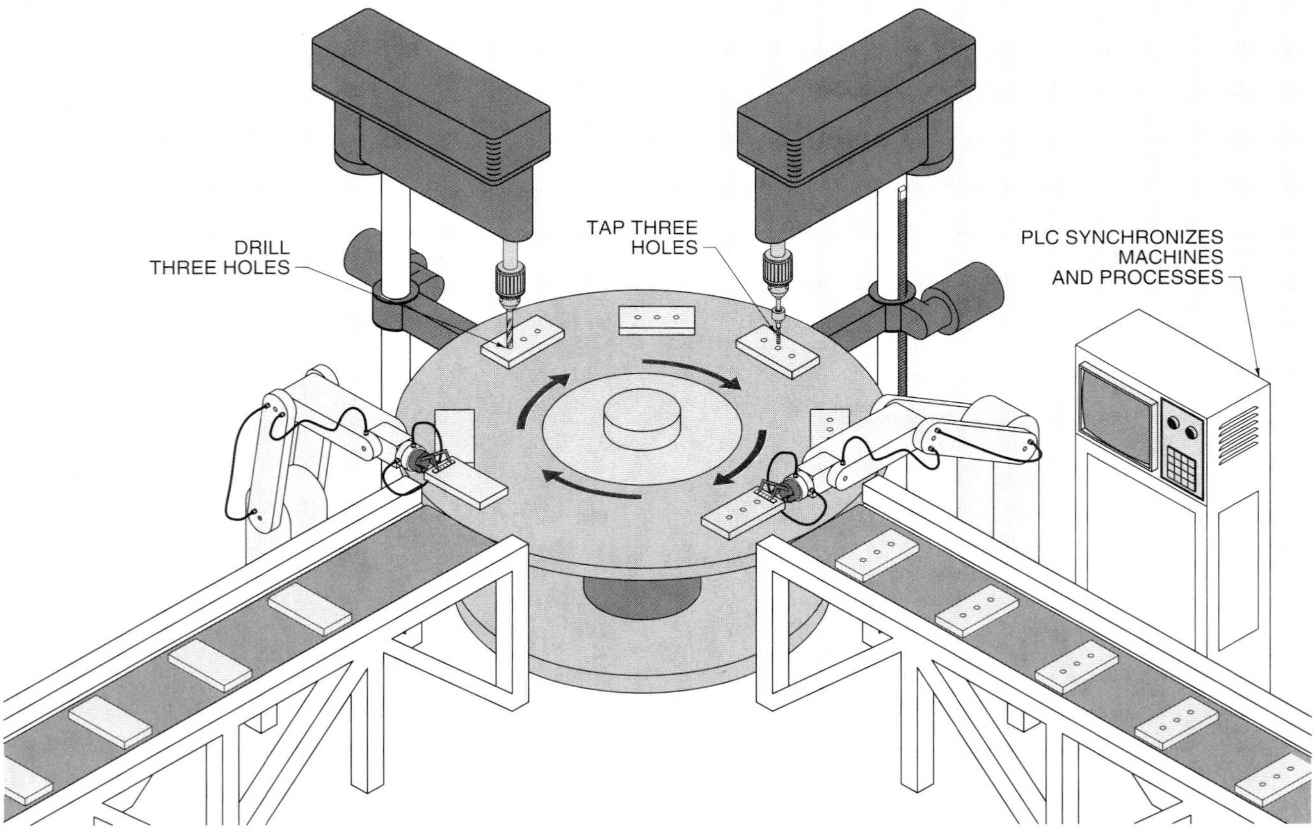

Figure 14-27. PLCs are used to control and synchronize individual machine operations with other machines.

Rockwell Automation, Allen-Bradley Company, Inc.

The output modules receive low-power digital signals from the processor and convert them into high-power signals to control the loads connected to the PLC.

Industrial Robot Control

PLCs are ideal devices for controlling any industrial robot. See Figure 14-28. The PLC can be used to control all operations such as rotate, grip, withdraw, extend, and lift. A PLC is recommended because most robots operate in an industrial environment.

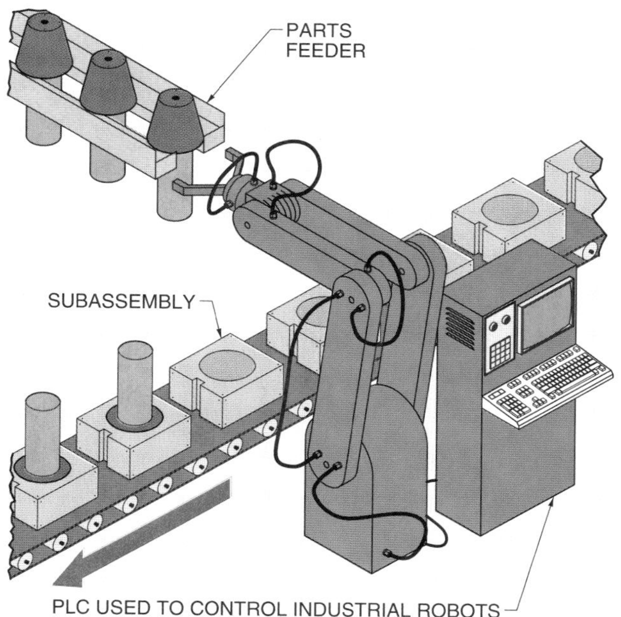

Figure 14-28. PLCs may be used to control the operations of an industrial robot.

Fluid Power Control

Fluid power cylinders are normally chosen when a linear movement is required in an automated application. Pneumatic cylinders are common because they are easy to install and most plants have compressed air. Pneumatics work well for most robot grippers, drives, positioning cylinders, machine loading and unloading, and tool-working applications. Hydraulic cylinders are used when a manufacturing process requires high forces. Hydraulic systems of several thousand psi are often used to punch, bend, form, and move components.

PLCs may be used to control linear and rotary actuators in an industrial fluid power circuit. See Figure 14-29. This system, as with any fluid power circuit, is ideal for control by a PLC. The controller's output module is connected to control the four solenoids. Solenoid A moves the cylinder in, solenoid B moves the cylinder out, solenoid C rotates the rotary actuator in the forward direction, and solenoid D rotates the rotary actuator in the reverse direction.

The PLC is used to control the energizing or de-energizing of the solenoids. Solenoids control the directional control valves, which control the actuators.

Siemens Corporation

The SIMATIC® S7 micro PLCs from Siemens may be used along with computer software programs to develop complex control systems for various robotic operations to increase flexibility and productivity.

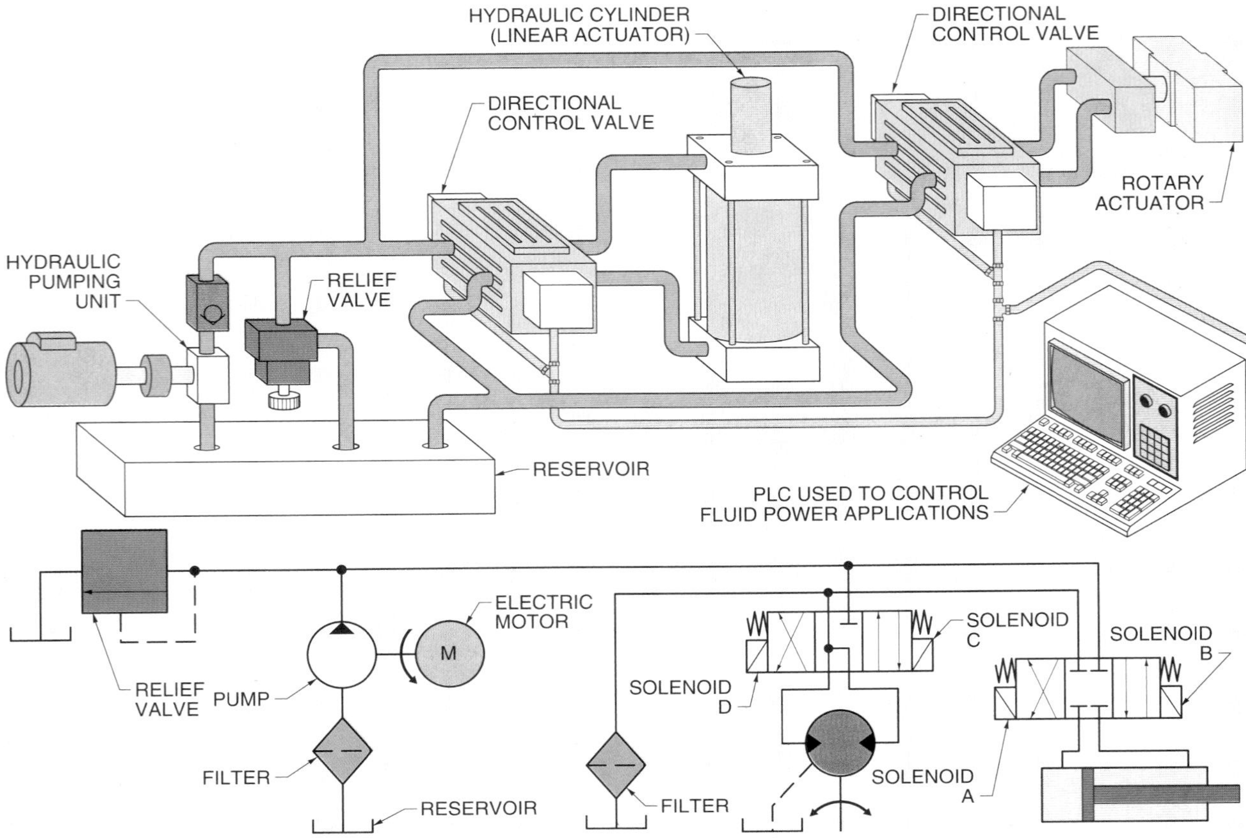

Figure 14-29. PLCs may be used to control linear and rotary actuators in an industrial fluid power circuit.

Industrial Drive Control

Motors have normally been connected directly to the power lines and operated at a set speed. As systems have become more automated, variable motor speed is required. Adjustable speed controls are available to control the speed of AC and DC motors. These controls are normally manually set for the desired speed, but many allow for automatic control of the set speed. A PLC may be used to control AC drives. See Figure 14-30. The drives can accept frequency and direction commands in a BCD format that the PLC can provide with a BCD output module.

Pulp and Paper Industries. Pulp and paper industries use PLCs to control each operation and diagnose a problem in the system. See Figure 14-31. This operation covers a large area. The control of this operation is ideal for a PLC because most control logic is start/stop, time delay, count sequential, and interlock functions.

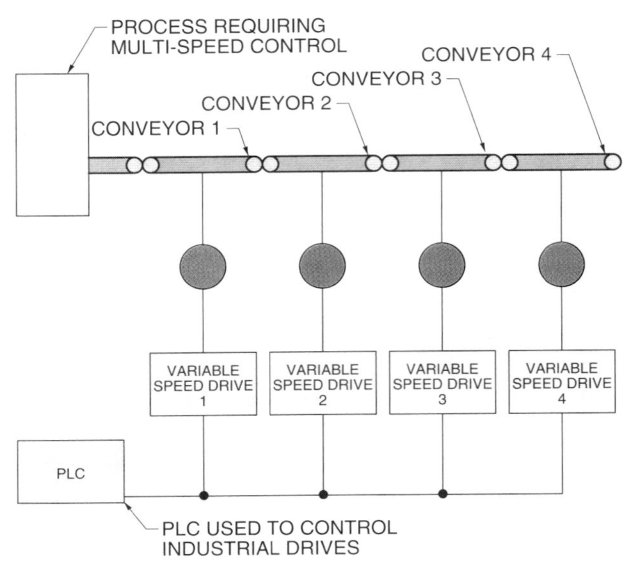

Figure 14-30. PLCs can be used to control and synchronize the speed of conveyors on an assembly line.

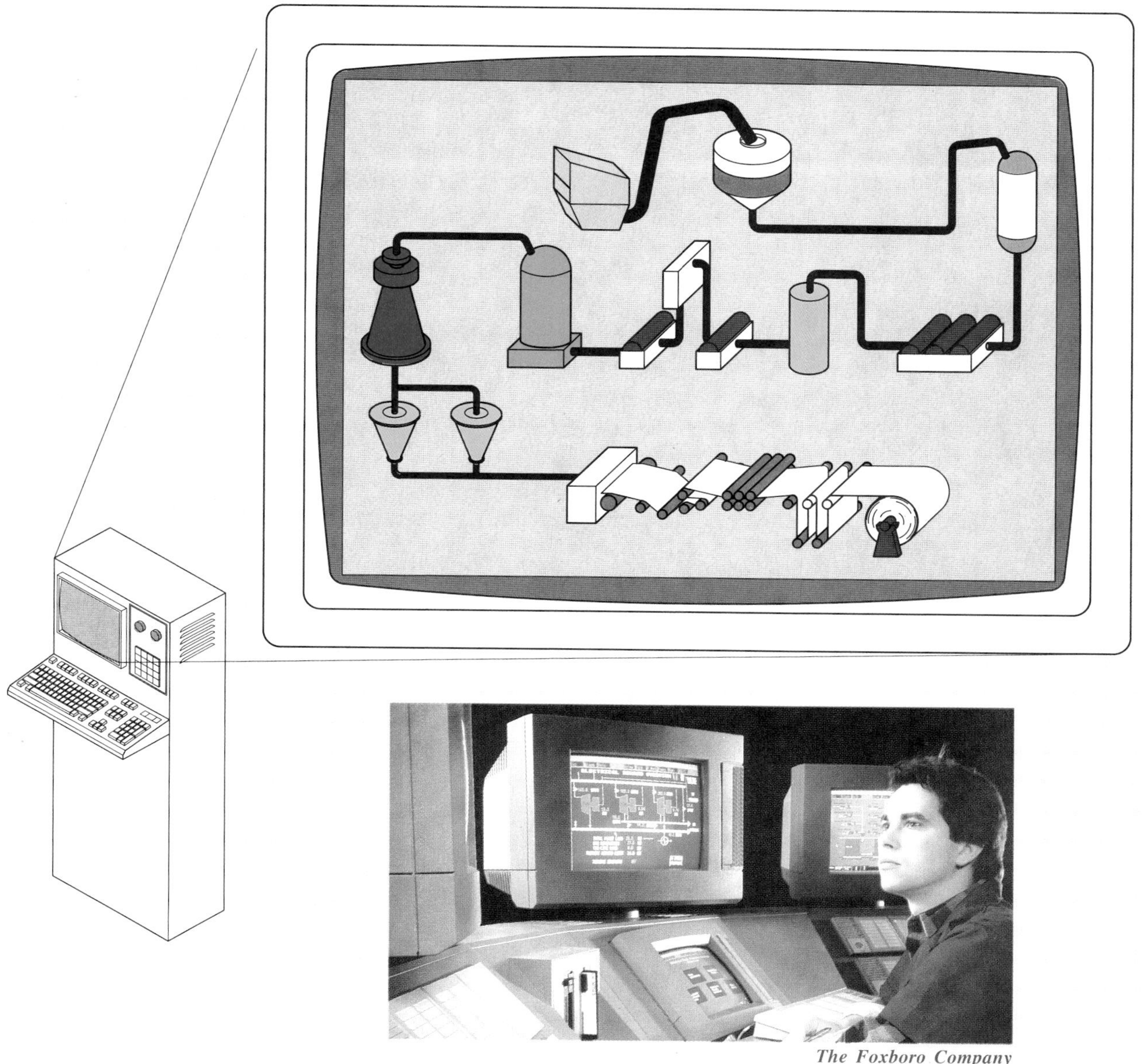

The Foxboro Company

Figure 14-31. In a paper mill, PLCs control each operation and diagnose problems in the system.

The PLC allows for required I/O which, when multiplexed, can transmit multiple signals over a single pair of wires. The basic operation of a paper mill is to receive raw material such as logs, pulpwood, or chips, and process, size, store, and deliver the material. This includes a large conveyor system that has diverter gates, overtravel switches, speed control, and interlocking. A break in any part of the system can shut down the entire system. Finding a fault is time-consuming because the system covers a large area. To solve this problem, a PLC with fault diagnostics can be used to analyze the system and give an alarm and printout of where the problem exists with suggested solutions.

Batch Process Control Systems. Batch processing blends sequential, step-by-step functions with continuous closed-loop control. Process control is systems control, and systems are made up of many parts. Individual PLCs can be used to control each part and step of the process, with additional PLCs and computers supervising the total operation.

In a batch process control system, an operator interface is used for instrumentation or other monitoring functions. See Figure 14-32. An operator interface is added as part of the system. This may be in the form of an instrumentation and process control station, a CRT (cathode-ray tube) terminal, or any other type of interface. To aid in interfacing and monitoring a programmable-based system, a serial port is used for monitoring and programming a system using a computer. Thus, the individual solenoids, motor starters, heating elements, etc. at each process step are directly controlled by the local PLCs with the host computer supervising all of the PLCs. See Figure 14-33.

PLC Circuits

Control circuits that do not use PLCs for control functions have been used for over 100 years. These control circuits do not allow for much flexibility or change. Today's electrical circuits are usually designed with change in mind. Changes include the way the circuit operates or additional safety features. With a PLC, changes can easily be made in an electrical circuit by changing the program.

A basic forward/reversing circuit is an example of a circuit that may require changes. In a basic forward/reversing circuit, very little circuit logic is required. See Figure 14-34.

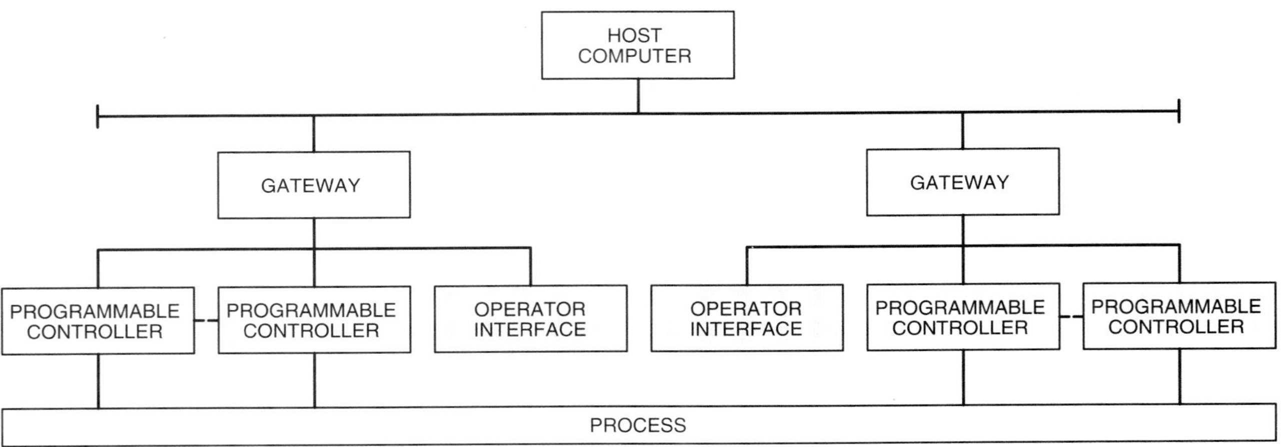

Figure 14-32. In a batch process control system, an operator interface is used for instrumentation or other monitoring functions.

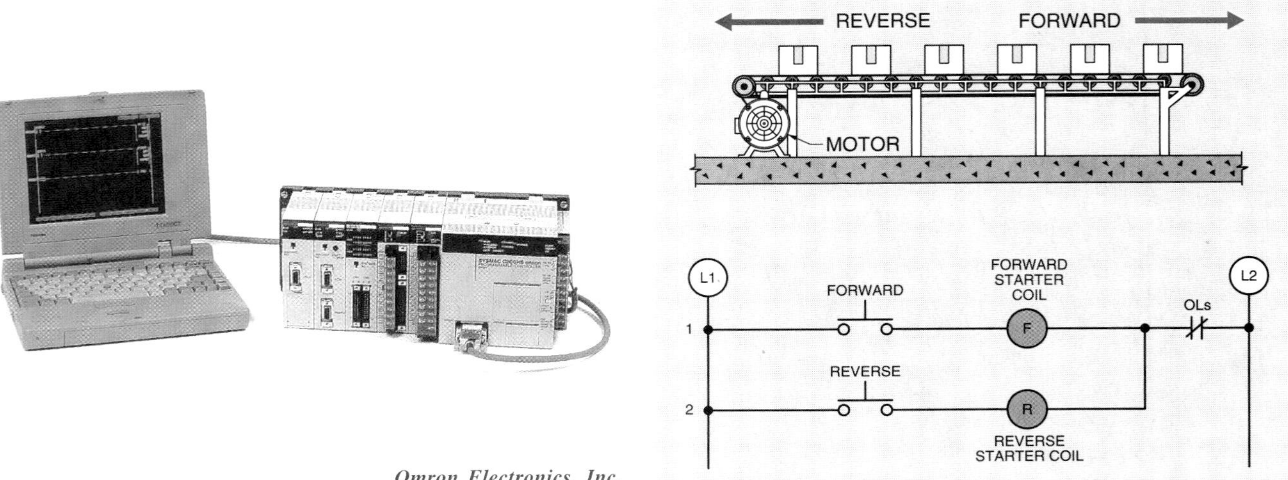

Omron Electronics, Inc.

Figure 14-33. A serial port is used for monitoring and programming a system using a computer.

Figure 14-34. In a basic forward/reversing circuit, very little circuit logic is required.

In this circuit, the forward pushbutton operates the forward starter coil, and the reverse pushbutton operates the reverse starter coil. This circuit operates satisfactorily if no operator error occurs. However, if an operator presses both pushbuttons at the same time, both starter coils energize. This causes a short circuit in the power circuit. Interlocking is added to solve this problem. Interlocking prevents the operator from energizing both starter coils at the same time. See Figure 14-35.

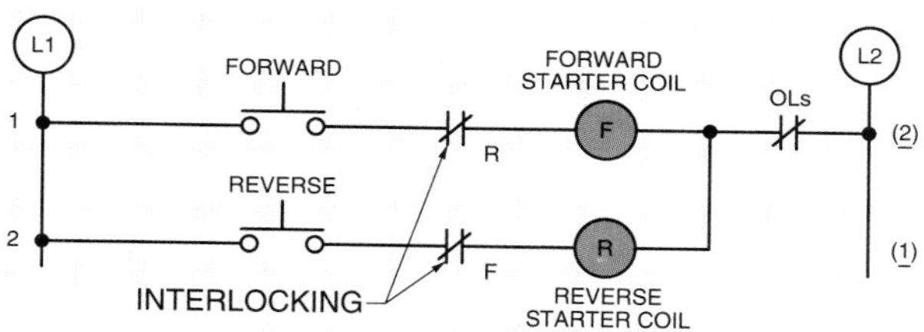

Figure 14-35. Interlocking is added to a circuit to prevent the operator from energizing both starter coils at the same time.

In a hard-wired circuit, auxiliary contacts must be added and wired to interlock the circuit. A PLC allows interlocking of the circuit with a simple change in the program and no additional components. See Figure 14-36.

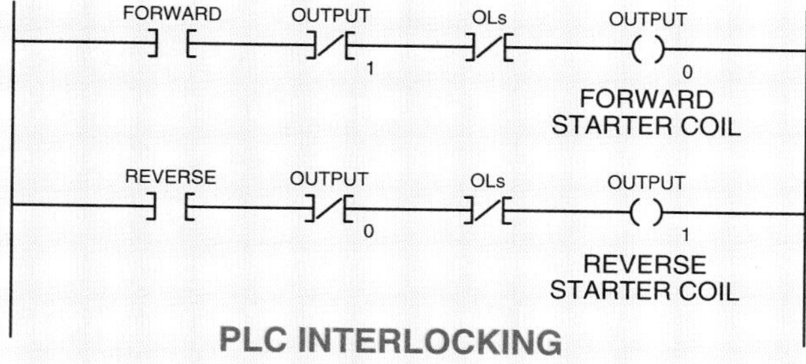

Figure 14-36. A PLC allows interlocking of the circuit with a simple change in the program and no additional components.

Another change that might be required is adding memory to the circuit. In a hard-wired circuit, auxiliary contacts are required. With a PLC, the program is changed. See Figure 14-37.

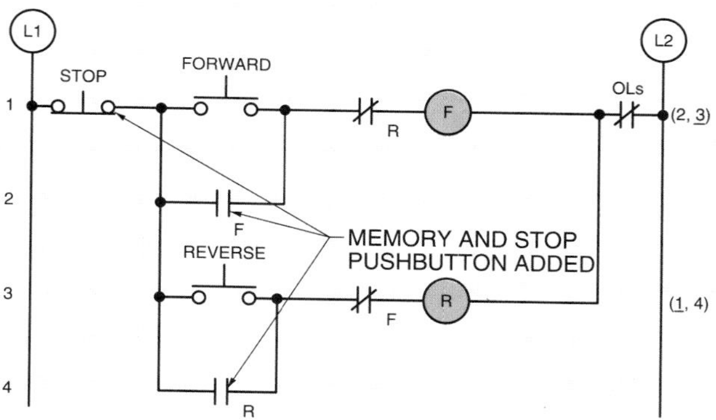

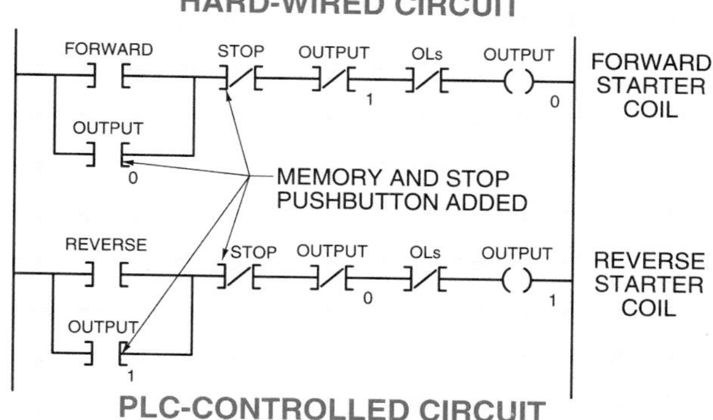

Figure 14-37. Adding memory to a circuit requires adding components and wiring if the circuit is hard wired or changing the program if the circuit is controlled by a PLC.

In a PLC circuit, many circuit changes can be programmed. However, additional inputs and outputs must be wired to the controller. For example, if a light is required to indicate the direction of motor (or product) travel, the light must be physically wired to the PLC. This is one of the similarities between hard-wired and PLC circuits. See Figure 14-38.

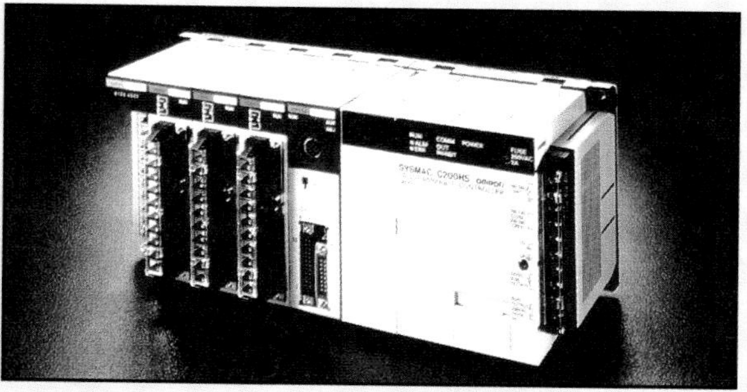

Omron Electronics, Inc.

Omron's C200H/HS Series mid-sized PLCs offer an expanded instruction set and data memory for high-speed machine control applications.

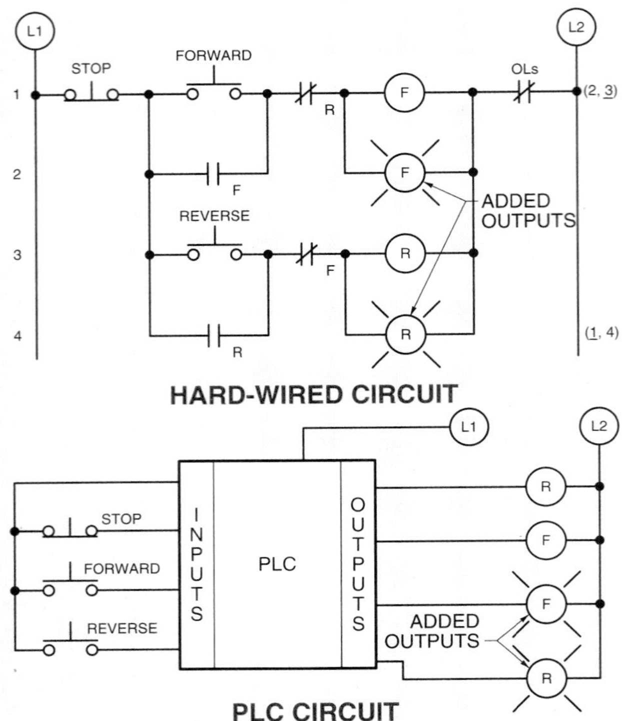

Figure 14-38. In a PLC circuit, additional inputs and outputs must be wired to the controller.

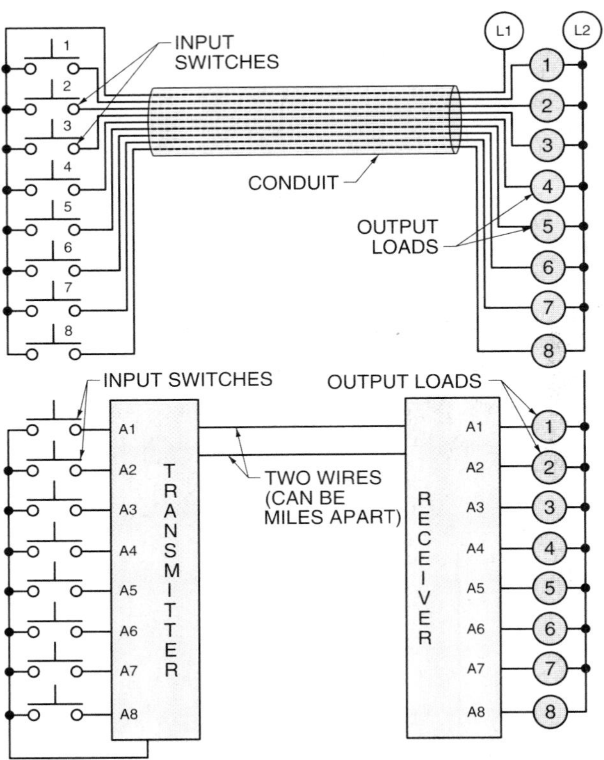

Figure 14-39. Multiplexing eliminates the need for costly hard wiring in a system.

MULTIPLEXING

Multiplexing is a method of transmitting more than one signal over a single transmission system. As the distance increases between any transmitting and receiving point, the cost of multi-conductor cable with separate wires for each signal becomes very expensive through installation, maintenance, and replacements. With multiplexing, two wires can serve multiple transmitters and receivers. A multiplexing system (2-wire system) is ideal when used with a PLC, as all inputs and outputs can be connected with just two wires.

One of the advantages in using a multiplexing system for control is the elimination of costly hard wiring. See Figure 14-39. In this circuit, eight control switches are hard wired to control eight loads. Two wires are required for each control switch. This means that time and money is wasted for even the shortest distance. As the distance between the control switches and loads increases, the cost of time and materials for the hard-wired circuit increases. The disadvantages of hard wiring include:

- Point-to-point wiring required between each switch and load
- Must pull dozens of wires for even small applications
- Number of wires and size required
- Conduit size required
- High conduit, wire, and labor costs
- Cost increases as distance between switches and loads increases

This same circuit can be connected using a multiplexing system. Only two wires are required between the eight control switches and eight loads. Additional control switches to be added require no additional transmission wires. Additional transmitters, receivers, displays, or PLCs can all be connected to the same two wires. The advantages of multiplexing include:

- No point-to-point wiring
- No conduit or multiple wires required
- Can easily expand with no additional transmission wires required
- Digital and analog signals can be transmitted on the same 2-wire system

Wiring a control circuit becomes more difficult as the circuit increases in size and function. A multiplexing system can send back a signal to indicate that the load is energized. A multiplexing system is much simpler than hard wiring and can be expanded to almost any number of inputs and outputs, all controlled by a PLC. See Figure 14-40. The PLC controls all inputs and outputs and makes timing, counting, sequencing, and any other required logic decisions.

A multiplexing system can be used to transmit both analog and digital signals on the same 2-wire system. This makes the system ideal for any instrumentation application including the transmission and control of temperatures, BCD signals, rpm, voltage and current levels, and counts.

In addition, a 24-hour clock and printer can be added to the system for documentation. This makes it possible to print out the time of day when a certain event has taken place on the multiplexing system.

Security Systems

A multiplexing system can be used in a security system for a plant or building. Each door and window can be connected to the 2-wire system. A display is located in a central control location to monitor the total system. See Figure 14-41. A clock and printer can be added to record the time each door or window is opened and closed. A PLC may be added to control all required circuit logic.

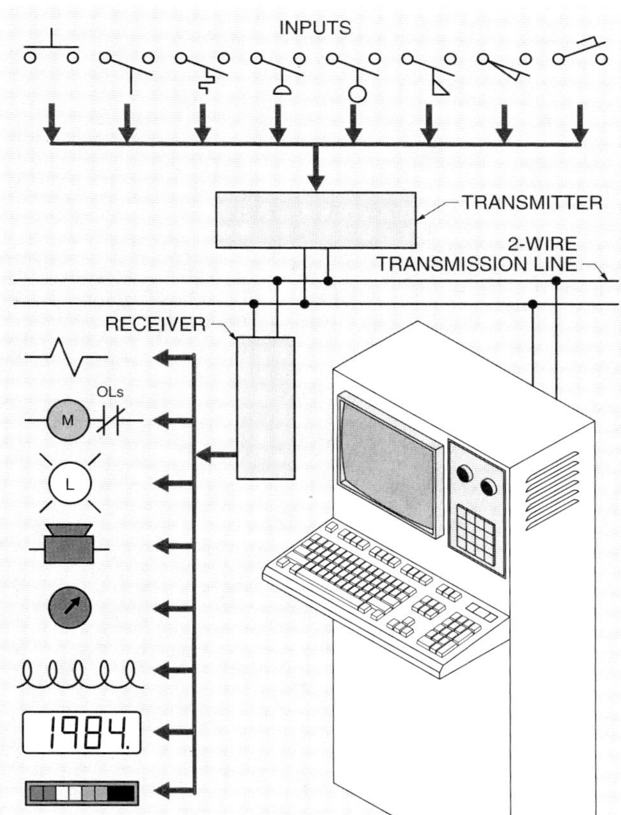

Figure 14-40. A multiplexing system is simpler than hard wiring and can be expanded to almost any number of inputs and outputs, all controlled by a PLC.

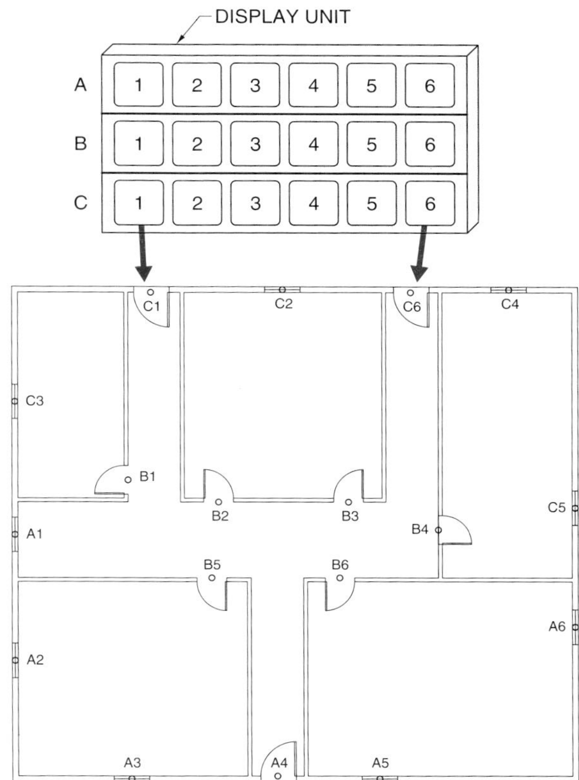

Figure 14-41. A multiplexing system can be used in a security system for a plant or building.

The PLC's controlling functions on the multiplexing system can be expanded as necessary. For example, if a security guard is to patrol a building, the controller can be programmed to monitor the guard as well as the building. As the guard moves through the building, the controller monitors the movement by recording when a door is opened and/or when the guard activates an assigned switch. The controller knows how long it should take the guard to move from station to station. If something happens to the guard, the controller detects this and takes corrective action, such as alarming a central control station.

Conveyor Systems

Conveyor systems are commonly used in industry for movement of materials. Additional control is required as industrial systems become more automated. Additional control requires additional wires to be connected from machine to machine. Multiplexing can be used to reduce the total wires required.

As in any assembly line application, a fault or breakdown at one station requires that all upstream machines be turned OFF to prevent a product jam. Multiplexing may be used to link the system together because this system may cover miles in many applications. See Figure 14-42.

A sensor may be used to detect a fault at one location and send a signal over the 2-wire system to stop all upstream machines and conveyors. This system may also be connected for total control of all functions using the multiplexing system and a PLC.

TROUBLESHOOTING PLCs

Troubleshooting PLCs normally involves finding a problem in the hardware or software. Most hardware problems are found in the input and output sections of the PLC and can usually be found using standard multimeters. Software problems require a knowledge of the specific program used and type of manufacturer equipment used.

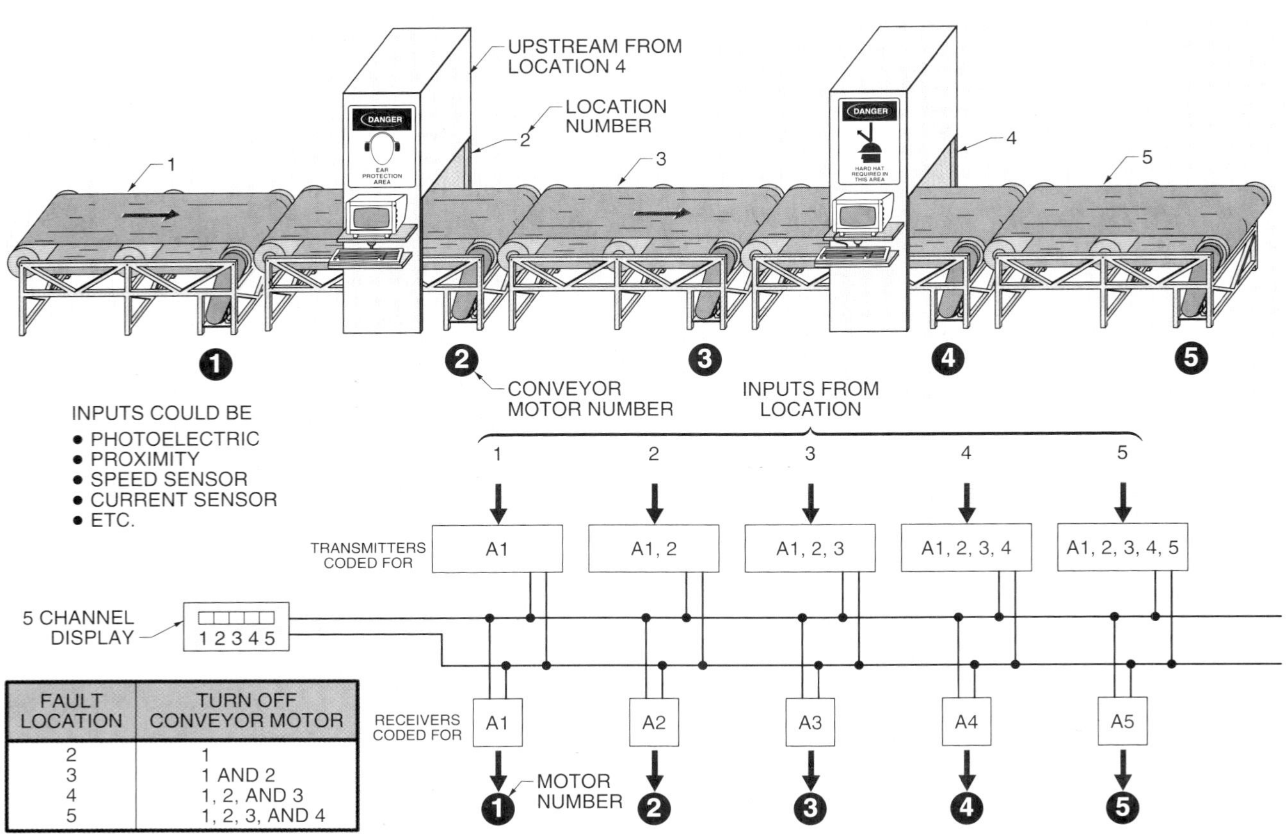

FAULT LOCATION	TURN OFF CONVEYOR MOTOR
2	1
3	1 AND 2
4	1, 2, AND 3
5	1, 2, 3, AND 4

Figure 14-42. An assembly line using several conveyors can be controlled by a 2-wire multiplexing system.

Troubleshooting Input Modules

Signals and information are sent to a PLC using input devices, such as pushbuttons, limit switches, level switches, and pressure switches.

The input devices are connected to the input module of the PLC. Input devices are connected to terminal screws at the back of the input module. The controller does not receive the proper information if the input device or input module is not operating correctly. See Figure 14-43.

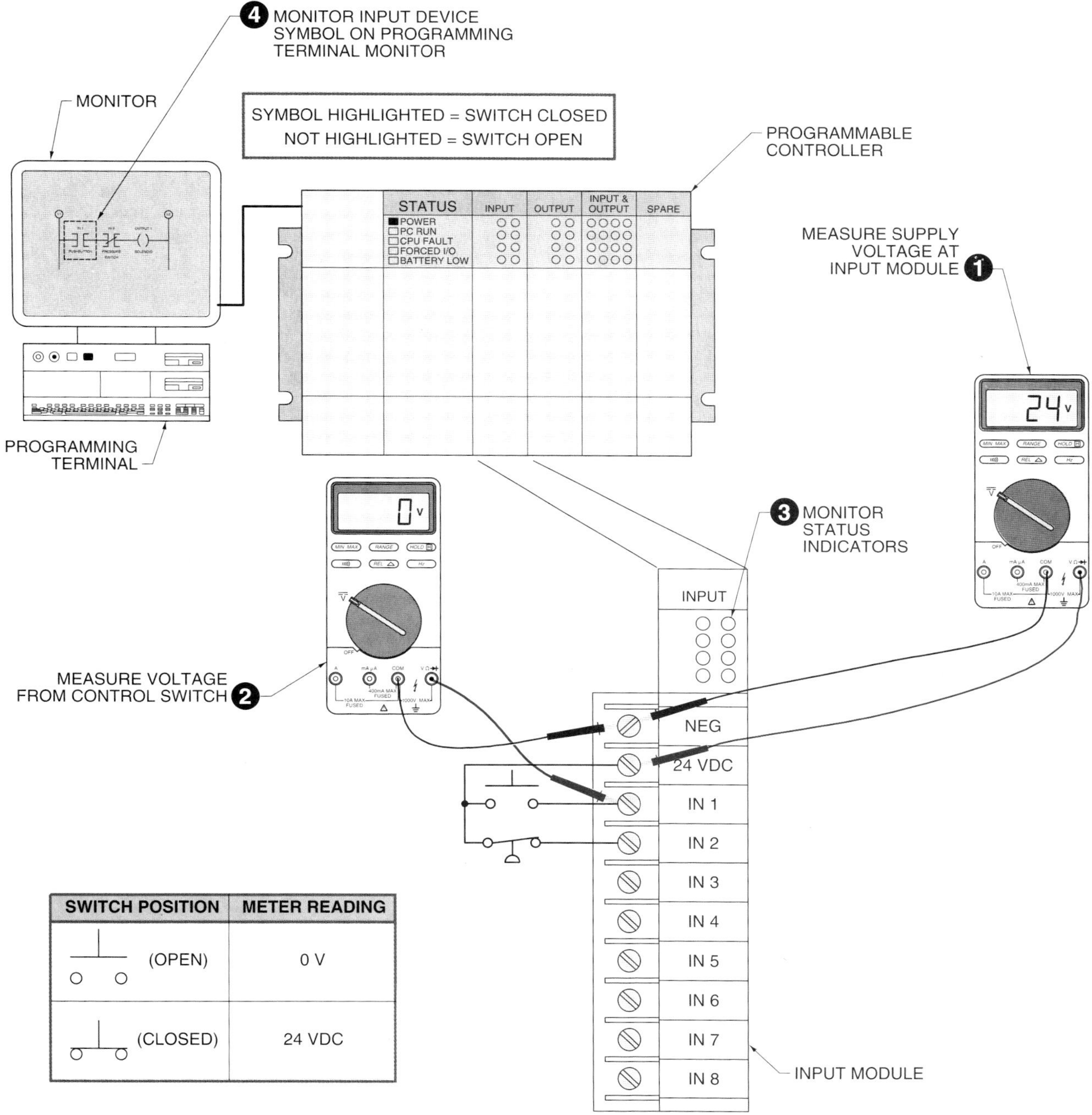

SWITCH POSITION	METER READING
(OPEN)	0 V
(CLOSED)	24 VDC

TROUBLESHOOTING INPUT MODULES

Figure 14-43. The controller does not receive the proper information if the input device or input module is not operating correctly.

To troubleshoot an input module of a PLC, apply the procedure:

1. Measure the supply voltage at the input module to ensure that there is power supplied to the input device(s). Test the main power supply of the controller when there is no power.
2. Measure the voltage from the control switch. Connect the meter directly to the same terminal screw to which the input device is connected.

The voltmeter should read the supply voltage when the control switch is closed. The voltmeter should read the full supply voltage when the control device uses mechanical contacts. The voltmeter should read nearly the full supply voltage when the control device is solid-state. Full supply voltage is not read because .5 V to 6 V is dropped across the solid-state control device. The voltmeter should read zero or little voltage when the control switch is open.

3. Monitor the status indicators on the input module. The status indicators should illuminate when the meter indicates the presence of supply voltage.
4. Monitor the input device symbol on the programming terminal monitor. The symbol should be highlighted when the meter indicates the presence of supply voltage.

Replace the control device if the control device does not deliver the proper voltage. Replace the input module if the control device delivers the correct voltage, but the status indicator does not illuminate.

Troubleshooting Input Devices

Input devices such as pushbuttons, limit switches, pressure switches, and temperature switches are connected to the input module(s) of a PLC. Input devices send information and data concerning circuit and process conditions to the controller. The processor receives the information from the input devices and executes the program. All input devices must operate correctly for the circuit to operate properly. See Figure 14-44.

To troubleshoot an input device of a PLC, apply the procedure:

1. Place the controller in the test or program mode. This step prevents the output devices from turning ON. Output devices are turned ON when the controller is placed in the run mode.
2. Monitor the input devices using the input status indicators (located on each input module), the programming terminal monitor, or the data file. A *data file* is a group of data values (inputs, timers, counters, and outputs) that is displayed as a group and whose status may be monitored.
3. Manually operate each input starting with the first input. Never reach into a machine when manually operating an input. Always use a wooden stick or other non-conductive device.

The input status indicator located on the input module should illuminate and the input symbol should be highlighted in the control circuit on the monitor screen when a normally-open input device is closed. The bit status on the programming terminal monitor screen should be set to 1 indicating a high or presence of voltage.

The input status indicator located on the input module should turn OFF and the input symbol should no longer be highlighted in the control circuit on the monitor screen when a normally-closed input device is open. The bit status on the programming terminal monitor screen should be set to 0 indicating a low or absence of voltage.

Select the next input device and test it when the status indicator and associated bit status match. Continue testing each input device until all inputs have been tested. Troubleshoot the input device and output device when the status indicator and associated bit status do not match.

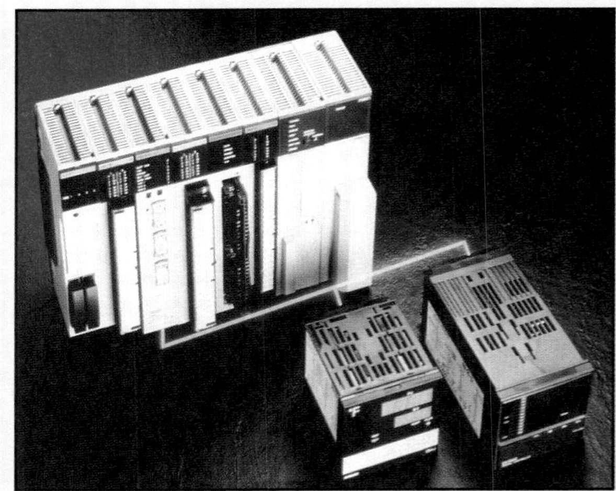

Omron Electronics, Inc.

Omron's CV Series PLCs are designed as an information-level supervisory control system that can link up to 32 temperature and process controllers.

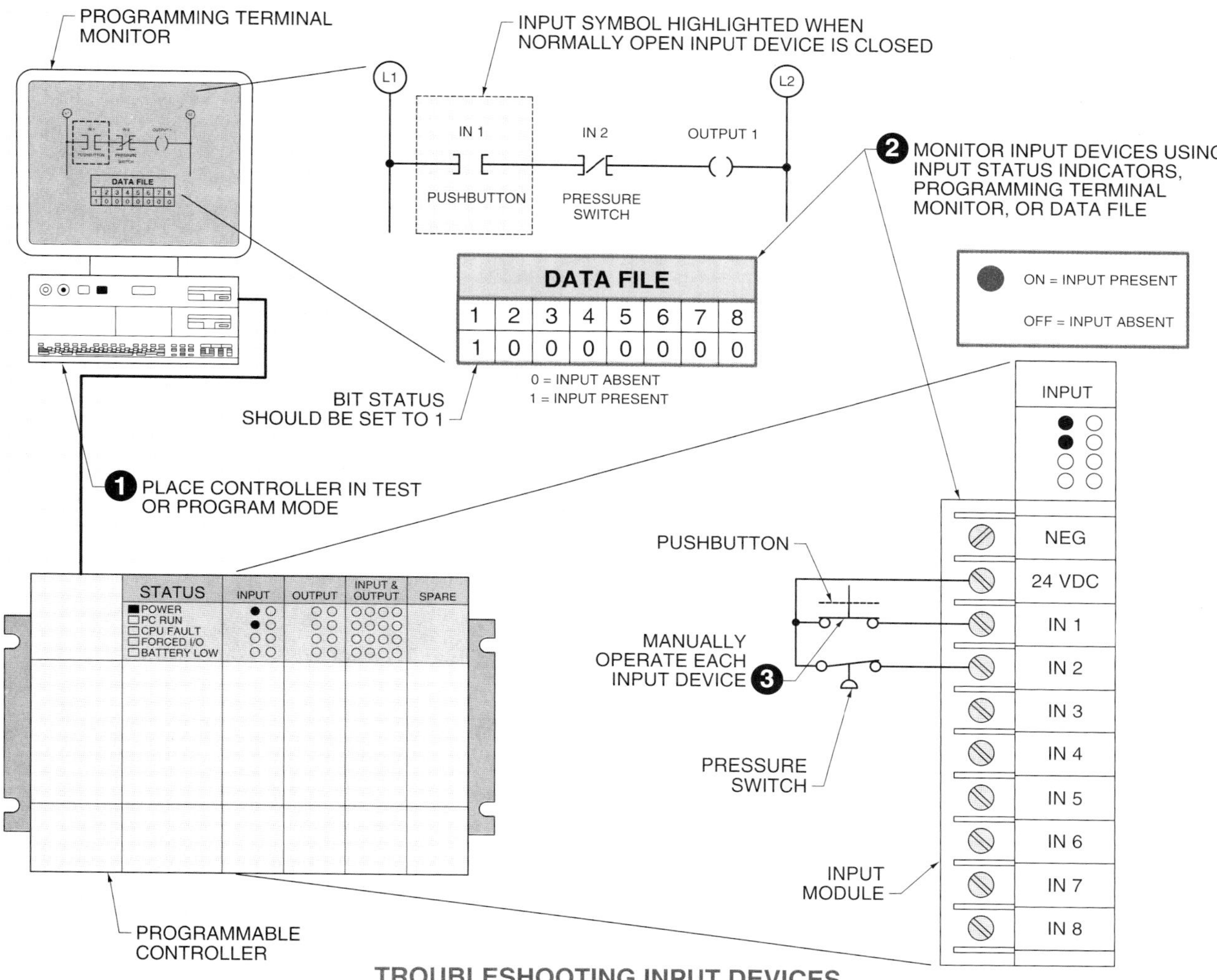

Figure 14-44. All input devices must operate correctly for the circuit to operate properly.

Troubleshooting Output Modules

A PLC turns ON and OFF the output devices (loads) in the circuit according to the program. The output devices are connected to the output module of the PLC. No work is produced in the circuit when the output devices or output module are not operating correctly. The problem may lie in the output module, output device, or controller when an output device does not operate. See Figure 14-45.

To troubleshoot an output module of a PLC, apply the procedure:

1. Measure the supply voltage at the output module to ensure that there is power supplied to the output devices. Test the main power supply of the controller when there is no power.
2. Measure the voltage delivered from the output module. Connect the meter directly to the same terminal screw to which the output device is connected.

The voltmeter should read the supply voltage when the program energizes the output device. The voltmeter should read full supply voltage when the output module uses mechanical contacts. The voltmeter should read almost full supply voltage when the output module uses a solid-state switch. Full voltage is not read because .5 V to 6 V is dropped across the solid-state switch. The voltmeter should read zero or little voltage when the program de-energizes the output device.

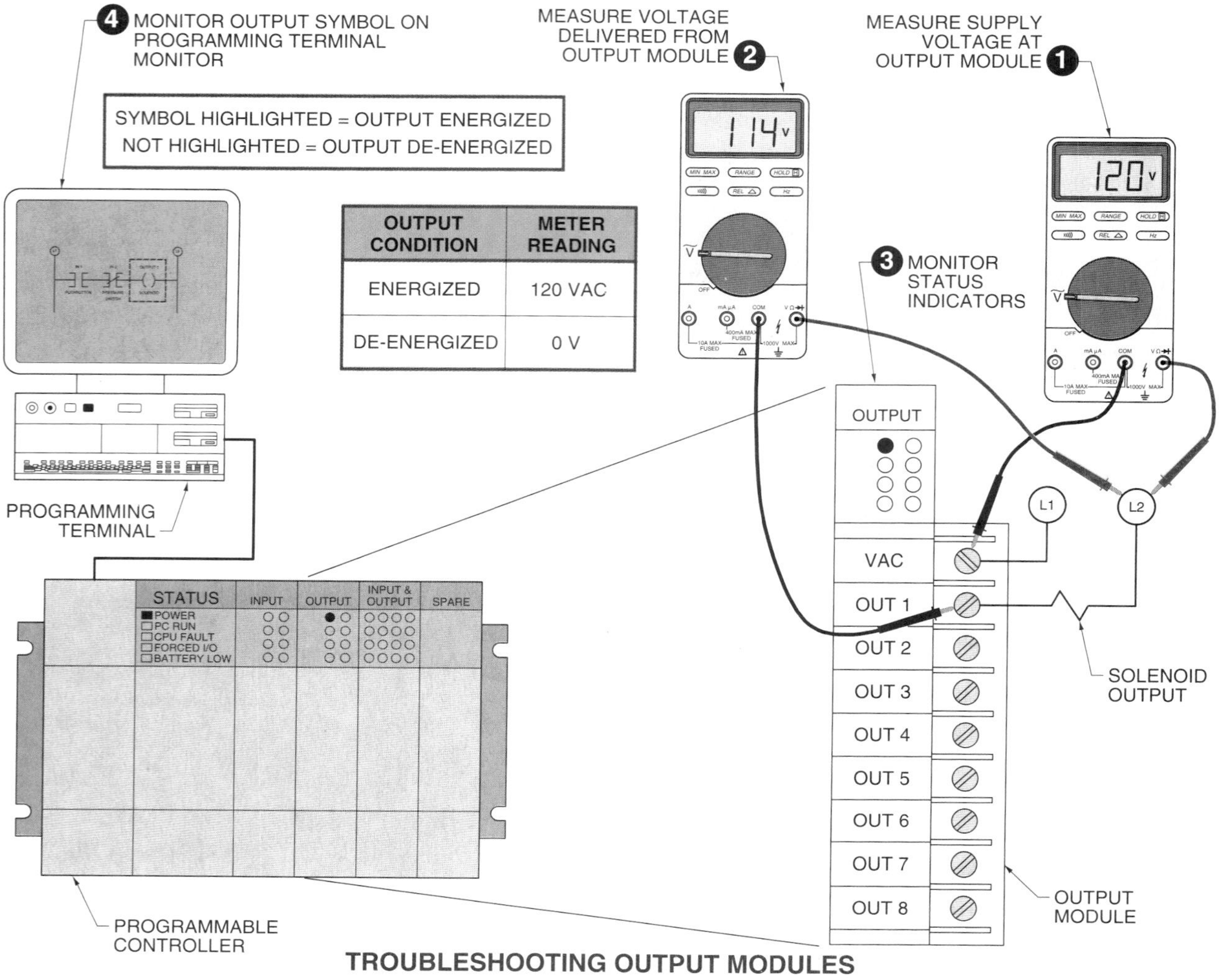

4 MONITOR OUTPUT SYMBOL ON PROGRAMMING TERMINAL MONITOR

MEASURE VOLTAGE DELIVERED FROM OUTPUT MODULE 2

MEASURE SUPPLY VOLTAGE AT OUTPUT MODULE 1

SYMBOL HIGHLIGHTED = OUTPUT ENERGIZED
NOT HIGHLIGHTED = OUTPUT DE-ENERGIZED

OUTPUT CONDITION	METER READING
ENERGIZED	120 VAC
DE-ENERGIZED	0 V

3 MONITOR STATUS INDICATORS

PROGRAMMING TERMINAL

PROGRAMMABLE CONTROLLER

SOLENOID OUTPUT

OUTPUT MODULE

TROUBLESHOOTING OUTPUT MODULES

Figure 14-45. No work is produced in the circuit when the output devices or output module are not operating correctly.

3. Monitor the status indicators on the output module. The status indicators should be energized when the voltmeter indicates the presence of supply voltage.

4. Monitor the output device symbol on the programming terminal monitor. The output device symbol should be highlighted when the voltmeter indicates the presence of supply voltage. Replace the output module when the output module does not deliver the proper voltage. Troubleshoot the output device when the output module does deliver the correct voltage but the output device does not operate.

Heidelberg Harris, Inc.

Monitoring the I/O status indicators helps locate any problem(s) when troubleshooting large systems.

Troubleshooting Output Devices

Output devices such as motor starters, solenoids, contactors, and lights are connected to the output modules of a PLC. An output device performs the work required for the application.

The processor energizes and de-energizes the output devices according to the program. All output devices must operate correctly for the circuit to operate properly. See Figure 14-46.

To troubleshoot an output device of a PLC, apply the procedure:

1. Place the controller in the test or program mode. Placing the controller in the test or program mode prevents the output devices from turning ON. Output devices turn ON when the controller is placed in the run mode.
2. Monitor the output devices using the output status indicators (located on each output module), the programming terminal monitor, or the data file.
3. Activate the input that controls the first output device. Check the program displayed on the monitor screen to determine which input activates which output device. Never reach into a machine to activate an input.

Select the next output device and test it when the status indicator and associated bit status match. Continue testing each output device until all output devices have been tested. Troubleshoot the input device and output device when the status indicator and associated bit status do not match.

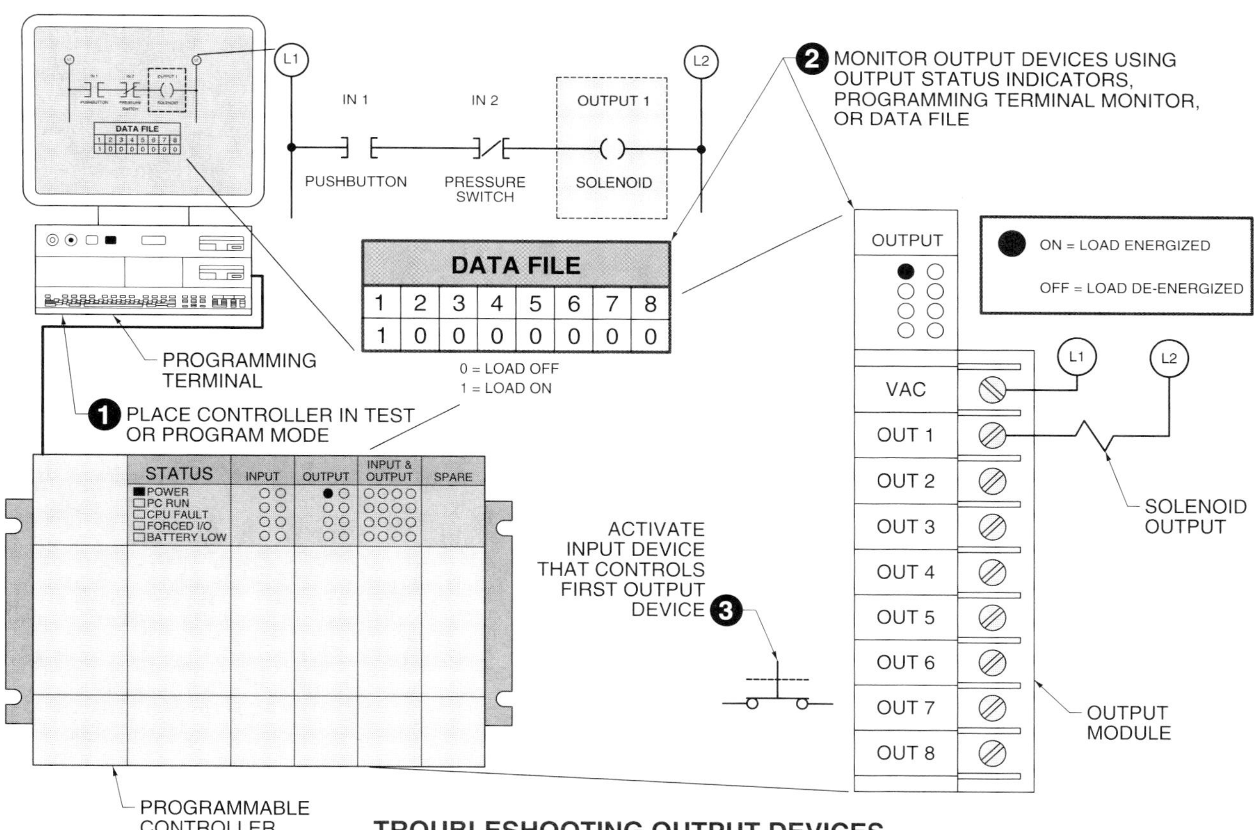

Figure 14-46. All output devices must operate correctly for the circuit to operate properly.

1. What are the two major categories of electrical systems designed to produce products?
2. Which manufacturing area represents durable good type products?
3. Which manufacturing area represents consumable good type products?
4. What are I/Os?
5. What type of diagrams are used to program a PLC?
6. What are the four basic parts of a PLC?
7. What are some examples of discrete type inputs?
8. What are some examples of data type inputs?
9. What is scan time?
10. What is a programming diagram?
11. Why does a PLC include a battery as part of its system?
12. What command is the opposite of the force command?
13. What is electrical noise?
14. How is interlocking added into a PLC controlled circuit?
15. What is multiplexing?

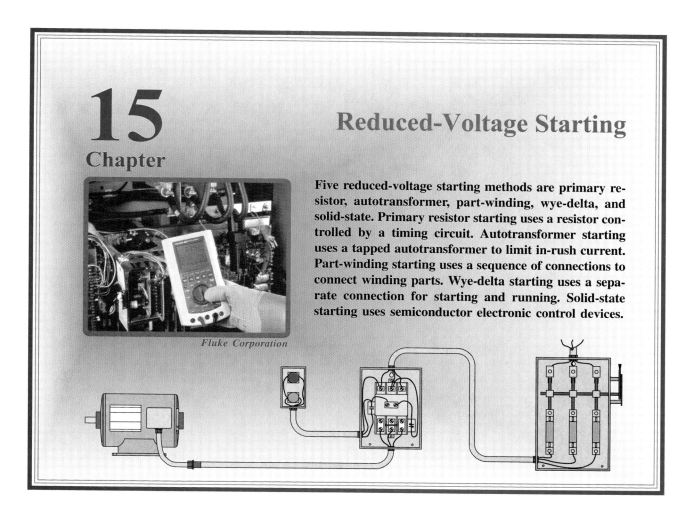

Chapter 15

Reduced-Voltage Starting

Five reduced-voltage starting methods are primary resistor, autotransformer, part-winding, wye-delta, and solid-state. Primary resistor starting uses a resistor controlled by a timing circuit. Autotransformer starting uses a tapped autotransformer to limit in-rush current. Part-winding starting uses a sequence of connections to connect winding parts. Wye-delta starting uses a separate connection for starting and running. Solid-state starting uses semiconductor electronic control devices.

Fluke Corporation

REDUCED-VOLTAGE STARTING

Full-voltage starting is the least expensive and most efficient means of starting a motor for applications involving small horsepower motors. Many applications involve large horsepower DC motors and AC motors that require reduced-voltage starting because full-voltage starting may create interference with other systems. Reduced-voltage starting reduces interference in the power source, the load, and the electrical environment surrounding the motor.

Power Source

Reduced-voltage starting is used to reduce the large current drawn from the power company lines by the across-the-line start of a large motor. An induction motor acts much like a short circuit in the secondary of a transformer when it is started. The current drawn by the motor is typically about two to six times the current rating found on the motor nameplate. This sudden demand for large current can reflect back into the power lines and create problems. Reduced-voltage starting reduces the amount of starting current a motor draws when starting. See Figure 15-1.

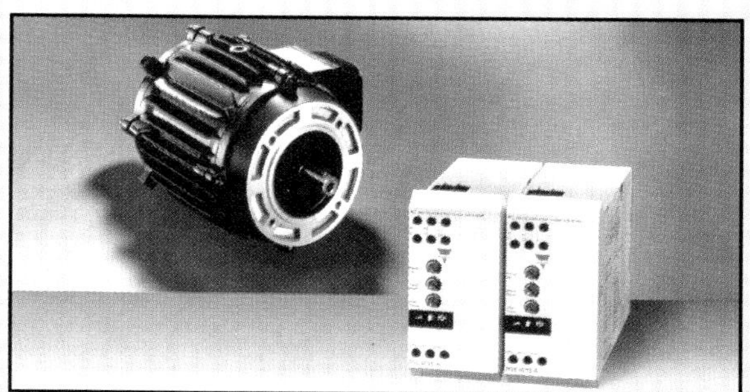

Carlo Gavazzi Inc. Electromatic Business Unit

Carlo Gavazzi Inc. manufactures 3ϕ soft starters for use in industrial applications that require reduced starting torque.

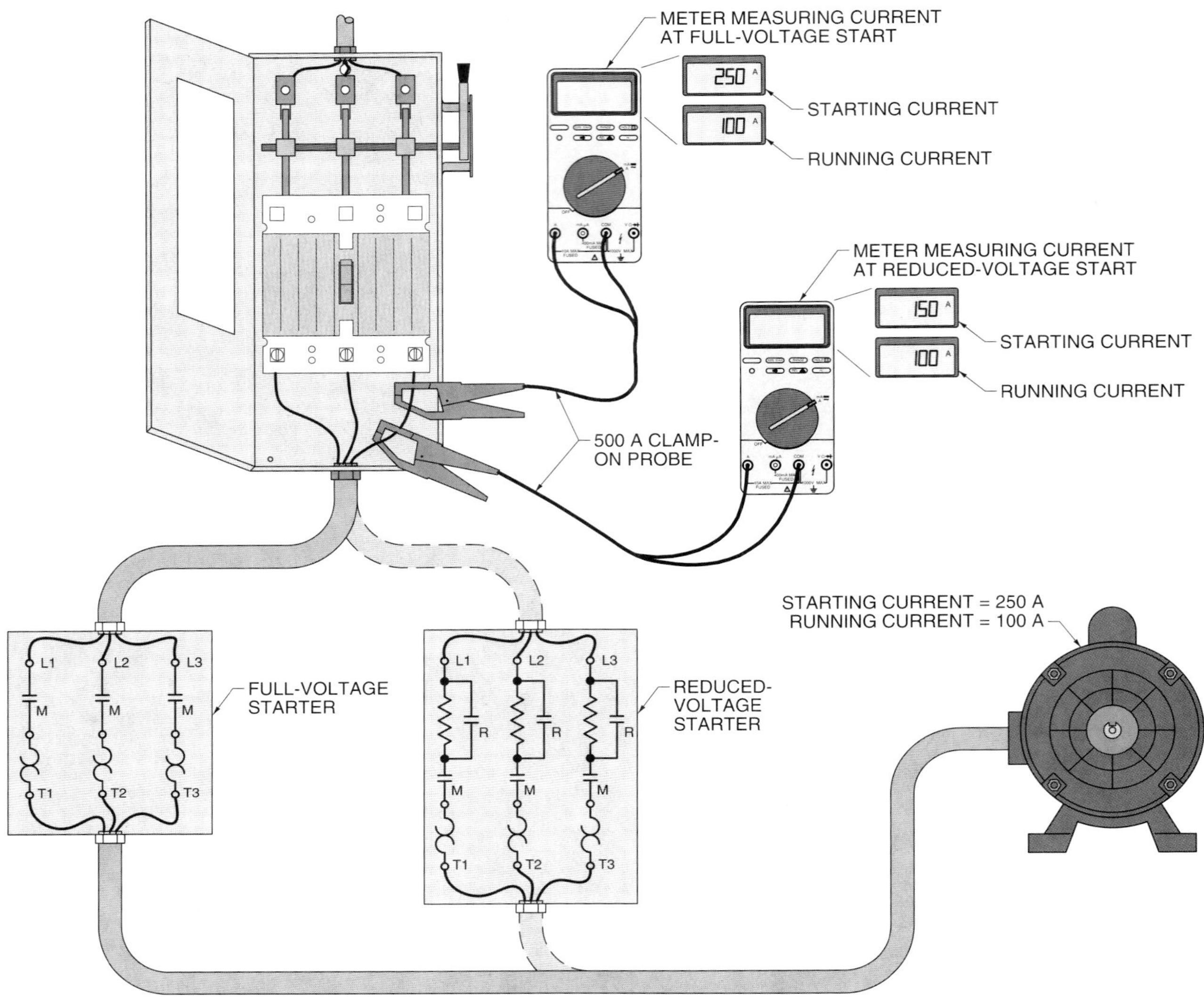

Figure 15-1. Reduced-voltage starting reduces the amount of starting current a motor draws when starting.

Electric utilities normally limit the inrush current drawn from their lines to a maximum amount for a specified period of time. Such limitations are necessary for smooth, steady power regulation and eliminating objectionable voltage disturbances such as annoying light flicker.

In these cases, the utility company is not limiting the total maximum amount of current that can be drawn, but rather dividing the amount of current into steps. This permits an incremental start that allows the utility voltage regulators sufficient time to compensate for the large current draw. *Increment current* is the maximum current permitted by the utility in any one step of an increment start. This increment current may be determined by checking with the local utility company. Reduced-voltage starting provides incremental current draw over a longer period of time.

Load Torque and Starting Requirements

In several industries, especially those dealing with paper and other delicate fabrics, care must be exercised to avoid sudden high starting torque (turning force). Such torque could stretch or tear the product.

To prevent product damage or damage to gears, belts, and chain drives, it is necessary to limit starting torque surges. Reduced-voltage starting is used to overcome excessive starting torque by providing a gentle start and smooth acceleration of a motor. See Figure 15-2.

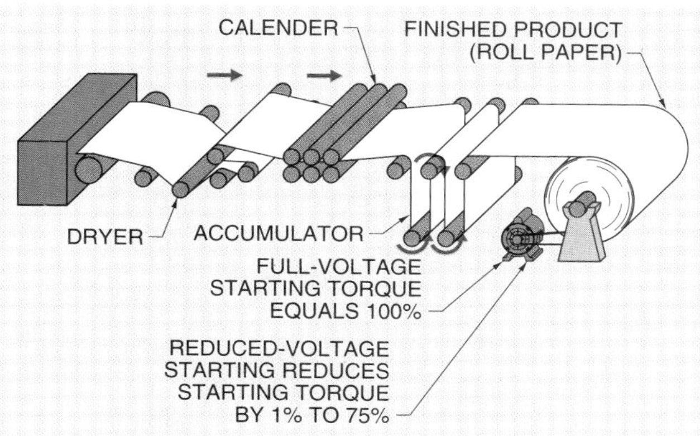

Figure 15-2. Reduced-voltage starting reduces the amount of starting torque produced on a load.

According to the formula $I = E \div R$, as voltage is reduced, current is reduced; and as current is reduced, torque is reduced because motor torque is proportional to current. A reduction in voltage reduces current, which reduces torque to provide a gentle start. As voltage is increased, current and torque is increased, providing smooth acceleration. Reduced-voltage starting is not speed control. Reduced-voltage starting acts as a buffer or shock absorber to the load when it is starting.

Reduced-voltage starting should not be considered for use on loads which are difficult to start. A load that is difficult to start at full voltage does not start at a reduced voltage.

Electrical Environment

A new electrical system should not create problems for the systems which are already installed and working properly. Electric current surges can cause disruptions. For example, a current surge may cause timers to reset or relays and starters to drop out. In buildings which are totally air conditioned, compressor motors have caused major computers and microcomputers to malfunction due to current surges. Reduced-voltage starting may be used to solve current surge problems even when not required by the utility companies.

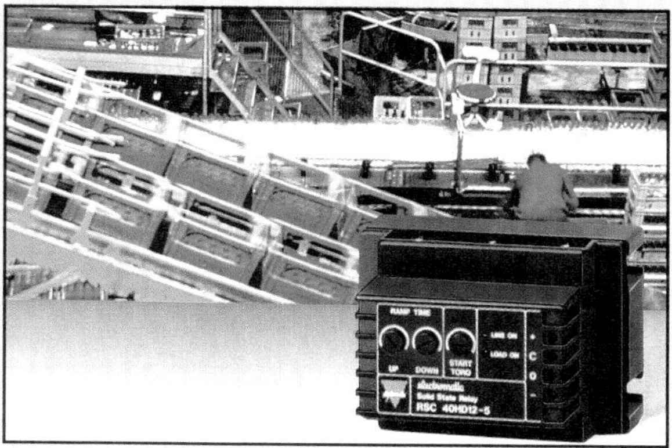

Carlo Gavazzi Inc. Electromatic Business Unit

The RSC 40HD12-6 control module from Carlo Gavazzi Inc. is used for soft starting and soft stopping of 3ϕ induction motors.

DC MOTOR REDUCED-VOLTAGE STARTING

A DC motor is used to convert electrical energy into a rotating mechanical force. Although AC and DC motors operate on the same fundamental principles of magnetism, they differ in the way the conversion of the electrical power to mechanical power is accomplished. This difference gives each motor its own operating characteristics. The two fundamental operating characteristics of DC motors that make them the choice for some applications are high torque outputs and good speed control.

Another factor in using DC motors is the available source of power. For applications such as an automobile starter, a DC motor is compatible to the power source (battery) which delivers only DC. DC motors run by batteries are also used in industrial applications using portable power equipment such as forklift trucks, dollies, and small locomotives used to move materials and supplies. In applications where the motor is to be connected to a power source other than a battery, the available power source may be either AC or DC.

DC Motors

Any current-carrying conductor has a magnetic field around it. In a DC motor, a magnetic field, caused by current flow in a conductor, interacts with another magnetic field. This interaction causes the conductor to move.

A current-carrying conductor moves when placed between the poles of a magnet. The direction of the movement depends on the direction of the current and the magnetic field. The electron flow motor rule is used to determine the motion of a current-carrying conductor in a magnetic field. See Figure 15-3.

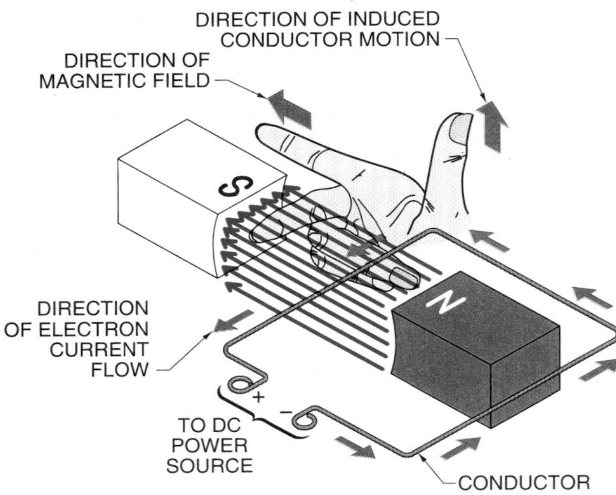

Figure 15-3. The electron flow motor rule is used to determine the motion of a current-carrying conductor in a magnetic field.

Rofin Sinar

Many industrial processes require electrical motors that must be designed to operate in hot and severe environments.

The index finger points in the direction of the magnetic field (N to S). The middle finger points in the direction of electron current flow in the conductor.

The conductor carries a current at right angles to the lines of the magnetic field. The force felt by the conductor is at right angles to both the current and the magnetic field. The amount of force depends on the intensity of the magnetic field, the current through the conductor, and the length of the conductor.

The intensity of the field and the amount of current are normally changed to increase force. The amount of force can be increased by increasing any of these three factors.

Torque is developed on a wire loop in a magnetic field. See Figure 15-4. The electron current flow is at a right angle to the magnetic field. This is required for induced motion because no force is felt by a conductor if the direction of electron current flow and magnetic field are the same (parallel).

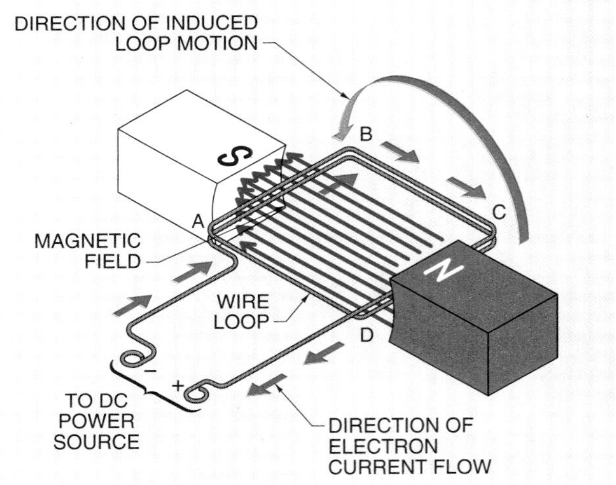

Figure 15-4. Torque is developed on a wire loop in a magnetic field.

Both sections of loop AB and CD have a force exerted on them because the direction of electron current flow in these segments is at right angles to the magnetic field. The exertion of force on AB and CD is opposite in direction because the current flow is opposite in each section.

The result of the two magnetic fields intersecting creates a turning force (torque) on the loop. The torque tends to rotate the loop in a counterclockwise direction. See Figure 15-5.

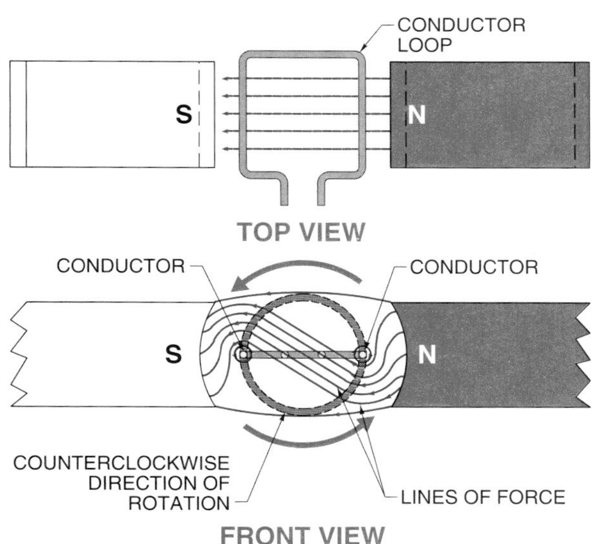

Figure 15-5. The distortion of the lines of force causes the conductor loop to rotate.

The interaction between the two magnetic fields causes a bending of the lines of force. The lines of force cause the loop to rotate when they straighten. The left conductor is forced downward, and the right conductor is forced upward, causing a counterclockwise rotation.

Motor Construction

The four main parts of a DC motor are its field, armature, brushes, and commutator. A rotating force is exerted on the armature when it is positioned so that the plane of the armature loop is parallel to the field, and the armature loop sides are at right angles to the magnetic field. See Figure 15-6.

No movement takes place if the armature loop is stopped in the vertical (neutral) position. In this position, no further torque is produced because the forces acting on the armature are upward on the top side and downward on the lower side.

The armature does not stop because of inertia. It continues to rotate for a short distance. As it rotates, the magnetic field in the armature is opposite that of the field. This pushes the conductor back in the direction it came, stopping the rotating motion.

A method is required to reverse the current in the armature every one-half rotation so that the magnetic fields work together. Brushes and a commutator are added to maintain a positive rotation.

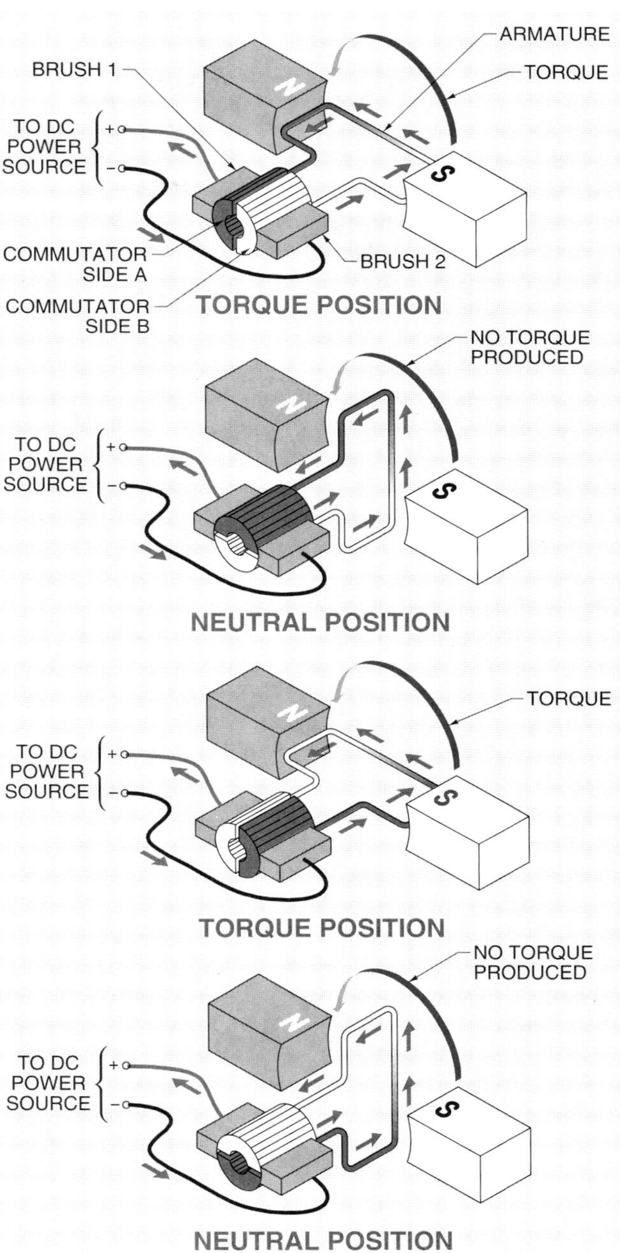

Figure 15-6. A rotating force is exerted on the armature when it is positioned so that the plane of the armature loop is parallel to the field, and the armature loop sides are at right angles to the magnetic field.

The commutator is used to reverse the direction of current flow in the armature. The commutator is split into two sections with each section connected to one side of the armature winding. The split-ring commutator is supplied voltage through brushes.

Each brush supplies a constant current from the power supply and does not change polarity.

Brush 1 begins in contact with side A of the commutator and brush 2 begins in contact with side B of the commutator. As the commutator rotates 90° through the magnetic field, brush 1 breaks contact with side A of the commutator and makes contact with side B of the commutator.

Brush 2 breaks contact with side B of the commutator and makes contact with side A. The flow of current through the commutator reverses because the flow of current is at the same polarity on the brushes at all times. This allows the commutator to rotate another 180° in the same direction. After the additional 180° rotation, brush 1 breaks contact with side B of the commutator and makes contact with side A of the commutator. Likewise, brush 2 breaks contact with side A of the commutator and makes contact with side B of the commutator. This reverses the direction of current in the commutator again and allows for another 180° of rotation. The armature continues to rotate as long as the commutator winding is supplied with current and there is a magnetic field.

A disadvantage of this type of motor is that no torque is produced each time the armature is in a neutral position. Maximum torque is produced when the armature is in the horizontal position. The motor torque and speed vary as the armature rotates. See Figure 15-7. Most practical devices require a motor to turn at a uniform speed.

Sprecher + Schuh

Reduced-voltage starting is required when starting large loads such as pump motors.

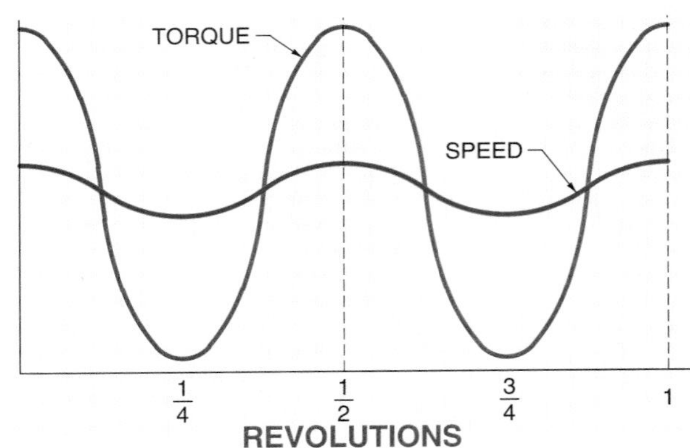

Figure 15-7. Torque and speed vary as the armature of a DC motor rotates.

Another disadvantage is that simple DC motors do not start easily. This is particularly true if the armature is in or near a neutral position. The armature must be moved out of the neutral position to start the motor. In DC motors, maximum torque is achieved by using an armature with more than one loop. See Figure 15-8.

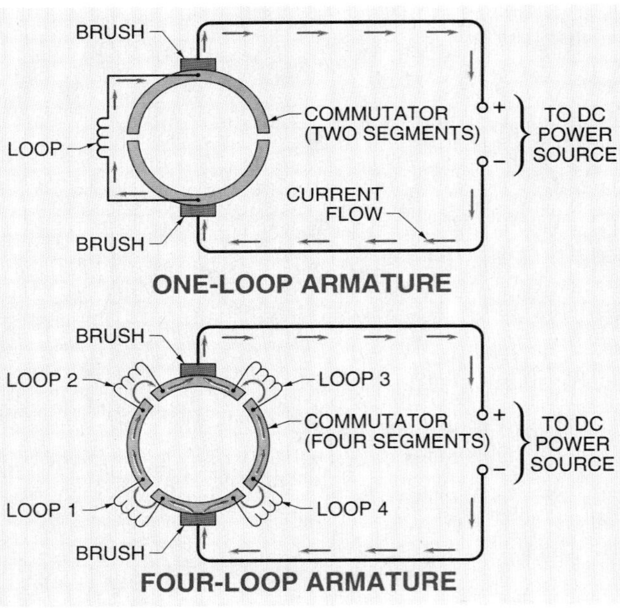

Figure 15-8. Torque in a DC motor is increased by using an armature with more than one loop.

Each loop of the armature is connected to a pair of commutator segments. A single pair of brushes makes contact with the commutator segments. The armature acts like two series circuits connected in parallel. See Figure 15-9.

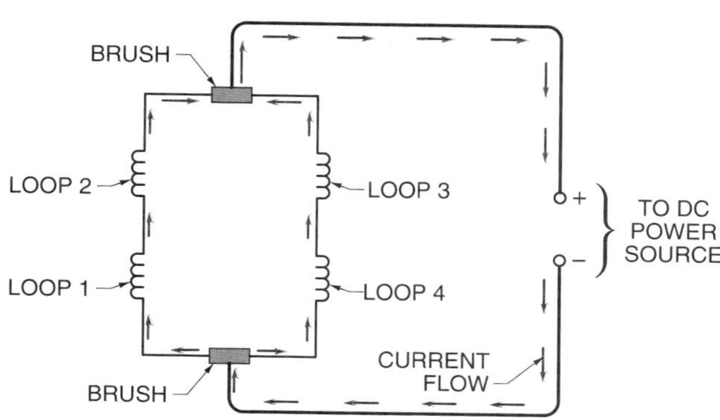

Figure 15-9. A four-loop armature acts like two series circuits connected in parallel.

All four loops act together and each loop adds to the total torque at all times when current flows through the brushes. There is no neutral armature position where torque is absent. The brushes are larger than the gaps between the commutator segments so contact with the commutator is maintained at every instant of rotation of the armature. A DC motor with a four-loop armature has uniform running and starting torque.

Field. A *field* is the stationary windings or magnets, of a DC motor. Many turns or windings are used in field circuits. See Figure 15-10. The larger the number of field windings, the smoother the motor runs. Field poles are used to concentrate the magnetic lines of force created by the field windings. The number of field poles used must always be an even number, with each set consisting of a north and south pole.

Baldor Electric Co.

DC motors provide high torque output and good speed control characteristics.

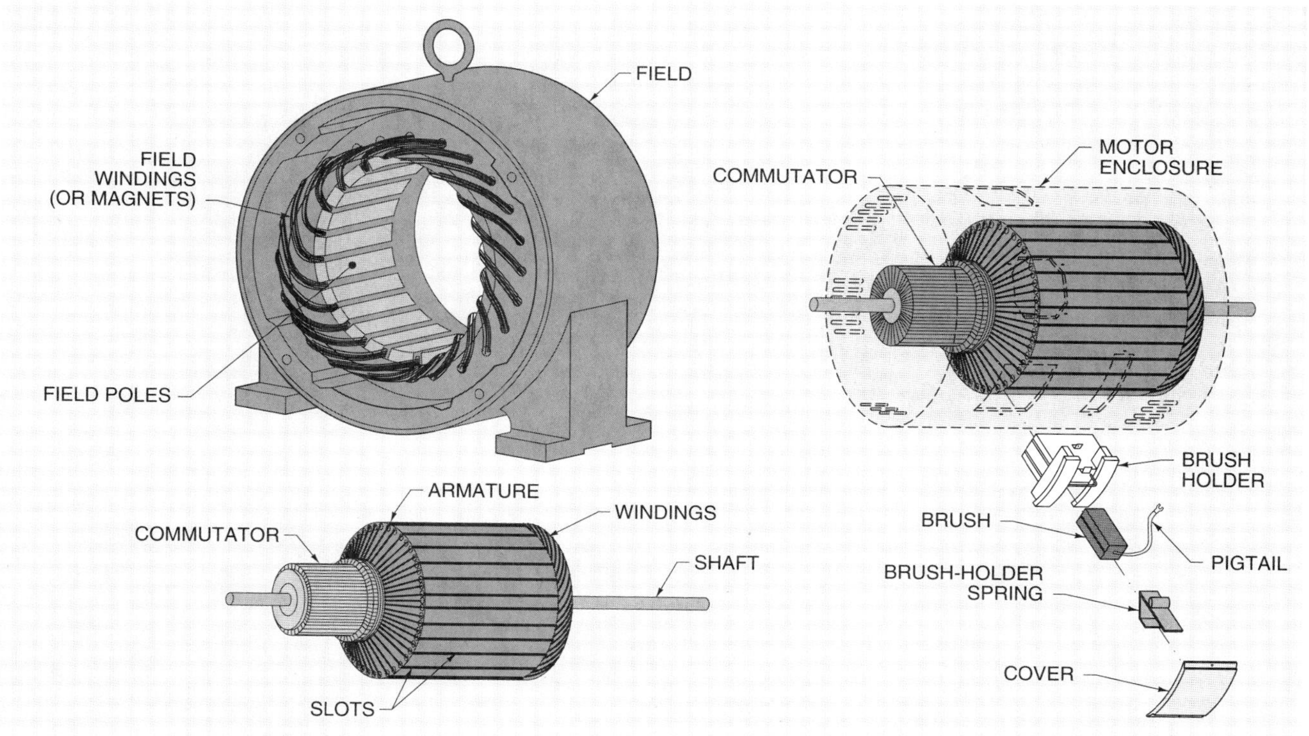

Figure 15-10. A field is the stationary windings or magnets of a DC motor.

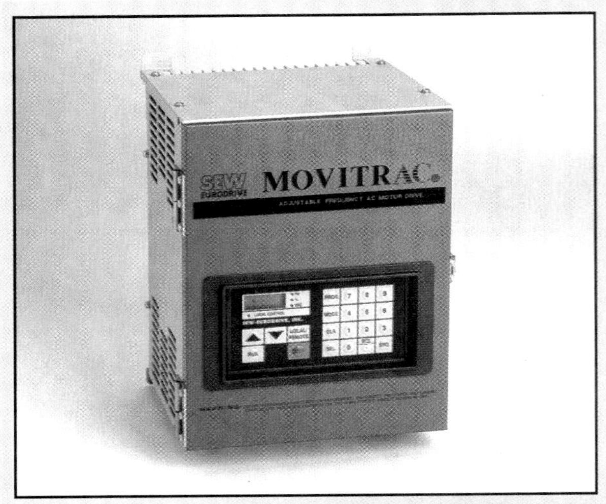

SEW-EURODRIVE, Inc.

The MOVITRAC® B-Series drive systems from SEW-EURODRIVE, Inc. have selectable acceleration/deceleration times, electrical braking, forward, reverse, jog, and multispeed run control, soft stall, automatic restart, and programmable run patterns for automatic operation.

Armature. The armature is constructed of steel laminations and is suspended at each end of the motor by bearings set in the motor frame. The commutator on the armature, along with the brushes, is used to supply the coil windings with current and reverse the current flow as needed. Commutators are normally constructed of drawn copper commutator bars which are insulated, one from the other. The armature coils are connected to each of the copper commutator bars.

Brushes. The brushes of a motor are used to provide the contact between the external power circuit and the commutator. Current is supplied to the commutator by brushes that ride on the commutator, making contact as it turns. The brushes are held in a stationary position by brush holders and are normally made from various grades of carbon. A flexible braided copper conductor (pigtail) is used to connect the brushes to the external circuit.

A brush must be positioned correctly for proper contact with the commutator. See Figure 15-11. Each brush is free to move up and down in the brush holder. This freedom allows the brush to follow irregularities in the surface of the commutator. A spring placed behind each brush forces the brush to make contact with the commutator. The spring pressure is normally adjustable, as is the entire brush-holder assembly. This allows shifting of the position of the brushes on the commutator.

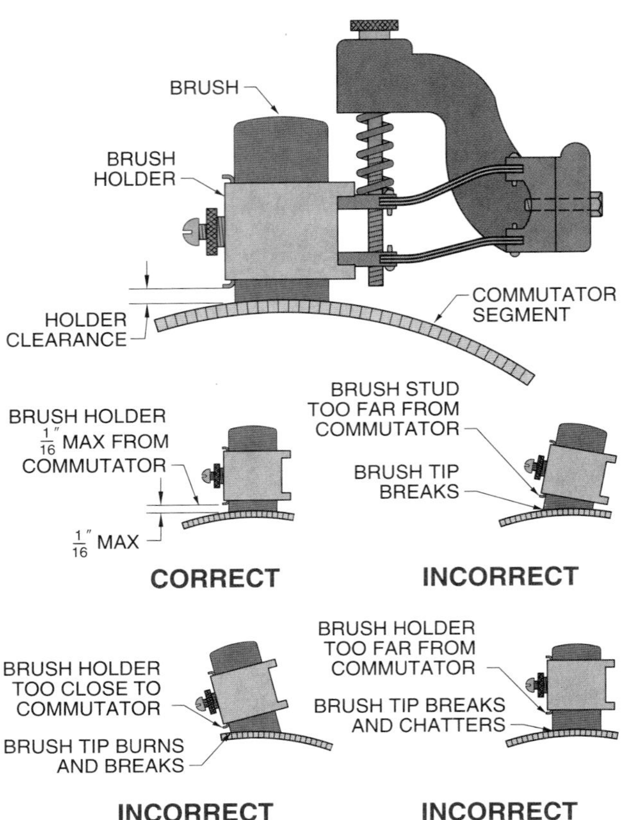

Figure 15-11. Brushes must be positioned correctly for proper contact with the commutator.

The brush must be pressed on the commutator with about 1.5 lb to 2 lb of pressure for each square inch of brush surface making contact with the armature. For this pressure to be applied, the spring must be allowed to move the brush up and down freely. There must also be very little space between the surface of the brush making contact with the commutator and the brush holder. Excessive space causes the brushes to chatter and break. A brush may become wedged in the brush holder. If this happens, the brush does not make good contact with the commutator and an open circuit exists.

Interpoles. *Interpoles (commutating field poles)* are auxiliary poles that are placed between the main field poles of the motor. See Figure 15-12. The interpoles are made with larger size wire than the main field poles to carry armature current. They are smaller in overall size than the main field poles because they have less windings. The interpoles are connected in series with the armature windings.

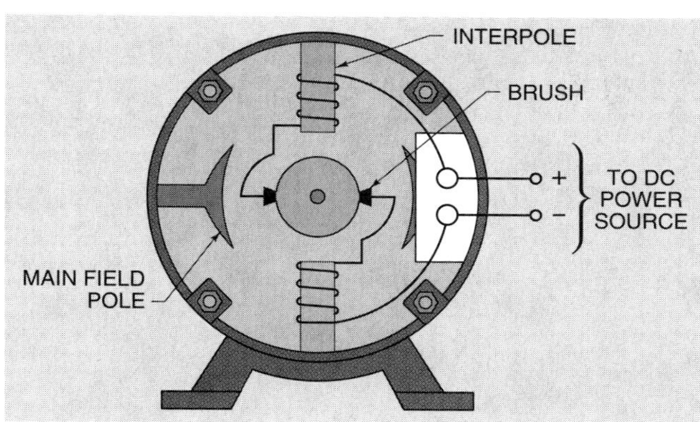

Figure 15-12. Interpoles are auxiliary poles that are placed between the main field poles of the motor.

The interpoles reduce sparking at the brushes of large DC motors. Interpoles are normally used with shunt and compound DC motors of ½ HP or more. Interpoles reduce sparking at the brushes by helping to overcome the effects of armature reaction. *Armature reaction* is the effect that the magnetic field of the armature coil has on the magnetic field of the main pole windings. See Figure 15-13.

The lines of magnetic force produced by the main field poles are directed through the armature's iron core from left to right when there is no current flow in the armature. Current flow through the armature creates a magnetic force in the main field windings. The magnetic force produced by the armature is directed through the armature's iron core and main field poles in an up-and-down movement. If both of these magnetic fields are combined, as they are in all DC motors, the produced magnetic field changes the angle of the neutral plane.

Leeson Electric Corporation

DC motor brushes must be accessible to aid in troubleshooting and preventive maintenance.

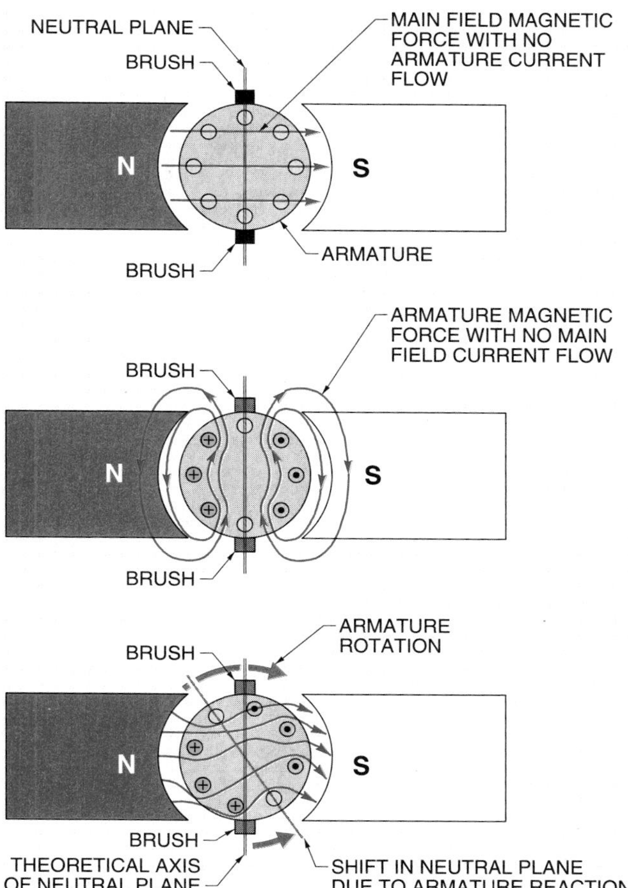

Figure 15-13. Interpoles reduce sparking at the brushes by helping to overcome the effects of armature reaction.

The shifting of the neutral plane has a direct effect on the commutator of the motor. This is because the brushes are best when connected to the commutator at the neutral plane. The current through the commutator and armature is minimum when the brushes make contact at the neutral plane. However, when the neutral plane moves, a current is produced by the magnetic field which allows current to flow through the brushes and commutator at the time when the brushes come in contact with the armature. This current causes sparking at the brushes and results in burning at the commutator.

To correct this sparking, the brushes can be shifted in a direction opposite the direction of motor rotation. This moves the brushes back to the neutral plane when they make contact with the armature. See Figure 15-14.

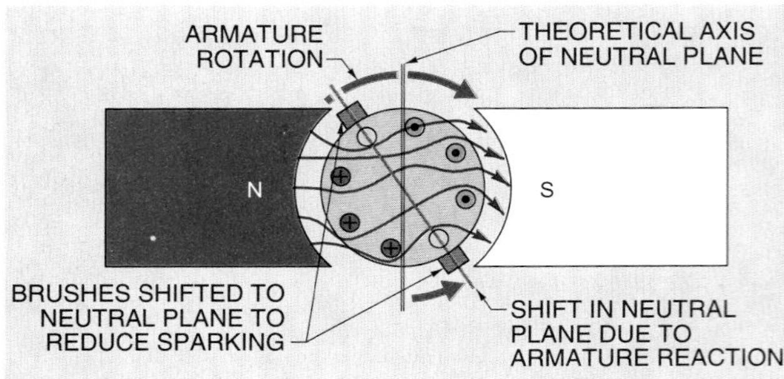

Figure 15-14. Brushes can be repositioned in a DC motor to reduce arcing.

This is not the best way to prevent sparking because the neutral plane changes angles with each change in the load connected to the motor. For a load that is constant, the brushes may be shifted to an angle that produces the least amount of sparking in an attempt to reduce the effect of armature reaction.

The best way to reduce sparking at the brushes is by using interpoles. The interpole's magnetic field opposes the distorted magnetic force caused by armature reaction. See Figure 15-15.

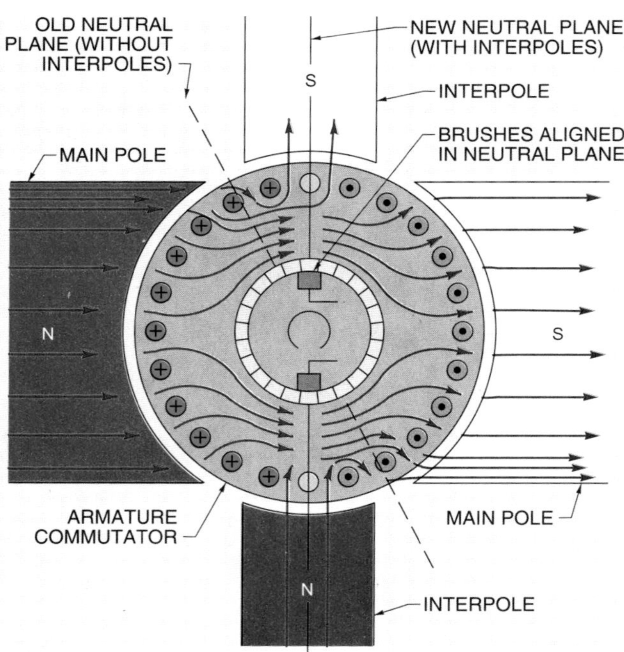

Figure 15-15. Interpoles may be used to automatically shift the neutral plane and reduce sparking at the brushes.

The result is that the combined magnetic force keeps the neutral plane at a fixed angle. Sparking is kept to a minimum and the motor is not influenced by a changing neutral plane because the neutral plane cannot be changed in motors that have interpoles.

All DC motors are supplied with current directly connected to the armature and field windings. During startup, current is limited by the resistance of the wire in the armature and field windings when current is connected directly to the motor. The larger the motor, the less the resistance and the larger the current. In DC motors, this starting current may be so high that it damages the motor. To prevent motor damage, reduced-voltage starting must be applied to DC motors larger than 1 HP.

Reduced-voltage starting of DC motors reduces the amount of current during starting. As the motor accelerates, the reduced voltage may be removed because the current in the motor decreases with an increase in motor speed. This decrease in current results from the motor generating a voltage that is opposite to the applied voltage as it accelerates. This opposing voltage, known as counter electromotive force (counter EMF), depends on the speed of the motor. Counter EMF is zero at standstill and increases with motor speed. Ohm's law is used to calculate motor starting current. To calculate motor starting current, apply the formula:

$$I = \frac{E}{R}$$

where

I = starting current (in A)

E = applied voltage (in V)

R = resistance (in Ω)

Example: Calculating Motor Starting Current

What is the starting current of a DC motor with an armature resistance of 1 Ω that is connected to a 200 V supply?

$$I = \frac{E}{R}$$

$$I = \frac{200}{1}$$

I = **200 A**

As the motor accelerates, a counter EMF is generated. The counter EMF reduces the current in the motor. To calculate the current drawn by a motor during starting, apply the formula:

$$I = \frac{E - C_{EMF}}{R}$$

where

I = starting current (in A)

C_{EMF} = generated counter electromotive force (in V)

E = applied voltage (in V)

R = resistance (in Ω)

Example: Calculating Current During Starting

What is the current during starting of a DC motor with an armature resistance of 1 Ω that is connected to a 200 V supply and is generating 100 V of counter EMF?

$$I = \frac{E - C_{EMF}}{R}$$

$$I = \frac{200 - 100}{1}$$

$$I = \frac{100}{1}$$

$$I = \mathbf{100\ A}$$

A motor at full speed generates an even higher counter EMF. The higher counter EMF further reduces the current in the motor. It is the 200 A (or starting current of any large DC motor) that the motor must be protected from to prevent damage.

Example: Calculating Motor Running Current

What is the running current of a DC motor with an armature resistance of 1 Ω that is connected to a 200 V supply and is generating 180 V of counter EMF?

$$I = \frac{E - C_{EMF}}{R}$$

$$I = \frac{200 - 180}{1}$$

$$I = \frac{20}{1}$$

$$I = \mathbf{20\ A}$$

REDUCED-VOLTAGE STARTING AND SQUIRREL-CAGE INDUCTION MOTORS

The majority of industrial applications normally use squirrel-cage induction motors. Squirrel-cage induction motors are normally chosen over other types of motors because of their simplicity, ruggedness, and reliability. Squirrel-cage induction motors have become the standard for AC, all-purpose, constant-speed motor applications.

Motor Construction

A squirrel-cage induction motor has certain advantages over a DC motor. A squirrel-cage induction motor has only two points (the two bearings) of mechanical wear. There are no brushes to wear because the motor does not have a commutator. For this reason, maintenance is minimal. No sparks are generated to create a hazard in the presence of flammable material. See Figure 15-16.

GE Motors & Industrial Systems

Figure 15-16. The majority of industrial applications normally use squirrel-cage induction motors because of their simplicity, ruggedness, and reliability.

The motor consists of a rotor and stator connected by bearings. A *rotor* is the rotating part of an AC motor. A *stator* is the stationary part of an AC motor. The motor is named for its rotor, which resembles a squirrel cage, and because current flowing in the stator induces AC current in the rotor similarly to a transformer. See Figure 15-17.

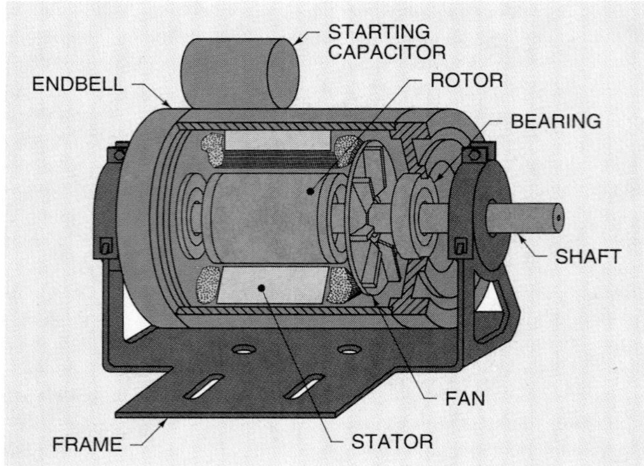

Figure 15-17. A squirrel cage induction motor consists of a rotor and stator connected by bearings.

AC Motor Operation

In a 3ϕ induction motor, there are three sets of windings on the stator frame arranged to produce a revolving magnetic field when connected to a 3ϕ power source. As phases A, B, and C change amplitude, the electromagnet interaction is turned into mechanical rotation of the rotor. See Figure 15-18.

Fluke Corporation

A Fluke Model 36 clamp meter is used to measure the amount of current a motor draws.

The rotor consists of steel laminations mounted rigidly on the motor shaft. Copper or aluminum bars placed or cast in the slots of the laminated steel core form the rotor (in contrast to the copper wire coils found in many other motor types). See Figure 15-19. These bars are designed to extend a sufficient distance beyond the end of the core so that they can all be interconnected by short-circuited end rings. All the conducting bars are connected into a short-circuited closed loop.

A voltage is induced into the rotor by transformer action, resulting in a heavy current flow in the rotor when the lines of magnetic force originating in the stator cross into the short-circuited bars of the rotor. Any induced voltages from transformer action are of reverse polarity to the voltage creating it. This results in a magnetic field opposite to that of the stator. The combined electromagnetic effects of the stator and rotor currents and their magnetic fields produce the torque (force) which create rotation. The air gap between the rotor and stator is kept extremely small to increase efficiency.

Motor Speed

A squirrel-cage induction motor is a constant speed device. It cannot operate for any length of time at speeds below those shown on the nameplate without danger of burning out. To calculate the synchronous speed of an induction motor, apply the formula:

$$S_{rpm} = \frac{120 \times F}{P}$$

where

S_{rpm} = synchronous revolutions per minute

120 = constant

F = supply frequency (in cycles/sec)

P = number of motor winding poles

Example: Calculating Motor Synchronous Speed

What is the synchronous speed of a motor having 4 poles connected to a 60 Hz power supply?

$$S_{rpm} = \frac{120 \times F}{P}$$

$$S_{rpm} = \frac{120 \times 60}{4}$$

$$S_{rpm} = \frac{7200}{4}$$

$$S_{rpm} = \mathbf{1800\ rpm}$$

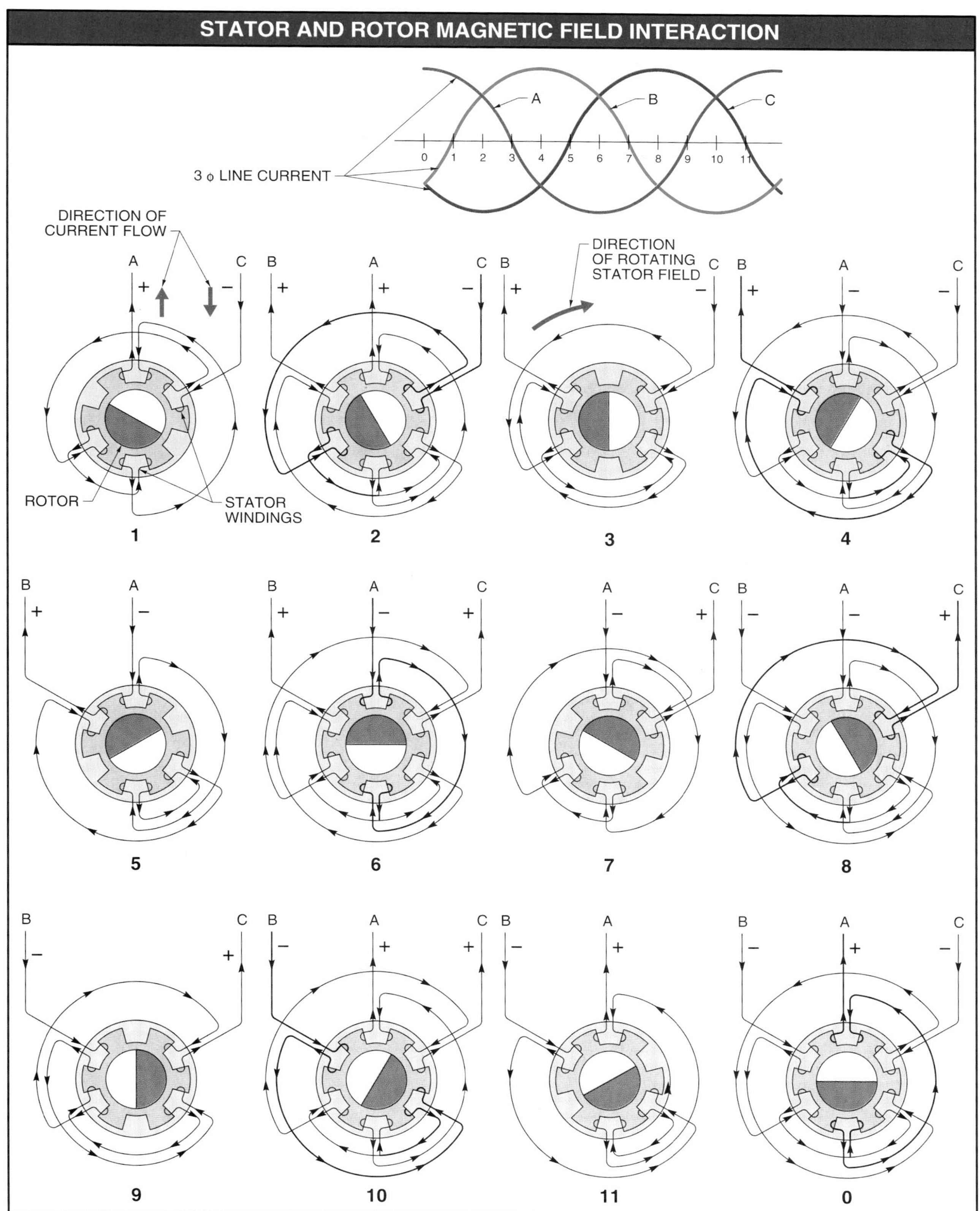

Figure 15-18. The rotating magnetic field in the stator windings causes the rotor to follow the rotating field.

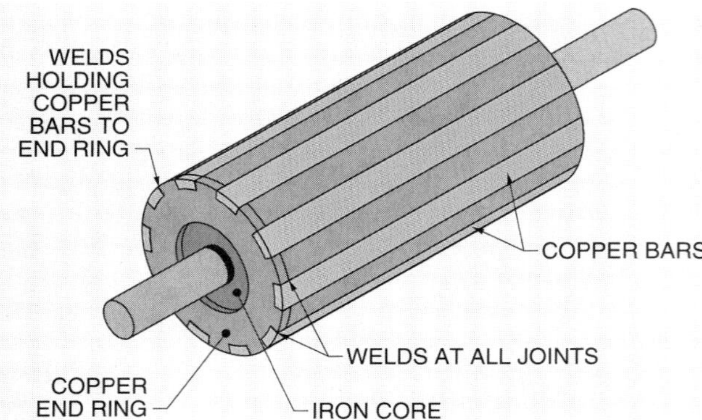

Figure 15-19. The rotor of a squirrel cage induction motor is the rotating portion of the motor.

Motors designed to operate on 60 Hz power (standard in the U.S.) have synchronous speeds. The speeds are based on the number of poles and line frequency. See Figure 15-20.

SYNCHRONOUS MOTOR SPEEDS	
Poles	**rpm**
2	3600
4	1800
6	1200
8	900
10	720
12	600
14	514
16	450

Figure 15-20. Motors designed to operate on 60 Hz power have synchronous speeds based on their number of poles and line frequency.

All induction motors have a full-load speed somewhat below the synchronous speed. *Percent slip* is the percentage reduction in speed below synchronous speed. A normal motor with an 1800 rpm synchronous speed and 2.8% slip has a slip of 50 rpm (1800 – 1750 = 50 rpm) and a full-load speed of 1750 rpm (1800 – 50 = 1750 rpm). It is this full-load speed that is listed on the motor's nameplate.

The speed of an induction motor is determined by the number of poles and the frequency of the power supply (not the supply voltage). It is for this reason that a reduced-voltage starter is not a speed controller and must not be used as one.

AC Motor Reduced-Voltage Starting

A heavy current is drawn from the power lines when an induction motor is started. This sudden demand for large current can reflect back into the power lines and create problems.

The revolving field of the stator induces this large current in the short-circuited rotor bars. The current is highest when the rotor is at a standstill and decreases as the motor increases speed. The current when starting is excessive because of a lack of counter EMF at the instant of starting. Once rotation begins, counter EMF is built up in proportion to speed, and the current decreases. See Figure 15-21.

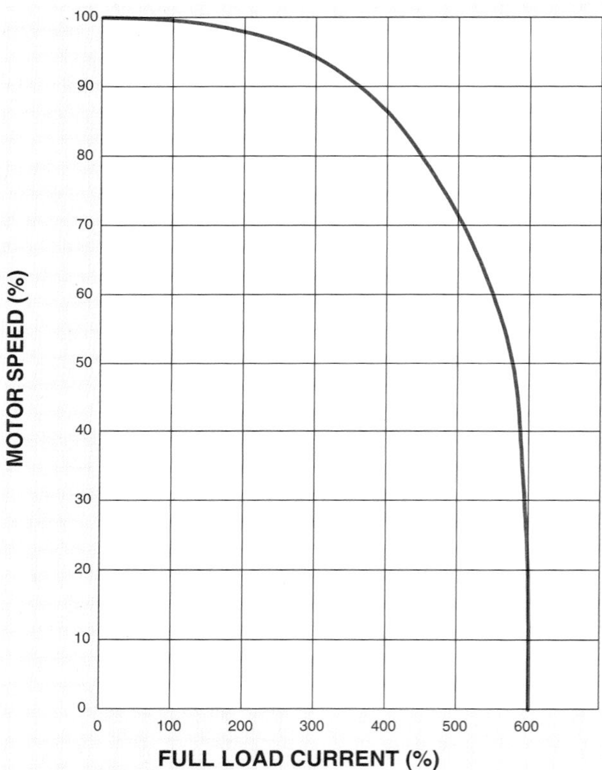

Figure 15-21. The current when starting is excessive because of a lack of counter EMF at the instant of starting.

The percent of full-load current is marked on the horizontal scale and the percent of motor speed is marked on the vertical scale. The starting current is quite high compared to the running current. The starting current remains fairly constant at this high value as the speed of the motor increases, but then drops sharply during the last few percentages up to 100%.

This illustrates that the heating rate is quite high during acceleration because the heating rate is a function of current. A motor may be considered to be in the locked condition during nearly all of the accelerating period.

Locked rotor current (LRC) is the steady-state current taken from the power line with the rotor locked (stopped) and with the voltage applied. LRC and the resulting torque produced in the motor shaft (in addition to load requirements) determine whether the motor can be connected across the line or whether the current has to be reduced through a reduced-voltage starter.

Full-load current (FLC) is the current required by the motor to produce full-load torque at the motor's rated speed. Full-load current is the current given on the motor's nameplate. The load current is less than what is given on the nameplate if the motor is not required to deliver full torque. This information is required when testing a motor (running a motor without a load). The only torque a motor must produce when not connected to a load is the torque that is needed to overcome its own internal friction and winding losses. For this reason, the current is less than the value given on the motor's nameplate.

PRIMARY RESISTOR STARTING

Primary resistor starting is a reduced-voltage starting method which uses a resistor connected in each motor line (in one line in a 1ϕ starter) to produce a voltage drop. This reduces the motor starting current as it passes through the resistor. See Figure 15-22. A timer is provided in the control circuit to short the resistors after the motor accelerates to a specified point. The motor is started at reduced voltage but operates at full-line voltage.

Primary resistor starters provide extremely smooth starting due to increasing voltage across the motor terminals as the motor accelerates. Standard primary resistor starters provide two-point acceleration (one step of resistance) with approximately 70% of line voltage at the motor terminals at the instant of motor starting. Multiple-step starting is possible by using additional contacts and resistors when extra smooth starting and acceleration is needed. This additional resistance stepping may be required in paper or fabric applications where even a small jolt in starting may tear the paper or snap the fabric.

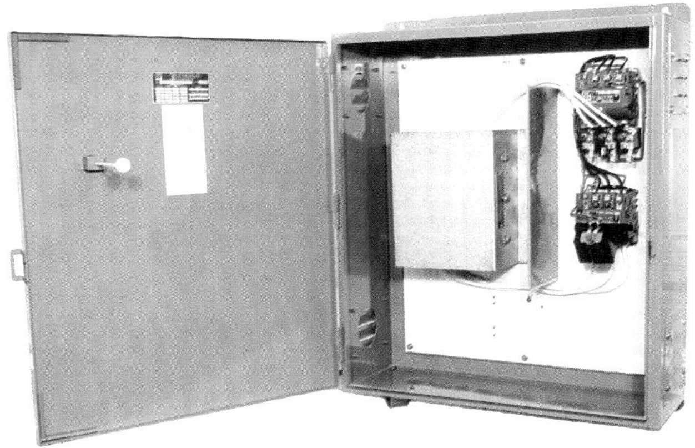

Furnas Electric Co.

Figure 15-22. Primary resistor starting uses a resistor connected in each motor line and a timer to short out the resistors after the motor accelerates to a specified point.

Primary Resistor Starting Circuits

In a primary resistor starting circuit, external resistance is added to and taken away from the motor circuit. See Figure 15-23. The control circuit consists of the motor starter coil M, ON-delay timer TR1, and contactor coil C. Coil M controls the motor starter which energizes the motor and provides overload protection. The timer provides a delay from the point where coil M energizes until contacts C close, shorting resistors R1, R2, and R3. Coil C energizes the contactor, which provides a short circuit across the resistors.

Baldor Electric Co.

Reduced-voltage starting is used with large industrial motors to reduce the damaging effect of a large starting current.

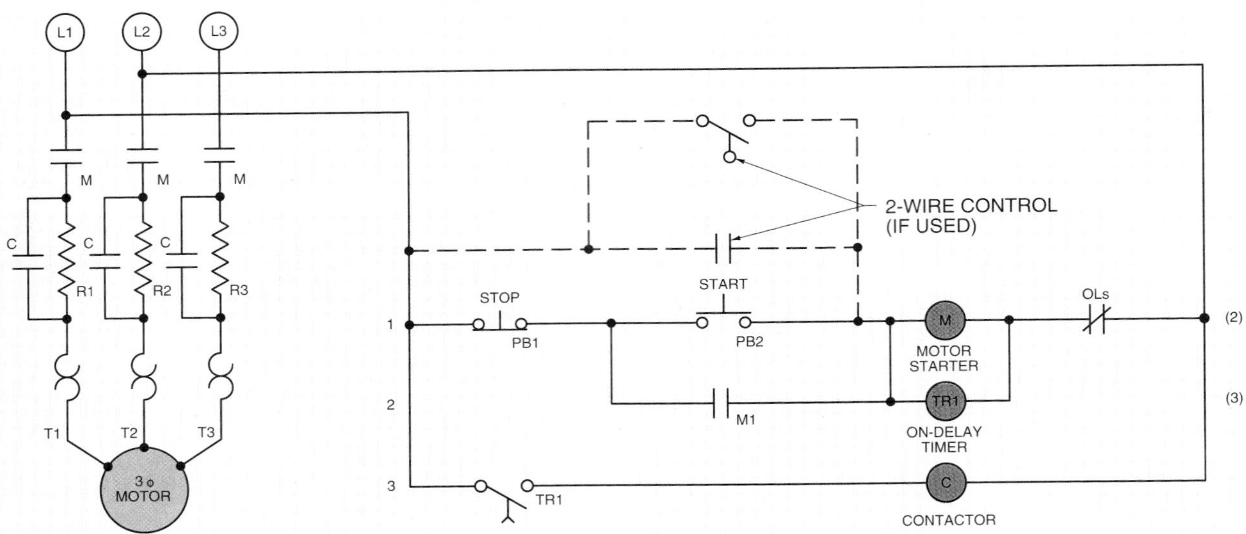

Figure 15-23. In a primary resistor starting circuit, external resistance is added to and taken away from the motor circuit.

Pressing start pushbutton PB2 energizes motor starter coil M and the ON-delay timer coil TR1. Motor starter coil M closes contacts M to create memory. ON-delay timer coil TR1 causes contacts TR1 to remain open during reset, stay open during timing, and close after timing out. Once timed out, the contactor coil C energizes, causing contacts C to close and the resistors to short.

This circuit is a common reduced-voltage starting circuit. Changes are often made in the values of resistance and wattage to accommodate motors of different horsepowers.

Furnas Electric Co.

Figure 15-24. Autotransformer starting uses a tapped 3ϕ autotransformer to provide reduced-voltage starting.

AUTOTRANSFORMER STARTING

Autotransformer starting uses a tapped 3ϕ autotransformer to provide reduced-voltage starting. See Figure 15-24. Autotransformer starting is one of the most effective methods of reduced-voltage starting. It is preferred over primary resistor starting when the starting current drawn from the line must be held to a minimum value, yet the maximum starting torque per line ampere is required.

The motor terminal voltage does not depend on the load current. The current to the motor may change because of the motor's changing characteristics, but the voltage to the motor remains relatively constant.

Autotransformer starting may use its turns ratio advantage to provide more current on the load side of the transformer than on the line side. In autotransformer starting, transformer motor current and line current are not equal as they are in primary resistor starting.

For example, a motor has a full-voltage starting torque of 120% and a full-voltage starting current of 600%. The power company has set a limitation of 400% current draw from the power line. This limitation is set only for the line side of the controller.

Because the transformer has a step-down ratio, the motor current on the transformer's secondary is larger than the line current as long as the primary of the transformer does not exceed 400%.

In this example, with the line current limited to 400%, 80% voltage can be applied to the motor, generating 80% motor current. The motor draws only 64% line current ($0.8 \times 80 = 64\%$) due to the 1:0.8 turns ratio of the transformer. The advantage is that the starting torque is 77% (0.8×80 of 120%) instead of the 51% obtained in primary resistor starting. This additional percentage may be sufficient accelerating energy to start a load which may have been difficult to start otherwise.

The two types of autotransformer connections or control schemes are open circuit transition and closed circuit transition. In open circuit transition, the motor may be temporarily disconnected when moving from one incremental voltage to another. In closed circuit transition, the motor is never removed from a source of voltage when moving from one incremental voltage to another.

Although both systems are used and function on the same principle, closed circuit transition poses the least amount of interference to the electrical environment. Closed circuit transition is the more expensive of the two.

Autotransformer Starting Circuits

In an autotransformer starting circuit, the various windings of the transformer are added to and taken away from the motor circuit to provide reduced voltage when starting. See Figure 15-25.

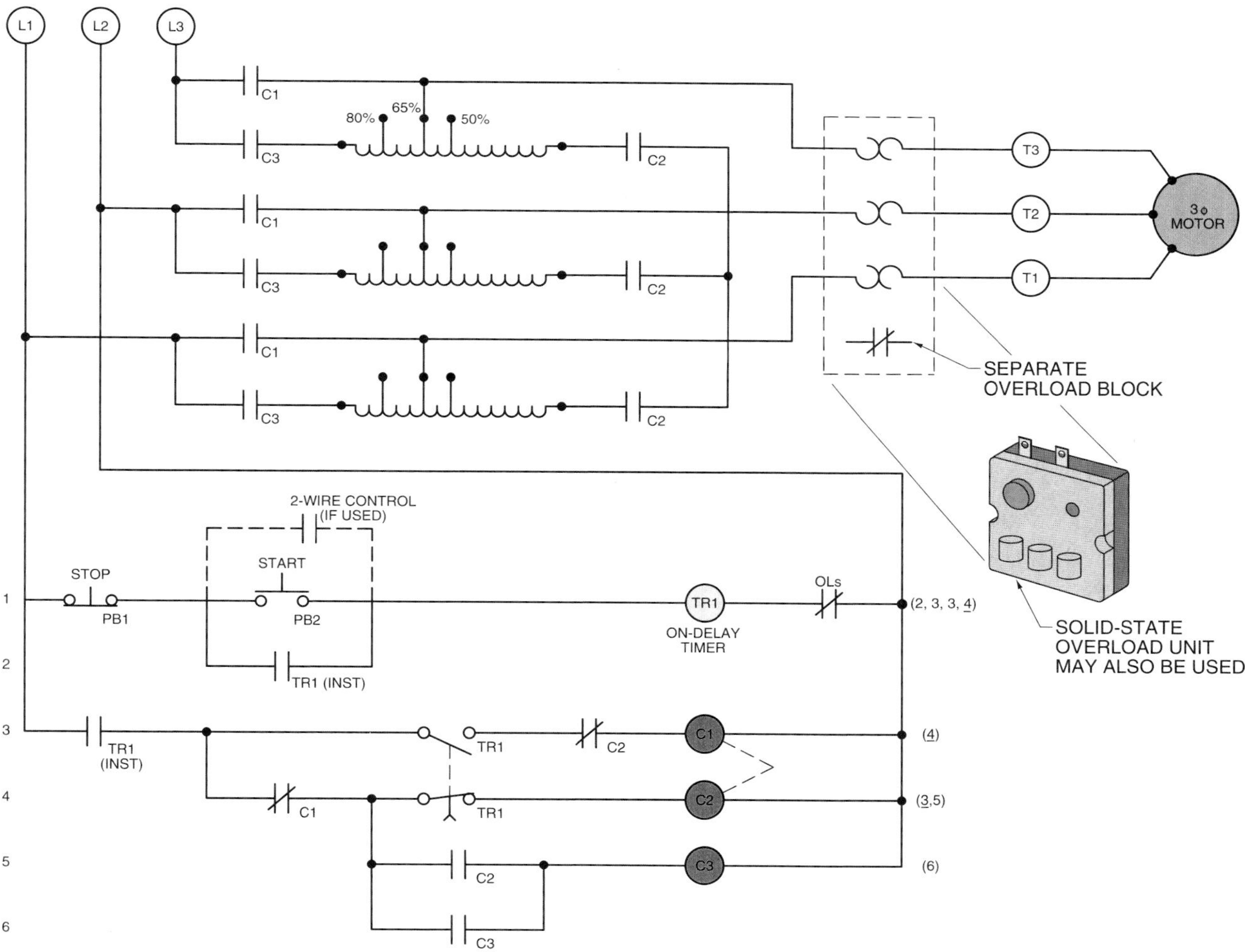

Figure 15-25. In an autotransformer starting circuit, the various windings of the transformer are added to and taken away from the motor circuit to provide reduced voltage when starting.

Giddings & Lewis, Inc.

Reduced-voltage starting may be used in some machining applications to reduce high-starting torque that can cause damage to the product.

The control circuit consists of an ON-delay timer TR1 and contactor coils C1, C2, and C3. Pressing start pushbutton PB2 energizes the timer, causing instantaneous contacts TR1 in line 2 and 3 of the line diagram to close. Closing the normally open (NO) timer contacts in line 2 provides memory for timer TR1, while closing NO timer contacts in line 3 completes an electrical path through line 4, energizing contactor coil C2. The energizing of coil C2 causes NO contacts C2 in line 5 to close, energizing contactor coil C3. The normally closed (NC) contacts in line 3 also provide electrical interlocking for coil C1 so that they may not be energized together. The NO contacts of contactor C2 close connecting the ends of the autotransformers together when coil C2 energizes. When coil C3 energizes, the NO contacts of contactor C3 close and connect the motor through the transformer taps to the power line, starting the motor at reduced inrush current and starting torque. Memory is also provided to coil C3 by contacts C3 in line 6.

After a predetermined time, the ON-delay timer times out and the NC timer contacts TR1 open in line 4, de-energizing contactor coil C2, and NO timer contacts TR1 close in line 3, energizing coil C1. In addition, NC contacts C1 provide electrical interlock in line 4, and NC contacts C2 in line 3 return to their NC position. The net result of de-energizing C2 and energizing C1 is the connecting of the motor to full-line voltage.

Note that during the transition from starting to full-line voltage, the motor was not disconnected from the circuit, indicating closed circuit transition. As long as the motor is running in the full-voltage condition, timer TR1 and contactor C1 remain energized. Only an overload or pressing the stop pushbutton stops the motor and resets the circuit. Overload protection is provided by a separate overload block.

In this circuit, pushbuttons are used to control the motor. However, any NO and/or NC device may be used to control the motor. Thus, in an air conditioning system, the pushbuttons would be replaced with a temperature switch, and the circuit would be connected for two-wire control.

PART-WINDING STARTING

Part-winding starting is a method of starting a motor by first applying power to part of the motor's coil windings for starting and then applying power to the remaining coil windings for normal running. The motor's stator windings must be divided into two or more equal parts for a motor to be started using part-winding starting. Each equal part must also have its terminal available for external connection to power. In most applications, a wye-connected motor is used, but a delta-connected motor can also be started using part-winding starting.

Wye-Connected Motors

Part-winding starting requires the use of a part-winding motor. A part-winding motor has two sets of identical windings which are intended to be used in parallel. These windings produce reduced-starting current and reduced-starting torque when energized in sequence. Most dual-voltage 230/460 V motors are suitable for part-winding starting at 230 V. See Figure 15-26.

Part-winding starters are available in either two- or three-step construction. The more common two-step starter is designed so that when the control circuit is energized, one winding of the motor is connected directly to the line. This winding draws about 65% of normal LRC and develops approximately 45% of normal motor torque. After about one second, the second winding is connected in parallel with the first winding in such a way that the motor is electrically complete across the line and develops its normal torque.

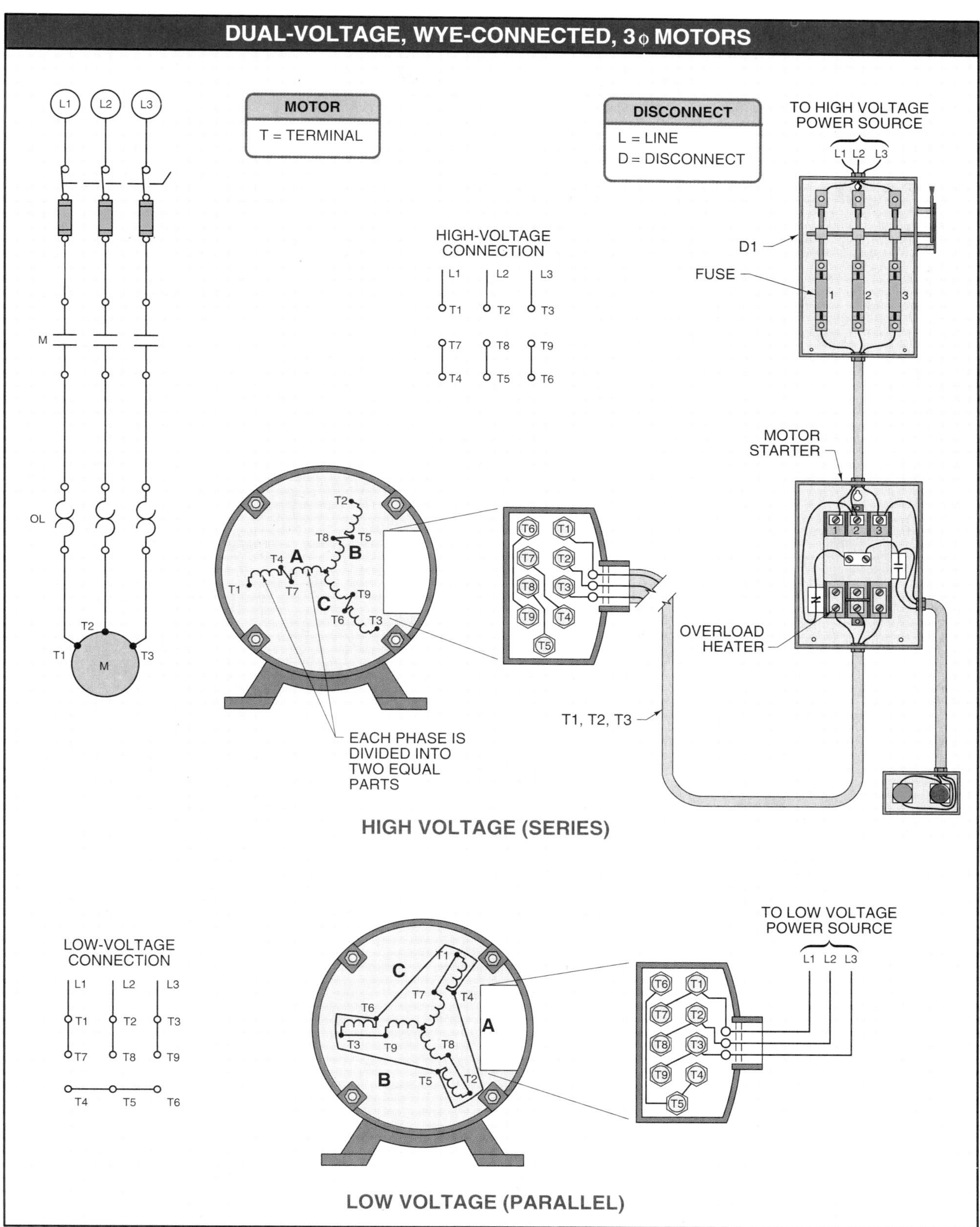

Figure 15-26. A part-winding motor has two sets of identical windings which are intended to be used in parallel.

Part-winding starting is not truly a reduced-voltage starting means. It is usually classified as reduced-voltage starting because of the resulting reduced current and torque.

Delta-Connected Motors

When a dual-voltage, delta-connected motor is operated at 230 V from a part-winding starter having a three-pole starting and a three-pole running contactor, an unequal current division occurs during normal operation, resulting in overloading the starting contactor. To overcome this defect, some part-winding starters are furnished with a four-pole starting contactor and a two-pole running contactor. This arrangement eliminates the unequal current division obtained with a delta-wound motor and enables wye-connected part-winding motors to be given either a one-half or two-thirds part-winding start.

Advantages and Disadvantages of Part-Winding Starting

Part-winding starting is less expensive than most other methods because it requires no voltage-reducing elements such as transformers or resistors, and it uses only two one-half size contactors. Also, its transition is inherently closed circuit.

Part-winding starting has poor starting torque because it is fixed. In addition, the starter is almost always an increment start device. Not all motors should be part-winding started. Consult the manufacturer' specifications before applying part-winding starting to a motor. Some motors are wound sectionally with part-winding starting in mind. Indiscriminate application to any dual-voltage motor can lead to excessive noise and vibration during starting, overheating, and extremely high-transient currents on switching.

The fuses in a part-winding starter must be proportionally smaller to protect the smaller contactors and overload elements allowed because of the low-current requirements in part-winding starters. Dual-element fuses are normally required.

Part-Winding Starter Circuits

Part-winding reduced-voltage starting is less expensive than other methods and produces less starting torque. See Figure 15-27.

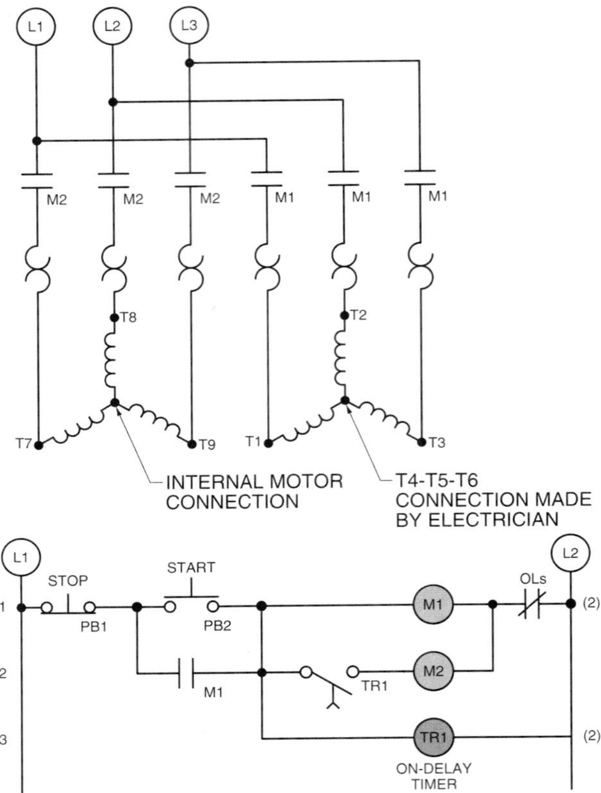

Figure 15-27. Part-winding reduced-voltage starting is less expensive than other methods and produces less starting torque.

The control circuit consists of motor starter M1, ON-delay timer TR1, and motor starter M2. Pressing start pushbutton PB2 energizes starter M1 and the timer TR1. M1 energizes the motor, and closes contacts M1 in line 2 to provide memory. With the motor starter M1 energized, L1 is connected to T1, L2 to T2, and L3 to T3, starting the motor at reduced current and torque through one-half of the wye windings.

The ON-delay NO contacts of ON-delay timer TR1 in line 2 remain open during timing and close after timing out, energizing coil M2. When M2 energizes, L1 is connected to T7, L2 to T8, and L3 to T9, applying voltage to the second set of wye windings. The motor now has both sets of windings connected to the supply voltage for full current and torque. The motor may normally be stopped by pressing stop pushbutton PB1 or by an overload in any line. Each magnetic motor starter need be only half-size because each one controls only one-half of the winding. Overloads must be sized accordingly.

WYE-DELTA STARTING

Wye-delta starting accomplishes reduced-voltage starting by first connecting the motor leads into a wye configuration for starting. A motor started in the wye configuration receives approximately 58% of the normal voltage and develops approximately 33% torque.

Wye-delta motors are specially wound with six leads extending from the motor to enable the windings to be connected in either a wye or delta configuration. When a wye-delta starter is energized, two contactors close, with one contactor connecting the windings in a wye configuration, and the second contactor connecting the motor-to-line voltage. After a time delay, the wye contactor opens (momentarily de-energizing the motor), and the third contactor closes to reconnect the motor to the lines with the windings connected in a delta configuration. A wye-delta starter is inherently an open transition system because the leads of the motor are disconnected and then reconnected to the power supply.

This starting method does not require any accessory voltage-reducing equipment such as resistors and transformers. Wye-delta starting gives a higher starting torque per line ampere than part-winding starting, with considerably less noise and vibration.

Wye-delta starters have the disadvantage of being open transition. Closed transition versions are available at additional cost. In closed transition, the motor windings are kept energized for the few cycles required to transfer the motor windings from wye to delta. Such starters are provided with one additional contactor plus a resistor bank.

Wye-Delta Motors

Windings of a wye-delta motor may be joined to form a wye or delta configuration. See Figure 15-28. There are no internal connections on this motor as there are on standard wye and delta motors. This allows the electrician to connect the motor leads into a wye-connected motor or into a delta-connected motor.

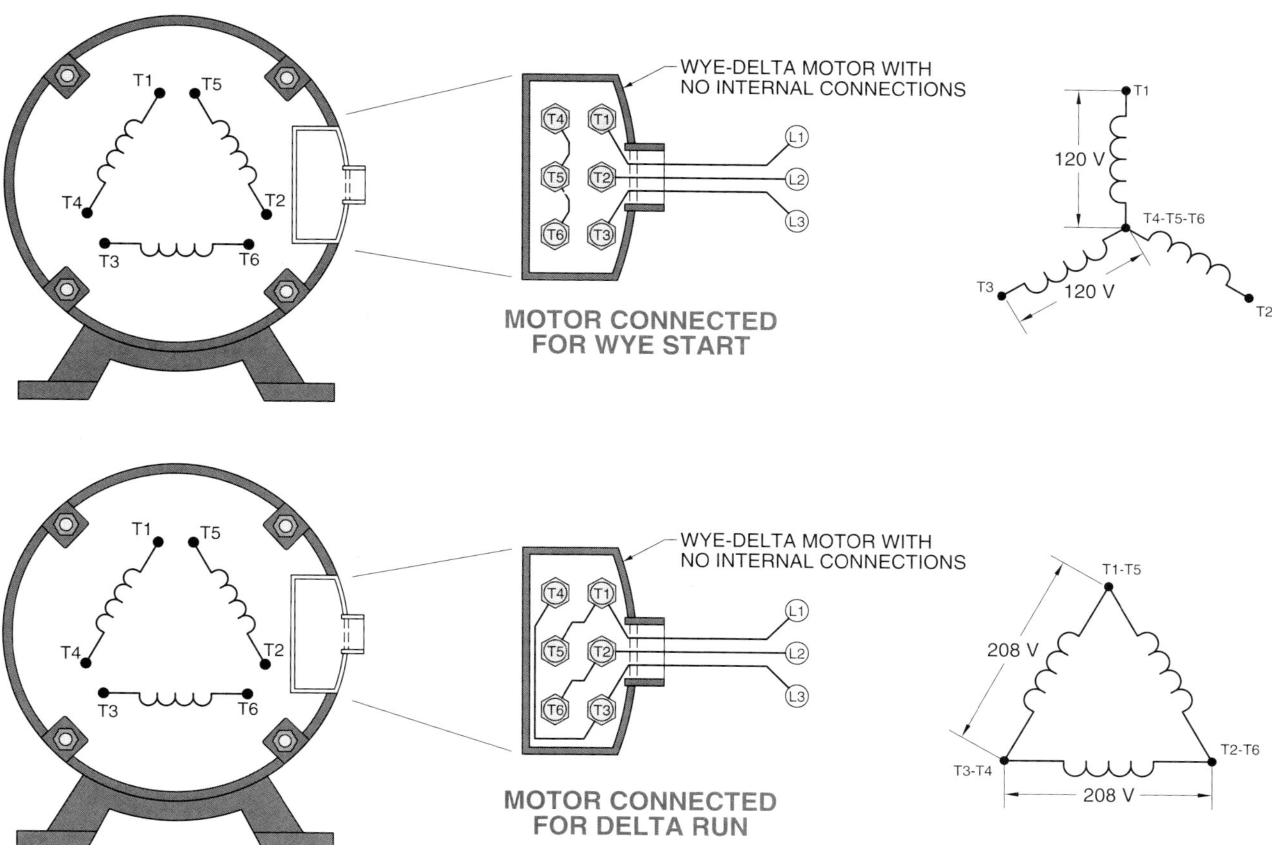

Figure 15-28. Windings on a wye-delta motor may be joined to form a wye or delta configuration.

Each coil winding in the motor receives 208 V if a delta-connected motor is connected across a 208 V, 3ϕ power line. This is because each coil winding in the motor is connected directly across two power lines. See Figure 15-29.

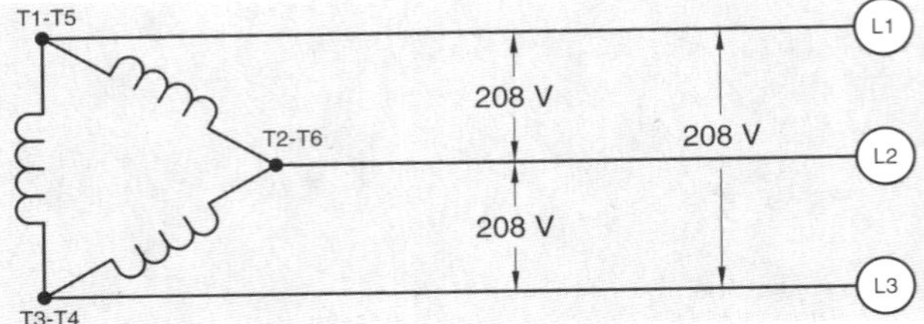

Figure 15-29. A delta-connected motor has each coil winding directly connected across two power lines so each winding receives the entire source voltage of 208 V.

Each coil winding in a wye-connected motor receives 120 V if it is connected across a 208 V, 3ϕ power line. This is because there are two coils connected in series across any pair of power lines. See Figure 15-30.

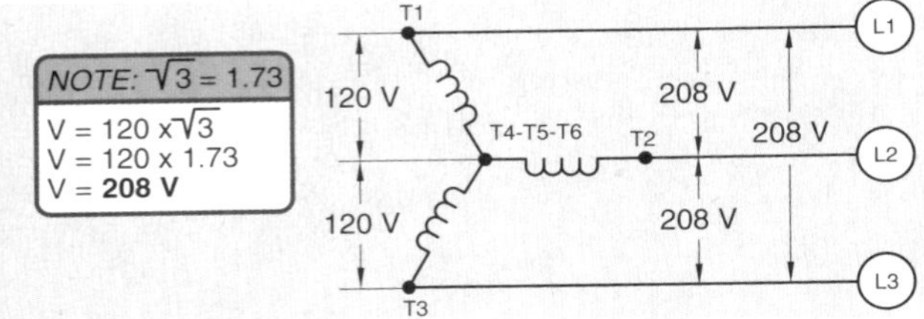

Figure 15-30. A wye-connected motor has two power lines connected across two sets of windings.

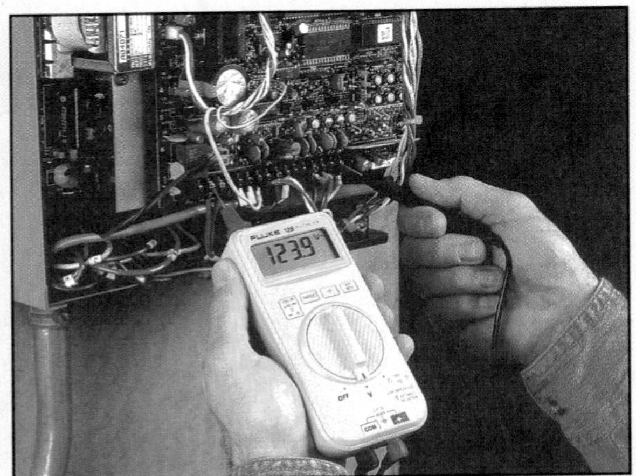

Fluke Corporation

A Fluke Model 12B multimeter is used to check reduced-voltage starting circuits and measure start and run voltages.

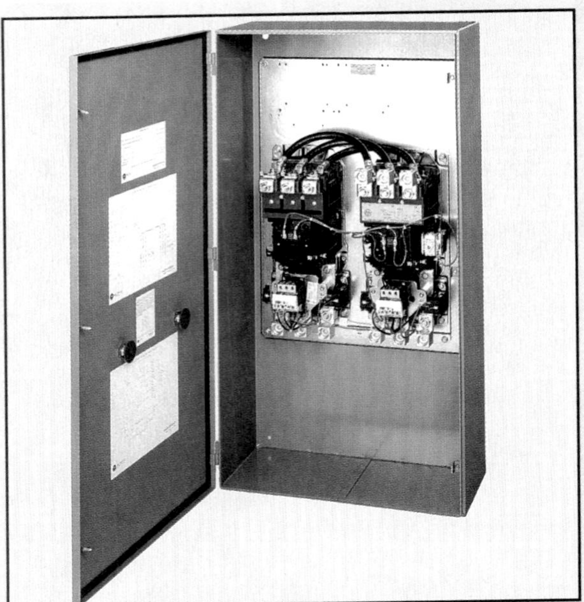

Rockwell Automation, Allen-Bradley Company, Inc.

Reduced-voltage starting circuits require two (or more) interlocked motor starters.

When calculating the voltage in the coil for a wye-connected circuit, the voltage is equal to the line voltage divided by the square root of 3 (1.73). The coil voltage is equal to 120 V ($^{208}/_{1.73}$) because the line voltage is equal to 208 V. A wye-delta motor connected to a line voltage of 208 V starts with 120 V (wye) and runs with 208 V (delta) across the motor windings, thus reducing starting voltage.

Wye-Delta Starting Circuits

The control circuit of a typical wye-delta starting circuit consists of motor starter coils M1 and M2, contactor C1, and ON-delay timer TR1. See Figure 15-31. Pressing start pushbutton PB2 energizes coil M1, which provides memory in line 2 and connects the power lines L1 to T1, L2 to T2, and L3 to T3. Contactor coil C1 in line 3 is energized, providing electrical interlock in line 2 and connecting motor terminals T4 and T5 to T6 so the motor starts in a wye configuration. TR1 in line 3 is also energized, and after a preset time the ON-delay timer times out, causing the NO TR1 contacts in line 2 to close and the NC TR1 contacts in line 3 to open. The opening of the NC contacts in line 3 disconnects contactor C1, and an instant later the NO contacts in line 2 energize the second motor starter through coil M2.

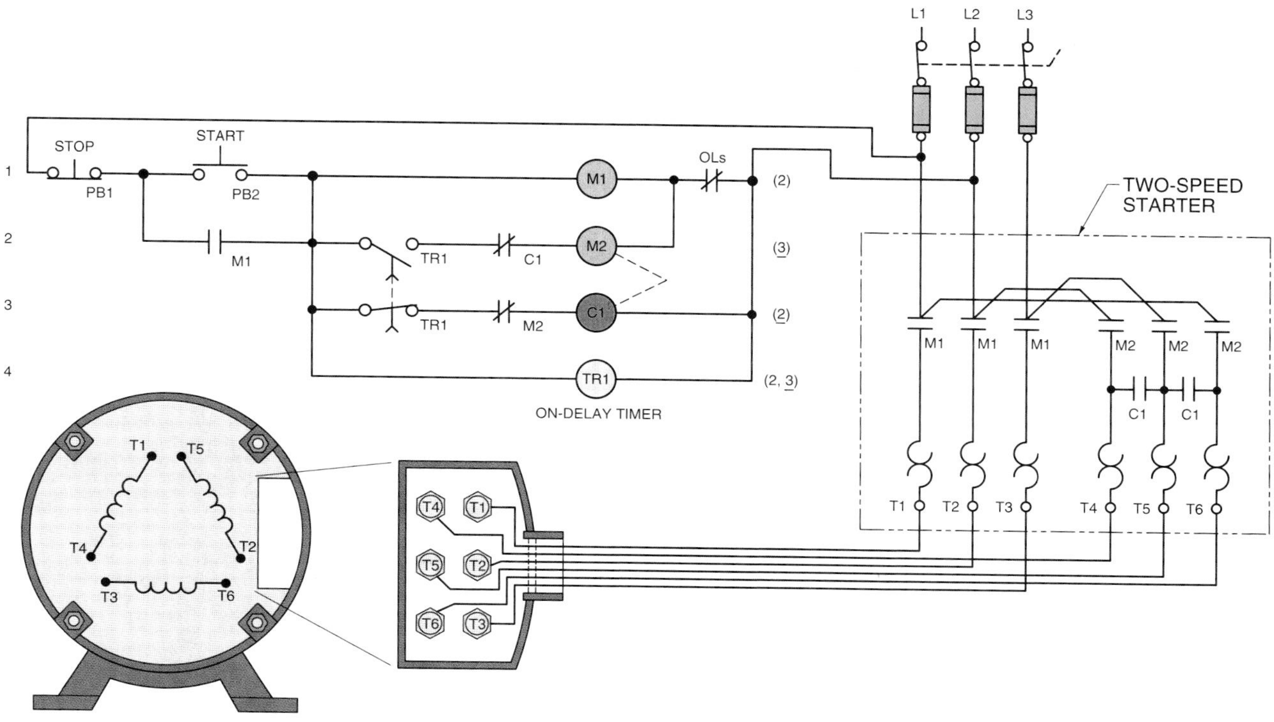

Figure 15-31. The control circuit of a typical wye-delta starting circuit consists of two motor starters, a contactor, and an ON-delay timer.

The short time delay between M2 and C1 is necessary to prevent a short circuit in the power lines and is provided through the NC auxiliary contacts of C1 in line 2. With contactor C1 de-energized, terminals T5, T6, and T4 are connected to power lines T1, T2, and T3 because L1, L2, and L3 are still connected to run in a delta configuration. The circuit can normally be stopped only by an overload in any line or by pressing the stop pushbutton PB1.

SOLID-STATE STARTING

A solid-state starter is the newest method of reduced-voltage starting for standard squirrel-cage motors. The heart of the solid-state system is the silicon-controlled rectifier (SCR) which controls motor voltage, current, and torque during acceleration.

An *SCR* is a solid-state rectifier with the ability to rapidly switch heavy currents. This type of starting provides a smooth, stepless acceleration in applications such as starting conveyors, compressors, pumps, and a wide range of other industrial applications because of this unique switching capability.

The advantage of SCRs is that they are small in size, rugged, and have no contacts. Unlimited life can be expected when SCRs are operated within specifications. The major disadvantage of solid-state starting is its relatively high cost in relation to other systems.

Electronic Control Circuitry

A solid-state controller determines to what degree the SCRs should be triggered ON to control the voltage, current, and the torque applied to a motor. A solid-state controller also includes current-limiting fuses and current transformers for protection of the unit. The current-limiting fuses are used to protect the SCRs from excess current. The current transformers are used to feed information back to the controller. Heat sinks and thermostat switches are also used to protect the SCRs from high temperatures.

The controller also provides the sequential logic necessary for interfacing other control functions of the starter, such as line loss detection during acceleration. The controller is turned OFF if any voltage is lost or too low on any one line. This may happen if one line opens or a fuse blows.

SCR Operation

An SCR includes an anode, cathode, and gate. See Figure 15-32. The anode and cathode of an SCR are similar to the anode and cathode of a diode rectifier. The gate gives the SCR added control not possible with an ordinary diode.

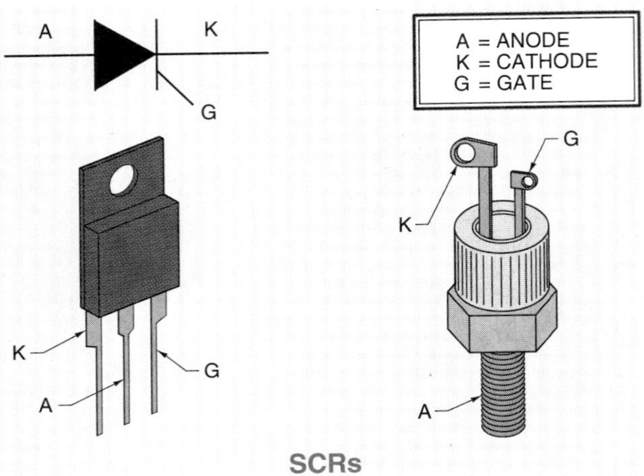

Figure 15-32. An SCR includes an anode, cathode, and gate.

With the gate, an SCR can be made to operate as an OFF/ON switch controlled by a voltage signal to the gate. Unlike an ordinary diode, an SCR does not pass current from cathode to anode unless an appropriate signal is applied to the gate.

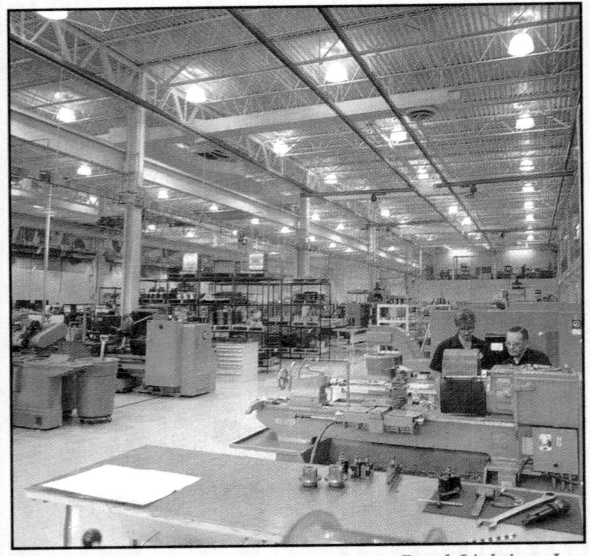

Ruud Lighting, Inc.

A variable output voltage is required in most industrial manufacturing applications.

When the signal is applied to the gate, the SCR is triggered ON, and the anode resistance decreases sharply, such that the resulting current flow through the SCR is only limited by the resistance of the load. The advantage of this device is its ability to turn ON at any point in the half-cycle. See Figure 15-33.

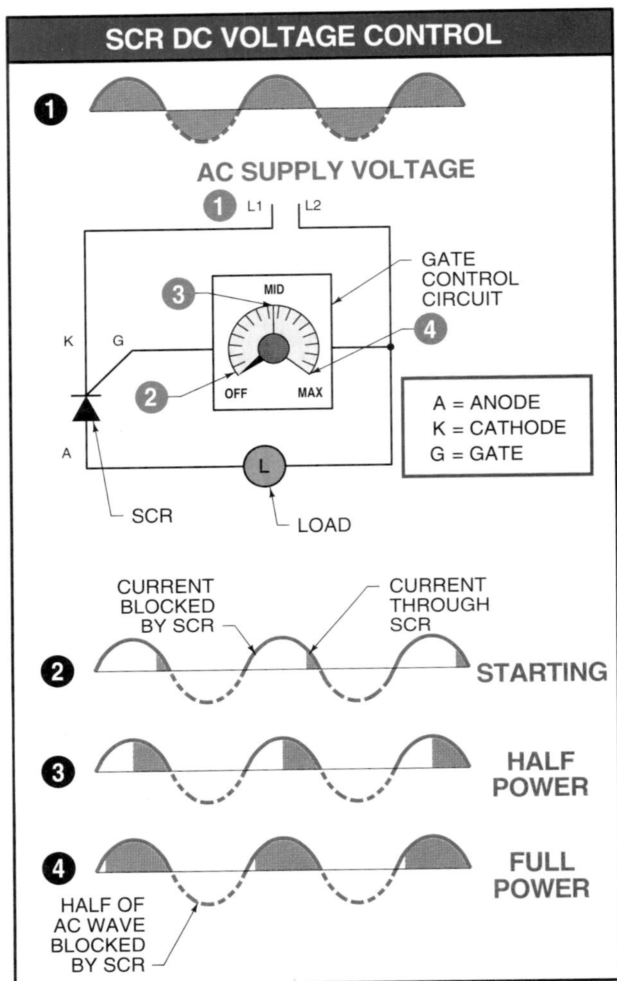

Figure 15-33. When the signal is applied to the gate, the SCR is triggered ON and the anode resistance decreases sharply, such that the resulting current flow through the SCR is only limited by the resistance of the load.

The average amount of voltage and current can be reduced or increased by the triggering of the SCRs because the amount of conduction can be varied. SCRs may be used alone in a circuit to provide one-way current control or may be wired in reverse parallel circuits to control AC line current in both directions. See Figure 15-34.

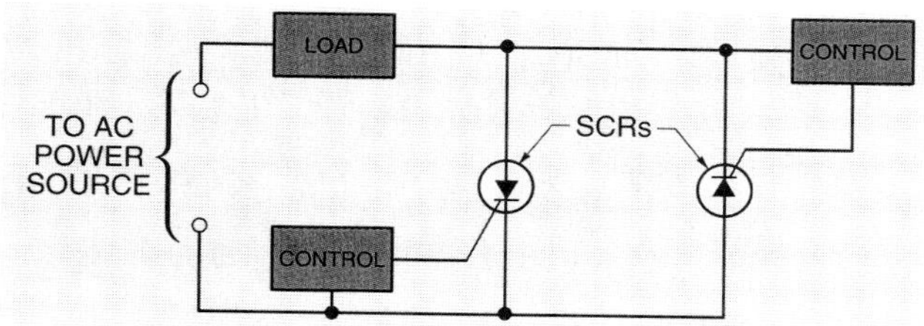

Figure 15-34. SCRs may be used alone in a circuit to provide one-way current control, or may be wired in reverse parallel circuits to control AC line current in both directions.

Solid-State Starting Circuits

A typical solid-state starting circuit consists of both start and run contactors connected in the circuit. The start contactor contacts C1 are in series with the SCRs and the run contactor contacts C2 are in parallel with the SCRs. See Figure 15-35.

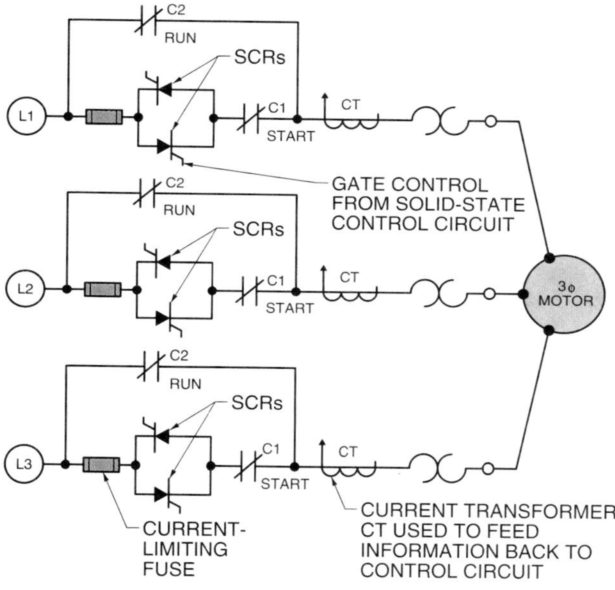

Figure 15-35. An SCR circuit with reverse parallel wiring of SCRs provides maximum control of an AC load.

The start contacts C1 close and the acceleration of the motor is controlled by triggering ON the SCRs when the starter is energized. The SCRs control the motor until it approaches full speed, at which time the run contacts C2 close, connecting the motor directly across the power line. At this point, the SCRs are turned OFF, and the motor runs with full power applied to the motor terminals.

STARTING METHOD COMPARISON

Several methods are available when an industrial application calls for using reduced-voltage starting. The amount of reduced current, the amount of reduced torque, and the cost of each starting method must be considered when selecting the method of starting.

The selection is not simply a matter of selecting the starting method that reduces the most amount of current. The motor does not start and the motor overloads trip if the starting torque is reduced too much.

A general comparison can be made of the amount of reduced current for each type of starting method compared to across-the-line starting. See Figure 15-36. The amount of reduced current is adjustable when using solid-state or autotransformer starting. Autotransformer starting uses taps so the amount of reduced current is somewhat adjustable. Solid-state starting is adjustable through its range. Some primary resistor starters are adjustable, others are not. Part-winding and wye-delta starting are not adjustable.

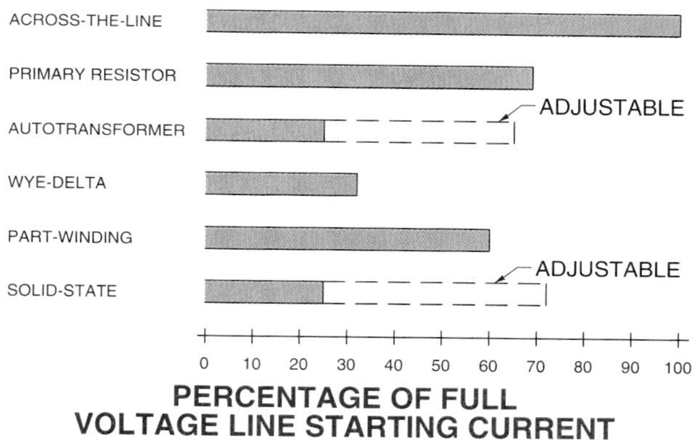

Figure 15-36. The different methods of reduced-voltage starting produce different percentages of full voltage current.

A general comparison can be made of the amount of reduced torque for each type of starting method compared to across-the-line starting. See Figure 15-37. The amount of reduced torque is adjustable when using the solid-state or autotransformer starting methods. The autotransformer starting method has taps, so the amount of reduced torque is somewhat adjustable. Solid-state starting is adjustable through its range. The motor overloads trip if the load requires more torque than the motor can deliver. The torque requirements of the load must be taken into consideration when selecting a starting method.

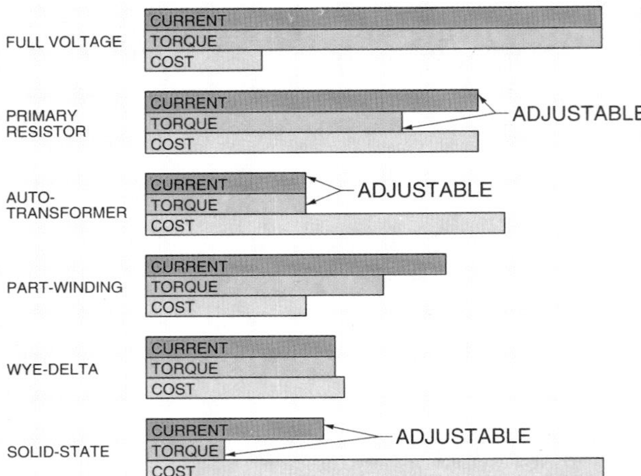

Figure 15-37. A general comparison can be made of the amount of reduced torque for each type of starting method compared to across-the-line starting.

A general comparison can also be made of the costs for each type of starting method compared to across-the-line starting. Although reducing the amount of starting current or starting torque in comparison to the load requirements is the primary consideration for selecting a starting method, cost may also have to be considered. The costs vary for each starting method. See Figure 15-38.

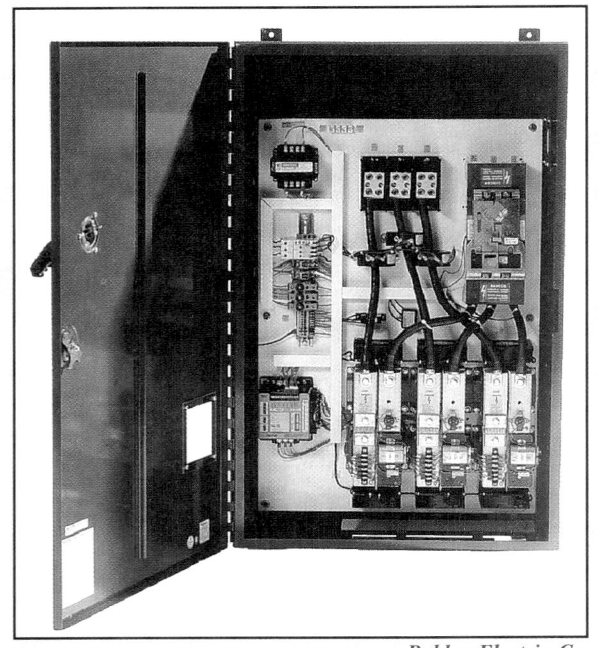

Baldor Electric Co.

Control circuit components and power circuit components are often housed in the same enclosure for protection and to provide one convenient place for troubleshooting.

The primary resistor starting method is used when it is necessary to restrict inrush current to predetermined increments. Primary resistors can be built to meet almost any current inrush limitation. They also provide smooth acceleration and can be used where it is necessary to control starting torque. Primary resistor starting may be used with any squirrel-cage motor.

The autotransformer starting method provides the highest possible starting torque per ampere of line current and is the most effective means of motor starting for applications where the inrush current must be reduced with a minimum sacrifice of starting torque. Three taps are provided on the transformers, making it field adjustable. Cost must be considered because the autotransformer is the most expensive type of transformer. Autotransformer starting can be used with any squirrel-cage motor.

The part-winding starting method is simple in construction and economical in cost. It provides a simple method of accelerating fans, blowers, and other loads involving low-starting torque. The part-winding starting method requires a nine-lead wye motor. The cost is less because no external resistors or transformers are required.

The wye-delta starting method is particularly suitable for applications involving long accelerating times or frequent starts. It is commonly used for high inertia loads such as centrifugal air conditioning units, although they can be used in applications where low-starting torque is necessary or where low-starting current and low-starting torque is permissible. The wye-delta starting method requires a special six-lead motor.

The solid-state starting method provides smooth, stepless acceleration in applications such as starting of conveyors, compressors, and pumps. It uses a solid-state controller which uses SCRs to control motor voltage, current, and torque during acceleration. Although this starting method offers the most control over a wide range, it is also the most expensive.

TROUBLESHOOTING REDUCED-VOLTAGE STARTING CIRCUITS

As with all motor circuits, the two main sections that must be considered when troubleshooting reduced-voltage starting circuits are the power circuit and control circuit. The power circuit connects the motor to the main power supply.

MOTOR STARTING METHOD COMPARISON										
Starter Type	Starting Characteristics			Standard Motor	Transition	Extra Acceleration Steps Available	Installation Cost	Advantages	Disadvantages	Applications
	Volts at Motor	Line Current	Starting Torque							
Across-The-Line	100%	100%	100%	Yes	None	None	Lowest	Inexpensive, readily available, simple to maintain, maximum starting torque	High inrush, high-starting torque	Many and various
Primary Resistor	65%	65%	42%	Yes	Closed	Yes	High	Smooth acceleration, high-power factor during start, less expensive than autotransformer starter in low HPs, available with as many as 5 accelerating points	Low-torque efficiency, resistors give off heat, starting time in excess of 5 sec requires expensive resistors, difficult to change starting torques under varying conditions	Belt and gear drives, conveyors, textile machines
Auto-Transformer	80% 65% 50%	64% 42% 25%	64% 42% 25%	Yes	Closed	No	High	Provides highest torque per ampere of line current, 3 different starting torques available through autotransformer taps, suitable for relatively long starting periods, motor current is greater than line current during starting	Is most expensive design in lower HP ratings, low power factor, large physical size	Blowers, pumps, compressors, conveyors,
Part-Winding	100%	65%	48%	*	Closed	Yes**	Low	Least expensive reduced-voltage starter, most dual-voltage motors can be started part-winding on lower voltage, small physical size	Unsuited for high inertia, long-starting loads, requires special motor design for voltage higher than 230 V, motor does not start when torque demanded by load exceeds that developed by motor when first half of motor is energized, first step of acceleration must not exceed 5 sec or motor overheats	Reciprocating compressors, pumps, blowers, fans
Wye-Delta	100%	33%	33%	No	Open***	No	Medium	Suitable for high inertia, long acceleration loads, high-torque efficiency, ideal for especially stringent inrush restrictions, ideal for frequent starts	Requires special motor, low-starting torque, momentary inrush occurs during open transition when delta contactor is closed	Centrifugal compressors, centrifuges
Solid-State	Adjust	Adjust	Adjust	Yes	Closed	Adjust	Highest	Energy-saving features available, voltage gradually applied during starting for a soft start condition, adjustable acceleration time, usually self-calibrating, adjustable built-in braking features included	High cost, requires specialized maintenance and installation, electrical transients can damage unit, requires good ventilation	Machine tools, hoists, packaging equipment, conveyor systems

* standard dual-voltage 230/460 V motor can be used on 230 V systems

** very uncommon

*** closed transition available for about 30% more in price

Figure 15-38. Several factors must be considered when selecting reduced-voltage starting systems.

In addition to including the main switching contacts and overload detection device (heaters or solid-state), the power circuit also includes the power resistors (primary resistor starting) and autotransformers (autotransformer starting).

The control circuit determines when and how the motor starts. The control circuit includes the motor starter (mechanical or solid-state), overload contacts, and timing circuit. To troubleshoot the control circuit, the same troubleshooting procedure is used as when troubleshooting any other motor control circuit.

Voltage and current readings are taken when troubleshooting power circuits. Current readings can be taken at the incoming power leads or the motor leads. The current reading during starting should be less than the current reading when starting without reduced-voltage starting. The amount of starting current varies by the starting method. See Figure 15-39.

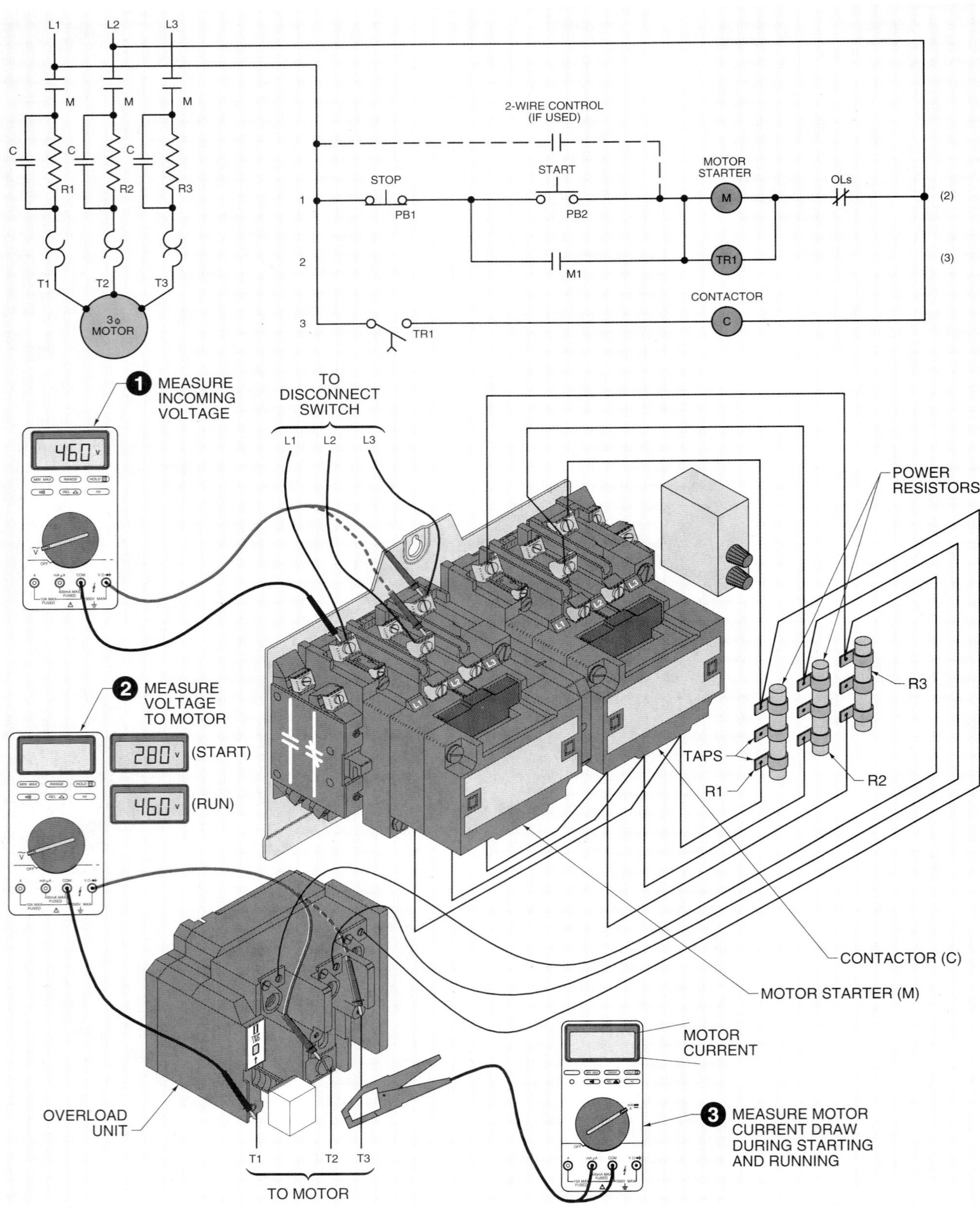

Figure 15-39. Voltage and current readings are taken when troubleshooting power circuits.

When troubleshooting the power circuit, voltage and current readings are taken. Current readings can be taken at the incoming power leads or the motor leads, since the current draw is the same at either point. With each starting method, there should be a reduction in starting current, as compared to a full voltage start.

1. Measure the incoming voltage coming into the power circuit. The voltage should be within 10% of the voltage rating listed on the motor nameplate. If the voltage is not within 10%, the problem is up-stream from the reduced-voltage power circuit.
2. Measure the voltage delivered to the motor from the reduced-voltage power circuit during starting and running. For primary resistor starting, the voltage during starting should be 10% to 50% less than the incoming measured voltage. The exact amount depends on the resistance added into the circuit. The resistance is set by using the resistor taps or adding resistors in series/parallel.

For autotransformer starting, the voltage during starting should be 50%, 65%, or 80% less than the incoming measured voltage. The exact amount depends on which tap connection is used on the autotransformer. For part-winding starting, the voltage during starting should be equal to the incoming measured voltage.

For wye-delta starting, the voltage during starting should equal the incoming measured voltage. For solid-state starting, the voltage during starting should be 15% to 50% less than the incoming measured voltage. The exact amount depends on the setting of the solid-state starting control switch.

The voltage measured after the motor is started should equal the incoming voltage with each method of reduced-voltage starting. There is a problem in the power circuit or control circuit if the voltage out of the starting circuit is not correct.

3. Measure the motor current draw during starting and after the motor is running. In each method of reduced-voltage starting, the starting current should be less than the current that the motor draws when connected for full-voltage starting. The current should normally be about 40% to 80% less. After the motor is running, the current should equal the normal running current of the motor. This current value should be less than the current rating listed on the motor nameplate.

1. What are three reasons for using reduced-voltage starting?
2. What is the difference in starting current between full-voltage starting and reduced-voltage starting?
3. What is the difference in starting torque between full-voltage starting and reduced-voltage starting?
4. What are the four main parts of a DC motor?
5. Why does a DC motor use many loops in the armature instead of just one loop?
6. What is the name of the stationary windings (magnets) in a DC motor?
7. How is the external power delivered to the commutator coils in a DC motor?
8. What are interpoles used for in DC motors?
9. Why is reduced-voltage starting used with large DC motors?
10. What is the name of the rotating part of an AC motor?
11. What is the name of the stationary part of an AC motor?
12. Do induction motors normally run at their synchronous speed?
13. What is locked-rotor current?
14. What is full-voltage current?
15. How can an additional reduction in current be accomplished in primary resistor starting?
16. What is open transition?
17. What is closed transition?
18. What is the advantage of part-winding starting?
19. What is the disadvantage of part-winding starting?
20. How many external leads does a wye/delta motor have?
21. Is wye/delta starting an open transition starting method or closed transition starting method?
22. In a solid-state reduced-voltage starter, what device is used to reduce the voltage applied to the motor?
23. Is a solid-state reduced-voltage starter limited to a preset number of steps?

16 Chapter

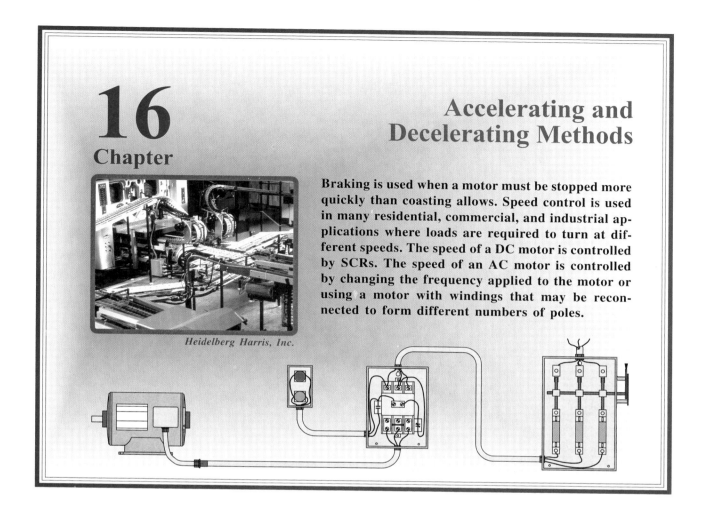

Accelerating and Decelerating Methods

Heidelberg Harris, Inc.

Braking is used when a motor must be stopped more quickly than coasting allows. Speed control is used in many residential, commercial, and industrial applications where loads are required to turn at different speeds. The speed of a DC motor is controlled by SCRs. The speed of an AC motor is controlled by changing the frequency applied to the motor or using a motor with windings that may be reconnected to form different numbers of poles.

BRAKING

A motor coasts to a stop when disconnected from the power supply. The time taken by the motor to come to rest depends on the inertia of the moving parts (motor and motor load) and friction. Braking is used when it is necessary to stop a motor more quickly than coasting allows.

Braking is accomplished by different methods having advantages and disadvantages. The braking method used depends on the application, available power, circuit requirements, cost, and desired results.

Braking applications vary greatly. For example, braking may be applied to a motor every time the motor is stopped, or it may be applied to a motor only in an emergency. In the first application, the braking action requires a method which is reliable with repeated use. In the second application, the method of stopping the motor may give little or no consideration to the damage braking may do to the motor or motor load.

DoALL Company

The C-650 Series cut-off saws from DoALL include an AC inverter which controls the speed of the band motor and provides infinitely variable band speeds from 40 fpm to 360 fpm.

Hazard braking may be required to protect an operator (hand in equipment, etc.) even if braking is not part of the normal stopping method. Brakes exist in industry that have never been used.

Friction Brakes

Friction brakes normally consist of two friction surfaces (shoes or pads) that come in contact with a wheel mounted on the motor shaft. Spring tension holds the shoes on the wheel and braking occurs as a result of the friction between the shoes and the wheel. Friction brakes (magnetic or mechanical) are the oldest motor stopping method. They are similar to brakes on automobiles. See Figure 16-1.

Heidelberg Harris, Inc.

Figure 16-1. Friction brakes normally consist of two friction surfaces that come in contact with a wheel mounted on the motor shaft.

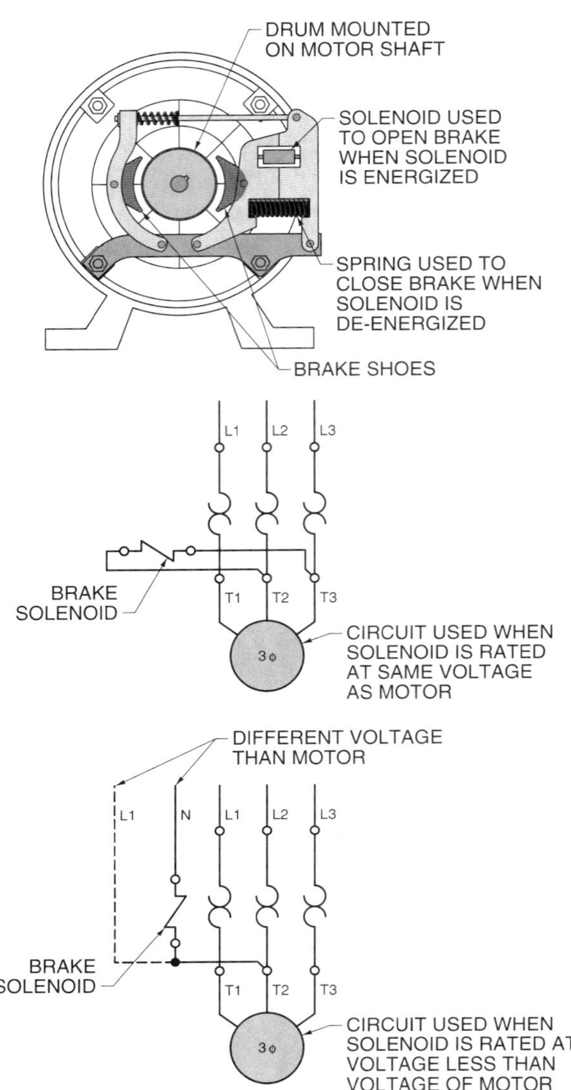

Figure 16-2. A friction brake may be connected to full-line voltage or to a voltage equal to that produced between L1 and the neutral.

Solenoid Operation. Friction brakes are normally controlled by a solenoid which activates the brake shoes. The solenoid is energized when the motor is running. This keeps the brake shoes from touching the drum mounted on the motor shaft. The solenoid is de-energized and the brake shoes are applied through spring tension when the motor is turned OFF.

Two methods are used to connect the solenoid into the circuit so that it activates the brake whenever the motor is turned ON and OFF. See Figure 16-2.

The first circuit is used if the solenoid has a voltage rating equal to the motor voltage rating. The second circuit is used if the solenoid has a voltage rating equal to the voltage between L1 and the neutral. Always connect the brake solenoid directly into the motor circuit, not into the control circuit. This eliminates improper activation of the brake.

Brake Shoes. In friction braking, the braking action is applied to a wheel mounted on the shaft of the motor rather than to the shaft directly. The wheel provides a much larger braking surface than could be obtained from the shaft alone. This permits the use of large brake shoe linings and low shoe pressure. Low shoe pressure, equally distributed over a large area, results in even wear and braking torque. The braking torque developed is directly proportional to surface area and spring pressure. The spring pressure is adjustable on nearly all friction brakes.

Determining Braking Torque. Full-load motor torque is calculated to determine the required braking torque of a motor. To calculate braking torque, apply the formula:

$$T = \frac{5252 \times HP}{rpm}$$

where

T = full-load motor torque (in lb-ft)

5252 = constant $(\frac{33{,}000}{\pi \times 2} = 5252)$

HP = motor horsepower

rpm = speed of motor shaft

Example: Calculating Braking Torque

What is the braking torque of a 60 HP, 240 V motor rotating at 1725 rpm?

$$T = \frac{5252 \times HP}{rpm}$$

$$T = \frac{5252 \times 60}{1725}$$

$$T = \frac{315{,}120}{1725}$$

T = **182.7 lb-ft**

The torque rating of the brake selected should be equal to or greater than the full-load motor torque. Manufacturers of electric brakes list the torque ratings (in lb-ft) for their brakes.

Braking torque may also be determined using a horsepower-to-torque conversion chart. See Figure 16-3. A line is drawn from the horsepower to rpm of the motor. The point at which the line crosses the torque values is the full-load and braking torque of the motor. For example, a 50 HP motor which rotates at 900 rpm requires a braking torque of 300 lb-ft.

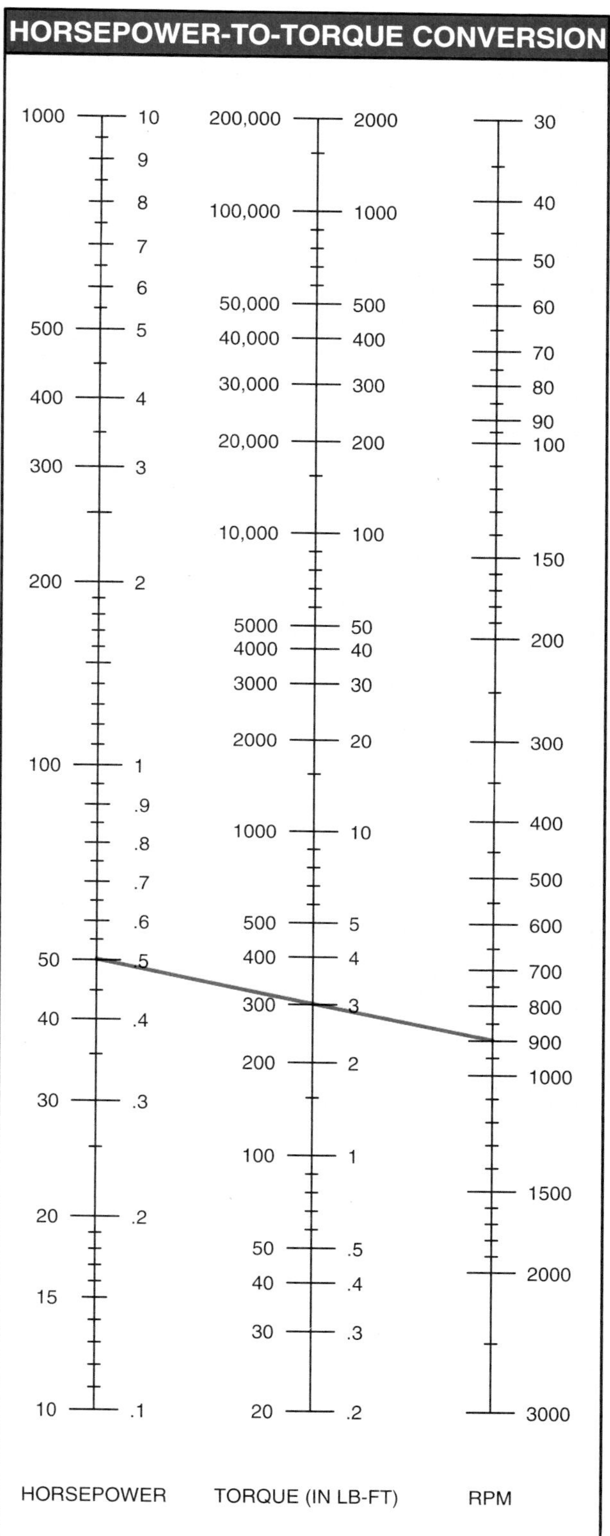

Figure 16-3. Braking torque may be determined using a horsepower-to-torque conversion chart.

Advantages and Disadvantages of Friction Brakes. The advantage of using friction brakes is lower initial cost and simplified maintenance. Friction brakes are less expensive to install than other braking methods because there are less expensive electrical components required. Maintenance is simplified because it is easy to see whether the shoes are worn and if the brake is working. Friction brakes are available in both AC and DC designs to meet almost any application. The disadvantage of friction brakes is that they require more maintenance than other braking methods. The maintenance consists of replacing the shoes which depends on the number of times the motor is stopped. A motor that is stopped often needs more maintenance than a motor that is almost never stopped. Friction brake applications include printing presses, small cranes, overhead doors, hoisting equipment, and machine tool control.

Plugging

Plugging is a method of motor braking in which the motor connections are reversed so that the motor develops a counter torque which acts as a braking force. The counter torque is accomplished by reversing the motor at full speed with the reversed motor torque opposing the forward inertia torque of the motor and its mechanical load. Plugging a motor allows for a very rapid stop. Although manual and electromechanical controls can be used to reverse the direction of a motor, a plugging switch is normally used in plugging applications. See Figure 16-4.

A plugging switch is connected mechanically to the shaft of the motor or driven machinery. The rotating motion of the motor is transmitted to the plugging switch contacts either by a centrifugal mechanism or by a magnetic induction arrangement. The contacts on the plugging switch are NO, NC, or both, and actuate at a given speed. The primary function of a plugging switch is to prevent the reversal of the load once the counter torque action of plugging has brought the load to a standstill. The motor and load would continue to run in the opposite direction without stopping if the plugging switch were not present.

Plugging Switch Operation. Plugging switches are designed to open and close sets of contacts as the shaft speed on the switch varies. As the shaft speed increases, the contacts are set to change at a given rpm. As the shaft speed decreases, the contacts return to their normal condition. As the shaft speed increases, the contact setpoint (point at which the contacts operate) reaches a higher rpm than the point at which the contacts reset (return to their normal position) on decreasing speed. The difference in these contact operating values is the differential speed or rpm.

In plugging, the continuous running speed must be many times the speed at which the contacts are required to operate. This provides high contact holding force and reduces possible contact chatter or false operation of the switch.

Continuous Plugging. A plugging switch may be used to plug a motor to a stop each time the motor is stopped. See Figure 16-5.

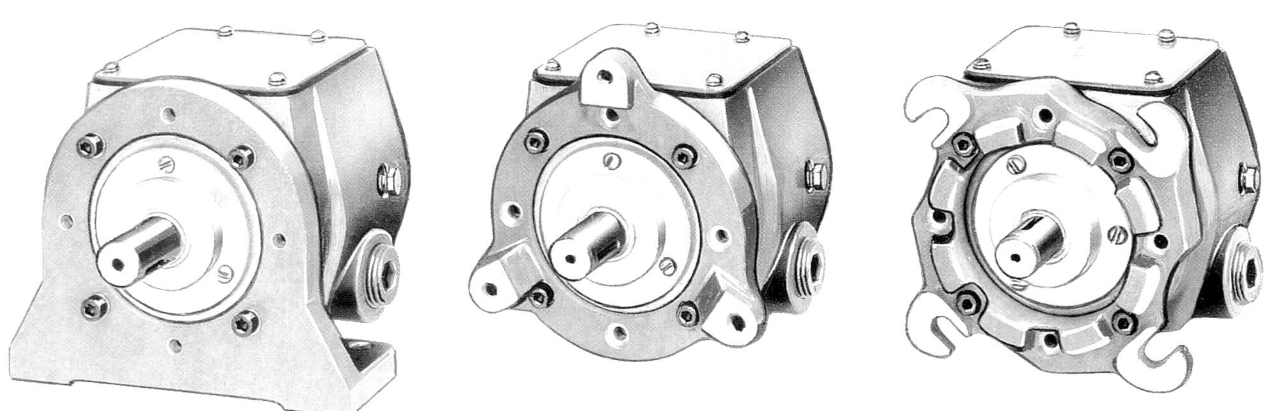

Rockwell Automation, Allen-Bradley Company, Inc.

Figure 16-4. Plugging switches prevent the reversal of the controlled load after the load has stopped.

In this circuit, the NO contacts of the plugging switch are connected to the reversing starter through an interlock contact. Pushing the start pushbutton energizes the forward starter, starting the motor in forward and adding memory to the control circuit. As the motor accelerates, the NO plugging contacts close. The closing of the NO plugging contacts do not energize the reversing starter because of the interlocks. Pushing the stop pushbutton drops out the forward starter and interlocks. This allows the reverse starter to immediately energize through the plugging switch and the NC forward interlock. The motor is reversed and the motor brakes to a stop. After the motor is stopped, the plugging switch opens to disconnect the reversing starter before the motor is actually reversed.

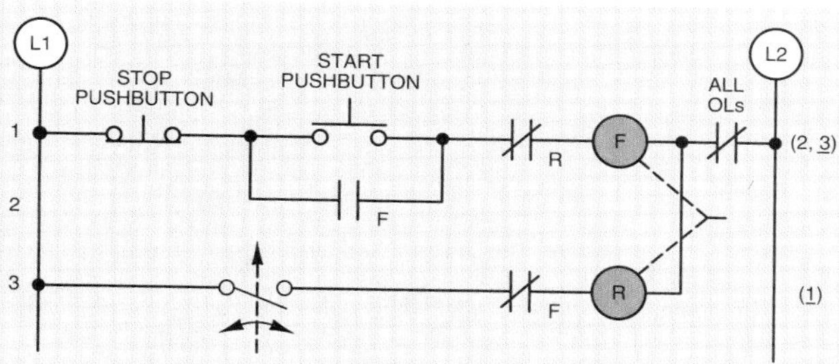

Figure 16-5. A plugging switch may be used to plug a motor to a stop each time the motor is stopped.

Plugging for Emergency Stops. A plugging switch may be used in a circuit where plugging is required only in an emergency. See Figure 16-6. In this circuit, the motor is started in the forward direction by pushing the run pushbutton. This starts the motor and adds memory to the control circuit. As the motor accelerates, the NO plugging contacts close. Pushing the stop pushbutton de-energizes the forward starter but does not energize the reverse starter. This is because there is no path for the L1 power to reach the reverse starter so the motor coasts to a stop.

Pushing the emergency stop pushbutton de-energizes the forward starter and simultaneously energizes the reversing starter. Energizing the reversing starter adds memory in the control circuit and plugs the motor to a stop. When the motor is stopped, the plugging switch opens to disconnect the reversing starter before the motor is actually reversed. The de-energizing of the reversing starter also removes the memory from the circuit.

Limitations of Plugging

Plugging is one method of braking that may not be applied to all motors and/or applications. Braking a motor to a stop using plugging requires that the motor be a reversible motor and that the motor can be reversed at full speed. Even if the motor can be reversed at full speed, the damage that plugging may do may outweigh its advantages.

Baldor Electric Co.

Capacitor-start motors cannot be plugged to a stop because they cannot be reversed at full speed.

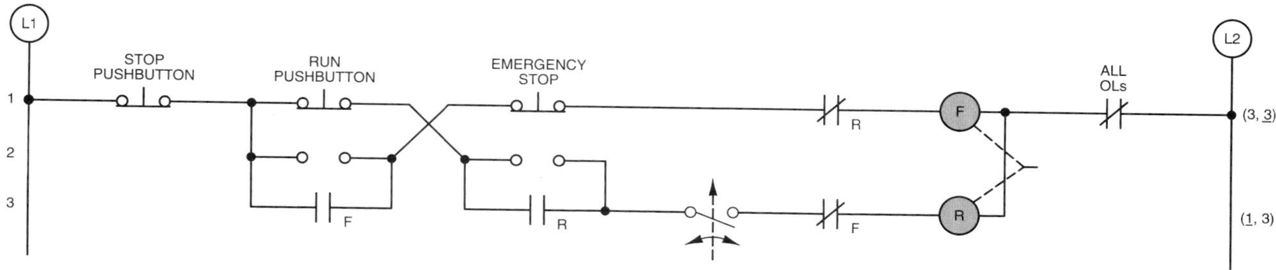

Figure 16-6. A plugging switch may be used in a circuit where plugging is required only in an emergency.

Reversing. A motor cannot be used for plugging if it cannot be reversed at full speed. For example, a 1ϕ shaded-pole motor cannot be reversed at any speed. Thus, a 1ϕ shaded-pole motor cannot be used in a plugging circuit. Likewise, most 1ϕ split-phase and capacitor-start motors cannot be plugged because their centrifugal switches remove the starting windings when the motor accelerates. Without the starting winding in the circuit, the motor cannot be reversed.

Heat. All 3ϕ motors, and most 1ϕ and DC motors, can be used for plugging. However, high current and heat result from plugging a motor to a stop. A motor is connected in reverse at full speed when plugging a motor. The current may be three or more times higher during plugging than during normal starting. For this reason, a motor designated for plugging or with a high service factor should be used in all cases except emergency stops. The service factor (SF) should be 1.35 or more for plugging applications.

Heidelberg Harris, Inc.

The amount of braking force and braking time is adjustable on applications such as printing presses.

Plugging Using Timing Relays

Plugging can also be accomplished by using a timing relay. The advantage of using a timing relay is normally in cost because a timer is much less expensive and does not have to be connected mechanically to the motor shaft or driven machine. The disadvantage is that, unlike a plugging switch, the timer does not compensate for a change in the load condition (which affects stop time) once the timer is preset.

An OFF-delay timer may be used in applications where the time needed to decelerate the motor is constant and known. See Figure 16-7. In this circuit, the NO contacts of the timer are connected into the circuit in the same manner as a plugging switch. The coil of the timer is connected in parallel with the forward starter.

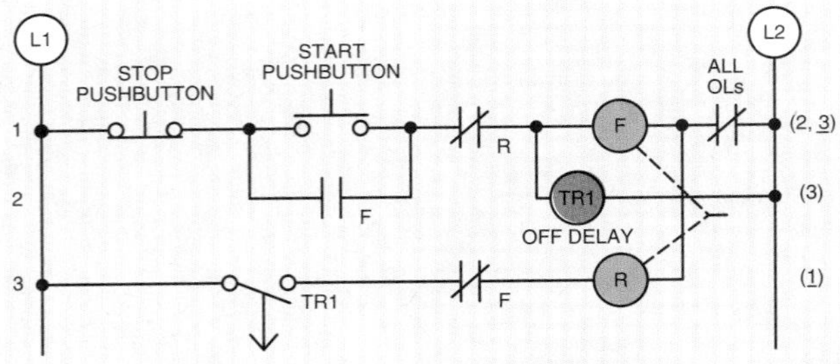

Figure 16-7. An OFF-delay timer may be used in applications where the time needed to decelerate the motor is constant.

The motor is started and memory is added to the circuit when the start pushbutton is pressed. In addition to energizing the forward starter, the OFF-delay timer is also energized. The energizing of the OFF-delay timer immediately closes the NO timer contacts. The closing of these contacts does not energize the reverse contacts because of the interlocks.

The forward starter and timer coil are de-energized when the stop pushbutton is pressed. The NO timing contact remains held closed for the setting of the timer. The holding closed of the timing contact energizes the reversing starter for the period of time set on the timer. This plugs the motor to a stop. The timer's contact must reopen before the motor is actually reversed. The motor reverses direction if the time setting is too long.

An OFF-delay timer may also be used for plugging a motor to a stop during emergency stops. See Figure 16-8. In this circuit, the timer's contacts are connected in the same manner as the plugging switch. The motor is started and memory is added to the circuit when the start pushbutton is pressed. The forward starter and timer are de-energized if the stop pushbutton is pressed. Although the timer's NO contacts are held closed for the time period set on the timer, the reversing starter is not energized. This is because no power is applied to the reversing starter from L1.

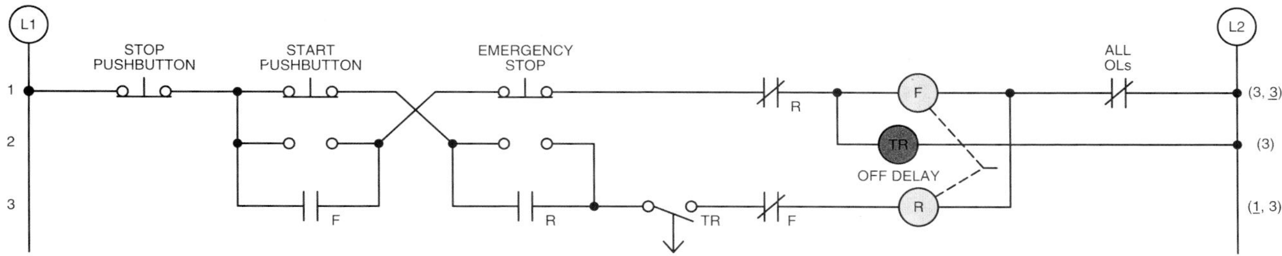

Figure 16-8. An OFF-delay timer may also be used for plugging a motor to a stop during emergency stops.

If the emergency stop pushbutton is pressed, the forward starter and timer are de-energized and the reversing starter is energized. The energizing of the reversing starter adds memory to the circuit and stops the motor. The opening of the timing contacts de-energizes the reversing starter and removes the memory.

Electric Braking

Electric braking is a method of braking in which a DC voltage is applied to the stationary windings of a motor after the AC voltage is removed. See Figure 16-9. This is an efficient and effective method of braking most AC motors. Electric braking provides a quick and smooth braking action on all types of loads including high-speed and high-inertia loads. Maintenance is minimum because there are no parts that come in physical contact during braking.

Electric Braking Operating Principles. Unlike magnetic poles attract each other and like magnetic poles repel each other. This principle, when applied to AC and DC motors, is the reason why the motor shaft rotates. The method in which the magnetic fields are created does change from one type of motor to another.

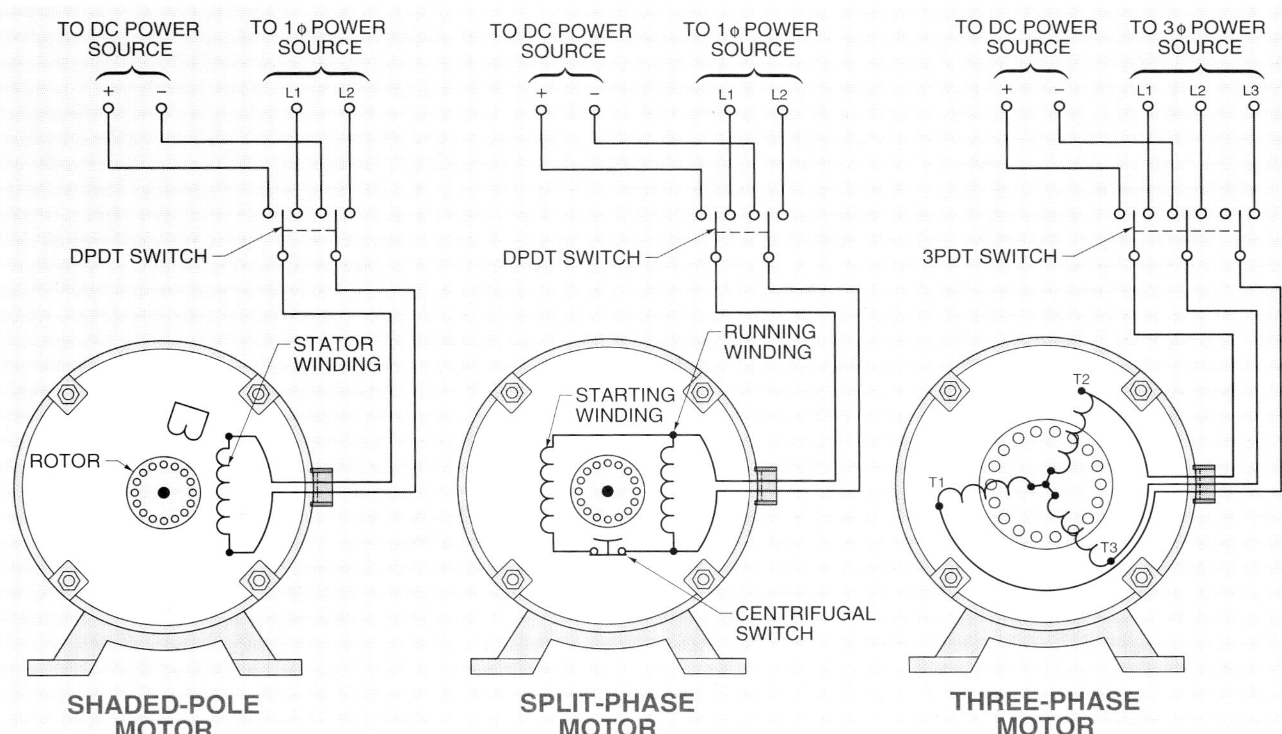

Figure 16-9. Electric braking is achieved by applying DC voltage to the stationary windings once the AC is removed.

In AC induction motors, the opposing magnetic fields are induced from the stator windings into the rotor windings by transformer action. The motor continues to rotate as long as the AC voltage is applied. The motor coasts to a standstill over a period of time when the AC voltage is removed because there is no induced field to keep it rotating.

Electric braking can be used to provide an immediate stop if the coasting time is unacceptable, particularly in an emergency situation. Electric braking is accomplished by applying a DC voltage to the stationary windings once the AC is removed. The DC voltage creates a magnetic field in the stator that does not change polarity.

The constant magnetic field in the stator creates a magnetic field in the rotor. Because the magnetic field of the stator does not change in polarity, it attempts to stop the rotor when the magnetic fields are aligned (N to S and S to N). See Figure 16-10. The only force that can keep the rotor from stopping with the first alignment is the rotational inertia of the load connected to the motor shaft. However, because the braking action of the stator is present at all times, the motor brakes quickly and smoothly to a standstill.

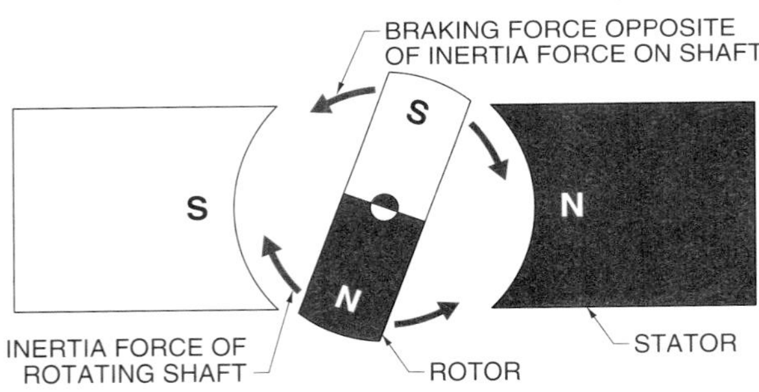

Figure 16-10. The DC voltage applied during electric braking creates a magnetic field in the stator that does not change polarity.

DC Electric Braking Circuits. DC is applied after the AC is removed to bring the motor to a stop quickly. See Figure 16-11. This circuit, like most DC braking circuits, uses a bridge rectifier circuit to change the AC into DC. In this circuit, a 3ϕ AC motor is connected to 3ϕ power by a magnetic motor starter.

The magnetic motor starter is controlled by a standard stop/start pushbutton station with memory. An OFF-delay timer is connected in parallel with the magnetic motor starter. The OFF-delay timer controls a NO contact that is used to apply power to the braking contactor for a short period of time after the stop pushbutton is pressed. The timing contact is adjusted to remain closed until the motor comes to a stop.

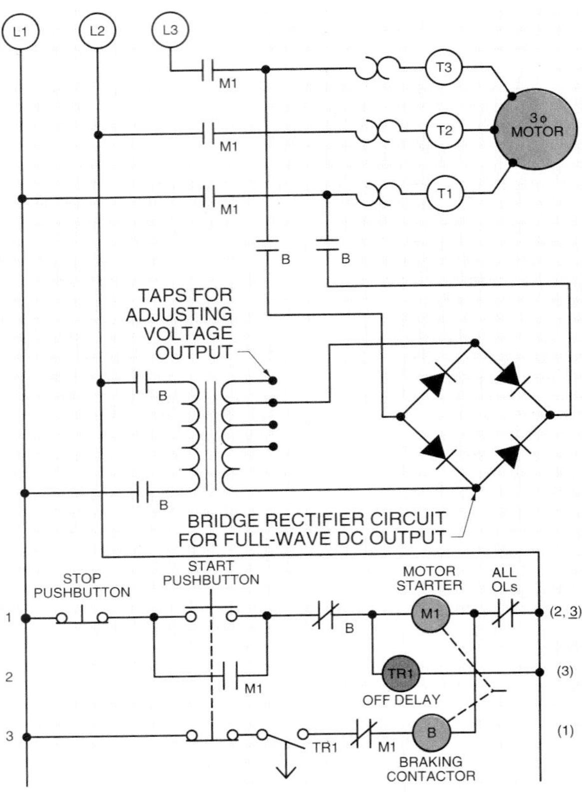

Figure 16-11. DC is applied after AC is removed to bring the motor to a stop quickly.

The braking contactor connects two motor leads to the DC supply. A transformer with tapped windings is used to adjust the amount of braking torque applied to the motor. Current-limiting resistors could be used for the same purpose. This allows for a low- or high-braking action depending on the application. The larger the applied DC voltage, the greater the breaking force.

The interlock system in the control circuit prevents the motor starter and braking contactor from being energized at the same time. This is required because the AC and DC power supplies must never be connected to the motor simultaneously. Total interlocking should always be used on electrical braking circuits. Total interlocking is the use of mechanical, electrical, and pushbutton interlocking. A standard forward and reversing motor starter can be used in this circuit, as it can with most electric braking circuits.

Dynamic Braking

Dynamic braking is a method of motor braking in which the motor is reconnected to act as a generator immediately after it is turned OFF. Connecting the motor in this way makes the motor act as a loaded generator that develops a retarding torque which rapidly stops the motor. The generator action converts the mechanical energy of rotation to electrical energy that can be dissipated as heat in a resistor.

Dynamic braking is normally applied to DC motors because there must be access to the rotor windings to reconnect the motor to act as a generator. Access is accomplished through the brushes on DC motors. See Figure 16-12. Dynamic braking of a DC motor may be needed because DC motors are often used for lifting and moving heavy loads that may be difficult to stop.

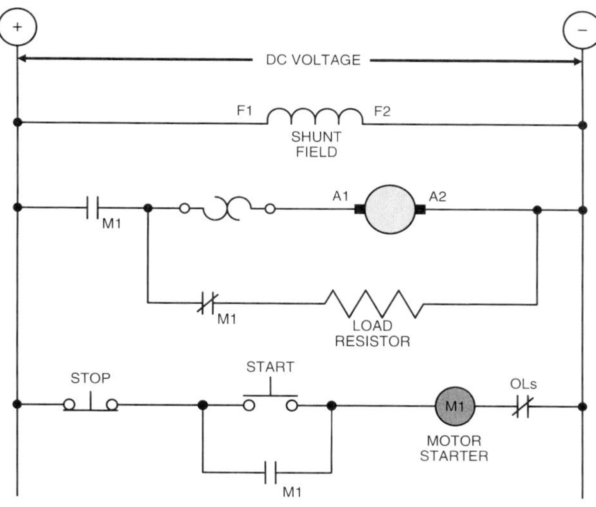

Figure 16-12. In dynamic braking, the motor is reconnected to act as a generator immediately after it is turned OFF.

In this circuit, the armature terminals of the DC motor are disconnected from the power supply and immediately connected across a resistor which acts as a load. The smaller the resistance of the resistor, the greater the rate of energy dissipation and the faster the motor comes to rest. The field windings of the DC motor are left connected to the power supply. The armature generates a counter electromotive force (CEMF). The CEMF causes current to flow through the resistor and armature.

The current causes heat to be dissipated in the resistor in the form of electrical watts. This removes energy from the system and slows the motor rotation.

The generated CEMF decreases as the speed of the motor decreases. As the motor speed approaches 0 rpm, the generated voltage also approaches 0 V. The braking action lessens as the speed of the motor decreases. As a result, a motor cannot be braked to a complete stop using dynamic braking. Dynamic braking also cannot hold a load once it is stopped because there is no braking action.

Electromechanical friction brakes are often used along with dynamic braking in applications that require the load to be held. A combination of dynamic braking and friction braking can also be used in applications where a large heavy load is to be stopped. In this application, the force of the load wears the friction shoe brakes too fast, so dynamic braking can be used to slow the load before the friction brakes are applied. This is similar to using a parachute to slow a race car before applying the brakes.

SPEED CONTROL

Speed control is essential in many residential, commercial, and industrial applications. For example, motors found in washing machines, commercial furnaces and air conditioners, as well as electrical appliances such as mixers and blenders, are required to turn the load at different speeds.

Industrial applications of speed control include mining machines, printing presses, cranes and hoists, elevators, assembly line conveyors, food processing equipment, and metalworking or woodworking lathes.

AC and DC Motors

The motor is the main consideration when choosing speed control for an application. Some motors offer excellent speed control through a total range of speeds, while others may offer only two or three different speeds. Other motors offer only one speed that cannot be changed except by external means such as gears, pulley drives, or changes of power source frequency.

The two basic motors used in speed control applications are AC and DC motors. Each motor has different ranges of speed control, applications, and cost effectiveness.

Load Requirements

The loads that are connected to and controlled by motors vary considerably. Each motor has its own ability to control different loads at different speeds. For example, certain motors are rated at high starting torque with low running torque and others are rated at poor starting torque with high running torque. Load requirements must be determined to select the correct motor for a given application. This is especially true in applications that require speed control. The requirements a motor must meet in controlling the load include force, work, torque, and horsepower in relation to speed.

Work

Work is applying a force over a distance. See Figure 16-13. *Force* is any cause that changes the position, motion, direction, or shape of an object. Work is done when a force overcomes a resistance. *Resistance* is any force that tends to hinder the movement of an object. If an applied force does not cause motion, no work is produced.

Sprecher + Schuh

DETERMINING WORK

How much work is performed when lifting a 30 lb motor from the floor to the top of a 3′ high workbench?

$W = F \times D$

$W = 30 \times 3$

$W =$ **90 lb-ft**

W = WORK
F = FORCE
D = DISTANCE

Figure 16-13. Work equals force times distance.

The amount of work (W) produced is determined by multiplying the force (F) that must be overcome by the distance (D) through which it acts. Thus, work is measured in pound-feet (lb-ft). To calculate the amount of work produced, apply the formula:

$$W = F \times D$$

where

W = work (in lb-ft)

F = force (in lb)

D = distance (in ft)

Example: Calculating Work

How much work is required to carry a 25 lb bag of groceries vertically from street level to the 4th floor of a building 30′ above street level?

$$W = F \times D$$

$$W = 25 \times 30$$

$$W = \textbf{750 lb-ft}$$

Resistance must be overcome to perform work. More work would be required if the groceries were heavier, the distance longer, or a combination of the two. For example, 1000 lb-ft of work is required to carry a 25 lb bag of groceries to a floor 40′ above the street ($25 \times 40 = 1000$ lb-ft). Less work would be required if the groceries were lighter, the distance were shorter, or a combination of the two. For example, 180 lb-ft of work is required to carry a 12 lb bag of groceries vertically from street level to the 2nd floor of a building 15′ above street level ($12 \times 15 = 180$ lb-ft).

Torque

Torque is the force that produces rotation. It causes an object to rotate. Torque (T) consists of a force (F) acting on a distance (D). See Figure 16-14. Torque, like work, is measured in pound-feet (lb-ft). However, torque, unlike work, may exist even though no movement occurs. To calculate torque, apply the formula:

$$T = F \times D$$

where

T = torque (in lb-ft)

F = force (in lb)

D = distance (in ft)

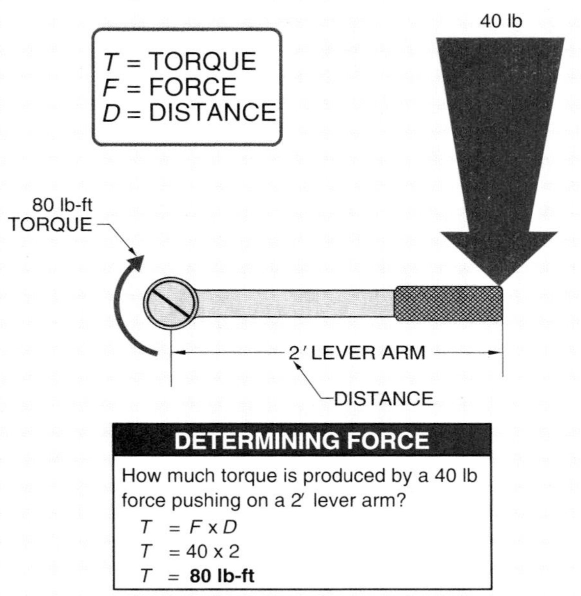

Figure 16-14. Torque is the force that produces or tends to produce rotation.

Example: Calculating Torque

What is the torque produced by a 60 lb force pushing on a 3′ lever arm?

$T = F \times D$

$T = 60 \times 3$

$T = \mathbf{180\ lb\text{-}ft}$

Work is done if the amount of torque produced is large enough to cause movement. More torque would be produced if the force were larger, the lever arm longer, or a combination of the two. Less torque would be produced if the force were smaller, the lever arm shorter, or a combination of the two. For example, 105 lb-ft of torque is produced by a 70 lb force pushing on a $1\frac{1}{2}'$ lever arm ($70 \times 1\frac{1}{2} = 105$ lb-ft).

Motor Torque

Motor torque is the force that produces or tends to produce rotation in a motor. A motor must produce enough torque to start and keep the load moving for the motor to operate the load connected to it. A motor connected to a load produces four types of torque. These four types of torque are full-load torque (FLT), locked rotor torque (LRT), pull-up torque (PUT), and breakdown torque (BDT). See Figure 16-15.

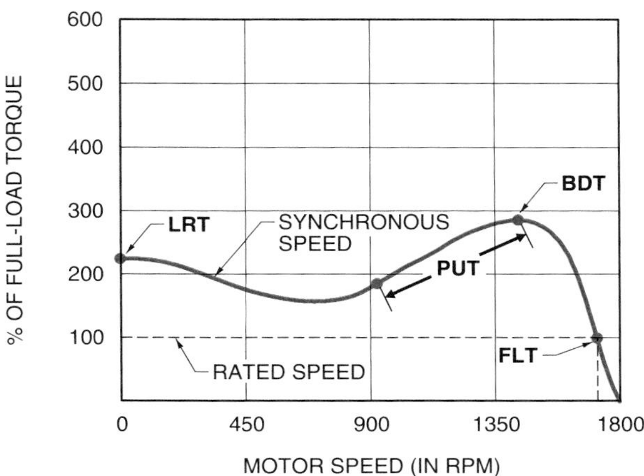

MOTOR TORQUE

FLT = FULL-LOAD TORQUE – PRODUCES RATED POWER AT FULL SPEED OF MOTOR

LRT = LOCKED ROTOR TORQUE – PRODUCED WHEN ROTOR IS STATIONARY AND FULL POWER IS APPLIED TO MOTOR

PUT = PULL-UP TORQUE – REQUIRED TO BRING MOTOR UP TO CORRECT SPEED

BDT = BREAKDOWN TORQUE – MAXIMUM TORQUE MOTOR CAN PROVIDE WITHOUT REDUCTION IN MOTOR SPEED

Figure 16-15. Motor torque is the force that produces or tends to produce rotation in a motor.

Full-load Torque. *Full-load torque (FLT)* is the torque required to produce the rated power at full speed of the motor. See Figure 16-16. The amount of torque a motor produces at rated power and full speed (full-load torque) can be found by using a horsepower-to-torque conversion chart. When using the conversion chart, place a straightedge along the two known quantities and read the unknown quantity on the third line. To calculate motor full-load torque, apply the formula:

$$T = \frac{HP \times 5252}{rpm}$$

where

T = torque (in lb-ft)

HP = horsepower

5252 = constant $\left(\frac{33{,}000}{\pi \times 2} = 5252\right)$

rpm = revolutions per minute

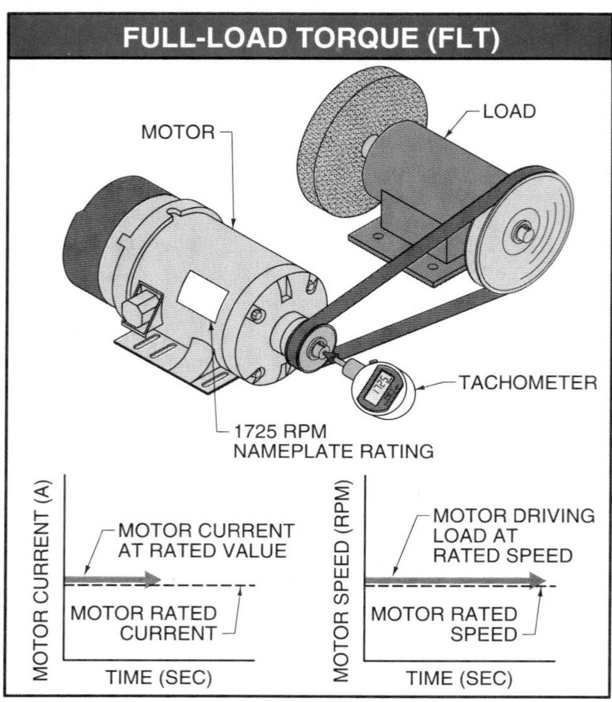

Figure 16-16. Full-load torque is the torque required to produce the rated power at full speed of the motor.

Cincinnati Milacron

Torque requirements on many machine processes vary with the type of material being machined.

Example: Calculating FLT

What is the FLT of a 30 HP motor operating at 1725 rpm?

$$T = \frac{HP \times 5252}{rpm}$$

$$T = \frac{30 \times 5252}{1725}$$

$$T = \frac{157{,}560}{1725}$$

$$T = \mathbf{91.34\ lb\text{-}ft}$$

If a motor is fully loaded, it produces full-load torque. If a motor is underloaded, it produces less than full-load torque. If a motor is overloaded, it must produce more than full-load torque to keep the load operating at the motor's rated speed. See Figure 16-17.

For example, the 30 HP motor operating at 1725 rpm can develop 91.34 lb-ft of torque at full speed. If the load requires 91.34 lb-ft at 1725 rpm, the 30 HP motor produces an output of 30 HP. However, if the load to which the motor is connected requires only half as much torque (45.67 lb-ft) at 1725 rpm, the 30 HP motor produces an output of 15 HP. The 30 HP motor draws less current (and power) from the power lines and operates at a lower temperature when producing 15 HP.

However, if the 30 HP motor is connected to a load that requires twice as much torque (182.68 lb-ft) at 1725 rpm, the motor must produce an output of 60 HP. The 30 HP motor draws more current (and power) from the power lines and operates at a higher temperature. If the overload protection device is correctly sized, the 30 HP motor automatically disconnects from the power line before any permanent damage is done to the motor.

Locked Rotor Torque. All motors can safely produce a higher torque output than the rated full-load torque for short periods of time. Since many loads require a higher torque to start them moving than to keep them moving, the motor must produce a higher torque when starting the load. *Locked rotor torque (LRT)* is the torque a motor produces when the rotor is stationary and full power is applied to the motor. See Figure 16-18. Locked rotor torque is also referred to as breakaway or starting torque. Starting torque is the torque required to start a motor. Starting torque is normally expressed as a percentage of full-load torque.

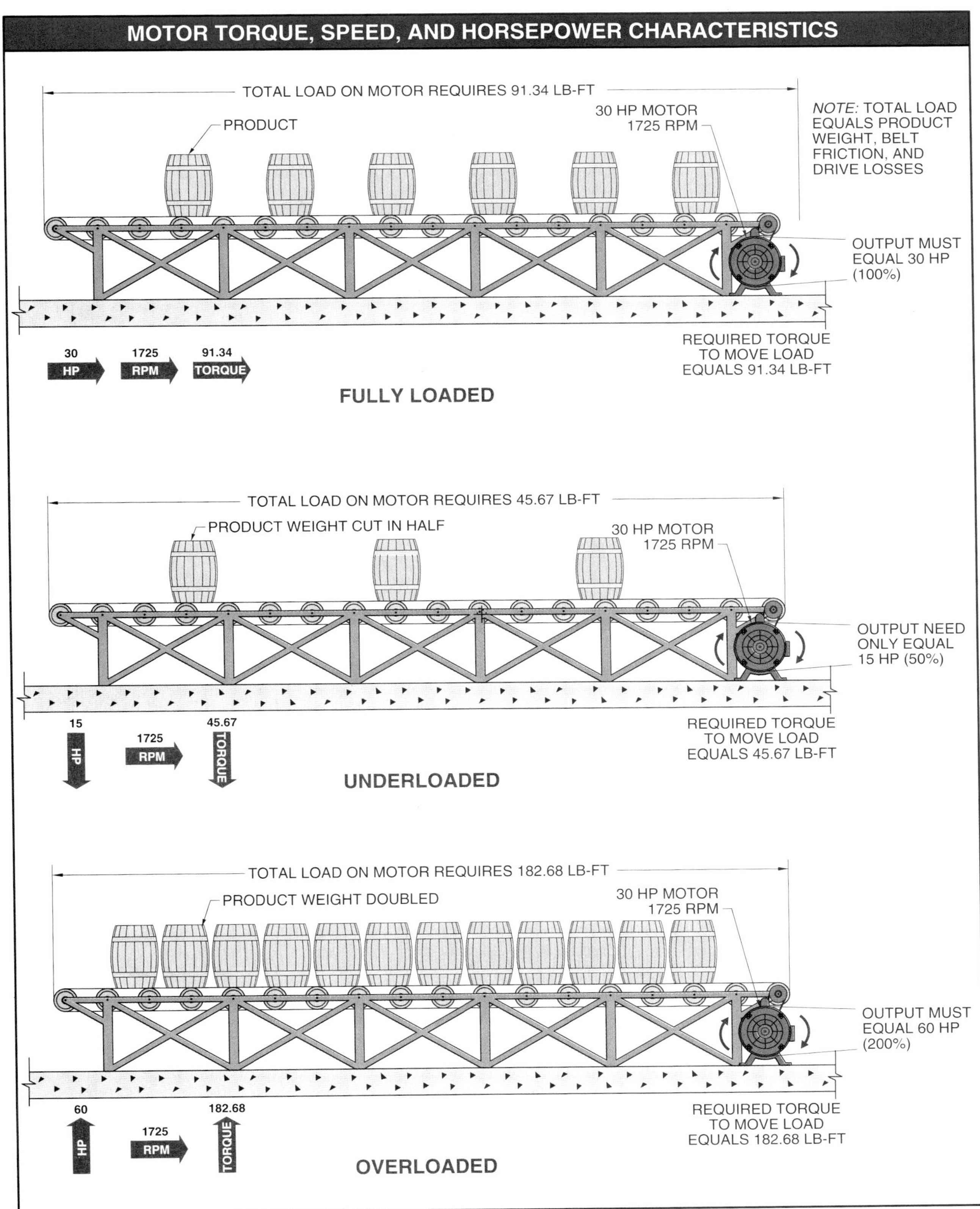

Figure 16-17. A motor may be fully loaded, underloaded, or overloaded.

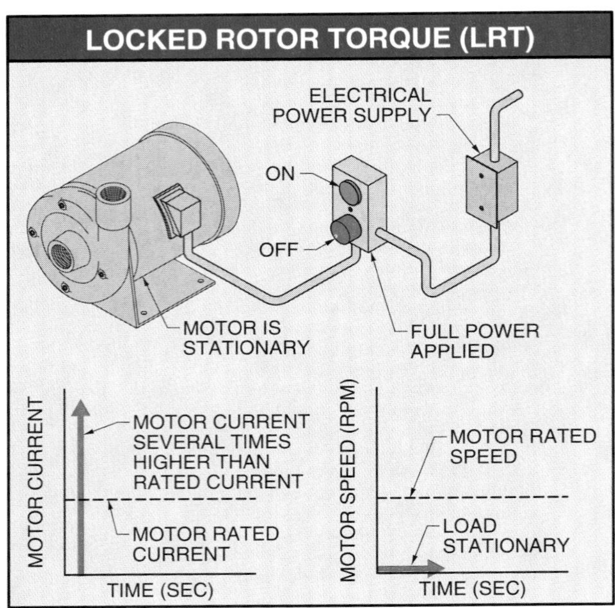

Figure 16-18. Locked rotor torque is the torque a motor produces when the rotor is stationary and full power is applied to the motor.

Baldor Electric Co.

Hoisting applications require motors that can produce very high starting torque.

Pull-Up Torque. *Pull-up torque (PUT)* is the torque required to bring a load up to its rated speed. See Figure 16-19. If a motor is properly sized to the load, pull-up torque is brief. If a motor does not have sufficient pull-up torque, the locked rotor torque may start the load turning but the pull-up torque cannot bring it up to rated speed. Once the motor is up to rated speed, full-load torque keeps the load turning. Pull-up torque is also referred to as accelerating torque.

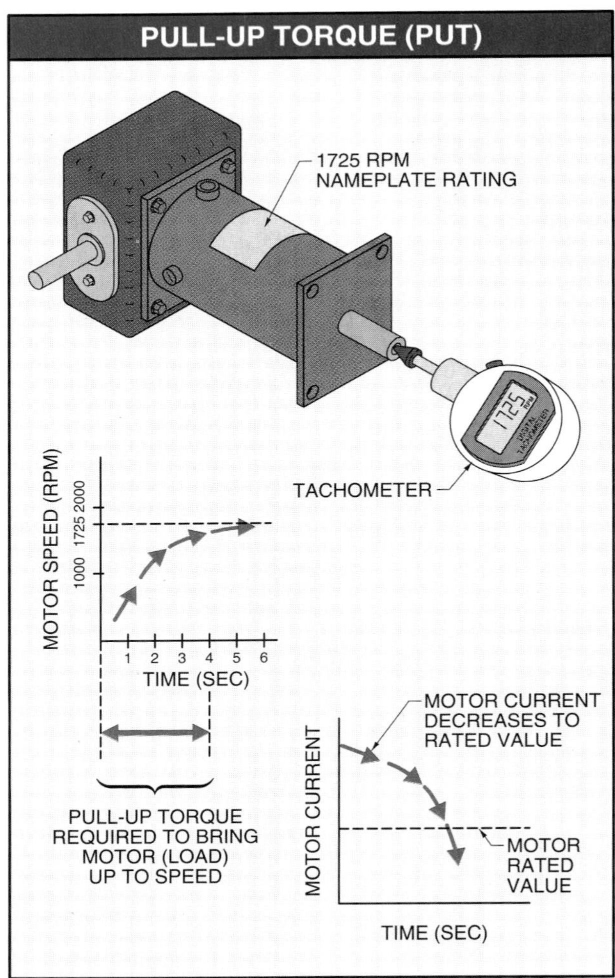

Figure 16-19. Pull-up torque is the torque required to bring a load up to its rated speed.

Breakdown Torque. *Breakdown torque (BDT)* is the maximum torque a motor can provide without an abrupt reduction in motor speed. See Figure 16-20. As the load on a motor shaft increases, the motor produces more torque. As the load continues to increase, the point at which the motor stalls is reached. This point is the breakdown torque.

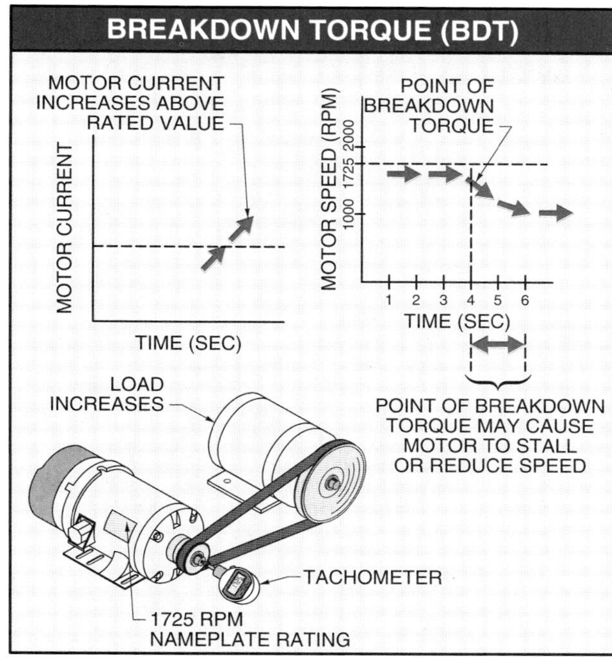

Figure 16-20. Breakdown torque is the maximum torque a motor can provide without an abrupt reduction in motor speed.

Horsepower

Electrical power is rated in horsepower or watts. A *horsepower (HP)* is a unit of power equal to 746 W or 33,000 lb-ft per minute (550 lb-ft per second). A *watt (W)* is a unit of measure equal to the power produced by a current of 1 A across a potential difference of 1 V. It is $\frac{1}{746}$ of 1 HP. The watt is the base unit of electrical power. Motor power is rated in horsepower and watts. See Figure 16-21.

Horsepower is used to measure the energy produced by an electric motor while doing work. To calculate the horsepower of a motor when current, efficiency, and voltage are known, apply the formula:

$$HP = \frac{V \times I \times Eff}{746}$$

where

HP = horsepower

V = voltage (in V)

I = current (in A)

Eff = efficiency

Example: Calculating Horsepower Using Voltage, Current, and Efficiency

What is the horsepower of a 230 V motor pulling 4 A and having 82% efficiency?

$$HP = \frac{V \times I \times Eff}{746}$$

$$HP = \frac{230 \times 4 \times .82}{746}$$

$$HP = \frac{754.4}{746}$$

$$HP = \mathbf{1\ HP}$$

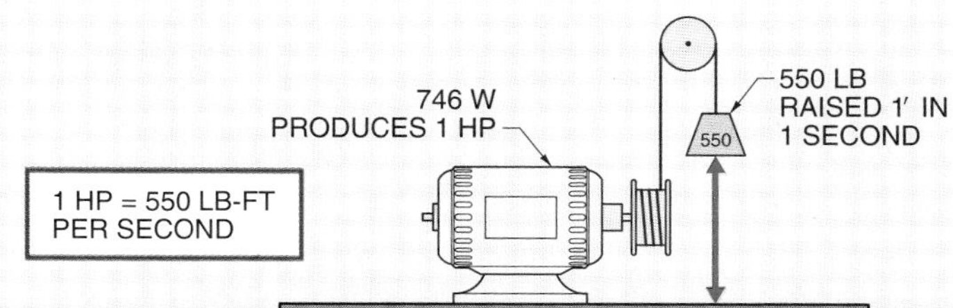

MECHANICAL ENERGY	ELECTRICAL ENERGY
$\frac{1}{2}$ HP	373 W
1 HP	746 W
2 HP	1492 W
5 HP	3730 W
100 HP	74,600 W

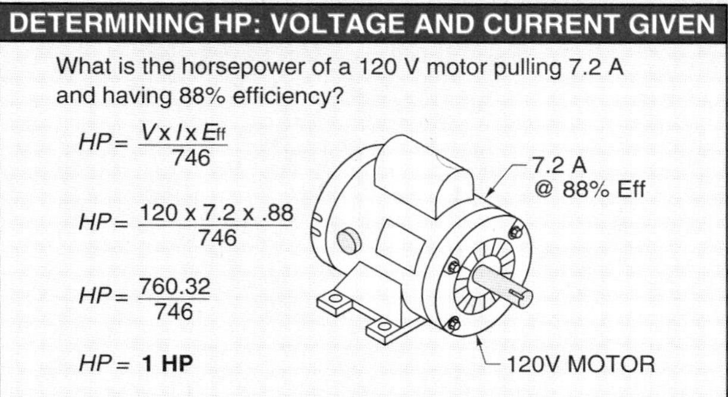

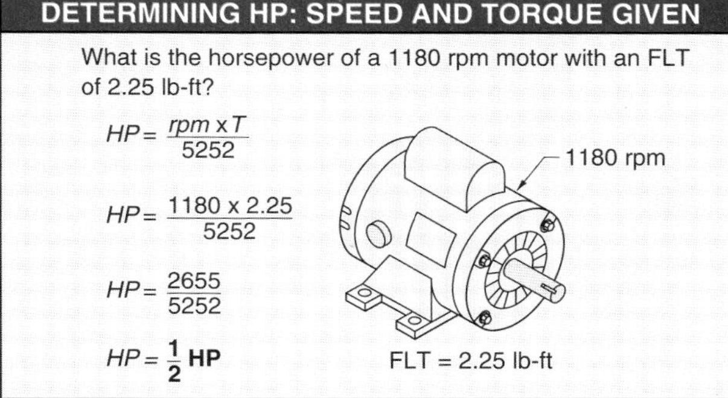

Figure 16-21. Motor power is rated in horsepower and watts.

The horsepower of a motor determines what size load a motor can operate and how fast the load turns. To calculate the horsepower of a motor when the speed and torque are known, apply the formula:

$$HP = \frac{rpm \times T}{5252}$$

where

HP = horsepower

rpm = revolutions per minute

T = torque (lb-ft)

5252 = constant $(\frac{33{,}000}{\pi \times 2} = 5252)$

Example: Calculating Horsepower Using Speed and Torque

What is the horsepower of a 1725 rpm motor with an FLT of 3.1 lb-ft?

$$HP = \frac{rpm \times T}{5252}$$

$$HP = \frac{1725 \times 3.1}{5252}$$

$$HP = \frac{5347.5}{5252}$$

$$HP = \mathbf{1\ HP}$$

Formulas for determining torque and horsepower are for theoretical values. When applied to specific applications, an additional 15% to 40% capability may be required to start a given load. Loads that are harder to start require the higher rating. To increase the rating, multiply the calculated theoretical value by 1.15 (115%) to 1.4 (140%). For example, what is the horsepower of a 1725 rpm motor with an FLT of 3.1 lb-ft with an added 25% output capability?

$$HP = \frac{rpm \times T}{5252} \times \%$$

$$HP = \frac{1725 \times 3.1}{5252} \times 1.25$$

$$HP = \frac{5347.5}{5252} \times 1.25$$

$$HP = 1.02 \times 1.25$$

$$HP = \mathbf{1.27\ HP}$$

Relationship Between Speed, Torque, and Horsepower

A motor's operating speed, torque, and horsepower rating determine the work the motor can produce. These three factors are interrelated when applied to driving a load. See Figure 16-22. If the torque remains constant, speed and horsepower are proportional. (A) If the speed increases, the horsepower must increase to maintain a constant torque. (B) If the speed decreases, the horsepower must decrease to maintain a constant torque.

If speed remains constant, torque and horsepower are proportional. (C) If the torque increases, the horsepower must increase to maintain a constant speed. (D) If the torque decreases, the horsepower must decrease to maintain a constant speed.

If torque and speed vary simultaneously but in opposite directions, the horsepower remains constant. (E) If the torque is increased and the speed is reduced, the horsepower remains constant. (F) If the torque is reduced and the speed is increased, the horsepower remains constant.

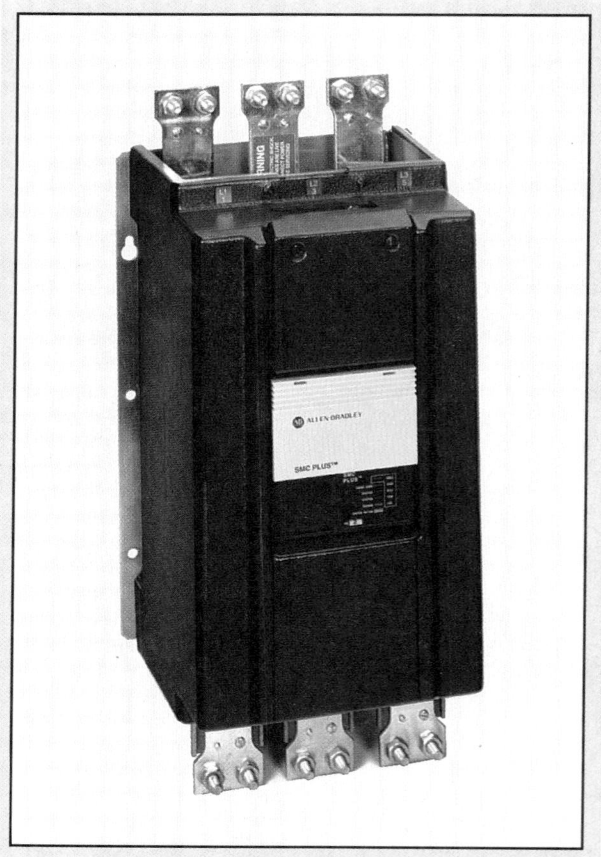

Rockwell Automation, Allen-Bradley Company, Inc.

SMC PLUS™ smart motor controllers from Allen-Bradley are available with ratings from 24 A to 1000 A and options that include pump control, SMB™ smart motor braking, slow speed with braking, Accu-Stop™, preset slow speed, and soft stop.

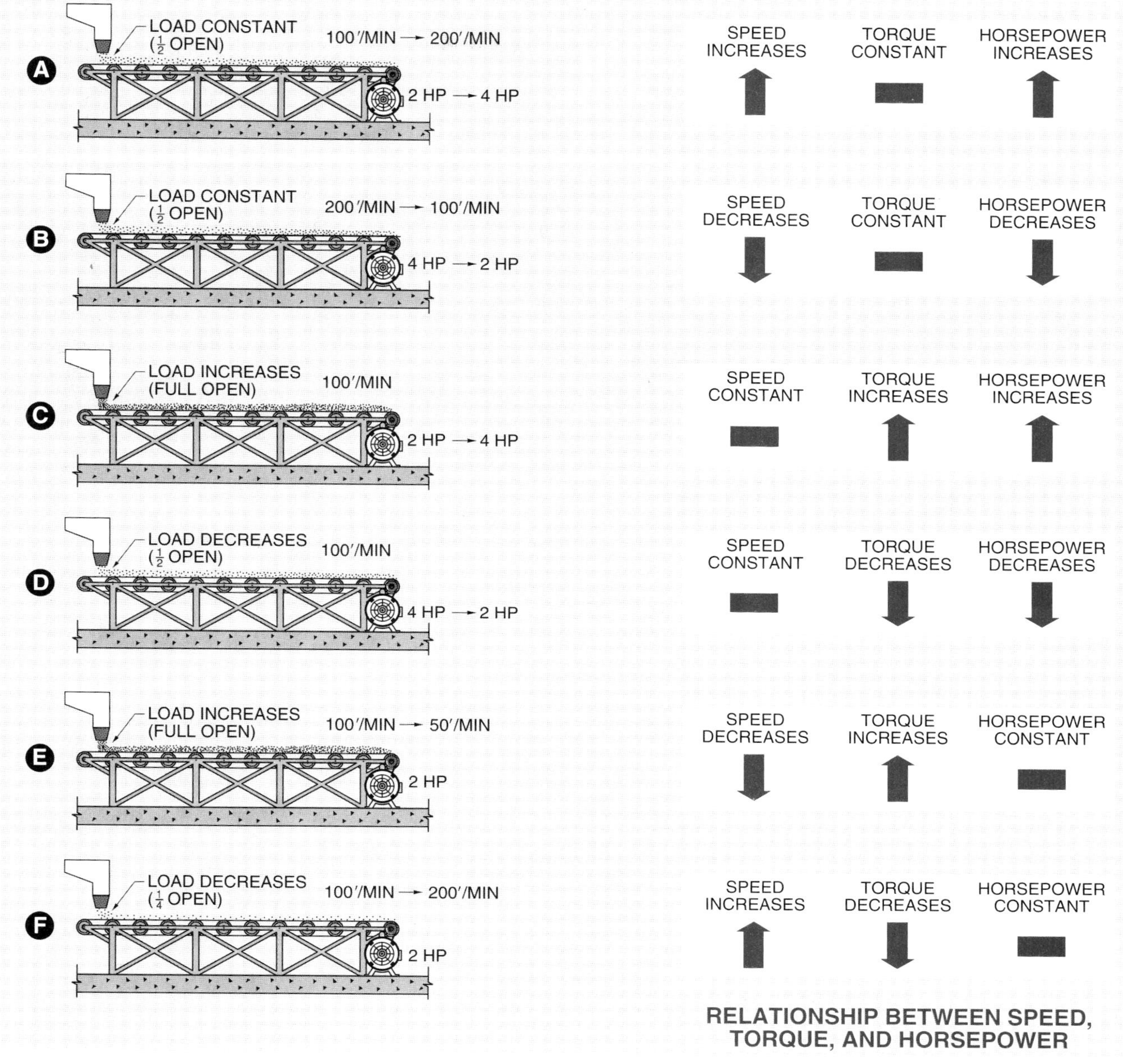

Figure 16-22. Operating speed, torque, and horsepower rating determine the work a motor can produce.

Motor Loads

Motor loads may require a constant torque, variable torque, constant horsepower, or variable horsepower when operating at different speeds. Each motor type has its own ability to control different loads at different speeds.

The best type of motor to use for a given application depends on the type of load the motor must drive. Loads are generally classified as constant torque/variable horsepower (CT/VH), constant horsepower/variable torque (CH/VT), or variable torque/variable horsepower (VT/VH). See Figure 16-23.

MOTOR LOADS				
Load	Motor Torque*		Classification	NEMA Motor Design
	LRT	PUT		
Ball Mill (mining)	125-150	175-200	CT/VH	C-D
Band Saws				
Production	50-80	175-225	CT/VH	C
Small	40	150	CT/VH	B
Car Pullers				
Automobile	150	200-225	CH/VT	C
Railroad	175	250-300	CH/VT	D
Chipper	60	225	CT/VH	B
Compressor (air)	60	150	VT/VH	B
Conveyors				
Unloaded at start	50	125-150	CT/VH	B
Loaded at start	125-175	200-250	CT/VH	C
Screw	100-125	50-175	CT/VH	C
Crushers				
Unloaded at start	75-100	150-175	CT/VH	B
With flywheel	125-150	175-200	CT/VH	D
Dryer, Industrial (loaded rotary drum)	150-175	175-225	CT/VH	D
Fan and blower	40	150	VT/VH	B
Machine Tools				
Drilling	40	150	CT/VH	B
Lathe	75	150	CT/VH	B
Press (with flywheel)	50-100	250-350	CH/VT	D
Pumps				
Centrifugal	50	150	VT/VH	B
Positive displacement	60	175	CT/VH	B
Propeller	50	150	VT/VH	B
Vacuum pumps	60	150	CT/VH	B-C

* in % of FLT

Figure 16-23. Loads are generally classified as constant torque/variable horsepower, constant horsepower/variable torque, or variable torque/variable horsepower.

Constant Torque/Variable Horsepower (CT/VH). *Constant torque/variable horsepower (CT/VH) loads* are loads in which the torque requirement remains constant. Any change in operating speed requires a change in horsepower. See Figure 16-24. CT/VH loads are the most common of all load types. They include loads that produce friction. Examples include conveyors, gear pumps and machines, metal-cutting tools, load-lifting equipment, and other loads that operate at different speeds.

Although the operating speed may change, a CT/VH load requires the same torque at low speeds as at high speeds. Since the torque requirement remains constant, an increase in speed requires an increase in horsepower.

Ruud Lighting, Inc.

A conveyor is a constant torque/variable horsepower load because the torque remains constant and any change in operating speed requires a change in horsepower.

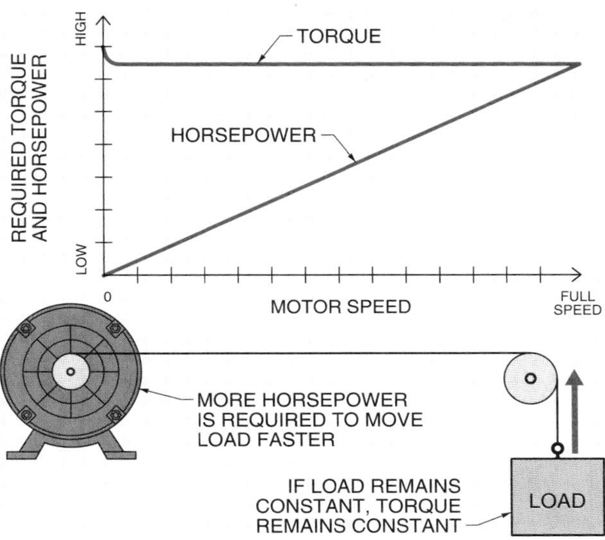

Figure 16-24. Constant torque/variable horsepower loads are loads in which the torque requirement remains constant.

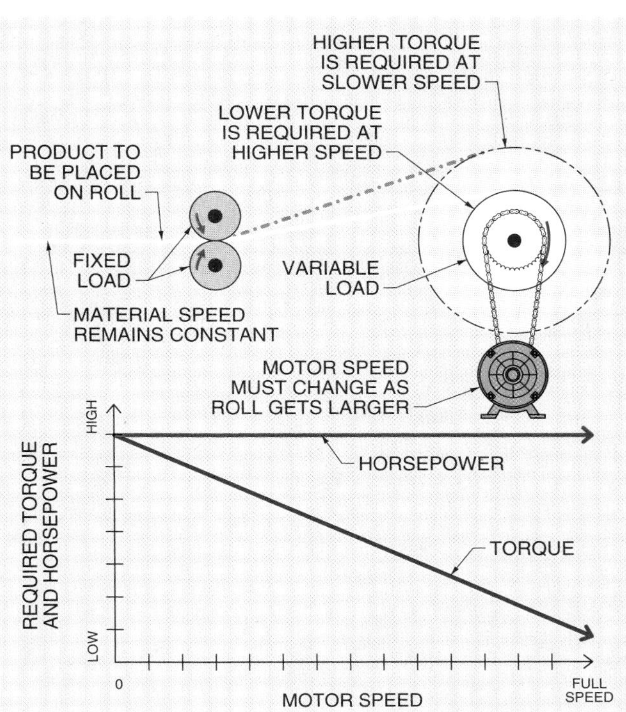

Figure 16-25. Constant horsepower/variable torque loads are loads that require high torque at low speeds and low torque at high speeds.

Constant Horsepower/Variable Torque (CH/VT). *Constant horsepower/variable torque (CH/VT) loads* are loads that require high torque at low speeds and low torque at high speeds. Since the torque requirement decreases as speed increases, the horsepower remains constant. Speed and torque are inversely proportional in CH/VT loads. See Figure 16-25.

Heidelberg Harris, Inc.

Printing presses require an increase in torque and horsepower when the press speed is increased.

An example of a CH/VT load is a center-driven winder used to roll and unroll material such as paper or metal. Since the work is done on a varying diameter with tension and linear speed of the material constant, horsepower must also be constant. Although the speed of the material is kept constant, the motor speed is not. The diameter of the material on the roll that is driven by the motor is constantly changing as material is added. At the start, the motor must run at high speed to maintain the correct material speed while torque is kept at a minimum. As material is added to the roll, the motor must deliver more torque at a lower speed. As the material is rolled, both torque and speed are constantly changing while the motor horsepower remains the same.

Variable Torque/Variable Horsepower (VT/VH). *Variable torque/variable horsepower (VT/VH) loads* are loads that require a varying torque and horsepower at different speeds. See Figure 16-26. With this type of load, the motor must work harder to deliver more output at a faster speed. Both torque and horsepower are increased with increased speed. Examples of VT/VH loads include fans, blowers, centrifugal pumps, mixers, and agitators.

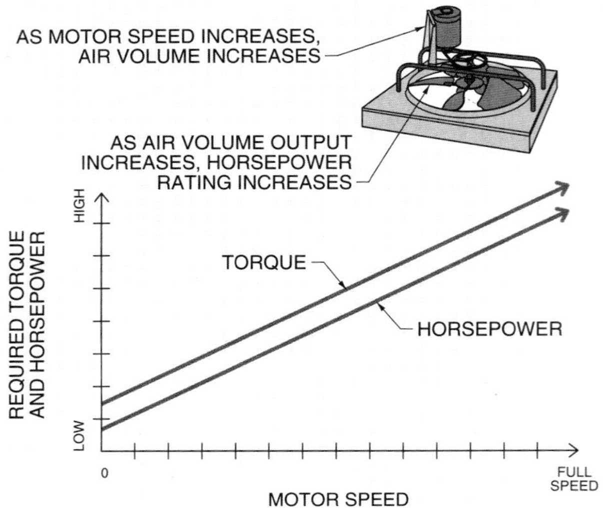

Figure 16-26. Variable torque/variable horsepower loads are loads that require a varying torque and horsepower at different speeds.

NEMA Design

Different motors are more suited for particular applications because each motor has its own characteristics of horsepower, torque, and speed. The basic characteristics of each motor are determined by the design of the motor and the supply voltage used. These designs are classified and given a letter designation which can be found on the nameplate of some motors listed as NEMA Design. See Figure 16-27.

MULTISPEED MOTORS

A motor's torque or HP characteristics change with a change in speed when motors are required to run at different speeds. The motor chosen depends on the application that the motor must perform. Once this selection is made, the motor is connected into the circuit. Common motor connection arrangements conforming to NEMA standards are used when connecting motors. See Figure 16-28.

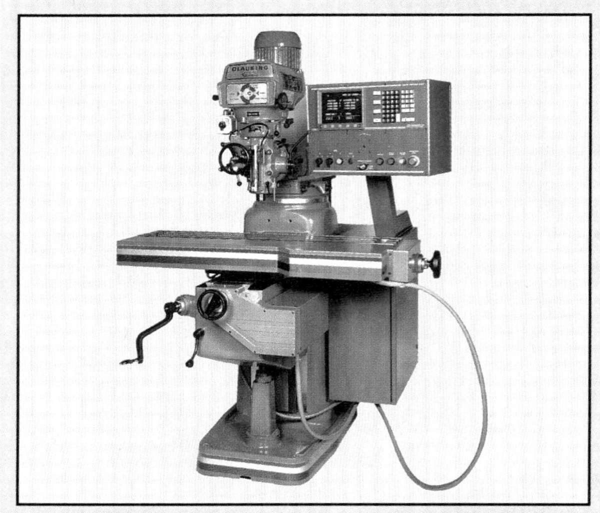

Clausing Industrial

Many machining processes require multiple speeds to machine different materials with different sizes and types of tooling.

MOTOR DESIGN CHARACTERISTICS					
NEMA Design	**Starting Torque**	**Starting Current**	**Breakdown Torque**	**Full-Load Slip**	**Typical Applications**
A	Normal	Normal	High	Low	Machine tools, fans, centrifugal pumps
B	Normal	Low	High	Low	Machine tools, fans, centrifugal pumps
C	High	Low	Normal	Low	Loaded compressors, loaded conveyors
D	Very High	Low	—	High	Punch presses

Figure 16-27. Motor design characteristics are classified and given a letter designation which can be found on the nameplate of some motors listed as NEMA Design.

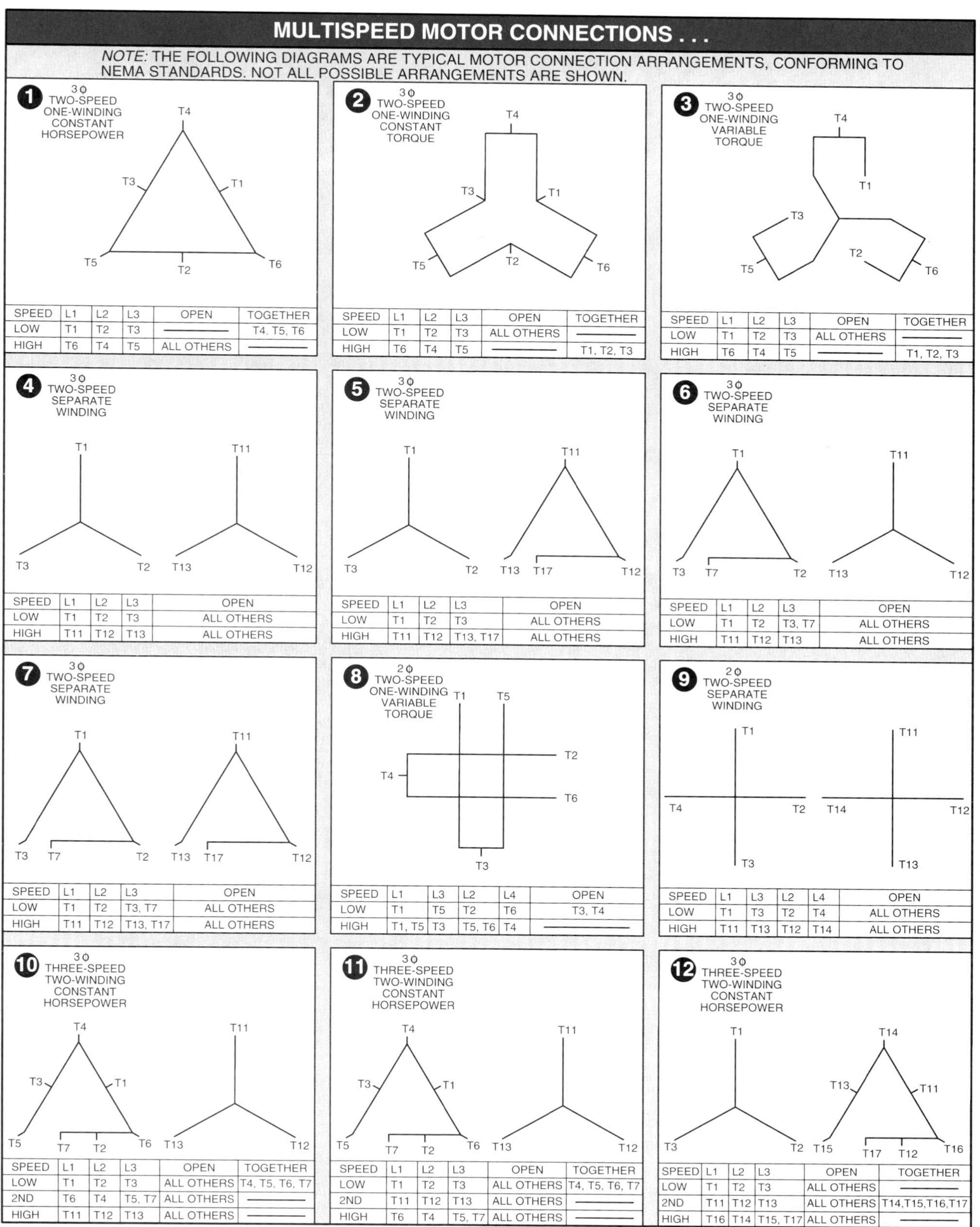

SPEED	L1	L2	L3	OPEN	TOGETHER
LOW	T1	T2	T3	—	T4, T5, T6
HIGH	T6	T4	T5	ALL OTHERS	—

SPEED	L1	L2	L3	OPEN	TOGETHER
LOW	T1	T2	T3	ALL OTHERS	—
HIGH	T6	T4	T5	—	T1, T2, T3

SPEED	L1	L2	L3	OPEN	TOGETHER
LOW	T1	T2	T3	ALL OTHERS	—
HIGH	T6	T4	T5	—	T1, T2, T3

SPEED	L1	L2	L3	OPEN
LOW	T1	T2	T3	ALL OTHERS
HIGH	T11	T12	T13	ALL OTHERS

SPEED	L1	L2	L3	OPEN
LOW	T1	T2	T3	ALL OTHERS
HIGH	T11	T12	T13, T17	ALL OTHERS

SPEED	L1	L2	L3	OPEN
LOW	T1	T2	T3, T7	ALL OTHERS
HIGH	T11	T12	T13	ALL OTHERS

SPEED	L1	L2	L3	OPEN
LOW	T1	T2	T3, T7	ALL OTHERS
HIGH	T11	T12	T13, T17	ALL OTHERS

SPEED	L1	L3	L2	L4	OPEN
LOW	T1	T5	T2	T6	T3, T4
HIGH	T1, T5	T3	T5, T6	T4	—

SPEED	L1	L3	L2	L4	OPEN
LOW	T1	T3	T2	T4	ALL OTHERS
HIGH	T11	T13	T12	T14	ALL OTHERS

SPEED	L1	L2	L3	OPEN	TOGETHER
LOW	T1	T2	T3	ALL OTHERS	T4, T5, T6, T7
2ND	T6	T4	T5, T7	ALL OTHERS	—
HIGH	T11	T12	T13	ALL OTHERS	—

SPEED	L1	L2	L3	OPEN	TOGETHER
LOW	T1	T2	T3	ALL OTHERS	T4, T5, T6, T7
2ND	T11	T12	T13	ALL OTHERS	—
HIGH	T6	T4	T5, T7	ALL OTHERS	—

SPEED	L1	L2	L3	OPEN	TOGETHER
LOW	T1	T2	T3	ALL OTHERS	—
2ND	T11	T12	T13	ALL OTHERS	T14,T15,T16,T17
HIGH	T16	T14	T15, T17	ALL OTHERS	—

Figure 16-28 continued...

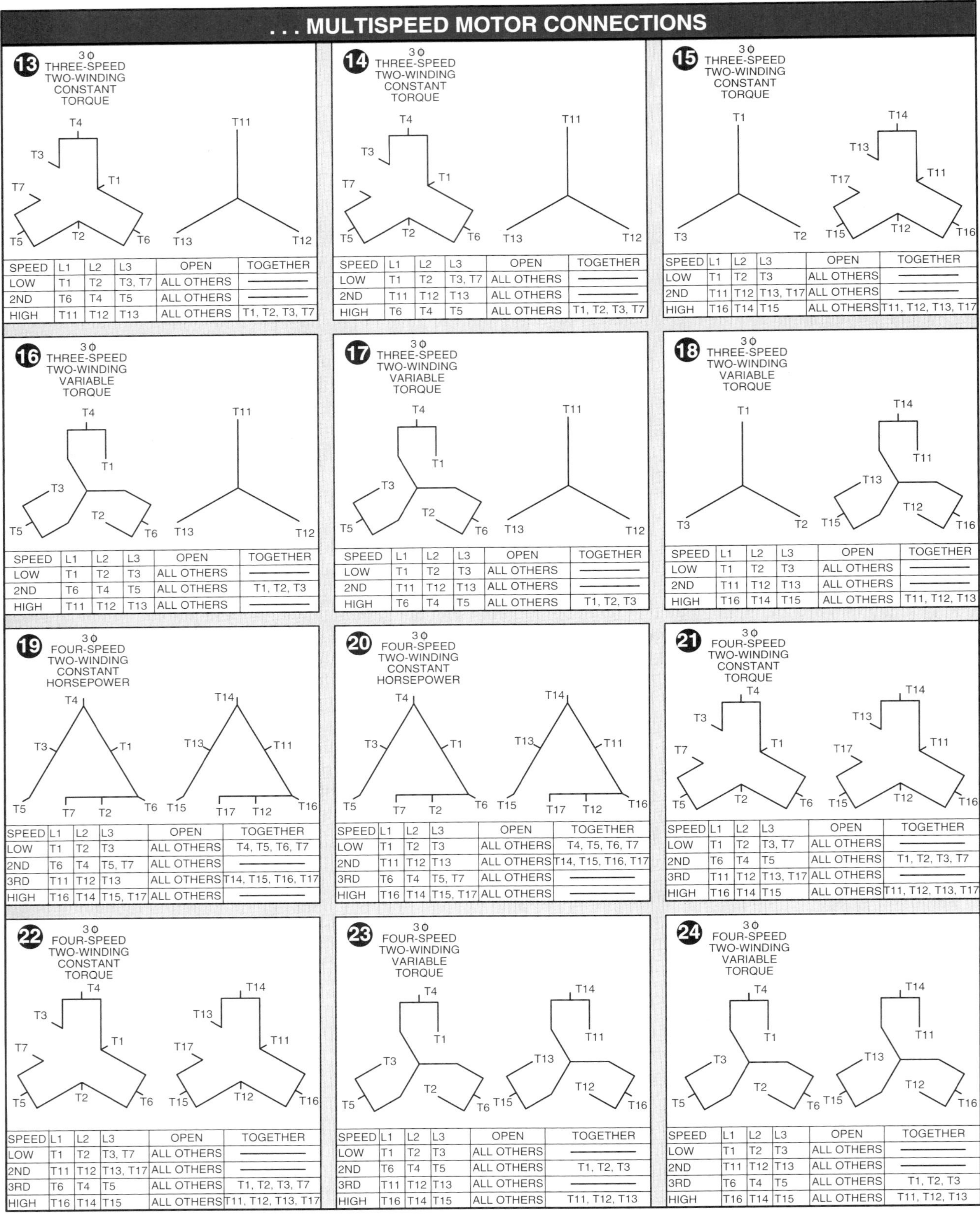

13

SPEED	L1	L2	L3	OPEN	TOGETHER
LOW	T1	T2	T3, T7	ALL OTHERS	—
2ND	T6	T4	T5	ALL OTHERS	—
HIGH	T11	T12	T13	ALL OTHERS	T1, T2, T3, T7

14

SPEED	L1	L2	L3	OPEN	TOGETHER
LOW	T1	T2	T3, T7	ALL OTHERS	—
2ND	T11	T12	T13	ALL OTHERS	—
HIGH	T6	T4	T5	ALL OTHERS	T1, T2, T3, T7

15

SPEED	L1	L2	L3	OPEN	TOGETHER
LOW	T1	T2	T3	ALL OTHERS	—
2ND	T11	T12	T13, T17	ALL OTHERS	—
HIGH	T16	T14	T15	ALL OTHERS	T11, T12, T13, T17

16

SPEED	L1	L2	L3	OPEN	TOGETHER
LOW	T1	T2	T3	ALL OTHERS	—
2ND	T6	T4	T5	ALL OTHERS	T1, T2, T3
HIGH	T11	T12	T13	ALL OTHERS	—

17

SPEED	L1	L2	L3	OPEN	TOGETHER
LOW	T1	T2	T3	ALL OTHERS	—
2ND	T11	T12	T13	ALL OTHERS	—
HIGH	T6	T4	T5	ALL OTHERS	T1, T2, T3

18

SPEED	L1	L2	L3	OPEN	TOGETHER
LOW	T1	T2	T3	ALL OTHERS	—
2ND	T11	T12	T13	ALL OTHERS	—
HIGH	T16	T14	T15	ALL OTHERS	T11, T12, T13

19

SPEED	L1	L2	L3	OPEN	TOGETHER
LOW	T1	T2	T3	ALL OTHERS	T4, T5, T6, T7
2ND	T6	T4	T5, T7	ALL OTHERS	—
3RD	T11	T12	T13	ALL OTHERS	T14, T15, T16, T17
HIGH	T16	T14	T15, T17	ALL OTHERS	—

20

SPEED	L1	L2	L3	OPEN	TOGETHER
LOW	T1	T2	T3	ALL OTHERS	T4, T5, T6, T7
2ND	T11	T12	T13	ALL OTHERS	T14, T15, T16, T17
3RD	T6	T4	T5, T7	ALL OTHERS	—
HIGH	T16	T14	T15, T17	ALL OTHERS	—

21

SPEED	L1	L2	L3	OPEN	TOGETHER
LOW	T1	T2	T3, T7	ALL OTHERS	—
2ND	T6	T4	T5	ALL OTHERS	T1, T2, T3, T7
3RD	T11	T12	T13, T17	ALL OTHERS	—
HIGH	T16	T14	T15	ALL OTHERS	T11, T12, T13, T17

22

SPEED	L1	L2	L3	OPEN	TOGETHER
LOW	T1	T2	T3, T7	ALL OTHERS	—
2ND	T11	T12	T13, T17	ALL OTHERS	—
3RD	T6	T4	T5	ALL OTHERS	T1, T2, T3, T7
HIGH	T16	T14	T15	ALL OTHERS	T11, T12, T13, T17

23

SPEED	L1	L2	L3	OPEN	TOGETHER
LOW	T1	T2	T3	ALL OTHERS	—
2ND	T6	T4	T5	ALL OTHERS	T1, T2, T3
3RD	T11	T12	T13	ALL OTHERS	—
HIGH	T16	T14	T15	ALL OTHERS	T11, T12, T13

24

SPEED	L1	L2	L3	OPEN	TOGETHER
LOW	T1	T2	T3	ALL OTHERS	—
2ND	T11	T12	T13	ALL OTHERS	—
3RD	T6	T4	T5	ALL OTHERS	T1, T2, T3
HIGH	T16	T14	T15	ALL OTHERS	T11, T12, T13

Figure 16-28. Common motor connection arrangements conforming to NEMA standards are used when connecting motors.

DC MOTOR SPEED CONTROL

DC motors are used in industrial applications that require variable speed control, high torque, or both. DC motors are used in many acceleration and deceleration applications because the speed of most DC motors can be controlled smoothly and easily from zero to full speed.

In addition to having excellent speed control, DC motors are ideal in applications that call for momentarily high torque outputs. This is because a DC motor can deliver three to five times its rated torque for short periods of time. Most AC motors stall with a load that requires twice the rated torque. Good speed control and high torque are the reasons DC motors are used in running cranes, hoists, and large machine tools found in the mining industry.

Baldor Electric Co.

Conveyor applications are normally constant torque/variable horsepower loads and may require a NEMA Design C motor or higher, if the conveyor is loaded when started.

DC Series Motors

A DC series motor produces high starting torque. See Figure 16-29. The field coil (series field) of the motor is connected in series with the armature. Although speed control is poor, a DC series motor produces very high starting torque and is ideal for applications in which the starting load is large. Applications include cranes, hoists, electric buses, streetcars, railroads, and other heavy-traction applications.

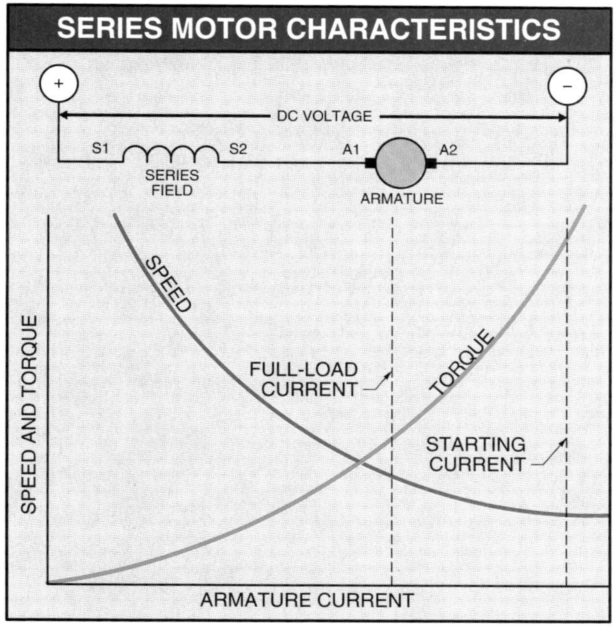

Figure 16-29. A DC series motor produces high starting torque.

The torque that is produced by a motor depends on the strength of the magnetic field in the motor. The strength of the magnetic field depends on the amount of current that flows through the series field. The amount of current that flows through a motor depends on the size of the load. The larger the load, the greater the current flow. Any increase in load increases current in both the armature and series field because the armature and field are connected in series. This increased current flow is what gives the series motor a high torque output.

In DC series motors, speed changes rapidly when torque changes. When torque is high, speed is low, and when speed is high, torque is low. This occurs because as the increased current (created by the load) flows through the series field, there is a large flux increase. This increased flux produces a large CEMF which greatly decreases the speed of the motor. As the load is removed, the motor rapidly increases speed. Without a load, the motor would gain speed uncontrollably. In certain cases, the speed may become great enough to damage the motor. For this reason, a DC series motor should always be connected directly (not through belts, chains, etc.) to the load.

The speed of a DC series motor is controlled by varying the applied voltage. Speed control of a series motor is poor when compared to the speed control of a shunt motor. Although the speed regulation of a series motor is not as good as the speed regulation of a shunt motor, not all applications require good speed regulation. The advantage of a high torque output outweighs good speed regulation in certain applications, such as the starter motor in automobiles.

DC Shunt Motors

In a DC shunt motor, the field coil (shunt field) is connected in parallel (shunt) with the armature. See Figure 16-30. DC shunt motors have good speed regulation and are used in applications that require a constant speed at any control setting, even with a changing load. The DC shunt motor is considered a constant speed motor for all reasonable loads.

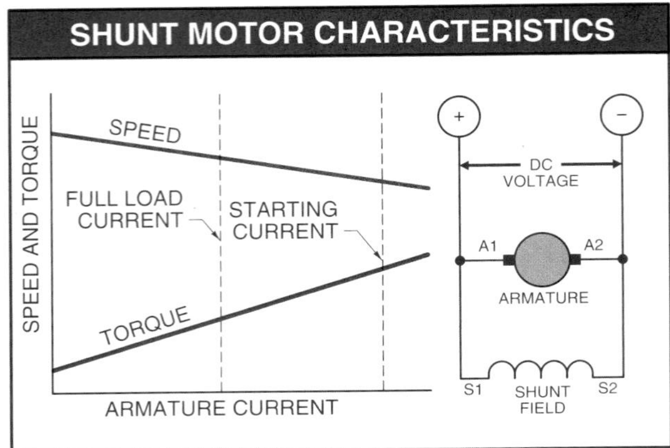

Figure 16-30. In a DC shunt motor, the field coil is connected in parallel with the armature.

Speed is determined by the voltage across the armature and the strength of the shunt field. If the voltage to the armature is reduced, the speed is also reduced. If the strength of the magnetic field is reduced, the motor speeds up. The motor speeds up with a reduction in shunt field strength because with less field strength, there is less CEMF developed in the armature. When the CEMF is lowered, the armature current increases, producing increased torque and speed. To control the speed of a DC shunt motor, the voltage to the armature is varied or the shunt field current is varied.

A field rheostat or armature rheostat is used to adjust the speed of a DC shunt motor. See Figure 16-31. The rheostat is used to increase or decrease the strength of the field or armature. Once the strength of the field is set, it remains constant regardless of changes in armature current. As the load is increased on the armature, the armature current and torque of the motor increase. This slows the armature, but the reduction of CEMF simultaneously allows a further increase in armature current and thus returns the motor to the set speed. The motor runs at a fairly constant speed at any control setting.

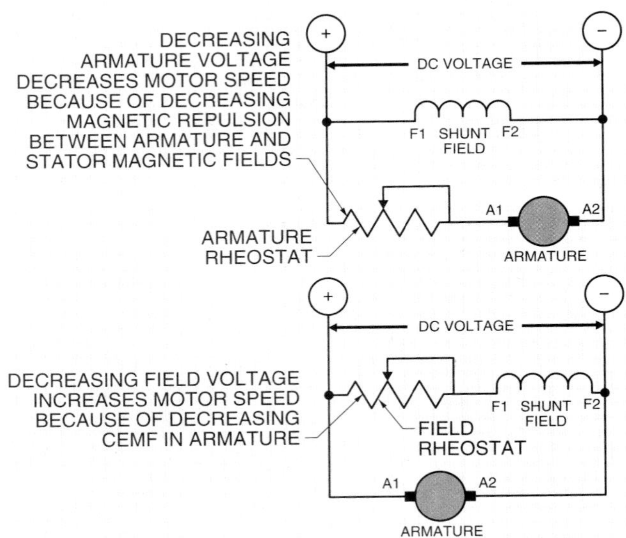

Figure 16-31. A field rheostat or armature rheostat is used to adjust the speed of a DC shunt motor.

A DC shunt motor has relatively high torque at any speed. The motor torque is directly proportional to the armature current. As armature current is increased, so is motor torque, with only a slight drop in motor speed.

DC Compound Motors

A DC compound motor has both a series coil (series field) and shunt coil (shunt field) connected in relationship to the armature. See Figure 16-32. A DC compound motor combines the operating characteristics of both the series and shunt motors. This produces high starting torque characteristics of a series motor and good speed control characteristics of a shunt motor. However, a DC compound motor does not have as high starting torque as a series motor and not as good speed regulation as a shunt motor. DC compound motors are used in applications where the load changes and precise speed control is not required. Applications include elevators, hoists, cranes, and conveyors.

Speed control is obtained in a DC compound motor by changing the shunt field current strength or changing the voltage applied to the armature. This is accomplished by using a controller that uses resistors to reduce the applied voltage or by using a variable voltage supply.

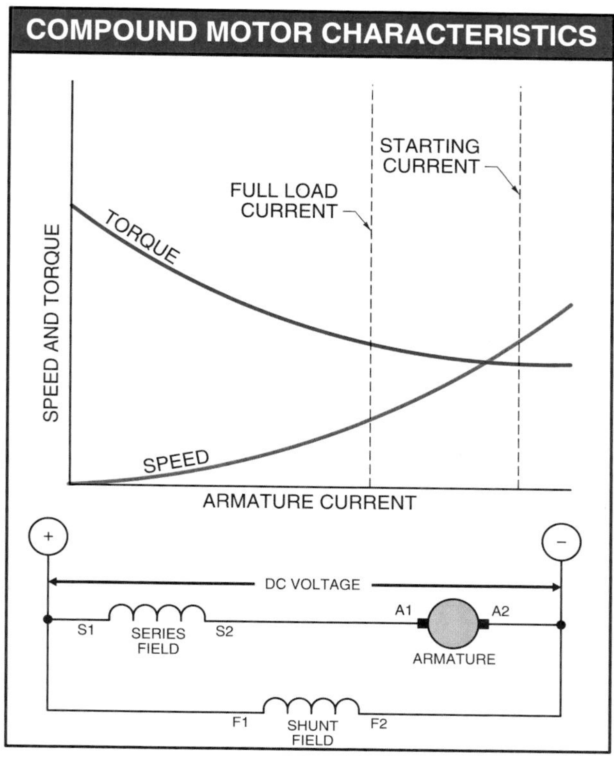

Figure 16-32. A DC compound motor combines the operating characteristics of series and shunt motors.

DC Motor Solid-State Speed Control

An SCR can perform the same function as resistance in controlling the voltage applied to a load. An SCR is similar to a rheostat (variable resistor) because it can be adjusted throughout its range. An SCR has some advantages over a rheostat. An SCR is smaller in size for the same rating, energy-efficient in not wasting power within itself, and less expensive than a rheostat. For these reasons, most DC speed controls produced today use SCRs instead of rheostats.

Controlling DC Motor Base Speed

DC motor control is one of the best industrial applications of an SCR. In DC motor control applications, an SCR can be used to control the speed of DC motors below the base speed by changing the amount of current that flows through the armature circuit. Speed below the base speed is controlled by changing the voltage applied to the field. *Base speed* of a DC motor is the speed (in rpm) listed on the motor's nameplate and the speed at which the motor runs with full-line voltage applied to the armature and field.

The speed of a DC motor is controlled by varying the applied voltage across the armature, the field, or both. When armature voltage is controlled, the motor delivers a constant torque characteristic. When field voltage is controlled, the motor delivers a constant horsepower characteristic. See Figure 16-33.

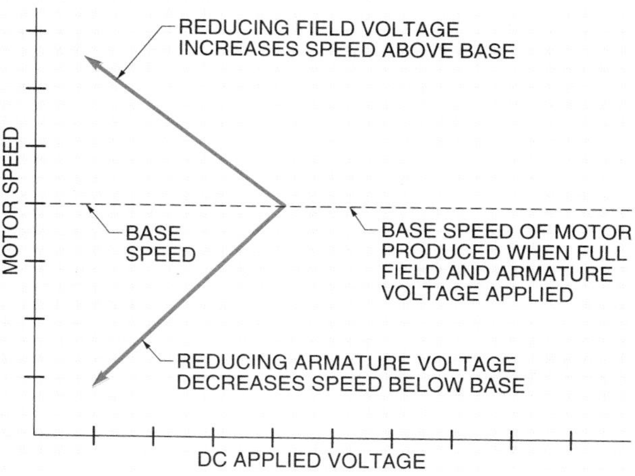

Figure 16-33. The speed of a DC motor is controlled by varying the applied voltage across the armature, the field, or both.

An SCR is used to control the speed of a DC motor. See Figure 16-34. In this circuit, the speed is controlled from zero to the base speed using an SCR.

American Precision Industries, Motion Technologies Group

Solid-state motor controllers are used for precision speed control applications.

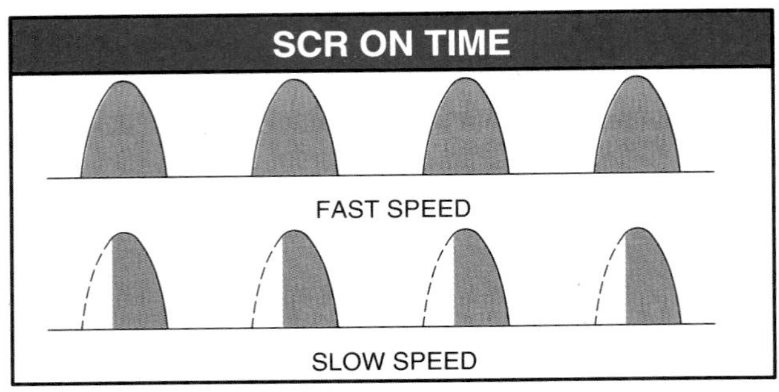

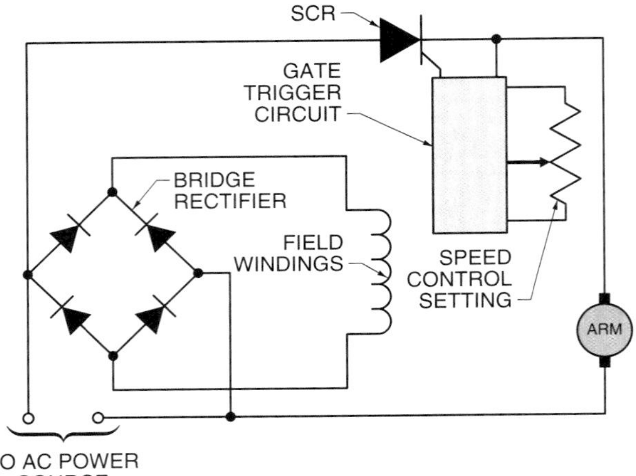

Figure 16-34. An SCR is used to control the speed of a DC motor.

The SCR is controlled by the setting of the gate trigger circuit, which varies the ON time of the SCR per cycle. This varies the amount of average current flow to the armature.

The voltage applied to the SCR is AC because the SCR rectifies (as well as controls) AC voltage. A rectifier circuit is required for the field circuit because the field circuit must be supplied with DC. If speed control above the base is required, the rectifier circuit in the field can also be changed to an SCR control.

AC MOTOR SPEED CONTROL

AC motors are considered constant speed motors. This is because the synchronous speed of an induction motor is based on the supply frequency and the number of poles in the motor winding. Motors designed for 60 Hz use have synchronous speeds of 3600, 1800, 1200, 900, 720, 600, 514, and 450 rpm. To calculate synchronous speed of an induction motor, apply the formula:

$$rpm_{syn} = \frac{120 \times f}{N_p}$$

where

rpm_{syn} = synchronous speed (in rpm)

f = supply frequency (in $^{cycles}/_{sec}$)

N_p = number of motor poles

Example: Calculating Synchronous Speed

What is the synchronous speed of a four-pole motor operating at 50 Hz?

$$rpm_{syn} = \frac{120 \times 50}{4}$$

$$rpm_{syn} = \frac{6000}{4}$$

$$rpm_{syn} = \mathbf{1500\ rpm}$$

All induction motors have a full-load speed somewhat below their synchronous speed. *Percent slip* is the percentage reduction in speed below synchronous speed. Most motors run from 2% to 10% slower than their synchronous speed at no load. As the load increases, the percentage of slip increases.

Supply frequency and number of poles are the only variables that determine the speed of an AC motor. Unlike the speed of a DC motor, the speed of an AC motor should not be changed by varying the applied voltage. Damage may occur to an AC motor if the supply voltage is varied more than 10% above or below the rated nameplate voltage. This is because in an induction motor, the starting torque and breakdown torque vary as the square of the applied voltage. For example, with 90% of rated voltage, the torque is 81% ($.9^2 = .81$ or 81%) of its rated torque.

Voltage drop compensation is required in applications where a motor is located at the end of a long line. The line voltage drop at the motor may be great enough to keep the motor from starting the load or developing the required torque to operate satisfactorily. The two methods of speed control available for AC motors are changing the frequency applied to the motor or using a motor with windings that may be reconnected to form different numbers of poles.

Multispeed AC motors, designed to be operated at constant frequency, are provided with stator windings that can be reconnected to provide a change in the number of poles and thus a change in the speed. These multispeed motors are available in two or more fixed speeds which are determined by the connections made to the motor. Two-speed motors normally have one winding that may be connected to provide two speeds, one of which is half the length of the other. Motors with more than two speeds normally include many windings that are connected and reconnected to provide different speeds by changing the number of poles.

In multispeed motors, the different speeds are determined by connecting the external winding leads to a multispeed starter. Although this starter may be manual, a magnetic motor starter is the most common means used for AC motor speed control. One starter is required for each speed of the motor. Each starter must be interlocked (mechanical, auxiliary contact, or pushbutton) to prevent more than one starter from being ON at the same time. The motor can run at only one speed at a time. For two-speed motors, a standard forward/reverse starter is normally used because it provides mechanical interlocking.

Basic Speed Control Circuits

Several control circuits can be developed to control a multispeed motor depending on the requirements of the circuit. In a two-speed motor control circuit, the motor can be started in the low or high speed. See Figure 16-35. In this circuit, the operator may start the motor from rest at either speed. The stop pushbutton must be pressed before changing from low to high speed or from high to low speed.

GE Motors & Industrial Systems

Pump motors must operate at different speeds to increase or decrease the amount of product flow.

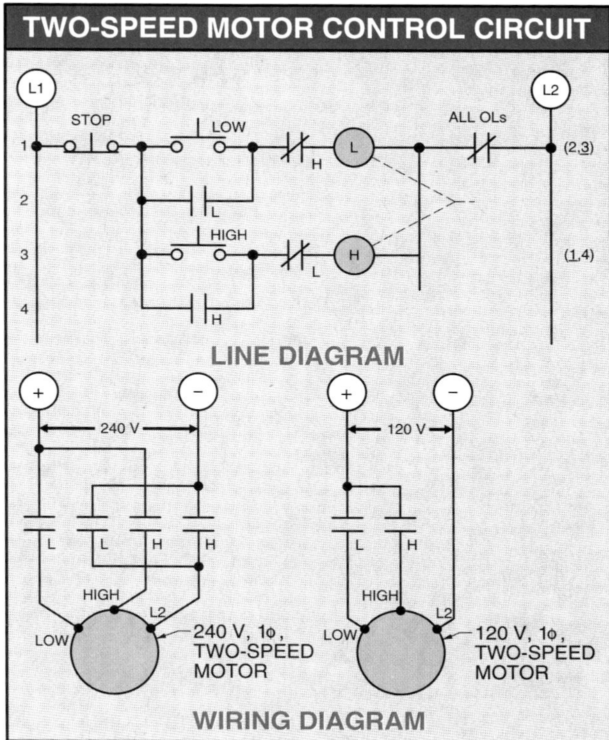

Figure 16-35. In a two-speed motor control circuit, the motor can be started in the low or high speed.

In a modified control circuit, the motor can be changed from low speed to high speed without first stopping the motor. See Figure 16-36. In this circuit, the operator can start the motor from rest at either speed or change from the low speed to the high speed without pressing the stop pushbutton. The stop pushbutton must be pressed before it is possible to change from high to low speed. This high-to-low arrangement prevents excessive line current and shock to the motor. In addition, the machinery, which is driven by the motor, is protected from shock that could result from connecting a motor at high speed to a low speed.

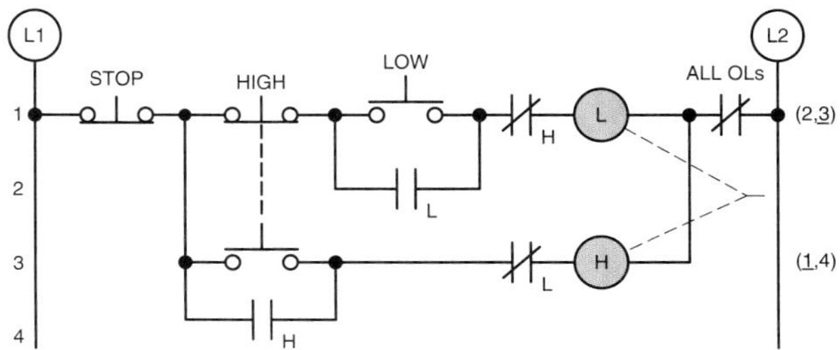

Figure 16-36. In a modified control circuit, the motor can be changed from low speed to high speed without first stopping the motor.

Compelling Circuit Logic. In many speed control applications, the motor must always be started at low speed before it can be changed to high speed. *Compelling circuit logic* is a control function that requires the operator to start and operate a motor in a predetermined order. See Figure 16-37.

This circuit does not allow the operator to start the motor at high speed. The circuit compels the operator to first start the motor at low speed before changing to high speed. This arrangement prevents the motor and driven machinery from starting at high speed. The motor and driven machinery are allowed to accelerate to low speed before accelerating to high speed. The circuit also compels the operator to press the stop pushbutton before changing speed from high to low.

Accelerating Circuit Logic. In many applications, a motor must be automatically accelerated from low to high speed even if the high pushbutton is pressed first. *Accelerating circuit logic* is a control function that permits the operator to select a high motor speed and the control circuit automatically accelerates the motor to that speed. See Figure 16-38.

A circuit with accelerating circuit logic allows the operator to select the desired speed by pressing either the low or high pushbutton. If the operator presses the low pushbutton, the motor starts and runs at low speed. If the operator presses the high pushbutton, the motor starts at low speed and runs at high speed only after the predetermined time set on the timer in the control circuit. This arrangement gives the motor and driven machinery a definite time period to accelerate from low to high speed. The circuit also requires that the operator press the stop pushbutton before changing speed from high to low. In any control circuit, the pushbuttons may be replaced with any control device such as a pressure switch, photoelectric switch, etc. without changing the circuit logic.

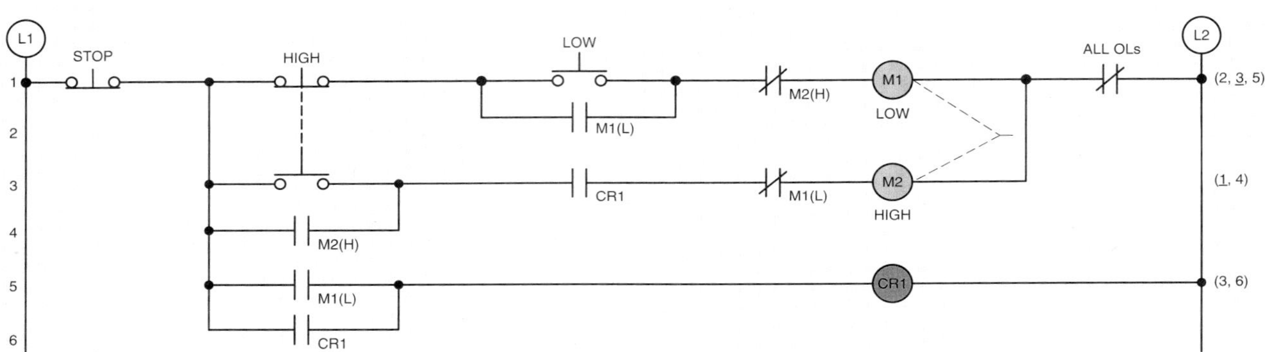

Figure 16-37. Compelling circuit logic is used where a motor must always be started at low speed before it can be changed to high speed.

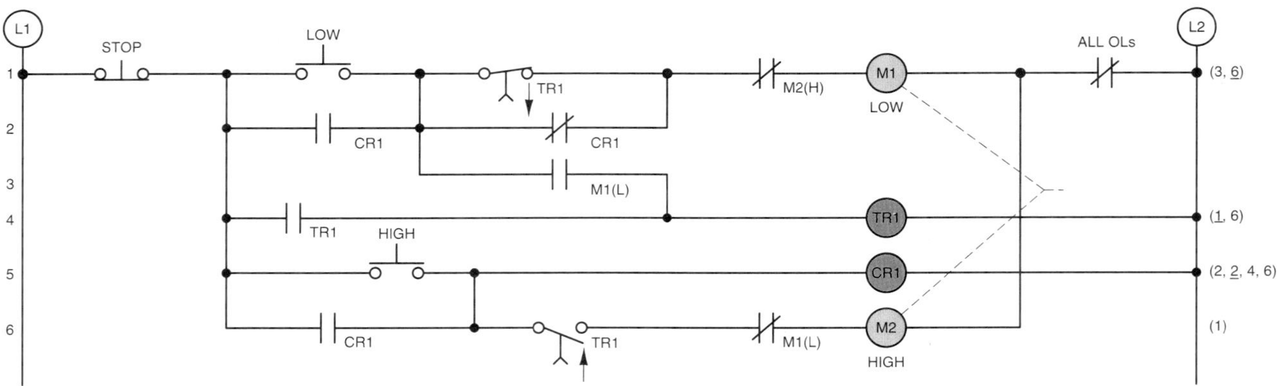

Figure 16-38. In accelerating circuit logic, a motor is automatically accelerated from low speed to high speed even if the high pushbutton is pressed first.

Decelerating Circuit Logic. In some applications, a motor or load cannot take the stress of changing from a high to a low speed without damage. In these applications, the motor must be allowed to decelerate by coasting or braking before being changed to a low speed. *Decelerating circuit logic* is a control function that permits the operator to select a low motor speed and the control circuit automatically decelerates the motor to that speed. See Figure 16-39.

In this circuit, the adjustable speed AC drive rectifies the 3ϕ AC voltage and delivers the power to the inverter circuit. The inverter circuit changes the DC power to an adjustable frequency AC output that controls the speed of the motor. The frequency is controlled by the setting of the rheostat. The rheostat controls the firing rate of the SCRs in the inverter circuit, and thus the frequency of the AC output.

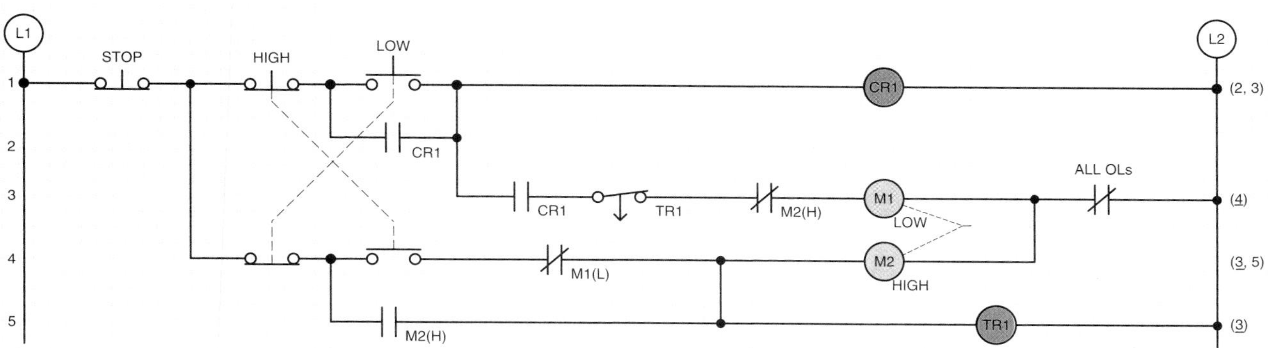

Figure 16-39. In decelerating circuit logic, a motor decelerates before being changed to a low speed.

This circuit allows the operator to select the desired speed by pressing either the low or high pushbutton. If the low pushbutton is pressed, the motor starts and runs at low speed. If the high pushbutton is pressed, the motor starts and runs at high speed. If the operator changes from high to low speed, the motor changes to low speed only after a predetermined time. This gives the motor and driven machinery a time period to decelerate from a high speed to a low speed.

Changing Frequency

Speed control of AC squirrel-cage induction motors can be accomplished if the frequency of the voltage applied to the stator is varied to change the synchronous speed. The change in synchronous speed causes the motor speed to change. The two methods used to vary the frequency of the AC voltage applied to a motor is by using an inverter or converter. An *inverter* is a machine, device, or system that changes DC voltage into AC and can be designed to produce variable frequency. A *converter* is a machine or device that changes AC power to DC power or vice versa, or from one frequency to another. A converter changes the standard 60 Hz AC power into almost any desired frequency. Inverters and converters use solid-state SCRs for control. See Figure 16-40.

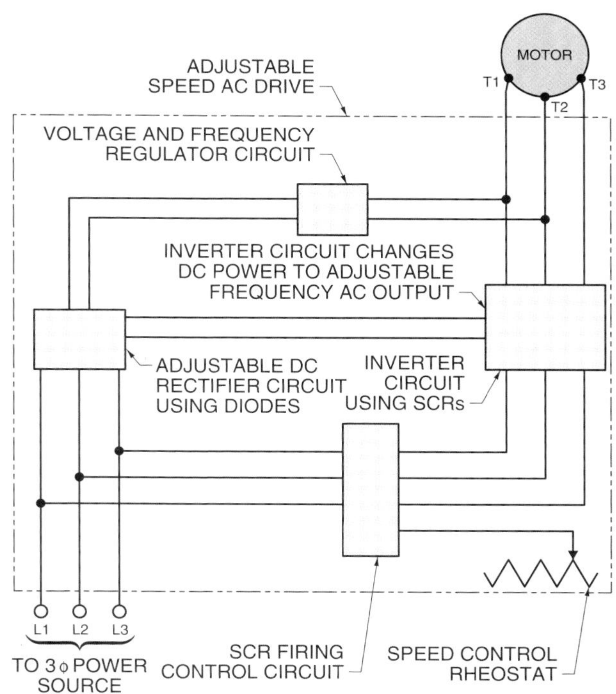

Figure 16-40. An inverter circuit changes DC power to an adjustable frequency AC output that controls the speed of a motor.

Each SCR is fired in sequence, one immediately following the other. The rate of firing and length of time the SCR is allowed to remain ON determines the frequency. Inverters are normally manufactured for the 5 HP to 150 HP range.

In a converter circuit, the frequency of the output voltage is determined by the length of time the SCRs are allowed to remain ON. See Figure 16-41. The operation of a variable frequency converter is basically the same as the inverter. Full-wave converters require a minimum of 36 SCRs with a complicated control circuit. For this reason, their cost is high. However, in applications that require several AC motors to have precise speed synchronization, the cost becomes less important. Converters are normally manufactured up to the 10,000 HP range and are often used to control large groups of motors simultaneously.

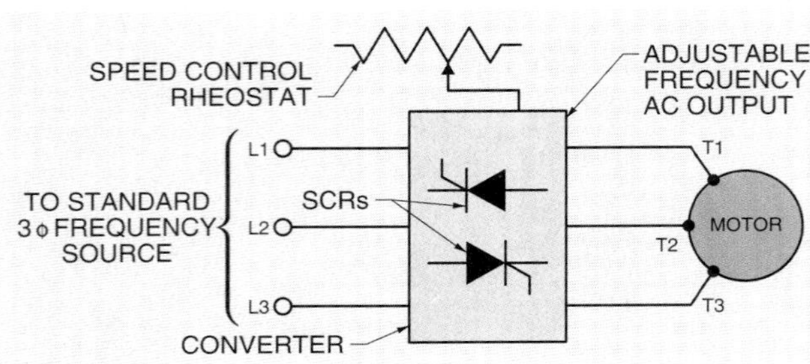

Figure 16-41. In a converter circuit, the frequency of the output voltage is determined by the length of time the SCRs are allowed to remain ON.

Changing Applied Voltage

The speed of standard AC squirrel-cage induction motors is normally varied by changing the number of poles or the applied frequency. Although the speed of an AC motor is determined by the number of poles and applied frequency, it is possible in some applications to control the speed of the load by varying the applied voltage to the motor. This method is not a standard method of speed control and caution must be taken in applying it. By varying the voltage applied to the motor, the torque that the motor can deliver to the load is varied. The torque of a squirrel-cage induction motor varies with the square of the applied voltage.

The greater the torque, the faster the acceleration time. If the torque of a motor is reduced, the speed at which the motor performs the work is also reduced. Although it is possible to reduce the speed of a large motor by reducing the applied voltage, this method could damage the motor.

This damage may come from excess heat buildup in the motor. Most AC motors are not designed to have their voltage varied more than 10% from the nameplate rating. However, manufacturers that know the load requirements and the motor type and size in advance may install this type of speed control. The advantage of less cost for control with a large motor is the determining factor. This type of speed control is limited to applications of soft-start light loads. Fan motors are sometimes controlled this way. Shaded-pole or permanent-magnet motors are normally used in these applications. Except in applications that are specifically designed for this type of speed control, it should not be considered a standard method.

Most variable voltage control circuits use a full-wave triac output to vary the voltage. See Figure 16-42. The triac varies the voltage by adjusting the point on the AC sine wave at which the triac is turned ON.

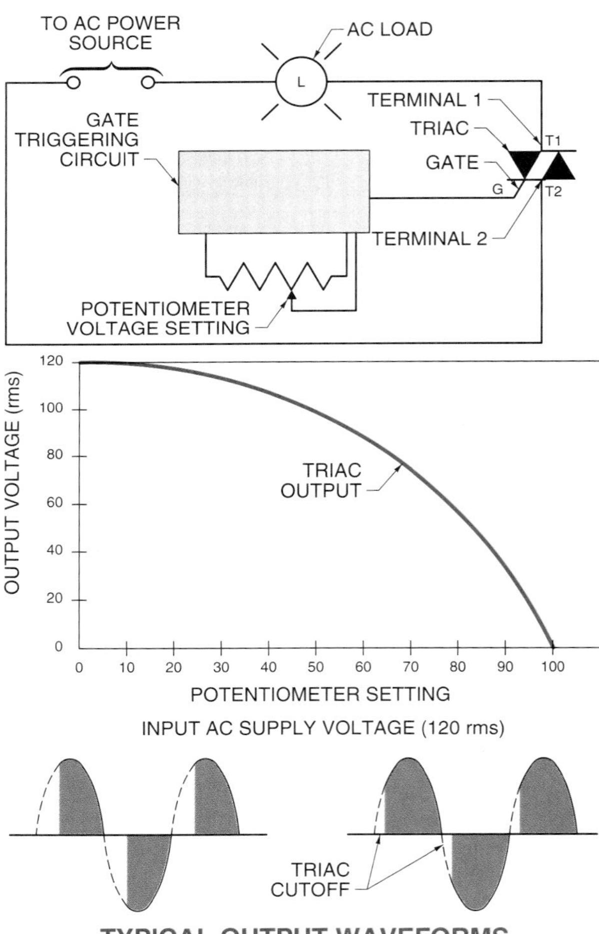

Figure 16-42. Most variable voltage control circuits use a full-wave triac output to vary the voltage.

The triac in this application is similar to the SCR used to control the speed of a DC motor. A triac is two reverse-parallel connected SCRs connected to allow the AC sine wave to pass in both directions at a controlled level. The triac's output is controlled by varying a potentiometer in the gate triggering circuit. This same basic triac circuit is also used to control the heat or light output of heating elements and incandescent lamps.

Motor Drives

A standard squirrel-cage induction motor runs at a constant speed for a given frequency and number of poles. The most common and economical running speed of a squirrel-cage motor is about 1800 rpm. Lower speeds require the addition of poles to reduce the speed. Because there are many applications that require some speed other than 1800 rpm, but not a variable speed control, some means must be provided to match the motor output speed to the lower or higher speed required by the load without changing the running speed of the motor. This is accomplished with belts, chains, or gear drives. These belts, chains, or gear drives are used for smooth speed changes between the motor drive and machine drive.

A pulley can be used to change the output speed of a motor, provided the manufacturer's limits are not exceeded. To determine the pulley size needed, the speed of the motor (drive rpm), the speed of the machine that is driven (driven rpm), and the diameter of the pulley on both the drive motor and driven machine must be considered. See Figure 16-43. To calculate the driven machine pulley diameter, apply the formula:

$$PD_m = \frac{PD_d \times N_d}{N_m}$$

where

PD_m = driven machine pulley diameter (in in.)

PD_d = drive pulley diameter (in in.)

N_d = motor drive speed (in rpm)

N_m = driven machine speed (in rpm)

Note: This formula may be rewritten to solve for any unknown value if the other three values are known.

Example: Calculating Pulley Diameter

What is the required driven machine pulley diameter if a motor running at 1800 rpm has a 6″ pulley, and the driven machine is run at 900 rpm?

$$PD_m = \frac{PD_d \times N_d}{N_m}$$

$$PD_m = \frac{6 \times 1800}{900}$$

$$PD_m = \frac{10{,}800}{900}$$

$$PD_m = \mathbf{12''}$$

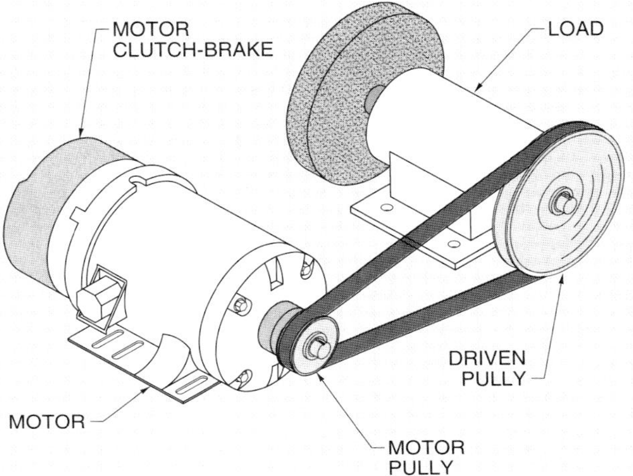

Figure 16-43. The pulley diameter for a driven machine is obtained from the correct motor rpm, driven rpm, and motor pulley diameter.

If the drive motor or driven machine does not have a pulley, a common pulley size can be selected for one or the other and the equation used to solve for the unknown size.

For very low speeds that are required in some applications, a gearmotor or motor connected to a gear drive is used. Gearmotors are designed with output speeds as low as 1 rpm. Gearmotors and geardrives work on the same gear reduction principles of most clocks and watches.

When selecting and using gearmotors, consideration must be given to the application of the motor. This is because mechanical overloads are more likely to cause failure of gears than failure of the motor. Gears are sized for maximum peak loads. Gears cannot take overloads, even for short periods of time. To solve this problem, select gearmotors that are one size larger than the application requires and/or use a motor that is designed to withstand the overload shock conditions.

1. What is the basic operating principle of friction brakes?
2. Why are large brake shoe linings required in friction brakes?
3. What is the formula used to determine braking torque?
4. What is plugging?
5. When do the contacts of a plugging switch change?
6. What are two reasons some motors cannot be plugged?
7. What type of timer is used when a timing relay is used for plugging?
8. What is the basic operating principle of electric braking?
9. Why is interlocking important in electrical braking?
10. What is the basic operating principle of dynamic braking?
11. What is work?
12. What is torque?
13. What is full-load torque?
14. What is locked-rotor torque?
15. What is pull-up torque?
16. What is breakdown torque?
17. What does a NEMA design letter indicate?
18. What is the relationship between speed and torque in a DC series motor?
19. What is the relationship between speed and torque in a DC shunt motor?
20. What is the relationship between speed and torque in a DC compound motor?
21. How is the speed of a DC motor changed?
22. What determines the speed of an AC motor?
23. What is the synchronous speed of an AC motor with eight poles?
24. What is compelling circuit logic?
25. What is accelerating circuit logic?
26. What is decelerating circuit logic?

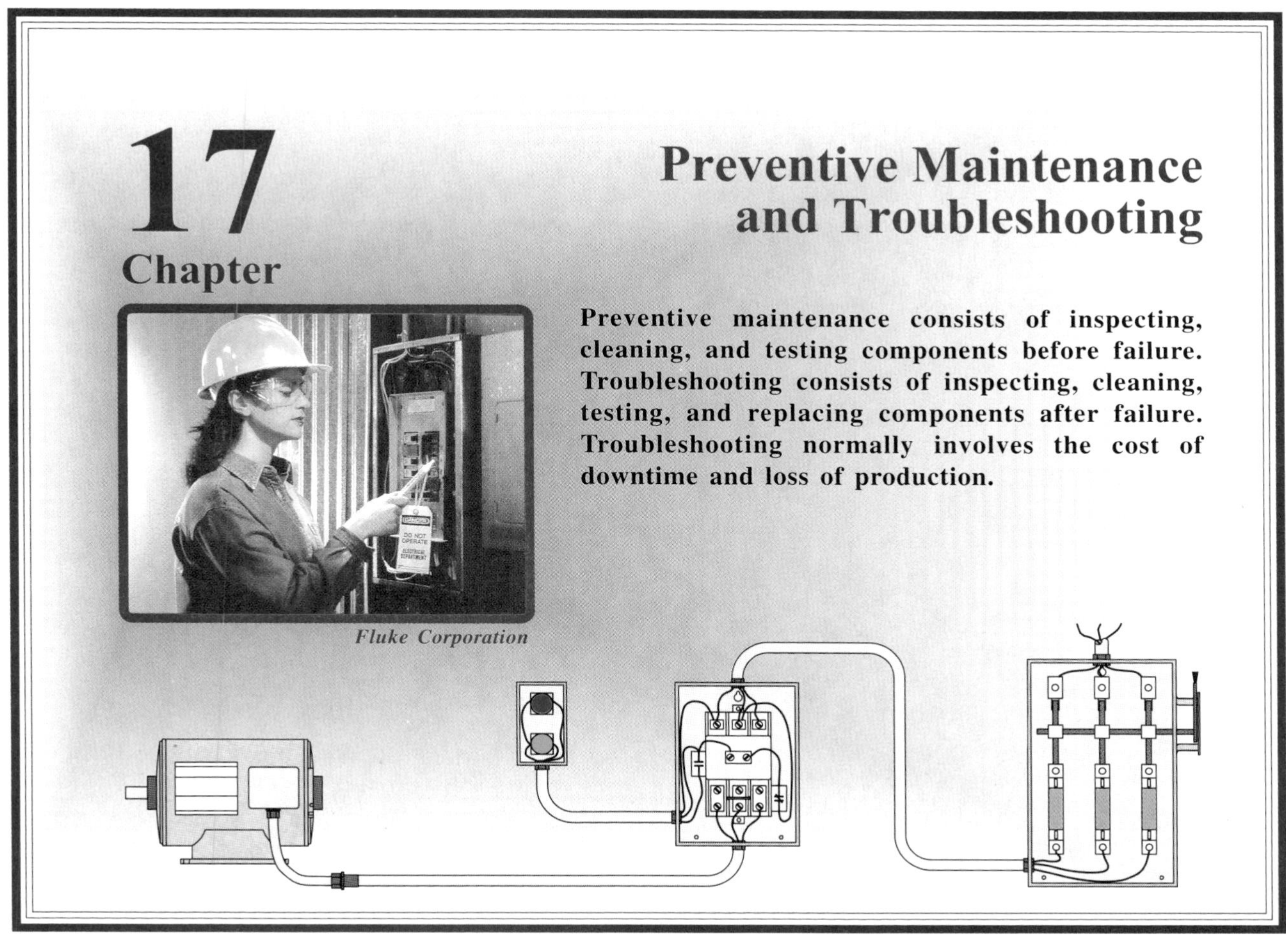

Chapter 17 Preventive Maintenance and Troubleshooting

Fluke Corporation

Preventive maintenance consists of inspecting, cleaning, and testing components before failure. Troubleshooting consists of inspecting, cleaning, testing, and replacing components after failure. Troubleshooting normally involves the cost of downtime and loss of production.

PREVENTIVE MAINTENANCE

Today's industrial plants and assembly lines turn out products faster and more economically than in any time in the past. Shutdowns of even short periods of time are extremely costly. Lower productivity and less profit occurs without preventive maintenance.

Preventive maintenance is maintenance performed to keep machines, assembly lines, production operations, and plant operations running with little or no downtime. In the past, the job of the maintenance department was almost always to repair broken equipment and install new equipment.

Today, preventive maintenance does not take the time and personnel it once did. This is due to the introduction of a number of different, inexpensive monitors that can detect and react to almost any problem before it becomes a major problem.

For example, inexpensive monitors are available that can monitor an electrical system for voltage or phase unbalances, voltage losses, phase reversals, over- or undervoltages, currents, temperatures, loss of a pump's prime, and other conditions that may be signs of major problems.

These monitors are easy to install and operate. They can be installed on new or old equipment and are designed to signal the maintenance department of trouble and take preventive measures until the maintenance personnel arrive. The monitors take measurements 24 hours a day.

A preventive maintenance program includes inspection, cleaning, tightening, adjusting and lubricating, keeping equipment dry, and electronically monitoring power circuits. The purpose of a preventive maintenance program is to:

- Maintain equipment in such condition as to ensure uninterrupted operations for as long as possible.
- Maintain the equipment in such condition that it always operates at the highest possible efficiency.

- Protect the equipment from dirt, dust, moisture, corrosion, and electrical and mechanical overloads.
- Maintain good records of all maintenance work to establish maintenance needs and priorities.

Inspection

Inspection of all equipment normally uncovers evidence of a problem before it causes downtime. In most cases, time can be saved if problems are corrected before they lead to major breakdowns. Inspection consists of observation for signs of overheating, dirt, loose parts, noise, and any other signs of abnormalities.

Cleaning

Keeping equipment clean helps eliminate overheating, high-voltage leakage, and breakdowns. Most equipment can be cleaned by blowing the dust and dirt away with a low-pressure, dry air stream. Care must be taken to remove the power if possible before cleaning. Cleaning should always be done on any equipment that is serviced for any reason.

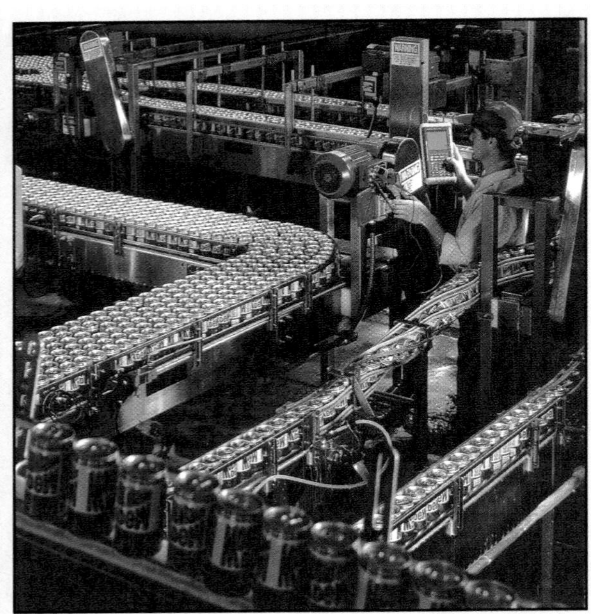

Fluke Corporation

Fluke scopemeters are used during preventive maintenance and troubleshooting to check for phase unbalance, voltage unbalance, single-phasing, improper phase sequence, voltage variations, and frequency variations.

Tightening

Vibration results in loose connections that eventually cause problems. All connections should be tightened firmly, but not beyond the pressure for which the connection is intended. Always use the correct tools of the proper size.

Adjusting and Lubricating

Routine maintenance such as adjusting and lubricating equipment is part of a good preventive maintenance program. Lubrication of bearings in motors and other rotating equipment helps to eliminate wear and heat. Adjustments in equipment that has been in operation for some time ensures that the equipment operates properly. Always follow the manufacturer's recommendations when making adjustments and adding lubrication.

Keeping Equipment Dry

Electrical equipment operates best in a dry atmosphere. Moisture on copper and other metal surfaces used in electrical equipment can cause corrosion and rust which lead to higher resistance and heating. The moisture often causes leakage currents or short circuits in the equipment. Always use the correct enclosure for the application.

Electronically Monitoring Power Circuits

Prevention of major motor failure and downtime can be accomplished by detecting problems before they can cause any damage. Problems such as phase unbalance, voltage unbalance, single-phasing, improper phase sequence, voltage surges, frequency variations, overcycling, improper ventilation, and improper motor mounting can all cause motor failure.

Electronic power monitors are available and can be installed easily to monitor phase unbalance, voltage unbalance, single-phasing, improper phase sequence, voltage surges, voltage variations, and frequency variations. Surge protectors can be used to protect against voltage surges. Properly designed control circuits can prevent overcycling and proper motor installation and maintenance can prevent improper ventilation and mounting problems.

Phase Unbalance

Phase unbalance is the unbalance that occurs when power lines are out-of-phase. Phase unbalance of a 3ϕ power system occurs when 1ϕ loads are applied, causing one or two of the lines to carry more or less of the load. The loads of 3ϕ power systems are balanced by electricians during installation. An unbalance begins to occur as additional 1ϕ loads are added to the system. This unbalance causes the 3ϕ lines to move out-of-phase so the lines are no longer 120 electrical degrees apart. See Figure 17-1.

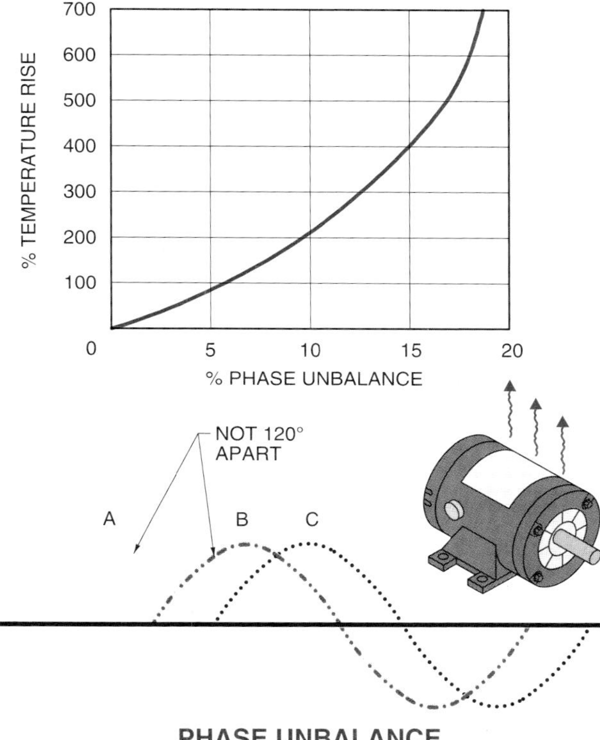

Figure 17-1. Phase unbalance is the unbalance that occurs when power lines are out-of-phase.

Phase unbalance causes 3ϕ motors to run at temperatures higher than their listed ratings. The greater the phase unbalance, the greater the temperature rise. High temperatures produce insulation breakdown and other related problems.

A 3ϕ motor operating in an unbalanced circuit cannot deliver its rated horsepower. For example, a phase unbalance of 3% causes a motor to work at 90% of its rated power. This requires the motor to be derated. See Figure 17-2.

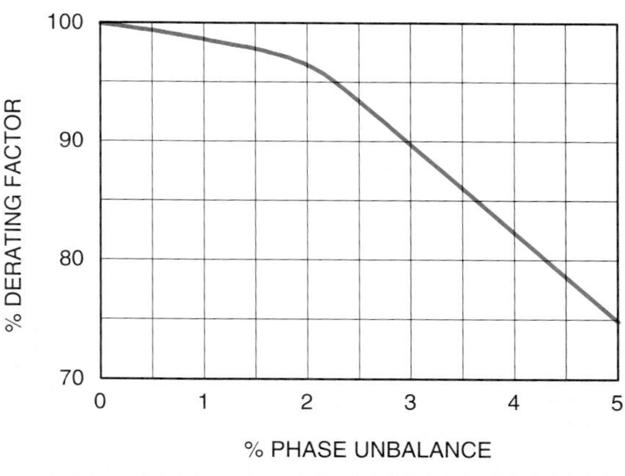

Figure 17-2. Motors operating on a circuit which has phase unbalance must be derated.

Voltage Unbalance

Voltage unbalance is the unbalance that occurs when the voltages at different motor terminals are not equal. One winding overheats, causing thermal deterioration of the winding if the voltage is not balanced. Voltage unbalance results in a current unbalance. Line voltage should be checked for voltage unbalance periodically and during all service calls. Whenever more than 2% voltage unbalance is measured, take the following steps:

- Check the surrounding power system for excessive loads connected to one line.
- Adjust the load or motor rating by reducing the load on the motor or oversizing the motor if the voltage unbalance cannot be corrected.
- Notify the power company.

To find voltage unbalance, apply the procedure:

1. Measure the voltage between each incoming power line. The readings are taken from L1 to L2, L1 to L3, and L2 to L3.
2. Add the voltages.
3. Divide by 3 to find the voltage average.
4. Subtract the voltage average from the voltage with the largest deviation to find the voltage deviation.

5. To find voltage unbalance, apply the formula:

$$V_u = \frac{V_d}{V_a} \times 100$$

where

V_u = voltage unbalance (%)

V_d = voltage deviation (in V)

V_a = voltage average (in V)

100 = constant

Example: Calculating Voltage Unbalance

Calculate the voltage unbalance of a feeder system with the following voltage readings: L1 to L2 = 442 V; L1 to L3 = 474 V; L2 to L3 = 456 V. See Figure 17-3.

1. Measure incoming voltage. Incoming voltage is 442 V, 474 V, and 456 V.
2. Add voltages. 442 V + 474 V + 456 V = 1372 V
3. Find voltage average.

$$V_a = \frac{V}{3}$$

$$V_a = \frac{1372}{3}$$

$$V_a = \mathbf{457\ V}$$

4. Find voltage deviation.

$$V_d = V - V_a$$

$$V_d = 474 - 457$$

$$V_d = \mathbf{17\ V}$$

5. Find voltage unbalance.

$$V_u = \frac{V_d}{V_a} \times 100$$

$$V_u = \frac{17}{457} \times 100$$

$$V_u = .0372 \times 100$$

$$V_u = \mathbf{3.72\%}$$

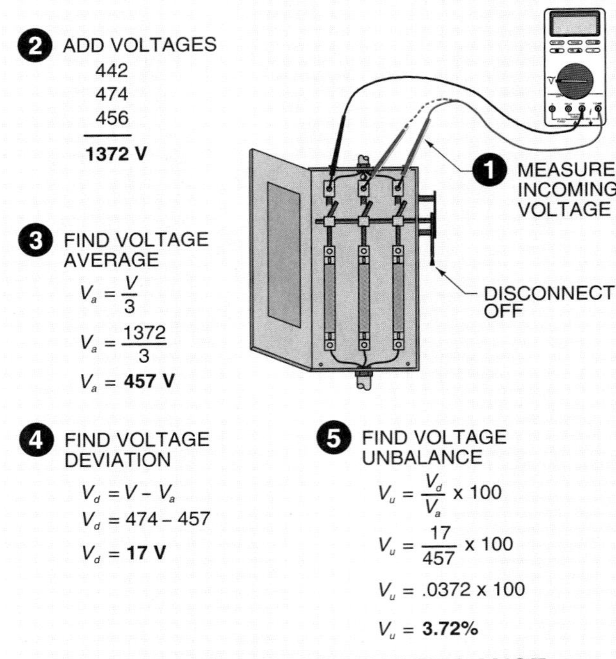

Figure 17-3. Voltage unbalance is the unbalance that occurs when the voltages at different motor terminals are not equal.

A troubleshooter can observe the blackening of one or two of the stator windings which occurs when a motor has failed due to voltage unbalance. The winding with the largest voltage unbalance is the darkest. See Figure 17-4.

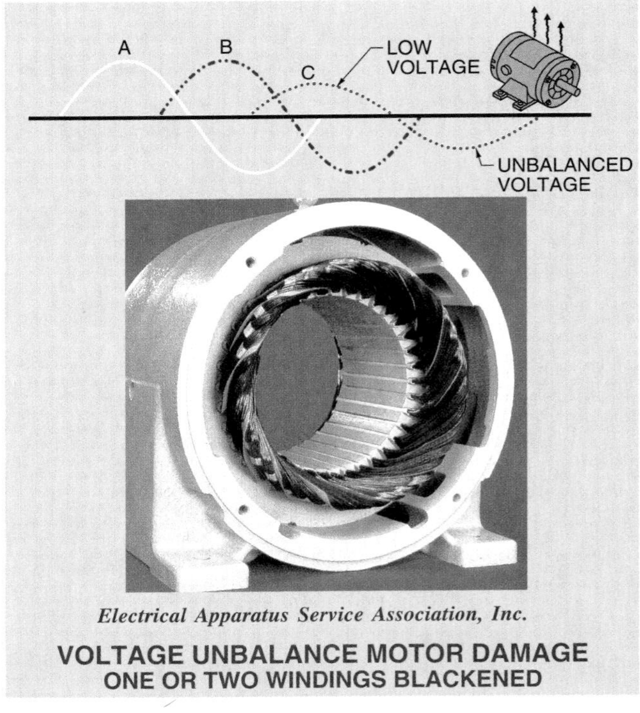

Figure 17-4. Voltage unbalance causes blackening of one or two of the stator windings.

Single-Phasing

Single-phasing is the operation of a motor that is designed to operate on three phases, but is only operating on two phases because one phase is lost. Single-phasing occurs when one of the 3ϕ lines leading to a 3ϕ motor does not deliver voltage to the motor. Single-phasing is the maximum condition of voltage unbalance.

Single-phasing occurs when one phase opens on either the primary or secondary power distribution system. This occurs when one fuse blows, when there is a mechanical failure within the switching equipment, or when lightning takes out one of the lines.

Single-phasing can go undetected on most systems because a 3ϕ motor running on 2ϕ continues to run in most applications. The motor usually runs until it burns out. When single-phasing, the motor draws all its current from two lines.

Measuring the voltage at a motor does not normally detect a single-phasing condition. The open winding in the motor generates a voltage almost equal to the phase voltage that is lost. In this case, the open winding acts as the secondary of a transformer, and the two windings connected to power act as the primary.

Single-phasing is reduced by using the proper size dual-element fuse and by using the correct heater sizes. In motor circuits, or other types of circuits in which a single-phasing condition cannot be allowed to exist for even a short period of time, an electronic phase-loss monitor is used to detect phase loss. The monitor activates a set of contacts to drop out the starter coil when a phase loss is detected.

Fluke Corporation

Fluke multimeters are used to check for single-phasing because a 3ϕ motor running on 2ϕ may continue to run with no signs of trouble and burn out the motor.

A troubleshooter can observe the severe blackening of one delta winding or two wye windings of the three 3ϕ windings which occurs when a motor has failed due to single phasing. The coil(s) that experienced the voltage loss indicate obvious and fast damage, which includes the blowing out of the insulation on the one (or two) windings. See Figure 17-5.

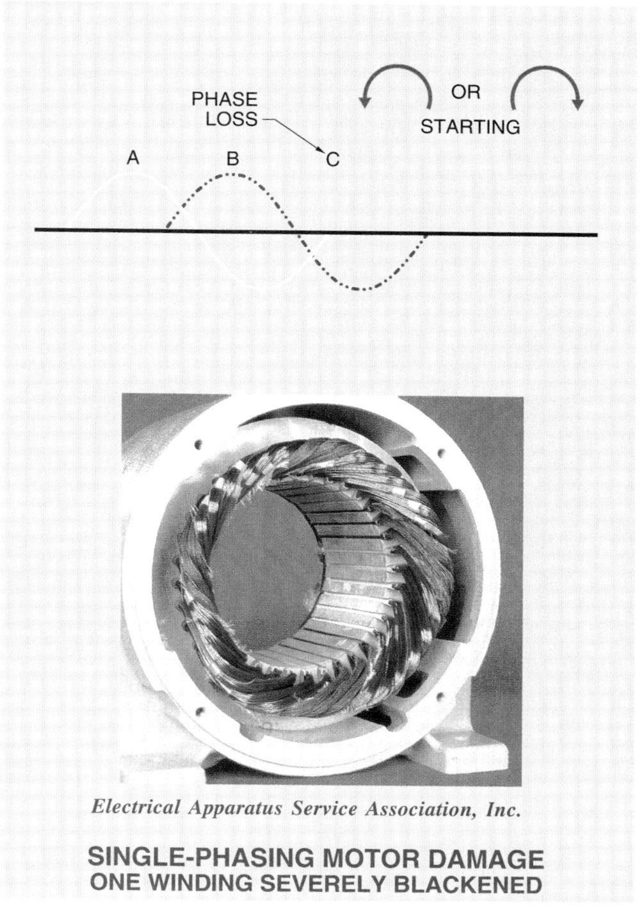

Figure 17-5. Single-phasing causes severe burning and distortion to one phase coil.

Single-phasing is distinguished from voltage unbalance by the severity of the damage. Voltage unbalance causes less blackening (but normally over more coils) and little or no distortion. Single-phasing causes burning and distortion to one phase coil.

Improper Phase Sequence

Improper phase sequence is the changing of the sequence of any two phases (phase reversal) in a 3ϕ motor control circuit. Improper phase sequence reverses the motor rotation. Reversing motor rotation can damage driven machinery or injure personnel.

Phase reversal can occur when modifications are made to a power distribution system or when maintenance is performed on electrical conductors or switching equipment. The NEC® requires phase reversal protection on all personnel transportation equipment such as moving walkways, escalators, and ski lifts. See Figure 17-6.

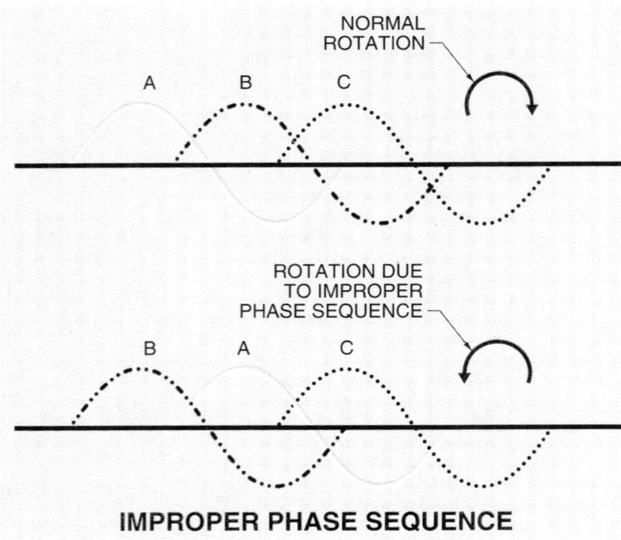

Figure 17-6. Improper phase sequence is the changing of the sequence of any two phases (phase reversal) in a 3ϕ motor control circuit.

Voltage Surges

A *voltage surge* is any higher-than-normal voltage that temporarily exists on one or more of the power lines. Lightning is a major cause of large voltage surges. A lightning surge on a power line comes from a direct hit or induced voltage. The lightning energy moves in both directions on the power lines, much like a rapidly moving wave.

This traveling surge causes a large voltage rise in a short period of time. The large voltage is impressed on the first few turns of the motor windings, destroying the insulation and burning out the motor.

A troubleshooter can observe the burning and opening of the first few turns of the windings which occur when a motor has failed due to a voltage surge. The rest of the windings appear normal, with little or no damage. See Figure 17-7.

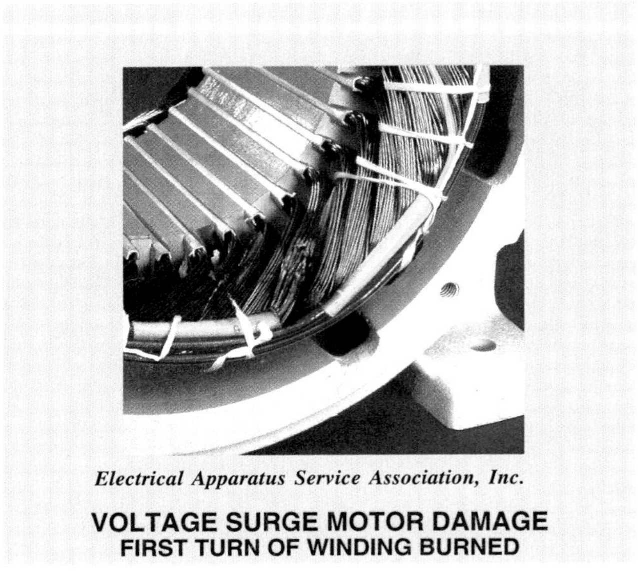

Electrical Apparatus Service Association, Inc.
VOLTAGE SURGE MOTOR DAMAGE
FIRST TURN OF WINDING BURNED

Figure 17-7. A voltage surge causes burning and opening of the first few turns of the windings.

Lightning arresters with the proper voltage rating and connection to an excellent ground assure maximum voltage surge protection. Surge protectors are also available. Surge protectors are placed on the equipment or throughout the distribution system.

Voltage surges can also occur from normal switching of high-rated power circuits. Voltage surges occurring from switching high-rated power circuits are a lesser magnitude than lightning strikes and normally do not cause any motor problems. A surge protector should be used on computer equipment circuits to protect sensitive electronic components.

AC Voltage Variations

Motors are rated for operation at specific voltages. Motor performance is affected when the supply voltage varies from a motor's rated voltage. A motor operates satisfactorily with a voltage variation of ±10% from the voltage rating listed on the motor nameplate. See Figure 17-8.

VOLTAGE VARIATION CHARACTERISTICS		
Performance Characteristics	**10% above Rated Voltage**	**10% below Rated Voltage**
Starting current	+10% to +12%	−10% to −12%
Full-load current	−7%	+11%
Motor torque	+20% to +25%	−20% to −25%
Motor efficiency	Little change	Little change
Speed	+1%	−1.5%
Temperature rise	−3°C to −4°C	+6°C to +7°C

Figure 17-8. A motor operates satisfactorily with a voltage variation of ±10% from the voltage rating listed on the motor nameplate.

AC Frequency Variations

Motors are rated for operation at specific frequencies. Motor performance is affected when the frequency varies from a motor's rated frequency. A motor operates satisfactorily with a frequency variation of ±5% from the frequency rating listed on the motor nameplate. See Figure 17-9.

FREQUENCY VARIATION CHARACTERISTICS		
Performance Characteristics	**5% above Rated Frequency**	**5% below Rated Frequency**
Starting current	−5% to −6%	+5% to +6%
Full-load current	−1%	+1%
Motor torque	−10%	+11%
Motor efficiency	Slight increase	Slight decrease
Speed	+5%	−5%
Temperature rise	Slight decrease	Slight increase

Figure 17-9. A motor operates satisfactorily with a frequency variation of ±5% from the frequency rating listed on the motor nameplate.

DC Voltage Variations

DC motors should be operated on pure DC power. *Pure DC power* is power obtained from a battery or DC generator. DC power is also obtained from rectified AC power. Most industrial DC motors obtain power from a rectified AC power supply. DC power obtained from a rectified AC power supply varies from almost pure DC power to half-wave DC power.

Half-wave rectified power is obtained by placing a diode in one of the AC power lines. Full-wave rectified power is obtained by placing a bridge rectifier (four diodes) in the AC power line. Rectified DC power is filtered by connecting a capacitor in parallel with the output of the rectifier circuit. See Figure 17-10.

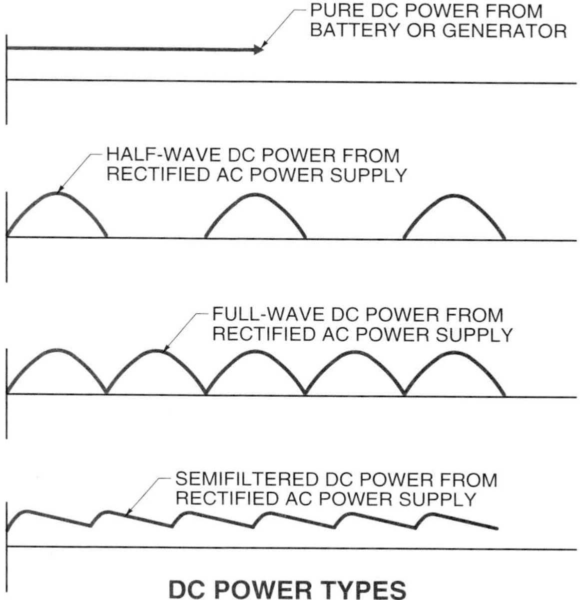

Figure 17-10. DC power obtained from a rectified AC power supply varies from almost pure DC power to half-wave DC power.

DC motor operation is affected by a change in voltage. The change may be intentional as in a speed-control application, or the change may be caused by variations in the power supply.

The power supply voltage normally should not vary by more than 10% of a motor's rated voltage. Motor speed, current, torque, and temperature are affected if the DC voltage varies from the motor rating. See Figure 17-11.

Allowable Motor Starting Time

A motor must accelerate to its rated speed within a limited time period. The longer a motor takes to accelerate, the higher the temperature rise in the motor. The larger the load, the longer the acceleration time. The maximum recommended acceleration time depends on the motor's frame size. Large motor frames dissipate heat faster than small motor frames. See Figure 17-12.

DC MOTOR PERFORMANCE CHARACTERISTICS				
Performance Characteristics	Voltage 10% below Rated Voltage		Voltage 10% above Rated Voltage	
	Shunt	Compound	Shunt	Compound
Starting torque	–15%	–15%	+15%	+15%
Speed	–5%	–6%	+5%	+6%
Current	+12%	+12%	–8%	–8%
Field temperature	Decreases	Decreases	Increases	Increases
Armature temperature	Increases	Increases	Decreases	Decreases
Commutator temperature	Increases	Increases	Decreases	Decreases

Figure 17-11. Motor speed, current, torque, and temperature are affected if the DC voltage varies from the motor rating.

MAXIMUM ACCELERATION TIME	
Frame Number	Maximum Acceleration Time (in sec)
48 and 56	8
143 – 286	10
324 – 326	12
364 – 505	15

Figure 17-12. A motor must accelerate to its rated speed within a limited time period.

Overcycling

Overcycling is the process of turning a motor ON and OFF repeatedly. Motor starting current is usually 6 to 8 times the full-load running current of the motor. Most motors are not designed to start more than 10 times per hour. Overcycling occurs when a motor is at its operating temperature and still cycles ON and OFF. This further increases the temperature of the motor, destroying the motor insulation. See Figure 17-13.

Totally enclosed motors withstand overcycling better than open motors because they are designed to dissipate heat faster without damaging the motor. When a motor application requires a motor to be cycled often, take the following steps:

- Use a motor with a 50°C rise instead of the standard 40°C.
- Use a motor with a 1.25 or 1.35 service factor instead of a 1.00 or 1.15 service factor.
- Provide additional cooling by forcing air over the motor.

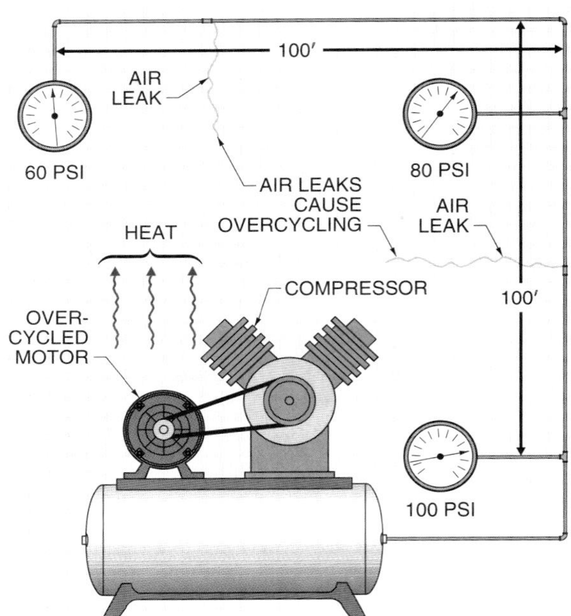

Figure 17-13. Overcycling is the process of turning a motor ON and OFF repeatedly, increasing the temperature of the motor, and destroying the motor insulation.

Heat Problems

Excessive heat is a major cause of motor failure and other motor problems. Heat destroys motor insulation which shorts the windings. The motor is not functional when motor insulation is destroyed.

The life of the insulation is shortened as the heat in a motor increases beyond the temperature rating of the insulation. The higher the temperature, the sooner the insulation fails. The temperature rating of motor insulation is listed as the insulation class. See Figure 17-14.

MOTOR INSULATION CLASS		
Class	**°C**	**°F**
A	105	221
B	130	266
F	155	311
H	180	356

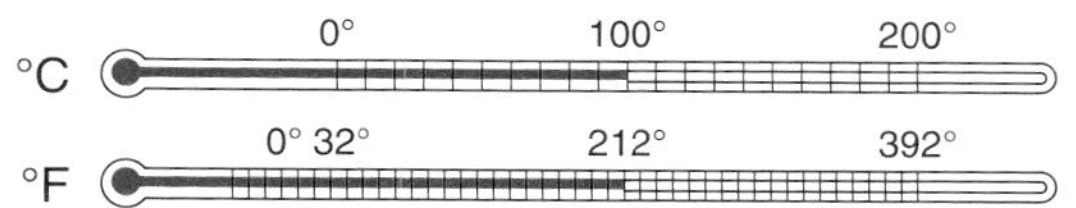

Figure 17-14. The temperature rating of motor insulation is listed as the insulation class.

The insulation class is given in Celsius (°C) and/or Fahrenheit (°F). A motor nameplate normally lists the insulation class of the motor. Heat buildup in a motor can be caused by the following conditions:

- Incorrect motor type or size for the application
- Improper cooling, normally from dirt buildup
- Excessive load, normally from improper use
- Excessive friction, normally from misalignment or vibration
- Electrical problems, normally voltage unbalance, phase loss, or a voltage surge

Improper Ventilation. All motors produce heat as they convert electrical energy to mechanical energy. This heat must be removed to prevent destruction of motor insulation. Motors are designed with air passages that permit a free flow of air over and through the motor. This airflow removes the heat from the motor. Anything that restricts airflow through a motor causes the motor to operate at a higher temperature than for which it is designed.

Airflow through a motor may be restricted by the accumulation of dirt, dust, lint, grass, pests, rust, etc. Airflow is restricted much faster if a motor becomes coated with oil from leaking seals or from overlubrication. See Figure 17-15.

Overheating can also occur if a motor is placed in an enclosed area. A motor overheats due to the recirculation of heated air when a motor is installed in a location that does not permit the heated air to escape. Vents added at the top and bottom of the enclosed area allow a natural flow of heated air.

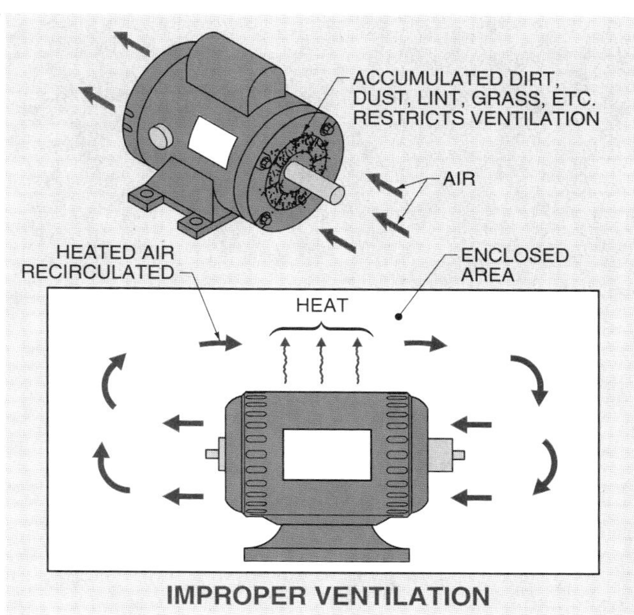

Figure 17-15. All motors produce heat which must be removed to prevent destruction of motor insulation.

Overloads. An *overload* is the application of excessive load to a motor. Motors attempt to drive the connected load when the power is ON. The larger the load, the more power required. All motors have a limit to the load they can drive. For example, a 5 HP, 460 V, 3ϕ motor should draw no more than 7.6 A. See NEC® Table 430-150.

Fluke Corporation

Fluke's Model 33 true rms clamp meter is used to check circuits for overloads without opening the circuit.

Overloads should not harm a properly-protected motor. Any overload present longer than the built-in time delay of the protection device is detected and removed. Properly-sized heaters in the motor starter assure that an overload is removed before any damage is done. See Figure 17-16.

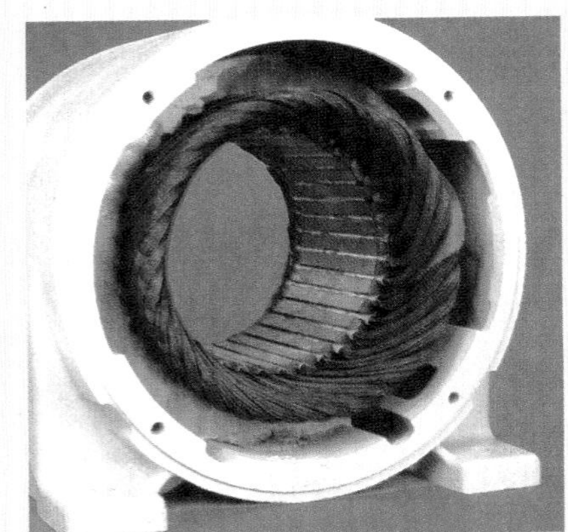

Electrical Apparatus Service Association, Inc.

MOTOR OVERLOADING
ALL WINDINGS EVENLY BLACKENED

Figure 17-16. Overloading causes an even blackening of all motor windings.

A troubleshooter can observe the even blackening of all motor windings, which occurs when a motor has failed due to overloading. The even blackening is caused by the motor's slow destruction over a long period of time. No obvious damage or isolated areas of damage to the insulation are visible.

Current readings are taken at a motor to determine an overload problem. A motor is working to its maximum if it is drawing rated current. A motor is overloaded if it is drawing more than rated current. The motor size may be increased or the load on the motor decreased if overloads are a problem. See Figure 17-17.

Altitude Correction

Temperature rise of motors is based on motor operation at altitudes of 3300′ or less. A motor with a service factor of 1.0 is derated when it operates at altitudes above 3300′. A motor with a service factor above 1.0 is derated based on the altitude and service factor. See Figure 17-18.

RATED CURRENT OF MOTOR	METER READING		
	MOTOR UNDERLOADED	MOTOR FULLY LOADED	MOTOR OVERLOADED
20 A	12 A	20 A	22 A
NAMEPLATE LISTED VALUE	0 TO 95% OF LISTED VALUE	95 TO 105% OF LISTED VALUE	105% + OF LISTED VALUE

20 A

TO POWER SOURCE

MOTOR CURRENT READINGS

Figure 17-17. Current readings are taken at a motor to determine an overload problem.

MOTOR ALTITUDE DERATINGS				
Altitude Range (in ft)	**Service Factor**			
	1.0	**1.15**	**1.25**	**1.35**
3300 – 9000	93%	100%	100%	100%
9000 – 9900	91%	98%	100%	100%
9900 – 13,200	86%	92%	98%	100%
13,200 – 16,500	79%	85%	91%	94%
Over 16,500	Consult manufacturer			

Figure 17-18. A motor with a service factor of 1.0 is derated when it operates at altitudes above 3300′.

Motor Mounting and Positioning

Motors that are not mounted properly are more likely to fail from mechanical problems. A motor must be mounted on a flat, stable base. This helps reduce vibration and misalignment problems. An adjustable motor base aids in proper mounting and alignment.

To ensure a long life span, a motor should be mounted so that it is kept as clean as possible. To reduce the chance of damaging material reaching a motor, a belt cleaner should be used in any application in which the belts are likely to bring damaging material to the motor.

Adjustable Motor Bases

An adjustable motor base makes the installation, tensioning, maintenance, and replacement of belts easier. An *adjustable motor base* is a mounting base that allows a motor to be easily moved over a short distance. See Figure 17-19. An adjustable motor base simplifies the installation of the motor and the tightening of belts and chains.

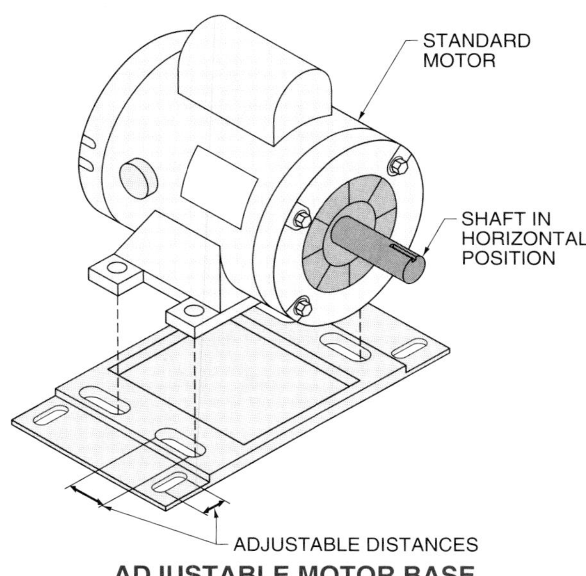

Figure 17-19. An adjustable motor base is a mounting base that allows a motor to be easily moved over a short distance.

Mounting Direction

The position of the driven machine normally determines whether a motor is installed horizontally or vertically. Standard motors are designed to be mounted with the shaft horizontal. The horizontal position is the best operating position for the motor bearings. A specially-designed motor is used for vertical mounting. Motors designed to operate vertically are more expensive and require more preventive maintenance.

Motor Belt Tension and Mounting

Belt drives provide a quiet, compact, and durable form of power transmission and are widely used in industrial applications. A belt must be tight enough not to slip, but not so tight as to overload the motor bearings.

Belt tension is normally checked by placing a straightedge from pulley to pulley and measuring the amount of deflection at the midpoint, or by using a tension tester. As a rule of thumb, belt deflection should equal $\frac{1}{64}''$ per inch of span. For example, if the span between the center of a drive pulley and the center of a driven pulley is 16″, the belt deflection is $\frac{1}{4}''$ ($16 \times \frac{1}{64} = \frac{1}{4}''$). Belt tension is normally adjusted by moving the drive component away from or closer to the driven component. See Figure 17-20.

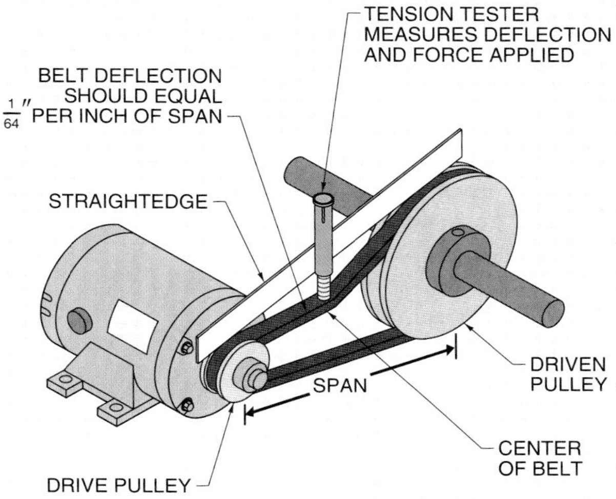

Figure 17-20. A belt must be tight enough not to slip, but not so tight as to overload the motor bearings.

TROUBLESHOOTING

Troubleshooting is the systematic elimination of the various parts of a system or process to locate a malfunctioning part. In most cases, troubleshooting is straight forward and simple. This is because in most cases only one problem exists. Proper tools and test equipment are essential to help troubleshoot problems quickly. The basic rules followed when using test instruments include:

- Always read the manufacturer's instructions. Save the instructions in a file or safe place for future reference.
- Always start with the highest scale available on a test instrument to prevent overloading due to unknown values.
- Always remove the component to be tested or disconnect the line voltage from the circuit before making any resistance measurements.

- Never try to use a test instrument beyond its rated capacity.
- Always close clamp-on instrument jaws tightly. All clamp-on instruments are designed to be clamped around one conductor. Clamping around two conductors neutralizes the fields and no reading can be taken.
- All leads must be insulated.
- All connections must be tight for accurate readings.
- Always check to ensure that any instrument fuses or batteries are in working condition. A new battery is needed if the needle on an ohmmeter cannot be zeroed on the ohm scale.
- The needle on a clamp-on instrument should read in the upper half of the scale for greatest accuracy.
- Always apply basic rules of electrical theory when testing any circuit.

Testing Control Transformers

A control transformer is used in a circuit to step down the supply voltage to provide a safe voltage level for the control circuit. The control transformer should be checked if there is a problem in the control circuit that may be related to the power supply. See Figure 17-21.

All transformers are capable of delivering a limited current output at a given voltage. The power limit of a transformer equals the current times the voltage. This power limit is listed on the nameplate of the transformer as kVA rating. This rating indicates the apparent power the transformer can deliver. The transformer overheats and the control circuit does not function properly if this limit is exceeded. The transformer comes closer to reaching its limit when loads are added.

To test a control transformer, apply the procedure:

1. Check the input and output voltages of the transformer with the power supply energized. The input and output voltages should be within 5% of the transformer's nameplate rating (10% max). The transformer is good if the voltage is within the rating or proportionally low.
2. Measure the current drawn by the transformer with an ammeter. The apparent power drawn by the control circuit is determined by multiplying the current reading by the voltage reading. A larger transformer is required if the VA drawn is greater than the rating of the transformer.
3. Check the transformer ground. A ground test should be performed on new transformer installations or if a ground problem is suspected. Connect one lead of the test circuit to the transformer. Do not connect to a painted or varnished surface. Connect the second lead of the test circuit to each lead of the transformer on both the primary and secondary. The transformer is defective and must be replaced if the light bulb lights on any lead.

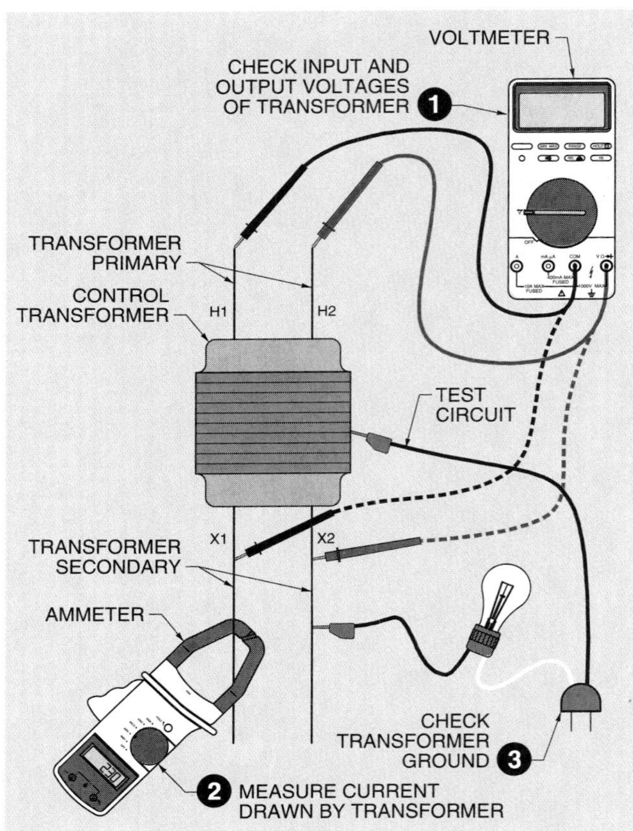

Figure 17-21. The control transformer should be checked if there is a problem in the control circuit that may be related to the power supply.

Troubleshooting Electric Motors

Electric motors are reliable and require little maintenance. It is not uncommon for motors to run satisfactorily for 20 or more years without repair. For example, refrigerators, air conditioners, and heating systems often last more than 20 years without any work to the motors. Any repair that is required almost always involves the related components.

Today, it is more common to replace a motor that has failed than it is to repair it. Small motors generally cost more to repair than to replace. Most large motors can be replaced with a more energy-efficient motor that justifies the extra expenditure.

The cost involved with a motor that has failed is almost always in downtime of the operation and maintenance time involved. For this reason, motor problems must be located as quickly as possible and the reason the motor failed must be determined to eliminate the cause and prevent the problem from returning.

Megohmmeter Tests. A *megohmmeter* is a device that detects insulation deterioration by measuring high resistance values under high test voltage conditions. A megohmmeter detects motor insulation deterioration before a motor fails. A megohmmeter is an ohmmeter capable of measuring very high resistances by using high voltages. Megohmmeter test voltages range from 50 V to 5000 V. A megohmmeter is used to perform motor insulation tests to prevent electrical shock and other causes of motor insulation failure, which include excessive moisture, dirt, heat, cold, corrosive vapors or solids, vibration, and aging.

A megohmmeter measures the resistance of different windings or the resistance from a winding to ground. An ohmmeter measures the resistance of common windings and components in a motor circuit. See Figure 17-22.

Several megohmmeter readings should be taken over a long period of time because the resistance of good insulation varies greatly. Megohmmeter readings are normally taken when the motor is installed and semiannually thereafter. A motor is in need of service if the megohmmeter reading is below the minimum acceptable resistance. See Figure 17-23.

RECOMMENDED MINIMUM RESISTANCE*	
Motor Voltage Rating (from nameplate)	**Minimum Acceptable Resistance**
Less than 208	100,000 Ω
208 – 240	200,000 Ω
240 – 600	300,000 Ω
600 – 1000	1 MΩ
1000 – 2400	2 MΩ
2400 – 5000	3 MΩ

* values for motor windings at 40°C

Figure 17-23. A motor is in need of service if the megohmmeter reading is below the minimum acceptable resistance.

Note: A motor with good insulation may have readings of 10 to 100 times the minimum acceptable resistance. Service the motor if the resistance reading is less than the minimum value.

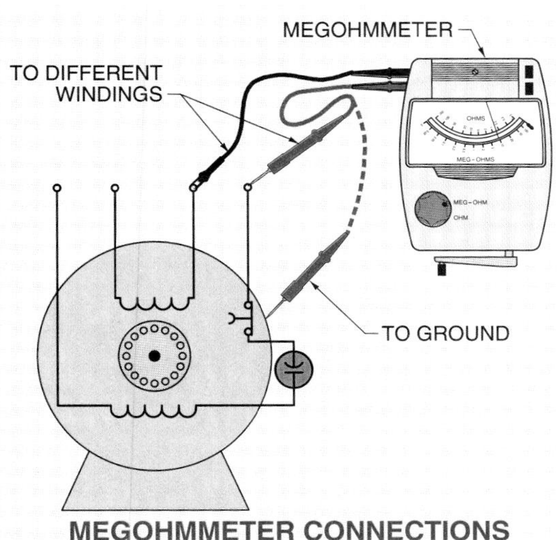

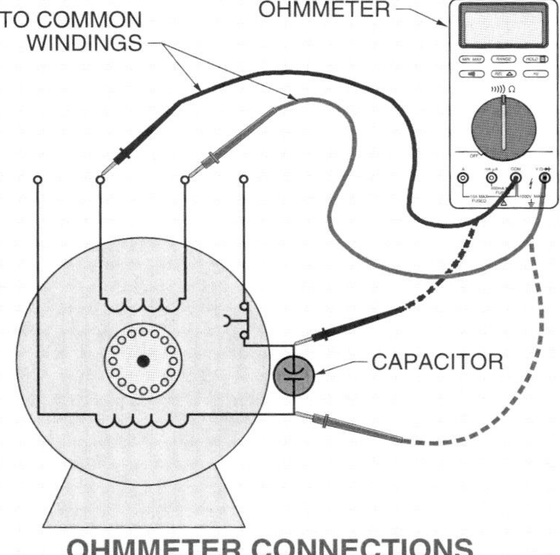

Figure 17-22. A megohmmeter measures the resistance of different windings or the resistance from a winding to ground. An ohmmeter measures the resistance of common windings and components in a motor circuit.

Caution: A megohmmeter uses high voltage for testing. Avoid touching the meter leads to the motor frame. Follow the manufacturer's recommended procedures and safety rules. After performing insulation tests with a megohmmeter, connect the motor windings to ground through a 5 kΩ, 5 Ω resistor. The winding should be connected for 10 times the motor testing time to discharge energy stored in the insulation.

Insulation Spot Tests

An *insulation spot test* is a test that checks motor insulation over the life of the motor. An insulation spot test is taken when the motor is placed in service and every six months thereafter. The test should also be taken after a motor is serviced. See Figure 17-24.

To perform a spot test, apply the procedure:

1. Connect a megohmmeter to measure the resistance of each winding lead to ground. Record the readings after 60 sec. Service the motor if a reading does not meet the minimum acceptable resistance. Record the lowest meter reading on an insulation spot test graph if all readings are above the minimum acceptable resistance. The lowest reading is used because a motor is only as good as its weakest point.
2. Discharge the motor windings.
3. Repeat Steps 1 and 2 every six months.

Interpret the results of the test to determine the condition of the insulation. Point A represents the motor insulation condition when the motor was placed in service. Point B represents effects of aging, contamination, etc., on the motor insulation. Point C represents motor insulation failure. Point D represents motor insulation condition after being rewound.

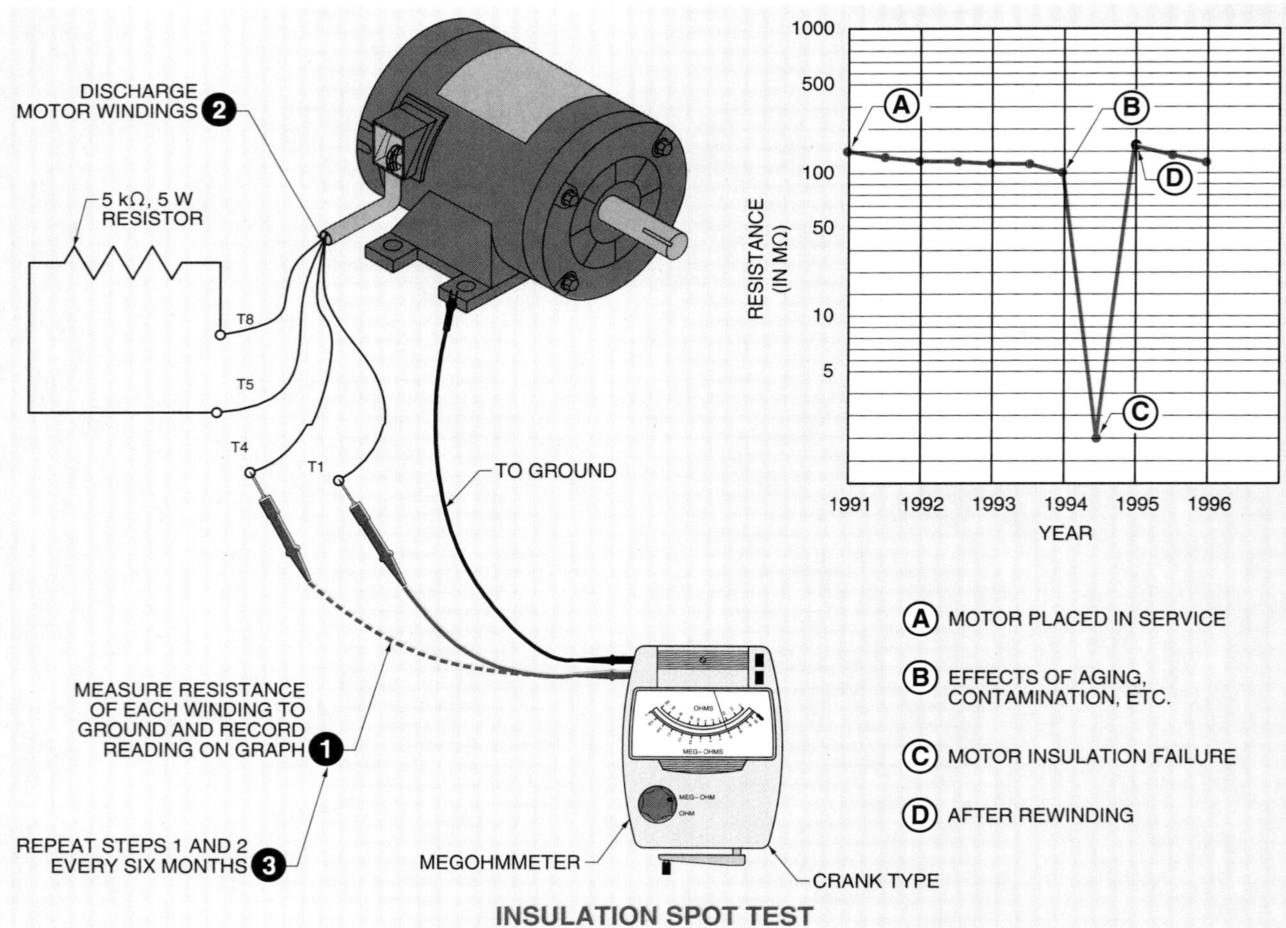

Figure 17-24. An insulation spot test checks motor insulation over the life of the motor.

Dielectric Absorption Tests

A *dielectric absorption test* is a test that checks the absorption characteristics of humid or contaminated insulation. The test is performed over a 10-minute period. See Figure 17-25. To perform a dielectric absorption test, apply the procedure:

1. Connect a megohmmeter to measure the resistance of each winding lead to ground.

Service the motor if a reading does not meet the minimum acceptable resistance. Record the lowest meter reading on a dielectric absorption test graph if all readings are above the minimum acceptable resistance. Record the readings every 10 sec for the first minute and every minute thereafter for 10 min.

2. Discharge the motor windings.

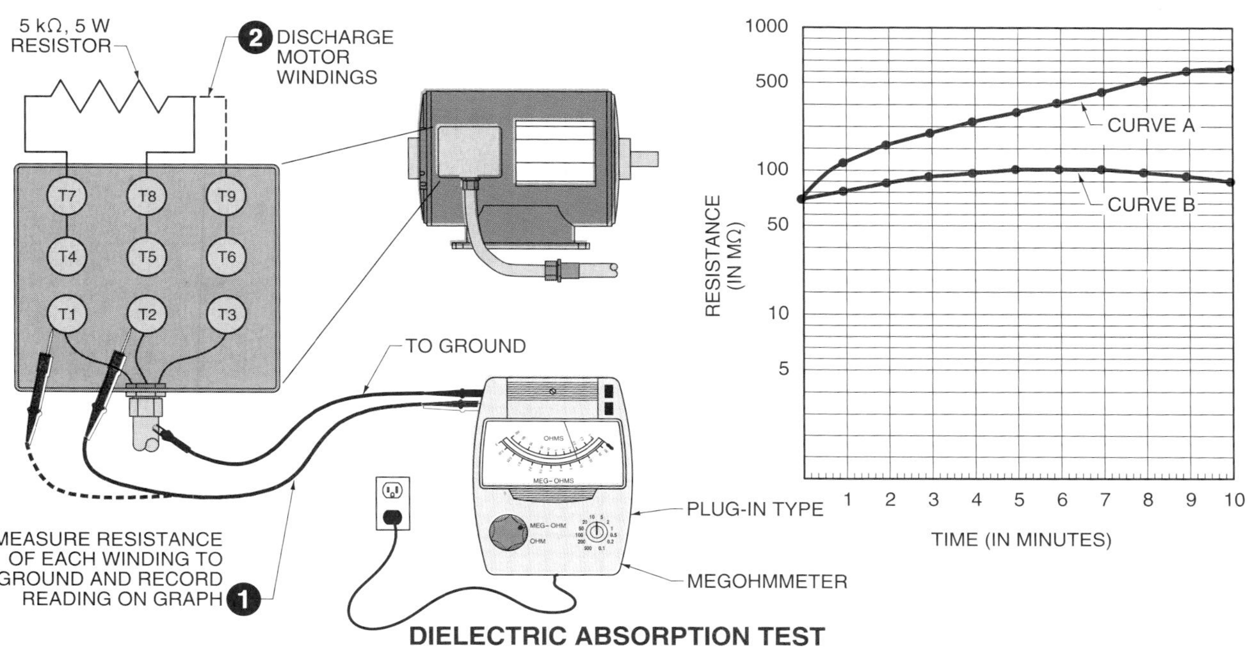

Figure 17-25. A dielectric absorption test checks the absorption characteristics of humid or contaminated insulation.

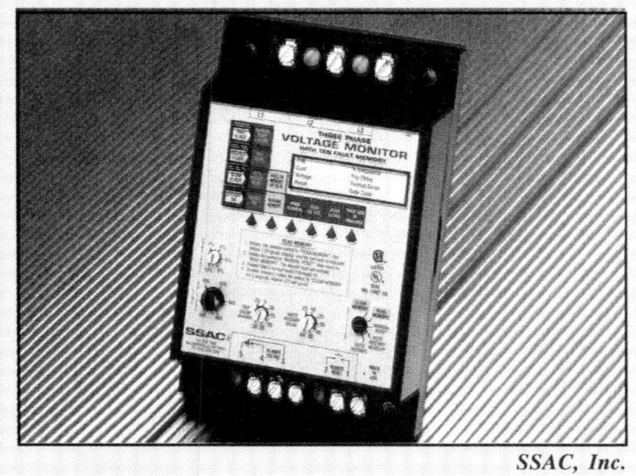

SSAC, Inc.

The WVM Series 3ϕ voltage monitor manufactured by SSAC, Inc. provides protection against premature equipment failure caused by voltage faults on the 3ϕ line.

Interpret the results of the test to determine the condition of the insulation. The slope of the curve shows the condition of the insulation. Good insulation (Curve A) shows a continual increase in resistance. Moist or cracked insulation (Curve B) shows a relatively constant resistance.

A polarization index is obtained by dividing the value of the 10-min reading by the value of the 1-min reading. The polarization index is an indication of the condition of the insulation. A low polarization index indicates excessive moisture or contamination. See Figure 17-26.

For example, if the 1-min reading of Class B insulation is 80 MΩ and the 10-min reading is 90 MΩ, the polarization index is 1.125 ($^{90}/_{80} = 1.125$). The insulation contains excessive moisture or contamination.

MINIMUM ACCEPTABLE POLARIZATION INDEX VALUES	
Insulation	**Value**
Class A	1.5
Class B	2.0
Class F	2.0

Figure 17-26. The polarization index is an indication of the condition of the insulation. A low polarization index indicates excessive moisture or contamination.

Insulation Step Voltage Tests

An *insulation step voltage test* is a test that creates electrical stress on internal insulation cracks to reveal aging or damage not found during other motor insulation tests. The insulation step voltage test is performed only after an insulation spot test. See Figure 17-27.

To perform an insulation step voltage test, apply the procedure:

1. Set the megohmmeter to 500 V and connect to measure the resistance of each winding lead to ground. Take each resistance reading after 60 sec. Record the lowest reading.
2. Place the meter leads on the winding that has the lowest reading.
3. Set the megohmmeter on increments of 500 V starting at 1000 V and ending at 5000 V. Record each reading after 60 sec.
4. Discharge the motor windings.

Interpret the results of the test to determine the condition of the insulation. The resistance of good insulation that is thoroughly dry (Curve A) remains approximately the same at different voltage levels. The resistance of deteriorated insulation (Curve B) decreases substantially at different voltage levels.

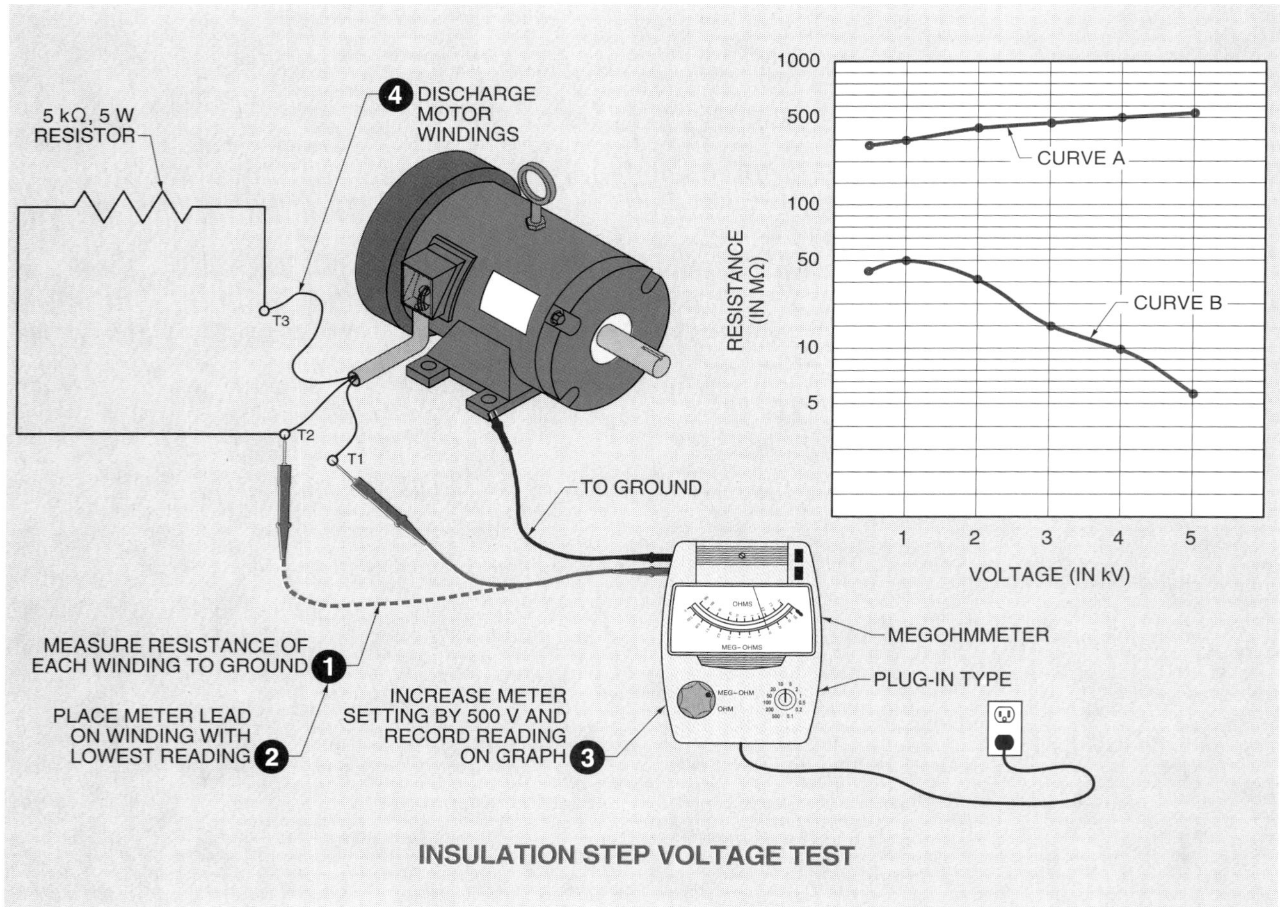

Figure 17-27. An insulation step voltage test is a test that creates electrical stress on internal insulation cracks to reveal aging or damage not found during other motor insulation tests.

Re-marking 3ϕ Induction Motor Connections

Three-phase induction motors are the most common motors used in industrial applications. Three-phase induction motors operate for many years with little or no required maintenance. It is not uncommon to find 3ϕ induction motors that have been in operation for 10 to 20 years in certain applications. The length of time a motor is in operation may cause the markings of the external leads to become defaced. This may also happen to a new or rebuilt motor that has been in the maintenance shop for some time. To ensure proper operation, each motor lead must be re-marked before troubleshooting and re-connecting the motor to a power source.

The two most common 3ϕ motors are the single-voltage, 3ϕ, three-lead motor and the dual-voltage, 3ϕ, nine-lead motor. Both may be internally connected in a wye or delta configuration.

The three leads of a single-voltage, 3ϕ, three-lead motor can be marked as T1, T2, and T3 in any order. The motor can be connected to the rated voltage and allowed to run. Any two leads may be interchanged if the rotation is in the wrong direction. The industry standard is to interchange T1 and T3.

Wye or Delta Connection Determination

A standard dual-voltage motor has nine leads extending from the motor and may be internally connected as a wye or delta motor. The internal connections must be determined when re-marking the motor leads. A multimeter is used to measure resistance or a test light circuit is used to determine whether a dual-voltage motor is internally connected in a wye or delta configuration.

A dual-voltage, wye-connected motor has four separate circuits. A dual-voltage, delta-connected motor has three separate circuits. See Figure 17-28. A wye-connected motor has three circuits of two leads each (T1-T4, T2-T5, and T3-T6) and one circuit of three leads (T7-T8-T9). A delta-connected motor has three circuits of three leads each (T1-T4-T9, T2-T5-T7, and T3-T6-T8).

A multimeter is used to determine the winding circuits (T1-T4, T2-T5, etc.) on an unmarked motor by connecting one meter lead to any motor lead and temporarily connecting the other meter lead to each remaining motor lead. See Figure 17-29. *Note:* ensure that the motor is disconnected from the power supply. A resistance reading other than infinity indicates a complete circuit.

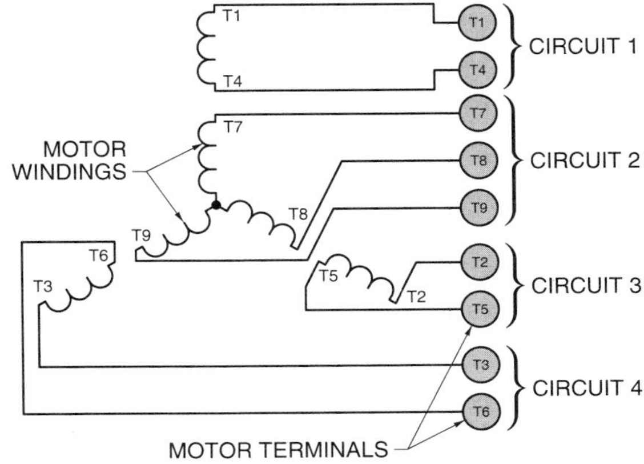

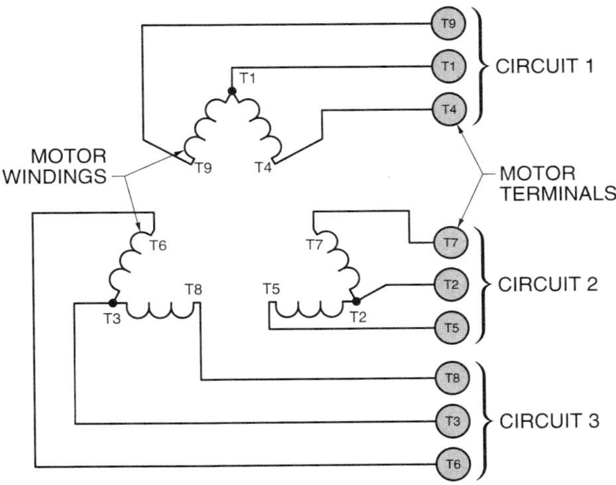

Figure 17-28. The internal connections of a motor must be determined when re-marking the motor leads.

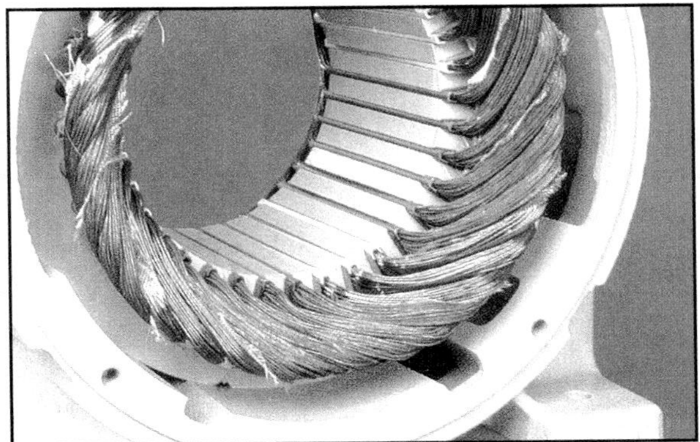

Electrical Apparatus Service Association, Inc.

A single-phased winding failure occurs when one phase of the power supply to the motor is open.

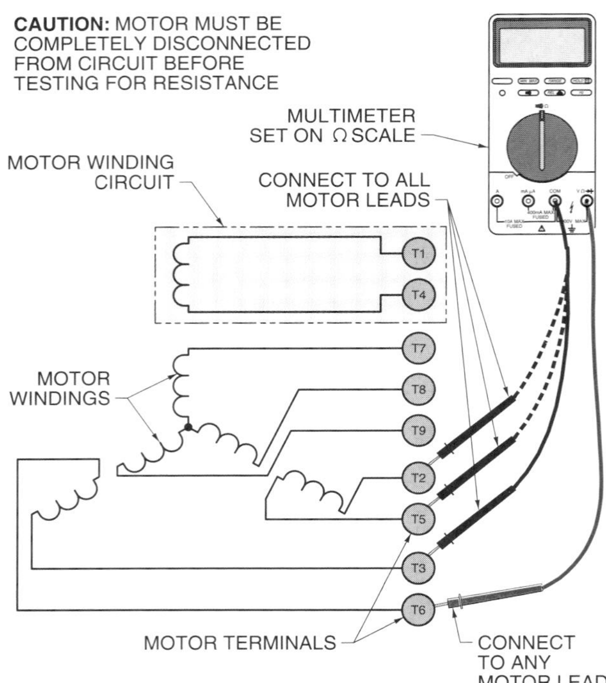

Figure 17-29. A multimeter is used to determine a winding circuit on an unmarked motor by connecting one meter lead to any motor lead and temporarily connecting the other meter lead to each of the remaining motor leads.

A test light circuit is used to determine the winding circuits on an unmarked motor by connecting one side of the test light circuit to any motor lead and temporarily connecting the other test light circuit lead to each remaining motor lead. See Figure 17-30.

Fluke Corporation

A Fluke Model 39 power meter is used to measure the kW, kVA, and power factor of a circuit.

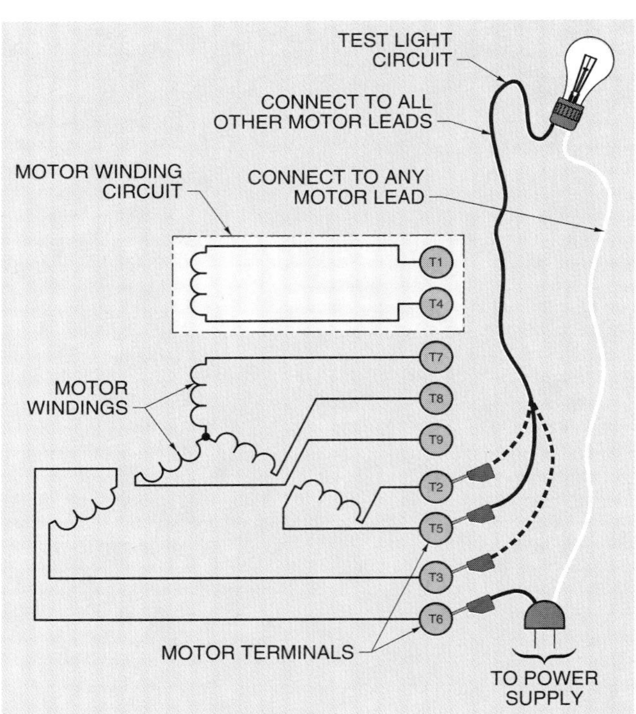

Figure 17-30. A test light circuit is used to determine a winding circuit on an unmarked motor by connecting one side of the test light circuit to any motor lead and temporarily connecting the other test light circuit lead to each of the remaining motor leads.

Apply power to the test light circuit. A glowing lamp indicates a complete circuit. Mark each connection that indicates a complete circuit by taping or pairing the leads together. Check all pairs of leads with all the remaining motor leads to determine if the circuit is a two- or three-lead circuit. The motor is a wye-connected motor if three circuits of two leads and one circuit of three leads are found. The motor is a delta-connected motor if three circuits of three leads are found.

Re-marking Dual-Voltage, Wye-Connected Motors

To re-mark a dual-voltage, wye-connected motor with no load, apply the procedure:

1. Determine the winding circuits using a meter or test light circuit. See Figure 17-31.
2. Mark the leads of the one three-lead circuit T7, T8, and T9 in any order. Separate the other motor leads in pairs, making sure none of the wires touch.

3. Connect the motor to the correct supply voltage. Connect T7 to L1, T8 to L2, and T9 to L3. The correct supply voltage is the lowest voltage rating of the dual-voltage rating given on the motor nameplate. The low voltage is normally 220 V because the standard dual-voltage motor operates on 220/440 V. For any other voltage, all test voltages should be changed in proportion to the motor rating.
4. Turn ON the supply voltage and let the motor run. The motor should run with no apparent noise or problems. The starting voltage should be reduced through a reduced-voltage starter if the motor is too large to be started by connecting it directly to the supply voltage.
5. Measure the voltage across each of the three open circuits while the motor is running, using a multimeter set on at least the 440 VAC scale. Care must be taken when measuring the high voltage of the running motor. Insulated test leads must be used. Connect only one test lead at a time. The voltage measured should be about 127 V (or slightly less) and should be the same on all three circuits.

The voltage is read on all circuits even though the two-wire circuits are not connected to the power lines because the voltage applied to the three-lead circuit induces a voltage into the two-wire circuits.

Draw the wiring diagram for the dual-voltage, wye-connected motor and mark the voltage readings on the wiring diagram. See Figure 17-32. Connect one lead of any two-wire circuit to T7 and connect the other lead of the circuit to one side of a multimeter. Temporarily mark the lead connected to T7 as T4 and the lead connected to the multimeter as T1. Connect the other lead of the multimeter to T8 and then to T9. Mark T1 and T4 permanently if the two voltages are the same and are approximately 335 V. Perform the same procedure on another two-wire circuit if the voltages are unequal. Mark the new terminals T1 and T4 if the new circuit gives the correct voltage (335 V). T1, T7, and T4 are found by this first test.

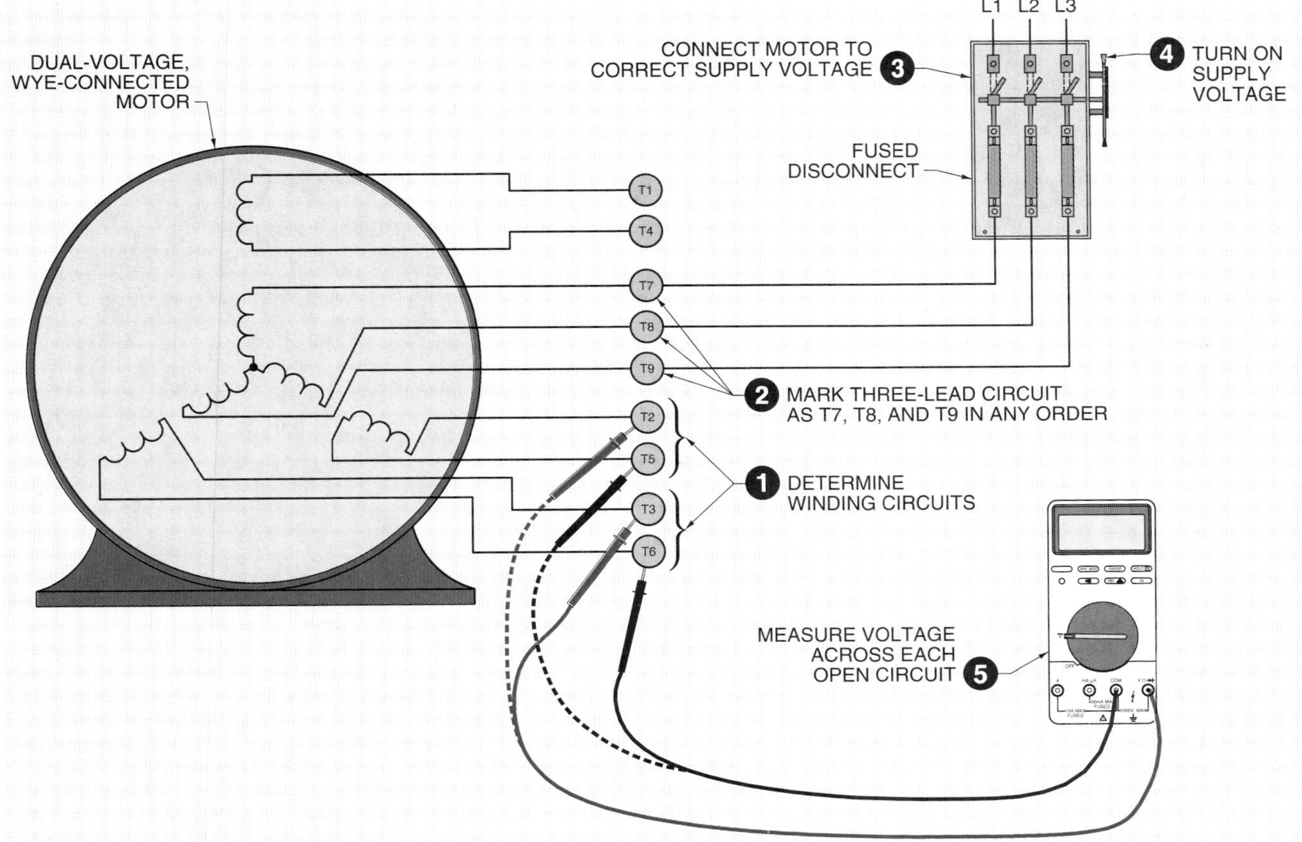

Figure 17-31. The three-lead circuit is connected to the correct supply voltage and the voltage across each of the three open circuits is measured when remarking a dual-voltage, wye-connected motor.

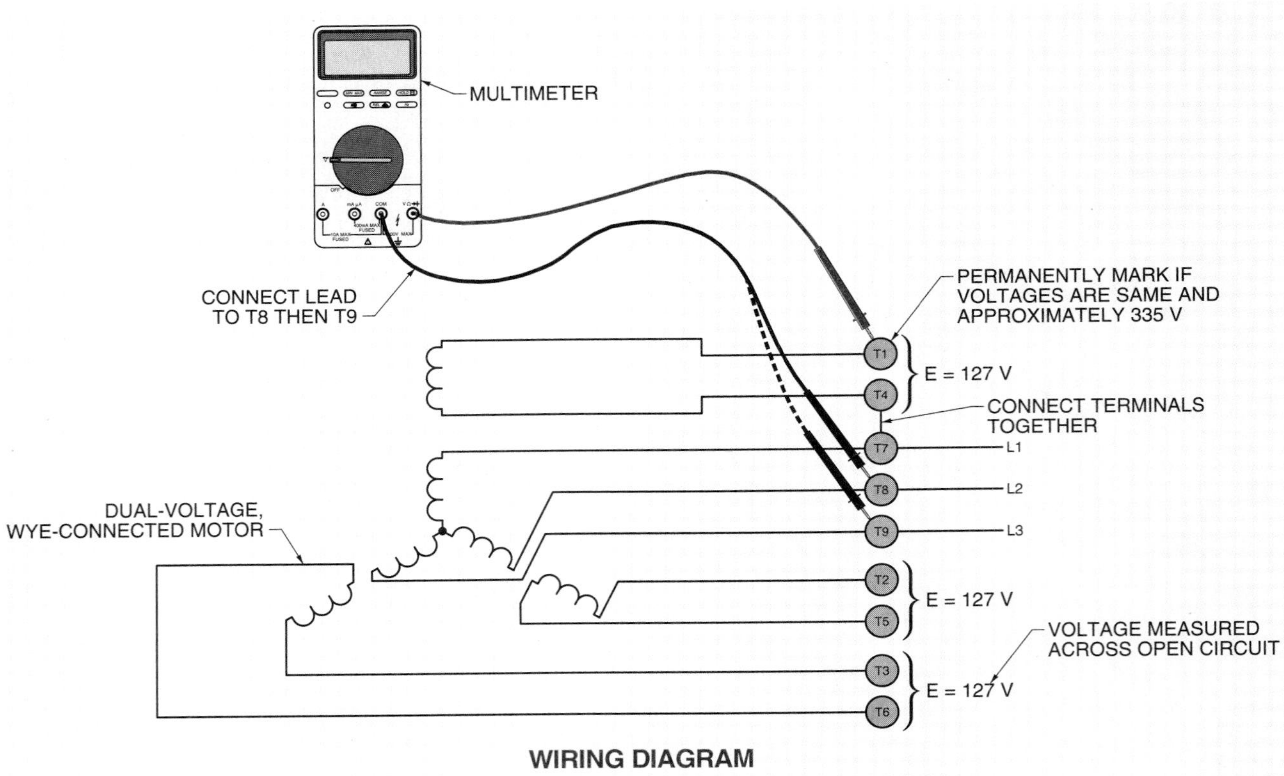

Figure 17-32. A wiring diagram is drawn when re-marking a dual-voltage, wye-connected motor to clarify the internal winding circuits.

Connect one lead of the two remaining unmarked two-wire circuits to T8 and the other lead to one side of the multimeter. Temporarily mark the lead connected to T8 as T5 and the lead connected to the multimeter as T2. Connect the other side of the multimeter to T7 and T9 and measure the voltage. Measurements and changes should be made until a position is found at which both voltages are the same and approximately 335 V. T2, T5, and T8 are found by this second test.

Check the third circuit in the same way until a position is found at which both voltages are the same and approximately 335 V. T3, T6, and T9 are found by this third test.

After each motor lead is found and marked, turn OFF the motor and connect L1 to T1 and T7, L2 to T2 and T8, L3 to T3 and T9, and connect T4, T5, and T6 together. Start the motor and let it run. Check the current on each power line with a clamp-on ammeter. The markings are correct and may be marked permanently if the current is approximately equal on all three power lines.

Re-marking Dual-Voltage, Delta-Connected Motors

A dual-voltage, delta-connected motor has nine leads grouped into three separate circuits. Each circuit has three motor leads connected which make the circuits (T1-T4-T9, T2-T5-T7, and T3-T6-T8). To re-mark a dual-voltage, delta-connected motor with no load, apply the procedure:

1. Determine the winding circuits using a multimeter or test light circuit. See Figure 17-33.
2. Measure the resistance of each circuit to find the center terminal. The resistance from the center terminal to the other two terminals is ½ the resistance between the other two terminals. Separate the three circuits and mark the center terminal for each circuit as T1, T2, and T3. Temporarily mark the two leads in the T1 group as T4 and T9, the two leads in the T2 group as T5 and T7, and the two leads in the T3 group as T6 and T8. Disconnect the multimeter.
3. Connect the group marked T1, T4, and T9 to L1, L2, and L3 of a 220 V power supply. This

should be the low-voltage rating on the nameplate of the motor. The other six leads should be left disconnected and must not touch because a voltage is induced into these leads even though these leads are not connected to power.

4. Turn the motor ON and let it run with the power applied to T1, T4, and T9.
5. Connect T4 (which is also connected to L2) to T7 and measure the voltage between T1 and T2. Set the multimeter on at least a 460 VAC range. Use insulated test leads and connect one meter lead at a time. The lead markings for T4 and T9, and T7 and T5 are correct if the measured voltage is approximately 440 V. Interchange T5 with T7 or T4 with T9 if the measured voltage is approximately 380 V. Interchange both T5 with T7 and T4 with T9 if the new measured voltage is approximately 220 V. T4, T9, T7, and T5 may be permanently marked if the voltage is approximately 440 V.

To correctly identify T6 and T8, connect T6 and T8 and measure the voltage from T1 and T3. The measured voltage should be approximately 440 V. Interchange leads T6 and T8 if the voltage does not equal 440 V. T6 and T8 may be permanently marked if the voltage is approximately 440 V.

Turn OFF the motor and reconnect the motor to a second set of motor leads. Connect L1 to T2, L2 to T5, and L3 to T7. Restart the motor and observe the direction of rotation. The motor should rotate in the same direction as with the previous connection. Turn OFF the motor and reconnect the motor to the third set of motor leads (L1 to T3, L2 to T6, and L3 to T8) after the motor has run and the direction is determined.

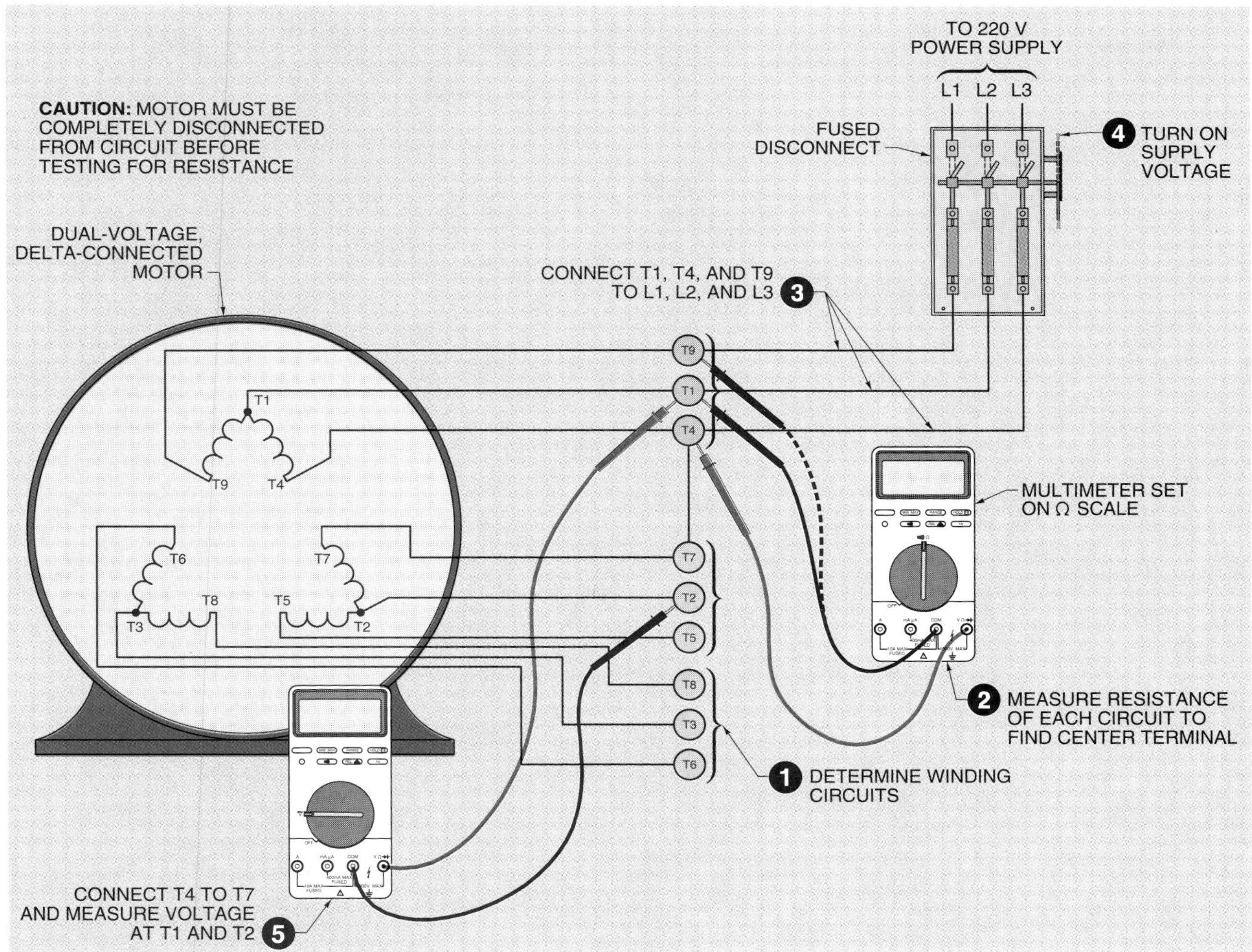

Figure 17-33. A dual-voltage, delta-connected motor has nine leads grouped into three separate circuits.

Restart the motor and observe the direction of rotation. The motor should rotate in the same direction as the first two connections. Start over carefully, re-marking each lead if the motor does not rotate in the same direction for any set of leads.

Turn OFF the motor and reconnect the motor for the low-voltage connection. Connect L1 to T1-T6-T7, L2 to T2-T4-T8, and L3 to T3-T5-T9. Restart the motor and take current readings on L1, L2, and L3 with a clamp-on ammeter. The markings are correct if the motor current is approximately equal on each line.

Re-marking DC Motor Connections

The three basic types of DC motors are the series, shunt, and compound motor. See Figure 17-34. All three types may have the same armature and frame but differ in the way the field coil and armature are connected. For all DC motors, terminal markings A1 and A2 always indicate the armature leads. Terminal markings S1 and S2 always indicate the series field leads. Terminal markings F1 and F2 always indicate the shunt field leads.

DC motor terminals can be re-marked using a multimeter. Measure the resistance of each pair of wires. A pair of wires must have a resistance reading or they are not a pair.

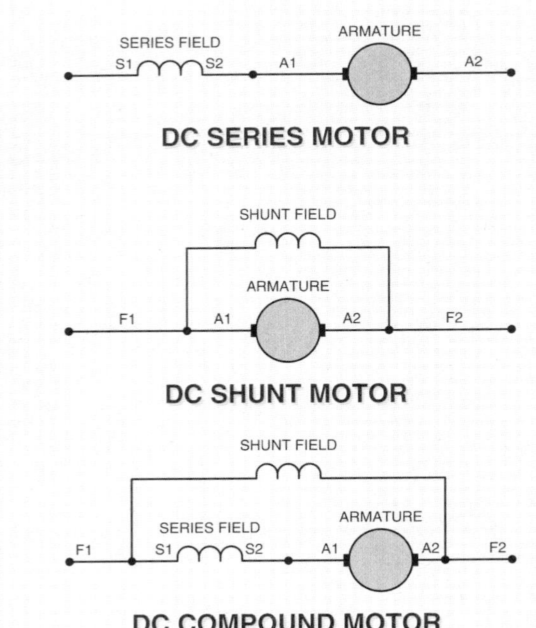

Figure 17-34. The three basic types of DC motors may have the same armature and frame, but differ in the way the field coil and armature are connected.

The field reading can be compared to the armature reading because each DC motor must have an armature. The series field normally has a reading less than the armature. The shunt field has a reading considerably larger than the armature. The armature can be easily identified by rotating the shaft of the motor when taking the readings. The armature moves the multimeter needle as it makes and breaks different windings. One final check can be made by lifting one of the brushes or placing a piece of paper under the brush. The multimeter moves to the infinity reading.

From this information, a motor is either a DC series or DC shunt motor if it has two pairs of leads (four wires) coming out. A coil is the series field if the reading of that coil is less than the armature coil resistance. A coil is the shunt field if the reading is considerably larger than the armature resistance.

Terminal Connections

A loose terminal connection can cause a load to burn out from undervoltage. A loose connection causes an increase in resistance. A voltage drop occurs at the resistance point and heat develops when current passes through resistance of any kind.

The heat at a terminal can be carried by the wire to the thermal overload inside a circuit breaker if the loose terminal is on a circuit breaker. The heat from the loose contact, added to the current in the overload, may cause the circuit breaker to trip on a current far below its rating. This may lead a troubleshooter to incorrectly suspect an overloaded circuit or faulty breaker.

Loads may burn out, coils may drop out (solenoids, starters, etc.), and timers and counters may reset if a loose terminal develops a high enough voltage drop across it. The heat developed at a loose terminal may also destroy the insulation around the terminal, leaving the possibility of a short circuit. This heat may also destroy any device that is connected to or near the loose connection.

To avoid loose connections, ensure that lugs clamp wires tightly. This is especially true with aluminum wire because the aluminum is softer than copper and does not hold its shape as well. Aluminum also expands and contracts more than copper, which may cause a loose connection. Ensure that both wires fit tightly if two wires are used in the same lug. Always check possible problem areas.

Locating Circuits in Switchboards, Panelboards, or Load Centers

A troubleshooter must often locate one circuit in a switchboard, panelboard, or load center to turn OFF the power before troubleshooting or working on the circuit. Switchboards, panelboards, and load centers are often crowded with wires that are not marked or that are mismarked. A troubleshooter cannot start turning OFF each circuit until the correct circuit is found because this disconnects all loads connected to that circuit. Timers, counters, clocks, starters, and other control devices must be reset, or critical equipment such as alarms and safety circuits may be stopped.

A flashing lamp and a clamp-on ammeter may be used to isolate a particular circuit. See Figure 17-35.

The flashing lamp is plugged into any receptacle on the circuit that is to be disconnected. As the lamp is flashing ON and OFF, a clamp-on ammeter is used to check each circuit. Each circuit displays a constant current reading except the one with the flashing lamp. The circuit with the flashing lamp displays a varying value on the ammeter equal to the flashing time of the lamp. This circuit may then be turned OFF for troubleshooting.

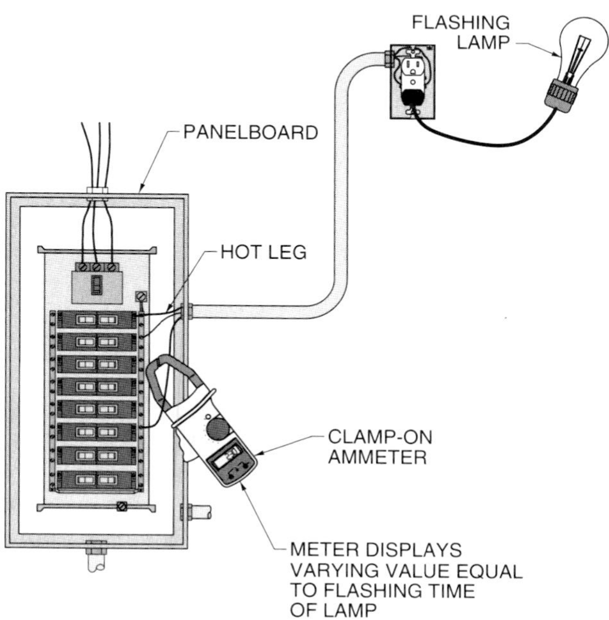

Figure 17-35. A flashing lamp and a clamp-on ammeter may be used to isolate a particular circuit.

Checking Capacitors

Capacitors are used on capacitor-start and capacitor-run motors as well as on other equipment. Capacitors have a limited life and should be checked whenever a problem is suspected. See Figure 17-36. To check a capacitor, apply the procedure:

1. Visually check the capacitor for any sign of leakage, cracks, or bulges. Replace if any of these signs are present.
2. Remove the capacitor from the circuit and discharge it. To safely discharge a capacitor, place a 20 kΩ, 5 W resistor across the terminals for 5 sec.
3. Connect a multimeter across the two leads of the capacitor. The capacitor is good, shorted, or open based on the resistance reading of the multimeter.

Good Capacitor. The multimeter value changes to zero resistance and slowly changes to infinity. Remove one of the leads and wait 30 sec. The value should change back to the previous value and continue to infinity when the lead is reconnected. The capacitor is not holding a charge and should be replaced if the value changes back to zero.

Shorted Capacitor. The multimeter value changes to zero and does not move. The capacitor is bad and must be replaced.

Open Capacitor. The value does not change from infinity. The capacitor is bad and must be replaced.

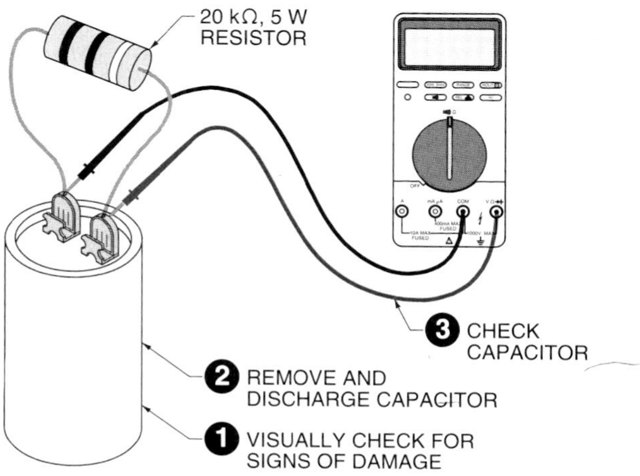

Figure 17-36. Capacitors have a limited life and should be checked whenever a problem is suspected.

1. What is the purpose of preventive maintenance in industrial plants?
2. How many electrical degrees apart is each power line in a balanced 3ϕ system?
3. As little as a 3% phase imbalance can cause how much of a power loss in a motor?
4. Why does measuring the voltage at a motor that is single-phasing give a voltage measurement on the power line that is open?
5. What is a voltage surge?
6. A motor should operate satisfactorily, as long as the voltage is within what percent of the motor's rated voltage?
7. What is overcycling?
8. Does the nameplate power rating of a transformer indicate the amount of apparent power or true power the transformer can deliver?
9. What is a megohmmeter?
10. Why should extra caution be taken when using a megohmmeter?
11. What is an insulation spot test?
12. What is a dielectric absorption test?
13. What is an insulation step voltage test?
14. How many external leads does a standard dual-voltage wye motor have?
15. How many external leads does a standard dual-voltage delta motor have?
16. How many separate circuits does a dual-voltage wye motor have?
17. How many separate circuits does a dual-voltage delta motor have?
18. What do terminal markings A1 and A2 indicate on a DC motor?
19. What do terminal markings S1 and S2 indicate on a DC motor?
20. What do terminal markings F1 and F2 indicate on a DC motor?

Electrical Tools, Instruments, and Safety

1. Three different methods used to organize electrical tools include pegboards, tool pouches, toolboxes, chests, and cabinets.
2.
- Know and understand all manufacturer's safety recommendations
- Ensure that all safety guards are properly in place and in working order
- Wear safety goggles and a dust mask when required
- Ensure that the material to be worked is free of obstructions and securely clamped
- Ensure that the switch is in the OFF position before connecting a tool to a power source
- Keep attention focused on the work
- A change in sound during tool operation normally indicates trouble. Investigate immediately
- Power tools should be inspected and serviced by a qualified repair person at regular intervals as specified by the manufacturer or by OSHA
- Inspect electrical cords to see that they are in good condition
- Shut OFF the power when work is completed. Wait until all movement of the tool stops before leaving a stationary tool or laying down a portable tool
- Clean and lubricate all tools after use
- Remove all defective power tools from service. Alert others to the situation
- Take extra precautions when working on damp or wet surfaces. Use additional insulation to prevent any body part from coming into contact with a wet or damp surface
- Work with at least one other coworker in hazardous or dangerous locations

3. The ground ensures that any short circuit trips the circuit breaker or blows the fuse.
4. They conduct electricity.
5. No, when using wrenches, never use a pipe extension or other form of "cheater" to increase the leverage of the wrench.
6. Long-nose pliers
7. Diagonal-cutting pliers
8. Combination wire stripper, bolt cutter, and crimper knife
9. Fish tape
10. Rigid conduit hickey
11. Fuse puller
12. An electromechanical device that indicates readings on a meter by the mechanical motion of a pointer
13. An electronic device that displays meter readings as numerical values
14. Range switch
15. By a light-emitting diode (LED) or a liquid crystal display (LCD)
16. A graph composed of segments that function as an analog pointer
17. A voltage that appears on a meter that is not connected to a circuit
18. Any output device that displays data about the circuit or operation
19. Voltmeter, ammeter, wattmeter, frequency meter, ohmmeter, and conductivity meter
20. Pressure gauge, tachometer, temperature meter, anemometer, manometer, hygrometer, pH meter, vibration meter, flowmeter, and counter
21. Voltmeter or multimeter
22. Clamp-on ammeter or multimeter
23. Set the function switch to AC voltage. Plug the black test lead into the common jack, plug the red test lead into the voltage jack, connect the test leads to the circuit, and read the voltage displayed on the meter.
24. To protect personnel by detecting potentially hazardous ground faults and quickly disconnecting power from the circuit
25. Inspection, servicing, or repair of electrical equipment
26. When in damp locations
27. Ordinary combustibles
28. Flammable liquids
29. Electrical equipment
30. Combustible metals

Electrical Symbols and Line Diagrams

1. To provide the information necessary to understand the operation of any electrical control system
2. By black nodes
3. Between the motor and L2
4. NC contacts
5. By a zigzag symbol
6. By a circle
7. Contacts which may be added to a contactor
8. NO
9. An electrically-operated switch (contactor) that includes motor overload protection

10. By pressing the stop pushbutton, a power failure, or an overload sensing a problem in the power circuit
11. Any circuit that requires a person to initiate an action for it to operate
12. The frame, plunger, and coil
13. Identical except for the overloads attached to them
14. Communicating with line diagrams
15. From left to right

Logic Applied to Line Diagrams

1. 1
2. In parallel
3. L2
4. Between L1 and the operating coil (OR load)
5. By numbering
6. By numbers which are underlined
7. Top left to bottom right
8. Dashed-line method and numerical cross-reference method
9. Signal(s), decision(s), and action
10. Section of circuit which starts or stops the flow of current by closing or opening the contacts
11. Section of circuit which determines what work is to be done and in what order the work is to occur
12. Section of the circuit which causes action (work) to take place
13. Load is ON only if all control signal contacts are closed
14. Load is ON if one or the other control signal contacts is closed
15. Load is ON if control signal is OFF
16. Two or more NC contacts are connected in series
17. Two or more NC contacts are connected in parallel
18. Retaining signal inputs to keep the load energized even after signals are removed
19. In series (NOR logic)
20. In parallel (OR logic)
21. Frequent starting and stopping a motor for short periods of time

AC Manual Contactors and Motor Starters

1. Exposed (live) parts, speed of opening and closing contacts is determined solely by operator, and soft copper knife switches require replacement after repeated arcing, heat generation, and mechanical fatigue
2. Electrical disconnects
3. Contacts that break the electrical circuit in two places
4. To protect the gap between the set of fixed contacts as the contacts make or break the circuit
5. Good conductivity, good mechanical strength, and oxide formed is an excellent conductor of electricity
6. A wiring diagram shows the connection of an installation or its component devices/parts. A wiring diagram shows, as closely as possible, the actual connection and placement of all component devices/parts in a circuit, including power circuit wiring.
7. Power circuit
8. To assure that both sets of contacts cannot be closed at the same time
9. A manual starter is a contactor with an added overload protection device and is used only in electrical motor circuits.
10. The NEC® requires that the control device shall not only turn a motor ON and OFF, but shall also protect the motor from destroying itself under an overload situation, such as a locked rotor.
11. Temperature of the air surrounding electrical equipment
12. To not open the circuit while the motor is starting, and to open the circuit if the motor gets overloaded and the fuses do not blow
13. The overload relay must indirectly monitor the temperature conditions of the motor because the overload relay is normally located at some distance from the motor.
14. The reset button is pushed, forcing the pawl across the ratchet wheel until the contacts are closed and the spring and ratchet wheel are returned to their original condition.
15. Single-pole
16. Double-pole
17. Three-pole
18. To provide mechanical and electrical protection for the operator and the starter
19. To provide a degree of protection against human contact with enclosed equipment in locations without unusual service conditions
20. Manual

Magnetism and Magnetic Solenoids

1. Permanent and temporary
2. Because of the loose molecular structure of soft iron
3. By increasing the current by increasing the voltage, increasing the number of coils, and inserting an iron core through the coils
4. Clapper, bell-crank, horizontal-action, vertical-action, and plunger
5. To reduce eddy currents produced by transformer action in the metal
6. To break the magnetic field and allow the armature to drop away freely when de-energized
7. Sets up an auxiliary magnetic field which is out of phase with the main coil magnetic field to help hold in the armature as the main coil magnetic field drops to zero
8. Because the large amount of iron in the magnetic circuit greatly increases the impedance of the coil, decreasing current through the coil
9. The coil draws more than its rated current
10. Low-coil current and reduced magnetic pull
11.

- Obtain full data on load requirements
- Allow for possible low-voltage conditions of the power supply
- Use shortest possible stroke
- Never use an oversized solenoid

12. Push or pull, length of stroke, required force, duty cycle, and uniform force curve
13. Paint mixing, filling processes, doorbells, washing machines, and kitchen appliances

14. Failure to operate when energized, noisy operation, and erratic operation

15. 50 Hz or 60 Hz

16. Suppresses noise and high voltage on power lines

17. Amount of voltage applied to the coil and amount of current allowed to pass through the coil

18. Coil burnout and mechanical damage

19. Copper or aluminum

20. Insulated copper wire

AC/DC Magnetic Contactors and Motor Starters

1. A control device that uses a small control current to energize or de-energize the load connected to it and has no overload protection

2. Two wires leading from the control device to the starter or contactor

3. In the event of a power loss in the control circuit, the contactor de-energizes, but also re-energizes if the control device remains closed when power is restored.

4. Three-wire control has three wires leading from the control device to the starter or contactor. A momentary contact stop pushbutton (NO) is wired in series with a momentary contact start pushbutton (NO) which is wired in parallel to a set of contacts forming a holding circuit interlock (memory).

5. Coil drops out at low or no voltage and cannot be reset unless voltage returns and the operator presses the start pushbutton.

6. AC contactor assemblies may have several sets of contacts and DC contactor assemblies have only one set.

7. Without arc suppression, contactors and motors require maintenance prematurely and result in excessive downtime.

8. Because the continuous DC supply causes current to flow constantly and with great stability across a much wider gap than does an AC supply of equal voltage

9. An arc chute confines and extinguishes arcs drawn between contacts opened under load. When a DC circuit carrying large amounts of current is interrupted, the collapsing magnetic field of the circuit current may induce a voltage which helps sustain the arc.

10. By their size and type of load by NEMA

11. A contactor that includes overload protection

12. Melting alloy, magnetic, and bimetallic

13. To open contacts to the magnetic coil, de-energizing the coil and disconnecting the power

14. To determine the amount of current going to the motor and reduce the current to a lower value for the overload relay

15. Each motor must be sized according to its own unique operating characteristics and applications

16. The percentage of extra demand that can be placed on a motor for short intervals without damaging the motor

17. Temperature of the air surrounding electrical equipment

18. Using a multiplier to size heaters as ambient temperature changes

19. An overload device located directly on or in a motor to provide overload protection

20. Additional electrical contacts, power poles, pneumatic timers, transient suppression modules, and control circuit fuse holders

Time Delay and Logic

1. Dashpot, synchronous clock, and solid-state

2. By controlling how fast air or liquid passes into or out of a container

3. By the speed at which a synchronous clock motor operates clock hands

4. By a solid-state timing circuit

5. A device that has a preset time period that must pass after the timer has been energized before any action occurs on the timer contacts

6. Delay ON operate

7. A device that does not start its timing function until the power is removed from the timer

8. Delay ON release

9. A device in which the contacts change position immediately and remain changed for the set period of time after the timer has received power

10. Interval

11. A device in which the contacts cycle open and closed repeatedly once the timer has received power

12. In series with the timer coil

13. The control switch can be wired using low-voltage wiring and switch contacts can be rated at a lower current level.

14. Yes

15. A sensor-controlled timer

16. With an O

17. With an X

18. A symmetrical recycle timer

19. The relay ON periods

20. Contact failure

Control Devices

1. One NO and one NC contact

2. Flush, half-shrouded, extended, jumbo mushroom, and illuminated

3. Mushroom

4. All basic NEMA enclosure types

5. To select or determine one of several different circuit conditions

6. To illustrate each switch position

7. To select one of eight different circuit conditions from one main control stick

8. 1–8

9. A circle with a dot and a truth table

10. NO or NC

11. The part of a limit switch that transfers the mechanical force of the moving part to the electrical contacts

12. Lever, fork-lever, push-roller, and wobble-stick

13. A switch that automatically turns lamps ON at dusk and OFF at dawn

14. A switch that responds to pressure changes and activates electrical contacts when the set force is reached

15. NO

16. The setting that adjusts the required pressure which must be removed before the contacts reset back to normal
17. The contacts may chatter
18. To maintain a predetermined tank pressure and accomplish sequencing
19. A switch that responds to the intensity of heat
20. Bimetallic, capillary tube, thermistor, and thermocouple control(s)
21. A switch used to sense the movement of a fluid
22. Yes
23. No
24. Pump and sump
25. 2
26. To measure wind speed (velocity)
27. Side-to-side movement

Reversing Motor Circuits

1. Stator
2. Rotor
3. One end of each of the three phases is internally connected to the other phases
4. Each winding is wired end-to-end to form a completely closed loop circuit
5. Because, the higher the voltage, the lower the current
6. L1 to T1 and T7, L2 to T2 and T8, L3 to T3 and T9, and tie T4, T5, and T6 together
7. L1 to T1, L2 to T2, L3 to T3, and tie T4 to T7, T5 to T8, and T6 to T9
8. L1 to T1, T6, and T7, L2 to T2, T4, and T8, and L3 to T9 and T5
9. L1 to T1, L2 to T2, L3 to T3, and tie T4 to T7, T5 to T8, and T6 to T9
10. By interchanging any two power lines
11. A 1ϕ AC motor that includes a running winding and a starting winding
12. To disconnect the starting winding from the circuit when the motor reaches approximately 75% of full speed
13. By interchanging the leads of the starting winding
14. To give the motor more torque
15. To provide a higher starting torque and higher running torque
16. By reversing the connections to the starting winding
17. Series, shunt, compound, and permanent-magnet
18. By reversing the direction of current through the armature
19. To ensure that both switching starters cannot be energized at the same time
20. Auxiliary contact
21. One NO and one NC
22. To give a visual indication of the direction of motor rotation
23. NC
24. Because it does not provide overload protection
25. The part of an electrical circuit that connects the loads to the main power lines

Power Distribution Systems

1. Alternators
2. When conductors are moved through magnetic fields, voltages are induced into the conductors
3. So that the external load may be easily attached to the rotor without interfering with its rotation
4. A decrease of the required speed of the rotor
5. 3ϕ power is smoother and more economical
6. Phase-to-neutral, phase-to-phase, and phase-to-phase-to-phase
7. 360 V
8. 240 V
9. An electrical interface that changes AC voltage levels
10. Primary and secondary
11. To provide a controlled path for the magnetic flux generated in the transformer by the current flowing through the windings
12. Step-down
13. Current is doubled
14. H1 and H2
15. X1 and X2
16. In kVA
17. If one transformer is damaged or removed from service, the other two transformers can be connected in an open-delta connection
18. Primary switchgear, transformer, and secondary switchgear
19. To allow for a variable output
20. Switchboard
21. A service-entrance switchboard has space and mounting provisions required by the local power company for metering equipment, as well as overcurrent protection and disconnect means for the service conductors. A distribution switchboard contains the protective devices and feeder circuits required to distribute the power throughout a building.
22. A wall-mounted distribution cabinet containing a group of overcurrent devices for lighting, appliance, or power distribution branch circuits
23. The portion of a distribution system between the final OCPDs and the outlet or load connected to them
24. To receive incoming power and deliver it to the control circuit and motor loads

Solid-State Electronic Control Devices

1. Valence electrons
2. By doping
3. An insulating material such as fiberglass or phenolic with conducting paths secured to one or both sides
4. Forward bias is the application of the proper polarity to a diode. Reverse bias is the application of the opposite polarity to a diode.
5. A circuit containing a diode which permits only the positive half-cycles of the AC sine wave to pass
6. In the reverse breakdown region
7. A thermally-sensitive resistor whose resistance changes with a change in temperature
8. A photoconductive cell conducts current when exposed to light. A photovoltaic cell converts light energy to electrical energy.
9. A sensor which produces a voltage depending on the strength of the magnetic field applied to the sensor

10. A transducer that changes resistance with a corresponding pressure change
11. A diode which produces light when current flows through it
12. A three-terminal device that controls current through the device depending on the amount of voltage applied to the base
13. By its transistor outline (TO) number
14. Changing AC into DC
15. A ratio of the amplitude of the output signal to the amplitude of the input signal
16. Common-emitter, common-base, and common-collector
17. The voltage required to switch an SCR into a conductive state.
18. A three-electrode AC semiconductor switch
19. A triac operates much like a pair of SCRs connected in a reverse parallel arrangement. The triac conducts if the appropriate signal is applied to the gate.
20. A diac acts much like two zener diodes that are series-connected in opposite directions. The diac is used primarily as a triggering device.
21. Thousands of semiconductors providing a complete circuit function in one small semiconductor package
22. By a standardized pin numbering system
23. A very high gain, directly-coupled amplifier that uses external feedback to control response characteristics
24. Analog and digital
25. AND, OR, NAND, and NOR

12 Electromechanical and Solid-State Relays

1. EMR and SSR
2. A reed relay
3. No
4. No
5. Yes
6. Yes
7. Single-pole, double-throw
8. Three-pole, double-throw
9. 2 (single-break or double-break)
10. Double-pole contact
11. 1
12. Mechanical contacts that can be placed in either a NO or NC position
13. A higher mechanical life rating
14. The formation of film on contact surfaces
15. Zero switching
16. Instant-ON
17. Voltage at the load starts at a low level and is increased over a period of time.
18. SCRs
19. Triacs
20. Transistors
21. It increases
22. The ability of a device to impede the flow of heat
23. The amount of current that leaks through an SSR when the switch is turned OFF
24. Temporary unwanted voltages in an electrical circuit
25. When a relay fails to turn OFF because the current and voltage in the circuit reach zero at different times

Photoelectric and Proximity Controls

1. A method of scanning in which the transmitter and receiver are placed opposite each other so the light beam from the transmitter shines directly at the receiver
2. When detecting very small targets
3. A method of scanning in which the transmitter and receiver are housed in the same enclosure and the transmitted light beam is reflected back to the receiver from a reflector
4. Retroreflective scan
5. A method of scanning in which the transmitter and receiver are placed at equal angles from a highly-reflective surface
6. A method of scanning in which the transmitter and receiver are housed in the same enclosure, and a small percentage of the transmitted light beam is reflected back to the receiver from the target
7. A method of scanning which simultaneously focuses and converges a light beam to a fixed focal point in front of the photoreceiver
8. Modulated
9. When the object to be detected moves at a high speed or the object to be detected is not much bigger than the effective beam of the controller
10. When the target is present
11. When the target is missing
12. Inductive and capacitive
13. A magnetic field
14. A thyristor output
15. An NPN or PNP transistor
16. The amount of current drawn by a load when energized
17. Residual or leakage current
18. The minimum amount of current required to keep a sensor operating
19. 3
20. NPN

Programmable Controllers

1. Discrete parts manufacturing and process manufacturing
2. Discrete parts manufacturing
3. Process manufacturing
4. Inputs and outputs
5. Ladder (line) diagrams
6. Power supply, input/output section, processor section, and programming section
7. Pushbuttons, temperature switches, and limit switches
8. Potentiometers, rheostats, and encoders
9. The time it takes a PLC to make a sweep of the program
10. A line diagram that better matches the PLC's language
11. To provide back-up power for the processor memory in case of an external power failure

12. Disable command
13. Unwanted signals that are present on a power line
14. Through programming
15. A method of transmitting more than one signal over a single transmission system

Reduced-Voltage Starting

1. To reduce interference in the power source, load, and the electrical environment surrounding the motor
2. Reduced-voltage starting reduces the amount of starting current a motor draws when starting.
3. Torque is reduced.
4. Field, armature, brushes, and commutator
5. To increase motor torque
6. Field
7. Through the brushes
8. To reduce sparking at the brushes of large DC motors
9. To reduce starting current
10. Rotor
11. Stator
12. No
13. The amount of current a motor draws when the rotor is stopped and power is applied
14. The amount of current required by the motor to produce full-load torque at the motor's rated speed
15. By using additional contacts and resistors during starting
16. When the motor is temporarily disconnected when moving from one incremental voltage to another
17. When the motor is not disconnected when moving from one incremental voltage to another
18. Less expensive
19. Poor starting torque
20. 6
21. Open transition
22. SCRs
23. No

16 Accelerating and Decelerating Methods

1. A friction surface comes in contact with a wheel mounted onto the motor shaft
2. Lower shoe pressure is permitted
3. $T = \frac{5252 \times HP}{rpm}$
4. A method of motor braking in which the motor connections are reversed so that the motor develops a counter torque which acts as a braking force
5. At a given rpm
6. Some motors cannot be reversed and some motors have centrifugal switches.
7. OFF-delay
8. To connect a DC voltage after the AC is removed.
9. To prevent both the DC and AC from being connected at the same time
10. To reconnect the motor and make it act as a generator immediately after it is turned OFF
11. Applying a force over a distance
12. The force that produces rotation
13. The torque required to produce the rated power at full speed of a motor
14. The torque a motor produces when the rotor is stationary and full power is applied to the motor
15. The torque required to bring a load up to its rated speed
16. The maximum torque a motor can provide without an abrupt reduction in motor speed
17. The motor's characteristics of HP, torque, and speed
18. Speed changes rapidly with torque changes.
19. The DC shunt motor has fairly high torque at any speed.
20. The DC compound motor has a higher starting torque than the shunt motor and a higher running torque than the series motor.
21. By changing the voltage across the armature or field
22. The supply frequency and the number of poles
23. 900 rpm
24. A control function that requires the operator to start and operate a motor in a predetermined order
25. A control function that permits the operator to select a high motor speed and the control circuit to automatically accelerate the motor to that speed
26. A control function that permits the operator to select a low motor speed and the control circuit to automatically decelerate the motor to that speed

Preventive Maintenance and Troubleshooting

1. To keep machines, assembly lines, production operations, and plant operations running with little or no downtime
2. 120 electrical degrees apart
3. A 10% loss
4. Because the open winding in the motor generates a voltage almost equal to the phase voltage that is lost
5. Any higher than normal voltage that temporarily exists on one or more of the power lines
6. Within 10%
7. The process of turning a motor ON and OFF repeatedly
8. Apparent power
9 An instrument that measures high resistance values under high test voltage conditions
10. Because a megohmmeter uses high voltage for testing
11. A test that checks motor insulation over the life of the motor
12. A test that checks the absorption characteristics of humid or contaminated insulation
13. A test that creates electrical stress on internal insulation cracks to reveal aging or damage
14. 9
15. 9
16. 4
17. 3
18. Armature leads
19. Series field leads
20. Shunt field leads

APPENDIX

ENGLISH SYSTEM				
		Unit	**Abbreviation**	**Equivalents**
LENGTH		mile	mi	5280′, 320 rd, 1760 yd
		rod	rd	5.50 yd, 16.5′
		yard	yd	3′, 36″
		foot	ft *or* ′	12″, .333 yd
		inch	in. *or* ″	.083′, .028 yd
AREA A = l x w		square mile	sq mi *or* mi^2	640 a, 102,400 sq rd
		acre	A	4840 sq yd, 43,560 sq ft
		square rod	sq rd *or* rd^2	30.25 sq yd, .00625 A
		square yard	sq yd *or* yd^2	1296 sq in., 9 sq ft
		square foot	sq ft *or* ft^2	144 sq in., .111 sq yd
		square inch	sq in. *or* in^2	.0069 sq ft, .00077 sq yd
VOLUME V = l x w x t		cubic yard	cu yd *or* yd^3	27 cu ft, 46,656 cu in.
		cubic foot	cu ft *or* ft^3	1728 cu in., .0370 cu yd
		cubic inch	cu in. *or* in^3	.00058 cu ft, .000021 cu yd
CAPACITY WATER, FUEL, ETC.	*U.S. liquid measure*	gallon	gal.	4 qt (231 cu in.)
		quart	qt	2 pt (57.75 cu in.)
		pint	pt	4 gi (28.875 cu in.)
		gill	gi	4 fl oz (7.219 cu in.)
		fluidounce	fl oz	8 fl dr (1.805 cu in.)
		fluidram	fl dr	60 min (.226 cu in.)
		minim	min	$\frac{1}{6}$ fl dr (.003760 cu in.)
VEGETABLES, GRAIN, ETC.	*U.S. dry measure*	bushel	bu	4 pk (2150.42 cu in.)
		peck	pk	8 qt (537.605 cu in.)
		quart	qt	2 pt (67.201 cu in.)
		pint	pt	$\frac{1}{2}$ qt (33.600 cu in.)
DRUGS	*British imperial liquid and dry measure*	bushel	bu	4 pk (2219.36 cu in.)
		peck	pk	2 gal. (554.84 cu in.)
		gallon	gal.	4 qt (277.420 cu in.)
		quart	qt	2 pt (69.355 cu in.)
		pint	pt	4 gi (34.678 cu in.)
		gill	gi	5 fl oz (8.669 cu in.)
		fluidounce	fl oz	8 fl dr (1.7339 cu in.)
		fluidram	fl dr	60 min (.216734 cu in.)
		minim	min	$\frac{1}{60}$ fl dr (.003612 cu in.)
MASS AND WEIGHT COAL, GRAIN, ETC.	*avoirdupois*	ton		2000 lb
		short ton		2000 lb
		long ton		2240 lb
		pound	lb *or* #	16 oz, 7000 gr
		ounce	oz	16 dr, 437.5 gr
		dram	dr	27.344 gr, .0625 oz
		grain	gr	.037 dr, .002286 oz
GOLD, SILVER, ETC.	*troy*	pound	lb	12 oz, 240 dwt, 5760 gr
		ounce	oz	20 dwt, 480 gr
		pennyweight	dwt *or* pwt	24 gr, .05 oz
		grain	gr	.042 dwt, .002083 oz
DRUGS	*apothecaries'*	pound	lb ap	12 oz, 5760 gr
		ounce	oz ap	8 dr ap, 480 gr
		dram	dr ap	3 s ap, 60 gr
		scruple	s ap	20 gr, .333 dr ap
		grain	gr	.05 s, .002083 oz, .0166 dr ap

METRIC SYSTEM			
LENGTH	**Unit**	**Abbreviation**	**Number of Base Units**
	kilometer	km	1000
	hectometer	hm	100
	dekameter	dam	10
	***meter**	m	1
	decimeter	dm	.1
	centimeter	cm	.01
	millimeter	mm	.001
AREA	square kilometer	sq km *or* km^2	1,000,000
	hectare	ha	10,000
	are	a	100
	square centimeter	sq cm *or* cm^2	.0001
VOLUME	cubic centimeter	cu cm, cm^3, *or* cc	.000001
	cubic decimeter	dm^3	.001
	***cubic meter**	m^3	1
CAPACITY	kiloliter	kl	1000
	hectoliter	hl	100
	dekaliter	dal	10
	***liter**	l	1
	cubic decimeter	dm^3	1
	deciliter	dl	.10
	centiliter	cl	.01
	milliliter	ml	.001
MASS AND WEIGHT	metric ton	t	1,000,000
	kilogram	kg	1000
	hectogram	hg	100
	dekagram	dag	10
	***gram**	g	1
	decigram	dg	.10
	centigram	cg	.01
	milligram	mg	.001

* base units

METRIC PREFIXES			
Multiples and Submultiples	**Prefixes**	**Symbols**	**Meaning**
$1,000,000,000,000 = 10^{12}$	tera	T	trillion
$1,000,000,000 = 10^{9}$	giga	G	billion
$1,000,000 = 10^{6}$	mega	M	million
$1000 = 10^{3}$	kilo	k	thousand
$100 = 10^{2}$	hecto	h	hundred
$10 = 10^{1}$	deka	d	ten
Unit $1 = 10^{0}$			
$.1 = 10^{-1}$	deci	d	tenth
$.01 = 10^{-2}$	centi	c	hundredth
$.001 = 10^{-3}$	milli	m	thousandth
$.000001 = 10^{-6}$	micro	μ	millionth
$.000000001 = 10^{-9}$	nano	n	billionth
$.000000000001 = 10^{-12}$	pico	p	trillionth

METRIC CONVERSIONS												
Initial Units	**Final Units**											
	giga	**mega**	**kilo**	**hecto**	**deka**	**base unit**	**deci**	**centi**	**milli**	**micro**	**nano**	**pico**
giga		3R	6R	7R	8R	9R	10R	11R	12R	15R	18R	21R
mega	3L		3R	4R	5R	6R	7R	8R	9R	12R	15R	18R
kilo	6L	3L		1R	2R	3R	4R	5R	6R	9R	12R	15R
hecto	7L	4L	1L		1R	2R	3R	4R	5R	8R	11R	14R
deka	8L	5L	2L	1L		1R	2R	3R	4R	7R	10R	13R
base unit	9L	6L	3L	2L	1L		1R	2R	3R	6R	9R	12R
deci	10L	7L	4L	3L	2L	1L		1R	2R	5R	8R	11R
centi	11L	8L	5L	4L	3L	2L	1L		1R	4R	7R	10R
milli	12L	9L	6L	5L	4L	3L	2L	1L		3R	6R	9R
micro	15L	12L	9L	8L	7L	6L	5L	4L	3L		3R	6R
nano	18L	15L	12L	11L	10L	9L	8L	7L	6L	3L		3R
pico	21L	18L	15L	14L	13L	12L	11L	10L	9L	6L	3L	

COMMON PREFIXES		
Symbol	**Prefix**	**Equivalent**
G	giga	1,000,000,000
M	mega	1,000,000
k	kilo	1000
base unit	—	1
m	milli	0.001
μ	micro	0.000001
n	nano	0.000000001
p	pico	0.000000000001
Z	impedance	ohms — Ω

CHEMICAL ELEMENTS

Name	Symbol	Atomic Weight*	Atomic Number	Name	Symbol	Atomic Weight*	Atomic Number
Actinium	Ac	[227]	89	Neon	Ne	20.183	10
Aluminum	Al	26.9815	13	Neptunium	Np	[237]	93
Americium	Am	[243]	95	Nickel	Ni	58.71	28
Antimony	Sb	121.75	51	Niobium	Nb	92.906	41
Argon	Ar	39.948	18	Nitrogen	N	14.0067	7
Arsenic	As	74.9216	33	Nobelium	No	[255]	102
Astatine	At	[210]	85	Osmium	Os	190.2	76
Barium	Ba	137.34	56	Oxygen	O	15.9994	8
Berkelium	Bk	[247]	97	Palladium	Pd	106.4	46
Beryllium	Be	9.0122	4	Phosphorus	P	30.9738	15
Bismuth	Bi	208.980	83	Platinum	Pt	195.09	78
Boron	B	10.811	5	Plutonium	Pu	[244]	94
Bromine	Br	79.909	35	Polonium	Po	[210]	84
Cadmium	Cd	112.40	48	Potassium	K	39.102	19
Calcium	Ca	40.08	20	Praseodymium	Pr	140.907	59
Californium	Cf	[251]	98	Promethium	Pm	[145]	61
Carbon	C	12.01115	6	Protactinium	Pa	[231]	91
Cerium	Ce	140.12	58	Radium	Ra	[226]	88
Cesium	Cs	132.905	55	Radon	Rn	[222]	86
Chlorine	Cl	35.453	17	Rhenium	Re	186.2	75
Chromium	Cr	51.996	24	Rhodium	Rh	102.905	45
Cobalt	Co	58.9332	27	Rubidium	Rb	85.47	37
Copper	Cu	63.54	29	Ruthenium	Ru	101.07	44
Curium	Cm	[247]	96	Samarium	Sm	150.35	62
Dysprosium	Dy	162.50	66	Scandium	Sc	44.956	21
Einsteinium	Es	[254]	99	Selenium	Se	78.96	34
Erbium	Er	167.26	68	Silicon	Si	28.086	14
Europium	Eu	151.96	63	Silver	Ag	107.870	47
Fermium	Fm	[257]	100	Sodium	Na	22.9898	11
Fluorine	F	18.9984	9	Strontium	Sr	87.62	38
Francium	Fr	[223]	87	Sulfur	S	32.064	16
Gadolinium	Gd	157.25	64	Tantalum	Ta	180.948	73
Gallium	Ga	69.72	31	Technetium	Tc	[97]	43
Germanium	Ge	72.59	32	Tellurium	Te	127.60	52
Gold	Au	196.967	79	Terbium	Tb	158.924	65
Hafnium	Hf	178.49	72	Thallium	Tl	204.37	81
Helium	He	4.0026	2	Thorium	Th	232.038	90
Holmium	Ho	164.930	67	Thulium	Tm	168.934	69
Hydrogen	H	1.00797	1	Tin	Sn	118.69	50
Indium	In	114.82	49	Titanium	Ti	47.90	22
Iodine	I	126.9044	53	Tungsten	W	183.85	74
Iridium	Ir	192.2	77	Unnilennium	Une	[266]	109
Iron	Fe	55.847	26	Unnilhexium	Unh	[263]	106
Krypton	Kr	83.80	36	Unniloctium	Uno	[265]	108
Lanthanum	La	138.91	57	Unnilpentium	Unp	[262]	105
Lawrencium	Lr	[256]	103	Unnilquadium	Unq	[261]	104
Lead	Pb	207.19	82	Unnilseptium	Uns	[262]	107
Lithium	Li	6.939	3	Uranium	U	238.03	92
Lutetium	Lu	174.97	71	Vanadium	V	50.942	23
Magnesium	Mg	24.312	12	Xenon	Xe	131.30	54
Manganese	Mn	54.9380	25	Ytterbium	Yb	173.04	70
Mendelevium	Md	[258]	101	Yttrium	Y	88.905	39
Mercury	Hg	200.59	80	Zinc	Zn	65.37	30
Molybdenum	Mo	95.94	42	Zirconium	Zr	91.22	40
Neodymium	Nd	144.24	60				

* a number in brackets indicates the mass number of the most stable isotope

THREE-PHASE VOLTAGE VALUES
For 208 V × 1.732, use 360
For 230 V × 1.732, use 398
For 240 V × 1.732, use 416
For 440 V × 1.732, use 762
For 460 V × 1.732, use 797
For 480 V × 1.732, use 831

POWER FORMULA ABBREVIATIONS AND SYMBOLS	
P = Watts	V = Volts
I = Amps	VA = Volt Amps
A = Amps	ϕ = Phase
R = Ohms	√ = Square Root
E = Volts	

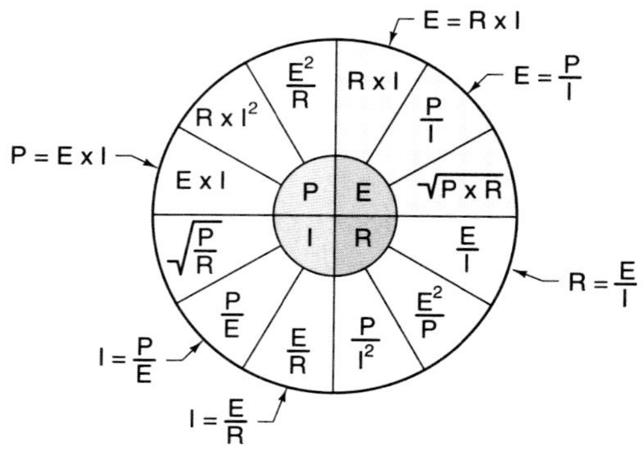

OHM'S LAW AND POWER FORMULA

POWER FORMULAS – 1ϕ, 3ϕ					
Phase	To Find	Use Formula	Example		
			Given	Find	Solution
1ϕ	I	$I = \frac{VA}{V}$	32,000 VA, 240 V	I	$I = \frac{VA}{V}$ $I = \frac{32{,}000\text{ VA}}{240\text{ V}}$ $I = \mathbf{133\ A}$
1ϕ	VA	$VA = I \times V$	100 A, 240 V	VA	$VA = I \times V$ $VA = 100\text{ A} \times 240\text{ V}$ $VA = \mathbf{24{,}000\ VA}$
1ϕ	V	$V = \frac{VA}{I}$	42,000 VA, 350 A	V	$V = \frac{VA}{I}$ $V = \frac{42{,}000\text{ VA}}{350\text{ A}}$ $V = \mathbf{120\ V}$
3ϕ	I	$I = \frac{VA}{V \times \sqrt{3}}$	72,000 VA, 208 V	I	$I = \frac{VA}{V \times \sqrt{3}}$ $I = \frac{72{,}000\text{ VA}}{360\text{ V}}$ $I = \mathbf{200\ A}$
3ϕ	VA	$VA = I \times V \times \sqrt{3}$	2 A, 240 V	VA	$VA = I \times V \times \sqrt{3}$ $VA = 2 \times 416$ $VA = \mathbf{832\ VA}$

AC/DC FORMULAS

To Find	DC	AC		
		1φ, 115 or 220 V	1φ, 208, 230, or 240 V	3φ – All Voltages
I, HP known	$\frac{HP \times 746}{E \times E_{ff}}$	$\frac{HP \times 746}{E \times E_{ff} \times PF}$	$\frac{HP \times 746}{E \times E_{ff} \times PF}$	$\frac{HP \times 746}{1.73 \times E \times E_{ff} \times PF}$
I, kW known	$\frac{kW \times 1000}{E}$	$\frac{kW \times 1000}{E \times PF}$	$\frac{kW \times 1000}{E \times PF}$	$\frac{kW \times 1000}{1.73 \times E \times PF}$
I, kVA known		$\frac{kVA \times 1000}{E}$	$\frac{kVA \times 1000}{E}$	$\frac{kVA \times 1000}{1.763 \times E}$
kW	$\frac{I \times E}{1000}$	$\frac{I \times E \times PF}{1000}$	$\frac{I \times E \times PF}{1000}$	$\frac{I \times E \times 1.73 \times PF}{1000}$
kVA		$\frac{I \times E}{1000}$	$\frac{I \times E}{1000}$	$\frac{I \times E \times 1.73}{1000}$
HP (output)	$\frac{I \times E \times E_{ff}}{746}$	$\frac{I \times E \times E_{ff} \times PF}{746}$	$\frac{I \times E \times E_{ff} \times PF}{746}$	$\frac{I \times E \times 1.73 \times E_{ff} \times PF}{746}$

E_{ff} = efficiency

HORSEPOWER FORMULAS

To Find	Use Formula	Example		
		Given	Find	Solution
HP	$HP = \frac{I \times E \times E_{ff}}{746}$	240 V, 20 A, 85% E_{ff}	HP	$HP = \frac{I \times E \times E_{ff}}{746}$ $HP = \frac{240\ V \times 20\ A \times 85\%}{746}$ $HP = \mathbf{5.5}$
I	$I = \frac{HP \times 746}{E \times E_{ff} \times PF}$	10 HP, 240 V, 90% E_{ff}, 88% PF	I	$I = \frac{HP \times 746}{E \times E_{ff} \times PF}$ $I = \frac{10\ HP \times 746}{240\ V \times 90\% \times 88\%}$ $I = \mathbf{39\ A}$

VOLTAGE DROP FORMULAS – 1φ, 3φ

Phase	To Find	Use Formula	Example		
			Given	Find	Solution
1φ	VD	$VD = \frac{2 \times R \times L \times I}{CM}$	240 V, 40 A, 60′ L, 16,510 CM, 12 R	VD	$VD = \frac{2 \times R \times L \times I}{CM}$ $VD = \frac{2 \times 12 \times 60 \times 40}{16{,}510}$ $VD = \mathbf{3.5}$
3φ	VD	$VD = \frac{2 \times R \times L \times I}{CM}$	208 V, 110 A, 75′ L, 66,360 CM, 12 R, .866 multiplier	VD	$VD = \frac{2 \times R \times L \times I}{CM}$ $VD = \frac{2 \times 12 \times 75 \times 110}{66{,}360}$ $VD = 2.98 \times .866$ $VD = \mathbf{2.58}$

$* \frac{\sqrt{3}}{2} = .866$

CAPACITORS			
Connected in Series		Connected in Parallel	Connected in Series/Parallel
Two Capacitors	Three or More Capacitors		
$C_T = \frac{C_1 \times C_2}{C_1 + C_2}$ where C_T = total capacitance (in μF) C_1 = capacitance of capacitor 1 (in μF) C_2 = capacitance of capacitor 2 (in μF)	$\frac{1}{C_T} = \frac{1}{C_1} + \frac{1}{C_2} + \ldots$	$C_T = C_1 + C_2 + \ldots$	1. Calculate capacitance of parallel branch. 2. Calculate capacitance of series combination. $C_T = \frac{C_1 \times C_2}{C_1 + C_2}$

SINE WAVES		
Frequency	Period	Peak-to-Peak Value
$f = \frac{1}{T}$ where f = frequency (in hertz) 1 = constant T = period (in seconds)	$T = \frac{1}{f}$ where T = period (in seconds) 1 = constant f = frequency (in hertz)	$V_{p-p} = 2 \times V_{max}$ where 2 = constant V_{p-p} = peak-to-peak value V_{max} = peak value

Average Value	rms Value
$V_{avg} = V_{max} \times .637$ where V_{avg} = average value (in volts) V_{max} = peak value (in volts) .637 = constant	$V_{rms} = V_{max} \times .707$ where V_{rms} = rms value (in volts) V_{max} = peak value (in volts) .707 = constant

CONDUCTIVE LEAKAGE CURRENT

$$I_L = \frac{V_A}{R_I}$$

where
I_L = leakage current (in microamperes)
V_A = applied voltage (in volts)
R_I = insulation resistance (in megohms)

TEMPERATURE CONVERSIONS	
Convert °C to °F	Convert °F to °C
$°F = (1.8 \times °C) + 32$	$°C = \frac{(°F - 32)}{1.8}$

FLOW RATE

$$Q = \frac{N \times V_d}{231}$$

where
Q = flow rate (in gpm)
N = pump drive speed (in rpm)
V_d = pump displacement (in cu in./rev)
231 = constant

BRANCH CIRCUIT VOLTAGE DROP

$$\%V_D = \frac{V_{NL} - V_{FL}}{V_{FL}} \times 100$$

where
$\%V_D$ = percent voltage drop (in volts)
V_{NL} = no-load voltage drop (in volts)
V_{FL} = full-load voltage drop (in volts)
100 = constant

POWERS OF 10

1×10^{4}	=	10,000	=	$10 \times 10 \times 10 \times 10$		Read ten to the fourth power
1×10^{3}	=	1000	=	$10 \times 10 \times 10$		Read ten to the third power or ten cubed
1×10^{2}	=	100	=	10×10		Read ten to the second power or ten squared
1×10^{1}	=	10	=	10		Read ten to the first power
1×10^{0}	=	1	=	1		Read ten to the zero power
1×10^{-1}	=	.1	=	1/10		Read ten to the minus first power
1×10^{-2}	=	.01	=	$1/(10 \times 10)$ or 1/100		Read ten to the minus second power
1×10^{-3}	=	.001	=	$1/(10 \times 10 \times 10)$ or 1/1000		Read ten to the minus third power
1×10^{-4}	=	.0001	=	$1/(10 \times 10 \times 10 \times 10)$ or 1/10,000		Read ten to the minus fourth power

UNITS OF ENERGY

Energy	Btu	ft lb	J	kcal	kWh
British thermal unit	1	777.9	1.056	0.252	2.930×10^{-4}
Foot-pound	1.285×10^{-3}	1	1.356	3.240×10^{-4}	3.766×10^{-7}
Joule	9.481×10^{-4}	0.7376	1	2.390×10^{-4}	2.778×10^{-7}
Kilocalorie	3.968	3.086	4.184	1	1.163×10^{-3}
Kilowatt-hour	3.413	2.655×10^{6}	3.6×10^{6}	860.2	1

UNITS OF POWER

Power	W	ft lb/s	HP	kW
Watt	1	0.7376	1.341×10^{-3}	0.001
Foot-pound/sec	1.356	1	1.818×10^{-3}	1.356×10^{-3}
Horsepower	745.7	550	1	0.7457
Kilowatt	1000	736.6	1.341	1

STANDARD SIZES OF FUSES AND CBs

NEC® 240-6(a) lists standard ampere ratings of fuses and fixed-trip CBs as follows:
15, 20, 25, 30, 35, 40, 45,
50, 60, 70, 80, 90, 100, 110,
125, 150, 175, 200, 225,
250, 300, 350, 400, 450,
500, 600, 700, 800,
1000, 1200, 1600,
2000, 2500, 3000, 4000, 5000, 6000

VOLTAGE CONVERSIONS

To Convert	To	Multiply By
rms	Average	.9
rms	Peak	1.414
Average	rms	1.111
Average	Peak	1.567
Peak	rms	.707
Peak	Average	.637
Peak	Peak-to-peak	2

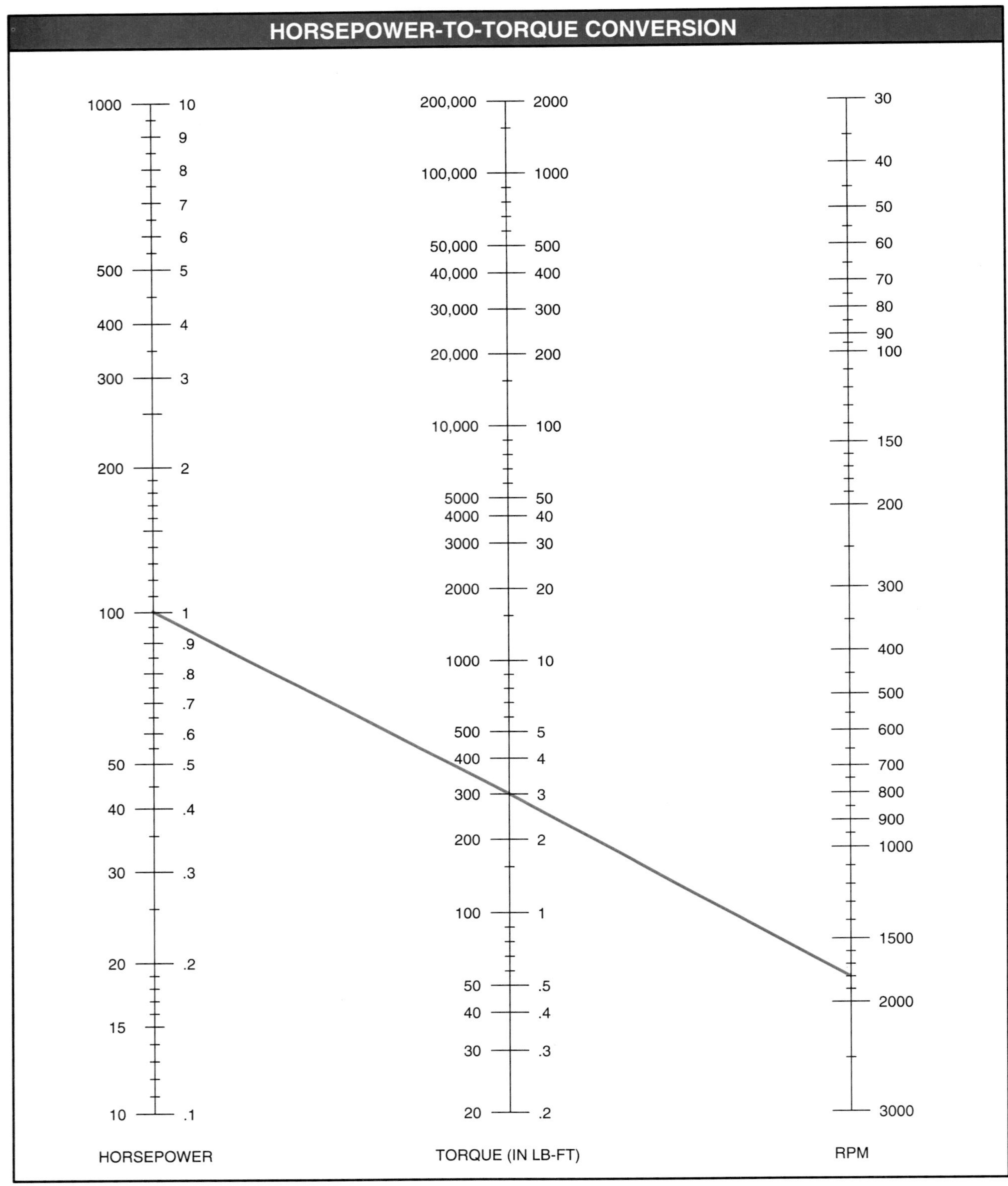
HORSEPOWER-TO-TORQUE CONVERSION
1000
10
9
8
7
6
500
5
400
4
300
3
200
2
100
1
.9
.8
.7
.6
50
.5
40
.4
30
.3
20
.2
15
10
.1
HORSEPOWER
200,000
2000
100,000
1000
50,000
500
40,000
400
30,000
300
20,000
200
10,000
100
5000
50
4000
40
3000
30
2000
20
1000
10
500
5
400
4
300
3
200
2
100
1
50
.5
40
.4
30
.3
20
.2
TORQUE (IN LB-FT)
30
40
50
60
70
80
90
100
150
200
300
400
500
600
700
800
900
1000
1500
2000
3000
RPM

ELECTRICAL ABBREVIATIONS			
Abbreviation	**Term**	**Abbreviation**	**Term**
AC	Alternating current	MB	Magnetic brake
ALM	Alarm	MCS	Motor circuit switch
AM	Ammeter	MEM	Memory
ARM	Armature	MTR	Motor
AU	Automatic	MN	Manual
BAT	Battery (electric)	NEG	Negative
BR	Brake relay	NEUT	Neutral
CAP	Capacitor	NC	Normally closed
CB	Circuit breaker	NO	Normally open
CEMF	Counter electromotive force	OHM	Ohmmeter
CKT	Circuit	OL	Overload relay
CONT	Control	PB	Pushbutton
CR	Control relay	PH	Phase
CRM	Control relay master	PLS	Plugging switch
CT	Current transformer	POS	Positive
D	Down	PRI	Primary switch
DB	Dynamic braking contactor or relay	PS	Pressure switch
DC	Direct current	R	Reverse
DIO	Diode	REC	Rectifier
DISC	Disconnect switch	RES	Resistor
DP	Double-pole	RH	Rheostat
DPDT	Double-pole, double-throw	S	Switch
DPST	Double-pole, single-throw	SCR	Silicon controlled rectifier
DS	Drum switch	SEC	Secondary
DT	Double-throw	1PH	Single-phase
EMF	Electromotive force	SOC	Socket
F	Forward	SOL	Solenoid
FLS	Flow switch	SP	Single-pole
FREQ	Frequency	SPDT	Single-pole, double-throw
FS	Float switch	SPST	Single-pole, single-throw
FTS	Foot switch	SS	Selector switch
FU	Fuse	SSW	Safety switch
GEN	Generator	T	Transformer
GRD	Ground	TB	Terminal board
IC	Integrated circuit	3PH	Three-phase
INTLK	Interlock	TD	Time delay
IOL	Instantaneous overload	THS	Thermostat switch
JB	Junction box	TR	Time delay relay
LS	Limit switch	U	Up
LT	Lamp	UCL	Unclamp
M	Motor starter	UV	Undervoltage

AC MOTOR CHARACTERISTICS

Motor Type	Typical Voltage	Starting Ability (Torque)	Size (HP)	Speed Range (rpm)	Cost*	Typical Uses
1φ						
Shaded-pole	115 V, 230 V	Very low 50% to 100% of full load	Fractional ½ HP to ⅓ HP	Fixed 900, 1200, 1800, 3600	Very low 75% to 85%	Light-duty applications such as small fans, hair dryers, blowers, and computers
Split-phase	115 V, 230 V	Low 75% to 200% of full load	Fractional ⅓ HP or less	Fixed 900, 1200, 1800, 3600	Low 85% to 95%	Low-torque applications such as pumps, blowers, fans, and machine tools
Capacitor-start	115 V, 230 V	High 200% to 350% of full load	Fractional to 3 HP	Fixed 900, 1200, 1800	Low 90% to 110%	Hard-to-start loads such as refrigerators, air compressors, and power tools
Capacitor-run	115 V, 230 V	Very low 50% to 100% of full load	Fractional to 5 HP	Fixed 900, 1200, 1800	Low 90% to 110%	Applications that require a high running torque such as pumps and conveyors
Capacitor start-and-run	115 V, 230 V	Very high 350% to 450% of full load	Fractional to 10 HP	Fixed 900, 1200, 1800	Low 100% to 115%	Applications that require both a high starting and running torque such as loaded conveyors
3φ Induction	230 V, 460 V	Low 100% to 175% of full load	Fractional to over 500 HP	Fixed 900, 1200, 3600	Low 100%	Most industrial applications
Wound rotor	230 V, 460 V	High 200% to 300% of full load	½ HP to 200 HP	Varies by changing resistance in rotor	Very high 250% to 350%	Applications that require high torque at different speeds such as cranes and elevators
Synchronous	230 V, 460 V	Very low 40% to 100% of full load	Fractional to 250 HP	Exact constant speed	High 200% to 250%	Applications that require very slow speeds and correct power factors

* based on standard 3φ induction motor

DC AND UNIVERSAL MOTOR CHARACTERISTICS

Motor Type	Typical Voltage	Starting Ability (Torque)	Size (HP)	Speed Range (rpm)	Cost*	Typical Uses
DC Series	12 V, 90 V, 120 V, 180 V	Very high 400% to 450% of full load	Fractional to 100 HP	Varies 0 to full speed	High 175% to 225%	Applications that require very high torque such as hoists and bridges
Shunt	12 V, 90 V, 120 V, 180 V	Low 125% to 250% of full load	Fractional to 100 HP	Fixed or adjustable below full speed	High 175% to 225%	Applications that require better speed control than a series motor such as woodworking machines
Compound	12 V, 90 V, 120 V, 180 V	High 300% to 400% of full load	Fractional to 100 HP	Fixed or adjustable	High 175% to 225%	Applications that require high torque and speed control such as printing presses, conveyors, and hoists
Permanent-magnet	12 V, 24 V, 36 V, 120 V	Low 100% to 200% of full load	Fractional	Varies from 0 to full speed	High 150% to 200%	Applications that require small DC-operated equipment such as automobile power windows, seats, and sun roofs
Stepping	5 V, 12 V, 24 V	Very low** .5 to 5000 oz/in.	Size rating is given as holding torque and number of steps	Rated in number of steps per sec (maximum)	Varies based on number of steps and rated torque	Applications that require low torque and precise control such as indexing tables and printers
AC/DC Universal	115 VAC, 230 VAC, 12 VDC, 24 VDC, 36 VDC, 120 VDC	High 300% to 400% of full load	Fractional	Varies 0 to full speed	High 175% to 225%	Most portable tools such as drills, routers, mixers, and vacuum cleaners

* based on standard 3φ induction motor
** torque is rated as holding torque

OVERCURRENT PROTECTION DEVICES

Motor Type	Code Letter	FLC (%)				
		Motor Size	TDF	NTDF	ITB	ITCB
AC*	—	—	175	300	150	700
AC*	A	—	150	150	150	700
AC*	B – E	—	175	250	200	700
AC*	F – V	—	175	300	250	700
DC	—	⅛ to 50 HP	150	150	150	150
DC	—	Over 50 HP	150	150	150	175

* full voltage and resistor starting

FULL-LOAD CURRENTS – DC MOTORS		
Motor rating (HP)	Current (A)	
	120 V	240 V
¼	3.1	1.6
⅓	4.1	2.0
½	5.4	2.7
¾	7.6	3.8
1	9.5	4.7
1½	13.2	6.6
2	17	8.5
3	25	12.2
5	40	20
7½	58	29

FULL-LOAD CURRENTS – 1ϕ, AC MOTORS		
Motor rating (HP)	Current (A)	
	115 V	230 V
⅙	4.4	2.2
¼	5.8	2.9
⅓	7.2	3.6
½	9.8	4.9
¾	13.8	6.9
1	16	8
1½	20	10
2	24	12
3	34	17
5	56	28
7½	80	40

FULL-LOAD CURRENTS – 3ϕ, AC INDUCTION MOTORS				
Motor rating (HP)	Current (A)			
	208 V	230 V	460 V	575 V
¼	1.11	.96	.48	.38
⅓	1.34	1.18	.59	.47
½	2.2	2.0	1.0	.8
¾	3.1	2.8	1.4	1.1
1	4.0	3.6	1.8	1.4
1½	5.7	5.2	2.6	2.1
2	7.5	6.8	3.4	2.7
3	10.6	9.6	4.8	3.9
5	16.7	15.2	7.6	6.1
7½	24.0	22.0	11.0	9.0
10	31.0	28.0	14.0	11.0
15	46.0	42.0	21.0	17.0
20	59	54	27	22
25	75	68	34	27
30	88	80	40	32
40	114	104	52	41
50	143	130	65	52
60	169	154	77	62
75	211	192	96	77
100	273	248	124	99
125	343	312	156	125
150	396	360	180	144
200	—	480	240	192
250	—	602	301	242
300	—	—	362	288
350	—	—	413	337
400	—	—	477	382
500	—	—	590	472

TYPICAL MOTOR EFFICIENCIES					
HP	Standard Motor (%)	Energy-Efficient Motor (%)	HP	Standard Motor (%)	Energy-Efficient Motor (%)
1	76.5	84.0	30	88.1	93.1
1.5	78.5	85.5	40	89.3	93.6
2	79.9	86.5	50	90.4	93.7
3	80.8	88.5	75	90.8	95.0
5	83.1	88.6	100	91.6	95.4
7.5	83.8	90.2	125	91.8	95.8
10	85.0	90.3	150	92.3	96.0
15	86.5	91.7	200	93.3	96.1
20	87.5	92.4	250	93.6	96.2
25	88.0	93.0	300	93.8	96.5

MOTOR FRAME DIMENSIONS . . .

Frame No.	Shaft U	Shaft V	Key W	Key T	Key L	Dimensions – Inches A	B	D	E	F	BA
48	1/2	1 1/2*	flat	3/64	—	5 5/8*	3 1/2*	3	2 1/8	1 3/8	2 1/2
56	5/8	1 7/8*	3/16	3/16	1 3/8	6 1/2*	4 1/4*	3 1/2	2 7/16	1 1/2	2 3/4
143T	7/8	2	3/16	3/16	1 3/8	7	6	3 1/2	2 3/4	2	2 1/4
145T	7/8	2	3/16	3/16	1 3/8	7	6	3 1/2	2 3/4	2 1/2	2 1/4
182	7/8	2	3/16	3/16	1 3/8	9	6 1/2	4 1/2	3 3/4	2 1/4	2 3/4
182T	1 1/8	2 1/2	1/4	1/4	1 3/4	9	6 1/2	4 1/2	3 3/4	2 1/4	2 3/4
184	7/8	2	3/16	3/16	1 3/8	9	7 1/2	4 1/2	3 3/4	2 3/4	2 3/4
184T	1 1/8	2 1/2	1/4	1/4	1 3/4	9	7 1/2	4 1/2	3 3/4	2 3/4	2 3/4
203	3/4	2	3/16	3/16	1 3/8	10	7 1/2	5	4	2 3/4	3 1/8
204	3/4	2	3/16	3/16	1 3/8	10	8 1/2	5	4	3 1/4	3 1/8
213	1 1/8	2 3/4	1/4	1/4	2	10 1/2	7 1/2	5 1/4	4 1/4	2 3/4	3 1/2
213T	1 3/8	3 1/8	5/16	5/16	2 3/8	10 1/2	7 1/2	5 1/4	4 1/4	2 3/4	3 1/2
215	1 1/8	2 3/4	1/4	1/4	2	10 1/2	9	5 1/4	4 1/4	3 1/2	3 1/2
215T	1 3/8	3 1/8	5/16	5/16	2 3/8	10 1/2	9	5 1/4	4 1/4	3 1/2	3 1/2
224	1	2 3/4	1/4	1/4	2	11	8 3/4	5 1/2	4 1/2	3 3/8	3 1/2
225	1	2 3/4	1/4	1/4	2	11	9 1/2	5 1/2	4 1/2	3 3/4	3 1/2
254	1 1/8	3 1/8	1/4	1/4	2 3/8	12 1/2	10 3/4	6 1/4	5	4 1/8	4 1/4
254U	1 3/8	3 1/2	5/16	5/16	2 3/4	12 1/2	10 3/4	6 1/4	5	4 1/8	4 1/4
254T	1 5/8	3 3/4	3/8	3/8	2 7/8	12 1/2	10 3/4	6 1/4	5	4 1/8	4 1/4
256U	1 3/8	3 1/2	5/16	5/16	2 3/4	12 1/2	12 1/2	6 1/4	5	5	4 1/4
256T	1 5/8	3 3/4	3/8	3/8	2 7/8	12 1/2	12 1/2	6 1/4	5	5	4 1/4
284	1 1/4	3 1/2	1/4	1/4	2 3/4	14	12 1/2	7	5 1/2	4 3/4	4 3/4
284U	1 5/8	4 5/8	3/8	3/8	3 3/4	14	12 1/2	7	5 1/2	4 3/4	4 3/4
284T	1 7/8	4 3/8	1/2	1/2	3 1/4	14	12 1/2	7	5 1/2	4 3/4	4 3/4
284TS	1 5/8	3	3/8	3/8	1 7/8	14	12 1/2	7	5 1/2	4 3/4	4 3/4
286U	1 5/8	4 5/8	3/8	3/8	3 3/4	14	14	7	5 1/2	5 1/2	4 3/4
286T	1 7/8	4 3/8	1/2	1/2	3 1/4	14	14	7	5 1/2	5 1/2	4 3/4
286TS	1 5/8	3	3/8	3/8	1 7/8	14	14	7	5 1/2	5 1/2	4 3/4
324	1 5/8	4 5/8	3/8	3/8	3 3/4	16	14	8	6 1/4	5 1/4	5 1/4
324U	1 7/8	5 3/8	1/2	1/2	4 1/4	16	14	8	6 1/4	5 1/4	5 1/4
324S	1 5/8	3	3/8	3/8	1 7/8	16	14	8	6 1/4	5 1/4	5 1/4
324T	2 1/8	5	1/2	1/2	3 7/8	16	14	8	6 1/4	5 1/4	5 1/4
324TS	1 7/8	3 1/2	1/2	1/2	2	16	14	8	6 1/4	5 1/4	5 1/4
326	1 5/8	4 5/8	3/8	3/8	3 3/4	16	15 1/2	8	6 1/4	6	5 1/4
326U	1 7/8	5 3/8	1/2	1/2	4 1/4	16	15 1/2	8	6 1/4	6	5 1/4
326S	1 5/8	3	3/8	3/8	1 7/8	16	15 1/2	8	6 1/4	6	5 1/4
326T	2 1/8	5	1/2	1/2	3 7/8	16	15 1/2	8	6 1/4	6	5 1/4
326TS	1 7/8	3 1/2	1/2	1/2	2	16	15 1/2	8	6 1/4	6	5 1/4
364	1 7/8	5 3/8	1/2	1/2	4 1/4	18	15 1/4	9	7	5 5/8	5 7/8
364S	1 5/8	3	3/8	3/8	1 7/8	18	15 1/4	9	7	5 5/8	5 7/8
364U	2 1/8	6 1/8	1/2	1/2	5	18	15 1/4	9	7	5 5/8	5 7/8

. . . MOTOR FRAME DIMENSIONS											
Frame No.	Shaft		Key			Dimensions – Inches					
	U	V	W	T	L	A	B	D	E	F	BA
364US	1⅞	3½	½	½	2	18	15¼	9	7	5⅝	5⅞
405	2⅛	6⅛	½	½	5	20	17¾	10	8	6⅞	6⅝
405S	1⅞	3½	½	½	2	20	17¾	10	8	6⅞	6⅝
405U	2⅜	6⅞	⅝	⅝	5½	20	17¾	10	8	6⅞	6⅝
405US	2⅛	4	½	½	2¾	20	17¾	10	8	6⅞	6⅝
405T	2⅞	7	¾	¾	5⅝	20	17¾	10	8	6⅞	6⅝
405TS	2⅛	4	½	½	2¾	20	17¾	10	8	6⅞	6⅝
444	2⅜	6⅞	⅝	⅝	5½	22	18½	11	9	7¼	7½
444S	2⅛	4	½	½	2¾	22	18½	11	9	7¼	7½
444U	2⅞	8⅜	¾	¾	7	22	18½	11	9	7¼	7½
444US	2⅛	4	½	½	2¾	22	18½	11	9	7¼	7½
444T	3⅜	8¼	⅞	⅞	6⅞	22	18½	11	9	7¼	7½
444TS	2⅜	4½	⅝	⅝	3	22	18½	11	9	7¼	7½
445	2⅜	6⅞	⅝	⅝	5½	22	20½	11	9	8¼	7½
445S	2⅛	4	½	½	2¾	22	20½	11	9	8¼	7½
445U	2⅞	8⅜	¾	¾	7	22	20½	11	9	8¼	7½
445US	2⅛	4	½	½	2¾	22	20½	11	9	8¼	7½
445T	3⅜	8¼	⅞	⅞	6⅞	22	20½	11	9	8¼	7½
445TS	2⅜	4½	⅝	⅝	3	22	20½	11	9	8¼	7½
504U	2⅞	8⅜	¾	¾	7¼	25	21	12½	10	8	8½
504S	2⅛	4	½	½	2¾	25	21	12½	10	8	8½
505	2⅞	8⅜	¾	¾	7¼	25	23	12½	10	9	8½
505S	2⅛	4	½	½	2¾	25	23	12½	10	9	8½

* not NEMA standard dimensions

MOTOR FRAME LETTERS	
LETTER	DESIGNATION
G	Gasoline pump motor
K	Sump pump motor
M and N	Oil burner motor
S	Standard short shaft for direct connection
T	Standard dimensions established
U	Previously used as frame designation for which standard dimensions are established
Y	Special mounting dimensions required from manufacturer
Z	Standard mounting dimensions except shaft extension

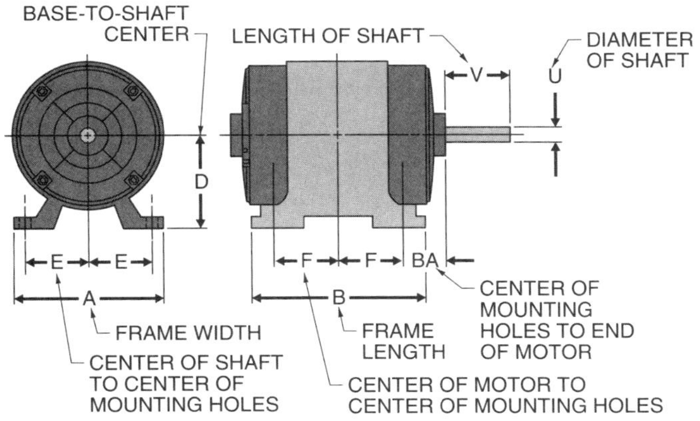

MOTOR FRAME TABLE

Frame No. Series	Third/Fourth Digit of Frame No.							
	D	1	2	3	4	5	6	7
140	3.50	3.00	3.50	4.00	4.50	5.00	5.50	6.25
160	4.00	3.50	4.00	4.50	5.00	5.50	6.25	7.00
180	4.50	4.00	4.50	5.00	5.50	6.25	7.00	8.00
200	5.00	4.50	5.00	5.50	6.50	7.00	8.00	9.00
210	5.25	4.50	5.00	5.50	6.25	7.00	8.00	9.00
220	5.50	5.00	5.50	6.25	6.75	7.50	9.00	10.00
250	6.25	5.50	6.25	7.00	8.25	9.00	10.00	11.00
280	7.00	6.25	7.00	8.00	9.50	10.00	11.00	12.50
320	8.00	7.00	8.00	9.00	10.50	11.00	12.00	14.00
360	9.00	8.00	9.00	10.00	11.25	12.25	14.00	16.00
400	10.00	9.00	10.00	11.00	12.25	13.75	16.00	18.00
440	11.00	10.00	11.00	12.50	14.50	16.50	18.00	20.00
500	12.50	11.00	12.50	14.00	16.00	18.00	20.00	22.00
580	14.50	12.50	14.00	16.00	18.00	20.00	22.00	25.00
680	17.00	16.00	18.00	20.00	22.00	25.00	28.00	32.00

Frame No. Series	Third/Fourth Digit of Frame No.								
	D	8	9	10	11	12	13	14	15
140	3.50	7.00	8.00	9.00	10.00	11.00	12.50	14.00	16.00
160	4.00	8.00	9.00	10.00	11.00	12.50	14.00	16.00	18.00
180	4.50	9.00	10.00	11.00	12.50	14.00	16.00	18.00	20.00
200	5.00	10.00	11.00	—	—	—	—	—	—
210	5.25	10.00	11.00	12.50	14.00	16.00	18.00	20.00	22.00
220	5.50	11.00	12.50	—	—	—	—	—	—
250	6.25	12.50	14.00	16.00	18.00	20.00	22.00	25.00	28.00
280	7.00	14.00	16.00	18.00	20.00	22.00	25.00	28.00	32.00
320	8.00	16.00	18.00	20.00	22.00	25.00	28.00	32.00	36.00
360	9.00	18.00	20.00	22.00	25.00	28.00	32.00	36.00	40.00
400	10.00	20.00	22.00	25.00	28.00	32.00	36.00	40.00	45.00
440	11.00	22.00	25.00	28.00	32.00	36.00	40.00	45.00	50.00
500	12.50	25.00	28.00	32.00	36.00	40.00	45.00	50.00	56.00
580	14.50	28.00	32.00	36.00	40.00	45.00	50.00	56.00	63.00
680	17.00	36.00	40.00	45.00	50.00	56.00	63.00	71.00	80.00

MOTOR RATINGS

Classification	Rating	Size
Milli	W	1, 1.5, 2, 3, 5, 7.5, 10, 15, 25, 35
Fractional	HP	1/20, 1/12, 1/8, 1/6, 1/4, 1/3, 1/2, 3/4
Full	HP	1, 1½, 2, 3, 5, 7½, 10, 15, 20, 25, 30, 40, 50, 60, 75, 100, 125, 150, 200, 250, 300
Full-Special Order	HP	350, 400, 450, 500, 600, 700, 800, 900, 1000, 1250, 1500, 1750, 2000, 2250, 2500, 3000, 3500, 4000, 4500, 5000, 5500, 6000, 7000, 8000, 9000, 10,000, 11,000, 12,000, 13,000, 14,000, 15,000, 16,000, 17,000, 18,000, 19,000, 20,000, 22,500, 30,000, 32,500, 35,000, 37,500, 40,000, 45,000, 50,000

3ϕ, 230 V MOTORS AND CIRCUITS – 240 V SYSTEM

1		2		3	4	5				6	
Size of motor		Motor overload protection: Low-peak or Fusetron®				Controller termination temperature rating				Minimum size of copper wire and trade conduit	
						60°C		75°C			
HP	Amp	Motor less than 40°C or greater than 1.15 SF (Max fuse 125%)	All other motors (Max fuse 115%)	Switch 115% minimum or HP rated or fuse holder size	Minimum size of starter	TW	THW	TW	THW	Wire size (AWG or kcmil)	Conduit (inches)
½	2	2½	2¼	30	00	•	•	•	•	14	½
¾	2.8	3½	3 2/10	30	00	•	•	•	•	14	½
1	3.6	4½	4	30	00	•	•	•	•	14	½
1½	5.2	6¼	5 6/10	30	00	•	•	•	•	14	½
2	6.8	8	7½	30	0	•	•	•	•	14	½
3	9.6	12	10	30	0	•	•	•	•	14	½
5	15.2	17½	17½	30	1	•	•	•	•	14	½
7½	22	25	25	30	1	•	•	•	•	10	½
10	28	35	30	60	2	•	•	•		8	¾
									•	10	½
15	42	50	45	60	2	•	•	•	•	6	1
										6	¾
20	54	60	60	100	3	•	•	•	•	4	1
25	68	80	75	100	3	•	•			3	1¼
								•		3	1
									•	4	1
30	80	100	90	100	3	•	•	•		1	1¼
									•	3	1¼
40	104	125	110	200	4	•	•	•		2/0	1½
									•	1	1¼
50	130	150	150	200	4	•	•	•		3/0	2
									•	2/0	1½
75	192	225	200	400	5	•	•	•		300	2½
									•	250	2½
100	248	300	250	400	5	•	•	•		500	3
									•	350	2½
150	360	450	400	600	6	•	•	•		300-2/ϕ*	2-2½*
									•	4/0-2/ϕ*	2-2*

* two sets of multiple conductors and two runs of conduit required

3ϕ, 460 V MOTORS AND CIRCUITS – 480 V SYSTEM											
1		2		3	4	5				6	
Size of motor		Motor overload protection Low-peak or Fusetron®				Controller termination temperature rating				Minimum size of copper wire and trade conduit	
						60°C		75°C			
HP	Amp	Motor less than 40°C or greater than 1.15 SF (Max fuse 125%)	All other motors (Max fuse 115%)	Switch 115% minimum or HP rated or fuse holder size	Minimum size of starter	TW	THW	TW	THW	Wire size (AWG or kcmil)	Conduit (inches)
½	1	1¼	1⅛	30	00	•	•	•	•	14	½
¾	1.4	1 6/10	1 6/10	30	00	•	•	•	•	14	½
1	1.8	2¼	2	30	00	•	•	•	•	14	½
1½	2.6	3 2/10	2 8/10	30	00	•	•	•	•	14	½
2	3.4	4	3½	30	00	•	•	•	•	14	½
3	4.8	5 6/10	5	30	0	•	•	•	•	14	½
5	7.6	9	8	30	0	•	•	•	•	14	½
7½	11	12	12	30	1	•	•	•	•	14	½
10	14	17½	15	30	1	•	•	•	•	14	½
15	21	25	20	30	2	•	•	•	•	10	½
20	27	30	30	60	2	•	•	•		8	¾
									•	10	½
25	34	40	35	60	2	•	•	•		6	1
									•	8	¾
30	40	50	45	60	3	•	•	•		6	1
									•	8	¾
40	52	60	60	100	3	•	•	•		4	1
									•	6	1
50	65	80	70	100	3	•	•	•		3	1¼
									•	4	1
60	77	90	80	100	4	•	•	•		1	1¼
									•	3	1¼
75	96	110	110	200	4	•	•	•		1/0	1½
									•	1	1¼
100	124	150	125	200	4	•	•	•		3/0	2
									•	2/0	1½
125	156	175	175	200	5	•	•	•		4/0	2
									•	3/0	2
150	180	225	200	400	5	•	•	•		300	2½
									•	4/0	2
200	240	300	250	400	5	•	•	•		500	3
									•	350	2½
250	302	350	325	400	6	•	•	•		4/0-2/ϕ*	2-2*
									•	3/0-2/ϕ*	2-2*
300	361	450	400	600	6	•	•	•		300-2/ϕ*	2-1½ *
									•	4/0-2/ϕ*	2-2*

* two sets of multiple conductors and two runs of conduit required

ENCLOSURES

Type	Use	Service Conditions	Tests	Comments	Type
1	Indoor	No unusual	Rod entry, rust resistance		1
3	Outdoor	Windblown dust, rain, sleet, and ice on enclosure	Rain, external icing, dust, and rust resistance	Do not provide protection against internal condensation or internal icing	
3R	Outdoor	Falling rain and ice on enclosure	Rod entry, rain, external icing, and rust resistance	Do not provide protection against dust, internal condensation, or internal icing	
4	Indoor/outdoor	Windblown dust and rain, splashing water, hose-directed water, and ice on enclosure	Hosedown, external icing, and rust resistance	Do not provide protection against internal condensation or internal icing	4
4X	Indoor/outdoor	Corrosion, windblown dust and rain, splashing water, hose-directed water, and ice on enclosure	Hosedown, external icing, and corrosion resistance	Do not provide protection against internal condensation or internal icing	4X
6	Indoor/outdoor	Occasional temporary submersion at a limited depth			
6P	Indoor/outdoor	Prolonged submersion at a limited depth			
7	Indoor locations classified as Class I, Groups A, B, C, or D, as defined in the NEC®	Withstand and contain an internal explosion of specified gases, contain an explosion sufficiently so an explosive gas-air mixture in the atmosphere is not ignited	Explosion, hydrostatic, and temperature	Enclosed heat-generating devices shall not cause external surfaces to reach temperatures capable of igniting explosive gas-air mixtures in the atmosphere	7
9	Indoor locations classified as Class II, Groups E or G, as defined in the NEC®	Dust	Dust penetration, temperature, and gasket aging	Enclosed heat-generating devices shall not cause external surfaces to reach temperatures capable of igniting explosive gas-air mixtures in the atmosphere	9
12	Indoor	Dust, falling dirt, and dripping noncorrosive liquids	Drip, dust, and rust resistance	Do not provide protection against internal condensation	12
13	Indoor	Dust, spraying water, oil, and noncorrosive coolant	Oil explosion and rust resistance	Do not provide protection against internal condensation	

HAZARDOUS LOCATIONS

Class	Division	Group	Material
I	1 or 2	A	Acetylene
	1 or 2	B	Hydrogen, butadiene, ethylene oxide, propylene oxide
	1 or 2	C	Carbon monoxide, ether, ethylene, hydrogen sulfide, morpholine, cyclopropane
	1 or 2	D	Gasoline, benzene, butane, propane, alcohol, acetone, ammonia, vinyl chloride
II	1 or 2	E	Metal dusts
	1 or 2	F	Carbon black, coke dust, coal
	1 or 2	G	Grain dust, flour, starch, sugar, plastics
III	1 or 2	No groups	Wood chips, cotton, flax, and nylon

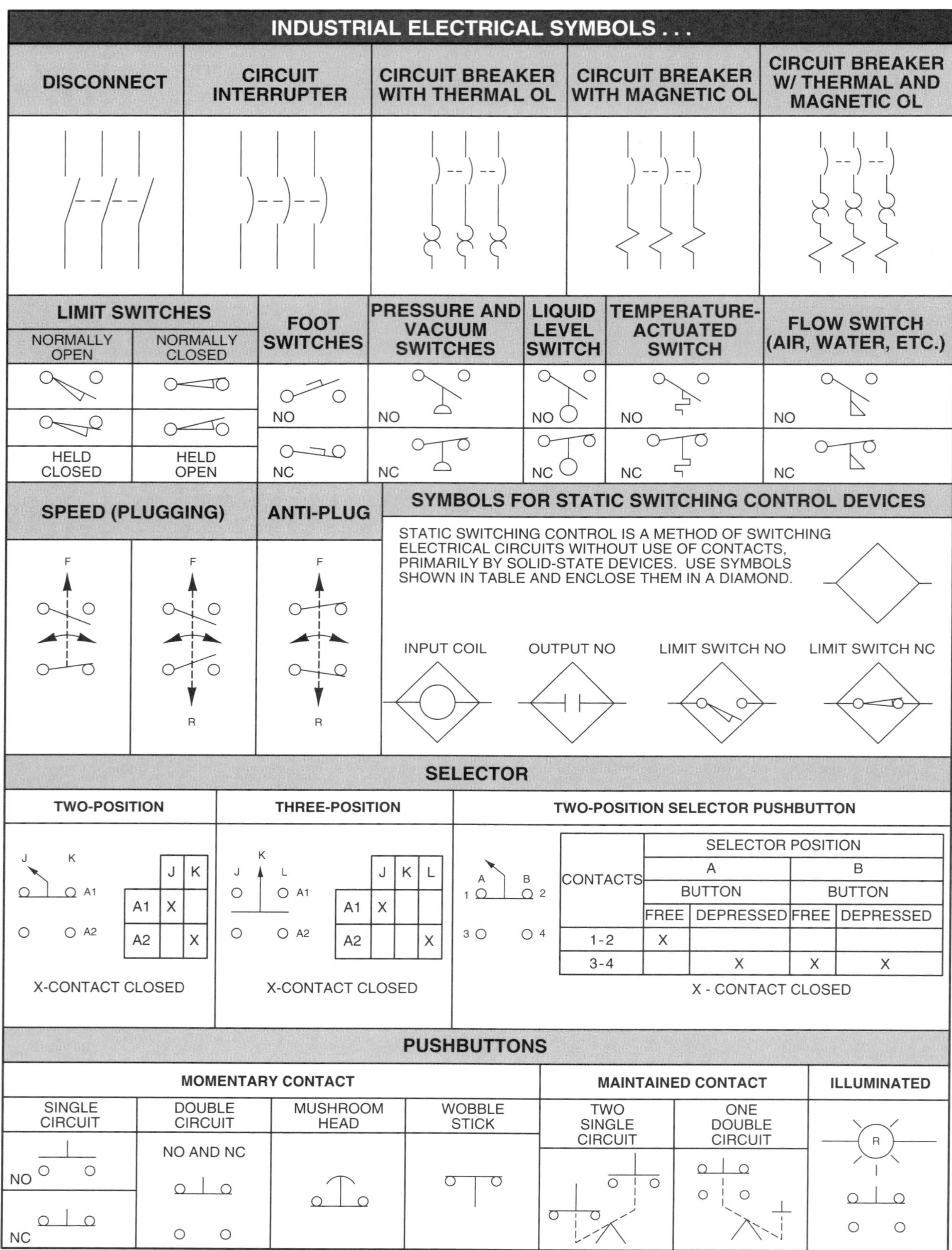
INDUSTRIAL ELECTRICAL SYMBOLS . . .
DISCONNECT
CIRCUIT INTERRUPTER
CIRCUIT BREAKER WITH THERMAL OL
CIRCUIT BREAKER WITH MAGNETIC OL
CIRCUIT BREAKER W/ THERMAL AND MAGNETIC OL
LIMIT SWITCHES
NORMALLY OPEN
NORMALLY CLOSED
HELD CLOSED
HELD OPEN
FOOT SWITCHES
PRESSURE AND VACUUM SWITCHES
LIQUID LEVEL SWITCH
TEMPERATURE-ACTUATED SWITCH
FLOW SWITCH (AIR, WATER, ETC.)
NO
NC
SPEED (PLUGGING)
ANTI-PLUG
F
R
SYMBOLS FOR STATIC SWITCHING CONTROL DEVICES
STATIC SWITCHING CONTROL IS A METHOD OF SWITCHING ELECTRICAL CIRCUITS WITHOUT USE OF CONTACTS, PRIMARILY BY SOLID-STATE DEVICES. USE SYMBOLS SHOWN IN TABLE AND ENCLOSE THEM IN A DIAMOND.
INPUT COIL
OUTPUT NO
LIMIT SWITCH NO
LIMIT SWITCH NC
SELECTOR
TWO-POSITION
J K
A1
A2
J | K
A1 | X |
A2 | | X
X-CONTACT CLOSED
THREE-POSITION
J K L
A1
A2
J | K | L
A1 | X | |
A2 | | | X
X-CONTACT CLOSED
TWO-POSITION SELECTOR PUSHBUTTON
A B
1 2
3 4
CONTACTS | SELECTOR POSITION
A | B
BUTTON | BUTTON
FREE | DEPRESSED | FREE | DEPRESSED
1-2 | X | | |
3-4 | | X | X | X
X - CONTACT CLOSED
PUSHBUTTONS
MOMENTARY CONTACT
SINGLE CIRCUIT
NO
NC
DOUBLE CIRCUIT
NO AND NC
MUSHROOM HEAD
WOBBLE STICK
MAINTAINED CONTACT
TWO SINGLE CIRCUIT
ONE DOUBLE CIRCUIT
ILLUMINATED
R

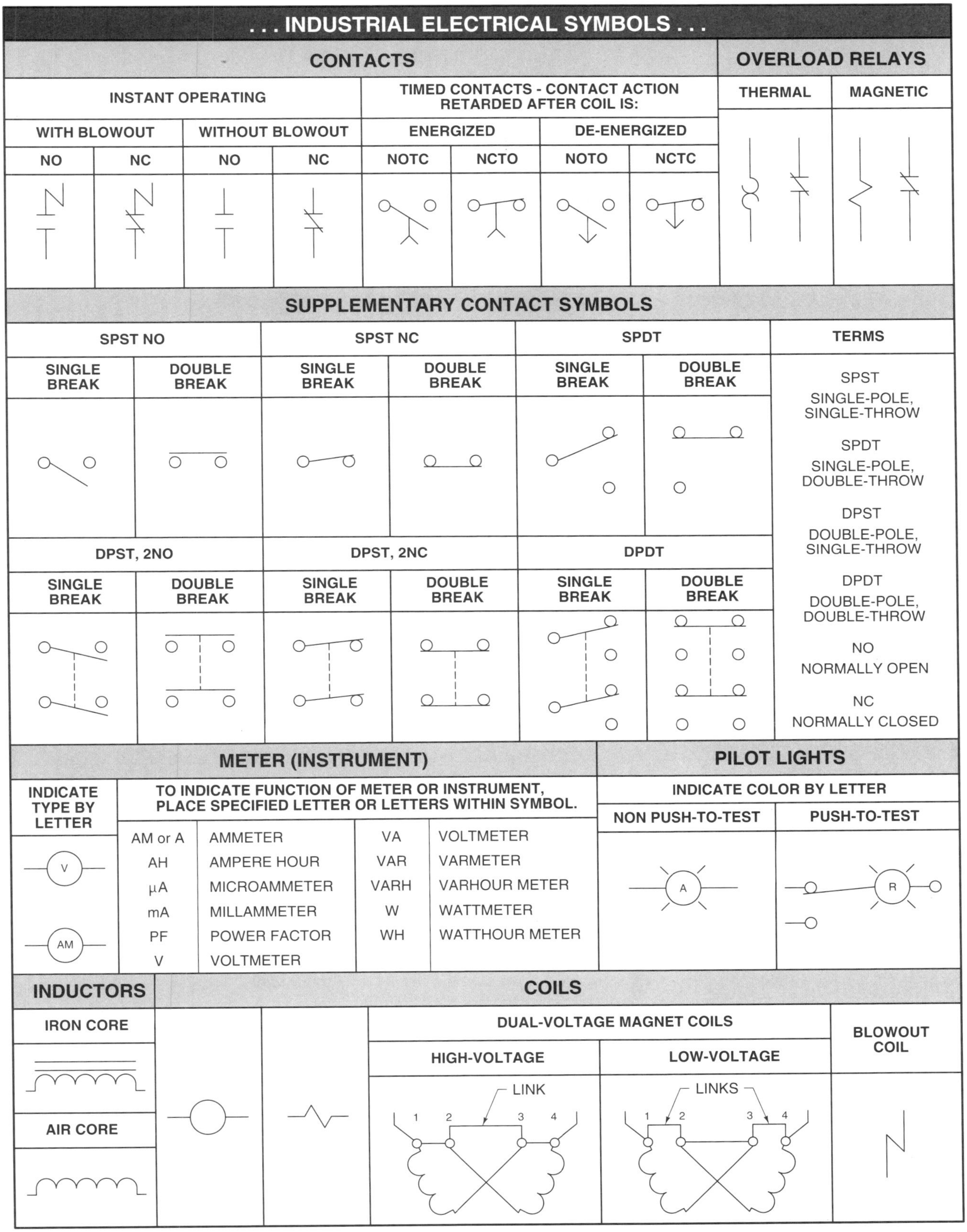

. . . INDUSTRIAL ELECTRICAL SYMBOLS . . .

CONTACTS

INSTANT OPERATING				TIMED CONTACTS - CONTACT ACTION RETARDED AFTER COIL IS:			
WITH BLOWOUT		WITHOUT BLOWOUT		ENERGIZED		DE-ENERGIZED	
NO	NC	NO	NC	NOTC	NCTO	NOTO	NCTC

OVERLOAD RELAYS

THERMAL	MAGNETIC

SUPPLEMENTARY CONTACT SYMBOLS

SPST NO		SPST NC		SPDT	
SINGLE BREAK	DOUBLE BREAK	SINGLE BREAK	DOUBLE BREAK	SINGLE BREAK	DOUBLE BREAK

DPST, 2NO		DPST, 2NC		DPDT	
SINGLE BREAK	DOUBLE BREAK	SINGLE BREAK	DOUBLE BREAK	SINGLE BREAK	DOUBLE BREAK

TERMS

SPST
SINGLE-POLE, SINGLE-THROW

SPDT
SINGLE-POLE, DOUBLE-THROW

DPST
DOUBLE-POLE, SINGLE-THROW

DPDT
DOUBLE-POLE, DOUBLE-THROW

NO
NORMALLY OPEN

NC
NORMALLY CLOSED

METER (INSTRUMENT)

INDICATE TYPE BY LETTER: V, AM

TO INDICATE FUNCTION OF METER OR INSTRUMENT, PLACE SPECIFIED LETTER OR LETTERS WITHIN SYMBOL.

AM or A	AMMETER	VA	VOLTMETER
AH	AMPERE HOUR	VAR	VARMETER
μA	MICROAMMETER	VARH	VARHOUR METER
mA	MILLAMMETER	W	WATTMETER
PF	POWER FACTOR	WH	WATTHOUR METER
V	VOLTMETER		

PILOT LIGHTS

INDICATE COLOR BY LETTER

NON PUSH-TO-TEST	PUSH-TO-TEST
A	R

INDUCTORS

IRON CORE

AIR CORE

COILS

DUAL-VOLTAGE MAGNET COILS

HIGH-VOLTAGE	LOW-VOLTAGE
LINK; 1 2 3 4	LINKS; 1 2 3 4

BLOWOUT COIL

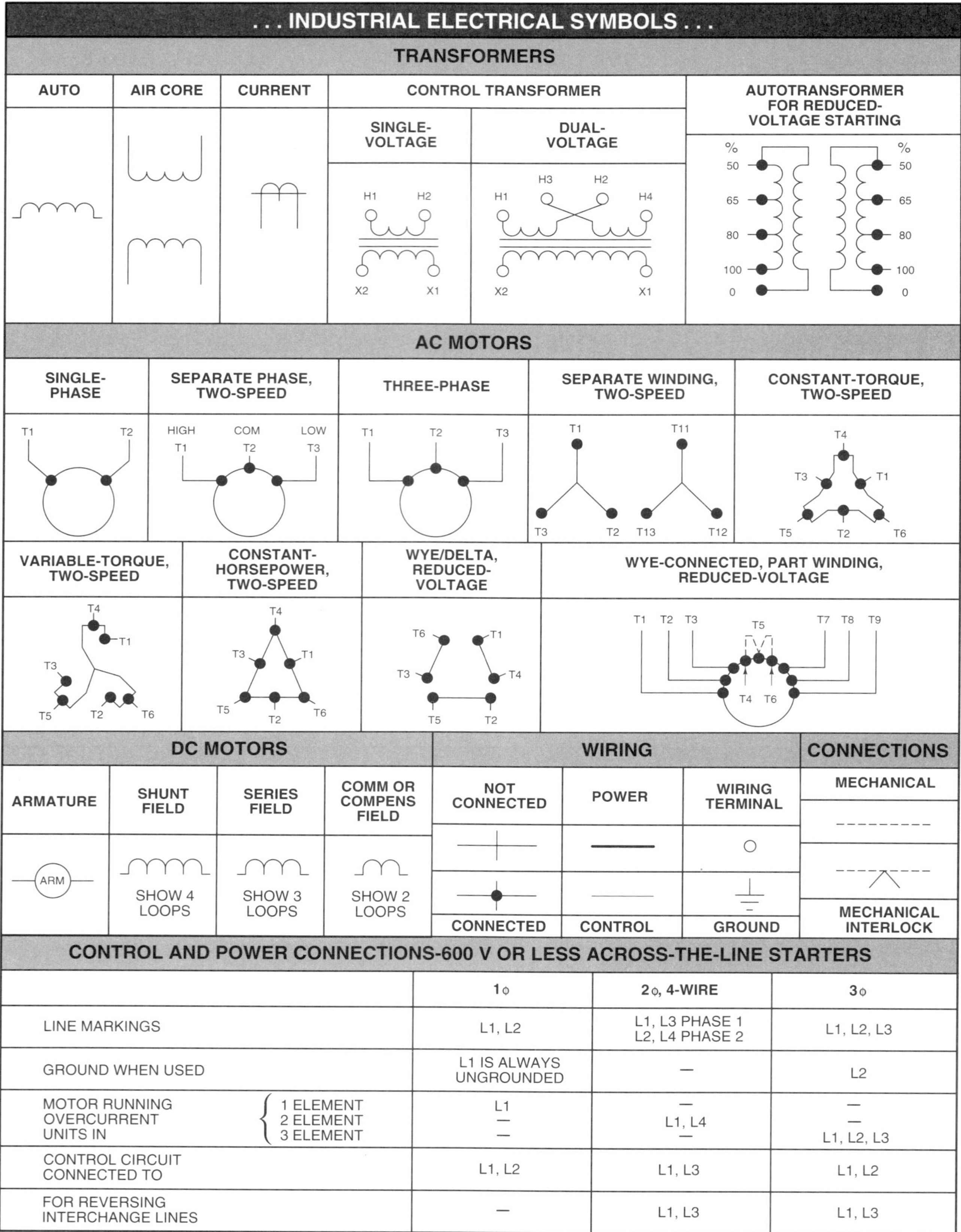

CONTROL AND POWER CONNECTIONS-600 V OR LESS ACROSS-THE-LINE STARTERS

	1φ	2φ, 4-WIRE	3φ
LINE MARKINGS	L1, L2	L1, L3 PHASE 1 L2, L4 PHASE 2	L1, L2, L3
GROUND WHEN USED	L1 IS ALWAYS UNGROUNDED	—	L2
MOTOR RUNNING OVERCURRENT UNITS IN { 1 ELEMENT 2 ELEMENT 3 ELEMENT	L1 — —	— L1, L4 —	— — L1, L2, L3
CONTROL CIRCUIT CONNECTED TO	L1, L2	L1, L3	L1, L2
FOR REVERSING INTERCHANGE LINES	—	L1, L3	L1, L3

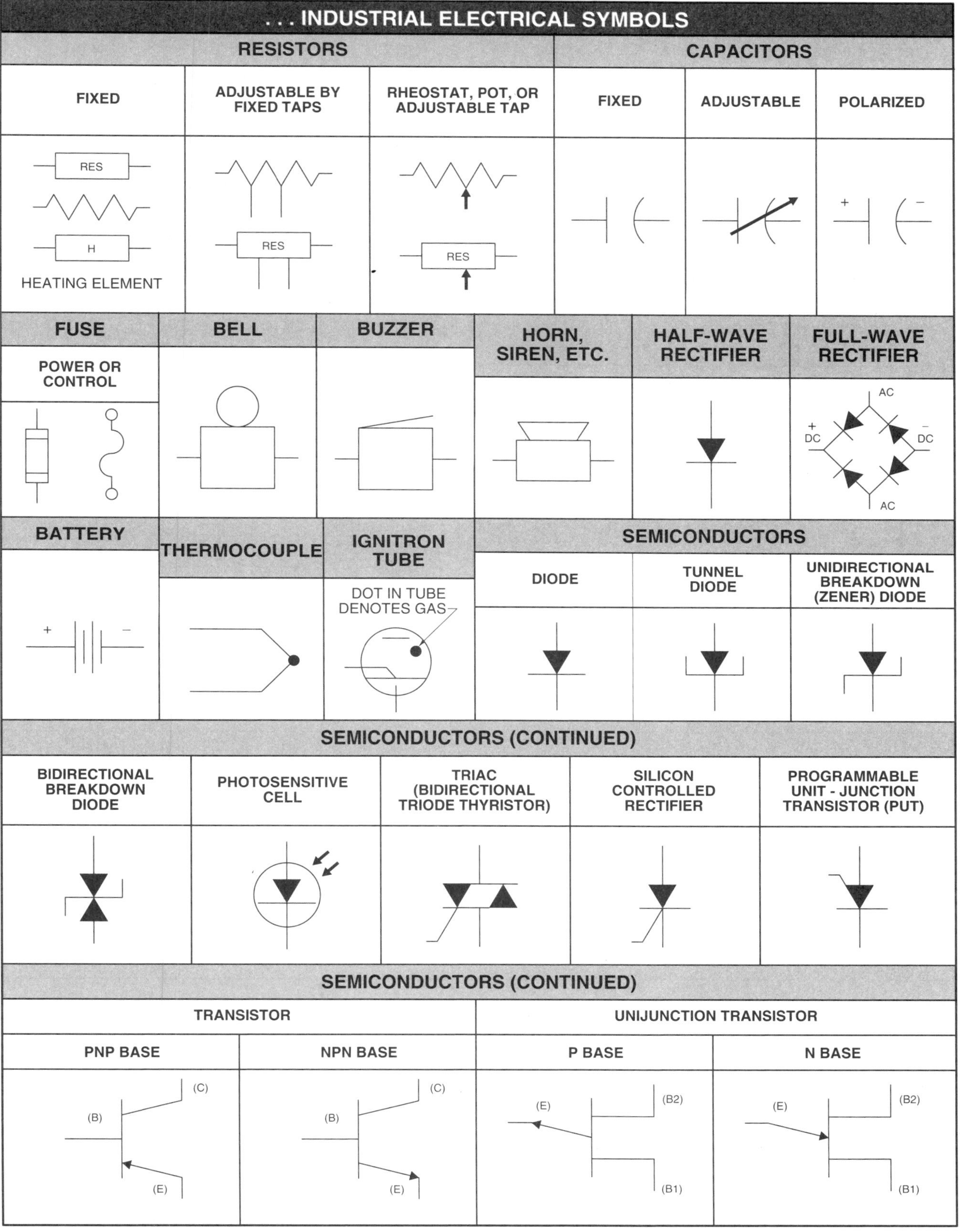
. . . INDUSTRIAL ELECTRICAL SYMBOLS
RESISTORS
FIXED
ADJUSTABLE BY FIXED TAPS
RHEOSTAT, POT, OR ADJUSTABLE TAP
RES
H
HEATING ELEMENT
CAPACITORS
FIXED
ADJUSTABLE
POLARIZED
FUSE
POWER OR CONTROL
BELL
BUZZER
HORN, SIREN, ETC.
HALF-WAVE RECTIFIER
FULL-WAVE RECTIFIER
AC
DC
BATTERY
THERMOCOUPLE
IGNITRON TUBE
DOT IN TUBE DENOTES GAS
SEMICONDUCTORS
DIODE
TUNNEL DIODE
UNIDIRECTIONAL BREAKDOWN (ZENER) DIODE
SEMICONDUCTORS (CONTINUED)
BIDIRECTIONAL BREAKDOWN DIODE
PHOTOSENSITIVE CELL
TRIAC (BIDIRECTIONAL TRIODE THYRISTOR)
SILICON CONTROLLED RECTIFIER
PROGRAMMABLE UNIT - JUNCTION TRANSISTOR (PUT)
SEMICONDUCTORS (CONTINUED)
TRANSISTOR
PNP BASE
NPN BASE
UNIJUNCTION TRANSISTOR
P BASE
N BASE
(B)
(C)
(E)
(B1)
(B2)

LOGIC SYMBOLS					
LOGIC ELEMENT	**AND**	**OR**	**NOT**	**NAND**	**NOR**
LOGIC ELEMENT FUNCTION	OUTPUT IF ALL CONTROL INPUT SIGNALS ARE ON	OUTPUT IF ANY ONE OF THE CONTROL INPUTS IS ON	OUTPUT IF SINGLE CONTROL INPUT SIGNAL IS OFF	OUTPUT IF ALL CONTROL INPUT SIGNALS ARE ON	OUTPUT IF ALL CONTROL INPUT SIGNALS ARE OFF
MIL-STD-806B AND ELECTRONIC LOGIC SYMBOL	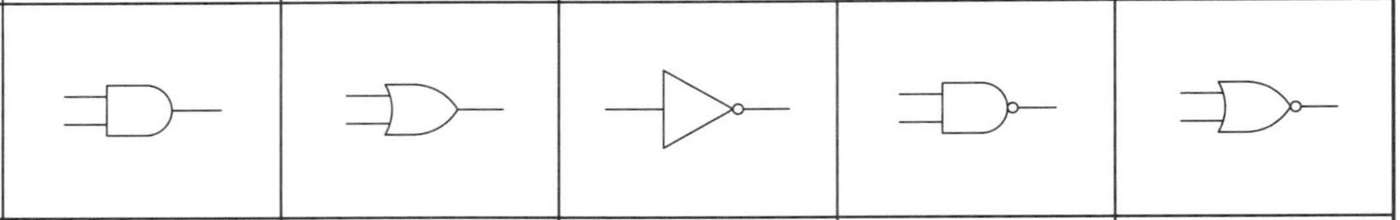				
ELECTRICAL RELAY LOGIC SYMBOL	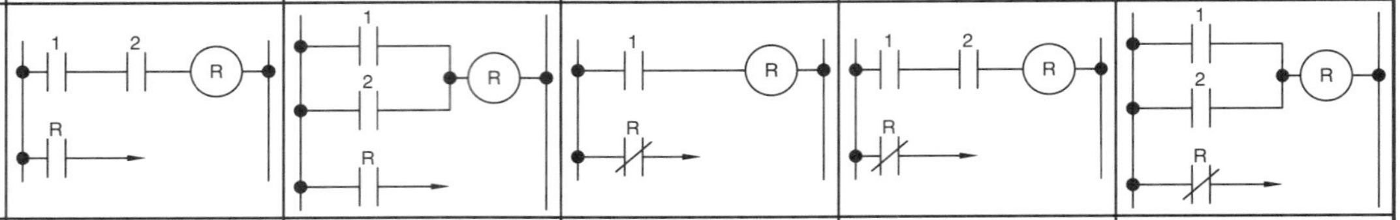				
ELECTRICAL SWITCH LOGIC SYMBOL	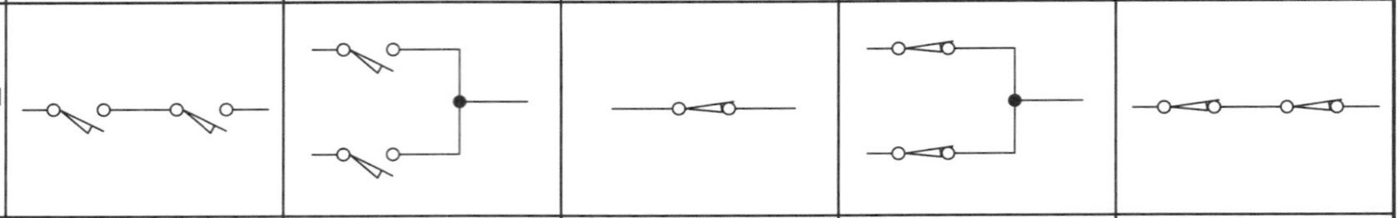				
ASA (JIC) VALVING SYMBOL	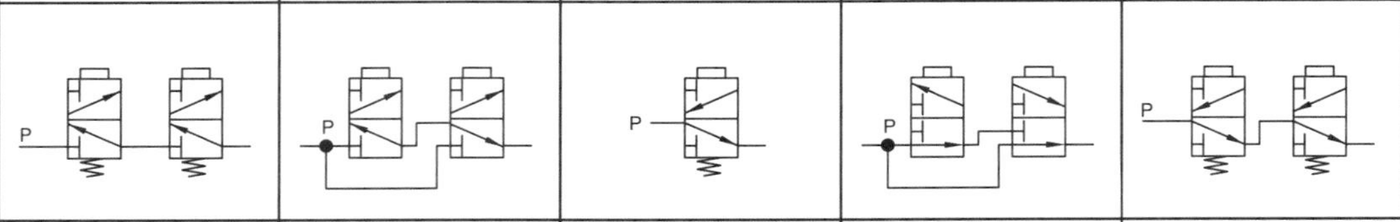				
ARO PNEUMATIC LOGIC SYMBOL	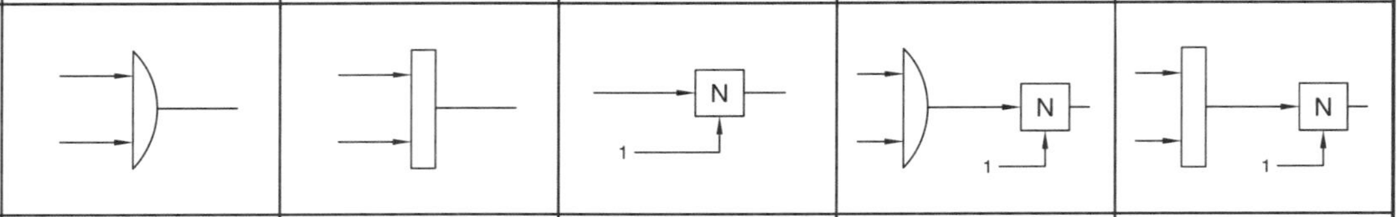				
NFPA STANDARD	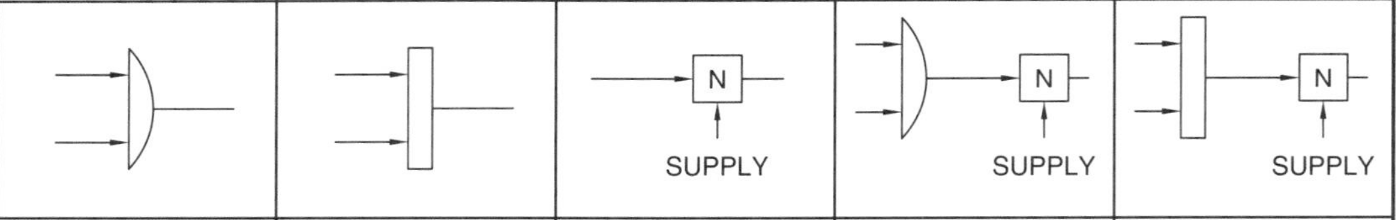		SUPPLY	SUPPLY	SUPPLY
BOOLEAN ALGEBRA SYMBOL	$(\ )\cdot(\ )$	$(\ )+(\ )$	$\overline{(\ )}$	$\overline{(\ )\cdot(\ )}$	$\overline{(\ )+(\ )}$
FLUIDIC DEVICE TURBULENCE AMPLIFIER	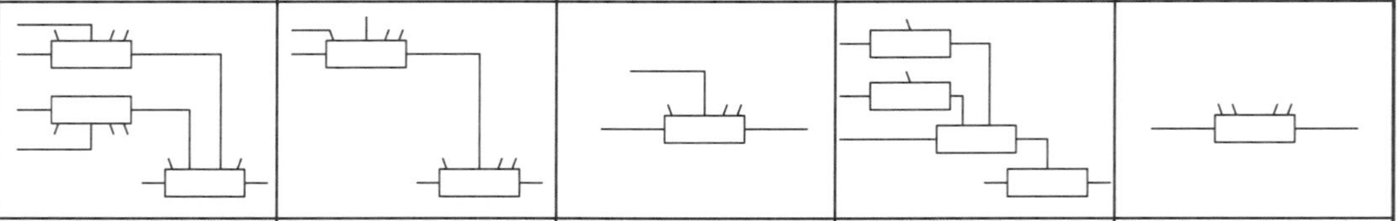				
NFPA STANDARD T3.7.68.2	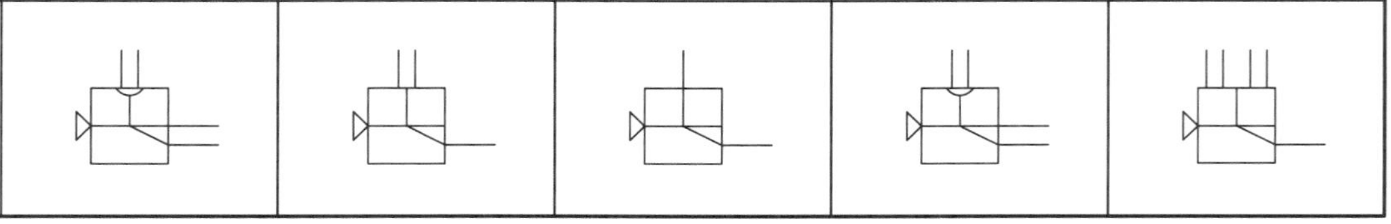				

NON-LOCKING WIRING DEVICES

2-POLE, 3-WIRE

WIRING DIAGRAM	NEMA ANSI	RECEPTACLE CONFIGURATION	RATING
	5-15 C73.11		15 A 125 V
	5-20 C73.12		20 A 125 V
	5-30 C73.45		30 A 125 V
	5-50 C73.46		50 A 125 V
	6-15 C73.20		15 A 250 V
	6-20 C73.51		20 A 250 V
	6-30 C73.52		30 A 250 V
	6-50 C73.53		50 A 250 V
	7-15 C73.28		15 A 277 V
	7-20 C73.63		20 A 277 V
	7-30 C73.64		30 A 277 V
	7-50 C73.65		50 A 277 V

4-POLE, 4-WIRE

WIRING DIAGRAM	NEMA ANSI	RECEPTACLE CONFIGURATION	RATING
	18-15 C73.15		15 A 3ϕ Y 120/208 V
	18-20 C73.26		20 A 3ϕ Y 120/208 V
	18-30 C73.47		30 A 3ϕ Y 120/208 V
	18-50 C73.48		50 A 3ϕ Y 120/208 V
	18-60 C73.27		60 A 3ϕ Y 120/208 V

3-POLE, 3-WIRE

WIRING DIAGRAM	NEMA ANSI	RECEPTACLE CONFIGURATION	RATING
	10-20 C73.23		20 A 125/250 V
	10-30 C73.24		30 A 125/250 V
	10-50 C73.25		50 A 125/250 V
	11-15 C73.54		15 A 3ϕ 250 V
	11-20 C73.55		20 A 3ϕ 250 V
	11-30 C73.56		30 A 3ϕ 250 V
	11-50 C73.57		50 A 3ϕ 250 V

3-POLE, 4-WIRE

WIRING DIAGRAM	NEMA ANSI	RECEPTACLE CONFIGURATION	RATING
	14-15 C73.49		15 A 125/250 V
	14-20 C73.50		20 A 125/250 V
	14-30 C73.16		30 A 125/250 V
	14-50 C73.17		50 A 125/250 V
	14-60 C73.18		60 A 125/250 V
	15-15 C73.58		15 A 3ϕ 250 V
	15-20 C73.59		20 A 3ϕ 250 V
	15-30 C73.60		30 A 3ϕ 250 V
	15-50 C73.61		50 A 3ϕ 250 V
	15-60 C73.62		60 A 3ϕ 250 V

LOCKING WIRING DEVICES

2-POLE, 3-WIRE

WIRING DIAGRAM	NEMA ANSI	RECEPTACLE CONFIGURATION	RATING
	ML2 C73.44		15 A 125 V
	L5-15 C73.42		15 A 125 V
	L5-20 C73.72		20 A 125 V
	L6-15 C73.74		15 A 250 V
	L6-20 C73.75		20 A 250 V
	L6-30 C73.76		30 A 250 V
	L7-15 C73.43		15 A 277 V
	L7-20 C73.77		20 A 277 V
	L8-20 C73.79		20 A 480 V
	L9-20 C73.81		20 A 600 V

3-POLE, 3-WIRE

WIRING DIAGRAM	NEMA ANSI	RECEPTACLE CONFIGURATION	RATING
	ML3 C73.30		15 A 125/250 V
	L10-20 C73.96		20 A 125/250 V
	L10-30 C73.97		30 A 125/250 V
	L11-15 C73.98		15 A 3ϕ 250 V
	L11-20 C73.99		20 A 3ϕ 250 V
	L12-20 C73.101		20 A 3ϕ 480 V
	L12-30 C73.102		30 A 3ϕ 480 V
	L13-30 C73.103		30 A 3ϕ 600 V

3-POLE, 4-WIRE

WIRING DIAGRAM	NEMA ANSI	RECEPTACLE CONFIGURATION	RATING
	L14-20 C73.83		20 A 125/250 V
	L14-30 C73.84		30 A 125/250 V
	L15-20 C73.85		20 A 3ϕ 250 V
	L15-30 C73.86		30 A 3ϕ 250 V
	L16-20 C73.87		20 A 3ϕ 480 V
	L16-30 C73.88		30 A 3ϕ 480 V
	L17-30 C73.89		30 A 3ϕ 600 V

4-POLE, 4-WIRE

WIRING DIAGRAM	NEMA ANSI	RECEPTACLE CONFIGURATION	RATING
	L18-20 C73.104		20 A 3ϕ Y 120/208 V
	L18-30 C73.105		30 A 3ϕ Y 120/208 V
	L19-20 C73.106		20 A 3ϕ Y 277/480 V
	L20-20 C73.108		20 A 3ϕ Y 347/600 V

4-POLE, 5-WIRE

WIRING DIAGRAM	NEMA ANSI	RECEPTACLE CONFIGURATION	RATING
	L21-20 C73.90		20 A 3ϕ Y 120/208 V
	L22-20 C73.92		20 A 3ϕ Y 277/480 V
	L23-20 C73.94		20 A 3ϕ Y 347/600 V

CAPACITOR RATINGS

110 – 125 VAC, 50/60 Hz, Starting Capacitors

Typical Ratings*	Dimensions**		Model Number***
	Diameter	Length	
88 – 106	$1\frac{7}{16}$	$2\frac{3}{4}$	EC8815
108 – 130	$1\frac{7}{16}$	$2\frac{3}{4}$	EC10815
130 – 156	$1\frac{7}{16}$	$2\frac{3}{4}$	EC13015
145 – 174	$1\frac{7}{16}$	$2\frac{3}{4}$	EC14515
161 – 193	$1\frac{7}{16}$	$2\frac{3}{4}$	EC16115
189 – 227	$1\frac{7}{16}$	$2\frac{3}{4}$	EC18915A
216 – 259	$1\frac{7}{16}$	$3\frac{3}{8}$	EC21615
233 – 280	$1\frac{7}{16}$	$3\frac{3}{8}$	EC23315A
243 – 292	$1\frac{7}{16}$	$3\frac{3}{8}$	EC24315A
270 – 324	$1\frac{7}{16}$	$3\frac{3}{8}$	EC27015A
324 – 389	$1\frac{7}{16}$	$3\frac{3}{8}$	EC2R10324N
340 – 408	$1\frac{13}{16}$	$3\frac{3}{8}$	EC34015
378 – 454	$1\frac{13}{16}$	$3\frac{3}{8}$	EC37815
400 – 480	$1\frac{13}{16}$	$3\frac{3}{8}$	EC40015
430 – 516	$1\frac{13}{16}$	$3\frac{3}{8}$	EC43015A
460 – 553	$1\frac{13}{16}$	$4\frac{3}{8}$	EC5R10460N
540 – 648	$1\frac{13}{16}$	$4\frac{3}{8}$	EC54015B
590 – 708	$1\frac{13}{16}$	$4\frac{3}{8}$	EC59015A
708 – 850	$1\frac{13}{16}$	$4\frac{3}{8}$	EC70815
815 – 978	$1\frac{13}{16}$	$4\frac{3}{8}$	EC81515
1000 – 1200	$2\frac{1}{16}$	$4\frac{3}{8}$	EC100015A

220 – 250 VAC, 50/60 Hz, Starting Capacitors

Typical Ratings*	Diameter	Length	Model Number***
53 – 64	$1\frac{7}{16}$	$3\frac{3}{8}$	EC5335
64 – 77	$1\frac{7}{16}$	$3\frac{3}{8}$	EC6435
88 – 106	$1\frac{13}{16}$	$3\frac{3}{8}$	EC8835
108 – 130	$1\frac{13}{16}$	$3\frac{3}{8}$	EC10835A
124 – 149	$1\frac{13}{16}$	$4\frac{3}{8}$	EC12435
130 – 154	$1\frac{13}{16}$	$4\frac{3}{8}$	EC13035
145 – 174	$2\frac{1}{16}$	$3\frac{3}{8}$	EC6R22145N
161 – 193	$2\frac{1}{16}$	$3\frac{3}{8}$	EC6R2216N
216 – 259	$2\frac{1}{16}$	$4\frac{3}{8}$	EC21635A
233 – 280	$2\frac{1}{16}$	$4\frac{3}{8}$	EC23335A
270 – 324	$2\frac{1}{16}$	$4\frac{3}{8}$	EC27035A

* in µF
** in inches
*** model numbers vary by manufacturer

CAPACITOR RATINGS

270 VAC, 50/60 Hz, Running Capacitors

Typical Ratings*	Dimensions**		Model Number***
	Oval	Length	
2	$1\frac{5}{16} \times 2\frac{5}{32}$	$2\frac{1}{8}$	VH5502
3		$2\frac{1}{8}$	VH5503
4		$2\frac{1}{8}$	VH5704
5		$2\frac{1}{8}$	VH5705
6		$2\frac{5}{8}$	VH5706
7.5	$1\frac{5}{16} \times 2\frac{5}{32}$	$2\frac{7}{8}$	VH9001
10		$2\frac{7}{8}$	VH9002
12.5		$3\frac{7}{8}$	VH9003
15	$1\frac{29}{32} \times 2\frac{29}{32}$	$2\frac{1}{8}$	VH9121
17.5		$2\frac{7}{8}$	VH9123
20	$1\frac{29}{32} \times 2\frac{29}{32}$	$2\frac{7}{8}$	VH5463
25		$3\frac{7}{8}$	VH9069
30		$3\frac{7}{8}$	VH5465
35	$1\frac{29}{32} \times 2\frac{29}{32}$	$3\frac{7}{8}$	VH9071
40		$3\frac{7}{8}$	VH9073
45	$1\frac{31}{32} \times 3\frac{21}{32}$	$3\frac{7}{8}$	VH9115
50		$3\frac{7}{8}$	VH9075

440 VAC, 50/60 Hz, Running Capacitors

Typical Ratings*	Oval	Length	Model Number***
10	$1\frac{5}{16} \times 2\frac{5}{32}$	$3\frac{7}{8}$	VH5300
15	$1\frac{29}{32} \times 2\frac{29}{32}$	$2\frac{7}{8}$	VH5304
17.5	$1\frac{29}{32} \times 2\frac{29}{32}$	$3\frac{7}{8}$	VH9141
20	$1\frac{29}{32} \times 2\frac{29}{32}$	$3\frac{7}{8}$	VH9082
25	$1\frac{29}{32} \times 2\frac{29}{32}$	$3\frac{7}{8}$	VH5310
30	$1\frac{29}{32} \times 2\frac{29}{32}$	$4\frac{3}{4}$	VH9086
35		$4\frac{3}{4}$	VH9088
40		$4\frac{3}{4}$	VH9641
45	$1\frac{31}{32} \times 3\frac{21}{32}$	$3\frac{7}{8}$	VH5351
50		$3\frac{7}{8}$	VH5320
55		$4\frac{3}{4}$	VH9084

* in µF
** in inches
*** model numbers vary by manufacturer

THERMAL UNIT CURRENT RATINGS

Motor Full-Load Current*			Thermal Unit Number
1 Unit	2 Units	3 Units	
0.29 – 0.31	0.29 – 0.31	0.28 – 0.30	B0.44
0.32 – 0.34	0.32 – 0.34	0.31 – 0.34	B0.51
0.35 – 0.38	0.35 – 0.38	0.35 – 0.37	B0.57
0.39 – 0.45	0.39 – 0.45	0.38 – 0.44	B0.63
0.46 – 0.54	0.46 – 0.54	0.45 – 0.53	B0.71
0.51 – 0.61	0.51 – 0.61	0.54 –0.59	B0.81
0.62 – 0.66	0.62 – 0.66	0.60 – 0.64	B0.92
0.67 – 0.73	0.67 – 0.73	0.64 – 0.72	B1.03
0.74 – 0.81	0.74 – 0.81	0.73 – 0.80	B1.16
0.82 – 0.94	0.82 – 0.94	0.81 – 0.90	B1.30
0.95 – 1.05	0.95 – 1.05	0.91 – 1.03	B1.45
1.06 – 1.22	1.06 – 1.22	1.04 – 1.14	B1.67
1.23 – 1.34	1.23 – 1.34	1.15 – 1.27	B1.88
1.35 – 1.51	1.35 – 1.51	1.28 – 1.43	B2.10
1.52 – 1.71	1.52 – 1.71	1.44 – 1.62	B2.40
1.72 – 1.93	1.72 – 1.93	1.63 – 1.77	B2.65
1.94 – 2.14	1.94 – 2.14	1.78 – 1.97	B3.00
2.15 – 2.40	2.15 – 2.40	1.98 – 2.32	B3.30
2.41 – 2.72	2.41 – 2.72	2.33 – 2.51	B3.70
2.73 – 3.15	2.73 – 3.15	2.52 – 2.99	B4.15
3.16 – 3.55	3.16 – 3.55	3.00 – 3.42	B4.85
3.56 – 4.00	3.56 – 4.00	3.43 – 3.75	B5.50
4.01 – 4.40	4.01 – 4.40	3.76 – 3.98	B6.25
4.41 – 4.88	4.41 – 4.88	3.99 – 4.48	B6.90
4.89 – 5.19	4.89 – 5.19	4.49 – 4.93	B7.70
5.20 – 5.73	5.20 – 5.73	4.94 – 5.21	B8.20
5.74 – 6.39	5.74 – 6.39	5.22 – 5.84	B9.10
6.40 – 7.13	6.40 – 7.13	5.85 – 6.67	B10.2
7.14 – 7.90	7.14 – 7.90	6.68 – 7.54	B11.5
7.91 – 8.55	7.91 – 8.55	7.55 – 8.14	B12.8
8.56 – 9.53	8.56 – 9.53	8.15 – 8.72	B14.0
9.54 – 10.6	9.54 – 10.6	8.73 – 9.66	B15.5
10.7 – 11.8	10.7 – 11.8	9.67 – 10.5	B17.5
11.9 – 13.2	11.9 – 12.0	10.6 – 11.3	B19.5
13.3 – 14.9	—	11.4 – 12.0	B22.0
15.0 – 16.6	—	—	B25.0
16.7 – 18.0	—	—	B28.0
Following Selections for Size 1 Only			
—	11.9 – 13.2	—	B19.5
—	13.3 – 14.9	11.4 – 12.7	B22.0
—	15.0 – 16.6	12.8 – 14.1	B25.0
16.7 – 18.9	16.7 – 18.9	14.2 – 15.9	B28.0
19.0 – 21.2	19.0 – 21.2	16.0 – 17.5	B32.0
21.3 – 23.0	21.3 – 23.0	17.6 – 19.7	B36.0
23.1 – 25.5	23.1 – 25.5	19.8 – 21.9	B40.0
25.6 – 26.0	25.6 – 26.0	22.0 – 24.4	B45.0
—	—	24.5 – 26.0	B50.0

* in amperes

RESISTOR COLOR CODES

Color	Number		Multiplier	Tolerance (%)
	1st	2nd		
Black (BK)	0	0	1	0
Brown (BR)	1	1	10	—
Red (R)	2	2	100	—
Orange (O)	3	3	1000	—
Yellow (Y)	4	4	10,000	—
Green (G)	5	5	100,000	—
Blue (BL)	6	6	1,000,000	—
Violet (V)	7	7	10,000,000	—
Gray (GY)	8	8	100,000,000	—
White (W)	9	9	1,000,000,000	—
Gold (Au)	—	—	0.1	5
Silver (Ag)	—	—	0.01	10
None	—	—	0	20

GLOSSARY

A

absolute pressure: Pressure measured relative to a perfect vacuum, expressed as pounds per square inch absolute (psia).

accelerating circuit logic: A control function that permits the operator to select a high motor speed and the control circuit automatically accelerates the motor to that speed.

accuracy: Limits in which the stated value of a measurement might vary relative to its true value, normally expressed as a percent of full scale, such as +1%.

accuracy (range): Minimum and maximum limits in which the stated value of a measurement might vary relative to the total range setting.

actuator: The part of a limit switch that transfers the mechanical force of the moving part to the electrical contacts.

address: A reference number assigned to a unique memory location. Each memory location has an address and each address has a memory location.

adjustable motor base: A mounting base that allows a motor to be easily moved over a short distance.

alternating current (AC): Current that reverses its direction of flow twice per cycle.

alternator: An alternating current generator.

ambient temperature: The temperature of the air surrounding equipment.

ampere (A): The basic unit of measurement for electric current. One ampere is the result of 1 V applied across a resistance of 1 Ω.

amplification: The process of taking a small signal and increasing its size.

amplifier: A device whose output is a larger reproduction of the input signal.

amplitude: The distance that a vibrating object moves from its position of rest as it vibrates.

analog device: Apparatus that measures continuous information such as voltage or current. The measured analog signal has an infinite number of possible values.

analog display: An electromechanical device that indicates readings on a meter by the mechanical motion of a pointer.

analog input interface: An input circuit that employs an analog-to-digital (A/D) converter to convert an analog value to a digital value that can be used for processing.

analog output interface: An output circuit that employs a digital-to-analog (D/A) converter to convert a digital value to an analog value that controls a connected analog device.

analog signal: Signal having the characteristic of being continuous and changing smoothly over a given range.

analog switching relay: An SSR that has an infinite number of possible output voltages within the relay's rated range.

AND gate: A device with an output that is high only when both of its inputs are high.

AND logic: Operation yielding logical 1 if and only if all inputs are 1.

arc chute: A device that confines, divides, and extinguishes arcs drawn between contacts opened under load.

arcing: The discharge of an electric current across a gap, such as when an electric switch is opened.

arc suppressor: A device that dissipates the energy present across opening contacts.

armature reaction: The effect that the magnetic field of the armature coil has on the magnetic field of the main pole windings.

asymmetrical: In timer terminology, the ability to have different settings for when the timer turns ON and OFF.

asymmetrical recycle timer: A timer that has independent adjustments for ON and OFF time periods.

atom: The smallest building block of matter that cannot be divided into a smaller unit without changing its basic character.

automatic condition: Any input which responds to changes in a system.

avalanche current: Current passed when a diode breaks down.

B

bar graph: A graph composed of segments that function as an analog pointer.

base speed: The speed of a DC motor (in rpm) listed on the motor's nameplate and the speed the motor runs with full-line voltage applied to the armature and field.

bellows: A cylindrical device with several deep folds which expand or contract when pressure is applied.

bimetallic overload relay: An overload relay which resets automatically.

bimetallic sensor: A sensor that bends or curls when the temperature changes.

bimetallic thermometer: Two metals having different coefficients of expansion, bonded together normally in the form of a spiral or strip.

binary coded decimal (BCD): A binary number system in which each decimal digit from 0 to 9 is represented by four binary digits (bits). The four positions have a weighted value of 1, 2, 4, and 8 respectively, starting with the least significant bit.

binary number system: A number system that uses two numerals (binary digits), 0 and 1.

binary word: A related grouping of 1s and 0s having coded meaning assigned by position, or as a group has some numerical value.

bipolar device: A device in which both holes and electrons are used as internal carriers for maintaining current flow.

bit: One binary digit. The smallest unit of binary information.

Boolean algebra: Shorthand notation for expressing logic functions. Used to understand the logic of a circuit to simplify the circuit when working with PCs.

Boolean operators: Logical operators such as AND, OR, NAND, NOR, NOT, and exclusive OR that can be used alone or in combination to form logical statements or circuits.

branch circuit: The portion of a distribution system between the final overcurrent protection device and the outlet or load connected to it.

break: The number of separate places on a contact that open or close an electrical circuit.

breakdown torque (BDT): The maximum torque which a motor can provide without an abrupt reduction in motor speed.

breakover voltage: The voltage across the output of a solid-state switch in the OFF-state at which a breakover turn-ON occurs.

burden current: The operating current in a line-powered (3-wire) sensor. This current does not pass through the load.

bus: A large trace extending around the edge to provide conduction from several sources.

capacitance: The ability to store energy in the form of an electrical charge.

capacitance detector: A device with single or multiple probes that causes a change in the probe capacitance when an object comes within proximity of the detector.

capacitive level switch: A level switch that detects the dielectric variation when the product is in contact (proximity) with the probe and when the product is not in contact with the probe.

capacitive proximity sensor: A sensor that detects either conductive or nonconductive substances.

capacitor motor: A 1ϕ, AC motor that includes a capacitor in addition to the running and starting windings.

capillary tube sensor: A sensor that changes internal pressure with a change in temperature.

cascaded amplifiers: Two or more amplifiers connected to obtain the required gain.

central processing unit (CPU): The part of a PC that governs system activities, including the interpretation and execution of programmed instruction.

circuit analysis method: A method of SSR replacement in which a logical sequence is used to determine the reason for failure.

circuit breaker (CB): A reusable OCPD that opens a circuit automatically at predetermined overcurrent.

cladding: The first layer of protection for the glass or plastic core of optical fiber cable.

clear: To remove data from a single memory location or all memory locations, and return to a nonprogrammed state or some initial condition (normally 0 in a PC).

clock signal: A clock pulse that is periodically generated and used throughout a system to synchronize equipment operation.

compelling circuit logic: A control function that requires the operator to start and operate a motor in a predetermined order.

complementary metal oxide semiconductor (CMOS): A low-power IC in which almost no current flows when a gate is not switching.

conductive level detector: A device with single or multiple probes that completes an electrical circuit between the container and/or probes when a change in level occurs.

conductive probe level switch: A level switch that uses liquid to complete the electrical path between two conductive probes.

conductor: Any material that allows electricity to flow through it.

constant torque/variable horsepower (CT/VH) loads: Loads in which the torque requirement remains constant.

contact: The conducting part of a switch that operates with another conducting part to make or break a circuit.

contact block: The part of the pushbutton that is activated when the operator is pressed.

contact-controlled timer: A timer that does not require the control switch to be connected in line with the timer coil.

contact life: The number of times a relay's contact switches the load controlled by the relay before malfunctioning.

contactor: A control device that uses a small control current to energize or de-energize the load connected to it.

contact protection circuit: A circuit that protects contacts by providing a nondestructive path for generated voltage as a switch is opened.

control circuit: The part of the relay that determines when the output component is energized or de-energized.

controller: An electrical device that continuously monitors any inputs (switches, etc.) connected to it and automatically activates any outputs (solenoids, etc.) connected to it, according to the way the controller was programmed or wired.

control point: The level at which a system is to be maintained.

convergent beam scan: A method of scanning which simultaneously focuses and converges a light beam to a fixed focal point in front of the photoreceiver.

converter: A machine or device that changes AC power to DC power or vice versa, or from one frequency to another.

current: The amount of electrons flowing in a circuit.

current loop: A 2-wire communication link in which the presence of a 20 mA current level indicates a binary 1 (mark) and its absence indicates a binary 0 (no data).

cutoff condition: The point at which all collector current is stopped by the absence of base current.

cutoff region: The point at which the transistor is turned OFF and no current flows.

cycle: One complete wave of alternating voltage or current.

dark-operated (DO): Control operating mode in which the output is energized when the light is blocked (photosensor is dark).

dark-operated photoelectric control: A control that energizes the output switch when a target is present (breaks the beam).

dashpot timer: A timer that provides time delay by controlling how rapidly air or a liquid is allowed to pass into or out of a container through an orifice (opening) that is either fixed or variable in diameter.

data file: A group of data values (inputs, timers, counters, and outputs) that is displayed as a group and whose status may be monitored.

daylight switch: A switch that automatically turns lamps ON at dusk and OFF at dawn.

dead band: The amount of pressure that must be removed before the switch contacts reset for another cycle after the setpoint has been reached and the switch has been actuated.

debouncing: The act of removing intermediate noise spikes from a mechanical switch.

decelerating circuit logic: A control function that permits the operator to select a low motor speed and the control circuit automatically decelerates the motor to that speed.

decimal number system: A number system that uses ten numeral digits (decimal digits), 0, 1, 2, 3, 4, 5, 6, 7, 8, and 9.

delta connection: A connection that has each coil end connected end-to-end to form a closed loop.

device: The smallest subdivision of a system that has a recognizable function of its own.

diac: A three-layer, bidirectional device used primarily as a triggering device.

diaphragm: A deflecting mechanism that moves when a force (pressure) is applied.

dielectric: A nonconductor of direct electric current.

dielectric absorption test: A test that checks the absorption characteristics of humid or contaminated insulation.

dielectric strength: The maximum allowable AC rms voltage which may be applied between input and output, input to case, and output to case.

dielectric variation: The range at which a material can sustain an electric field with a minimum dissipation of power.

differential: In ON/OFF controllers, the temperature difference between the temperature at which the controller turns heat OFF and the temperature at which the heat is turned ON.

diffuse scan (proximity scan): A method of scanning in which the transmitter and receiver are housed in the same enclosure and a small percentage of the transmitted light beam is reflected back to the receiver from the target.

digital: The representation of data in the form of pieces (bits or digits).

digital display: An electronic device that displays readings on a meter as numerical values.

diodes: Electronic components that allow current to pass through them in only one direction.

direct current (DC): Current that flows in only one direction.

direct memory access (DMA): A process in which a direct transfer of data to or from the memory of a processor-based system can take place without involving the central processing unit.

direct scan: A method of scanning in which the transmitter and receiver are placed opposite each other so that the light beam from the transmitter shines directly at the receiver.

disconnect: A device used only periodically to remove electrical circuits from their supply source.

doping: The process by which the crystal structure is altered.

double-break contacts: Contacts that break the electrical circuit in two places.

drop-out voltage: The voltage which exists when the voltage has reduced sufficiently to allow the solenoid to open.

drum switch: A manual switch made up of moving contacts mounted on an insulated rotating shaft.

dual-voltage motor: A motor that operates at more than one voltage level.

dynamic braking: A method of motor braking in which the motor is reconnected to act as a generator immediately after it is turned OFF.

eddy current: Unwanted current induced in the metal structure of a device due to the rate of change in the induced magnetic field.

edge card: A PC board with multiple terminations (terminal contacts) on one end.

edge card connector: A connector that allows the edge card to be connected to the system's circuitry with the least amount of hardware.

electrical noise: Unwanted signals that are present on a power line.

electric braking: A method of braking in which a DC voltage is applied to the stationary windings of a motor after the AC voltage is removed.

electromagnetic actuation: A passive method of sensor activation in which a magnetic field produced by a coil of wire is used to activate a Hall effect sensor.

electromechanical relay (EMR): A switching device that has sets of contacts which are closed by a magnetic effect.

electron: Negatively-charged particle of an atom that orbits the nucleus of an atom.

electronic overload: A device that has built-in circuitry to sense changes in current and temperature.

electrons: Negatively charged particles whirling around the nucleus at great speeds in shells.

erasable programmable read only memory (EPROM): A ROM that can be erased with ultraviolet light and then reprogrammed.

eutectic alloy: A metal that has a fixed temperature at which it changes from a solid to a liquid state without going through a "mushy" condition.

exact replacement method: A method of SSR replacement in which a bad relay is replaced with a relay of the same type and size.

extended button operator: A pushbutton that has the button extended beyond the guard.

ferrous proximity shunt actuation: A passive method of sensor activation in which the magnetic induction around the Hall effect sensor is shunted with a gear tooth.

fiber optics: A technology that uses a thin flexible glass or plastic optical fiber to transmit light.

field: The stationary windings or magnets of a motor.

filter: Electrical device used to suppress undesirable electrical noise.

555 timer: An integrated circuit designed to output timing pulses for control of certain types of circuits.

flip-flop: An electronic circuit having two stable states or conditions normally designated "set" and "reset."

flow: The travel of fluid in response to a force caused by pressure or gravity.

flow switch: A control switch that detects the movement of a fluid.

flush button operator: A pushbutton with a guard ring surrounding the button which prevents accidental operation.

foot switch: A control switch that is operated by a person's foot.

force: Any cause that changes the position, motion, direction, or shape of an object.

fork-lever actuator: An actuator operated by either one of two roller arms.

forward-biased voltage (forward bias): The application of the proper polarity to a diode. Forward bias always results in forward current.

forward breakover voltage: The voltage required to switch the SCR into a conductive state.

frequency: Occurrence of a periodic function (with time as the independent variable), generally specified as a certain number of cycles per unit time.

full-load current (FLC): The current required by the motor to produce full-load torque at the motor's rated speed.

fuse: An overcurrent protection device (OCPD) with a fusible link that melts and opens the circuit on an overcurrent condition.

gain: A ratio of the amplitude of the output signal to the amplitude of the input signal.

gas switch (gas detector): A switch that detects a set amount of a specified gas and activates a set of electrical contacts.

general purpose relays: Electromechanical relays that include several sets (normally 2, 3, or 4) or nonreplaceable NO and NC contacts (normally rated at 5 A to 15 A) that are activated by a coil.

ghost voltage: A voltage that appears on a meter that is not connected to a circuit.

graph: A diagram that shows a variable in comparison to other variables.

ground fault: A condition in which current from a hot power line is flowing to ground.

ground fault circuit interrupter (GFCI): An electrical device which protects personnel by detecting potentially hazardous ground faults and quickly disconnecting power from the circuit.

grounding electrode: A conductor embedded in the earth to provide a good ground.

half-cycling: A false turn-ON of an SSR for a portion of one-half cycle normally caused by voltage transients appearing across the output that exceed OFF-state dv/dt or breakover voltage capabilities of the SSR.

half-shrouded button operator: A pushbutton with a guard ring which extends over the top half of the button.

half-wave rectifier: A circuit containing a diode which permits only the positive half-cycles of the AC sine wave to pass.

half-waving: A phenomenon that occurs when a relay fails to turn OFF because the current and voltage in the circuit reach zero at different times.

Hall effect sensor: A sensor that detects the proximity of a magnetic field.

Hall generator: A thin strip of semiconductor material through which a constant control current is passed.

hammer: A striking or splitting tool with a hardened head fastened perpendicular to the handle.

handshaking: Two-way communication between two devices to ensure successful data transfer. Handshaking can be accomplished through hardware using special lines or through software using special codes.

hard copy: A printed document of what is stored in memory.

head-on actuation: An active method of sensor activation in which a magnet is oriented perpendicularly to the surface of the sensor and is usually centered over the point of maximum sensitivity.

heater coil: A sensing device used to monitor the heat generated by excessive current and the heat created through ambient temperature rise.

heat sink: A piece of metal used to dissipate the heat of solid-state components mounted on it.

hertz (Hz): The international unit of frequency equal to one cycle per second.

holding current: The minimum current necessary for an SCR to continue conducting.

holes: The missing electrons in the crystal structure.

horsepower (HP): A unit of power equal to 746 W or 33,000 lb-ft per minute (550 lb-ft per second).

hybrid relay: A combination of electromechanical and solid-state technology used to overcome unique problems which cannot be resolved by one or other devices.

improper phase sequence: The changing of the sequence of any two phases (phase reversal) in a 3ϕ motor control circuit.

increment current: The maximum current permitted by the utility in any one step of an increment start.

index pin: A metal extension from the transistor case.

induction motor: A motor that has no physical electrical connection to the rotor.

inductive detector: A level measuring system incorporating an oscillator and electromagnetic field.

inductive proximity sensor: A sensor that detects conductive substances only.

infinity: An unlimited number or amount.

informational output: Any output device that displays data about the circuit or operation.

infrared (IR): The invisible radiation that certain LEDs emit. Infrared light is used with photoelectric controls.

infrared light: Light that is not visible to the human eye.

inherent motor protectors: Overload devices located directly on or in a motor to provide overload protection.

input circuit: The part of the relay to which the control component is connected.

inrush current: The current flowing in a load circuit immediately following turn-ON.

instant ON switching relay: An SSR that turns ON the load immediately when the control voltage is present.

insulation spot test: A test that checks motor insulation over the life of a motor.

insulation step voltage test: A test that creates electrical stress on internal insulation cracks to reveal aging or damage not found during other motor insulation tests.

insulator: Any material that resists the flow of electricity.

integrated circuit (IC): Circuit composed of thousands of semiconductors providing a complete circuit function in one small semiconductor package.

interface: A circuit that permits communication between the central processing unit and a field input or output device.

interference: Any object other than the object to be detected that is sensed by the sensor.

interpoles (commutating field poles): Auxiliary poles that are placed between the main field poles of the motor.

inversion: In relays, allowing the relay contacts to be changed from one state, such as normally open to closed contacts.

inverter: A machine, device, or system that changes DC voltage into AC and can be designed to produce variable frequency.

isolation: The value of insulation resistance measured between the input and output, input to case, and output to case.

jogging: The frequent starting and stopping of a motor for short periods of time.

joystick: An operator that selects one to eight different circuit conditions by shifting from the center position into one of the other positions.

jumbo mushroom button operator: A pushbutton that has a large curved operator extending beyond the guard.

kVA: Kilovolt-amperes (1000 volt amps).

kW: Kilowatts (1000 watts).

lag: A delay in output change following a change in input.

laser diode: A diode similar to an LED but has an optical cavity, which is required for lasing production (emitting coherent light).

latch: An instruction or component that retains its state even though the conditions that caused it to latch ON may go OFF. A latched output retains its last state (ON or OFF) if power is removed.

legend plate: The part of a switch that includes the written description of the switch's operation.

lever actuator: An actuator operated by means of a lever which is attached to the shaft of the limit switch.

light-activated silicon controlled rectifier (LASCR): An SCR that is activated by light.

light emitting diode (LED): A diode which produces light when current flows through it.

lightning arrester: A device which protects transformers and other electrical equipment from voltage surges caused by lightning.

light-operated (LO): Control operating mode in which the output is energized when the light beam is not blocked when the photosensor is illuminated (light).

light-operated photoelectric control: A control that energizes the output switch when the target is missing (removed from the beam).

limiting: A boundary imposed on the upper or lower range of a controller.

limit switch: A mechanical input that requires physical contact of the object with the switch actuator.

linear scale: A scale that is divided into equally spaced segments.

line diagram (ladder diagram): A diagram which shows, with single lines and symbols, the logic of an electrical circuit or system of circuits and components.

liquid crystal display (LCD): A display device consisting of a liquid crystal hermetically sealed between two glass plates.

load: The electrical device in the line diagram that uses the electrical power from L1 to L2.

load current: The amount of current drawn by the load when energized.

locked rotor: A condition when a motor is loaded so heavily that the motor shaft cannot turn.

locked rotor current (LRC): The steady-state current taken from the power line with the rotor locked (stopped and with the voltage applied).

locked rotor torque (LRT): The torque a motor produces when the rotor is stationary and full power is applied to the motor.

lockout: The process of removing the source of electrical power and installing a lock which prevents the power from being turned ON.

machine control relay: An electromechanical relay that includes several sets (usually 2 to 8) of NO and NC replaceable contacts (typically rated at 10 A to 20 A) that are activated by a coil.

machine language: A program written in binary form.

magnet: A substance that attracts iron and produces a magnetic field.

magnetic field: The invisible field produced by a current-carrying conductor (or coil), a permanent magnet, or the earth itself, that develops a north and a south polarity.

magnetic field interference: A form of interference induced into a circuit due to the presence of a magnetic field.

magnetic level switch: A switch that contains a float, a moving magnet, and a magnetically-operated reed switch to detect the level of a liquid.

magnetic motor starter: A contactor that includes motor overload protection.

manual condition: Any input into the circuit by a person.

manual contactor: A control device that uses pushbuttons to energize or de-energize the load connected to it.

manual control circuit: Any circuit that requires a person to initiate an action for the circuit to operate.

manual starter: A contactor with an added overload protective device.

maximum control voltage and/or current: The maximum control voltage and/or current intended to be applied to the input of an SSR.

maximum OFF-state dv/dt: The maximum rate of rise of OFF-state voltage to be applied to the output. Higher dv/dt may result in SSR turn-ON.

mechanical condition: Any input into the circuit by a mechanically moving part.

mechanical interlocking: The arrangement of contacts in such a way that both sets of contacts cannot be closed at the same time.

mechanical level switch: Level switch that uses a float which moves up and down with the level of the liquid and activates electrical contacts at a set height.

mechanical life: The number of times a relay's mechanical parts operate before malfunctioning.

mechanical relay: An electromechanical device that completes or interrupts a circuit by physically moving electrical contacts into contact with each other.

megohmmeter: A device that detects insulation deterioration by measuring high resistance values under high test voltage conditions.

memory: The part of a programmable controller where data and instructions are stored.

minimum holding current: The minimum amount of current required to keep a sensor operating.

minimum load current: The minimum ON-state load current that ensures proper operation of a load and sensor.

minimum ON-state holding current: The minimum current required to maintain a solid-state switch in the ON-state.

modulated light source (MLS) control: A photoelectric control that operates on modulated (pulsed) infrared light and responds to modulating frequency rather than light intensity.

molecular theory of magnetism: A theory that states that all substances are made up of an infinite number of molecular magnets that can be arranged in either an organized or disorganized manner.

motor torque: The force that produces, or tends to produce rotation in a motor.

multiplex: The act of channeling two or more signals to one source.

multiplexing: A method of transmitting more than one signal over a single transmission system.

NAND gate: A device that provides a low output when both inputs are high.

NAND logic: Operation yielding 0 if and only if all inputs are 1.

negative logic: The use of binary logic in such a way that 0 represents the voltage level normally associated with logic 1.

negative resistance characteristic: The characteristic that current decreases with an increase of applied voltage.

neutrons: Particles within an atom that have no electrical charge.

noise: Any condition that interferes with the desired signal in a circuit.

nominal control voltage and/or current: The normal control voltage and/or current intended to be applied to the input of an SSR.

nonlinear scale: A scale that is divided into unequally spaced segments.

nonrepetitive blocking voltage: The maximum voltage to be applied to the output of an SSR in the OFF-state. Higher voltages may result in breakover turn-ON.

nonrepetitive surge current: The maximum nonrepetitive peak surge current that may be conducted by the control switch for a specific duration. Control may be lost during and following the surge due to excessive heating.

NOR logic: Operation yielding logical 1 if and only if all inputs are 0.

NOT logic: Operation yielding logical 1 if the input is 0 and yielding 0 if the input is 1.

N-type material: Material created by doping a region of a crystal with atoms from an element that has more electrons in its outer shell than the crystal.

nucleus: The heavy, dense center of the atom that has a positive electrical charge.

OFF-delay (delay on release) timer: A device that does not start its timing function until the power is removed from the timer.

off-line programming: The use of a personal computer to program a PLC.

OFF-state leakage current: The amount of current that leaks through an SSR when the switch is turned OFF, normally about 2 mA to 10 mA.

ohm (Ω): The basic unit of measurement of resistance and impedance. One ohm is the result of 1 V applied across a resistance that allows 1 A to flow through it.

ON-delay (delay on operate) timer: A device that has a preset time period that must pass after the timer has been energized before any action occurs on the timer contacts.

one-shot (interval) timer: A device in which the contacts change position immediately and remain changed for the set period of time after the timer has received power.

opacity: The characteristic of an object that prevents light from passing through it.

op-amp: A very high gain, directly-coupled amplifier that uses external feedback to control response characteristics.

operating (residual or leakage) current: The amount of current a sensor draws from the power lines to develop a field that can detect the target.

operational diagram: A diagram that illustrates the function of a circuit or control module.

operator: The device that is pressed, pulled, or rotated by the individual operating the circuit.

optical isolation: Two circuits which are connected only through an LED transmitter and photoelectric receiver with no electrical continuity between the two circuits.

optical level switches: Level switches that use a photoelectric beam to sense the liquid.

optocoupler: A device that consists of an infrared emitting diode (IRED) as the input stage, and a silicon phototransistor as the output stage.

OR gate: A device with an output that is high when either or both inputs are high.

OR logic: Operation yielding logical 1 if one or any number of inputs is 1.

overcurrent protection device (OCPD): Disconnect switches with CBs or fuses added to provide overcurrent protection of the switched circuit.

overcycling: The process of turning a motor ON and OFF repeatedly.

overload: The application of excessive load to a motor.

overload relay: A relay that responds to electrical loads and operates at a preset value.

pads: Small round conductors to which component leads are soldered.

panelboard: A wall-mounted distribution cabinet containing a group of overcurrent and short-circuit protection devices for lighting and appliance or power distribution branch circuits.

part-winding starting: A method of starting a motor by first applying power to part of motor coil windings for starting and then applying power to the remaining coil windings for normal running.

PC board: An insulating material such as fiberglass or phenolic with conducting paths secured to one or both sides of the board.

peak inverse voltage (PIV): The maximum reverse bias voltage that a diode can withstand.

peak switching relay: An SSR that turns ON the load when the control voltage is present and the voltage at the load is at peak.

pendulum actuation: A method of sensor activation which is a combination of the head-on and the slide-by actuation methods.

percent slip: The percentage reduction in speed below synchronous speed.

permanent magnet: A magnet which can retain its magnetism after the magnetizing force has been removed.

personal computer (PC): A desktop or laptop computer intended for personal use in the home or office.

phase: The relationship between the current and voltage in an AC circuit with respect to their angular displacement. It is expressed in degrees, with 360° representing one complete cycle.

phase unbalance: The unbalance that occurs when power lines are out-of-phase.

photoconductive cell: A device which conducts current when energized by light.

photoconductive diode (photodiode): A diode which is switched ON and OFF by a light.

photoelectric switch: A solid-state sensor that can detect the presence of an object without touching the object.

photoreceiver: A unit consisting of a photosensor circuit that is designed to detect a predetermined light source.

phototransistor: A device that combines the effect of the photodiode and the switching capability of a transistor.

phototriac: A triac that is activated by light.

photovoltaic cell (solar cell): A device that converts solar energy to electrical energy.

pick-up voltage: The minimum control voltage which causes the armature to start to move.

PIN photodiode: A diode with a large intrinsic region sandwiched between P-type and N-type regions.

piston: A cylinder that is moved back and forth in a tight-fitting chamber by the pressure applied in the chamber.

pliers: Hand tools with opposing jaws for gripping and/or cutting.

plugging: A method of braking in which the motor connections are reversed so that the motor develops a counter torque which acts as a braking force.

PN junction: The area on a semiconductor material between the P-type and N-type material.

polarity: The particular state of an object, either positive or negative, which refers to the two electrical poles, north and south.

polarized scan: A method of scanning in which the receiver responds only to the depolarized reflected light from corner cube reflectors or polarized sensitive reflective tape.

pole: The number of completely isolated circuits that a relay can switch.

positive logic: The use of binary logic in such a way that 1 represents a positive logic level.

power distribution: The process of delivering electrical power to where it is needed.

pressure: A force exerted over surface divided by its area.

pressure switch: A control switch that detects a set amount of force and activates electrical contacts when the set amount of force is reached.

preventive maintenance: Maintenance performed in order to keep machines, assembly lines, production operations, and plant operations running with little or no downtime.

primary resistor starting: A reduced-voltage starting method which uses a resistor connected in each motor line (in one line in a 1ϕ starter) to produce a voltage drop.

primary winding: The coil that draws power from the source.

processor section: The section of a PLC that organizes all control activity by receiving inputs, performing logical decisions according to the program, and controlling the outputs.

programmable controller (PLC): A solid-state control device that is programmed and reprogrammed to automatically control an industrial process or machine.

programmable read-only memory (PROM): A digital storage device that can be written into only once, but can be read continually.

programming diagram: A line diagram that better matches the PLC's language.

programming section: The section of a PLC that allows input into the PLC through a keyboard.

protons: Particles in an atom which have a positive electrical charge of one unit.

proximity switch: A solid-state sensor that detects the presence of an object by means of an electronic sensing field.

P-type material: Material with empty spaces (holes) in its crystalline structure.

pull-up torque (PUT): The torque required to bring a load up to its rated speed.

pulse duration: The time that an input pulse must be present to correctly be registered by the control module. Given as a minimum time, such as 15 ms, or as a time range such as 15 ms to infinity.

pure DC power: Power obtained from a battery or DC generator.

pushbutton station: An enclosure that protects the pushbutton, contact block, and wiring from dust, dirt, water, or corrosive fluids.

push roller actuator: An actuator operated by direct forward movement into the limit switch.

radio frequency interference (RFI): Electromagnetic interference in the radio frequency range.

random turn-ON: Initial turn-ON that may occur at any point on the AC line voltage cycle.

RC circuit: A circuit in which resistance (R) and capacitance (C) are used to help filter the power in a circuit.

read-only memory (ROM): A digital storage device specified for a single function.

rectification: The changing of AC into DC.

rectifier: A circuit that converts AC to DC.

recycle timer: A device in which the contacts cycle open and close repeatedly once the timer has received power.

reed relay: A fast-operating, single-pole, single-throw switch with NO contacts hermetically sealed in a glass envelope.

reflective scan: A scan technique in which the light source is aimed at a reflecting surface to illuminate the photoreceiver.

relay: A device that controls one electrical circuit by opening and closing contacts in another circuit.

renewable fuses: OCPDs designed so that the fusible link can be replaced.

replacement method: Method in which a bad relay is replaced with a relay of the same type and size.

residual leakage current: A small amount of current that flows through a non-conducting (open) load-powered sensor.

resistance: Any force that tends to hinder the movement of an object.

resistive load: An electrical load characterized by not having any significant inrush current.

resistor: A device with a specific amount of electrical resistance.

retroreflective scan: A method of scanning in which the transmitter and receiver are housed in the same enclosure and the transmitted light beam is reflected back to the receiver from a reflector.

reversed-biased voltage (reverse bias): The application of the opposite polarity to a diode.

reverse polarity protection: Internal protective circuitry that prevents damage to the sensor in case of accidental reverse polarity connection.

response time: The number of pulses (objects) per second the controller can detect.

root mean square (rms): The value of AC voltage or current required to do the same amount of work as the same value of DC voltage or current. For a sine wave, the rms value is .707 times the peak value.

rotary actuation: An active method of sensor activation in which a multipolar ring magnet or a collection of magnets is used to produce an alternating magnetic pattern.

rotor: The rotating part of an AC motor.

RS232: An Electronics Industries Association (EIA) standard for the transmission of data over a twisted-wire pair. It defines pin assignments, signal levels, etc. for receiving and transmitting devices.

RS422: An Electronics Industries Association (EIA) standard for the electrical characteristics of balanced voltage digital interface. Unlike the RS232 connection, the RS422 allows data transmission to be received by multiple locations.

saturation region: The maximum current that can flow in the transistor circuit.

scan: The process of evaluating the input/output status, executing the program, and updating the system.

scanning: The process of using a light source and photosensor together to allow them to measure a change in light intensity when a target is present in, or absent from, the transmitted light beam.

scan time: The time it takes a PLC to make a sweep of the program.

screwdriver: A hand tool with a tip designed to fit into a screw head for turning.

seal-in voltage: The minimum control voltage required to cause the armature to seal against the pole faces of the magnet.

secondary winding: The coil that delivers the energy at a transformed or changed voltage to the load.

selector switch: A switch with an operator that is rotated (instead of pushed) to activate the electrical contacts.

semiconductor: A class of materials, such as silicon, whose electrical properties lie between those of conductors (copper, etc.) and insulators (glass, etc.)

semiconductor devices: Devices in which electrical conductivity is between that of a conductor (high conductivity) and that of an insulator (low conductivity).

sensor: A device that detects motion, light, sound, etc.

sensor-controlled timer: A timer controlled by an external sensor in which the timer supplies the power required to operate the sensor.

service factor (SF): A number designation that represents the percentage of extra demand that can be placed on the motor for short intervals without damaging the motor.

shading coil: A single turn of conducting material (normally copper or aluminum) mounted on the face of the magnetic laminate assembly or armature.

silicon controlled rectifier (SCR): A solid-state rectifier with the ability to rapidly switch heavy currents.

sine wave: A wave that can be expressed as the sine of a linear function of time and/or space.

single-phasing: The operation of a motor that is designed to operate on three phases, but is only operating on two phases because one phase is lost.

single-voltage motor: A motor that operates at only one voltage level.

slide-by actuation: An active method of sensor activation in which a magnet is moved across the face of the Hall effect sensor at a constant distance (gap).

smoke switch (smoke detector): A switch that detects a set amount of smoke caused by smoldering or burning material and activates a set of electrical contacts.

snubber circuit: A circuit that suppresses noise and high voltage on power lines.

solenoid: An electric output device that converts electrical energy into a linear mechanical force.

solid-state pressure sensor: A transducer that changes resistance with a corresponding change in pressure.

solid-state relay (SSR): A switching device that has no contacts and switches entirely by electronic means.

solid-state timer: A timer whose time delay is provided by solid-state electronic devices enclosed within the timing device.

specular scan: A method of scanning in which the transmitter and receiver are placed at equal angles from a highly reflective surface.

split-phase motor: A single-phase, AC motor that includes a running winding (main winding) and a starting winding (auxiliary winding).

stator: The stationary part of an AC motor.

sulfidation: The formation of film on the contact surface.

supply voltage timer: A timer that requires the control switch to be connected so that it controls power to the timer coil.

switch: An input device that detects and reacts to some type of manual, mechanical, or automatic condition.

switchboard: The piece of equipment in which a large block of electric power is delivered from a substation and broken down into smaller blocks for distribution throughout a building.

symmetrical: In timer terminology, having an identical and set length of time between when the timer turns ON and OFF.

symmetrical recycle timer: A timer that operates with equal ON and OFF time periods.

synchronous clock timer: A timer that opens and closes a circuit depending on the position of the hands of a clock.

tagout: The process of placing a danger tag on the source of electrical power which indicates that the equipment may not be operated until the danger tag is removed.

temperature switches: Control devices that react to heat intensity.

temporary magnet: A magnet which has extreme difficulty in retaining any magnetism after the magnetizing force has been removed.

thermal resistance (R_{TH}): The ability of a device to impede the flow of heat.

thermistor: A temperature-sensitive resistor that changes its electrical resistance with a change in temperature.

thermocouple: A temperature sensor of two dissimilar metals joined at the end where heat is to be measured, which produces a voltage output at the other end proportional to the measured temperature.

three-wire, line-powered sensor: A sensor that draws its operating current (burden current) directly from the line. The operating current does not flow through the load.

throw: The number of closed contact positions per pole.

thyristor: A bistable semiconductor device that can be switched from the OFF-state to the ON-state or vice versa.

torque: The force that produces rotation.

traces (foils): Conducting paths used to connect components on a PC board.

transducer: A device used to convert physical parameters, such as temperature, pressure, and weight, into electrical signals.

transformer: An electrical interface designed to change AC from one voltage level to another voltage level.

transient: A temporary current or voltage that occurs randomly and rides the AC voltage sine wave.

transient voltages: Temporary, unwanted voltages in an electrical circuit.

transistor-controlled timer: A timer that is controlled by an external transistor from a separately powered electronic circuit.

transistors: Three-terminal devices that control current through the device depending on the amount of voltage applied to the base.

transistor-transistor logic (TTL): A family of integrated circuit logic which normally uses 5 V as high and 0 V as low.

translucent: The quality of allowing light to pass through.

triac: A solid-state switching device used to switch alternating current. A three-electrode AC semiconductor switch.

troubleshooting: The systematic elimination of the various parts of a system or process to locate a malfunctioning part.

two-wire, load-powered sensor: A sensor that draws its operating current (residual current) through the load.

unijunction transistor (UJT): A transistor that consists of N-type material with a region of P-type material doped within the N-type material.

upper range value: The highest quantity that a device can adjust to or measure.

valence electrons: Electrons in the outermost shell of an atom.

vane actuation: A passive method of sensor activation in which an iron vane shunts or redirects the magnetic field in the air gap away from the Hall effect sensor.

vapor: A gas that can be liquified by compression without lowering the temperature.

variable torque/variable horsepower (VT/VH) loads: Loads that require a varying torque and horsepower at different speeds.

varistor: A resistor whose resistance is inversely proportional to the voltage applied to it.

vise: A portable or stationary clamping device used to firmly hold work in place.

volt (V): The unit of electrical potential difference and electromotive force.

volt amp (VA): Volts × amps.

voltage surge: Any higher-than-normal voltage that temporarily exists on one or more of the power lines.

voltage unbalance: The unbalance that occurs when the voltages at different motor terminals are not equal.

watt (W): A unit of measure equal to the power produced by a current of 1 A across a potential difference of 1 V.

wiring diagram: A diagram that shows the connection of an installation or its component devices or parts.

wobble stick actuator: An actuator operated by means of any movement into the switch, except a direct pull.

work: Applying a force over a distance.

wrap-around bar graph: A bar graph that displays a fraction of the full range on the graph.

wrench: A hand tool with jaws at one end that is designed to turn bolts, nuts, or pipes.

wye connection: A connection that has one end of each coil connected together and the other end of each coil left open for external connections.

yawing: A side-to-side movement.

zener diode: A silicon PN junction that differs from a rectifier diode in that it operates in the reverse breakdown region.

zero current turn-OFF: Turn-OFF at the zero crossing of the load current that flows through an SSR.

zero switching relay: An SSR that turns ON the load when the control voltage is applied and the voltage at the load crosses zero (or within a few volts of zero).

zero voltage turn-ON: Initial turn-ON that occurs at a point near the zero crossing of the AC line voltage.

B